Learning Resources
Brevard Community College
Cocoa, Florida

Direct and Alternating Currents

Second Edition

Samuel L. Oppenheimer
Founder and Former President, Ohio Technical Institute

F. Roger Hess, Jr.
Director
Central Institute of Technology

Jean Paul Borchers
President, Miami Plating Company

McGRAW-HILL BOOK COMPANY

New York
St. Louis
San Francisco
Düsseldorf
Johannesburg
Kuala Lumpur
London
Mexico
Montreal
New Delhi
Panama
Rio de Janiero
Singapore
Sydney
Toronto

Library of Congress Cataloging in Publication Data

Oppenheimer, Samuel L.
DIRECT AND ALTERNATING CURRENTS.

1. Electric circuits. 2. Electric currents.
I. Hess, F. Roger, joint author. II. Borchers,
1. Electric circuits. 2. Electric currents. I. Hess,
F. Roger, joint author. II. Borchers, Jean Paul, joint author
III. Title.
TK454.066 1972 621.319'12 73-39882
ISBN 0-07-047665-9

DIRECT AND ALTERNATING CURRENTS

Copyright © 1963, 1973 by McGraw-Hill, Inc. All rights reserved. Printed in the United States of America. No part of this publication may be reproduced, stored in a retrieval system, or transmitted, in any form or by any means, electronic, mechanical, photocopying, recording, or otherwise, without the prior written permission of the publisher.

1 2 3 4 5 6 7 8 9 0 K P K P 7 9 8 7 6 5 4 3

The editors for this book were Alan Lowe and Cynthia Newby, the designer was Marsha Cohen, and its production was supervised by James Lee. It was printed and bound by Kingsport Press, Inc.

Dedicated to the memory of

ROGER A. BARNETT

Chairman, Department of Physics
Ohio Institute of Technology

Contents

Preface

The vast growth in the body of science in the field of electrical engineering during the past decade has placed very heavy demands upon the engineer of today. No longer can the electrical engineer devote his time to the *practice* of his art; he is much too involved with the *science* of electrical engineering. Often, he tends to become more physical scientist than engineer. So that the practice of his art does not suffer, the electrical engineer needs support from the engineering technician, who must be capable of replacing the engineer in the many routine tasks that are part of any engineering function.

Just as the modern engineer has an education that is far more rigorous than in the past, so must today's engineering technician receive much more rigorous training. This book has been written to meet the need for a fundamental circuits text that will supply the necessary rigor for the training of engineering technicians capable of providing support for the electrical engineer of today.

Direct and Alternating Currents is a fundamental text in electric circuits and analysis. It is intended for use in technical institutes, junior and community colleges, and college courses in electricity for nonelectrical majors. The text may be used for home study or industrial training courses.

This text is not a compromise between vocational training and university training but is written for, and aimed at, the requirements of the electrical and electronics engineering technician. For this reason much attention is paid to areas where little, if any, was given in the past.

The emphasis throughout the text is on analysis and understanding of electric circuits. Electron flow convention is used. Numerous examples and figures supplement the theory in each chapter so that the student will gain a thorough understanding of circuit analysis.

Modern network analysis is covered in detail. Some of the network-analysis techniques explained in rigorous detail include voltage division, current division mesh currents, node voltages, superposition, Thévenin's theorem, Norton's theorem and reciprocity. Properties of magnetic circuits are presented using the rationalized mks system of units. Other systems of units and their conversion factors are discussed after the completion of magnetic fundamentals. Chapter 13 deals with coupled circuits and includes a thorough discussion of two-port network parameters, transformers, and filter circuits. The chapter on polyphase circuits is complete, including the use of impedance parameters in the analysis of unbalanced -y loads. This text contains some chapters which need not be taught in class but which will remain as useful references for the student long after the completion of electric circuit courses.

The contents of the text may be covered in two semesters or three quarters. In many schools algebra is taught simultaneously with direct current, and trigonometry is taught simultaneously with alternating current. The mathematical requirements of this text increase with the normal progress of a student in algebra and trigonometry.

The authors owe their deepest gratitude to Mr. Eugene F. Christopher and department chairmen James D. Ayres and March C. Fleming of the Ohio Institute of Technology for their many suggestions in the revision of the first edition. The authors thank the many students of Ohio Tech whose use of the first edition for over seven years helped to iron out the many wrinkles that any text has. Also the authors wish to thank Marilyn F. Hess for the excellent fashion in which she typed the manuscript of the second edition.

Samuel L. Oppenheimer
F. Roger Hess, Jr.
Jean Paul Borchers

1

FUNDAMENTALS, QUANTITIES, AND CONCEPTS

1-1 INTRODUCTION

The field of electronics deals with the use and control of electrons or other charge carriers in tubes, transistors, and related devices. A clear understanding of the behavior of moving charge (electric currents)—whether in a vacuum, a gas, or a solid—is essential for anyone studying electronics.

Electronics has diverse applications, which range from a simple device like a photoelectric relay to a complex system like a computer or the control system for a missile launch. Radio, television, radar, automatic control in industry, and guidance control systems for missiles are but a few of the many examples of electronics in action. Electronics is important in medical research, production control, quality control, and aircraft design and operation. Hardly an industry can be mentioned that does not apply electronics in some fashion. Through all these different applications of electronics one common factor stands out: all the devices use vacuum tubes, transistors, and interconnecting electric circuits that rely upon the flow of charge.

Before proceeding with the study of electronics, the laws and concepts governing charge flow, and the circuits in which this motion takes place, we must first pursue some basic physical quantities and concepts. We must develop our conceptual vocabulary—in other words, learn the language. In the remainder of Chap. 1 we discuss qualitatively the basic quantities and concepts to be used throughout the book. In Chap. 2 we deal with these quantities and concepts again but in a more quantitative manner.

There are four fundamental quantities with which we can begin: distance, time, mass, and charge, which are identified by symbols as follows: distance l, time t, mass m, and charge Q. These four fundamental quantities are also fundamental dimensions.

1-2 LENGTH, OR DISTANCE

Length is the dimension we are probably most familiar with. It is the quantity with which we describe space and, more specifically, position. The dimension of distance identifies the quantity separating two points. When we think of the distance from ourselves to some other point, we are using the quantity of length. It makes no difference whether the distance we identify is large (macroscopic), like the distance from Cape Kennedy to the moon, or infinitesimally small (microscopic), like the distance between the smallest particles of a substance.

There is a difference in meaning between *unit* and *dimension*. As previously pointed out, the dimension is the quantity being measured. The unit is just the specific size breakdown of the fundamental dimension. Then, using the familiar example of the dimension of length, the *unit* of length becomes the inch, foot, yard, mile, meter, micrometer, millimeter, kilometer, or centimeter. We could define a distance between two points as 1 yd, 3 ft, or 36 in. All three define the same fundamental quantity or dimension, length or distance. In Chap. 2 we deal with the dimension of length and its units quantitatively.

1-3 TIME

Time is a continuum best described by a sequence of events. Thus, time was first described by the continuing events of night and day, later by the phases of the moon and the position of the stars, and more recently by the rotation of the earth on its axis. The devices for measuring time began with instruments to measure the position of the sun and stars. An hourglass was developed to provide an artificial continuum to compare to nature's continuums. Mechanical clocks followed, and today we have electronic devices which provide extremely accurate artificial continuums.

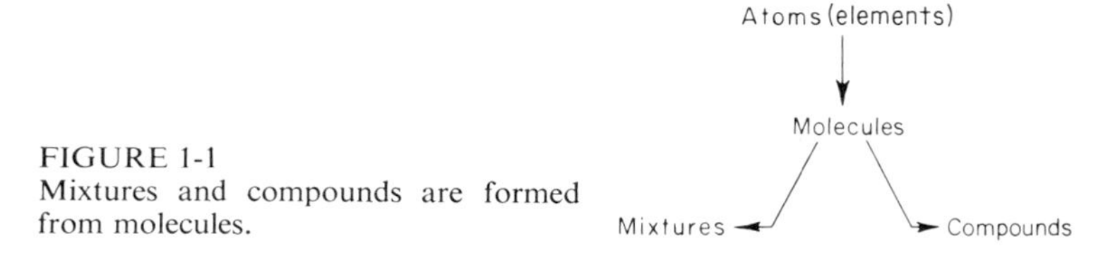

FIGURE 1-1
Mixtures and compounds are formed from molecules.

1-4 MASS AND MATTER

Mass is the quantity of inertia possessed by a substance which occupies space. All matter has the property of mass. The physical forms around us are examples of matter. It is possible for the physical forms of matter to take on a variety of physical states such as solid, liquid, or gas. The air we breathe is the most common example of matter in a gaseous state. Even though air is gaseous and invisible it still has weight and occupies space, thus meeting the defining conditions of mass.

The discussion of mass and matter requires some investigation of the structure of matter. Further, the structure of matter has a direct bearing on electronics. Matter can be divided into smaller and smaller portions until there remains the smallest portion or particle it can be divided into and still retain all the properties of the macroscopic portion we began with. A compound of matter can be divided into smaller and smaller particles until there remains a molecule of that compound (Fig. 1-1). An element of matter can be divided into smaller and smaller particles until there remains an atom of that element. The elements are thus the building blocks of the compounds of matter. In the Appendix Table A-1 lists the atomic number and atomic mass of the elements, and Table A-2 shows the periodic table of the elements.

We can take as a common and familiar example of a compound table salt (sodium chloride). We can begin with a grain of salt and divide that grain into smaller and smaller particles. Eventually, particles will be so small that they cannot be seen. As we continue (theoretically) dividing the invisible particle of salt into smaller particles, we finally reach a point where the particle we have can be divided no further if it is still to retain the properties of salt (sodium chloride). We now have one molecule of salt. This single molecule of salt, however, is made up of smaller particles of matter (elements) and can be divided into them. The two elements which make up a molecule of salt are sodium and chlorine (see Fig. 1-2).

When we divide the molecule of salt into smaller particles, neither particle has the properties of salt. Instead these particles have the properties of the individual elements, chlorine and sodium. These particles of elements are called *atoms*. An atom is thus the smallest particle into which an element can be divided and still retain the physical and chemical properties of the element. It should be further noted that

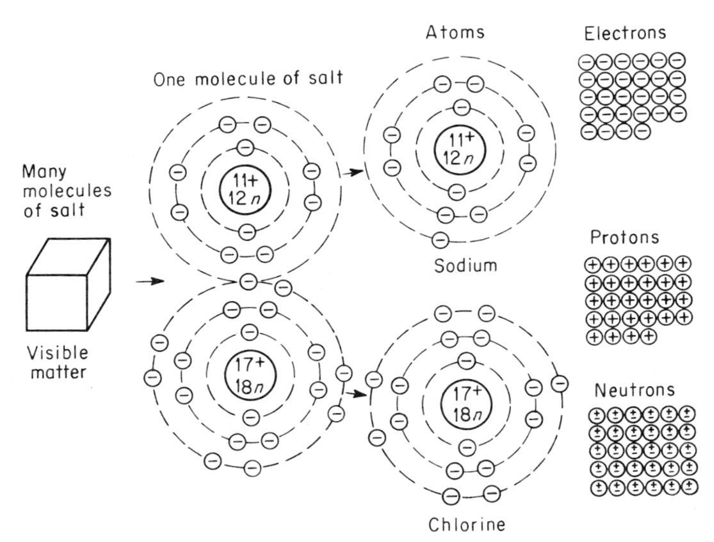

FIGURE 1-2
Breakdown of a salt crystal

an atom of an element has mass (see Appendix Table A-1). At this time there are 103 known elements (see Appendix Table A-2). These elements enter into chemical combinations with one another to form molecules of an endless number of substances.

Compounds may have properties very different from those of the elements that form them. Let us return to our example of salt, made up of chlorine and sodium. We are all familiar with the physical properties of salt commonly used in flavoring food. Some specific properties of salt are the following:

1 White solid
2 Soluble in water
3 Melting point 804°C

Chlorine has the following properties:

1 Pale yellow-green gas
2 A distinct odor
3 Poisonous; irritating to the eyes
4 Boiling point −34.6°C

Sodium has the following properties:

1 Silver-white metal
2 Reacts violently with water
3 Boiling point 880°C

All matter is made up of a single element or a combination of elements. One way in which elements or molecules combine is by chemical reaction. Another way is by a mixture. An example of this type of combination is salt water, where the salt and water do not enter into chemical combination but mix to a point where the molecules of salt and water are uniformly distributed in a solution. The definitive quality of a mixture is that the molecules making up the mixture can be separated from each other by evaporation or filtering. Atoms combined by chemical reaction cannot be separated from each other by evaporation or filtering.

1-5 CHARGE

Charge (or more descriptively electric charge) is the fourth fundamental quantity or dimension. The quantity of charge is defined on the basis of the property of fundamental atomic particles, the electron and the proton. A *negative* electric charge is the charge possessed by an electron; the proton possesses a *positive* charge. It should be noted that the positive proton and the negative electron are completely arbitrary. Thus the proton is positive and our electron is negative *by definition*. The symbol Q is used to identify the fundamental quantity charge in a general way. The symbol q_e stands for the charge of a single electron or proton.

1-6 THE STRUCTURE OF THE ATOM

It would seem that with breaking salt down into individual atoms the process of decomposition would be complete. Indeed, an atom is a small particle as far as physical size and weight are concerned. The weight of one atom of an element varies with the type of element. The weight of the chlorine atom is 58.2×10^{-27} kilogram (kg), while the weight of a sodium atom is 38×10^{-27} kg. Since 1 kg is equal to 2.205 lb we can convert the unfamiliar unit of kilograms to the more familiar pounds as follows:

1 The weight of chlorine in kilograms times pounds per kilogram equals weight of the chlorine atom in pounds. Thus

$$\begin{aligned}(58.2 \times 10^{-27})(2.205) &= 128.331 \times 10^{-27} \text{ lb} \\ &= 0.000000000000000000000000128331 \text{ lb}\end{aligned}$$

2 The weight of the sodium atom in pounds is

$$\begin{aligned}(38 \times 10^{-27})(2.205) &= 83.79 \times 10^{-27} \text{ lb} \\ &= 0.00000000000000000000000008379 \text{ lb}\end{aligned}$$

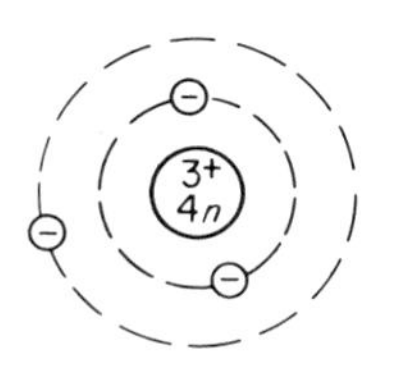

FIGURE 1-3
The lithium atom.

We can also make a relative comparison of the weight of chlorine and sodium. Thus, making a ratio of the weight of chlorine to the weight of sodium, we find that chlorine weighs 1.54 times as much as sodium. A more important comparison might be one between the weight of all the other elements and that of carbon (see the column of atomic weights listed in Appendix Table A-1). Despite the fact that atoms are rather small particles, they can be further divided into three main classes of particles, *protons*, *neutrons*, and *electrons*. The steps in the breakdown of salt from a visible crystal to electrons, neutrons, and protons are illustrated in Fig. 1-2.

The common representation of an atom is shown in Fig. 1-3, which pictures electrons circling the nucleus of the atom much as the planets circle the sun. The proton and the electron are unique particles of matter in that they possess electric charge Q in addition to the ordinary material properties such as mass and moment. The neutron has no electric charge. Other atomic particles have electric charge, but they are unstable, have short lives, and do not contribute to electric currents. Table 1-1 lists the atomic particles and their atomic mass. The mass unit (u) is one-twelfth the mass of the carbon atom. ($1\ u = 1.660 \times 10^{-27}$kg.) The electron has little weight. (9.1083×10^{-31} kg.) In fact, the weight of the electron is approximately 1/1,800 the weight of a proton.

The central portion of an atom, called the *nucleus*, consists primarily of protons and neutrons. The proton weighs 1.6724×10^{-27} kg. The polarity assigned to the electric charge of the proton is positive. The other nuclear particle of importance, the neutron, has no charge. It weighs approximately the same as the proton—more precisely, 1.6747×10^{-27} kg.

The atom of any particular element has a relatively heavy nucleus made up of protons and neutrons with the lighter electrons circling the nucleus.

Table 1-1 ATOMIC PARTICLES

Particle	Atomic mass, u	Mass kg	Charge
electron	5.4876×10^{-4}	9.1083×10^{-31}	$-1e$
proton	1.0076	1.6724×10^{-27}	$+1e$
neutron	1.0090	1.6747×10^{-27}	0

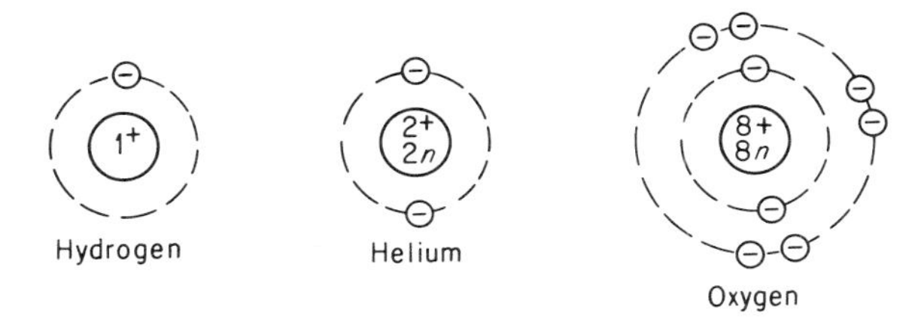

FIGURE 1-4
Various elements.

The nucleus has a net positive charge equal to the number of its protons $+q_e$. In the normal state, there is an equal number of electrons, negative charges $-q_e$, surrounding the nucleus. Since the net charge of both nucleus and the encircling electrons is zero, all elements are electrically neutral unless external forces remove electrons from their natural orbital shells or add electrons to them.

It is the total number of electrons, protons, and neutrons found in the atom of a given element that makes one element differ from another. Figure 1-4 shows this relationship for several elements.

The words "positive" and "negative" are relative terms, just like the words "higher" and "lower." When a particle is said to have a positive charge, we mean that it has the opposite charge to a particle with a negative charge. The assignment of negative polarity to the electron is purely arbitrary, but it is the accepted convention. The concept of negative and positive also holds for points or bodies in space. If we say that a point in space is positive, we mean that it is positive with respect to some other point in space. Conversely, a negative point in space can be negative only by comparison with some other point in space.

It is an observed physical fact that like charges repel each other and unlike charges attract each other. This fact was first demonstrated by rubbing a glass rod with a silk cloth. The charge thus induced could be transferred to a pith ball. If two pith balls are electrified in this manner and then brought into close proximity, they repel each other. If we rub a hard-rubber rod with a wool cloth, we can demonstrate the same force of repulsion as with the glass rod. However, if we take one pith ball charged with the glass rod and one charged with the hard-rubber rod and bring them into close proximity, we find that they are attracted to each other. The obvious conclusion is that the glass rod and the hard-rubber rod have taken on opposite charges. Further, it can be determined that the glass rod lost electrons and thus took on a positive charge. The hard-rubber rod picked up extra electrons and was thus negatively charged. The proton and electron exhibit an attractive force because they have unlike charges. The attractive force between electrons and the protons of the nucleus aids in holding the atom together.

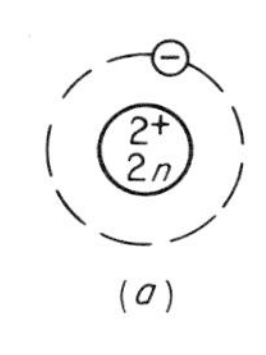

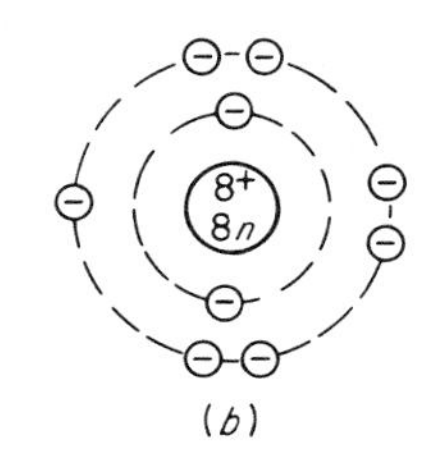

FIGURE 1-5
Ions: (*a*) positive, (*b*) negative.

1-7 IONIZATION

The graphic models of atoms in Fig. 1-4 indicate that there are distinct "shells" in which the electrons surrounding the nucleus travel. There are also physical relationships which determine the number of electrons required to complete a particular shell or subshell. Section 1-9 further discusses the relationship of the shells as energy levels; however, it is observable from our pith-ball experiment that as we move the pith balls closer together, the force of attraction or repulsion becomes greater. We can then assume that the force of attraction holding the electrons in orbit about the nucleus becomes less as they become more distant from the nucleus. Thus, the electrons in the outermost shell can most easily be removed from their orbit.

In some elements the electrons in the outermost shell are so loosely bound to the nucleus that they can move about from atom to atom at random. Electrons that engage in this random motion are called *free electrons.* Their random motion can be increased by the addition of small amounts of energy. For example, heating a material increases the random motion. Some types of material can be excited by light energy.

If an atom receives sufficient energy from an external source, it gives up an electron. This is basically what happened when we rubbed the glass rod with the silk cloth. The electrons in the glass rod were excited by the heat generated by the friction of rubbing with the silk cloth. An atom that has lost one of its electrons possesses a net positive charge because the number of protons now exceeds the number of electrons. With equal numbers of electrons and protons, an atom is electrically neutral. When it loses one or more electrons, an atom becomes a *positive ion.*

It is also possible for the reverse condition to occur. That is, an atom can have associated with it more electrons than there are protons in its nucleus. It then has a net negative charge and is a *negative ion.* An ionized atom behaves in accordance with the laws governing electric charges. Figure 1-5*a* shows a positive ion (helium) and Fig. 1-5*b* a negative ion (oxygen).

1-8 FORCE

Force is that which pushes, pulls, extends, compresses, or in any way acts to change the position or state of any mass. The force we are most commonly familiar with is the gravitational force which, when acting upon a mass, produces weight. An important relationship of force is that it has both magnitude and direction. The force of gravity pulls the mass in a direction with some magnitude. In the case of the force of gravity the letter **W** is used to symbolize the force, weight. For forces in general the symbol **F** is used. The boldface print symbolizes the fact that weight **W** and force **F** are vector quantities, that is, a quantity having both magnitude and direction (as opposed to a scalar quantity, or a quantity having magnitude only).

Force can further be defined as the product of mass and acceleration. Thus in equation form

$$\mathbf{F} = m \times a \tag{1-1}$$

In Eq. (1-1) mass is a fundamental quantity and acceleration is defined as the rate of change of velocity. Acceleration expressed in fundamental units reduces to l/t^2. Force can now be expressed in fundamental quantities:

$$\mathbf{F} = m \times \frac{l}{t^2} \tag{1-2}$$

We can now write the relationship defining the force of gravity, weight, as

$$\mathbf{W} = m \times g \tag{1-3}$$

In Eq. (1-3) g is the acceleration due to gravity and m is the mass of the body under the influence of the force of gravity.

An electric force is defined by Coulomb's law.

Coulomb's law The force of attraction or repulsion between two charges is inversely proportional to the square of the distance between the charges and directly proportional to the product of the charges.

This is further defined by the equation

$$\mathbf{F} = \frac{kQ_1Q_2}{d^2} \tag{1-4}$$

where

$\mathbf{F}$ = force of attraction or repulsion between two charged bodies
Q_1 = charge on body 1
Q_2 = charge on body 2
d = distance separating bodies
k = constant dependent upon medium separating charges and system of units being used

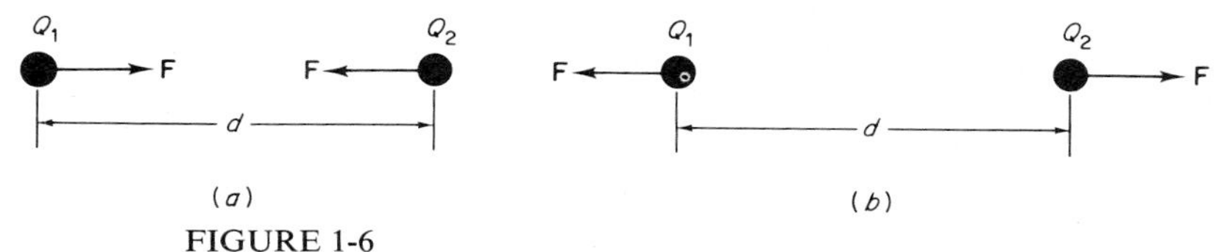

FIGURE 1-6
Two point charges separated by distance d: (a) opposite charges, (b) like charges.

Figure 1-6 shows the situation described above. We shall deal further with Coulomb's law in Chap. 2.

1-9 WORK, ENERGY, AND POWER

Work and energy are basically equivalent. Energy is the ability to do work, and both are equated to the work being done or the ability to do work. Energy and work are symbolized by W. It should be noted that, unlike force, work is a scalar quantity. Work and energy are described by the following equation

$$\begin{aligned} W &= \text{force} \times \text{distance} \\ &= F \times d \end{aligned} \tag{1-5}$$

When Eq. (1-5) is reduced to its fundamental dimensions, it yields

$$W = \frac{ml^2}{t^2}$$

Now that we have defined work, it becomes a simple matter to define power. Power is the rate of doing work, or, in equation form,

$$P = \frac{\text{work}}{\text{time}}$$

$$= \frac{\text{force} \times \text{length}}{\text{time}}$$

$$P = \frac{W}{t} \tag{1-6}$$

The dimensional descriptions for work and power are obtained as follows:

$$W = F \times d$$

Substituting the basic dimensions yields

$$W = \frac{m \times l}{t^2} \times l$$

$$= \frac{ml^2}{t^2}$$

$$P = \frac{W}{t}$$

Substituting the basic dimensions gives

$$P = \frac{ml^2/t^2}{t}$$

$$= \frac{ml^2}{t^3}$$

1-10 CURRENT

Current is defined as a directed movement of charge per unit of time. Current I is thus defined by

$$I = \frac{\text{charge}}{\text{time}}$$

Reduced to basic dimensions,

$$I = \frac{Q}{t}$$

From this definition of current it can be seen that a movement of either positive charges or negative charges in a common or net direction constitutes a current (or more descriptively an electric current).

The electron is the lightest of the charged particles we have discussed. It would be logical to expect this particle to be most easily forced into directed motion as an electric current. This is indeed the case. The direction of electron flow in any circuit is away from the most negative point in the circuit and toward the most positive point in the circuit. This type of current is referred to as electric current.

Even though a movement of electrons is the most common type of electric current, there are devices in which ions, usually positive ions, contribute to current. An example of simultaneous current flow of both types is a gas-filled tube used in industrial electronics. Here, electrons travel toward a plate inside the tube because the plate has been made positive with respect to a source of free electrons. A great

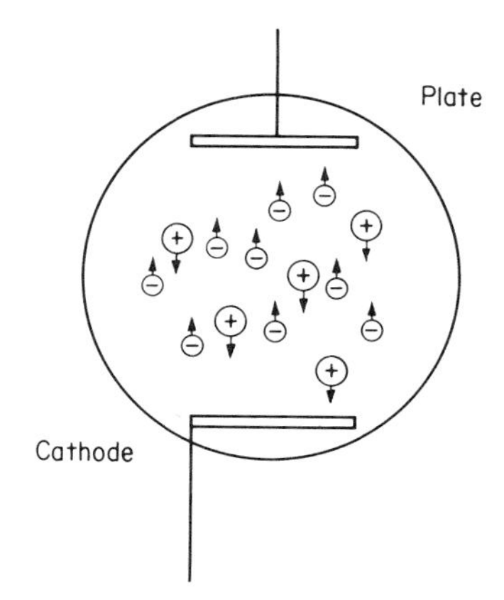

FIGURE 1-7
Charge movement in a gas-filled tube.

many of these electrons collide with gas molecules in the space between the plate and the source of the free electrons. The effect of these collisions is to tear other electrons loose from the gas molecules. The gas molecules become positive ions and travel toward the source of free electrons, since this is the most negative point in the tube. Figure 1-7 illustrates this action in a gas-filled tube. The source of free electrons is called the *cathode*. The *plate* is the electrode, which is positive with respect to the cathode.

1-11 CONDUCTORS, SEMICONDUCTORS, AND INSULATORS

In the discussion of atomic structure, reference was made to the movement of electrons in orbits around the nucleus of an atom. If a negative electric charge is applied to one side of a material that has large numbers of free electrons and a positive charge is applied to the opposite side, the free electrons will travel from atom to atom of the material. This drift of electrons is toward the side with positive charge and away from the side with negative charge. This type of matter is called a *conductor*, meaning that it permits relatively easy drift of free electrons. Not all conductors have the same degree of conductivity, or ability to allow free-electron drift. For one thing, different materials have larger or smaller quantities of free electrons. The degree of freedom of the free electrons also varies. Thus numerous materials may be classed as conductors even though the ability to conduct charge, or *conductivity*, of each is different.

Metals are the most common conductors. Their physical characteristics as well as their conductivity contribute to this choice. Even in metals there are different degrees of conductivity. For example, silver is a better conductor than copper; on the other hand, copper is a better conductor than aluminum. Table 1-2 lists some common conductors in order of decreasing conductivity, with silver used as a reference. Appendix Table A-5 lists more materials.

When the conductivity of a material is low, as in carbon or nichrome (a metal alloy), the material is referred to as a *resistor* rather than a conductor.

Some materials have practically no free electrons. Materials of this type prevent current from flowing when electric force is applied to them. They are called *insulators*. Materials used as insulators have electrons that cannot be easily moved from atom to atom, as they can in a conductor.

A measurable current can flow in an insulator only if the electric force applied to it is great enough to tear electrons loose from the atoms of the insulator. In short, no insulator is perfect. Insulators have varying degrees of insulating properties, just as conductors have varying degrees of conductivity. Insulators are rated in terms of the electric force necessary to break down the material and cause measurable conduction. The rating of the insulator's breakdown point is its *dielectric strength*. Table 1-3 lists some common insulating materials and their dielectric strengths.

Table 1-2

Conductor	Conductivity (compared with silver)
Silver	1.000
Copper	0.945
Aluminum	0.576
Tungsten	0.297
Iron	0.166
Nichrome	0.015
Carbon	0.0245–0.0089

Table 1-3

Material	Dielectric strength, V/mil of thickness
Air	20.6*
Pyrex glass	335.0
Mica	1,050.0*
Paper	305.0
Polystyrene	638.0*
Porcelain	150.0
Hard rubber	450.0

* Average.

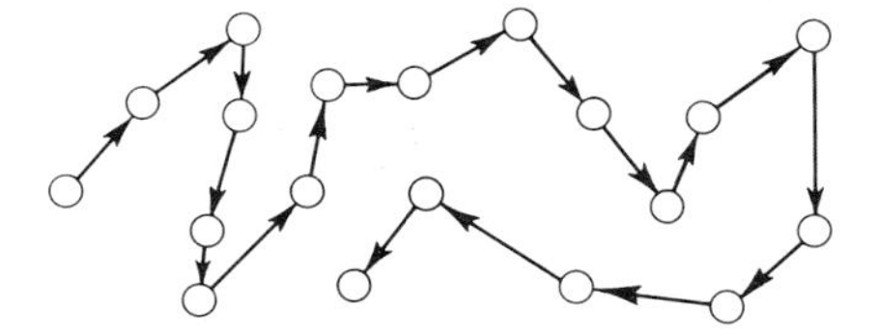

FIGURE 1-8
Random motion of electrons in a wire.

There is a class of material which falls between the conductor and insulator, called a *semiconductor*. Germanium, selenium, silicon, and titanium compounds are some examples of semiconductors. They are shown in the resistivity table (Table A-5). Semiconductors are the vital ingredient of transistors.

1-12 ELECTROMOTIVE FORCE

Current in a wire is a directed drift of electrons from one atom to another throughout the length of the wire. This does not mean that an electron starting at one end of the wire travels through the wire until it reaches the other end. What happens is that one electron is forced into an adjacent atom, and since like charges repel each other, the intruding electron forces another electron out of the atom. The dislodged electron forces itself into the next adjacent atom, dislodging another electron, and so on, until an electron at the other end of the wire is forced to leave its parent atom. This drift of electrons from atom to atom takes place with nearly the speed of light and is therefore considered to be instantaneous throughout a circuit. It should be pointed out that for simplicity we are using single electrons in our discussion while in the actual conductor tremendous numbers of electrons are engaged either in random movement in the conductor or directed motion, depending on the application of an external electric force.

The force that causes electrons in a wire to move in one general direction from atom to atom is a difference of charge. The negative side of a source of charge is connected at one end of the wire, and the positive source of charge is connected to the other end of the wire. Electron current flow in the wire is from the negative side of the source of charge, through the wire to the positive end of the source of charge, and through the charge source from the positive side to the negative side. If the source of charge remains connected to the wire, the current will be continuous. It is important to understand clearly that current is a movement of charges in one direction at any instant in time. Electrons are constantly in random motion within a conductor, but this is not a current (Fig. 1-8). Only when electrons travel from atom to atom in one general direction is there current in the conductor (Fig. 1-9).

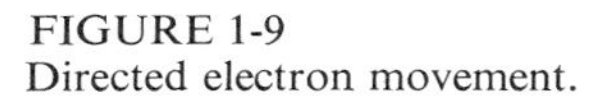

FIGURE 1-9
Directed electron movement.

Two conditions are necessary for electron flow (electric current): (1) a conductor permitting an easy drift of electrons from atom to atom and (2) a charge difference along the wire from one end to the other.

This difference of charge is called *voltage*. Voltage may be described as the electric pressure that causes current to flow. Voltage is also called *electric pressure*, *electric potential*, *difference of potential*, *voltage drop*, or *electromotive force* (emf).

Voltage exists between any two points in a circuit whenever the quantity of charge at one point differs from the quantity of charge at another point. In other words, voltage exists between two points when one point has a shortage of electrons (making it positive) while the other point has a surplus of electrons (making it negative).

1-13 SOURCES OF ELECTROMOTIVE FORCE OR VOLTAGE; BATTERIES

When two dissimilar elements such as carbon and zinc are immersed in a solution of sulfuric acid and water, the acid attacks the zinc more readily than the carbon and a potential difference appears between the two elements. This device is called a *voltaic cell*. A voltaic cell converts chemical energy into electric energy. The rods of zinc and carbon are called *electrodes*. The acid–water solution is called the *electrolyte*.

When the ends of a conductor are connected to the electrodes, electrons flow because of the potential difference that is developed chemically between the electrodes. In our example the zinc is the negative electrode, and the carbon is the positive electrode. The zinc and acid engage in a "burning" action that produces potential electric energy instead of heat.

The potential difference a given cell will develop depends on the materials making up the cell. The current that a cell can deliver depends on the opposition to the flow of current presented by the path through which the current must flow. Not only is there opposition to current in the wire connecting the two electrodes but also there is opposition to the movement of charges within the cell. The larger the area of the electrodes and the closer they are to each other (without touching), the smaller the opposition to charge movement within the cell.

When the current flows through a cell, the zinc gradually dissolves and the acid is neutralized. A chemical equation is sometimes used to illustrate the reaction that takes place. The symbols in the equation represent the different materials involved. The symbol for carbon is C, and the symbol for zinc is Zn. Consider the chemical

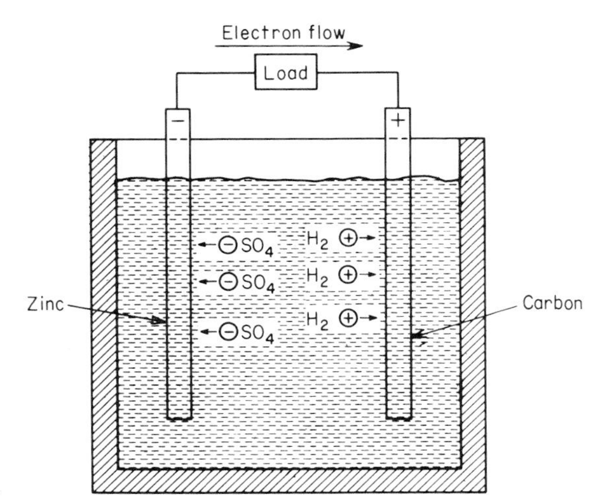

FIGURE 1-10
A wet-cell primary battery.

formula for water, H_2O, which states the chemical fact that one molecule of water is made up of two atoms of hydrogen and one atom of oxygen. Oxygen and hydrogen are both gases, but when combined they form water, a liquid. Sulfuric acid has the formula H_2SO_4, indicating that two atoms of hydrogen, one atom of sulfur, and four atoms of oxygen make up the molecule. In combining water, H_2O, and sulfuric acid, H_2SO_4, there is no chemical reaction. Therefore, we have a simple mixture.

When current flows through a cell having carbon and zinc electrodes immersed in a dilute solution of sulfuric acid and water, the reaction that takes place is expressed by the chemical equation

$$Zn + H_2SO_4 + H_2O \longrightarrow ZnSO_4 + H_2O + H_2 \tag{1-7}$$

This indicates that as current flows, an atom of zinc reacts with a molecule of sulfuric acid to yield a molecule of zinc sulfate, $ZnSO_4$, and a molecule of hydrogen, H_2. The zinc sulfate dissolves, and the hydrogen appears as gas bubbles around the carbon electrode. As current continues to flow, the zinc electrode is gradually consumed and the solution is changed to zinc sulfate and water. The carbon electrode does not enter into the chemical reaction but simply provides a return path for the current.

In the process of oxidizing the zinc, the solution breaks up into positive and negative ions that move in opposite directions through the solution. The positive ions are hydrogen ions that appear around the carbon (positive) electrode. They are attracted to it by the free electrons from the zinc that are returning to the cell by way of the external circuit and the positive carbon electrode. Negative SO_4 (sulfate) ions gather

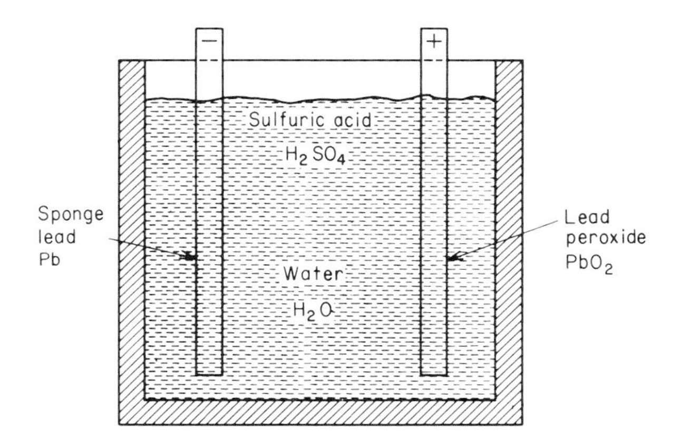

FIGURE 1-11
The lead-acid cell.

around the zinc electrode. Positive zinc ions enter the solution from the zinc electrode and combine with the negative sulfate ions to form zinc sulfate, which dissolves in the water. At the same time that the positive and negative ions are moving in opposite directions in the solution, electrons are moving through the external circuit from the negative zinc terminal back to the positive carbon terminal. When the zinc is used up, the potential difference of the cell drops to zero, and the battery is discharged. There is no way to reverse this process to recharge the battery. This type of cell is called a *primary cell.*

There are two types of primary cells, the wet cell, just discussed in detail, and the dry cell. Figure 1-10 illustrates a primary wet cell. In the dry cell, the electrolyte is in the form of a paste, and the container can be sealed so that there is no chance of spilling the electrolyte. A flashlight battery is a common example of a primary dry cell.

Another familiar type of battery is the automobile storage battery, which is a lead-acid type (Fig. 1-11). When this battery is fully charged, the active material of the positive plate is in the form of lead peroxide, PbO_2, and the negative plate is pure sponge lead. The lead peroxide has a lack of electrons while the sponge lead has an excess of electrons. The specific gravity of the electrolyte (ratio of its weight to the weight of an equal volume of water) is at its maximum.

If an external circuit is closed between the positive and negative terminals of the cell, electron flow begins because of the reaction of the electrolyte with the active material. Chemical energy is transformed into electric energy, and the cell is said to be *discharging*. The electrolyte reacts with the lead on the negative plate and the lead

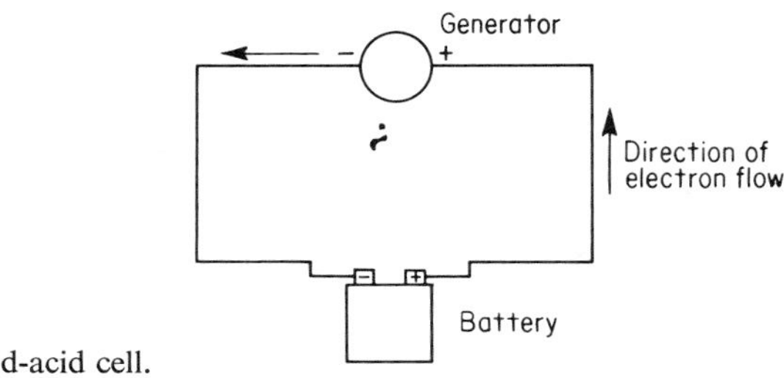

FIGURE 1-12
Charging the lead-acid cell.

peroxide on the positive plate to form lead sulfate on both plates. This reaction is expressed in the equation

$$Pb + PbO_2 + 2H_2SO_4 \underset{\text{charging}}{\overset{\text{discharging}}{\rightleftharpoons}} 2PbSO_4 + 2H_2O \tag{1-8}$$

The left side of Eq. (1-8) represents the cell in the charged state, and the right side shows the cell in a discharged condition. This type of cell is called a *secondary cell.*

When the battery is charged, the positive plate is lead peroxide, PbO_2, the negative plate is composed of sponge lead, Pb, and the solution is sulfuric acid, H_2SO_4, and water. In the discharged condition, both plates consist of lead sulfate, $PbSO_4$, and the solution is more dilute. As discharge progresses, the acidity of the electrolyte decreases because the sulfate ions are removed to form lead sulfate. The specific gravity of the electrolyte also decreases. A point is reached where so much of the active material has been converted into lead sulfate that the cell can no longer produce enough current to be of practical value. At this point the cell is said to be *discharged.* Since the removal of sulfate ions from the solution to the plates during discharge is directly proportional to the total current or electron flow delivered, the specific gravity of the electrolyte is a guide in determining the state of charge of the lead-acid cell.

If a discharged cell is properly connected to a source of voltage higher in electromotive force (emf) than the cell in question, current will flow through the cell in the direction opposite to the current flow when the cell is connected to a load. When the battery is connected to a source of voltage in this manner, the battery is said to be *charging.* The effect of the current is to change the lead sulfate on both the positive and negative plates back to its original active forms, lead peroxide and sponge lead, respectively. At the same time, the sulfate is restored to the electrolyte, with the result that the specific gravity of the electrolyte increases. When all the sulfate has been restored to the electrolyte, the specific gravity is maximum. The cell is then fully charged and is ready for service again. Figure 1-12 shows the discharged cell connected

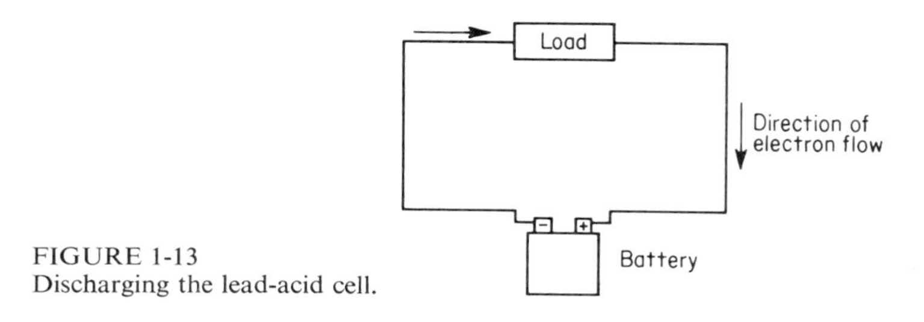

FIGURE 1-13
Discharging the lead-acid cell.

into a circuit to be recharged. Figure 1-13 shows the fully charged cell connected to a load and therefore discharging.

The most common source of voltage is the generator, a mechanical device that produces electric energy. Mechanical energy is used to rotate coils of wire within a magnetic field. The interaction between the coils and the magnetic field causes voltage to be developed at the output terminals of the generator. The voltage is thus the result of the conversion of mechanical energy into electric energy. Chapter 7 covers the theory in detail. In another type of generator, the coil is stationary and the magnetic field is rotated. Regardless of the method used, the interaction between the coils of wire and the magnetic field produces the output voltage.

Light energy can be used to develop a difference of potential, as in the photocell. Here light falls upon the photosensitive surface of the cell and causes electrons to leave one surface and accumulate on the other. This makes one side of the cell negative and the other side positive, resulting in a difference of potential. This phenomenon, the *photoelectric effect*, makes it possible to convert light energy into electric energy.

It is possible to convert mechanical energy into electric energy in several other ways. If we apply pressure to a thin crystal of quartz or rochelle salts, the mechanical stress in the crystalline structure develops a difference of potential across the surfaces of the crystal. One common application of this principle is the phonograph pickup. The needle in the pickup follows the grooves in the record and produces mechanical strains on the crystal. The mechanical strains on the crystal cause voltage to be developed at the terminals of the crystal. This voltage is proportional to the strain placed upon the crystal. The voltage at the output of the crystal is a representation of the sound recorded on the record (see Fig. 1-14).

This characteristic of certain crystals is called the *piezoelectric effect*, meaning that voltages are produced when mechanical stresses act upon the crystal, and conversely, motion by the crystal occurs when voltages are applied to the surfaces of the crystal.

When two dissimilar metals are connected so that small portions of their surfaces are in contact with each other, the device is capable of generating a potential

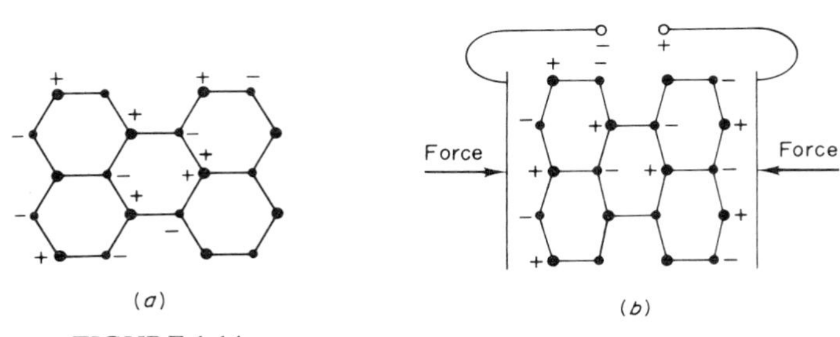

FIGURE 1-14
Crystalline structure producing voltage: (*a*) piezoelectric crystal, (*b*) crystal distorted under stress, producing voltage.

difference. This is called a *thermocouple*. The voltage developed by a thermocouple is a function of the heat at the junction of the metals. Further it is a function of the separation of the materials in the electromotive series (see Appendix Table A-6). Thermocouples have application as a means of measuring temperature and electric current.

QUESTIONS

1-1 Give an example of electronics applied to the medical profession.
1-2 Is sugar a compound or a mixture?
1-3 What is the important distinction between a compound and a mixture?
1-4 Can a mixture consist of compounds?
1-5 Must a molecule be made up of the atoms of different elements?
1-6 Define molecule.
1-7 What is meant by one atom of iron?
1-8 Define electron.
1-9 What are the two stable particles of the nucleus of an atom?
1-10 If the weight of a certain material is 3.1 lb, how many kilograms does it weigh?
1-11 How does the oxygen atom differ from the carbon atom?
1-12 The nucleus of an atom consists of 5 neutrons and 12 protons. Surrounding the nucleus are 11 electrons. Does this atom have charge, and if so, what is the polarity of the charge?

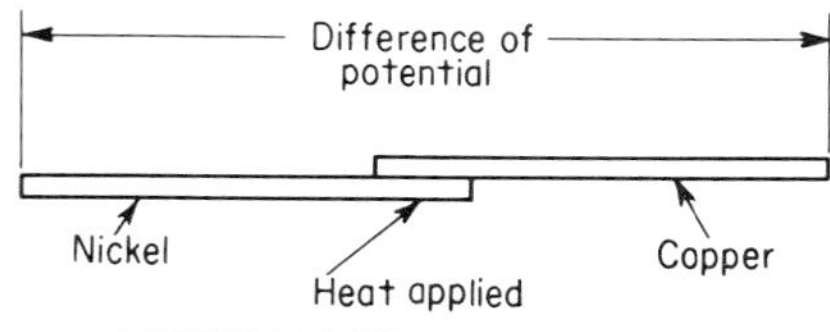

FIGURE 1-15
The thermocouple.

1-13 Describe the structure of a negative ion.

1-14 Define current.

1-15 What is meant by a conductor?

1-16 What characteristic of an element makes it a good conductor?

1-17 Explain what is meant by insulator.

1-18 Why is nickel not a good insulator?

1-19 Explain what is meant by electromotive force.

1-20 A battery converts ________ energy into electric energy.

1-21 The photo cell converts ________ energy into electric energy.

1-22 List two sources of emf that convert mechanical energy into electric energy.

1-23 Explain the difference between primary and secondary cells.

1-24 A thermocouple changes ________ energy into electric energy.

2

PHYSICAL AND ELECTRICAL UNITS

2-1 UNITS

In Chap. 1 we developed the concepts of the fundamental physical quantities length, time, mass, and charge. The fundamental dimensions are physical characteristics. Further, these fundamental physical characteristics can be used to define other physical quantities. Dimensions defined in terms of the fundamental quantities are called *secondary quantities.* Some examples of these secondary quantities were discussed in Chap. 1, namely, work, energy, power, current, voltage, and force.

The quantities we have defined, both fundamental and secondary, have magnitudes, and we need a standard, or reference, system to communicate the magnitudes of these quantities.

Units are simply a defined method whereby dimensions can be communicated. That is, by using units of measurement, a distance of, say, 2 ft can be measured in one place and this measurement can be accurately communicated to some other place without having to send a 2-ft string along to show the measured length. This is possible because the foot has been defined and accepted as a standard unit of length.

Just as the unit of length has been defined and standardized, so have the other quantities or dimensions. The basic unit of time is the second. Mass or matter is described in terms of the slug. Charge is defined by the coulomb.

Any unit may be too large or small to be convenient. For example, the distance from New York to Los Angeles would not be stated in feet because the number would be large and unwieldy. Therefore, another unit of length, the mile, is used. The mile, however, is still *defined* in terms of the fundamental unit of length, the foot. For similar reasons days are not ordinarily measured in seconds but in hours.

A system of units begins with basic units that describe the quantities of length, time, mass, and charge. The *English system* of units developed in a rather natural way based on handy physical objects used to convey the dimensional information. The foot, rod, and chain are examples of English units of length. The stone and slug are examples of weight and mass units. In the English system of measurement the basic unit of length is the foot, the unit of time the second, the unit of mass the slug, and the unit of charge the coulomb. These four units are the fundamental units of the English system. They can be divided into smaller units, such as 12 in. equal 1 ft. A smaller basic unit can be used to express a larger secondary unit; for example, 60 min makes 1 hr.

Combinations of these units can be used to describe other quantities. For example, the unit of a quantity of speed can be defined by feet per second or miles per hour. The unit of a quantity of density can be defined as pound mass per cubic foot. Units that are compounded of fundamental units are called secondary units.

Up to this point, the discussion has been confined to one system of fundamental units. There are two other sets of fundamental units, the mks and the cgs systems. The cgs system is the basic metric system, where the unit of length is the centimeter (cm), the unit of mass is the gram (g), the unit of time is the second (s), and the unit of charge is the coulomb (C). Much scientific work is done in the cgs system, primarily because the breakdown of multiples or submultiples of the fundamental units of length and mass is done decimally (by tens). For example, 1 meter equals 100 centimeters, and 1 kilogram equals 1,000 grams.

The mks system is simply a rationalization of the cgs system. The fundamental unit of length in the mks system is the meter, which is larger than the centimeter. This is also true of the kilogram, the fundamental unit of mass in the mks system. The units of time and charge are the same for all systems. The secondary units of the mks and cgs systems are formulated in the same manner as in the English system.

International physical standards of the fundamental units are maintained at the International Bureau of Standards in Sèvres, France. Replicas are maintained at the National Bureau of Standards, Washington, D.C.

2-2 LENGTH, OR DISTANCE

The *meter* is the unit of length in the mks system. The standard meter is the distance between two scribed lines on a platinum-iridium bar at atmospheric pressure and 0°C. The meter (m) was originally defined as one ten-millionth the distance from the equator to the pole at sea level. The international standards are the fundamental standards, and the standards maintained by our National Bureau of Standards are considered secondary to them.

The ratio of the yard to the standard meter is 3,600/3,937 by definition. From this defined ratio the conversion factors can be developed:

$$1 \text{ yd} = \frac{3{,}600}{3{,}937} \text{ m}$$

therefore,

$$\frac{3{,}937 \text{ yd}}{3{,}600 \text{ m}} = 1.094 \text{ yd/m}$$

or

$$\frac{3{,}600 \text{ m}}{3{,}937 \text{ yd}} = 0.914 \text{ m/yd}$$

thus with 36 in./yd

$$\frac{\text{Inches}}{\text{Meter}} = \frac{36 \text{ in./yd}}{0{\cdot}914\text{m/yd}} = 39.37 \text{ in./m}$$

or

$$\frac{1.00 \text{ m}}{39.37 \text{ in.}} = 0.0254 \text{ m/in.}$$

Since 12 in. = 1 ft by definition,

$$\frac{12 \text{ in.}}{1 \text{ ft}} 0.0254 \text{ m/in.} = 0.3048 \text{ m/ft}$$

From the above relationships two points are established: (1) The table of conversion factors can be mathematically derived, given the defined relationships between systems of units and multiples of units. Further, we can convert any number of units to an equivalent number of another unit. (2) The conversion is an algebraically supported conversion. Consider the following example.

EXAMPLE 2-1 Convert 47 m to (*a*) yards, (*b*) feet, and (*c*) inches.

SOLUTION

(*a*)

$$47 \text{ m} = 47 \text{ m} \times 1.092 \frac{\text{yd}}{\text{m}}$$

$$= 51.4 \text{ yd}$$

(*b*)
$$47 \text{ m} = 47 \text{ m} \times \frac{1}{0.3048 \text{ m/ft}}$$
$$= 47\text{m} \times \frac{1 \text{ ft}}{0.3048 \text{ m}}$$
$$= 154 \text{ ft}$$

(*c*)
$$47 \text{ m} = 47 \text{ m} \times 39.37 \frac{\text{in.}}{\text{m}}$$
$$= 1{,}850 \text{ in.}$$
////

Some discussion is in order concerning the problem of quantities with extremely large or small magnitude. For example, consider 745,000 m. We can express this large distance in powers of 10 as 7.45×10^5 m. We can also express it as 745 kilometers (km) or 0.745 megameters (Mm). The preceding example points out the practical advantage of the mks system. Conversion from one unit to another is primarily by decimal movement or powers of 10. Decimal prefixes for powers of 10 which are applied to the abbreviation for the unit are shown in Table 2-1. Thus 2 mm is read 2 millimeters, and 3 μm is read 3 micrometers.

The diameter of an atom is approximately 0.00000000001 m. For clarification and ease of expression we can convert the diameter of an atom as follows:

$$0.00000000001 \text{m} = 1 \times 10^{-11} \text{ m}$$

or
$$(1 \times 10^{-3})(10^{-8} \text{ m}) = 1 \times 10^{-8} \text{ mm}$$

or
$$(1 \times 10^{-6})(10^{-5} \text{ m}) = 1 \times 10^{-5} \ \mu\text{m}$$

or
$$(1 \times 10^{-9})(10^{-2} \text{ m}) = 1 \times 10^{-2} \text{ nm}$$
$$= 0.01 \text{ nm}$$

or
$$(1 \times 10^{-12})(10^{1} \text{ m}) = 1 \times 10^{1} \text{ pm}$$
$$= 10 \text{ pm}$$

Table 2-1 DECIMAL PREFIXES

Prefix	Abbreviation	Power of 10
pico	p	10^{-12}
nano	n	10^{-9}
micro	μ	10^{-6}
milli	m	10^{-3}
kilo	k	10^{3}
mega	M	10^{6}

2-3 TIME

Time as we defined in Chap. 1 is a continuum of events. Thus, our basic unit is the fractional part of a natural continuum. The mean solar second, the fundamental unit of time, is defined as 1/86,400 a mean solar day. The second is the fundamental unit of time in all three systems of units. We are all familiar with the larger secondary units, minutes, hours, and days. Appendix Table A-8 shows the units of time and the conversion factors. Consider the following example.

EXAMPLE 2-2 The speed of light is 186,000 mi/s. The diameter of an atom was given as approximately 1×10^{-11} m. How long does it take light energy to travel from one side of the atom to the other?

SOLUTION Converting 186,000 mi/s to meters per second gives

$$\begin{aligned} 186{,}000 \text{ mi/s} &= 186{,}000 \text{ mi/s} \times 1{,}609 \text{ m/mi} \\ &= (1.86 \times 10^5)(1.609 \times 10^3) \text{ m/s} \\ &= 2.99 \times 10^8 \text{ m/s} \end{aligned}$$

Recalling that speed × time = distance, we find

$$\begin{aligned} S \times t &= d \\ t &= \frac{d}{s} \\ &= \frac{1 \times 10^{-11} \text{ m}}{2.99 \times 10^8 \text{ m/s}} \\ &= 0.334 \times 10^{-19} \text{ s} \end{aligned}$$

////

2-4 MASS

The standard unit of mass in the mks system is the kilogram. In the English system the slug is the basic unit. The standard kilogram is maintained in the form of a platinum-iridium cylinder at the International Bureau of Standards.

EXAMPLE 2-3 The mass of a body is 475 kg. Convert this body's mass into (*a*) slugs and (*b*) grams (see Appendix Table A-8).

SOLUTION

(*a*)

$$\begin{aligned} 475 \text{ kg} &= 475 \text{ kg} \times 6.852 \times 10^{-2} \text{ sl/kg} \\ &= 475 \times 6.852 \times 10^{-2} \text{ sl} \\ &= 32.6 \text{ sl} \end{aligned}$$

(*b*)

$$\begin{aligned} 475 \text{ kg} &= 475 \text{ kg} \times 1{,}000 \text{ g/kg} \\ &= 475 \times 1{,}000 \text{ g} \\ &= 475{,}000 \text{ g} \\ &= 4.75 \times 10^5 \text{ g} \end{aligned}$$

////

Density is an important concept associated with mass. Density by definition is the mass per unit volume. The concept of density is best demonstrated by an example.

EXAMPLE 2-4 The density of water is 1.94 sl/ft^3. (*a*) What is the mass of 2.4 ft^3 of water? (*b*) What is the mass of 1 m^3 of water in slugs and kilograms?

SOLUTION

(*a*)

$$\text{Mass} = \text{density} \times \text{volume}$$

$$\begin{aligned} \text{Mass (sl)} &= \text{density (sl/ft}^3) \times \text{volume (ft}^3) \\ &= 1.94 \frac{\text{sl}}{\text{ft}^3} \times 2.4 \text{ ft}^3 \\ &= 4.65 \text{ sl} \end{aligned}$$

Notice that the units of the above equations are homogeneous. Just because an equation is homogeneous is no proof that it is correct, but if the equation is not homogeneous, it cannot be correct.

(*b*)

$$\text{Mass} = \text{density} \times \text{volume}$$

Convert cubic meters to cubic feet,

$$\begin{aligned} 1 \text{ m} &= 3.281 \text{ ft} \\ 1 \text{ m}^3 &= (3.281)^3 \text{ ft}^3 \\ &= 35.31 \text{ ft}^3 \\ \text{Mass} &= 1.94 \text{ sl/ft}^3 \times 35.31 \text{ ft}^3 \\ &= 68.5 \text{ sl} \end{aligned}$$

To convert to kilograms, from Table A-8, 1 kg = 6.852×10^{-2} sl

$$\begin{aligned} \text{Mass} &= (68.5 \text{ sl}) \left(\frac{1 \text{ kg}}{6.852 \times 10^{-2} \text{ sl}} \right) \\ &= 1{,}000 \text{ kg} \end{aligned}$$

In this example we have converted the density of water in slugs per cubic foot to kilograms per cubic meter. ////

2-5 CHARGE

The electric charge was defined in Chap. 1 as a property of the electron. However, the electron charge as a unit is extremely small, and the coulomb is a more practical unit. The coulomb (C) is defined as follows:

$$1\text{ C} = \frac{1}{1.602 \times 10^{-19}} \text{ electron charge}$$
$$= 0.624 \times 10^{19} \text{ electron charges}$$
$$= 6.24 \times 10^{18} \text{ electron charges}$$

EXAMPLE 2-5 Calculate the charge in coulombs on a sphere if the number of electron charges $-q_e$ is 3.756×10^{19}.

SOLUTION

$$Q(\text{C}) = \frac{-q_e \text{ electron charges}}{6.24 \times 10^{18} \text{ electron charges/C}}$$
$$= \frac{3.756 \times 10^{19}}{6.24 \times 10^{18}}$$
$$= 6.03 \text{ C}$$

////

Table A-8 lists the other units of charge and their conversion factors.

2-6 FORCE

The unit of force in the mks system is defined in Newton's second law:

> The acceleration of a body is directly proportional to the force causing the acceleration and inversely proportional to the mass of the body being accelerated. In equation form, Newton's second law becomes $a = F/m$.

Clearing the right side of the equation, we then have the equation for the force: $F = ma$. This equation stated with mks units becomes F (newtons) $= m$ (kilograms) $\times\ a$ (meters per second per second). In words, the unit of force in the mks system, the newton (N), is the force required to accelerate a one kilogram mass at the rate of one meter per second per second. This last unit is written m/s^2.

The weight of a body is by definition the force due to the gravitational interaction with the earth. If we express the acceleration of a body due to the gravitational field as g, we have

$$W = mg$$

where W = weight of force due to gravity, N
m = mass of body, kg
g = acceleration due to gravity, m/s^2

The pound is the weight of a one slug mass where the acceleration of gravity has the standard value $g = 32.174\ ft/s^2$.
The equation for Newton's second law can be applied:

$$W = mg$$

where W is in pounds, m is in slugs, and g is in feet per second per second.

EXAMPLE 2-6 Calculate the mass of a body in slugs and kilograms if 3.6 lb provides an acceleration of $128\ ft/s^2$.

SOLUTION

$$m = \frac{F}{a}$$
$$= \frac{3.6}{128} = \frac{3.6}{1.28 \times 10^2}$$
$$= 2.81 \times 10^{-2}\ \text{slug}$$

Convert slugs to kilograms according to Table A-8.

$$m(\text{kg}) = 2.81 \times 10^{-2}\ \text{slug} \times \frac{14.59\ \text{kg}}{1\ \text{slug}}$$
$$= (2.81 \times 10^{-2})(1.459 \times 10^1)\ \text{kg}$$
$$= 4.12 \times 10^{-1}\ \text{kg}$$
$$= 0.412\ \text{kg}$$

We can check our computations by converting the English units of the original problem to mks units and then apply Newton's second law.

$$F\,(\mathrm{N}) = 3.6\ \mathrm{lb} \times \frac{4.448\ \mathrm{N}}{\mathrm{lb}}$$

$$= 16.01\ \mathrm{N}$$

$$a\,(\mathrm{m/s^2}) = \frac{128\ \mathrm{ft}}{\mathrm{s}^2} \times \frac{0.3048\ \mathrm{m}}{\mathrm{ft}}$$

$$= 39.0\ \mathrm{m/s^2}$$

Applying Newton's second law gives

$$m = \frac{(F\ \mathrm{N})}{a\,(\mathrm{m/s^2})}$$

$$= \frac{16.01}{39} = 0.412\ \mathrm{kg} \qquad ////$$

EXAMPLE 2-7 A body having a mass of 4.32×10^{-3} kg is subjected to a force of 3.27 N. Calculate the acceleration in meters per second per second and in feet per second per second.

SOLUTION

$$F = ma$$

$$a = \frac{F}{m}$$

$$= \frac{3.27\ \mathrm{N}}{4.32 \times 10^{-3}\ \mathrm{kg}}$$

$$= 0.756 \times 10^3\ \mathrm{m/s^2}$$

Converting $\mathrm{m/s^2}$ to $\mathrm{ft/s^2}$,

$$a\,(\mathrm{ft/s^2}) = 7.56 \times 10^2\ \mathrm{m/s^2} \times 3.281\ \mathrm{ft/m}$$
$$= 24.8 \times 10^2\ \mathrm{ft/s^2} \qquad ////$$

With the previous discussion of force we are ready to examine Coulomb's law in a more quantitative way. In equation form Coulomb's law states

$$F = k\,\frac{Q_1 Q_2}{d^2}$$

We can now define the proportionally constant for the mks system as

$$k = 8.9878 \times 10^9\ \frac{\mathrm{N \cdot m^2}}{\mathrm{C^2}}\ \text{(in free space)}$$

where

$$F = \text{force, N}$$
$$Q_1, Q_2, = \text{charge, C}$$
$$d = \text{distance between } Q_1 \text{ and } Q_2, \text{ m}$$

EXAMPLE 2-8 Two charges are separated by 6.85 m. Q_1 is 1.75 C, and Q_2 is -12.2×10^{17} electron charges. Calculate the force resulting. Is this a force of attraction or repulsion?

SOLUTION Convert Q_2 to coulombs.

$$Q_2 = \frac{-12.2 \times 10^{17} \text{ electron charges}}{6.24 \times 10^{18} \text{ electron charges/C}}$$
$$= -1.957 \times 10^{-1} = -0.1957 \text{ C}$$

Apply Coulomb's law.

$$F = k\,\frac{Q_1 Q_2}{d^2}$$
$$= \frac{(8.9878 \times 10^9)(1.75)(-1.957 \times 10^{-1})}{(6.85)^2}$$
$$= -6.55 \times 10^7 \text{ N}$$

The negative sign indicates the force is attractive. Note that so long as the signs of the charges are different, the force will be attractive. If both charges are alike, the force will be repulsive. ////

EXAMPLE 2-9 Calculate the constant k for the English system using the quantities of Example 2-8.

SOLUTION Since the units of charge are the same in both systems,

$$Q_1 = 1.75 \text{ C}$$
$$Q_2 = -0.1957 \text{ C}$$
$$d = 6.85 \text{ m} \times 3.281 \text{ ft/m}$$
$$= 22.45 \text{ ft}$$
$$F = (-6.55 \times 10^7 \text{ N})(.2248 \text{ lb/N})$$
$$= -1.472 \times 10^7 \text{ lb}$$

Coulomb's law gives

$$F = k\,\frac{Q_1 Q_2}{d^2}$$

Clearing the right side of the equation, we have

$$\begin{aligned} k &= \frac{F d^2}{Q_1 Q_2} \\ &= \frac{(-1.472 \times 10^7)(22.45)^2}{(1.75)(-.1957)} \\ &= 2.17 \times 10^{10}\ \text{lb} \cdot \text{ft}^2/\text{C}^2 \end{aligned}$$

////

2-7 WORK, ENERGY, AND POWER

As defined in Chap. 1, work and energy are described by the equation

$$\begin{aligned} W &= \text{force} \times \text{distance} \\ &= F \times d \end{aligned}$$

In the mks system the work equation shows

$$W = F\,(\text{N}) \times d\,(\text{m})$$

Thus the unit of work or energy is the newton-meter (N · m), also called the joule (J). The work done in displacing a body one meter by a constant force of one newton is one joule. In the English system the force in pounds and the distance in feet yields the work in foot-pounds.

Some clarification should be made at this point with regard to the equivalency of work and energy. Work and energy have the same units. They are defined by the same equational relationships. However, energy is a measure of the capacity or ability of a body to perform work.

EXAMPLE 2-10 A force of 6.28 N is applied to a body. If the body is displaced 4.7 ft, how much work is done in (*a*) joules and (*b*) foot-pounds?

SOLUTION

(*a*) Convert feet to meters.

$$\begin{aligned} 4.7\ \text{ft} &= 4.7\ \text{ft} \times 0.3048\ \text{m/ft} \\ &= 1.432\ \text{m} \\ W &= (6.28\ \text{N})(1.432\ \text{m}) \\ &= 9.00\ \text{N} \cdot \text{m} \\ &= 9.00\ \text{J} \end{aligned}$$

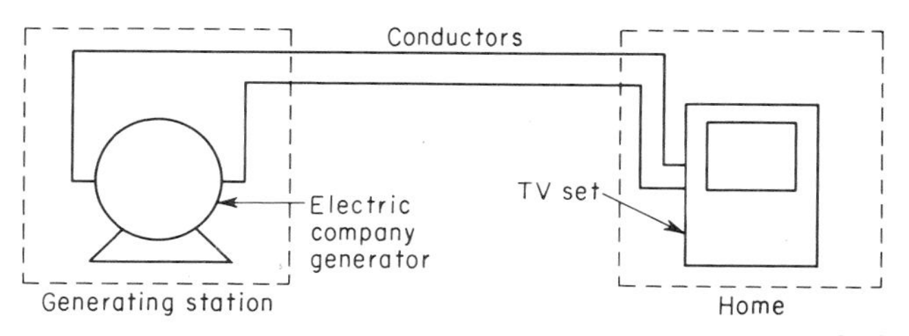

FIGURE 2-1
An electrical system.

(*b*) We can convert the answer of (*a*) directly by the conversion factor of Table A-8.

$$\begin{aligned} W &= 9.00 \text{ J} \times 0.7376 \text{ ft} \cdot \text{lb/J} \\ &= 6.64 \text{ ft} \cdot \text{lb} \end{aligned}$$

We can also convert newtons to pounds force and then apply the work equation.

$$\begin{aligned} F &= 6.28 \text{ N} \times 0.2248 \text{ lb/N} \\ &= 1.41 \text{ lb} \\ W &= F \times d \\ &= 1.41 \text{ lb} \times 4.7 \text{ ft} \\ &= 6.64 \text{ ft} \cdot \text{lb} \end{aligned}$$

////

Power is the rate of doing work and is defined by

$$P = \frac{\text{work}}{\text{time}} = \frac{\text{J}}{\text{s}} = \text{watts}$$

Thus work done at the rate of a joule per second is a watt (W).

2-8 THE ELECTRIC CIRCUIT

The purpose of an electric circuit is to deliver energy in the form of electricity from a source of electric energy to a load. The function of the load is to convert electric energy into some other form, such as heat, light, sound, or mechanical energy. Numerous examples of electric circuits can be found in daily life: electric appliances, the battery that turns over the engine of a car, radio, television, neon signs, lights, and motors, to mention a few. Electric energy supplied by a power company's generating station is distributed over wires to homes, where various loads are connected to the circuit, and the electric energy is converted to other forms of energy, thereby performing useful work. See Fig. 2-1.

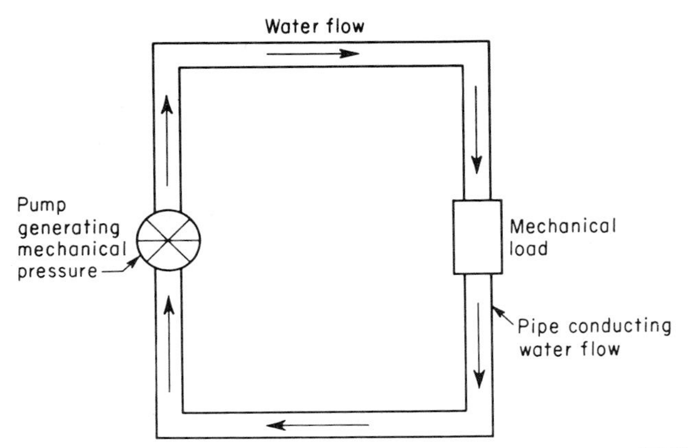

FIGURE 2-2
The hydraulic circuit.

The electric circuit may be compared to mechanical and hydraulic systems that deliver energy from a source of energy to loads (Fig. 2-2). If water is made to move under pressure developed by a pump, it can be used to deliver energy to a load and do useful work. The amount of work that can be performed by the water depends on the amount of flow of the water and the water pressure. In other words, the number of gallons of water flow will determine the amount of work done by the moving water.

What factors determine the amount of flow of the water? The pressure applied by the pump is a prime factor in determining the water flow. If the pump pressure is increased, the flow of the water will increase. If the pressure of the pump is decreased, flow is decreased. It can then be said that the flow is directly proportional to the pressure.

Another important factor governing the flow of water is the friction offered to the water flow by the pipe and by the load itself. If friction is increased by making the pipe smaller in diameter, by increasing its length, or by using a material having greater surface friction, then the water flow decreases. If, on the other hand, the pipe is increased in diameter, shortened, or replaced with a pipe of material having less surface friction, the water flow increases. From this discussion it can be seen that the flow is inversely proportional to the friction of the system.

The relationship between flow, pressure, and friction is expressed by the equation

$$\text{Flow} = \frac{\text{pressure}}{\text{friction}} \tag{2-1}$$

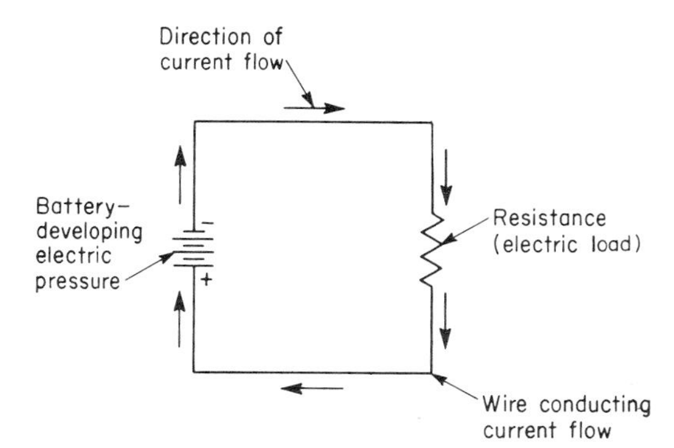

FIGURE 2-3
The electric circuit.

In an electric circuit, the pressure is the voltage, and the flow is the current. The property of an electric circuit that may be compared to friction is *resistance*. See Fig. 2-3.

2-9 RESISTANCE

Resistance is the property of an electric circuit that opposes movement of charge and dissipates energy. In mechanical systems friction causes energy losses and reduces speed. In electric circuits, resistance dissipates electric energy in the form of heat and reduces the current.

The unit of resistance is the *ohm* (Ω) named for George Simon Ohm, the German physicist who propounded the formula known as Ohm's law in 1826. Resistance is designated by the letter R.

The standard international ohm is defined as the resistance at zero degrees Celsius of a column of mercury of uniform cross section having a length of 106.3 cm and a mass of 14.4521 g. The precise specifications used in the definition of the standard ohm make it a reproducible standard. The factors that affect resistance are similar to those which determine friction. The longer a conductor, the greater its resistance; the larger the cross-sectional area of a conductor, the smaller its resistance.

If we reason intuitively and take into consideration the free electrons present in a material, the following factors become important. The larger the cross section of a material the greater the number of free electrons subject to movement. In mathematical form we say that the resistance is inversely proportional to the cross-sectional area ($R \propto 1/A$). As the length of the conductor becomes longer, the opposition to current flow becomes greater. This is reasonable when one considers the greater number of free electrons that must be moved to accomplish electron flow. In mathematical form the resistance is directly proportional to the length ($R \propto l$).

It should be noted that in the definition for the standard ohm the cross-sectional area of the column of mercury is specified in terms of the mass of mercury in a column 106.3 cm high.

The resistance of any material is a property of that material, just as the melting point, density, and mass are properties of the material.

2-10 RESISTIVITY

Resistance is a function of length l, cross-sectional area A, and a constant ρ:

$$R = \frac{\rho l}{A} \tag{2-2}$$

The Greek letter ρ (rho) is the *resistivity constant* for a material. In other words, ρ has a different value for each material. The resistivity constant of a material is its resistance times its unit cross-sectional area per unit length. One way of stating the resistivity of a material is $6\Omega \cdot \text{in.}^2/\text{ft}$.

Since most conductors are circular in cross section, resistivity usually is given in terms of circular mils (cmil) of area. The mil is a unit of length equal to 0.001 in. For a material of circular cross section, the cross-sectional area is found by the equation

$$A = \pi r^2$$

where r = radius, units of length
π = the constant 3.1416
A = area, units of length squared

EXAMPLE 2-11 For a circle 4 in. in radius, the area would be

$$A = \pi r^2 = (3.1416)(4^2) = 50.2 \text{ in.}^2 \qquad ////$$

The area of a circular cross section in square mils is determined by using the radius in mils.

EXAMPLE 2-12 The area of a circle 4 mils in radius would be

$$A = \pi r^2 = (3.1416)(4^2) = 50.2 \text{ sq mils} \qquad ////$$

Up to this point two examples have been given for determining the cross-sectional area of a conductor. The first area was in square inches and the second in square mils. There is another way of specifying the cross-sectional area of a circular conductor that is more convenient. This system uses the *circular mil* (cmil) as the unit of cross-sectional area. The area of a conductor in circular mils is found by simply squaring the diameter of the conductor.

EXAMPLE 2-13 For a circular cross section 3 mils in radius find the area in circular mils.

SOLUTION

$$\text{Diameter } d = 2 \times \text{radius} = (2)(3) = 6 \text{ mils}$$
$$\text{Area (cmils)} = d^2 = 6^2 = 36 \text{ cmils}$$

////

It is a simple matter to convert from circular-mil measurement to square-mil measurement when the area in square mils is needed. This is shown in Example 2-14.

EXAMPLE 2-14

$$\text{Area (sq mils)} = \pi r^2$$

but

$$r = \frac{d}{2}$$

Therefore,

$$r^2 = \frac{d^2}{4}$$

and

$$\text{Area (sq mils)} = \pi \frac{d^2}{4} = \frac{\pi}{4} d^2 = 0.7854 d^2$$

Area in circular mils $= d^2$. Therefore the area in square mils is $\pi/4$ times the area in circular mils. ////

To convert from square mils to circular mils use the equation

$$\text{Area (cmils)} = \frac{\text{Area (sq mils)}}{0.7854}$$

The following example demonstrates the use of these conversion factors.

EXAMPLE 2-15 What is the cross-sectional area of a conductor with a radius of 0.0055 in. (1) in circular mils and (2) in square mils?

SOLUTION

1

$$r(\text{ mils}) = 5.5$$
$$d\,(\text{mils}) = 11$$
$$\text{Area (cmils)} = d\,(\text{mils})^2 = 121 \text{ cmils}$$

2 To convert the area in circular mils to square mils,

$$\text{Area (sq mils)} = 0.7854 \times A \text{ (cmils)} = (0.7854)(121) = 95 \text{ sq mils}$$

To check the conversion,

$$\text{Area (sq mils)} = \pi r^2 = (3.1416)(5.5)^2 = 95 \text{ sq mils} \qquad ////$$

When the circular mil is used as the unit of cross-sectional area, the resistivity ρ of a material is the number of ohms in a piece of the material with a cross-sectional area of 1 cir mil and a length of 1 ft.

In Appendix Table A-5, annealed copper is shown to have a resistivity of 10.37 $\Omega \cdot$ cmil/ft. From this, the resistance of a copper conductor of any dimensions can be found.

EXAMPLE 2-16 Find the resistance of a copper wire 3 ft long and 2 mils in diameter.

SOLUTION

$$R = \frac{\rho l}{A} = \frac{(10.37)(3)}{2^2} = 7.77\ \Omega$$

where A = area, cmils
l = length, ft
ρ = resistivity, $\Omega \cdot$cmil/ft ////

EXAMPLE 2-17 Find the resistance of a piece of commercial iron wire with a radius of 0.009 in. and a length of 1.7 ft.

SOLUTION

$$d = 0.018 \text{ in.} = 18 \text{ mils}$$

$$R = \frac{\rho l}{A} = \frac{(75)(1.7)}{18^2} = 0.393\ \Omega \qquad ////$$

With Eq. (2-2) it is possible to determine unknown values of ρ, l, or A for a conductor.

EXAMPLE 2-18 A conductor of an unknown material is 4.24 ft long and 6 mils in diameter. The resistance of the conductor is 2 Ω. Find its resistivity.

SOLUTION

1

$$R = \frac{\rho l}{A}$$

2 Multiply both sides of the equation by A/l.

$$\frac{A}{l} R = \frac{A}{l} \frac{\rho l}{A}$$

$$\frac{AR}{l} = \rho$$

$$\rho = \frac{(36)(2)}{4.24} = 17\ \Omega \cdot \text{cmil/ft} \qquad ////$$

The value of resistivity obtained would indicate that the conductor is made of aluminum. Since the resistivity of a material is a definite physical property of the material, it can be used to identify the material (see Table A-5).

EXAMPLE 2-19 An annealed copper conductor 35 ft long has a resistance of 9 Ω. What is the diameter of the conductor in inches?

SOLUTION

1 From Eq. (2-2), $R = \rho l/A$.
2 Multiply both sides of the equation by A/R.

$$\frac{AR}{R} = \frac{\rho l A}{AR}$$

$$A = \frac{\rho l}{R} = \frac{(10.37)(35)}{9} = 40.2 \text{ cmils}$$

3

$$A \text{ (cmils)} = d^2$$

$$d^2 = 40.2 \text{ cmils}$$

$$d = \sqrt{40.2} = 6.34 \text{ mils}$$

$$d \text{ (in.)} = 0.001 \times d \text{ (mils)} = 0.00634 \text{ in.}$$

EXAMPLE 2-20 Given a tungsten conductor 0.002 in. in diameter with a resistance of 1.65 Ω, find its length.

SOLUTION

1

$$R = \frac{\rho l}{A}$$

2 Multiply both sides of the equation by A/ρ.

$$\frac{AR}{\rho} = \frac{A}{\rho}\frac{\rho l}{A}$$

$$\frac{AR}{\rho} = l$$

$$l = \frac{(4)(1.65)}{33}$$

$$= 0.2 \text{ ft}$$

////

In the final analysis, the resistance offered by a material is inversely proportional to the quantity of free electrons per unit volume. As noted in Chap. 1, addition of energy to the atom increases the energy level of the atom in some fashion. An increase in temperature represents an increase of heat energy. This increased energy may result in an increase in the energy level of the free electrons. The greater the energy level of the free electrons the more energy will be required to obtain directed movement. Thus, most materials increase in resistivity as the temperature is increased.

2-11 TEMPERATURE COEFFICIENT OF RESISTANCE

It must be noted that in the definition of the standard ohm the temperature was specified to be 0°C. In the table of resistivities the temperature is specified to be 20°C. These temperatures are clearly specified because resistance of a material changes with changes in temperature of the material. This property of a material varying in resistance with variation of temperature is defined by the *temperature coefficient* of the material. The temperature coefficient of resistance is the resistance change per degree Celsius per ohm at the standard temperature. The symbol for temperature coefficient is the Greek letter α. Equation (2-3) shows how resistance varies with change in temperature.

$$\Delta R = \alpha_s(R_s)(T_1 - T_s) \tag{2-3}$$

where T_s = standard temperature, usually 0°C
α_s = temperature coefficient of resistance at standard temperature
R_s = resistance of material at standard temperature
T_1 = any other temperature for which it is desired to know resistance of material, °C

The Greek letter Δ stands for a small change. ΔR is then a change in resistance. Change in temperature would be ΔT.

Table 2-1 gives a number of temperature coefficients of resistance.

EXAMPLE 2-21 An annealed copper conductor has a resistance of 6 Ω at 0°C. What is the resistance of the conductor at 35°C?

SOLUTION

1 By Eq. (2-3),

$$\Delta R = \alpha_s R_s(T_1 - T_s) = (0.00426)(6)(35 - 0) = 0.895\ \Omega$$

2 The resistance at 35°C is the sum of ΔR and R_s.

$$R_1 = R_s + \Delta R \tag{2-4}$$
$$= 6.0 + 0.895 = 6.895\ \Omega$$

////

Table 2-2 RESISTIVITY AND TEMPERATURE COEFFICIENT OF RESISTANCE

Material	Resistivity Ω · cmil/ft at 20°C	Temperature coefficient per °C at 0°C
Aluminum	17.0	0.00420
Antimony	251.0	0.00388
Brass, annealed	42.0	0.00208
Copper:		
Annealed	10.37	0.00426
Hard-drawn	10.7	0.00413
Pure	10.2	0.00410
German silver	199.0	0.00036
Gold	16.7	0.00365
Iron, commercial	66.4–81.4	0.00618
Lead	132.5	0.00466
Mercury	577.0	0.00088
Nickel	47.0	0.006
Platinum-iridium	148.0	0.0012
Platinum	60.2	0.0037
Silver	9.89–11.2	0.00411
Tungsten	33.2	0.0049
Zinc	37.4	0.0040
Nichrome	600.0	0.00044

Example 2-21 indicates that the resistance of a material at any temperature may be found if its resistance is known at standard temperature. Equations (2-3) and (2-4) indicate the required calculations. If Eq. (2-3) is substituted into Eq. (2-4), Eq. (2-5) results.

$$R_1 = R_s + \alpha_s(R_s)(T_1 - T_s) \tag{2-5}$$

Writing Eq. (2-5) in terms of another temperature T_2 yields

$$R_2 = R_s + \alpha_s(R_s)(T_2 - T_s) \tag{2-6}$$

where R_2 is the resistance at T_2.

Since we are using the same material in Eqs. (2-5) and (2-6), R_s is the same in both equations and α_s is the same in both. Dividing Eq. (2-5) by Eq. (2-6) results in another useful equation.

$$\frac{R_1}{R_2} = \frac{R_s + \alpha_s(R_s)(T_1 - T_s)}{R_s + \alpha_s(R_s)(T_2 - T_s)}$$

$$= \frac{R_s[1 + \alpha_s(T_1 - T_s)]}{R_s[1 + \alpha_s(T_2 - T_s)]}$$

$$R_2 \frac{R_1}{R_2} = \frac{1 + \alpha_s(T_1 - T_s)}{1 + \alpha_s(T_2 - T_s)} R_2$$

$$R_1 = \frac{1 + \alpha_s(T_1 - T_s)}{1 + \alpha_s(T_2 - T_s)} R_2 \tag{2-7}$$

If T_s is 0°C, Eq. (2-7) becomes

$$R_1 = \frac{1 + \alpha_s T_1}{1 + \alpha_s T_2} R_2 \tag{2-8}$$

Equation (2-8) makes it possible to find the resistance of a material at any temperature if its resistance at a certain temperature is known.

EXAMPLE 2-22 An aluminum conductor has a resistance of 20 Ω at 15°C. Find its resistance at (1) 65°C, (2) 0°C, (3) −120°C.

SOLUTION

1 Use Eq. (2-8).

$$R_1 = \frac{1 + \alpha_s t_1}{1 + \alpha_s t_2} R_2 = \frac{1 + (0.0042)(65)}{1 + (0.0042)(15)} \times 20 = 23.9\ \Omega$$

2

$$R_1 = \frac{1 + \alpha_s t_1}{1 + \alpha_s t_2} R_2 = \frac{1 + (0.0042)(0)}{1 + (0.0042)(15)} \times 20 = 18.8\ \Omega$$

3 $$R_1 = \frac{1 + \alpha_s t_1}{1 + \alpha_s t_2} R_2 = \frac{1 + (0.0042)(-120)}{1 + (0.0042)(15)} \times 20 = 9.32\ \Omega \qquad ////$$

Example 2-22 illustrates the use of Eq. (2-8) to find resistance at any temperature when the resistance is known at any other temperature. The temperature for the unknown resistance may be above or below the temperature at the known resistance. In Example 2-21, the resistance changed in the same direction as the temperature. When a material's resistance changes in the same direction as the temperature change, the material is said to have a *positive* temperature coefficient of resistance. Pure metals have positive temperature coefficients of resistance and are relatively close in value, as shown in Table 2-1. Metal alloys can be made that have very small temperature coefficients of resistance, in some cases almost zero.

If the resistance of a material decreases as temperature rises, the material is said to have a *negative* temperature coefficient of resistance. Carbon, porcelain, and glass have negative temperature coefficients. Metal alloys can be compounded that have negative temperature coefficients of resistance within certain temperature ranges.

It is possible to determine the temperature of a conductor if we know its resistance at a given temperature and at the unknown temperature.

EXAMPLE 2-23 A piece of nickel has a resistance of 95 Ω at 10°C. The temperature changes, and the resistance is found to be 98 Ω. What is the new temperature?

SOLUTION

1 Start with Eq. (2-8).

$$R_1 = \frac{1 + \alpha_s T_1}{1 + \alpha_s T_2} R_2$$

2 Multiply both sides by $(1 + \alpha_s T_2)/R_2$.

$$\frac{1 + \alpha_s T_2}{R_2} R_1 = \frac{1 + \alpha_s T_2}{R_2} R_2 \frac{1 + \alpha_s T_1}{1 + \alpha_s T_2}$$

3 Subtract 1 from both sides of the equation.

$$-1 + \frac{R_1}{R_2}(1 + \alpha_s T_2) = 1 + \alpha_s T_1 - 1$$

$$\frac{R_1}{R_2}(1 + \alpha_s T_2) - 1 = \alpha_s T_1$$

4 Divide both sides by α_s.

$$\frac{R_1(1 + \alpha_s T_2)}{R_2 \alpha_s} - \frac{1}{\alpha_s} = T_1 \tag{2-9}$$

Let T_1 = unknown temperature, °C
$T_2 = 10°C$
$R_1 = 98\ \Omega$
$R_2 = 95\ \Omega$
$\alpha_s = 0.006$ (nickel)

5 Then

$$T_1 = \frac{98(1 + 0.006 \times 10)}{95 \times 0.006} - \frac{1}{0.006} = 15.9°C \qquad ////$$

A practical application of resistance change with temperature is a thermometer based upon this phenomenon.

By using Eq. (2-8), it is also possible to calculate the temperature coefficient of resistance for an unknown material if its resistance is known at two temperatures.

EXAMPLE 2-24 An unknown material has a resistance of 175 Ω at 60°C and a resistance of 200 Ω at 90°C. Find its temperature coefficient of resistance.

SOLUTION

1 Start with Eq. (2-8).

$$R_1 = \frac{1 + \alpha_s T_1}{1 + \alpha_s T_2} R_2$$

2 Multiply both sides by $1 + \alpha_s T_2$.

$$R_1 + R_1\alpha_s T_2 = R_2 + R_2 \alpha_s T_1$$

3 Collect terms containing α_s.

$$\alpha_s(R_1 T_2 - R_2 T_1) = R_2 - R_1$$

4 Solve for α_s.

$$\alpha_s = \frac{R_2 - R_1}{R_1 T_2 - R_2 T_1}$$

5 Let

$$R_1 = 175\ \Omega$$
$$T_1 = 60°\text{C}$$
$$R_2 = 200\ \Omega$$
$$T_2 = 90°\text{C}$$

Then

$$\alpha_s = \frac{200 - 175}{(1.75 \times 10^2)(9.0 \times 10^1) - (2.0 \times 10^2)(6.0 \times 10^1)}$$

$$= \frac{25}{3.75 \times 10^3} = 6.67 \times 10^{-3}\ °\text{C} \qquad ////$$

Determination of α_s, as in Example 2.23, indicates that an unknown material can be identified by its temperature coefficient of resistance.

2-12 WIRE GAUGES

The diameters of wires used for electric conductors are standardized by the American Wire Gauge (AWG) system. Appendix Table A-7 gives the gauge number and the corresponding diameter in mils. The AWG system is in general use in the United States, but there are other wire-size standards. The BWG (Birmingham Wire Gauge) system is in general use in Great Britain.

2-13 CONDUCTANCE

Conductance is defined as the reciprocal of resistance. The letter G (or g) denotes conductance. The reciprocal relationship between R and G is shown in Eqs. (2-10) and (2-11).

$$G = \frac{1}{R} \tag{2-10}$$

$$R = \frac{1}{G} \tag{2-11}$$

The unit of conductance is the *mho* (ohm spelled backward). The primary reason for having the quantity conductance is that in many types of circuit problems conductance is more convenient to work with than resistance. As resistance is indicative of opposition to current in a circuit, conductance is an indication of the ease or lack of opposition in the circuit. As conductance increases, current increases for a given voltage. Conversely, when conductance decreases, current decreases, provided the applied voltage does not change. It can then be stated that current is directly proportional to conductance. To find conductance when resistance is known, Eq. (2-10) is used.

EXAMPLE 2-25 Find the conductance of the wire used in Example 2-16.

$$G = \frac{1}{R} = \frac{1}{7.77} = 0.1288 \text{ mho} \qquad ////$$

EXAMPLE 2-26 Find the conductance of a 50-Ω resistor.

$$G = \frac{1}{R} = \frac{1}{50} = 0.02 \text{ mho} \qquad ////$$

Examples 2-25 and 2-26 show that conductance decreases as resistance increases.

2-14 ELECTRIC CURRENTS

In the hydraulic analog of the electric circuit, water flow is the hydraulic equivalent of electron flow. As water flow increases with an increase in pressure or a decrease in friction, so electron flow increases with an increase in electric pressure (voltage) or a decrease in resistance.

It has been stated that current is a directed movement of charges. It is now necessary to provide a unit for current. This unit is the *ampere* (A), named for the eighteenth-century French physicist André Marie Ampère. One ampere is defined as the movement of one coulomb of charge past a given point in one second. This is shown in Eq. (2-12).

$$I\,(\text{A}) = \frac{Q\,(\text{C})}{t\,(\text{s})} \tag{2-12}$$

I is the common symbol for current. The use of Eq. (2-12) is demonstrated in the following examples.

EXAMPLE 2-27 If 12.48×10^{18} electrons pass a point in a circuit every 1.5 s, what is the magnitude of the current in amperes?

SOLUTION

1 $I = Q/t$, where Q is in coulombs. Therefore, the charge in electrons is first converted to coulombs.

$$Q = \frac{\#\ \text{electrons}}{6.24 \times 10^{18}} = \frac{12.48 \times 10^{18}}{6.24 \times 10^{18}} = 2\ \text{C}$$

2 Then

$$I = \frac{2}{1.5} = 1.333\ \text{A} \qquad ////$$

EXAMPLE 2-28 If a movement of 6 C creates a current of 8 A, in what time does this movement of charge take place?

SOLUTION Since $I = \dfrac{Q}{t}$,

$$t = \frac{Q}{I} = \frac{6}{8} = 0.75\ \text{s} \qquad ////$$

EXAMPLE 2-29 If 2 A of current flows, how much charge passes a given point in 1.2 s?

SOLUTION Rearranging Eq. (2-12) gives

$$Q = I \times t = (2)(1.2) = 2.4\ \text{C} \qquad ////$$

The reproducible standard ampere is defined as the unvarying current which, when passed through a solution of silver nitrate in accordance with standard specifications, deposits silver at the rate of 0.001118 g/s.

2-15 TYPES OF CURRENT

There are three basic types of current, distinguished by the ways in which they vary in direction and magnitude.

1 Pure direct current flows in one direction only, and varies only slightly (if at all) in magnitude.

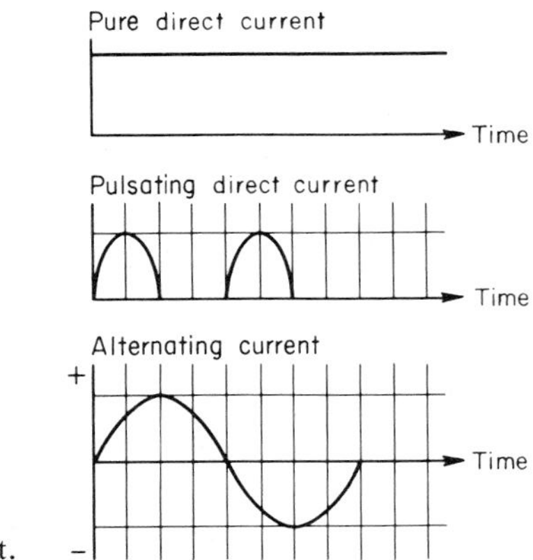

FIGURE 2-4
Three types of current.

2 Pulsating direct current is a current whose direction is constant but whose magnitude varies considerably in a short interval of time.

3 Alternating current varies in both magnitude and direction. That is, the direction of the current periodically reverses.

Examples of these three types of current are shown in Fig. 2-4.

2-16 ELECTROMOTIVE FORCE

The unit of electric pressure, or emf, is the volt. The reproducible international standard volt is 1/1.083 part of the potential difference developed by a Weston cell. The symbols for voltage are the letters E and V (either upper- or lowercase).

QUESTIONS

2-1 What is the purpose of units?
2-2 What four quantities are defined by the fundamental units?
2-3 What are secondary units?
2-4 What are the fundamental units of the English system?
2-5 What is the advantage of the mks system?
2-6 Why is the coulomb used as a unit of charge rather than an electron charge?
2-7 How is the standard kilogram defined?
2-8 What is the purpose of an electric circuit?
2-9 What is the purpose of an electric load?
2-10 Compare a hydraulic system to an electric circuit.

2-11 What are the properties of resistance?

2-12 Does temperature have any effect on resistance?

2-13 Does the diameter of a conductor have any effect on its resistance?

2-14 Can a material be identified by its resistance?

2-15 What is the definition of conductance?

2-16 What is the relation between resistance and current?

2-17 What is the relation between conductance and current?

2-18 What is the unit of current?

2-19 What is the definition of the unit of current?

2-20 What is the difference between pulsating direct current and alternating current?

2-21 What is the unit of potential?

PROBLEMS

2-1 Convert 3 in. to centimeters.

2-2 Convert 2,750 ft to meters.

2-3 Convert 6 km to miles.

2-4 Convert 0.0000026 slug to grams.

2-5 Convert 765 g to kilograms.

2-6 Convert 1.25 h to seconds.

2-7 What is the resistance at 20°C of an annealed copper conductor 1.5 ft long and 0.001 in. in diameter?

2-8 What is the resistance at 20°C of an aluminum conductor that is 30 ft long and has a cross-sectional area of 40×10^{-4} in.^{2}?

2-9 Which will have greater resistance, a nickel conductor 4 ft long with a radius of 0.032 in. or an aluminum conductor 4 ft long with a diameter of 0.032 in.? What is the resistance of each?

2-10 A copper conductor 9 in. long is found to have a resistance of 2 Ω. What is the diameter of the conductor?

2-11 A piece of wire is found to be 3 ft long. It has a cross-sectional area of 30 mils and has 6 Ω of resistance. What is the wire made of?

2-12 A heater coil is to be made from no. 15 nichrome wire. A resistance of 1,500 Ω is needed. How long must the wire be?

2-13 What is the difference in the resistances (at 20°C) of 100 ft of no. 10 copper wire and 100 ft of no. 10 aluminum wire?

2-14 A conductor having a resistance of 120 Ω at 20°C is 25 ft long and 0.00297 in. in diameter. What material is the conductor made of?

2-15 What is the change in resistance of an annealed copper conductor having 65 Ω resistance at 25°C if the temperature goes up to 40°C?

2-16 What is the resistance of the conductor of Prob. 2-8 if the temperature goes up to 75°C?

2-17 At 35°C, a nickel conductor has a resistance of 300 Ω. When the temperature is −40°C, what is its resistance?

2-18 At 70°C, a lead conductor has a resistance of 50 Ω. What is its temperature when its resistance is (*a*) 59 Ω, (*b*) 20 Ω, (*c*) 30 Ω, (*d*) 70 Ω?

2-19 Find α_s if $R_1 = 70$ Ω, $T_1 = 40$°C, $R_2 = 62$ Ω, and T_2 equals (*a*) 10°C, (*b*) 78°C, (*c*) −10°C, (*d*) 0°C.

2-20 A nichrome conductor has a resistance of 1,500 Ω at 35°C. What is its resistance at (*a*) −60°C, (*b*) −40°C, (*c*) 0°C, (*d*) 40°C, (*e*) 250°C?

3

OHM'S LAW, POWER, AND THE ELECTRIC CIRCUIT

3-1 THE ELECTRIC CIRCUIT

In this chapter the student will learn how to determine the various quantities of an electric circuit: current, voltage, resistance, and power.

Circuits are classified as *series*, *parallel*, or *compound* (*series-parallel*).

In a series circuit the current can follow only one path from the negative terminal of the source of voltage, through the various components of the circuit, and back to the positive terminal of the source. Figure 3-1 shows such a circuit. A familiar example of a series circuit is the type of Christmas-tree lighting where the bulbs are wired end to end (in the sockets). In this type of wire, if one bulb goes out, all bulbs go out. In a series circuit, the current is the same throughout the circuit. Electrons cannot leave a point in the circuit in any greater quantity than they enter that point. The hydraulic analog of the series electric circuit is shown in Fig. 3-2.

In a parallel circuit there are two or more separate paths for current to follow. Figure 3-3 shows a parallel hydraulic circuit. Note that the pump supplies pressure and this pressure moves water through the separate pipes. Note also that the rate of water flow in the mains must equal the sum of the rates of flow through the pipes

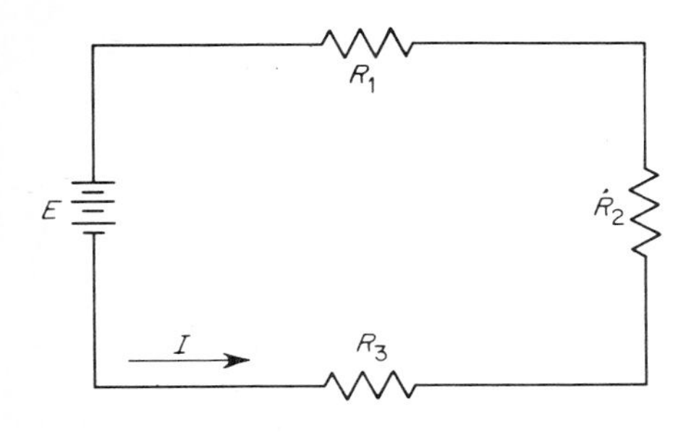

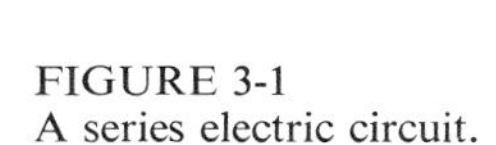

FIGURE 3-1
A series electric circuit.

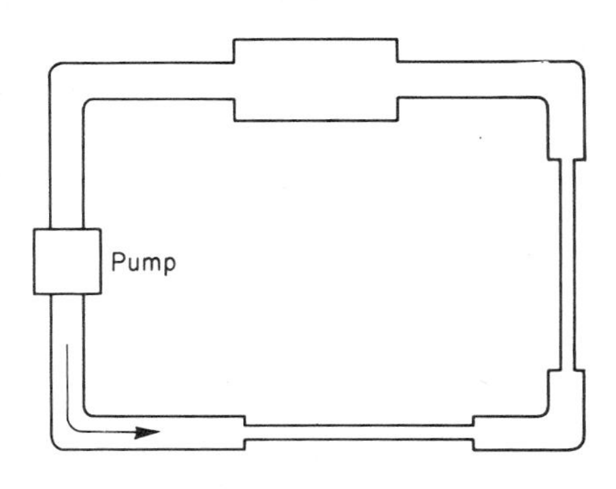

FIGURE 3-2
A series hydraulic circuit.

between the mains. Figure 3-4 is an example of a parallel electric circuit consisting of three branches, where a branch is defined as a part of a circuit containing just one component or source. Note that the total current leaving the negative terminal of the battery and returning at the positive terminal must be the sum of branch currents I_1 and I_2.

$$I_T = I_1 + I_2$$

The compound (series-parallel) circuit is any combination of series and parallel circuits. In reality, all parallel circuits are compound circuits since all sources of voltage have internal resistance. It is only in circuits where the external circuit resistance is far greater than the internal resistance of the source that we can treat the circuit as a parallel circuit. Figure 3-5 is a schematic diagram of a series-parallel circuit.

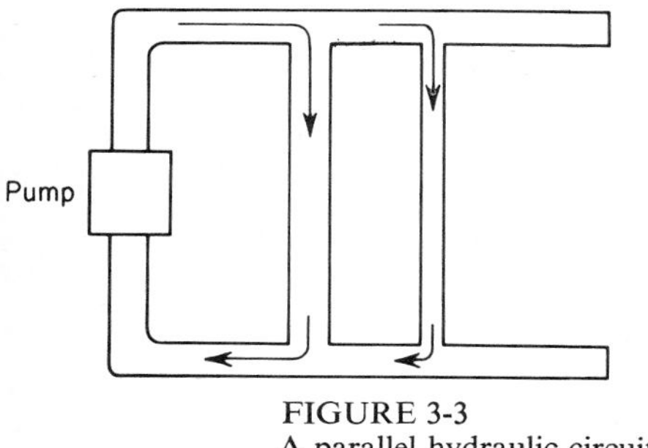

FIGURE 3-3
A parallel hydraulic circuit.

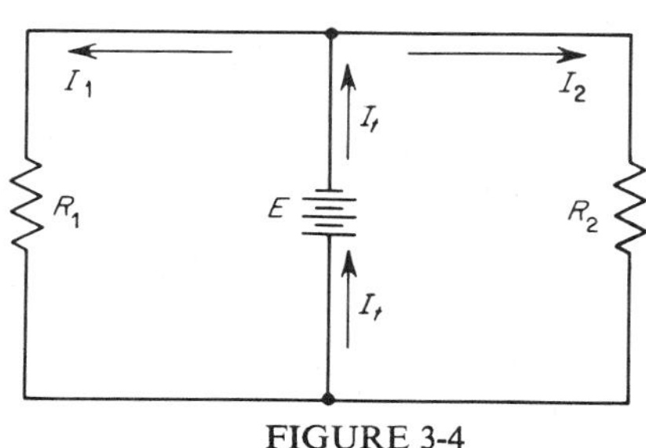

FIGURE 3-4
A parallel electric circuit.

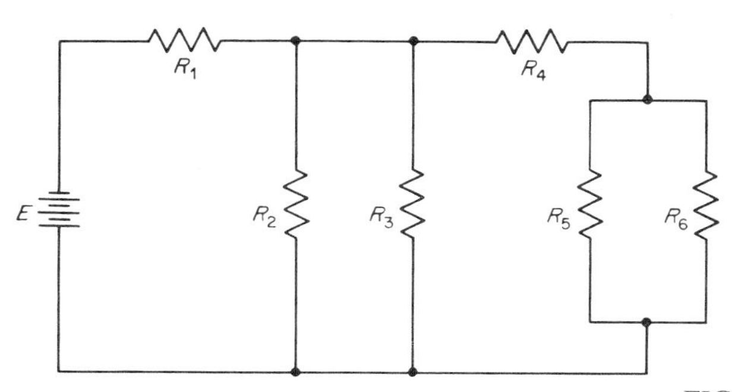

FIGURE 3-5
A series-parallel circuit.

3-2 WORK AND ENERGY

Work is performed when a force causes a body to be displaced. Energy is the ability to do work. A familiar example of work and energy involves a pump and an elevated storage tank. The pump does work when it lifts water to the level of the storage tank against the force of gravity. This situation is shown in Fig. 3-6. The water upon being lifted to the level of the storage tank and being stored at this higher level possesses energy. That is, the water stored at this higher level has the ability to do work. If a valve at the bottom of the tank were opened, the water would flow and could be used to drive a turbine, thus performing work.

What raises the water to the higher level of the storage tank against the force of gravity? The pump does the work. Thus, the pump must have had energy of some form because work was done by the pump. In this process, the ability to do work is transferred from the pump to the water. From this it can be seen that *work is simply a transformation of energy from one form into another.* The pump does work, and its mechanical energy is converted into the energy possessed by the water in the tank. The action involved in bringing about this conversion of energy is termed *work*. It follows that work and energy are measured in the same units.

An axiom of physics states that energy can be neither created nor destroyed, only changed in form. When this principle is applied to the pump, it appears that all the energy spent by the pump is converted into energy possessed by the water. However, some of the energy of the pump is converted into heat energy by friction as the water passes through the pipes. The water in the storage tank, therefore, does not acquire all the energy supplied by the pump. If the heat energy lost as a result of friction in the pipes were added to the energy of the water in the tank, the principle of conservation of energy would be satisfied.

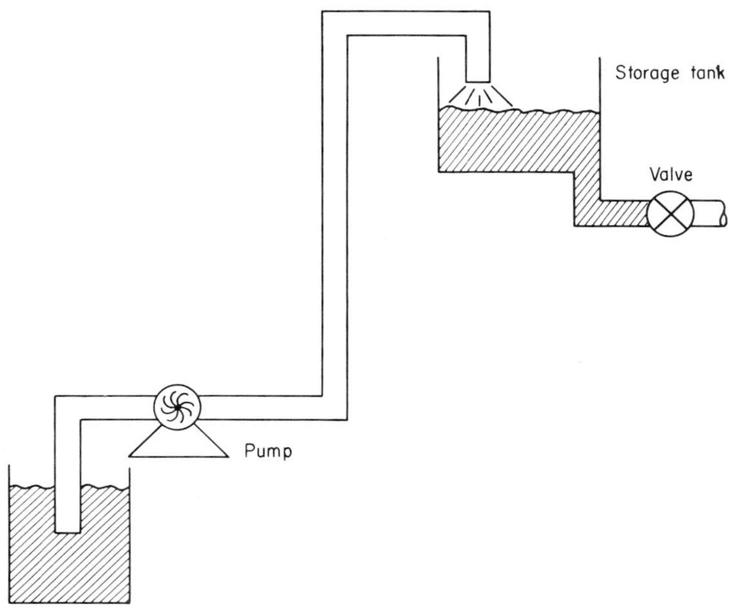

FIGURE 3-6
An example of hydraulic energy used to perform work.

Energy exists in several forms: chemical, mechanical, heat, light, and electric energy. There are also many devices that convert one form of energy into another. The pump in the previous example is an energy-converting device. Devices that convert electric energy into some other form of energy and devices that convert other forms of energy into electric energy are called *transducers*.

An energy-converting device of importance is the battery. A battery is a device that converts chemical energy into electric energy. The chemical energy causes work to be done when the negative charges are accumulated on one electrode of the battery and positive charges are accumulated on the other electrode. This storage of negative charges on one electrode can be accomplished only by overcoming the repelling force of the electrons already there. Thus a battery performs an action similar to that of the pump and storage tank, except in that situation energy was converted by elevating water, and in a battery charge is "elevated."

The voltage of a battery is a measure of the rise in electric energy per unit charge. This becomes clearer if it is remembered that voltage is actually a potential difference that opposes the "elevation" of charge by chemical action. Thus, the greater the potential difference through which a charge must be raised, the greater the energy needed. Also, the greater the amount of charge overcoming this potential difference, the greater the energy needed. This relationship is expressed mathematically in the equation

$$W = E \times Q \tag{3-1}$$

W represents the work done by the chemical action and the energy transferred to the "elevated" electrons in the form of electric energy. The unit of W is the joule when E is in volts and Q is in coulombs. E is the potential difference in volts through which Q coulombs are "elevated."

EXAMPLE 3-1 If a charge of 4.1 C is moved through a potential difference of 12 V, (1) how much work is done? (2) How much energy is acquired by this amount of charge?

SOLUTION

1

$$\begin{aligned} W &= E \times Q \\ &= (12)(4.1) = 49.2 \text{ J} \end{aligned} \tag{3-1}$$

2 Neglecting any loss of energy, the electric energy is identical to the work done. ////

EXAMPLE 3-2 If 4.5 J of work is done in moving a charge through a difference of potential of 90 V, (1) how much charge is moved? (2) How many electrons are moved?

SOLUTION

1

$$W = E \times Q \tag{3-1}$$

$$Q = \frac{W}{E}$$

$$= \frac{4.5}{90} = 0.05 \text{ C}$$

2

$$(6.24 \times 10^{18})(0.05) = 0.312 \times 10^{18} \text{ electrons}$$

////

3-3 OHM'S LAW

In 1826, George Simon Ohm, a German mathematician and physicist, related current to voltage and resistance. The relationship is known as *Ohm's law.*

Ohm's law The current through a resistance is directly proportional to the voltage across the resistance and inversely proportional to the resistance.

Equation (3-2) states Ohm's law in mathematical form.

$$I = \frac{E}{R} \tag{3-2}$$

It is possible to rearrange Eq. (3-2) to solve for voltage or resistance. For example, the voltage across a resistor equals the product of the current through the resistor and the resistance of the resistor.

$$E = IR \tag{3-3}$$

The magnitude of a resistance can be found if the current through it and the voltage across it are known. The resistance is proportional to the voltage and inversely proportional to the current.

$$R = \frac{E}{I} \tag{3-4}$$

3-4 APPLICATION OF OHM'S LAW

Ohm's law can be applied to an entire circuit or to any part of a circuit. When Ohm's law is applied to an entire circuit, the voltage used must be the applied voltage, the current must be the current flowing through the source of voltage, and the resistance must be the resistance of the entire circuit.

EXAMPLE 3-3 Find the current in the circuit of Fig. 3-7.

SOLUTION

$$I = \frac{E}{R} = \frac{25}{50} = 0.5 \text{ A}$$ ////

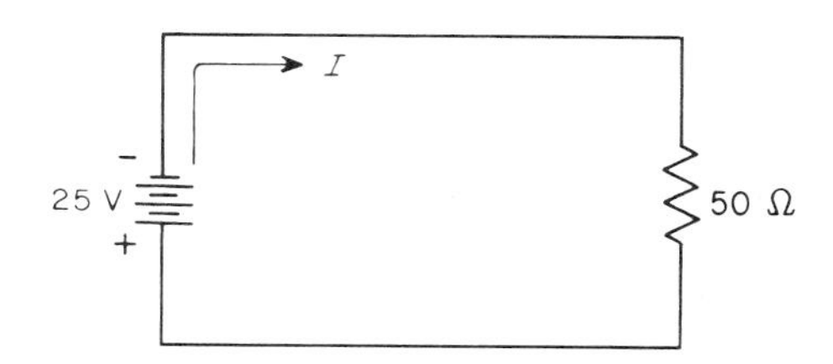

FIGURE 3-7
Circuit for Example 3-3.

EXAMPLE 3-4 Find the total resistance of the circuit of Fig. 3-8. The current measured at the source is 2.5 A.

SOLUTION

$$R_t = \frac{E}{I} = \frac{120}{2.5} = 48\ \Omega \qquad ////$$

EXAMPLE 3-5 Find the applied voltage in the circuit of Fig. 3-9 when the total resistance is 45 Ω.

SOLUTION

$$E = IR_t = (0.5)(45) = 22.5\ \text{V} \qquad ////$$

EXAMPLE 3-6 In the circuit of Fig. 3-10, find the value of resistor R_1. A voltmeter connected across R_1 reads 100 V.

SOLUTION

$$R_1 = \frac{E}{I} = \frac{100}{0.25} = 400\ \Omega \qquad ////$$

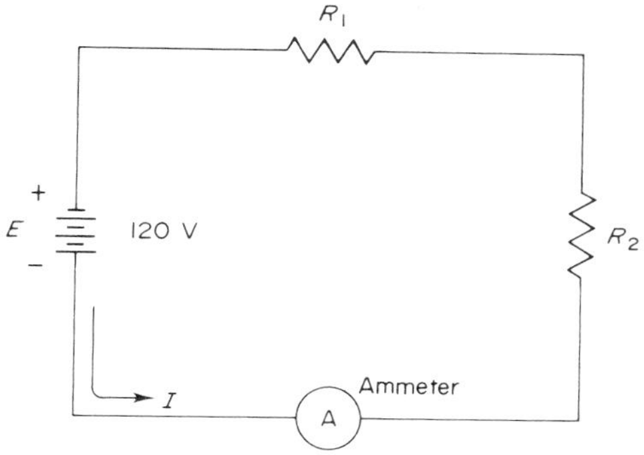

FIGURE 3-8
Circuit for Example 3-4

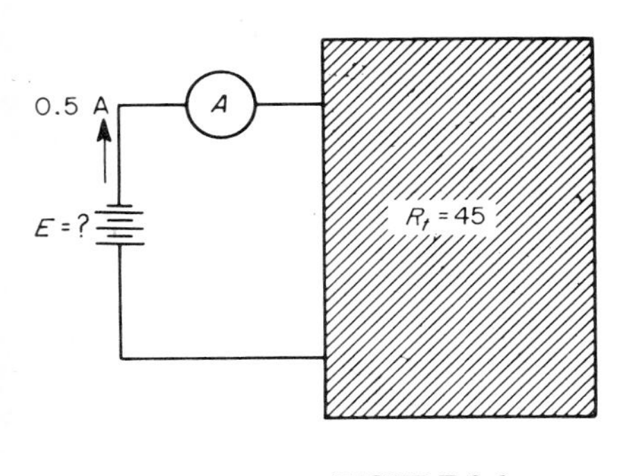

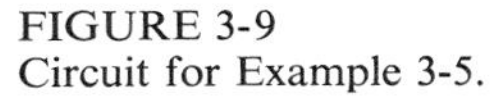
FIGURE 3-9
Circuit for Example 3-5.

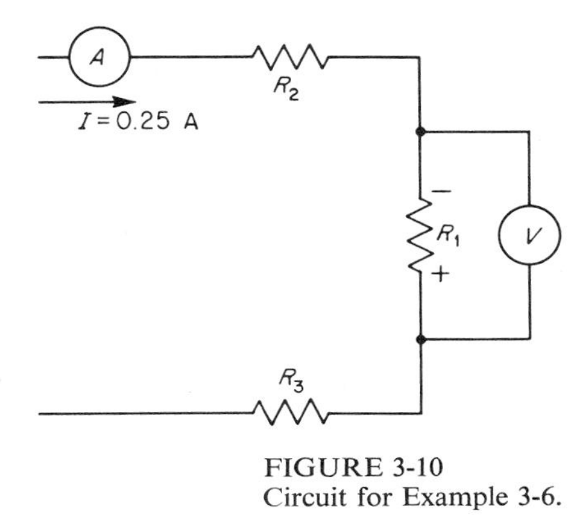

FIGURE 3-10
Circuit for Example 3-6.

EXAMPLE 3-7 The current through a resistance of 1,200 Ω is 0.1 A. What is the voltage drop across the resistor?

SOLUTION

$$E = IR = (0.1)(1{,}200) = 120 \text{ V} \qquad ////$$

The term *voltage drop* has the same meaning as potential difference between two points. Potential difference is measured by the energy required to move electrons between two points.

EXAMPLE 3-8 The voltage drop across a 250-Ω resistor is 10 V. What is the current through this resistor? (See Fig. 3-11.)

SOLUTION

$$I = \frac{E}{R} = \frac{10}{250} = 0.04 \text{ A} \qquad ////$$

Ohm's law can be expressed in terms of the *conductance* of a component or a circuit. Since resistance is a measure of the opposition to electron flow, conductance is a measure of the ease with which electrons flow.

Conductance G is the reciprocal of resistance.

$$G = \frac{1}{R} \text{ mho} \tag{3-5}$$

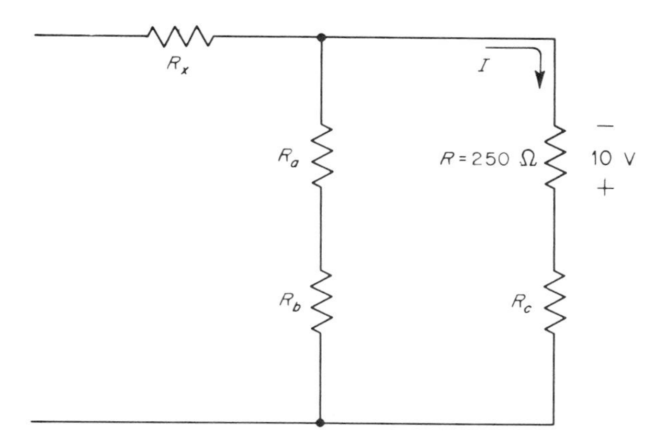

FIGURE 3-11
Circuit for Example 3-8.

The units of conductance would be $(1/\Omega)$, which is defined as a mho. As resistance increases conductance decreases, and vice versa. According to Eq. (3-2),

$$I = \frac{E}{R} = \frac{E}{1}\frac{1}{R}$$

But $1/R$ is conductance. Therefore,

$$I = EG \tag{3-6}$$

Equation (3-3) states that voltage is the product of current and resistance ($E = IR$). Since resistance is the reciprocal of conductance, this equation becomes

$$E = I\frac{1}{G} = \frac{I}{G} \tag{3-7}$$

Since R equals voltage divided by current, Eq. (3-4), conductance equals current divided by voltage.

$$R = \frac{E}{I}$$

$$G = \frac{1}{R} = \frac{1}{E/I} = \frac{I}{E} \tag{3-8}$$

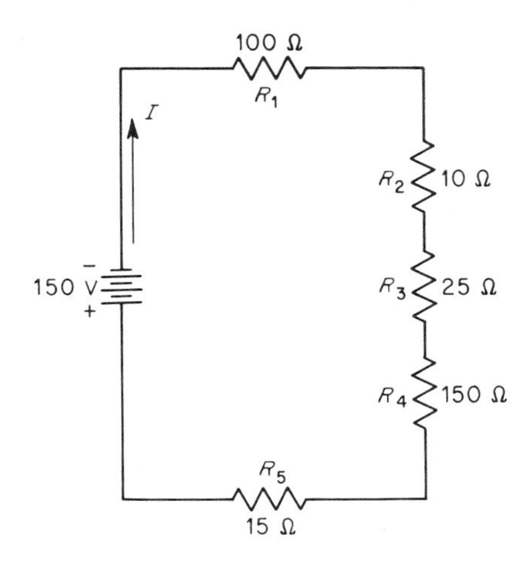

FIGURE 3-12
A series circuit of five resistors.

EXAMPLE 3-9 Given a circuit with a conductance of 0.005 mho, find the current if the applied voltage is 200 V.

SOLUTION

$$I = EG = (200)(0.005) = 1 \text{ A} \qquad ////$$

EXAMPLE 3-10 What is the voltage drop across a conductance of 0.0003 mho if the current through the conductance is 0.015 A?

SOLUTION

$$E = \frac{I}{G} = \frac{0.015}{0.0003} = \frac{15 \times 10^{-3}}{3 \times 10^{-4}} = 5 \times 10^{1} = 50 \text{ V} \qquad ////$$

3-5 THE SERIES CIRCUIT

There are three fundamental rules for series circuits:

1. The current is the same everywhere in the circuit.
2. The sum of the voltage drops around the circuit equals the applied voltage.
3. The total resistance of a series circuit equals the sum of the individual resistances in the circuit.

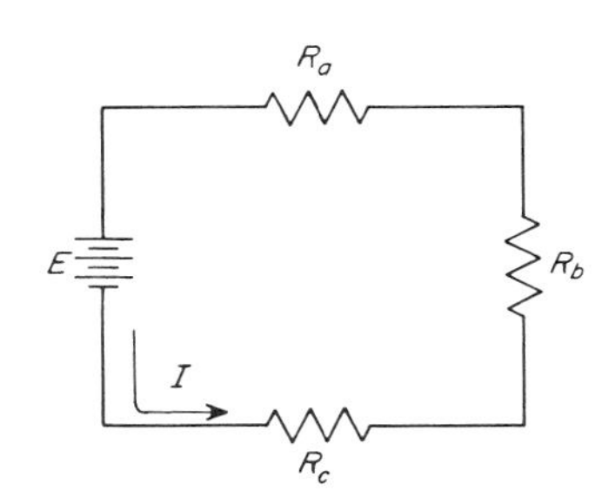

FIGURE 3-13
Circuit for Examples 3-11 and 3-12.

In the circuit of Fig. 3-12 it is apparent that the currents through all components are the same, since charges cannot enter a point in any greater quantity than they leave.

To find the current in Fig. 3-12, Ohm's law is used. The quantities used must pertain to the entire circuit, since no individual voltage drops are given. The solution is

$$R_t = R_1 + R_2 + R_3 + R_4 + R_5 = 300\ \Omega$$

$$I = \frac{E}{R_t} = \frac{150}{300} = 0.5\ \text{A}$$

The voltage drop across each resistor is found by Ohm's law.

$$\begin{aligned}
E &= IR \\
E_1 &= IR_1 = (0.5)(100) = 50\ \text{V} \\
E_2 &= IR_2 = (0.5)(10) = 5\ \text{V} \\
E_3 &= IR_3 = (0.5)(25) = 12.5\ \text{V} \\
E_4 &= IR_4 = (0.5)(150) = 75\ \text{V} \\
E_5 &= IR_5 = (0.5)(15) = 7.5\ \text{V}
\end{aligned}$$

Note that the sum of the voltage drops E_1, E_2, E_3, E_4, and E_5 equals the applied voltage.

The following examples illustrate applications of Ohm's law and the rules of series circuits.

EXAMPLE 3-11 In the circuit of Fig. 3-13, what is the voltage drop across R_c if $E = 50$ V, $R_a = 125\ \Omega$, $R_b = 225\ \Omega$, and $R_c = 150\ \Omega$?

SOLUTION

1 Find *I*. In a series circuit *I* is the same everywhere.

$$I = \frac{E}{R_t} = \frac{50}{500} = 0.1\ \text{A}$$

2 The voltage drop across R_c equals IR_c.

$$E = IR_c = (0.1)(150) = 15 \text{ V} \qquad ////$$

EXAMPLE 3-12 In the circuit of Fig. 3-13, $R_a = 200\ \Omega$, $R_c = 350\ \Omega$, and R_b is unknown. The applied voltage is 120 V, and the voltage drop across R_a is 40 V. What is the resistance of R_b?

SOLUTION

1 Solve for the current. (Note that in this case Ohm's law is applied to one part of the circuit, rather than to the entire circuit.)

$$I = \frac{E_a}{R_a} = \frac{40.0}{200} = 0.2 \text{ A}$$

2 Once the current is known, there are two ways in which R_b can be found.

a The total resistance of a series circuit is the ratio of the applied voltage to the current.

$$R_t = \frac{E}{I} = \frac{120}{0.2} = 600\ \Omega$$

$$R_b = R_t - R_a - R_c = 600 - 200 - 350 = 50\ \Omega$$

b The alternative solution for R_b is to find the voltage drop across R_b and then apply Ohm's law.

1 The drop across R_c equals IR_c.

$$E_c = IR_c = (0.2)(350) = 70 \text{ V}$$

2 The drop across R_b is the difference between the applied voltage and the drops across R_a and R_c.

$$E_b = 120 - 70 - 40 = 10 \text{ V}$$

3 By Ohm's law,

$$R_b = \frac{E_b}{I} = \frac{10}{0.2} = 50\ \Omega$$

EXAMPLE 3-13 What is the applied voltage in the circuit of Fig. 3-14? The voltmeter reads 12.1 V.

Note that electron current I must flow clockwise since the voltmeter indicates the polarity shown across R_1. Thus the negative terminal of the applied voltage must be attached to R_1 instead of R_5.

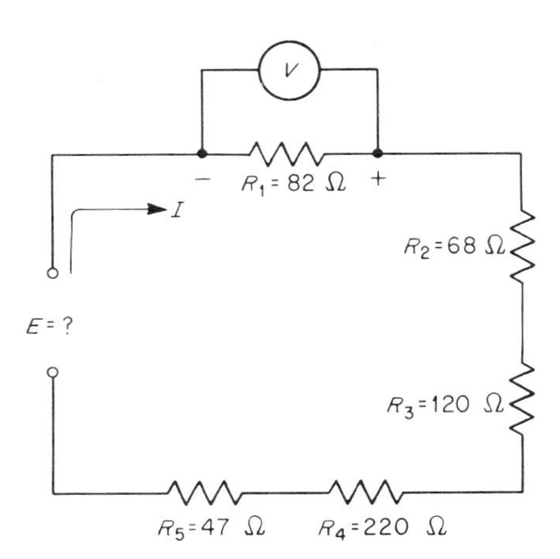

FIGURE 3-14
Circuit for Examples 3-13 and 3-14.

SOLUTION

1 $$E = IR_t$$

2 $$R_t = R_1 + R_2 + R_3 + R_4 + R_5 = 537\ \Omega$$

3 $$I = \frac{E_1}{R_1} = \frac{12.1}{82} = 0.148\ \text{A}$$

4 $$E = IR_t = (0.148)(537) = 79.5\ \text{V}$$ ////

EXAMPLE 3-14 What is the voltage drop across resistor R_4 in the circuit of Fig. 3-14?

SOLUTION

1 Find the current in the same manner as in Example 3-13.
2 The voltage drop across R_4 is then found by Ohm's law.

$$E_4 = IR_4 = (0.148)(220) = 32.6\ \text{V}$$ ////

PROBLEMS

3-1 What is the current through a 1,200-Ω resistor if the voltage drop across the resistor is 8.3 V?

3-2 A certain vacuum-tube filament has a voltage drop of 6.3 V when a current of 150 milliampere (mA) flows. What is the resistance of the filament? (1 mA = 0.001 A.)

3-3 The voltage drop across a 1,500-Ω resistor is 72 V. What is the current through the resistor?

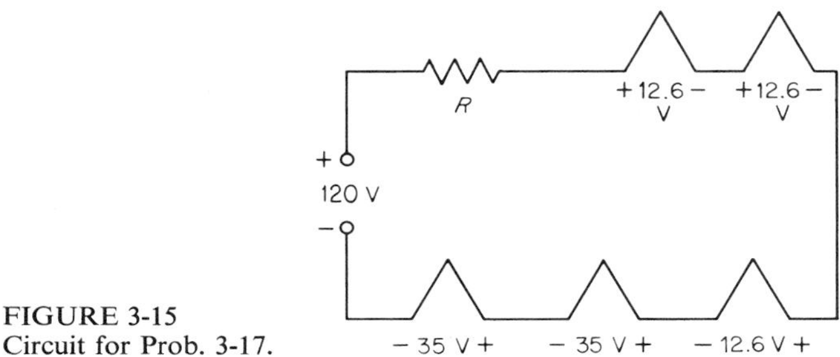

FIGURE 3-15
Circuit for Prob. 3-17.

3-4 The current through a 150-Ω resistor is 0.21 A. What is the voltage drop across the resistor?

3-5 The current through an unknown resistance is 2 mA. The voltage drop across the resistor is 125 V. What is the size of the unknown resistance?

3-6 An electric toaster draws a current of 7.2 A from a 120-V line. What is the resistance of the heater element?

3-7 An electric soldering iron has a resistance of 192 Ω. How much current will flow if it is connected to a 120-V line?

3-8 A pilot lamp is rated at 0.3 A, 6 V. What is the resistance of the filament?

3-9 A resistor used in a radio receiver has become so discolored that it is impossible to determine its rating by its color code. A technician measures the voltage drop across the resistor and the current through it. The voltage drop is 15 V and the current is 21 mA. What is the resistance of the resistor?

3-10 A certain relay coil has a resistance of 350 Ω. The voltage rating of the coil is 18 V. How much current flows through the coil?

3-11 A choke used in a radio power supply has a dc resistance of 30 Ω. The voltage drop across the coil is 12 V. How much current flows through the choke coil?

3-12 A certain heating element is rated at 220 V. If its resistance is 50 Ω, how much current flows through the heater when operating?

3-13 If a resistor rated at 510 Ω is used in a line with a current of 420 mA, what is the voltage drop across the resistor?

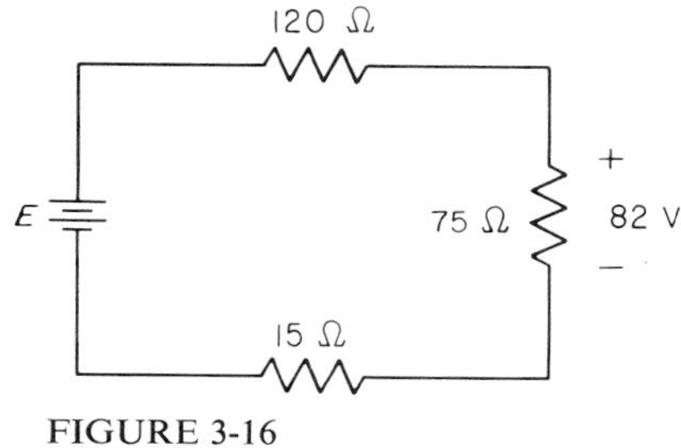

FIGURE 3-16
Circuit for Prob. 3-18.

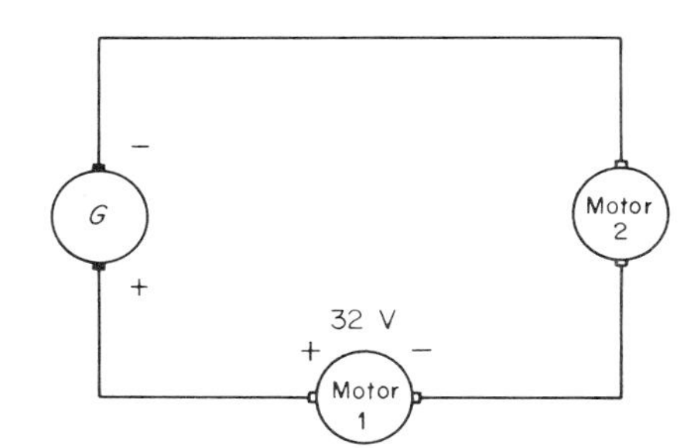

FIGURE 3-17
Circuit for Prob. 3-19.

3-14 A circuit consists of two series resistors. One is 82 Ω, and the other 150 Ω. If a potential of 32 V is applied to the circuit, what is the voltage drop across the 82-Ω resistor?

3-15 What voltage must be applied to the circuit in Prob. 3-14 for the current to be 85 mA?

3-16 Referring to Prob. 3-14, what is the applied voltage if the voltage drop across the 82-Ω resistor is 1.2 V?

3-17 An ac/dc radio uses the circuit arrangement shown in Fig. 3-15. What series resistance is needed to limit the current to 150 mA?

3-18 Given the circuit of Fig. 3-16, solve for the applied voltage.

3-19 A generator develops a terminal voltage of 150 V dc. It is connected to a load so that the current supplied is 2.1 A. Refer to the circuit of Fig. 3-17 and determine the resistance of motor 2.

3-20 What is the current in the circuit of Fig. 3-18?

3-21 What is the size of resistor R in the circuit of Fig. 3-18?

3-22 A series circuit consists of three resistors. The drop across the first is twice the drop across the second, and the drop across the second is twice the drop across the third. The applied voltage is 300 V and the current is 220 microamperes (μA). ($1 \mu A = 1 \times 10^{-6}$ A.) What are the sizes of the resistors?

3-23 A distribution system consists of a source whose terminal voltage is 220 V and a load located 3,100 yd from the source. The resistance of the load is 75 Ω, and the wires connecting source and load are no. 8 copper conductors (hard drawn) at an average temperature of 20°C. What is the voltage across the load, and what is the voltage drop in the connecting lines?

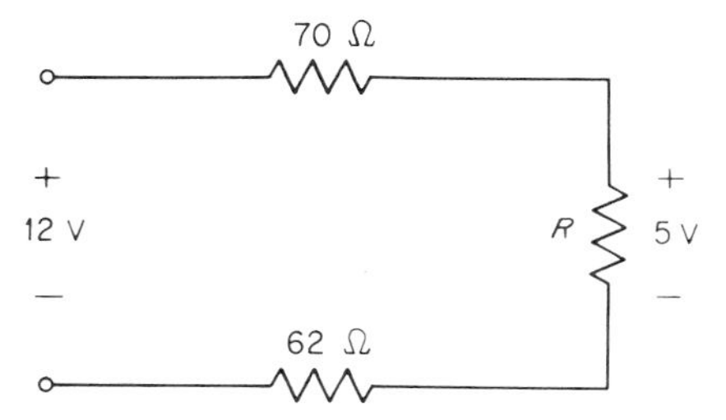

FIGURE 3-18
Circuit for Probs. 3-20, 3-21, and 3-24.

3-24 What is the conductance of the circuit of Fig. 3-18?

3-25 A series circuit consists of the following conductances: 0.0025, 0.05, and 0.0006 mho. If applied voltage to the circuit is 25 V, what is the current? Draw the circuit.

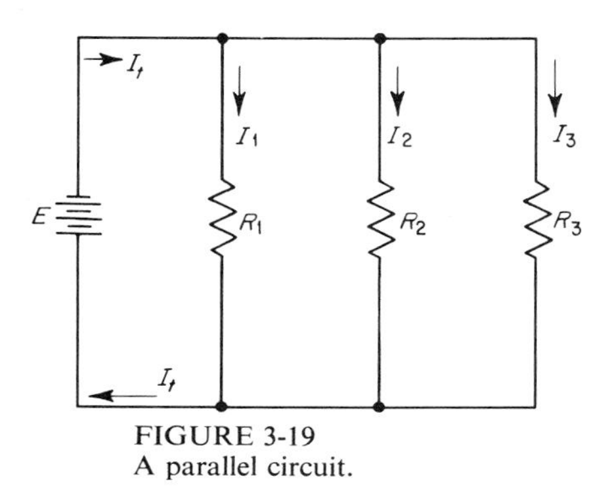

FIGURE 3-19
A parallel circuit.

3-6 THE PARALLEL CIRCUIT

In Sec. 3-1, the parallel circuit was described as a circuit in which there are two or more paths for current from the source to various loads. House wiring is a common example of a parallel circuit. All appliances, lights, radios, and TV sets are connected directly to the line, *not* to one another.

All circuit analyses in this book ignore the resistance of the connecting wires because the resistances of the loads are hundreds and even thousands of times larger than the resistances of the connecting wires. Therefore, no inaccuracy results from considering connectors to have negligible resistance.

We then have three simple laws for parallel circuits:

1 The voltages across all branches of a parallel circuit are the same.

2 The total current through a parallel circuit is the sum of the individual branch currents.

3 The total resistance of a parallel circuit is less than the resistance of any branch.

Examination of Fig. 3-19 serves to verify all three rules. Rule 1 is obvious because it has been agreed that the resistances of connecting wires are zero. Therefore the battery is connected directly across each branch.

Rule 2 is also self-evident. Each of the individual branch currents flows in and out of the source. Since the direction of the currents is determined by the polarity of

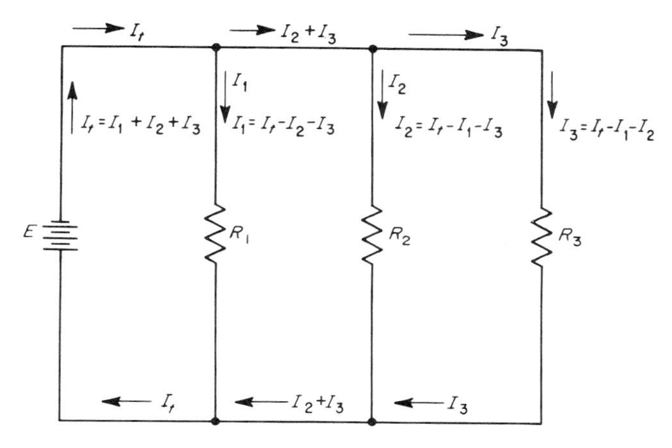

FIGURE 3-20
Current in a parallel circuit.

the source, the direction through all branches is the same. Hence, the total current through the source equals the sum of the individual branch currents.

$$I_t = I_1 + I_2 + I_3$$

The current can be followed in detail in the circuit of Fig. 3-19. The current I_t flows in and out of the battery and up to the branch containing R_1. Then an amount $(I_t - I_1)$ flows to the next branch, while I_1 flows through resistor R_1. Current I_2 flows through resistor R_2, while the remainder of the current $[I_t - (I_1 + I_2)]$ flows through R_3. This process is shown in detail in Fig. 3-20.

Rule 3 can be proved by rule 2. The total resistance of any circuit is the applied voltage divided by the total current.

$$R_t = \frac{E}{I_t} \tag{3-9}$$

Since the total current in a parallel circuit is greater than the current in any branch, it follows that the resistance of the entire circuit is less than the resistance of any branch.

The total resistance for the circuit can be found with Eq. (3-9) (refer to Fig. 3-21). Since I_t equals the sum of the branch currents, this quantity may be substituted in Eq. (3-9):

$$R_t = \frac{E}{I_1 + I_2 + I_3 + \cdots + I_n} \tag{3-10}$$

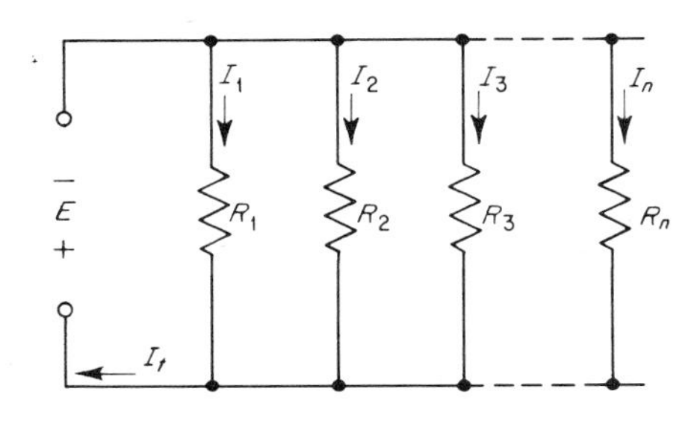

FIGURE 3-21
Circuit for the derivation of Eq. (3-10).

The branch currents are found by Ohm's law:

$$I_1 = \frac{E}{R_1}$$

$$I_2 = \frac{E}{R_2}$$

$$I_3 = \frac{E}{R_3}$$

$$\cdots\cdots\cdots$$

$$I_n = \frac{E}{R_n}$$

Equation (3-10) may now be written

$$\begin{aligned} R_t &= \frac{E}{\dfrac{E}{R_1} + \dfrac{E}{R_2} + \dfrac{E}{R_3} + \cdots + \dfrac{E}{R_n}} \\ &= \frac{E}{E\left(\dfrac{1}{R_1} + \dfrac{1}{R_2} + \dfrac{1}{R_3} + \cdots + \dfrac{1}{R_n}\right)} \\ &= \frac{1}{\dfrac{1}{R_1} + \dfrac{1}{R_2} + \dfrac{1}{R_3} + \cdots + \dfrac{1}{R_n}} \end{aligned} \tag{3-11}$$

Equation (3-11) may be used to determine the total resistance of any parallel circuit. It states that the total resistance equals the reciprocal of the sum of the reciprocals of the branch resistances.

Equation (3-11) can be derived in another manner. It is apparent that each branch of a parallel circuit is a separate conducting path for current. The total

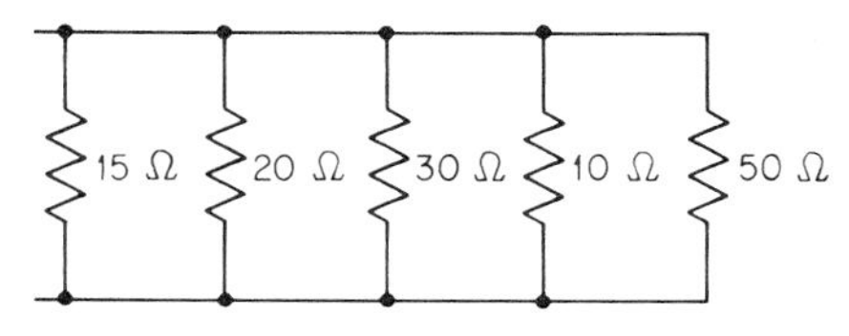

FIGURE 3-22
Circuit for Example 3-15.

conductance of the parallel circuit must then be the sum of the individual branch conductances.

$$G_t = G_1 + G_2 + G_3 + \cdots + G_n \tag{3-12}$$

$$G_1 = \frac{1}{R_1}$$

$$G_2 = \frac{1}{R_2}$$

$$G_3 = \frac{1}{R_3}$$

$$\cdots\cdots\cdots$$

$$G_n = \frac{1}{R_n}$$

$$G_t = \frac{1}{R_1} + \frac{1}{R_2} + \frac{1}{R_2} + \cdots + \frac{1}{R_n}$$

$$R_t = \frac{1}{G_t}$$

$$= \frac{1}{\dfrac{1}{R_1} + \dfrac{1}{R_2} + \dfrac{1}{R_3} + \cdots + \dfrac{1}{R_n}} \tag{3-11}$$

EXAMPLE 3-15 Find the total resistance of the circuit of Fig. 3-22.

SOLUTION

$$R_t = \frac{1}{\dfrac{1}{R_1} + \dfrac{1}{R_2} + \dfrac{1}{R_3} + \dfrac{1}{R_4} + \dfrac{1}{R_5}} = \frac{1}{\dfrac{1}{15} + \dfrac{1}{20} + \dfrac{1}{30} + \dfrac{1}{10} + \dfrac{1}{50}}$$

$$= \frac{1}{0.0666 + 0.05 + 0.0333 + 0.1 + 0.02} = \frac{1}{0.2699} = 3.704\ \Omega \quad ////$$

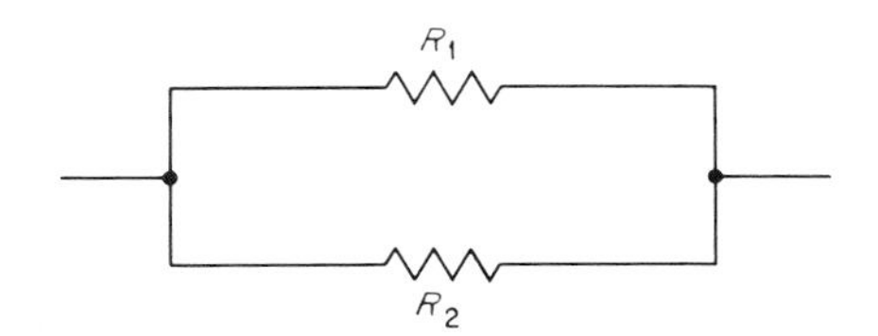

FIGURE 3-23
A two-branch parallel circuit.

A special form of Eq. (3-11) can be derived for a parallel circuit consisting of two branches. Figure 3-23 illustrates such a circuit.

From Eq. (3-11), the total resistance of the circuit is

$$R_t = \frac{1}{1/R_1 + 1/R_2}$$

When the denominator of this complex fraction is rationalized, the common denominator becomes R_1R_2. The equation then becomes

$$R_t = \frac{1}{(R_1 + R_2)/R_1R_2}$$

If we now solve this fraction, we come up with a useful formula for two-branch parallel circuits.

$$R_t = \frac{R_1R_2}{R_1 + R_2} \tag{3-13}$$

EXAMPLE 3-16 Find the total resistance of the circuit of Fig. 3-24 by two methods.

SOLUTION

1 By Eq. (3-11),

$$R_t = \frac{1}{1/R_1 + 1/R_2} = \frac{1}{\frac{1}{25} + \frac{1}{50}} = \frac{1}{0.04 + 0.02} = \frac{1}{0.06} = 16.66\ \Omega$$

2 By Eq. (3-13),

$$R_t = \frac{R_1R_2}{R_1 + R_2} = \frac{(25)(50)}{75} = 16.66\ \Omega$$ ////

The total resistance of parallel circuits can be found in still another manner. Since the total resistance of any circuit equals the applied voltage divided by the total

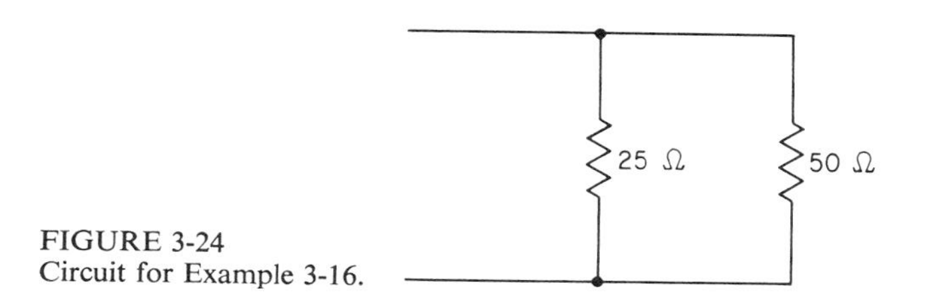

FIGURE 3-24
Circuit for Example 3-16.

current, it is possible to find the total resistance of a parallel circuit by calculating the total current that will flow if a certain voltage is applied.

1 Use the assumed voltage and solve for the branch currents.
2 Divide the assumed voltage by the sum of the branch currents. This yields the total resistance of the circuit.
3 To simplify the calculations, always assume a convenient voltage, usually an amount equal to the highest branch resistance, so that all currents will be one or more amperes.

EXAMPLE 3-17 Find the total resistance of the circuit of Fig. 3-25 by two methods.

SOLUTION By Eq. (3-11),

$$R_t = \frac{1}{\frac{1}{40} + \frac{1}{25} + \frac{1}{20}} = \frac{1}{0.025 + 0.04 + 0.05} = \frac{1}{0.115} = 8.7\ \Omega$$

ALTERNATIVE SOLUTION By the method outlined above:

1 Assume $E = 40$ V.
2 Solve for the currents.

$$I_1 = \tfrac{40}{40} = 1\text{ A} \qquad I_2 = \tfrac{40}{25} = 1.6\text{ A} \qquad I_3 = \tfrac{40}{20} = 2\text{ A}$$

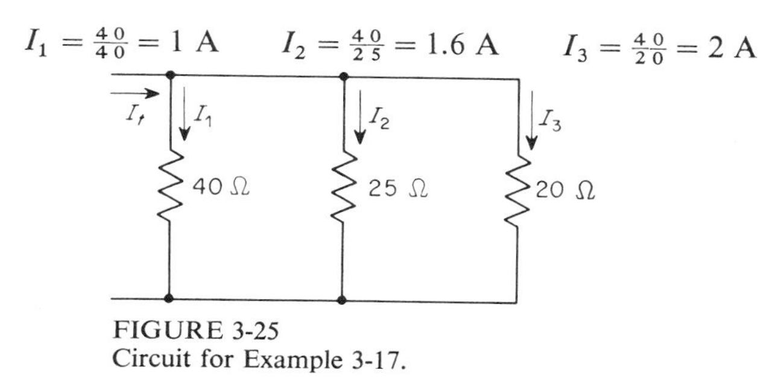

FIGURE 3-25
Circuit for Example 3-17.

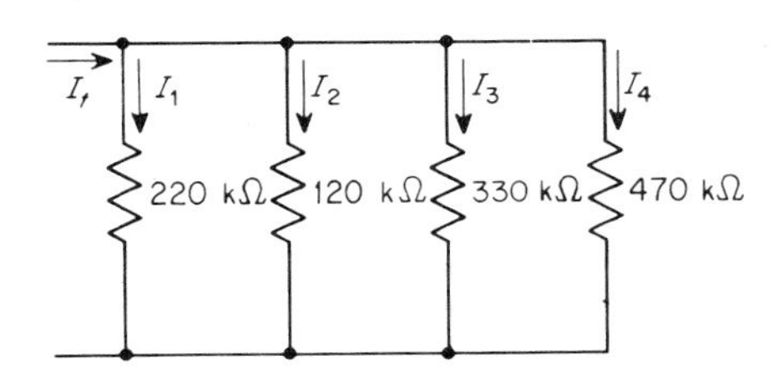

FIGURE 3-26
Circuit for Example 3-18.

3 Divide the assumed voltage E by the total current.

$$I_t = I_1 + I_2 + I_3 = 1 + 1.6 + 2 = 4.6 \text{ A}$$

$$R_t = \frac{E}{I_t} = \frac{40}{4.6} = 8.7 \ \Omega$$

////

The alternative method in Example 3-17 is very useful when the parallel circuit consists of many branches containing large resistances. This is further illustrated in Example 3-18.

EXAMPLE 3-18 Solve for the total resistance of the circuit of Fig. 3-26.

SOLUTION

1 Assume an applied voltage of 470,000 V.
2 Solve for the currents in the branches.

$$I_1 = \frac{470 \times 10^3}{220 \times 10^3} = 2.14 \text{ A}$$

$$I_2 = \frac{470 \times 10^3}{120 \times 10^3} = 3.91 \text{ A}$$

$$I_3 = \frac{470 \times 10^3}{330 \times 10^3} = 1.42 \text{ A}$$

$$I_4 = \frac{470 \times 10^3}{470 \times 10^3} = 1.00 \text{ A}$$

3 Compute the total current.

$$I_t = I_1 + I_2 + I_3 + I_4 = 8.47 \text{ A}$$

4 Solve for R_t.

$$R_t = \frac{E_t}{I} = \frac{470 \times 10^3}{847.} = 55.4 \text{ k}\Omega$$

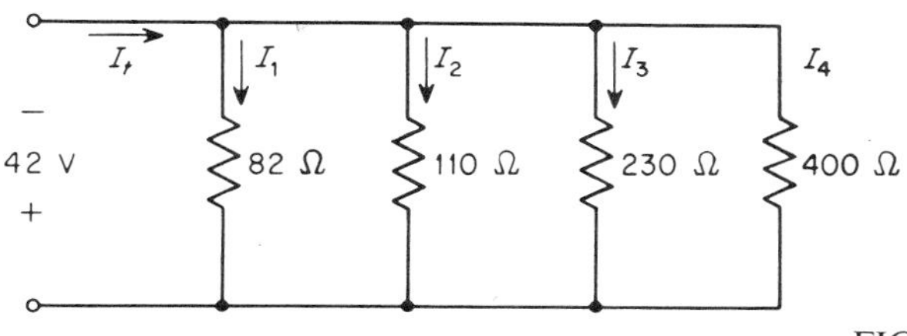

FIGURE 3-27
Circuit for Prob. 3-33.

PROBLEMS

3-26 A parallel circuit consists of two branches, one 27 Ω and the other 33 Ω. What is the total resistance of the circuit?

3-27 A circuit consists of three resistors in parallel. One is 125 Ω, the second is 15 Ω, and the third, 90 Ω. If 18 V is applied to the circuit, what is the total current? (Solve by two methods.)

3-28 A parallel circuit consists of two branches: one 280 Ω, and the other 120 Ω. What is the total resistance of the circuit?

3-29 A parallel circuit consists of two branches: one 56 Ω and the other 30 Ω. How much current will flow from a source of 20 V?

3-30 Referring to Prob. 3-29, how much current will flow in the 30-Ω branch if 12 V is applied? What is the voltage drop across the 56-Ω branch?

3-31 A certain radio has six tubes, all with 6.3-V, 300-mA heaters. If the heaters are wired in parallel, what is their total resistance? How much current will flow from the heater voltage source?

3-32 Derive an equation to show that if there are n equal resistors in parallel, the total resistance of the parallel circuit equals $1/n$ times any branch resistance.

3-33 Given the circuit of Fig. 3-27, find the current in each branch, the total current, and the total resistance of the circuit.

3-34 What is the value of resistor R_3 in the circuit of Fig. 3.28 when the ammeter reads

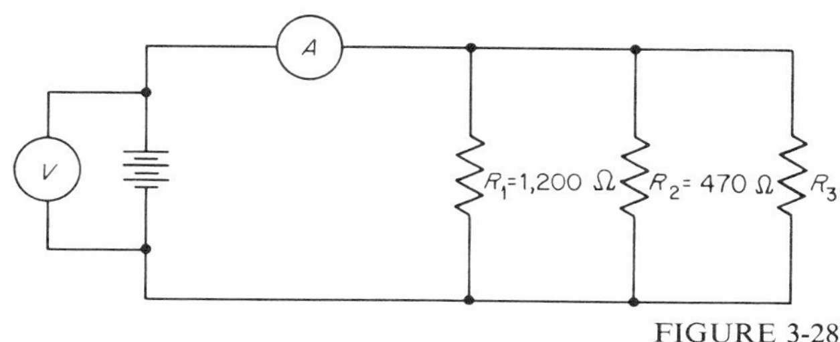

FIGURE 3-28
Circuit for Probs. 3-34 and 3-35.

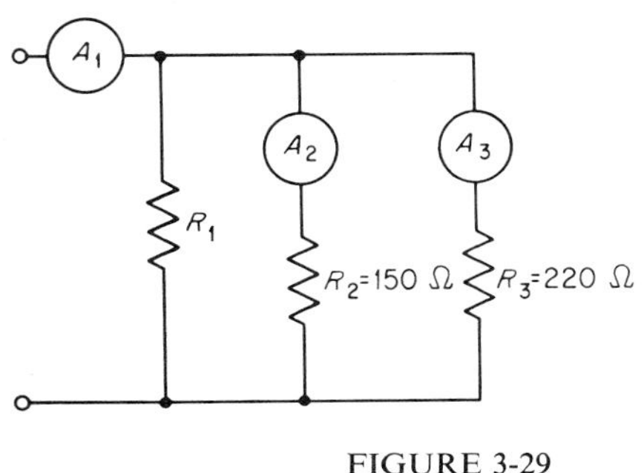

FIGURE 3-29
Circuit for Prob. 3-38.

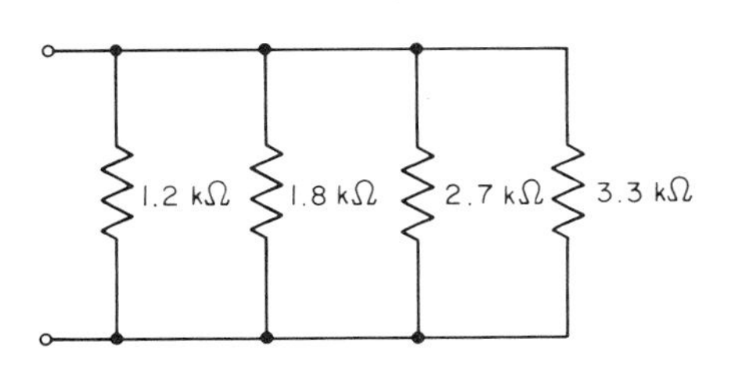

FIGURE 3-30
Circuit for Probs. 3-39 and 3-41.

380 mA and the voltmeter reads 80 V?

3-35 What is the total resistance of the circuit of Fig. 3.28?

3-36 A circuit consists of five equal resistors in parallel. When 70 V is applied to the circuit, the current is 1.75 A. What is the resistance of each branch?

3-37 The voltage drop across a parallel circuit is 125 V when a current of 0.166 A flows. The resistance of one branch is 1,200 Ω. What is the resistance of the other branch?

3-38 Given the circuit of Fig. 3.29, solve for resistor R_1 when A_1 reads 1.26 A, A_2 reads 0.733 A, and A_3 reads 0.5 A.

3-39 Solve for the total resistance of the circuit of Fig. 3-30.

3-40 Solve for the total resistance of the circuit of Fig. 3-31.

3-41 Given the circuit of Fig. 3-30, find the current between a 200-V source and the circuit. What are the branch currents?

3-42 In the circuit of Fig. 3-31, when a certain voltage is applied to the circuit, the total current is 0.5 A. What are the branch currents?

3-43 Given the circuit of Fig. 3-31 with a total current of 2 mA, find the applied voltage.

3-44 Solve for the total resistance of the circuit of Fig. 3-32.

3-45 What voltage must be applied to the circuit of Fig. 3-32 for the current to be 1 mA?

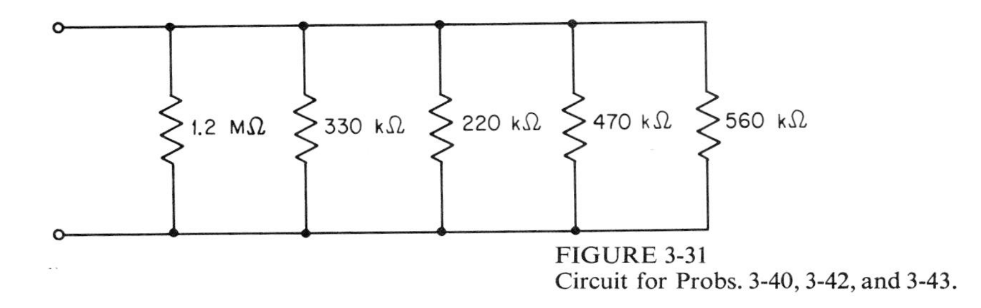

FIGURE 3-31
Circuit for Probs. 3-40, 3-42, and 3-43.

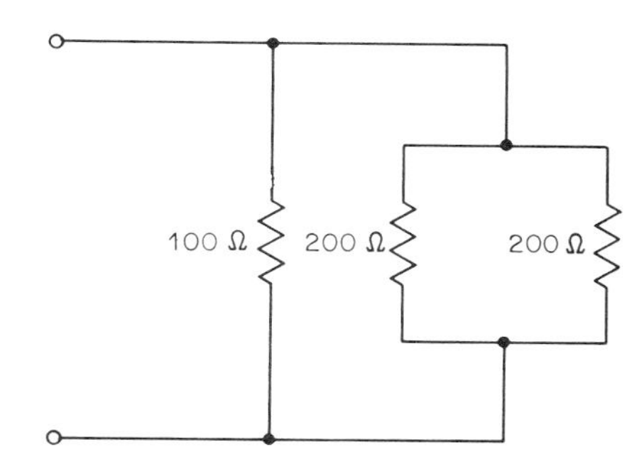

FIGURE 3-32
Circuit for Probs. 3-44 and 3-45.

3-7 THE SERIES-PARALLEL CIRCUIT

The solution of series-parallel circuits is often a source of considerable trouble to students. The trouble arises from difficulty in understanding the action of the current in this type of circuit. To make compound circuits as clear as possible, the analysis will be based upon separate analyses of both series and parallel circuits.

In any circuit, the source sees a resistance equal to the terminal voltage divided by the total current. The resistance that the source sees may be considered a lumped quantity. In other words, a circuit whose total resistance is 100 Ω might consist of a 100-Ω resistor, a pair of 50-Ω resistors in series, ten 10-Ω resistors in series, a 10-Ω resistor in series with a 90-Ω resistor, two 200-Ω resistors in parallel, five 500-Ω resistors in parallel, or any combination of resistors, in series, parallel, or both, that would equal 100 Ω. In short, if the circuit is enclosed in a box, as in Fig. 3-33, and the terminal voltage and line current are measured, we can calculate the resistance, but we cannot describe the circuit inside the box.

The resistance between terminals *A* and *B* in Fig. 3.33*a* is found by Ohm's law.

$$R_{AB} = \frac{250}{0.125} = 2{,}000\ \Omega$$

This 2,000 Ω is shown in Fig. 3-33*b*.

Any parallel circuit must be equivalent to a series circuit. This means the opposition to current that a generator sees in a parallel circuit can be duplicated by a simple series circuit. It does not mean that the series circuit can function as a parallel circuit but simply that as far as the source of voltage is concerned, the circuit has not been changed.

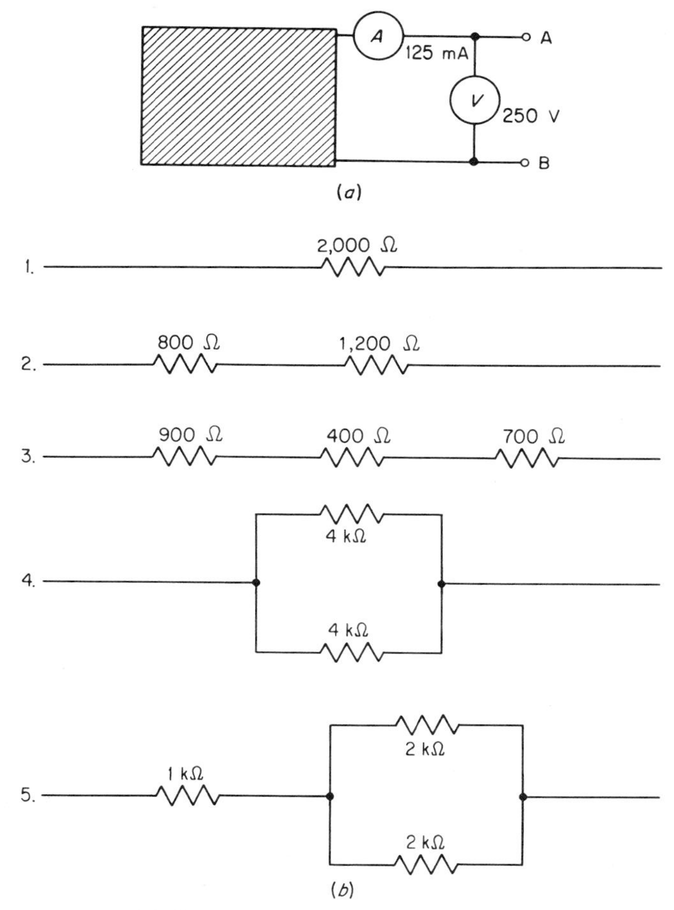

FIGURE 3-33
(*a*) A black-box circuit. (*b*) Some resistance combinations that are equivalent to 2kΩ.

EXAMPLE 3-19 Determine the series-equivalent resistance of the circuit of Fig. 3-34.

SOLUTION Solve for the total resistance of the parallel circuit. This resistance is the series equivalent of the parallel circuit. The most convenient method for finding the total resistance in this example would be to assume a voltage, find total current, and then solve for total resistance.

1 Assume 60 V.

2 The current through the 15-Ω branch is 4 A.

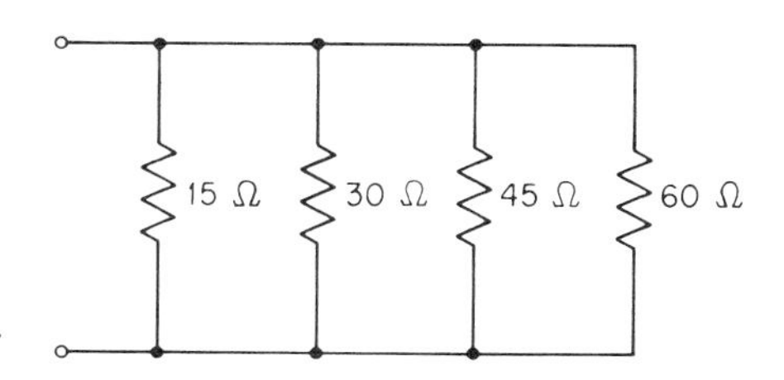

FIGURE 3-34
Circuit for Example 3-19.

3 The current through the 30-Ω branch is 2 A.
4 The current through the 45-Ω branch is 1.33 A.
5 The current through the 60-Ω branch is 1 A.
6 The total current is 8.33 A.
7 The total resistance is

$$R_t = \frac{60}{8.33} = 7.2\ \Omega$$

////

The series resistance that offers the same opposition to current as far as the source is concerned is known as the *series equivalent* of the parallel circuit. This idea of a series equivalent is very important. It is a concept that is used throughout the analysis of electric circuits.

EXAMPLE 3-20 Determine the series-equivalent resistances of the circuits in Fig. 3-35.

The proper solution for compound circuits involves a sequence of steps which, if followed in the proper order, make the solution much easier.

1 The first step is to solve the parallel circuits for the series-equivalent resistances.
2 Redraw the circuit, substituting the series equivalents for the parallel circuits. This yields a simple series circuit.
3 Solve the series circuit for current.
4 Calculate the voltage drops across the parallel circuits by using the series-equivalent resistances of the parallel circuits.
5 Determine the branch currents of the parallel circuits by using the voltage drop across the parallel circuit as found in step 4.

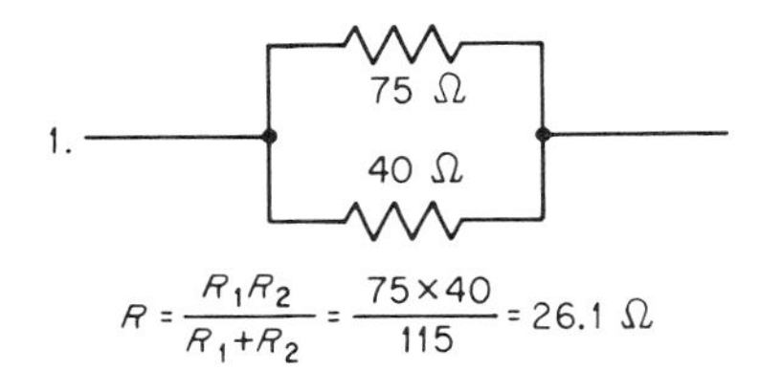

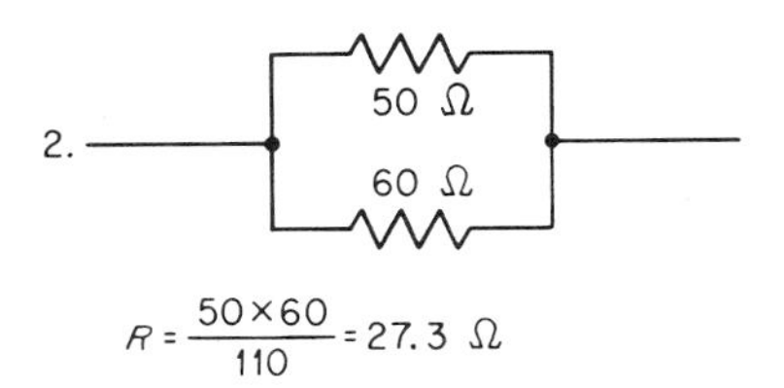

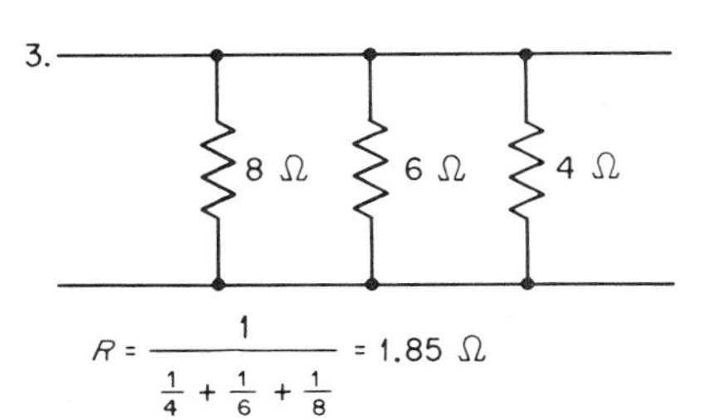

FIGURE 3-35
Circuits used in the solution of Example 3-20.

EXAMPLE 3-21 Find the currents I_1, I_2, I_3, and I_t in the circuit of Fig. 3-36.

SOLUTION

1 The series equivalent of the parallel section is found first.

$$R = \frac{1}{\dfrac{1}{R_1} + \dfrac{1}{R_2} + \dfrac{1}{R_3}} = \frac{1}{\dfrac{1}{200} + \dfrac{1}{300} + \dfrac{1}{100}}$$

$$= \frac{1}{0.005 + 0.00333 + 0.01} = \frac{1}{0.01833} = 54.5\ \Omega$$

2 Redraw the circuit, using the series-equivalent resistance calculated in step 1.

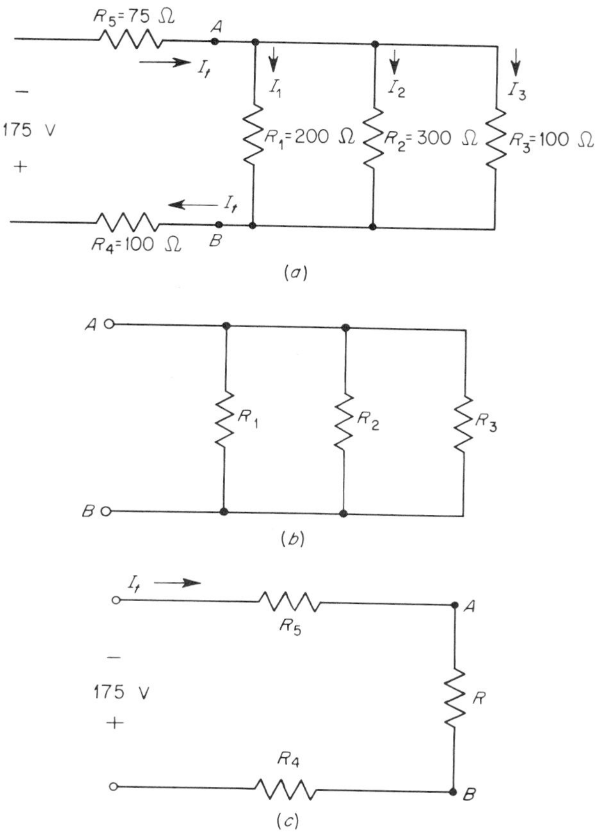

FIGURE 3-36
(*a*) Circuit for Example 3-21. (*b*) The parallel resistances in Example 3-21. (*c*) Series equivalent of the circuit.

3 Solve for the current.

$$R_t = 75 + 54.5 + 100 = 229.5\ \Omega$$

$$I_t = \frac{E}{R_t} = \frac{175}{229.5} = 0.763\ \text{A}$$

4 Calculate the voltage drop across the series-equivalent resistance of the parallel circuit.

$$E = I_t R = (0.763)(54.5) = 41.6\ \text{V}$$

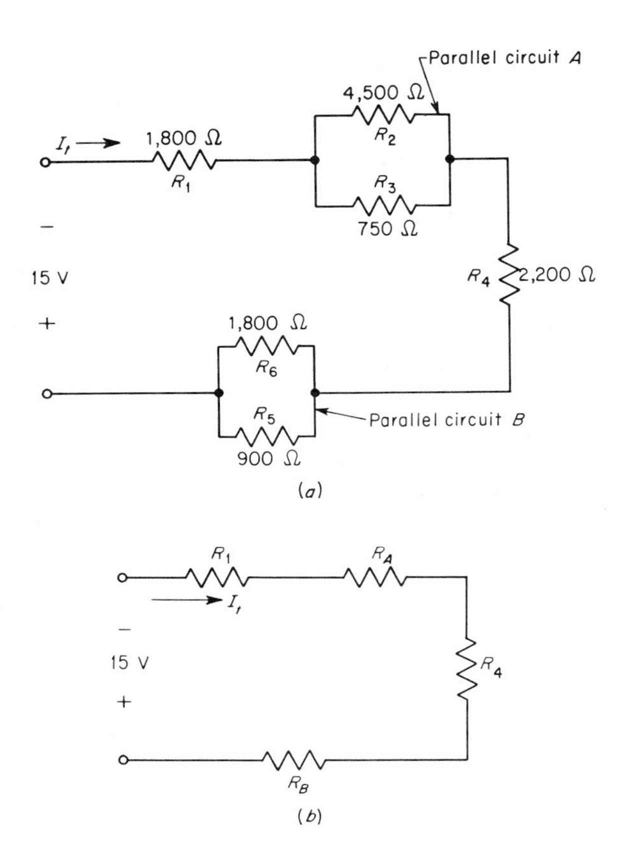

FIGURE 3-37
(*a*) Circuit for Example 3-22. (*b*) Series equivalent of actual circuit.

5 Calculate the branch currents.

$$I_t = \frac{E}{R_1} = \frac{41.6}{200} = 0.208 \text{ A}$$

$$I_2 = \frac{E}{R_1} = \frac{41.6}{300} = 0.139 \text{ A}$$

$$I_3 = \frac{E}{R_3} = \frac{41.6}{100} = 0.416 \text{ A}$$

PROOF

1 The sum of the branch currents must equal the total current.

$$I_t = 0.763 \text{ A} \qquad \text{from step 3}$$
$$= I_1 + I_2 + I_3 = 0.208 + 0.139 + 0.416 = 0.763 \text{ A}$$

2 The sum of the voltage drops must equal the applied voltage.

$$E = 41.6 \text{ V} \qquad \text{from step 4}$$
$$E_4 = I_t R_4 = (0.763)(100) = 76.3 \text{ V}$$
$$E_5 = I_t R_5 = (0.763)(75) = 57.1 \text{ V}$$
$$41.6 + 57.1 + 76.3 = 175 \text{ V}$$

////

EXAMPLE 3-22 Given the circuit of Fig. 3-37, find the voltage drop across R_4.

SOLUTION

1 Solve for the series-equivalent resistances of the parallel circuits.

$$R_A = \frac{R_2 R_3}{R_2 + R_3} = \frac{(4{,}500)(750)}{4{,}500 + 750} = 642\ \Omega$$

$$R_B = \frac{R_5 R_6}{R_5 + R_6} = \frac{(1{,}800)(900)}{1{,}800 + 900} = 600\ \Omega$$

2 Redraw the circuit, substituting the series equivalents R_A and R_B for the parallel circuits.

3 Solve for the total resistance of the circuit of step 2.

$$R_t = R_1 + R_A + R_4 + R_B = 1{,}800 + 642 + 2{,}200 + 600 = 5{,}242\ \Omega$$

4 Solve for the current.

$$I_t = \frac{E}{R_t} = \frac{15}{5{,}242} = 0.00296 \text{ A}$$

5 The voltage drop across R_4 is now found by Ohm's law.

$$E_4 = I_t R_4 = (0.00286)(2{,}200) = 6.29 \text{ V}$$

////

The solution of compound circuits outlined enables the student to solve complex circuits as readily as simple series circuits. The following problems will provide an opportunity to practice this technique.

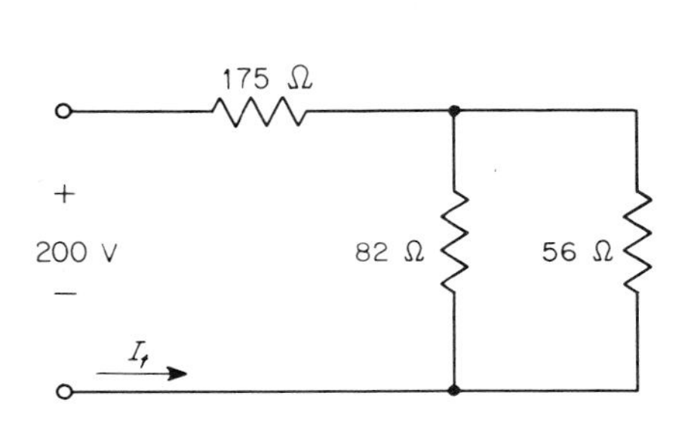

FIGURE 3-38
Circuit for Prob. 3-46.

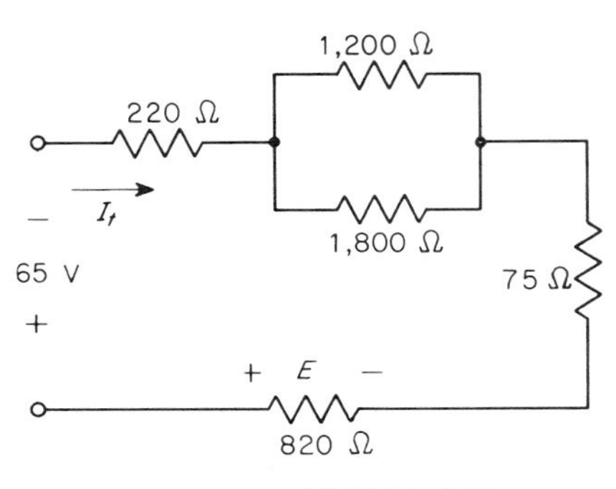

FIGURE 3-39
Circuit for Prob. 3-47.

PROBLEMS

3-46 Given the circuit of Fig. 3-38, solve for the current through the 56-Ω resistor.

3-47 What is the voltage drop across the 820-Ω resistor in the circuit of Fig. 3-39?

3-48 Find the voltage drop across the 1,800-Ω resistor in the circuit of Fig. 3-40.

3-49 Find the value of the unknown resistor in Fig. 3-41.

3-50 What is R if E_{out} is 1 V? (Use the circuit of Fig. 3-42.)

3-51 Given the circuit of Fig. 3-43, find the current through each resistance, the total current, the applied voltage, and the voltage drop across each resistance when the current through the 56-Ω resistor is 0.8 A.

3-52 The circuit of Fig. 3-44 represents an adjustable level control. As the contactor is moved from contacts 1-1 to 4-4 successively, the voltage across load R_L is one-tenth of what it was at each previous contact. The total resistance across the input terminals remains 1,500 Ω. What must be the values of resistances R_a, R_a', R_a'', R_b, R_b', and R_b''?

3-53 Solve the circuit of Fig. 3-45 for total resistance.

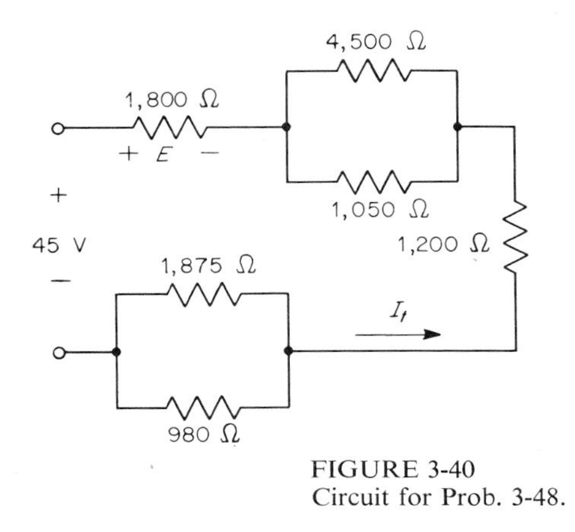

FIGURE 3-40
Circuit for Prob. 3-48.

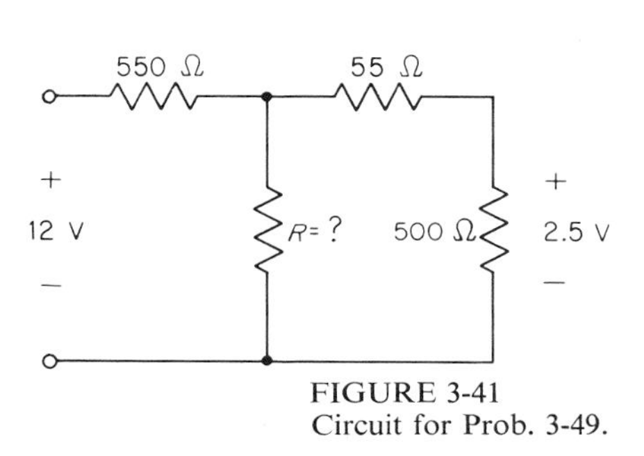

FIGURE 3-41
Circuit for Prob. 3-49.

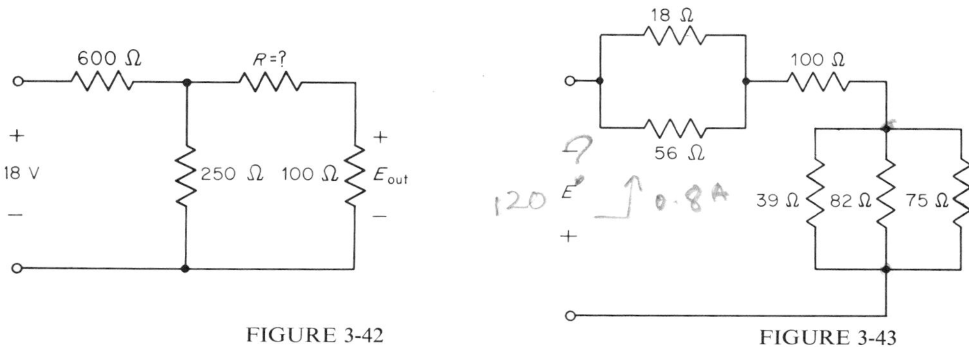

FIGURE 3-42
Circuit for Prob. 3-50.

FIGURE 3-43
Circuit for Prob. 3-51.

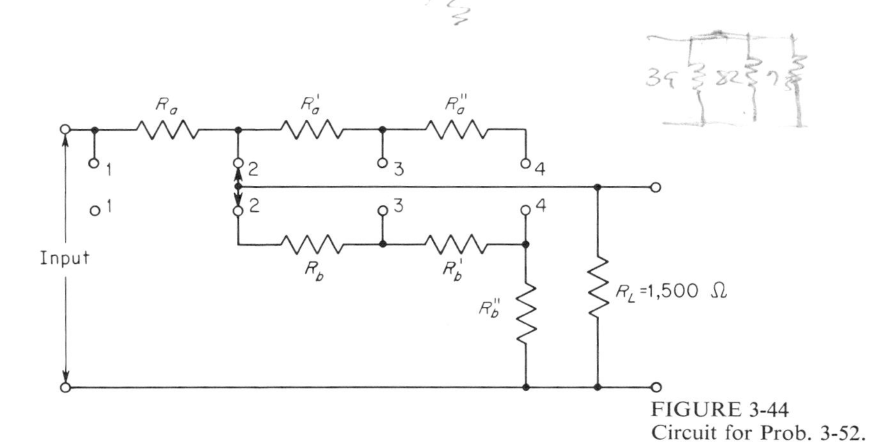

FIGURE 3-44
Circuit for Prob. 3-52.

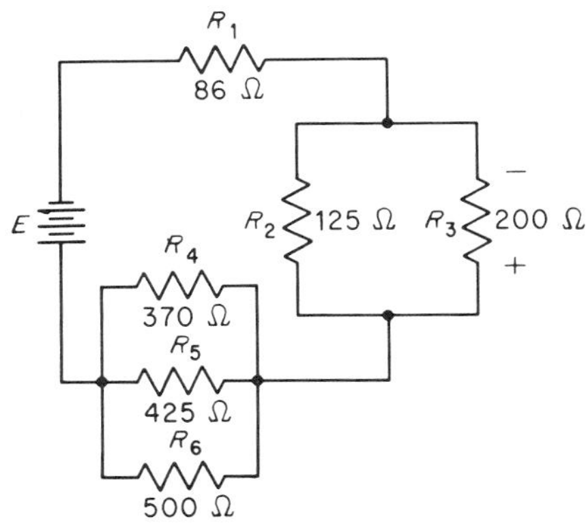

FIGURE 3-45
Circuit for Probs. 3-53 to 3-56.

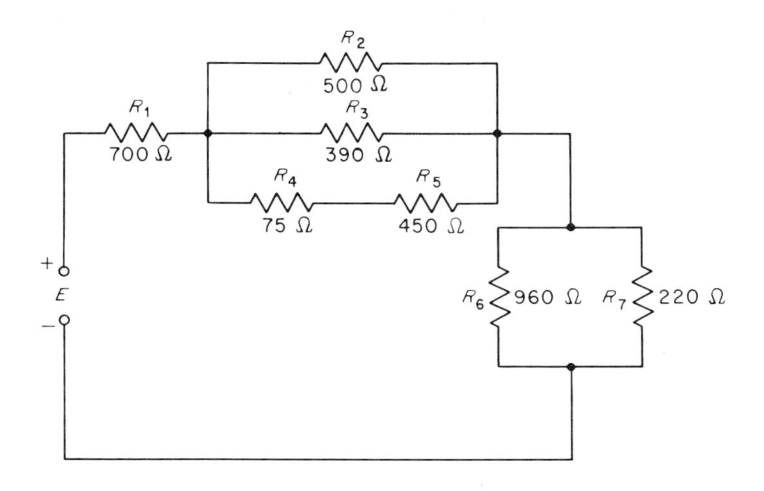

FIGURE 3-46
Circuit for Probs. 3-57 to 3-60.

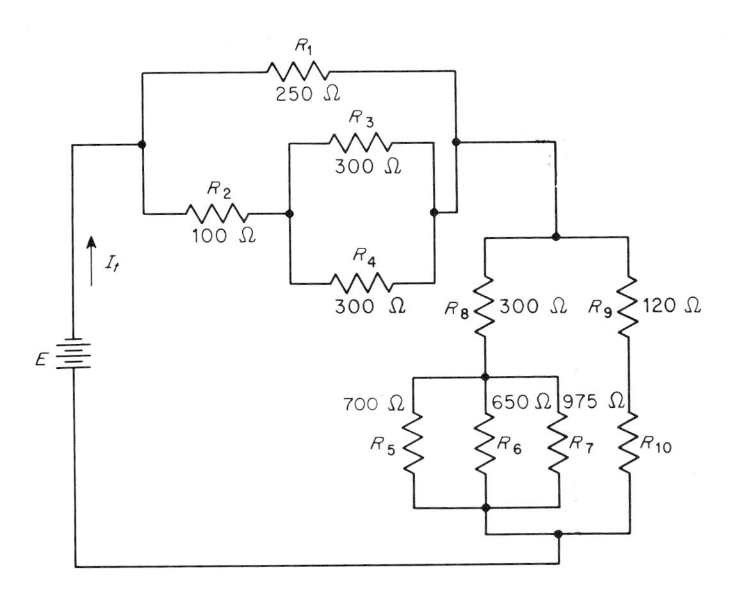

FIGURE 3-47
Circuit for Probs. 3-61 to 3-64.

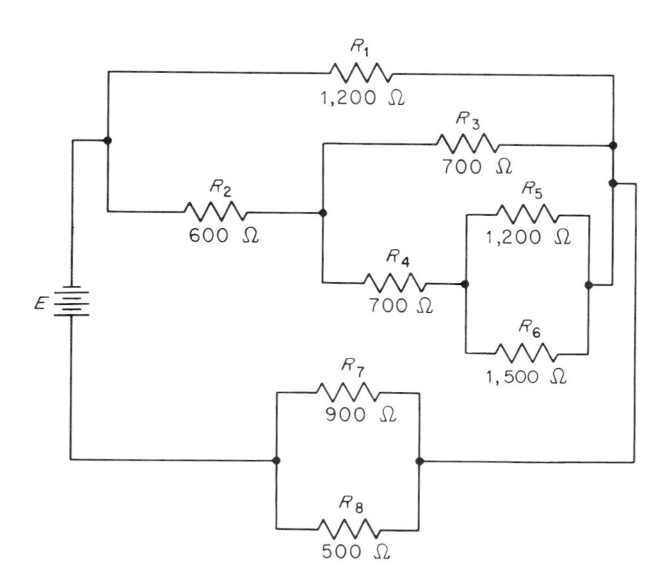

FIGURE 3-48
Circuit for Probs. 3-65 to 3-68.

3-54 In the circuit of Fig. 3-45, find the applied voltage E when the voltage across R_3 is 40 V.

3-55 In the circuit of Fig. 3-45, find the voltage drop across R_2 if a current of 65 mA is measured in the R_6 branch.

3-56 In the circuit of Fig. 3-45, find the current through R_5 if E is 275 V.

3-57 Solve the circuit of Fig. 3-46 for the total resistance.

3-58 In the circuit of Fig. 3-46, find the drop across R_5 if E is 350 V.

3-59 In the circuit of Fig. 3-46, find the applied voltage E if the drop across R_4 is 25 V.

3-60 In the circuit of Fig. 3-46, find the current through R_2 if the current through R_7 is 20 mA.

3-61 In the circuit of Fig. 3-47, find R_{10} if I_t is 95 mA and I_6 is 15 mA.

3-62 In the circuit of Fig. 3-47, find I_t if R_{10} is 400 Ω and E is 500 V.

3-63 In the circuit of Fig. 3-47, find I_5 if R_{10} is 200 Ω and E_9 is 45 V.

3-64 In the circuit of Fig. 3-47, find E_8 if R_{10} is 500 Ω and I_3 is 40 mA.

3-65 In the circuit of Fig. 3-48, find E if I_6 is 20 mA.

3-66 In the circuit of Fig. 3-48, find E_1 if E_8 is 90 V.

3-67 In the circuit of Fig. 3-48, find I_4 if E is 1,200 V.

3-68 What is the total conductance of the circuit of Fig. 3-48?

3-69 In the circuit of Fig. 3-49, find E_7 if E is 200 V.

3-70 In the circuit of Fig. 3-49, find E_5 if I_5 is 45 mA.

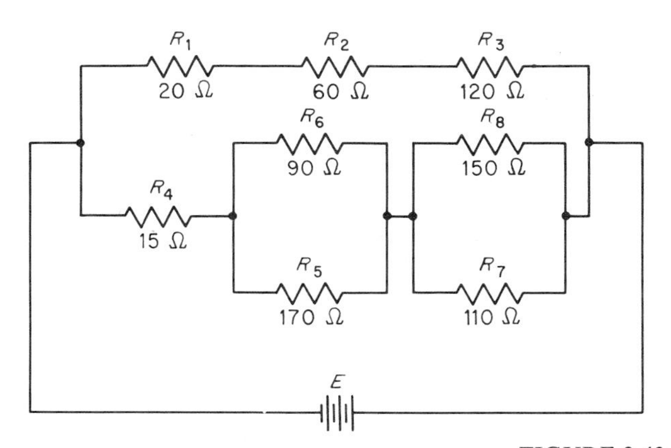

FIGURE 3-49
Circuit for Probs. 3-69 and 3-70.

3-8 POWER

Power is defined as the *rate of doing work* or the *rate of transforming energy*. In mathematical form,

$$P = \frac{W}{t} \tag{3-14}$$

where P = power, W
W = work, J
t = time, s

If Eq. (3-1) is substituted into Eq. (3-14), the following relation results:

$$W = EQ$$

$$P = \frac{W}{t} = \frac{EQ}{t} = E\frac{Q}{t}$$

As defined earlier, charge per unit time is current I. Therefore,

$$P = EI \tag{3-15}$$

Using Ohm's law and Eq. (3-15), other forms of the power equation can be derived.

$$P = EI = E\frac{E}{R} = \frac{E^2}{R} \tag{3-16}$$

$$P = EI = IR \times I = I^2 R \tag{3-17}$$

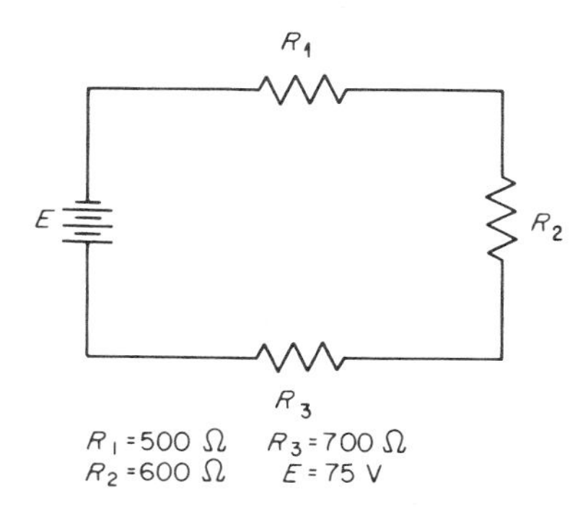

FIGURE 3-50
Circuit for Example 3-23.

In Eqs. (3-15) to (3-17), power P is in watts when current is in amperes, voltage is in volts, and resistance is in ohms.

It can be seen from the preceding that current flowing through a resistance causes energy to be transformed. In this case electric energy is transformed into heat energy. The rate at which energy is transformed is power. Many electric components are given power ratings, which indicate the rates at which they can dissipate heat energy without being destroyed.

Multiplying both sides of Eq. (3-14) by time t yields (3-18)

$$W = Pt$$

where W = energy, J
P = power, W
t = time, s

It is now apparent that 1 J is equivalent to 1 W·s.

EXAMPLE 3-23 In the circuit of Fig. 3-50, find the total power supplied by the battery and the power dissipated as heat in each resistor.

SOLUTION

1 To find the total power supplied by the battery, the current supplied must first be found.

$$I_t = \frac{E}{R_1 + R_2 + R_3} = \frac{75}{1{,}800} = 0.0417 \text{ A}$$

Equation (3-15) then yields

$$P = E \times I = (75)(.0417) = 3.13 \text{ W}$$

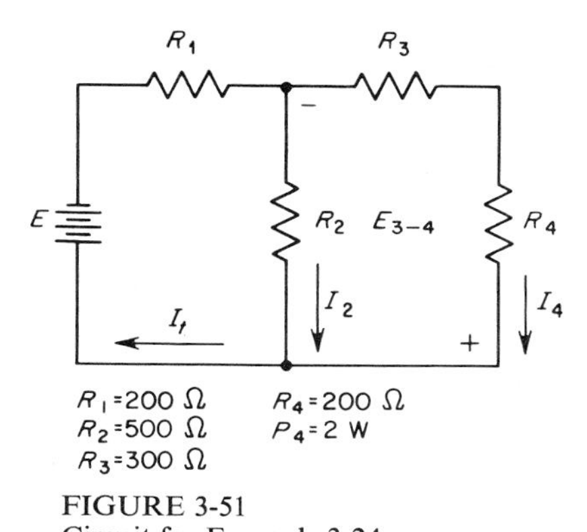

FIGURE 3-51
Circuit for Example 3-24.

2 To find the power dissipated in each resistor, Eq. (3-17) is applied.

$$P_1 = I^2R_1 = (0.0417)^2(500) = 0.870 \text{ W}$$
$$P_2 = I^2R_2 = (0.0417)^2(600) = 1.044 \text{ W}$$
$$P_3 = I^2R_3 = (0.0417)^2(700) = 1.216 \text{ W}$$

3 It should be noted that all the power supplied by the battery is dissipated in the form of heat by the resistors.

$$P_t = P_1 + P_2 + P_3 = 0.870 + 1.044 + 1.216 = 3.13 \text{ W} \qquad ////$$

EXAMPLE 3-24 In the circuit of Fig. 3-51, find the total power delivered by the source.

SOLUTION

1 The current through R_4 can be found

$$I_4 = \sqrt{\frac{P_4}{R_4}} = \sqrt{\frac{2}{200}} = \sqrt{.01} = 0.1 \text{ A}$$

2
$$E_{3\text{-}4} = (I_4)(R_3 + R_4) = (0.1)(500) = 50 \text{ V}$$

3 The voltage found in step 2 is also the voltage across R_2. Therefore, the current through R_2 can be found.

$$I_2 = \frac{E_{3\text{-}4}}{R_2} = \frac{50}{500} = 0.1 \text{ A}$$

4 The current through R_1 is the sum of the currents through R_2 and R_3.

$$I_t = I_2 + I_4 = 0.1 + 0.1 = 0.2 \text{ A}$$

5 The voltage across R_1 is

$$E_1 = I_t R_1 = (0.2)(200) = 40 \text{ V}$$

6 The applied voltage E is the sum of the drops across R_1 and the parallel circuit.

$$E = E_1 + E_{3\text{-}4} = 40 + 50 = 90 \text{ V}$$

7 The power supplied by the battery can now be found.

$$P = EI_t = (90)(0.2) = 18 \text{ W}$$

8 As a check on the solution in step 7, the sum of the power dissipated by the resistors should equal the power supplied by the battery.

$$\begin{aligned} P &= P_1 + P_2 + P_3 + P_4 \\ &= I_t^2 R_1 + I_2^2 R_2 + I_4^2 R_3 + I_4^2 R_4 \\ &= (0.2)^2(200) + (0.1)^2(500) + (0.1^2)(300) + (0.1^2)(200) \\ &= 8.0 + 5.0 + 3.0 + 2.0 = 18.0 \text{ W} \end{aligned}$$

////

PROBLEMS

3-71 When 75 V is applied to a circuit, there is a movement of 46 C of charge. Find the energy transformed.

3-72 If 91 J of work is performed when 2.5 C of charge move, what potential difference is the charge moved through?

3-73 If 2.5 J of work is performed while moving a charge through a potential of 62 millivolts (mV) how much charge is moved?

3-74 One electron is moved through a difference of potential of 1 V. How much energy is transformed?

3-75 If 4.9 C of charge is moved through 1.9 V difference of potential in 0.002 s, what is the rate of energy transformation in watts?

3-76 The rate of energy transformation is 90 W. How much energy is transformed in 0.059 s?

3-77 In the circuit of Fig. 3-52, how much power is dissipated in each resistor if $E = 100$ V? How much energy is transformed to heat energy in each resistor in 4.7 min if $I_3 = 0.2$ A?

3-78 When the power in R_2 of Fig. 3-52 is 2.1 W, find the total power delivered to the circuit.

3-79 In the circuit of Fig. 3-53, find the total power delivered when P_2 is 270 milliwatts (mW) and I_2 is 3.5 mA.

3-80 How much power is dissipated in R_5 of Fig. 3-53 when I_2 is 75 mA?

3-81 A load rated at 40 V and 75 W is to be operated from a 110-V source. Find the series resistor necessary. What must be the power rating of the series resistor?

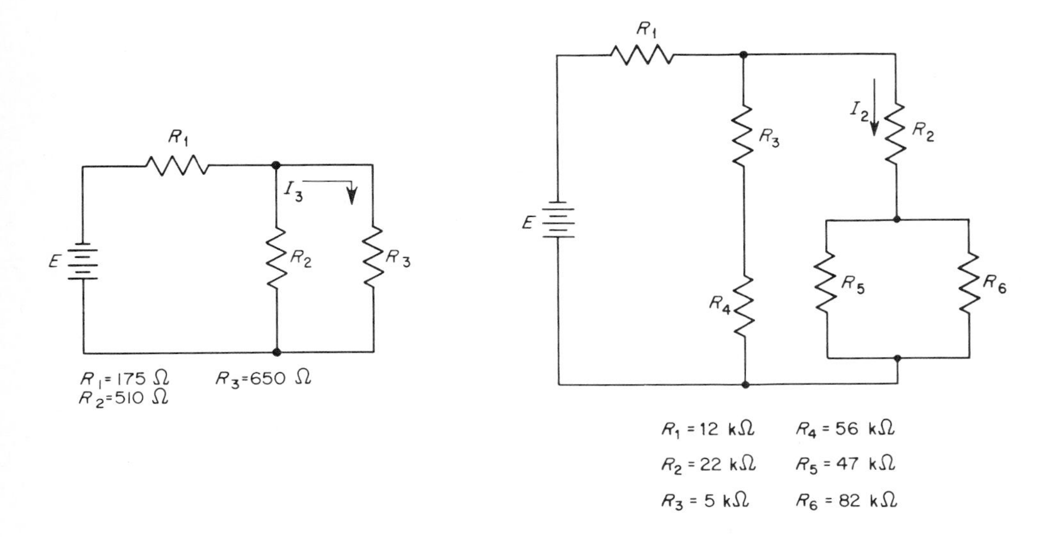

FIGURE 3-52
Circuit for Probs. 3-77 and 3-78.

FIGURE 3-53
Circuit for Probs. 3-79 and 3-80.

3-9 MAXIMUM POWER TRANSFER

In any device that converts another form of energy into electric energy there is internal resistance. This is another way of saying that losses occur within the device so that not all its original energy is available at its terminals. In Fig. 3-54, points A and B represent the terminals of the battery. E_s represents the chemical energy converted into electric energy by the battery. If no current is drawn from the battery by an external load, the voltage across terminals A-B will be the electromotive force E_s. E_s is also referred to as the *no-load voltage.*

If a load is connected to terminals A-B of Fig. 3-54, current will flow, and there will be a voltage drop within the battery due to R_I. Thus the voltage at terminals A-B will be less than the no-load voltage by the amount of the voltage drop across R_I. This condition is shown in Fig. 3-55. As R_L is decreased the current drawn will increase, and as a result the internal voltage drop IR_I will increase. As the internal voltage drop increases, the terminal voltage seen across the load decreases. This sequence of events shows that it is desirable for the internal resistance to be very small to ensure that changes in terminal voltage will be negligible over a wide range of load currents.

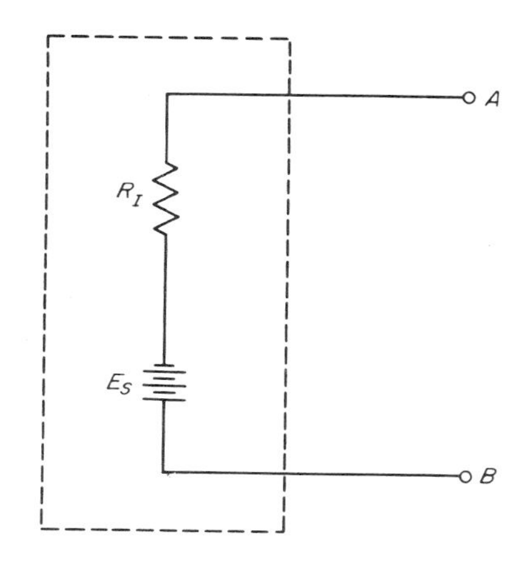

FIGURE 3-54
Battery under no-load conditions.

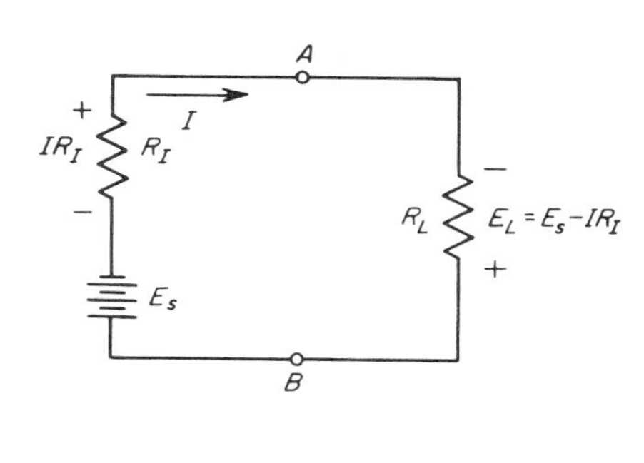

FIGURE 3-55
Battery under load conditions.

EXAMPLE 3-25 If the internal resistance of a battery is 10 Ω and the no-load voltage is 100 V, find the terminal voltage when (1) $R_L = 100\ \Omega$, (2) $R_L = 10\ \Omega$, (3) $R_L = 1\ \Omega$.

SOLUTION

1

$$I = \frac{E_s}{R_I + R_L} = \frac{100}{10 + 100} = \frac{100}{110} = 0.91\ \text{A}$$

$$E_L = IR_L = (0.91)(100) = 91\ \text{V}$$

2

$$I = \frac{E_s}{R_I + R_L} = \frac{100}{20} = 5\ \text{A}$$

$$E_L = IR_L = (5)(10) = 50\ \text{V}$$

3

$$I = \frac{E_s}{R_I + R_L} = \frac{100}{11} = 9.1\ \text{A}$$

$$E_L = IR_L = (9.1)(1) = 9.1\ \text{V}$$

EXAMPLE 3-26 If the internal resistance of a battery is 0.1 Ω and the no-load voltage is 100 V, find the terminal voltage when (1) $R_L = 100\ \Omega$, (2) $R_L = 10\ \Omega$, (3) $R_L = 1\ \Omega$.

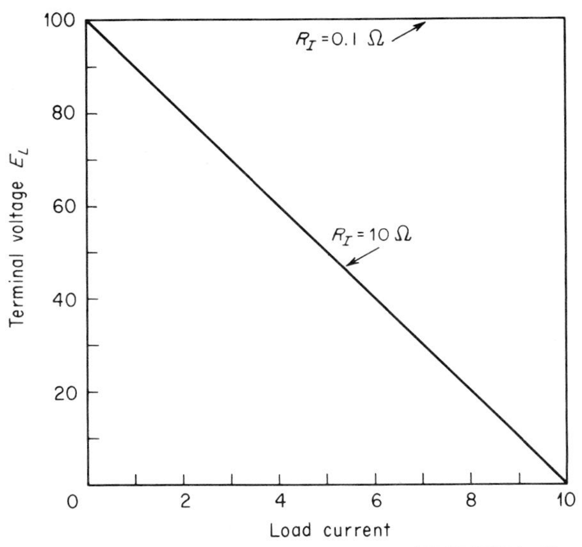

FIGURE 3-56
Terminal voltage versus load current.

SOLUTION

1

$$I = \frac{E_s}{R_I + R_L} = \frac{100}{100.1} \cong 1\ \text{A}$$

$$E_L = IR_L = (1)(100) = 100\ \text{V}$$

2

$$I = \frac{E_s}{R_I + R_L} = \frac{100}{10.1} = 9.9\ \text{A}$$

$$E_L = IR_L = (9.9)(10) = 99\ \text{V}$$

3

$$I = \frac{E_s}{R_I + R_L} = \frac{100}{1.1} = 91\ \text{A}$$

$$E_L = IR_L = (91)(1) = 91\ \text{V}$$

In Example 3-26, the battery has a large internal resistance compared with Example 3-25. To supply less than 10 A of current to the load, the voltage across the load drops from 100 to 9.1 V in Example 3-25. In Example 3-26, to supply a similar load current, the voltage across the load drops only 1 V. The current and voltage relationships of Examples 3-25 and 3-26 are plotted in Fig. 3-56. In this figure, load voltage E_L is plotted against load current I for the values of R_I given in these examples.

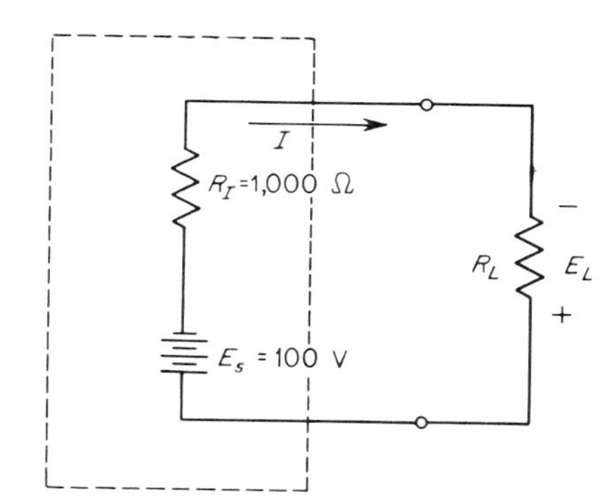

FIGURE 3-57
Circuit for Example 3-28.

EXAMPLE 3-27 The no-load voltage of a dc supply is 350 V, and a load connected to the supply draws 150 mA. What is the internal resistance of this supply? The terminal voltage when 150 mA is drawn is 320 V.

SOLUTION

1 Voltage not present across the load must be dropped across the internal resistance.

$$E_I = E_s - E_L = 350 - 320 = 30 \text{ V}$$

2 The internal resistance can now be found.

$$R_I = \frac{E_I}{I} = \frac{30}{0.150} = 200 \ \Omega$$

There is a relationship between load resistance and internal resistance of a source called the *maximum-power-transfer theorem.*

Maximum-power-transfer theorem Maximum power is delivered to a load when the load resistance equals the internal resistance of the source.

EXAMPLE 3-28 In the circuit of Fig. 3-57, find the power in the load when R_L is (1) 500 Ω, (2) 1,000 Ω, (3) 1,500 Ω.

SOLUTION

1

$$I = \frac{E_s}{R_I + R_L} = \frac{100}{1{,}500} = 0.0667 \text{ A}$$

$$E_L = IR_L = (0.0667)(500) = 33.4 \text{ V}$$

$$P_L = IE_L = (0.0667)(33.4) = 2.22 \text{ W}$$

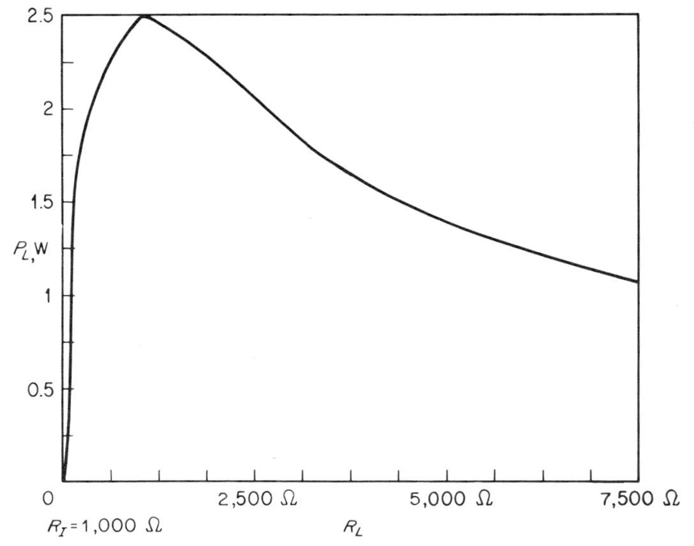

FIGURE 3-58
Power developed in the load versus load resistance.

2

$$I = \frac{E_s}{R_I + R_L} = \frac{100}{2{,}000} = 0.05\ \text{A}$$

$$E_L = IR_L = (0.05)(1{,}000) = 50\ \text{V}$$

$$P_L = IE_L = (0.05)(50) = 2.5\ \text{W}$$

3

$$I = \frac{E_s}{R_I + R_L} = \frac{100}{2{,}500} = 0.04\ \text{A}$$

$$E_L = IR_L = (0.04)(1{,}500) = 60\ \text{V}$$

$$P_L = IE_L = (0.04)(60) = 2.4\ \text{W}$$

The relationship between P_L and R_L is plotted in Fig. 3-58. The graph is plotted for the values of R_L and E_s given in Fig. 3-57. ////

EXAMPLE 3-29 In the circuit of Fig. 3-59, what must R_L be so that maximum power is developed in R_L? What is the maximum power in R_L?

SOLUTION

1 Determine the voltage drop across the load.

$$E_L = IR_L = (10 \times 10^{-3})(9.5 \times 10^{3}) = 95\ \text{V}$$

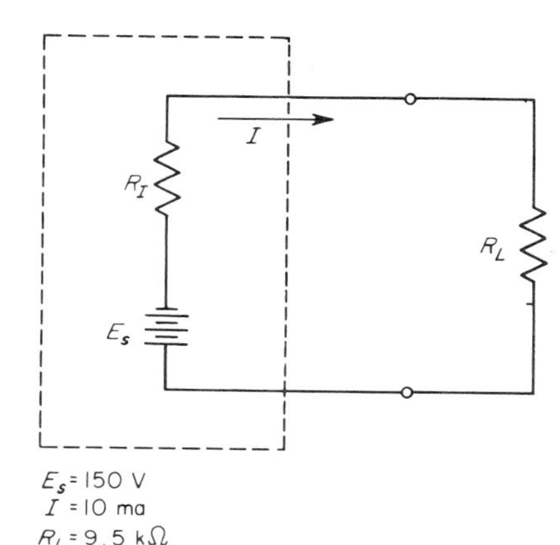

FIGURE 3-59
Circuit for Example 3-29.

2 The voltage dropped by the internal resistance is

$$E_I = E_s - E_L = 150 - 95 = 55 \text{ V}$$

3 The internal resistance can now be found.

$$R_I = \frac{E_I}{I} = \frac{55}{10 \times 10^{-3}} = 5.5 \times 10^3 \ \Omega$$

4 By the maximum-power-transfer theorem, maximum power is developed when load resistance equals internal resistance. Therefore $R_L = 5{,}500 \ \Omega$ for maximum power transfer.

$$I = \frac{E_s}{R_I + R_L} = \frac{150}{11{,}000} = 1.36 \times 10^{-2} \text{ A}$$

5 Power in the load can now be found.

$$P_L = I^2 R_L = (1.36 \times 10^{-2})^2(5{,}500) = 1.02 \text{ W}$$ ////

3-10 EFFICIENCY

It was pointed out in Sec. 3-2 that whenever energy is converted from one form into another or transmitted, there are losses of energy as a result of this transmission. This situation gives rise to the concept of efficiency. Efficiency is the ratio of power received by a load to the power supplied by a source of energy. This ratio is shown in Eq.

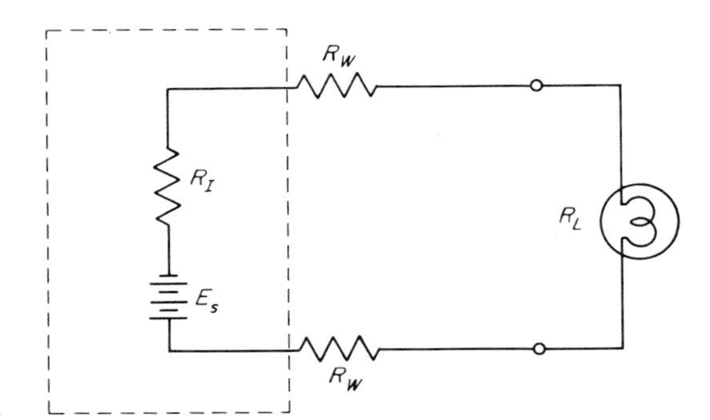

FIGURE 3-60
Battery with a light-bulb load.

(3-19). Efficiency has no units because it is simply a ratio of one quantity of power to another.

$$\text{Efficiency} = \frac{P_L}{P_s} \tag{3-19}$$

Efficiency is normally specified as a percentage:

$$\text{Percent efficiency} = \frac{P_L}{P_s} \times 100\% \tag{3-20}$$

Examples of efficiency are many and varied. Consider a light bulb connected to a source of voltage as in Fig. 3-60. The light bulb is connected to the source by wires that have some resistance, however small. This wire resistance is shown as R_W and the internal resistance of the source as R_I in Fig. 3-60. Any power dissipated in R_I or R_W is not being delivered to the useful load R_L. Thus the efficiency of the action taking place in Fig. 3-60 is the ratio of P_L, the power dissipated by the light bulb, to P_s, the power delivered by the source.

A mechanical analog of the preceding example is the gasoline engine. When gasoline is burned in the cylinders, its energy is converted to a form that sets the pistons in motion. The motion of the pistons, through a mechanical linkage, drives a vehicle. Not all the energy converted within the cylinders, however, is available to move the vehicle. Much of the energy is converted to heat by friction within the mechanical linkage. The result is a loss of efficiency due to frictional losses in the system.

In electric circuitry, many components function as electric "linkage." The function of these linkage components is to make available at the load the desired amount of voltage, and thus the power dissipated in these linkage components is lost as far as the load is concerned.

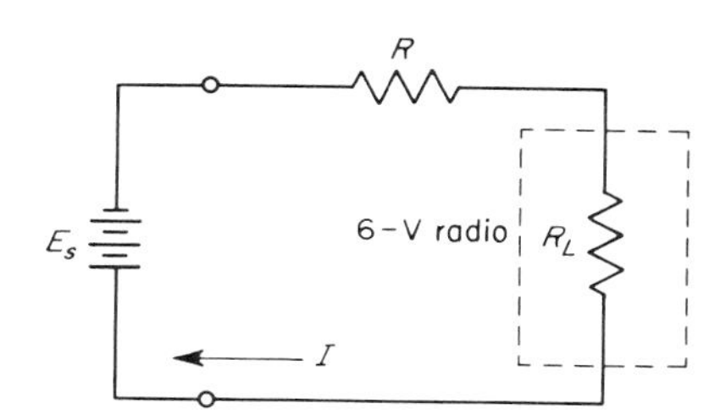

FIGURE 3-61
Circuit for Example 3-30.

EXAMPLE 3-30 A car is purchased with a 12-V electric system. A radio whose input is rated at 6 V and 6 A is also available. What amount of series resistance must be used to adapt the radio for use in the car? What is the percentage efficiency of this conversion? See Fig. 3-61.

SOLUTION

1 Sketch the circuit shown in Fig. 3-61.

2 The voltage to be dropped across R is

$$E_R = E_s - E_L = 12 - 6 = 6$$

3 The required series resistance can now be found.

$$R = \frac{E_R}{I} = \frac{6}{6} = 1\ \Omega$$

4 The power supplied by the source can be found.

$$P_s = E_s I = (12)(6) = 72\ \text{W}$$

5 Power in the load is now calculated.

$$P_L = E_L I = (6)(6) = 36\ \text{W}$$

6 The percent efficiency can be calculated by Eq. 3-20.

$$\text{Percent efficiency} = \frac{P_L}{P_s} \times 100\% = \tfrac{36}{72} \times 100\% = 50\% \qquad ////$$

It has already been mentioned that each resistor or other electric component is given a power rating that indicates the power the component can dissipate without being destroyed. The resistor to be used in Example 3-30 would have to be rated at 36 W or more to dissipate safely the power developed.

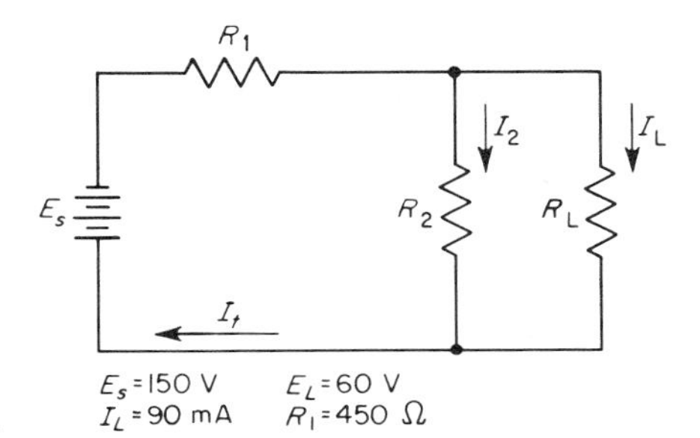

FIGURE 3-62
Circuit for Example 3-31.

EXAMPLE 3-31 In the circuit of Fig. 3-62, the desired load current and voltage are given. The only dc supply available is rated at 150 V. The 450-Ω resistor is available. (1) What resistance must be purchased for R_2? (2) What power rating is necessary for R_2? (3) What is the percent efficiency?

SOLUTION

1 The drop across R_1 is

$$E_1 = E_s - E_L = 150 - 60 = 90 \text{ V}$$

The current I_t through R_1 is

$$I_t = \frac{E_1}{R_1} = \frac{90}{450} = 0.2 \text{ A}$$

The current through R_2 can be calculated, since the sum of the parallel branch currents must equal the total current flowing into the parallel circuit.

$$I_t = I_2 + I_L$$
$$I_2 = I_t - I_L = 0.2 - 0.090 = 0.110 \text{ A}$$

Resistance R_2 can now be found.

$$R_2 = \frac{E_L}{I_2} = \frac{60}{0.110} = 545 \ \Omega$$

2 Power in R_2 is

$$P_2 = I_2 E_L = (0.110)(60) = 6.6 \text{ W}$$

Note: Resistors are available in certain standard ohm and watt values. For R_2 it would probably be necessary to purchase a 10-W resistor.

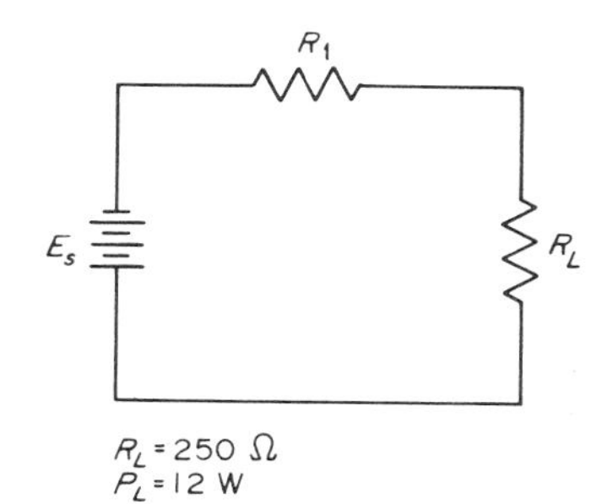

FIGURE 3-63
Circuit for Probs. 3-91 and 3-92.

3 The power in the load is

$$P_L = I_L E_L = (0.090)(60) = 5.4 \text{ W}$$

The power supplied by the source is

$$P_s = I_t E_s = (0.2)(150) = 30 \text{ W}$$

$$\text{Percent efficiency} = \frac{P_L}{P_s} \times 100\% = \frac{5.4}{30} \times 100\% = 18\% \quad ////$$

PROBLEMS

3-82 A source whose no-load voltage is 250 V delivers 100 W to a load when the current is 0.47 A. Find the internal resistance of the source.

3-83 What must the load resistance be for the source of Prob. 3-82 to deliver maximum power to the load? What is the maximum power that can be delivered?

3-84 The no-load voltage of a given source is 1,200 V. When a load draws 120 mA, the voltage across the load is 1,100 V. Calculate the internal resistance of the source.

3-85 If R_3 of Fig. 3-52 is considered the useful load, compute the efficiency when E is 190 V.

3-86 A given load develops 0.75 hp and is 92 percent efficient. How much current is drawn from a 30-V source? (1 hp = 746 W.)

3-87 A source of 325 V has an internal resistance of 75 Ω. What must the load resistance be for 65 percent efficiency?

3-88 In the circuit of Fig. 3-53, if R_5 and R_6 represent the useful load, find the percent efficiency.

3-89 In the circuit of Fig. 3-40, consider the 1,200-Ω resistor the useful load and calculate percent efficiency.

3-90 Find the power in each resistor of the circuit of Fig. 3-43.

3-91 In the circuit of Fig. 3-63, R_L is the useful load and the circuit operates at 70 percent efficiency. Find R_1 and E_s.

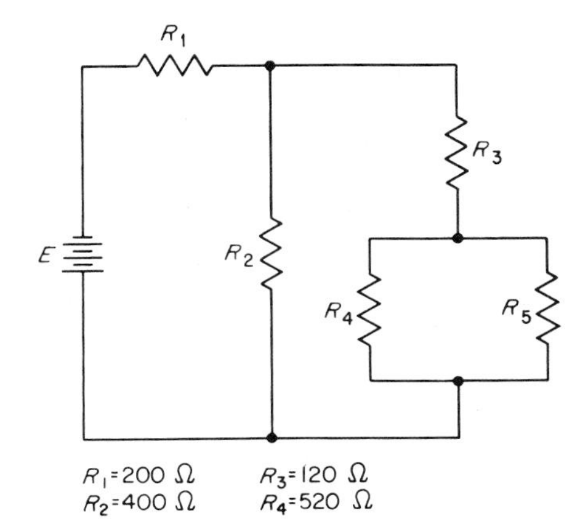

FIGURE 3-64
Circuit for Probs. 3-94 to 3-96.

3-92 In the circuit of Fig. 3-63, using the values of E_s and R_1 found in Prob. 3-91, find R_L for an efficiency of 82 percent.

3-93 Calculate the percentage efficiency for the conditions of Example 3-25.

3-94 In the circuit of Fig. 3-64, if I_5 is 0.2 A and P_5 is 4.8 W, find E.

3-95 Calculate the power in R_2 for the information given in Prob. 3-94.

3-96 In the circuit of Fig. 3-64, if E_5 is 35 V and P_5 is 7 W, find P_1.

3-11 DECIBELS

Soon after the telephone was invented by Alexander Graham Bell, investigations into various types of transmission or telephone lines were started. The ideal telephone line would have zero power loss, and thus the output power would equal the input power. However, all lines exhibited some power loss. A standard figure of measure was needed to permit comparisons between the power-loss properties of various telephone lines.

With a simple ratio such as output power divided by input power, efficiency, the resultant quantity, expressed the fraction of electric power leaving the mouthpiece that arrived as electric power at the earpiece. The problem with this type of measurement is the ear is not sensitive to sound intensity on a linear basis. Power which is doubled to a sound-producing device is not sensed by the ear as a sound with twice the intensity. Ears react to sound intensity on a *logarithmic basis*.

Since the unit of measure should include power levels at two points which reflect ear sensitivity, the bel (B) was adopted. The bel is defined as the common logarithm of the ratio of two powers.

$$\text{Bel} = \log \frac{P_1}{P_2} \tag{3-21}$$

In most practical applications, the bel represents a large power ratio which is usually not encountered in circuits. Thus the decibel (dB), which is one-tenth of a bel, is normally used.

$$\text{Decibel} = 10 \log \frac{P_1}{P_2} \tag{3-22}$$

Terms P_1 and P_2 in Eqs. (3-21) and (3-22) represent power levels at any two points in a circuit. Usually these two power levels occur at the output and input. Therefore, P_1 represents P_{out} and P_2 represents P_{in}. This orientation is adopted because in situations where P_{out} is less than P_{in}, power is lost; the decibel quantity is negative since the logarithm of a fraction is negative. Restating Eq. (3-22) using P_{out} and P_{in} yields

$$\text{dB} = 10 \log \frac{P_{out}}{P_{in}} \tag{3-23}$$

This equation expresses the power gain of a circuit.

In circuits incorporating amplifiers, the output power is larger than the input power, and the decibel quantity is positive. Positive decibels indicate a power increase, and negative decibels indicate a power loss.

EXAMPLE 3-32 An input power of 3 W is applied to the sending end of a telephone line. At the receiving end, 0.5 W of power is measured. Determine the decibel power gain in this telephone line. Also find the bel power gain in the line.

SOLUTION

1 By Eq. (3-23),

$$\text{dB} = 10 \log \frac{0.5}{3}$$

2 Obtaining the log of a fraction is quite easy by taking note of a log property.

$$10 \log X = -10 \log X^{-1} = -10 \log \frac{1}{X}$$

Thus,

$$10 \log \frac{0.5}{3} = -10 \log \frac{3}{0.5} = -10 \log 6$$

3 Taking log 6 and multiplying by -10,

$$\begin{aligned} \text{dB} &= (-10)(0.778) \\ &= -7.78 \end{aligned}$$

4 Expressing power gain in bels,

$$\text{B} = -0.778 \qquad ////$$

EXAMPLE 3-33 A transistor amplifier produces an output power of 2.5 W for an input power of 0.1 W. Determine the amplifier power gain in decibels.

SOLUTION By Eq. (3-23),

$$\begin{aligned} \text{dB} &= 10 \log \frac{2.5}{0.1} = 10 \log 25 \\ &= (10)(1.398) = 13.98 \end{aligned} \qquad ////$$

EXAMPLE 3-34 If 1 W of input power is applied to a circuit which has a power gain of -3 dB, calculate the resulting output power.

SOLUTION

1 By Eq. (3-23),

$$\text{dB} = 10 \log \frac{P_{\text{out}}}{P_{\text{in}}}$$

$$-3 = -10 \log \frac{P_{\text{in}}}{P_{\text{out}}}$$

$$\frac{P_{\text{in}}}{P_{\text{out}}} = \text{antilog } 0.30 = 2.0$$

2 Knowing that $P_{\text{in}} = 1$ W, we have

$$P_{\text{out}} = \frac{P_{\text{in}}}{2.0} = \frac{1}{2.0} = 0.5 \text{ W}$$

Thus a power gain of -3 dB corresponds to a 50 percent loss of input power.

The equations for decibels could be expressed in terms of voltages and currents. Since powers can be expressed

$$P_{\text{out}} = \frac{E_{\text{out}}^2}{R_{\text{out}}} = I_{\text{out}}^2 R_{\text{out}}$$

$$P_{\text{in}} = \frac{E_{\text{in}}^2}{R_{\text{in}}} = I_{\text{in}}^2 R_{\text{in}}$$

decibel power gain can be obtained from Eq. (3-24).

$$\text{dB} = 10 \log \frac{E_{\text{out}}^2/R_{\text{out}}}{E_{\text{in}}^2/R_{\text{in}}} = 10 \log \frac{I_{\text{out}}^2 R_{\text{out}}}{I^2_{\text{in}} R_{\text{in}}} \qquad (3\text{-}24)$$

If the resistances at output and input are equal, the resistive terms in Eq. (3-24) will cancel and (provided $R_{\text{out}} = R_{\text{in}}$) reduce to

$$\text{dB} = 20 \log \frac{E_{\text{out}}}{E_{\text{in}}} = 20 \log \frac{I_{\text{out}}}{I_{\text{in}}} \qquad (3\text{-}25)$$

Another use of the decibel quantity is in expressing the power value in a circuit compared to some standardized reference power level. The reference level is usually taken to be 1 mW (0.001 W) although a level of 6 mW is sometimes used. Referring to Eq. (3-22), P_2 is used as reference level at 1 mW, and P_1 is the power value at some point.

EXAMPLE 3-35 Calculate the decibel output power level of a microphone which produces an output of 0.08 μW.

SOLUTION By Eq. (3-22) and $P_2 = 1$ mW

$$\text{dB} = 10 \log \frac{8 \times 10^{-8}}{1 \times 10^{-3}} = -10 \log \frac{1 \times 10^{-3}}{8 \times 10^{-8}}$$

$$\text{decibel} = -10 \log 12{,}500 = (-10)(4.096)$$

$$\text{decibel} \approx -41$$

////

EXAMPLE 3-36 A mechanical-to-electrical transducer produces a decibel power level of -20 dB. Calculate the output power in watts.

SOLUTION

$$\text{dB} = 10 \log \frac{P_1}{P_2}$$

$$-20 = -10 \log \frac{P_2}{P_1}$$

$$2 = \log \frac{P_2}{P_1}$$

$$\frac{P_2}{P_1} = \text{antilog } 2 = 100$$

$$P_1 = \frac{P_2}{100} = \frac{1 \text{ mW}}{100} = 10\ \mu\text{W}$$

////

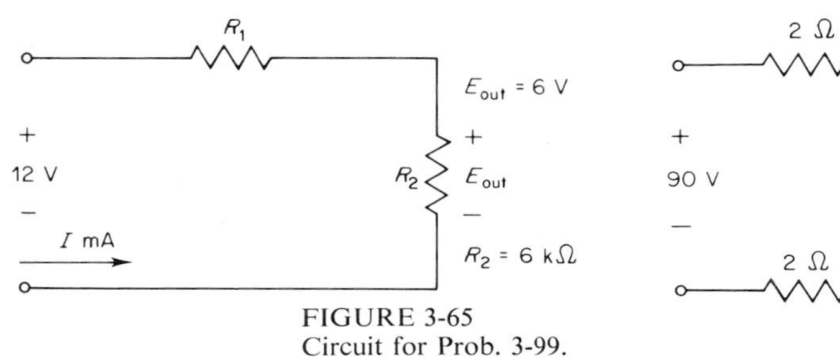

FIGURE 3-65
Circuit for Prob. 3-99.

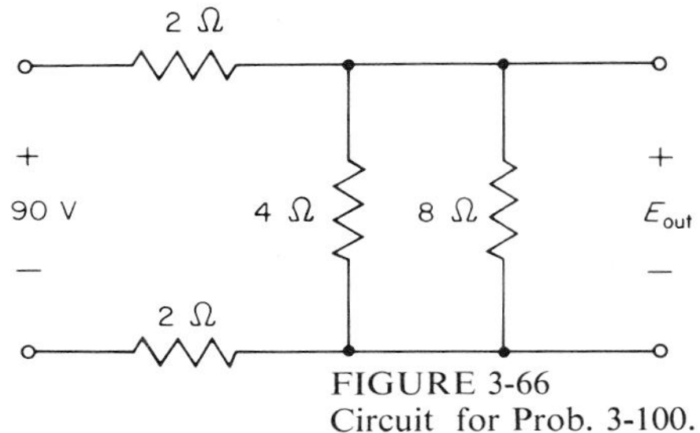

FIGURE 3-66
Circuit for Prob. 3-100.

PROBLEMS

3-97 A network yields an output power of 1 nW for an input power of 1.2 mW. Determine the decibel power gain of the network.

3-98 The percent efficiency of a circuit is 96 percent. Calculate the power gain of the circuit in decibels. If 10 W is desired at the output, how much input power must be supplied?

3-99 Find the decibel power gain for the circuit in Fig. 3-65.

3-100 For the circuit in Fig. 3-66, calculate the decibel power gain. Specify the input power and output power.

3-101 A mechanical transducer provides 8 mV to the input of an amplifier which has an input resistance of 100 kΩ. The amplifier output is 40 V across a 10-Ω resistor. Calculate the decibel power gain.

3-102 Determine the decibel power level the transducer in Prob. 3-101 provides to the input of the amplifier.

4

NETWORK RULES, LAWS, AND THEOREMS

4-1 INTRODUCTION

Network rules, laws, and theorems are necessary tools for analyzing electric circuits. Not only are they necessary but also very convenient. With the proper application of a rule, law, or theorem, a complex network problem can be reduced to a simple straightforward problem.

Though all the analyses will consider resistive networks with dc excitations, the techniques and procedures can be extended to include the impedance effects of capacitors and inductors in networks with ac excitations. The capacitor and inductor are discussed in following chapters.

Sources connected to circuits are of two types. One type is the voltage source. An ideal voltage source maintains the same output terminal voltage regardless of the terminal load resistance. This means, theoretically, that the source would provide a constant voltage in the range of 0 Ω (short circuit) to ∞ Ω (open circuit). Such a large resistance range is possible only if the internal resistance of the source is equal to 0 Ω (see Fig. 4-1). All actual voltage sources have some finite internal resistance (see Fig. 4-2). The ideal current source produces the same output terminal current regardless of terminal load resistance. Once again, load-resistance variations from

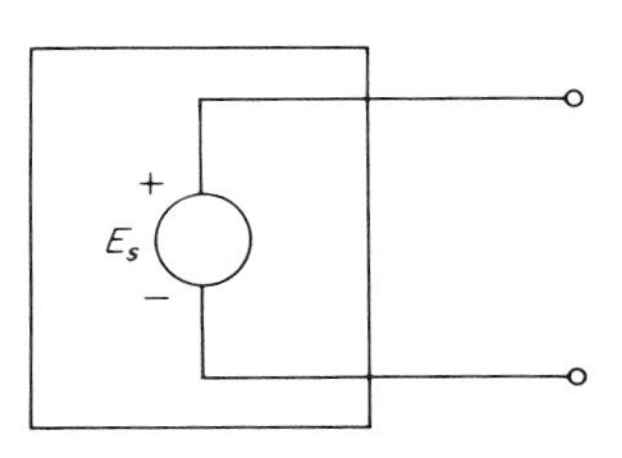

FIGURE 4-1
Ideal voltage source.

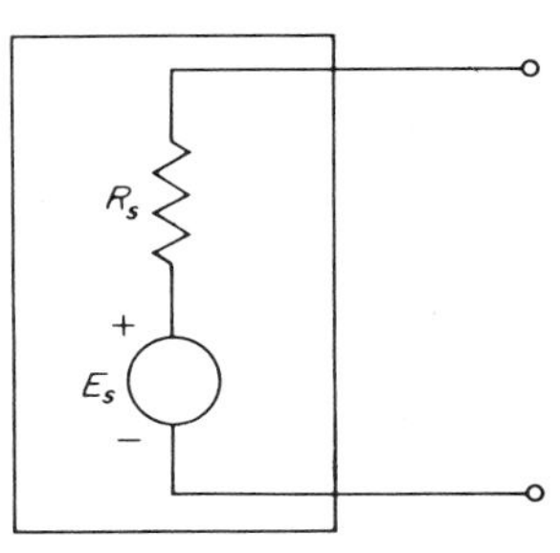

FIGURE 4-2
Actual voltage source.

0 Ω to ∞ Ω would result in the same output current. This implies that the internal resistance of the source must be infinite, and so all source current flows through the load resistance (see Fig. 4-3). Actual current sources have some internal resistance (see Fig. 4-4). The arrow inside the current source points in the direction of electron current flow.

A voltage source or current source is often called an *excitation* in a network, and the various voltage drops or currents throughout the network are called *responses*. The current convention adopted in this text is that of electron current, in other words, the charge movement of electrons from negative terminal to positive terminal on a voltage source. The potential or voltage drop established across a resistance (Fig. 4.5) is such that the negative reference of the voltage drop coincides with the location where electrons enter the resistance.

Conventional current flow is the other type of current convention. When the early discoveries of electricity were being made, scientists thought current flow was accomplished through the movement of positive charges from the positive terminal to the negative terminal of a voltage source. This implies a current direction opposite the electron current direction. The proposed charge direction, though in error, was

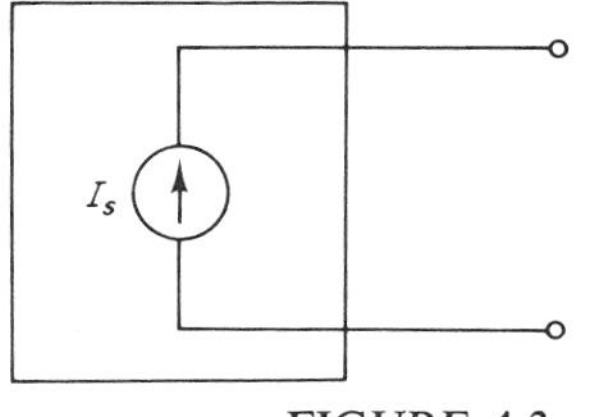

FIGURE 4-3
Ideal current source.

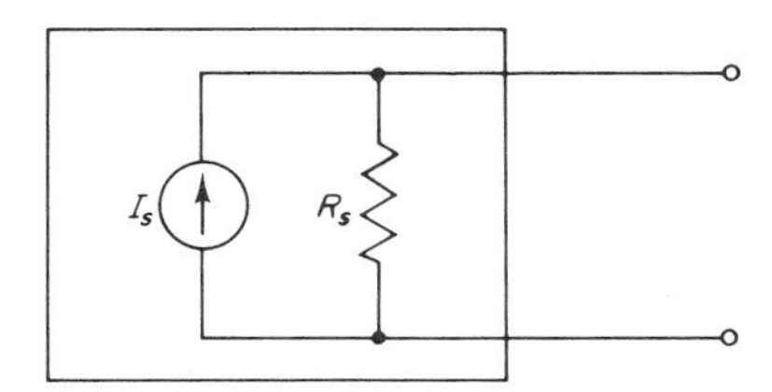

FIGURE 4-4
Actual current source.

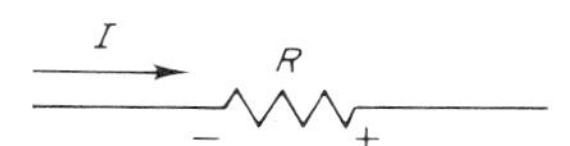

FIGURE 4-5
Voltage drop for given electron current direction.

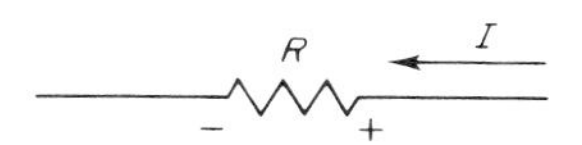

FIGURE 4-6
Voltage drop for given conventional current direction.

accepted as convention and thus the name conventional current. Conventional current flow entering a resistance produces a positive reference at that point with respect to the voltage drop (see Fig. 4-6).

All the network rules, laws, and theorems are valid for both electron current flow and conventional current flow. The only difference is the direction of current flow. Thus the voltage drop in Fig. 4-5 is the same for electron current from left to right or conventional current from right to left.

As mentioned previously, the text will use electron current flow. It will be referred to simply as current flow.

4-2 VOLTAGE DIVISION RULE

This rule yields the voltage drop across a single resistor or several resistors in a series circuit containing voltage sources.

Voltage division rule (VDR) The voltage drop across a resistance in a series circuit is equal to the applied voltage multiplied by the ratio of the resistance where the voltage drop occurs to the sum of the resistance in the series circuit.

PROOF Consider the series circuit containing a voltage source and n resistors in Fig. 4-7. Derive the voltage drop across R_3 using the VDR. Then use Ohm's law for determining the voltage drop.

Using the VDR,

$$E_3 = E\frac{R_3}{R_1 + R_2 + R_3 + \cdots + R_n}$$

Applying Ohm's law,

$$I = \frac{E}{R_1 + R_2 + R_3 + \cdots + R_n}$$

$$E_3 = IR_3$$

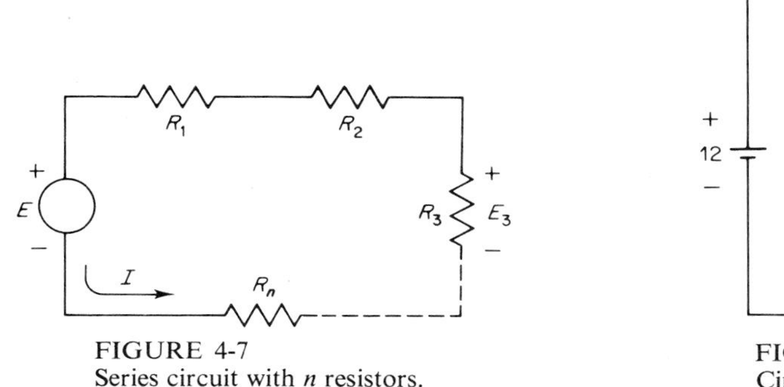

FIGURE 4-7
Series circuit with n resistors.

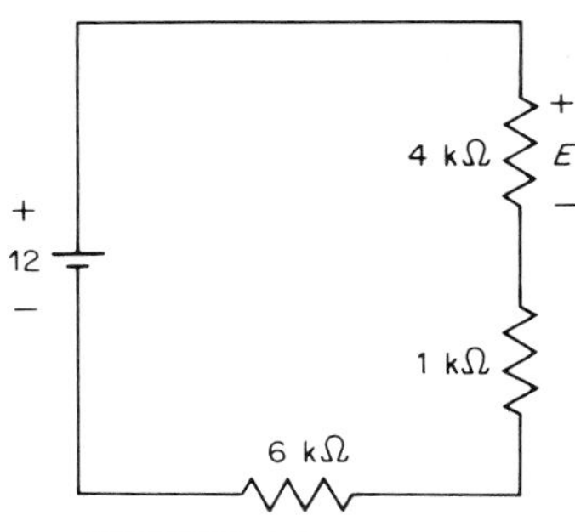

FIGURE 4-8
Circuit for Example 4-1.

Therefore,

$$E_3 = \frac{E}{R_1 + R_2 + R_3 + \cdots + R_n} R_3 = E \frac{R_3}{R_1 + R_2 + R_3 + \cdots + R_n}$$

Thus, E_3 is the same by both methods. ////

EXAMPLE 4-1 Determine the voltage drop across the 4-kΩ resistor in Fig. 4-8.

SOLUTION

$$E = 12 \frac{4\ \text{k}\Omega}{4\ \text{k}\Omega + 1\ \text{k}\Omega + 6\ \text{k}\Omega} = 12 \frac{4\ \text{k}\Omega}{11\ \text{k}\Omega}$$
$$= 4.36\ \text{V}$$

////

EXAMPLE 4-2 The series circuit in Fig. 4-9 contains two voltage sources. Find the combined voltage drop across the 3 kΩ and 15 kΩ resistors.

SOLUTION The 30 V and 40 V sources oppose each other. The result is a net applied voltage of 10 V in the series circuit. The polarity of the net applied voltage is determined by the 40 V source. The assumed polarity of E then agrees with the net applied voltage. If it did not agree, a minus sign would occur in the answer for E and would simply indicate that the actual polarity was opposite to that assumed for E.

$$E = 10 \frac{3\ \text{k}\Omega + 15\ \text{k}\Omega}{3\ \text{k}\Omega + 15\ \text{k}\Omega + 8\ \text{k}\Omega + 4\ \text{k}\Omega} = 10 \frac{18\ \text{k}\Omega}{30\ \text{k}\Omega}$$
$$= 6\ \text{V}$$

////

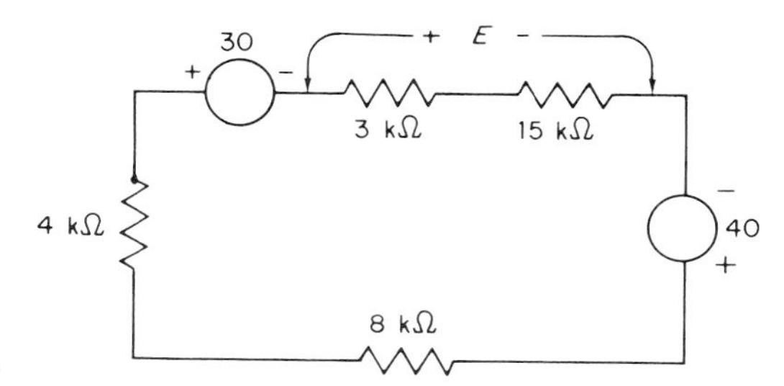

FIGURE 4-9
Circuit for Example 4-2.

EXAMPLE 4-3 Calculate the voltage drop E assumed in Fig. 4-10.

SOLUTION The polarity assumed for voltage drop E cannot be obtained with the given source orientation.

$$E = -100 \frac{20 + 30 + 80}{20 + 30 + 80 + 12} = -100 \frac{130}{142}$$

$$= -91.5 \text{ V}$$

This answer indicates that the magnitude of the drop is 91.5 volts and the actual polarity drop is opposite the polarity assumed. ////

4-3 KIRCHHOFF'S VOLTAGE LAW

Gustav Kirchhoff formulated two laws, one a relationship for voltages and the other a relationship for currents. These laws are fundamental in analyzing a circuit using mesh currents or node voltages. Mesh and node analysis will be discussed in later sections in this chapter.

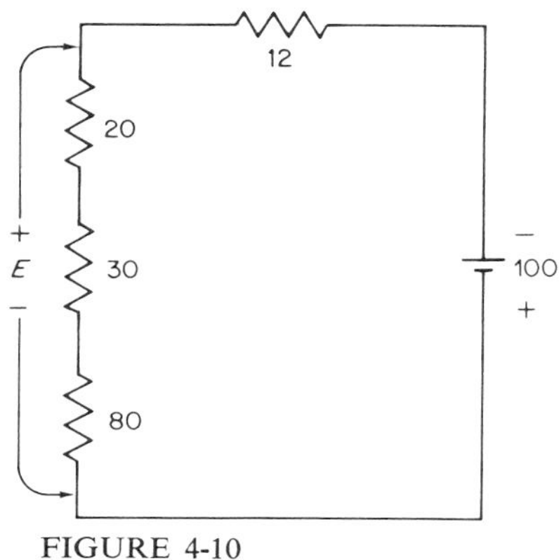

FIGURE 4-10
Circuit for Example 4-3.

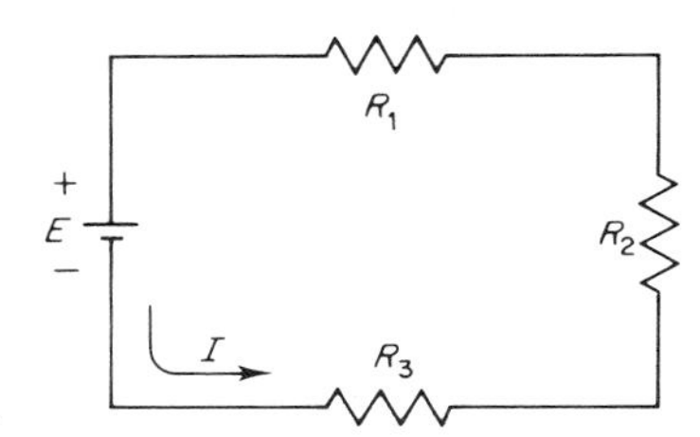

FIGURE 4-11
Circuit for the KVL equation.

Kirchhoff's voltage law (KVL) The algebraic sum of the voltages around any closed loop is zero.

A closed loop may be defined as a continuous connection of branches where current flow leaving a point in one direction and traced around the branches enters the same point from the opposite direction. Voltage sources producing the assumed current direction around the loop are positive algebraic terms. Voltage drops across resistors are negative algebraic terms in the sum. Applying the KVL to the circuit in Fig. 4-11,

$$E - IR_1 - IR_2 - IR_3 = 0$$

or

$$E = IR_1 + IR_2 + IR_3$$

Another way of stating Kirchhoff's voltage law is that the net sum of the voltage sources in the loop producing the assumed direction of current flow is equal to the sum of the voltage drops. This form is stated in the second equation above.

EXAMPLE 4-4 Write KVL equation for the circuit in Fig. 4-12. Use Ohm's law to check the validity of the equation.

SOLUTION

1 KVL equation

$$40 = 500I + (4\ \text{k}\Omega)I + (2\ \text{k}\Omega)I + (1\ \text{k}\Omega)I$$

2 Using Ohm's law,

$$I = \frac{40}{7.5\ \text{k}\Omega} = 5.34\ \text{mA}$$

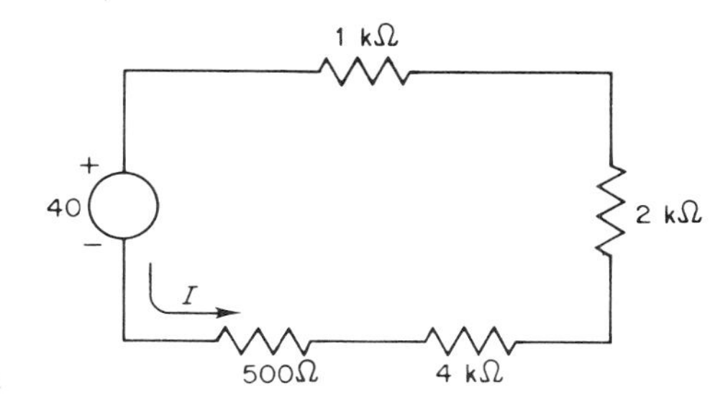

FIGURE 4-12
Circuit for Example 4-4.

3 Calculate each of the voltage drops.

$$500I = (500)(5.34\text{ mA}) = 2.66\text{ V}$$
$$(4\text{ k}\Omega)I = (4\text{ k}\Omega)(5.34\text{ mA}) = 21.34\text{ V}$$
$$(2\text{ k}\Omega)I = (2\text{ k}\Omega)(5.34\text{ mA}) = 10.66\text{ V}$$
$$(1\text{ k}\Omega)I = (1\text{ k}\Omega)(5.34\text{ m}\Omega) = 5.34\text{ V}$$

4 Substituting into the KVL equation gives

$$40 = 2.66 + 21.34 + 10.66 + 5.34$$ ////
$$40 = 40 \qquad \text{checks}$$

EXAMPLE 4-5 Write KVL equation for the circuit in Fig. 4-13.

SOLUTION The net applied loop voltage agrees with the assumed loop current direction. Thus the net applied voltage is positive.

KVL equation: $12 - 10 = 100I + 400I + 600I$

$$2 = (1.1\text{ k}\Omega)I$$ ////

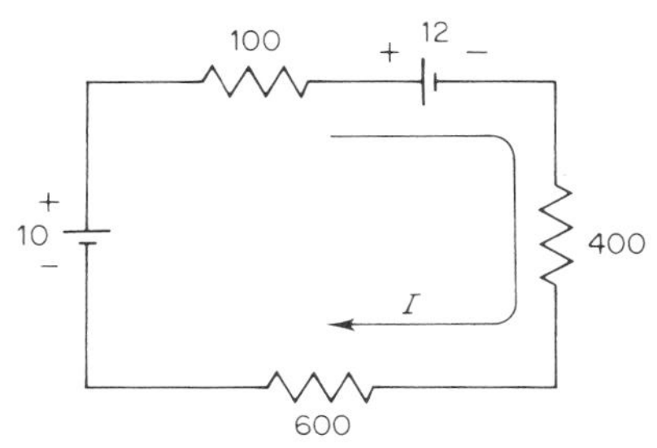

FIGURE 4-13
Circuit for Example 4-5.

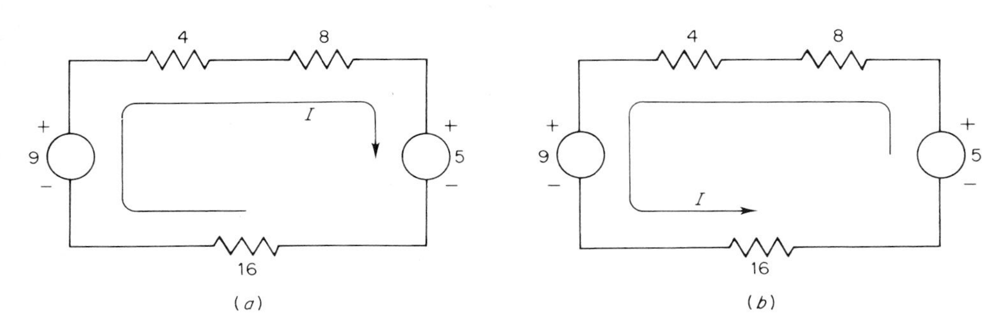

FIGURE 4-14
Circuits for Example 4-6: (*a*) assumed clockwise loop current, (*b*) assumed counter clockwise loop current.

EXAMPLE 4-6 For the circuits in Fig. 4-14, first write the KVL equation assuming clockwise loop current flow. Then write the KVL equation assuming counterclockwise current flow. Show that in solving for the loop current the magnitude and resultant direction will be the same in both cases.

SOLUTION

1 From Fig. 4-14*a* with clockwise current assumption:

KVL equation:
$$5 - 9 = 4I + 8I + 16I$$
$$-4 = 28I$$
$$I = -\tfrac{4}{28} = -143 \text{ mA}$$

Which means the actual direction is opposite the assumed direction and the magnitude is 143 mA. Thus the actual direction is counterclockwise.

2 From Fig. 4-14*b* with counterclockwise current assumption:

KVL equation:
$$9 - 5 = 4I + 8I + 16I$$
$$4 = 28I$$
$$I = \tfrac{4}{28} = 143 \text{ mA}$$

Since I is positive, the assumed direction and the actual direction are the same.

3 Actual results are the same in the previous steps. ////

As can be seen from Example 4-6, the assumed loop current direction is completely arbitrary. The actual direction will be revealed in the final solution. Remember that a negative sign on a current value simply indicates that the actual direction is just opposite the assumed direction.

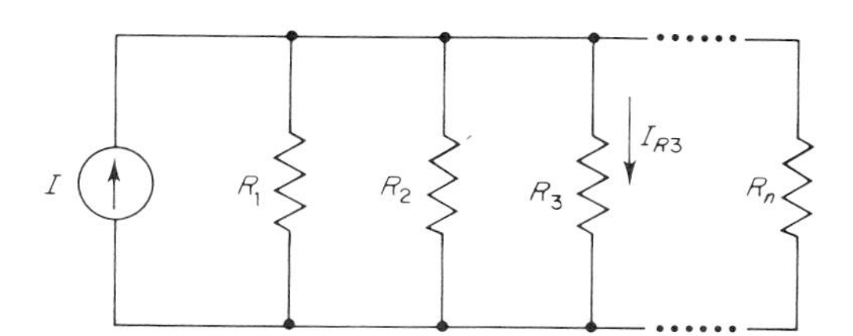

FIGURE 4-15
Parallel circuit with n resistors.

The real significance of Kirchhoff's voltage law will become apparent in the section on mesh currents.

4-4 CURRENT DIVISION RULE

Current flowing into a circuit consisting of parallel resistors divides and flows into each resistor. The amount of current through each resistor depends upon the resistor values in the circuit. Small values of resistance have larger currents, and large resistances have smaller currents. Application of the current division rule yields the current flowing in a branch resistance of a parallel circuit for some given input current.

Current division rule (CDR) The current in a branch of a parallel circuit is equal to the applied current multiplied by the ratio of the conductance where the current flow occurs to the sum of the conductances in the circuit.

PROOF Consider the parallel circuit containing a current source and n resistors in Fig. 4-15. Derive the current in R_3 using the CDR. Then use Ohm's law in determining the voltage across the parallel network and the resultant current in R_3.
Using the CDR,

$$I_{R_3} = I \frac{\dfrac{1}{R_3}}{\dfrac{1}{R_1} + \dfrac{1}{R_2} + \dfrac{1}{R_3} + \cdots + \dfrac{1}{R_n}}$$

Applying Ohm's law for voltage across network gives

$$E = IR_{\text{tot}} = I \frac{1}{\dfrac{1}{R_1} + \dfrac{1}{R_2} + \dfrac{1}{R_3} + \cdots + \dfrac{1}{R_n}}$$

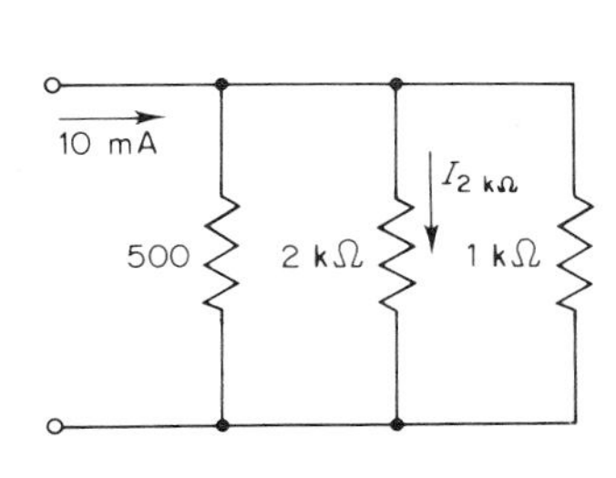

FIGURE 4-16
Circuit for Example 4-7.

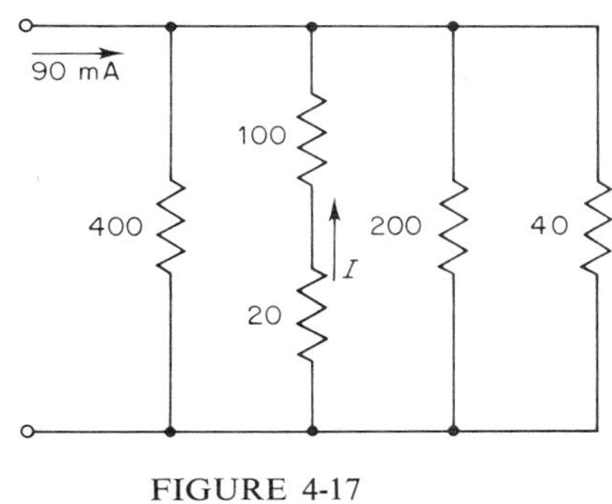

FIGURE 4-17
Circuit for Example 4-8.

Therefore,

$$I_{R_3} = \frac{E}{R_3} = I \frac{\dfrac{1}{R_3}}{\dfrac{1}{R_1} + \dfrac{1}{R_2} + \dfrac{1}{R_3} + \cdots + \dfrac{1}{R_n}}$$

Thus, I_{R_3} is the same by both methods. ////

EXAMPLE 4-7 Determine the current in the 2-kΩ resistor for the circuit in Fig. 4-16.

SOLUTION

$$I_{2\,k\Omega} = 10\text{ mA} \frac{\dfrac{1}{2\text{ k}\Omega}}{\dfrac{1}{0.5\text{ k}\Omega} + \dfrac{1}{2\text{ k}\Omega} + \dfrac{1}{1\text{ k}\Omega}}$$

$$= 10\text{ mA} \frac{0.5 \times 10^{-3}}{3.5 \times 10^{-3}}$$

$$= 1.43\text{ mA}$$ ////

In the preceding example, the assumed direction of $I_{2\,k\Omega}$ was from top to bottom. This assumption was correct since the input current enters from the top and leaves at the bottom. If the assumed direction is opposite the actual direction caused by the input current, a negative sign will appear on the branch current.

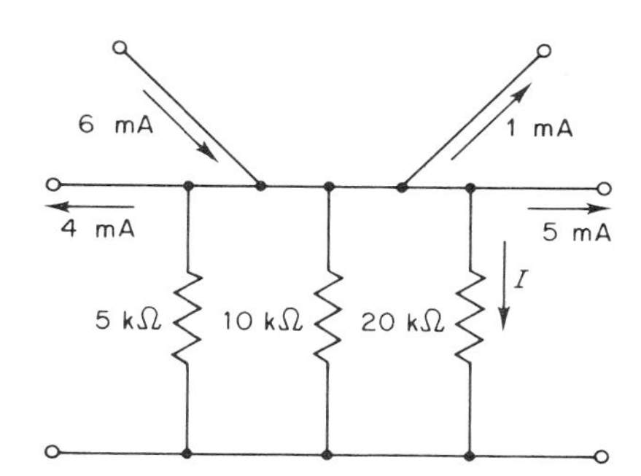

FIGURE 4-18
Circuit for Example 4-9.

EXAMPLE 4-8 Find the current in the two series resistors of the parallel circuit (see Fig. 4-17).

SOLUTION

$$I = -90\ \text{mA}\,\frac{\frac{1}{120}}{\frac{1}{400}+\frac{1}{120}+\frac{1}{200}+\frac{1}{40}}$$

$$= -90\ \text{mA}\,\frac{8.34 \times 10^{-3}}{40.84 \times 10^{-3}}$$

$$= -18.4\ \text{mA}$$

////

Parallel circuits may have more than a single input current. If so, the net current which produces the assumed branch-current direction is used in the CDR as the applied input current. Input currents that agree with the assumed direction are positive terms, and those which disagree are negative terms.

EXAMPLE 4-9 Calculate the indicated branch current for the parallel circuit in Fig. 4-18, which has four input currents.

SOLUTION

$$I = (-4\ \text{mA} + 6\ \text{mA} - 1\ \text{mA} - 5\ \text{mA})\,\frac{\frac{1}{20\ \text{k}\Omega}}{\frac{1}{5\ \text{k}\Omega}+\frac{1}{10\ \text{k}\Omega}+\frac{1}{20\ \text{k}\Omega}}$$

$$= -4\ \text{mA}\,\frac{5 \times 10^{-5}}{35 \times 10^{-5}}$$

$$= -0.571\ \text{mA}$$

The actual branch current of 0.571 mA is opposite the assumed direction. ////

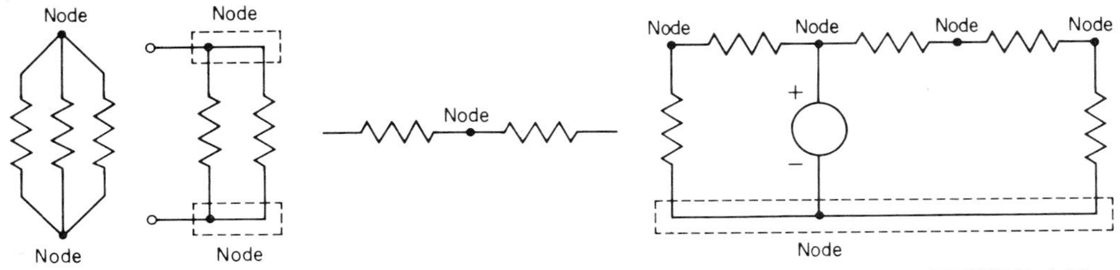

FIGURE 4-19
Node locations.

4-5 KIRCHHOFF'S CURRENT LAW

This law is basic to node voltage analysis. A node is defined as the junction point of two or more circuit elements. Some examples of node locations are shown in Fig. 4-19.

> **Kirchhoff's current law (KCL)** The algebraic sum of the currents at a node is zero.

Currents into a node are taken as positive algebraic terms in the sum. Those flowing out of a node are negative algebraic terms. Applying KCL to node 1 in the circuit of Fig. 4-20 gives

$$I - I_1 - I_2 - I_3 = 0$$

or

$$I = I_1 + I_2 + I_3$$

Kirchhoff's current law can also be stated as follows: the sum of the currents into a node is equal to the sum of the currents out of the same node. This form is given in the second equation above.

EXAMPLE 4-10 Write KCL equation for the circuit in Fig. 4-21. Use Ohm's law to check the validity of the equation.

SOLUTION

1 KCL equation

$$0.1 = I_1 + I_2 + I_3$$

$$0.1 = \tfrac{1}{10}E + \tfrac{1}{20}E + \tfrac{1}{4}E$$

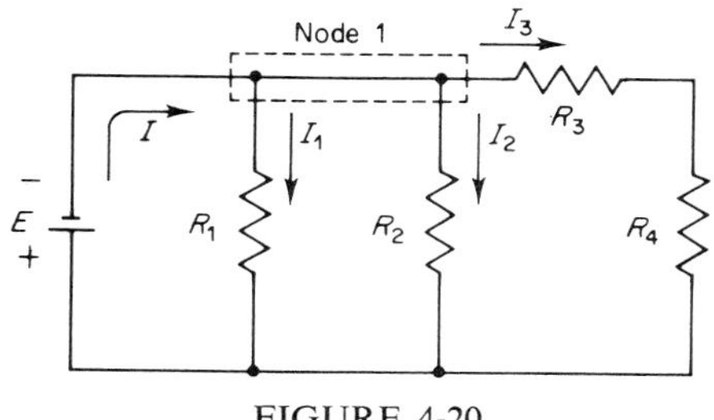

FIGURE 4-20
Circuit for the KCL equation.

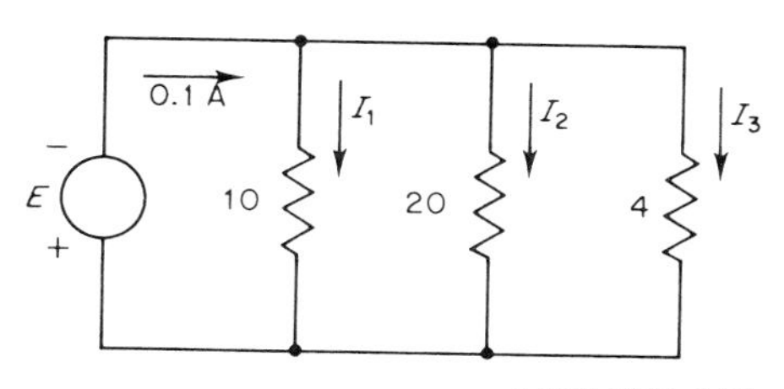

FIGURE 4-21
Circuit for Example 4-10.

2 Using Ohm's law,

$$E = 0.1R_{\text{tot}} = (0.1)(2.5)$$

$$E = 0.25 \text{ V}$$

3 Calculate each branch current

$$\tfrac{1}{10}E = 0.0250$$

$$\tfrac{1}{20}E = 0.0125$$

$$\tfrac{1}{4}E = 0.0625$$

4 Substituting into KCL equation gives

$$0.1 = 0.0250 + 0.0125 + 0.0625$$

$$0.1 = 0.1 \qquad \text{checks}$$

////

EXAMPLE 4-11 Determine the unknown currents in the circuit of Fig. 4-22 by writing and solving the KCL equations.

SOLUTION KCL equation for node 1

$$I_1 + I_3 + I_4 = I_2 + I_5$$

$$10 \text{ mA} + 4 \text{ mA} + 2 \text{ mA} = 5 \text{ mA} + I_5$$

$$I_5 = 11 \text{ mA}$$

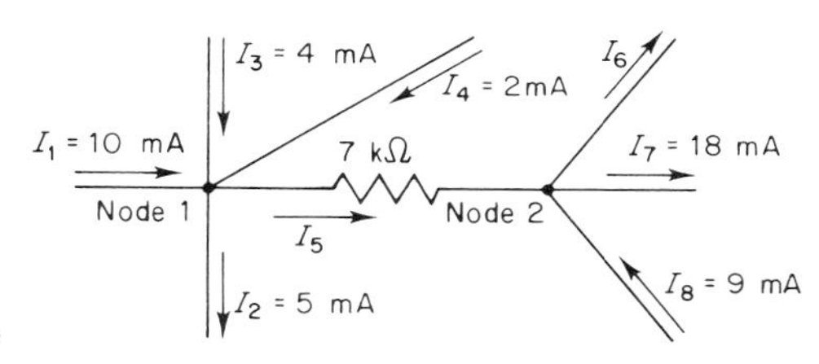

FIGURE 4-22
Circuit for Example 4-11.

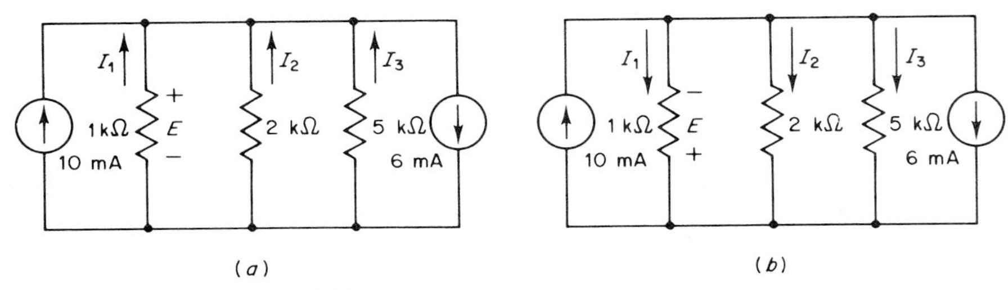

FIGURE 4-23
Circuits for Example 4-12: (*a*) assumed positive top node, (*b*) assumed negative top node.

KCL equation for node 2

$$I_5 + I_8 = I_7 + I_6$$

$$11 \text{ mA} + 9 \text{ mA} = 18 \text{ mA} + I_6$$

$$I_6 = 2 \text{ mA}$$

A negative numerical answer indicates that the actual current direction is opposite to the assumed current direction. ////

EXAMPLE 4-12 For the circuits in Fig. 4-23, first write the KCL equation assuming the top node is positive with respect to the bottom node. Then write the KCL equation assuming the top node is negative with respect to the bottom node. Show that in solving for the voltage between the top and bottom nodes the magnitude and resultant polarity will be the same in both cases.

SOLUTION

1 From Fig. 4-23*a* with positive-top-node assumption, the resistor currents would be upward to provide for the E drop indicated.
KCL equation for top node:

$$\text{Currents in} = \text{currents out}$$

$$10 \text{ mA} + I_1 + I_2 + I_3 = 6 \text{ mA}$$

$$10 \text{ mA} + \frac{1}{1 \text{ k}\Omega} E + \frac{1}{2 \text{ k}\Omega} E + \frac{1}{5 \text{ k}\Omega} E = 6 \text{ mA}$$

$$(1.7 \times 10^{-3})E = -4 \text{ mA}$$

$$E = -2.35 \text{ V}$$

Since E is negative, the actual polarity is opposite the assumed polarity. Thus, the top node is negative with respect to the bottom node.

2 From Fig. 4-23*b* with negative-top-node assumption, the resistor currents would be downward to provide for E drop indicated.
KCL equations for top node:

$$\text{Currents in} = \text{currents out}$$

$$10 \text{ mA} = I_1 + I_2 + I_3 + 6 \text{ mA}$$

$$4 \text{ mA} = \frac{1}{1 \text{ k}\Omega} E + \frac{1}{2 \text{ k}\Omega} E + \frac{1}{5 \text{ k}\Omega} E$$

$$E = 2.35 \text{ V}$$

Since E is positive, actual and assumed polarity are the same.

3 Results are the same in steps 1 and 2. ////

The assumed polarity between nodes is completely arbitrary. The actual polarity is revealed in the sign of the final solution.

4-6 MESH-CURRENT ANALYSIS

A mesh is nothing more than a closed loop in a circuit. Even the most complicated circuits can be represented by a collection of meshes. An equation is written for each mesh, and the resulting simultaneous equations are solved for the value of each mesh current. The equation for each mesh is basically the KVL equation. Several guidelines should be understood before writing mesh equations.

1 The circuit must contain only voltage sources. When current sources appear in the circuit, they must be converted into the equivalent voltage-source representation before analysis.

2 The direction of the assumed mesh current is completely arbitrary (the sign of the answer will indicate the actual direction), though eventually a clockwise direction will be adopted to facilitate writing the mesh equations.

3 Voltage sources that would produce the assumed mesh-current direction are considered positive. Those opposing the current direction are negative.

4 Unknown mesh currents in the simultaneous equations can be solved by addition-substitution methods or determinants.

Step 1 is necessary since the KVL deals with voltage sources and voltage drops around a closed loop. Equivalent source interchange is shown in Fig. 4-24. Note in the conversion that the current source indicates electron flow downward, the same direction the voltage source would force electrons to flow. Both sources considered

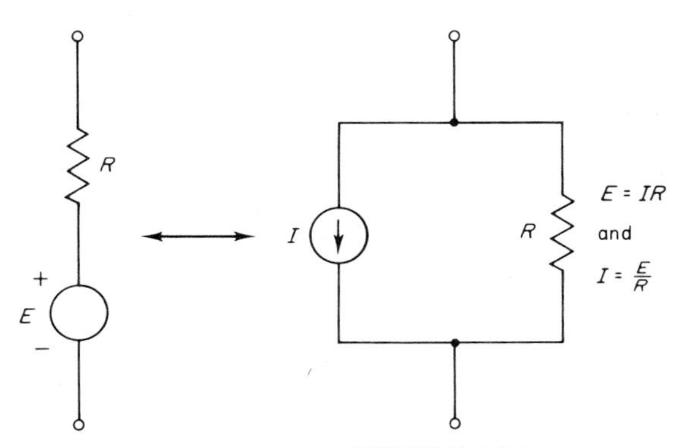

FIGURE 4-24
Conversion between voltage and current sources.

are of the nonideal variety. Thus the voltage source must have a series resistance, and the current source must have a parallel resistance. The validity of these conversions will be shown in following sections of this chapter.

EXAMPLE 4-13 Write the mesh equations for the circuit in Fig. 4-25. The direction of I_1 and I_2 was arbitrarily selected. Once the equations are obtained, solve for the unknown mesh currents.

SOLUTION First notice that the 4-V source would produce the mesh-current direction assumed for I_1. Thus this source is considered positive in the KVL mesh equation. The 6-V source would not produce the assumed direction for I_2. Therefore, this source is considered negative in the mesh equation.

Both mesh currents I_1 and I_2 flow through the 2-kΩ resistor common to both meshes. The resulting voltage drop is $(I_1 + I_2)(2\ \text{k}\Omega)$.

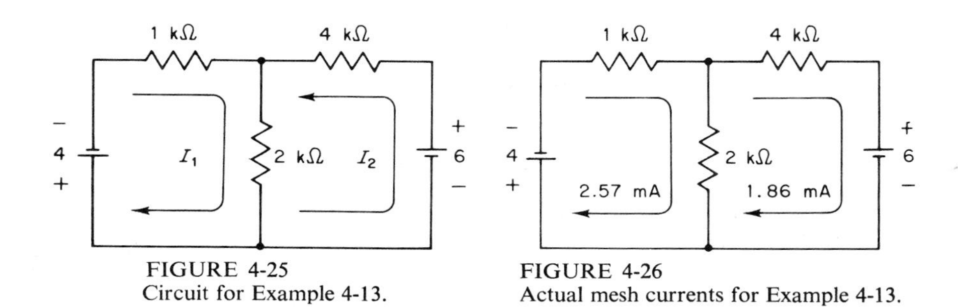

FIGURE 4-25
Circuit for Example 4-13.

FIGURE 4-26
Actual mesh currents for Example 4-13.

Recalling the format for KVL equations, we have

Net sum of sources producing current direction = voltage drops around loop

First mesh equation:

$$4 = (1\ \text{k}\Omega)I_1 + (2\ \text{k}\Omega)(I_1 + I_2)$$

$$4 = (3\ \text{k}\Omega)I_1 + (2\ \text{k}\Omega)I_2$$

Second mesh equation:

$$-6 = (4\ \text{k}\Omega)I_2 + (2\ \text{k}\Omega)(I_1 + I_2)$$

$$-6 = (2\ \text{k}\Omega)I_1 + (6\ \text{k}\Omega)I_2$$

Using determinants, the two mesh currents can be found from the two equations.

$$4 = (3\ \text{k}\Omega)I_1 + (2\ \text{k}\Omega)I_2$$

$$-6 = (2\ \text{k}\Omega)I_1 + (6\text{k}\ \Omega)I_2$$

$$I_1 = \frac{\begin{vmatrix} 4 & 2\ \text{k}\Omega \\ -6 & 6\ \text{k}\Omega \end{vmatrix}}{\begin{vmatrix} 3\ \text{k}\Omega & 2\ \text{k}\Omega \\ 2\ \text{k}\Omega & 6\ \text{k}\Omega \end{vmatrix}} = \frac{24\ \text{k}\Omega + 12\ \text{k}\Omega}{18 \times 10^6 - 4 \times 10^6}$$

$$= \frac{36 \times 10^3}{14 \times 10^6} = 2.57\ \text{mA}$$

$$I_2 = \frac{\begin{vmatrix} 3\ \text{k}\Omega & 4 \\ 2\ \text{k}\Omega & -6 \end{vmatrix}}{\begin{vmatrix} 3\ \text{k}\Omega & 2\ \text{k}\Omega \\ 2\ \text{k}\Omega & 6\ \text{k}\Omega \end{vmatrix}} = \frac{-18\ \text{k}\Omega - 8\ \text{k}\Omega}{14 \times 10^6}$$

$$= \frac{-26 \times 10^3}{14 \times 10^6} = -1.86\ \text{mA}$$

The results indicate that I_1 flows in the assumed direction and I_2 is reversed. Figure 4-26 shows the actual mesh-current directions and magnitudes. As a check, the values of I_1 and I_2 can be substituted into the two mesh equations.

$$4 = (3\ \text{k}\Omega)I_1 + (2\ \text{k}\Omega)I_2$$

$$= (3\ \text{k}\Omega)(2.57\ \text{mA}) + (2\ \text{k}\Omega)(-1.86\ \text{mA})$$

$$4 = 7.7 - 3.7 = 4 \qquad \text{checks}$$

$$-6 = (2\ \text{k}\Omega)I_1 + (6\ \text{k}\Omega)I_2$$

$$= (2\ \text{k}\Omega)(2.57\ \text{mA}) + (6\ \text{k}\Omega)(-1.86\ \text{mA})$$

$$-6 = 5.14 - 11.14 = -6 \qquad \text{checks}$$

The net current through the 2-kΩ resistor is 2.57 − 1.86 mA = 0.71 mA in a downward direction.

////

The next two examples analyze the same circuit as the previous example. The only difference is the assumed direction of mesh current.

EXAMPLE 4-14 Write the mesh equations for the circuit in Fig. 4-27. Solve for the unknown mesh currents.

SOLUTION The mesh currents oppose each other in the common 2-kΩ resistor.

First mesh equation: $-4 = (1\text{ k}\Omega)I_1 + (2\text{ k}\Omega)(I_1 - I_2)$

$$= (3\text{ k}\Omega)I_1 - (2\text{ k}\Omega)I_2$$

Second mesh equation: $-6 = (4\text{ k}\Omega)I_2 + (2\text{ k}\Omega)(I_2 - I_1)$

$$= (-2\text{ k}\Omega)I_1 + (6\text{ k}\Omega)I_2$$

Note that I_1 is assumed larger when writing the first mesh equation, $I_1 - I_2$, and I_2 is assumed larger for the second mesh equation, $I_2 - I_1$.

Solving the equations by determinants,

$$-4 = (3\text{ k}\Omega)I_1 - (2\text{ k}\Omega)I_2$$
$$-6 = (-2\text{ k}\Omega)I_1 + (6\text{ k}\Omega)I_2$$

$$I_1 = \frac{\begin{vmatrix} -4 & -2\text{ k}\Omega \\ -6 & 6\text{ k}\Omega \end{vmatrix}}{\begin{vmatrix} 3\text{ k}\Omega & -2\text{ k}\Omega \\ -2\text{ k}\Omega & 6\text{ k}\Omega \end{vmatrix}} = \frac{-24\text{ k}\Omega - 12\text{ k}\Omega}{18 \times 10^6 - 4 \times 10^6}$$

$$= \frac{-36 \times 10^3}{14 \times 10^6} = -2.57\text{ mA}$$

$$I_2 = \frac{\begin{vmatrix} 3\text{ k}\Omega & -4 \\ -2\text{ k}\Omega & -6 \end{vmatrix}}{\begin{vmatrix} 3\text{ k}\Omega & -2\text{ k}\Omega \\ -2\text{ k}\Omega & 6\text{ k}\Omega \end{vmatrix}} = \frac{-18\text{ k}\Omega - 8\text{ k}\Omega}{14 \times 10^6}$$

$$= \frac{-26 \times 10^3}{14 \times 10^6} = -1.86\text{ mA}$$

Thus the actual mesh-current directions are reversed. This is in complete agreement with the results of Example 4-13 shown in Fig. 4-26. ////

EXAMPLE 4-15 Write the mesh equations for the circuit in Fig. 4-28. Solve for the unknown mesh currents.

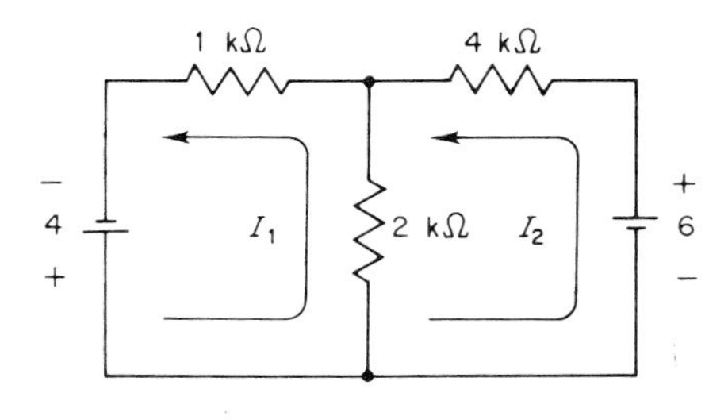

FIGURE 4-27
Circuit for Example 4-14.

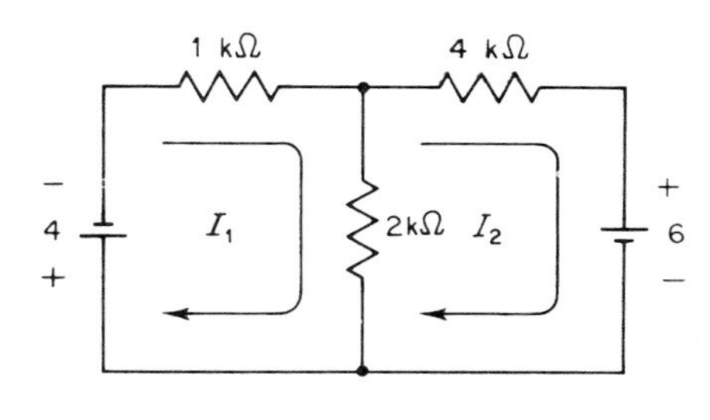

FIGURE 4-28
Circuit for Example 4-15.

SOLUTION The mesh currents oppose each other in the common 2-kΩ resistor.

First mesh equation:
$$4 = (1\ \text{k}\Omega)I_1 + (2\ \text{k}\Omega)(I_1 - I_2)$$
$$= (3\ \text{k}\Omega)I_1 - (2\ \text{k}\Omega)I_2$$

Second mesh equation:
$$6 = (4\ \text{k}\Omega)I_2 + (2\ \text{k}\Omega)(I_2 - I_1)$$
$$= (-2\ \text{k}\Omega)I_1 + (6\ \text{k}\Omega)I_2$$

Solving the equations by determinants,

$$4 = (3\ \text{k}\Omega)I_1 - (2\ \text{k}\Omega)I_2$$
$$6 = (-2\ \text{k}\Omega)I_1 - (6\ \text{k}\Omega)I_2$$

$$I_1 = \frac{\begin{vmatrix} 4 & -2\ \text{k}\Omega \\ 6 & 6\ \text{k}\Omega \end{vmatrix}}{\begin{vmatrix} 3\ \text{k}\Omega & -2\ \text{k}\Omega \\ -2\ \text{k}\Omega & 6\ \text{k}\Omega \end{vmatrix}} = \frac{24\ \text{k}\Omega + 12\ \text{k}\Omega}{18 \times 10^6 - 4 \times 10^6}$$

$$= \frac{36 \times 10^3}{14 \times 10^6} = 2.57\ \text{mA}$$

$$I_2 = \frac{\begin{vmatrix} 3\ \text{k}\Omega & 4 \\ -2\ \text{k}\Omega & 6 \end{vmatrix}}{\begin{vmatrix} 3\ \text{k}\Omega & -2\ \text{k}\Omega \\ -2\ \text{k}\Omega & 6\ \text{k}\Omega \end{vmatrix}} = \frac{18\ \text{k}\Omega + 8\ \text{k}\Omega}{14 \times 10^6}$$

$$= \frac{26 \times 10^3}{14 \times 10^6} = 1.86\ \text{mA}$$

Both currents are in the assumed direction. Once again, this agrees with Fig. 4-26 in Example 4-13. ////

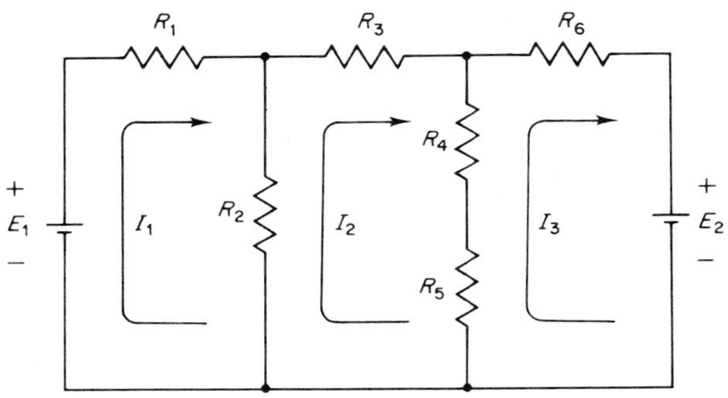

FIGURE 4-29
Three-mesh circuit.

The last three examples have shown that the selection of mesh current direction is completely arbitrary. From now on the text will adopt clockwise mesh-current directions for two reasons:

1 A consistent mesh-current assumption is provided for future analyses.
2 Mesh-current equations can be written quickly with greater ease.

Consider the circuit shown in Fig. 4-29 with assumed clockwise mesh currents. This circuit contains three meshes and two voltage sources. The mesh equations for the circuit would be:

First mesh equation:
$$-E_1 = R_1 I_1 + R_2(I_1 - I_2)$$
$$= (R_1 + R_2)I_1 - R_2 I_2$$

Second mesh equation:
$$0 = R_3 I_2 + R_2(I_2 - I_1) + (R_4 + R_5)(I_2 - I_3)$$
$$= -R_2 I_1 + (R_2 + R_3 + R_4 + R_5)I_2 - (R_4 + R_5)I_3$$

Third mesh equation:
$$E_2 = R_6 I_3 + (R_4 + R_5)(I_3 - I_2)$$
$$= -(R_4 + R_5)I_2 + (R_4 + R_5 + R_6)I_3$$

Notice that in the first mesh equation, the resistive coefficient for the current in the mesh I_1 is equal to the sum of the resistors in that mesh. For the second mesh, the resistive coefficient on I_2 is the sum of the resistors in that mesh. Also the sum of the resistors in the third mesh is the resistive coefficient for I_3. Coefficients for the other current terms are simply the negative of the resistance common between meshes. For example, I_2 has a coefficient $(-R_2)$ in the first mesh equation, where R_2 is the common resistance between the second and first meshes. Zero is the coefficient of I_3 in the first mesh equation. This is valid since there is no resistance common to the third and first meshes

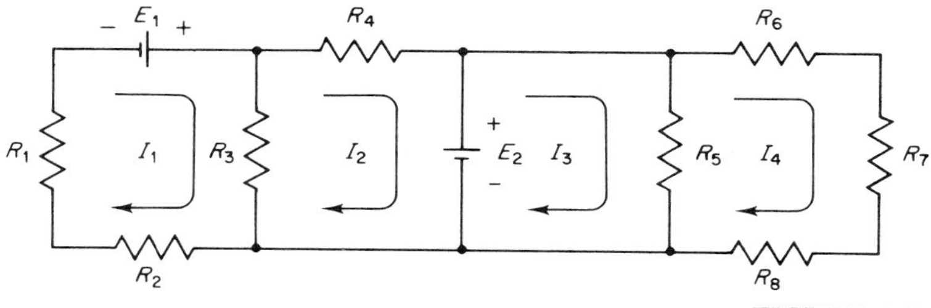

FIGURE 4-30
Circuit for Example 4-16.

In conclusion:

1 The resistive coefficient for the current I_k, when writing the kth mesh equation, is the sum of resistances in the kth mesh

2 The resistive coefficients for the other currents I_l, I_m, I_n, etc., when writing the kth mesh equation, are the negative of the resistance between the kth and lth mesh, kth and mth mesh, kth and nth mesh, etc.

EXAMPLE 4-16 Write the four mesh equations for the circuit in Fig. 4-30.

SOLUTION Applying the procedure for resistive coefficients, we have:

First mesh equation: $-E_1 = (R_1 + R_2 + R_3)I_1 - R_3 I_2$

Second mesh equation: $E_2 = -R_3 I_1 + (R_3 + R_4)I_2$

Third mesh equation: $-E_2 = R_5 I_3 - R_5 I_4$

Fourth mesh equation: $0 = -R_5 I_3 + (R_5 + R_6 + R_7 + R_8)I_4$ ////

EXAMPLE 4-17 Write mesh equations for the circuit in Fig. 4-31. Solve for the mesh currents. Resketch the original circuit and show the actual direction of mesh currents, current through each resistor, and voltage drop across each resistor. Also, calculate the power dissipation in every resistor.

SOLUTION

First mesh equation: $-12 = (6\ \text{k}\Omega)I_1 - (5\ \text{k}\Omega)I_2$

Second mesh equation: $0 = (-5\ \text{k}\Omega)I_1 + (11\ \text{k}\Omega)I_2 - (2\ \text{k}\Omega)I_3$

Third mesh equation: $0 = (-2\ \text{k}\Omega)I_2 + (11\ \text{k}\Omega)I_3$

We express the equations for determinant solution as

$$\begin{aligned} -12 &= (6\ \text{k}\Omega)I_1 - (5\ \text{k}\Omega)I_2 \\ 0 &= (-5\ \text{k}\Omega)I_1 + (11\ \text{k}\Omega)I_2 - (2\ \text{k}\Omega)I_3 \\ 0 &= -(2\ \text{k}\Omega)I_2 + (11\ \text{k}\Omega)I_3 \end{aligned}$$

$$I_1 = \frac{\begin{vmatrix} -12 & -5\ \text{k}\Omega & 0 \\ 0 & 11\ \text{k}\Omega & -2\ \text{k}\Omega \\ 0 & -2\ \text{k}\Omega & 11\ \text{k}\Omega \end{vmatrix}}{\begin{vmatrix} 6\ \text{k}\Omega & -5\ \text{k}\Omega & 0 \\ -5\ \text{k}\Omega & 11\ \text{k}\Omega & -2\ \text{k}\Omega \\ 0 & -2\ \text{k}\Omega & 11\ \text{k}\Omega \end{vmatrix}} = \frac{-14.04 \times 10^8}{4.27 \times 10^{11}} = -3.29\ \text{mA}$$

$$I_2 = \frac{\begin{vmatrix} 6\ \text{k}\Omega & -12 & 0 \\ -5\ \text{k}\Omega & 0 & -2\ \text{k}\Omega \\ 0 & 0 & 11\ \text{k}\Omega \end{vmatrix}}{4.27 \times 10^{11}} = \frac{-6.60 \times 10^8}{4.27 \times 10^{11}} = -1.55\ \text{mA}$$

$$I_3 = \frac{\begin{vmatrix} 6\ \text{k}\Omega & -5\ \text{k}\Omega & -12 \\ -5\ \text{k}\Omega & 11\ \text{k}\Omega & 0 \\ 0 & -2\ \text{k}\Omega & 0 \end{vmatrix}}{4.27 \times 10^{11}} = \frac{-1.20 \times 10^8}{4.27 \times 10^{11}} = -0.28\ \text{mA}$$

See Fig. 4-32 for currents and voltages in the circuit.
Power dissipation in resistors:

$$P = IE$$

$$\begin{aligned} 1\ \text{k}\Omega:&\quad P = (3.29\ \text{mA})(3.29) = 10.8\ \text{mW} \\ 2\ \text{k}\Omega:&\quad P = (1.27\ \text{mA})(2.54) = 3.22\ \text{mW} \\ 3\ \text{k}\Omega:&\quad P = (0.28\ \text{mA})(0.84) = 0.236\ \text{mW} \\ 4\ \text{k}\Omega:&\quad P = (1.55\ \text{mA})(6.2) = 9.6\ \text{mW} \\ 5\ \text{k}\Omega:&\quad P = (1.74\ \text{mA})(8.7) = 15.1\ \text{mW} \\ 6\ \text{k}\Omega:&\quad P = (0.28\ \text{mA})(1.68) = 0.47\ \text{mW} \end{aligned}$$

////

EXAMPLE 4-18 Solve for the three mesh currents indicated in the circuit of Fig. 4-33.

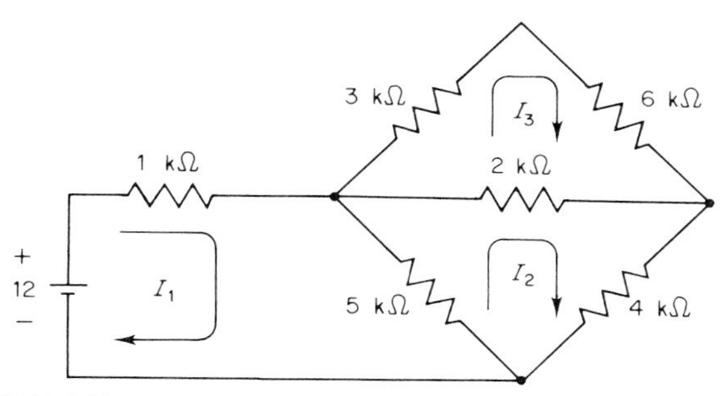

FIGURE 4-31
Circuit for Example 4-17.

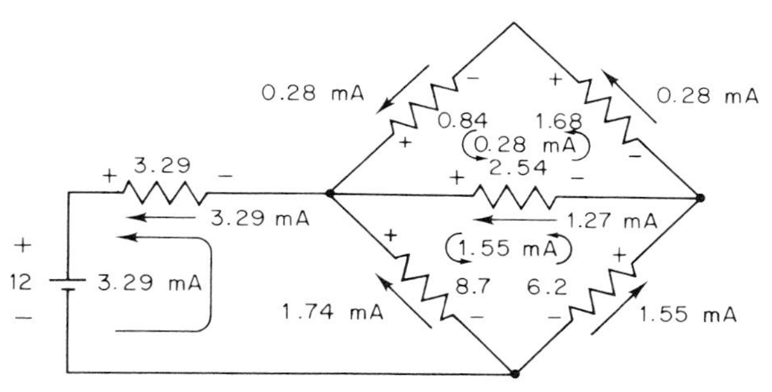

FIGURE 4-32
Currents and voltages for Example 4-17.

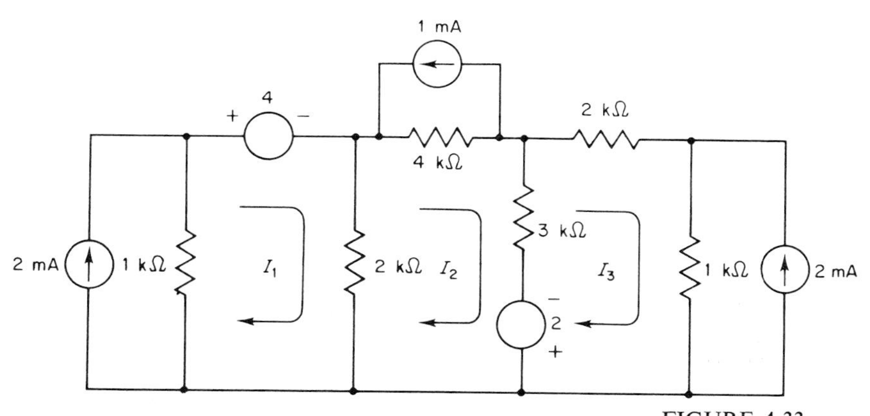

FIGURE 4-33
Circuit for Example 4-18.

SOLUTION Before mesh equations are written, the current sources must be converted into equivalent voltage sources. Figure 4-34 represents the circuit after conversion.

First mesh equation: $6 = (3\ \text{k}\Omega)I_1 - (2\ \text{k}\Omega)I_2$

Second mesh equation: $-6 = (-2\ \text{k}\Omega)I_1 + (9\ \text{k}\Omega)I_2 - (3\ \text{k}\Omega)I_3$

Third mesh equation: $0 = (-3\ \text{k}\Omega)I_2 + (6\ \text{k}\Omega)I_3$

Solving for mesh currents,

$$I_1 = \frac{\begin{vmatrix} 6 & -2\ \text{k}\Omega & 0 \\ -6 & 9\ \text{k}\Omega & -3\ \text{k}\Omega \\ 0 & -3\ \text{k}\Omega & 6\ \text{k}\Omega \end{vmatrix}}{\begin{vmatrix} 3\ \text{k}\Omega & -2\ \text{k}\Omega & 0 \\ -2\ \text{k}\Omega & 9\ \text{k}\Omega & -3\ \text{k}\Omega \\ 0 & -3\ \text{k}\Omega & 6\ \text{k}\Omega \end{vmatrix}} = \frac{98 \times 10^6}{111 \times 10^9} = 0.883\ \text{mA}$$

$$I_2 = \frac{\begin{vmatrix} 3\ \text{k}\Omega & 6 & 0 \\ -2\ \text{k}\Omega & -6 & -3\ \text{k}\Omega \\ 0 & 0 & 6\ \text{k}\Omega \end{vmatrix}}{111 \times 10^9} = \frac{-36 \times 10^6}{111 \times 10^9} = -0.324\ \text{mA}$$

$$I_3 = \frac{\begin{vmatrix} 3\ \text{k}\Omega & -2\ \text{k}\Omega & 6 \\ -2\ \text{k}\Omega & 9\ \text{k}\Omega & -6 \\ 0 & -3\ \text{k}\Omega & 0 \end{vmatrix}}{111 \times 10^9} = \frac{-18 \times 10^6}{111 \times 10^9} = -0.162\ \text{mA}$$ ////

If a current source does not have resistance in parallel with it, the conversion to an equivalent voltage source cannot be accomplished. Then the current source remains in the circuit during the analysis. Obviously, the current in the branch containing the current source is equal to the value of the source. This situation is presented in the following example.

EXAMPLE 4-19 Determine the mesh currents in the circuit of Fig. 4-35.

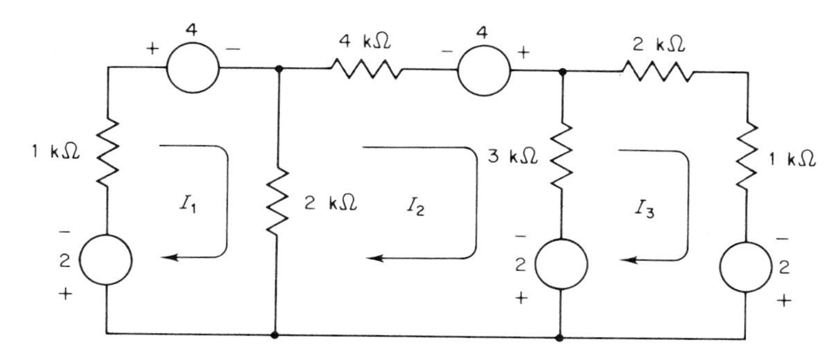

FIGURE 4-34
Circuit after conversions for Example 4-18.

SOLUTION The 10-mA current source remains for analysis. By inspection

$$I_1 = 10 \text{ mA}$$

Thus the only unknowns are I_2 and I_3. These can be found by writing mesh equations.

Second mesh equation:

$$10 = (-8 \text{ k}\Omega)I_1 + (22 \text{ k}\Omega)I_2 - (10 \text{ k}\Omega)I_3$$

$$10 = (-8 \text{ k}\Omega)(10 \text{ mA}) + (22 \text{ k}\Omega)I_2 - (10 \text{ k}\Omega)I_3$$

$$10 = -80 + (22 \text{ k}\Omega)I_2 - (10 \text{ k}\Omega)I_3$$

$$90 = (22 \text{ k}\Omega)I_2 - (10 \text{ k}\Omega)I_3$$

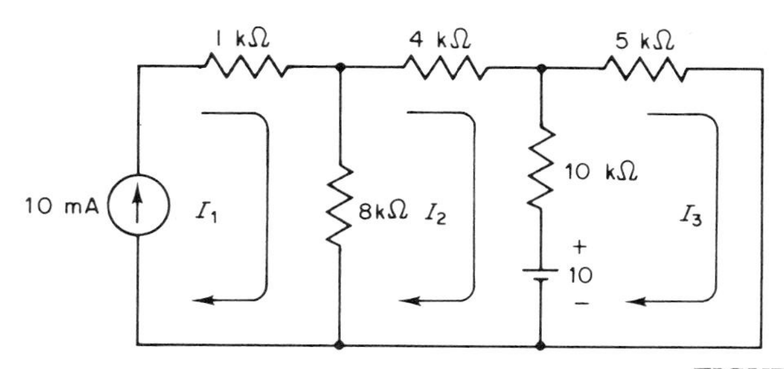

FIGURE 4-35
Circuit for Example 4-19.

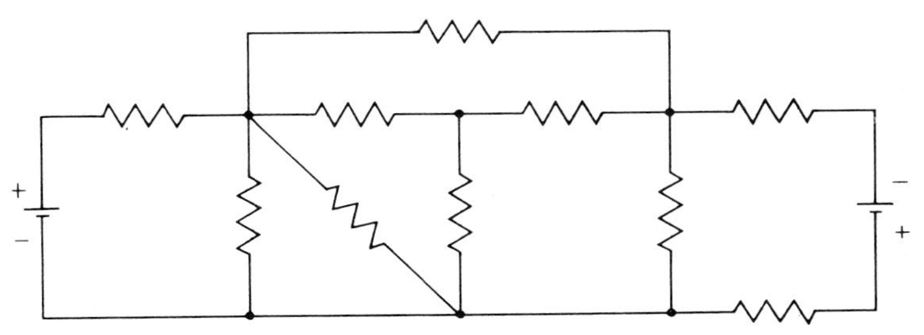

FIGURE 4-36

Third mesh equation: $-10 = (-10\ \text{k}\Omega)I_2 + (15\ \text{k}\Omega)I_3$

$$I_2 = \frac{\begin{vmatrix} 90 & -10\ \text{k}\Omega \\ -10 & 15\ \text{k}\Omega \end{vmatrix}}{\begin{vmatrix} 22\ \text{k}\Omega & -10\ \text{k}\Omega \\ -10\ \text{k}\Omega & 15\ \text{k}\Omega \end{vmatrix}} = \frac{1.25 \times 10^6}{2.3 \times 10^8} = 5.44\ \text{mA}$$

$$I_3 = \frac{\begin{vmatrix} 22\ \text{k}\Omega & 90 \\ -10\ \text{k}\Omega & -10 \end{vmatrix}}{2.3 \times 10^8} = \frac{6.8 \times 10^5}{2.3 \times 10^8} = 2.96\ \text{mA} \qquad ////$$

The choice of using mesh-current analysis or node-voltage analysis on a given circuit depends upon the number of equations required for each method. The approach requiring the fewest equations is usually selected. With fewer equations, the determinants are simpler.

A node was defined previously as a point in a circuit common to two or more elements. Principal nodes are nodes joining three or more circuit elements. Branches are electrical paths containing one or more elements. These paths must also connect the principal nodes. The number of required mesh equations in a circuit with voltage sources, either initially or through current-to-voltage conversions, is

Number of mesh equations = number of branches − (number of principal nodes − 1)

$$M = B - (PN - 1) \tag{4-1}$$

The circuit in Fig. 4-36 contains nine branches and four principal nodes. Therefore,

$$M = 9 - (4 - 1) = 6 \text{ mesh equations}$$

4-7 NODE-VOLTAGE ANALYSIS

When we know the mesh currents in a circuit, the current, voltage, and power dissipation in each resistance can easily be determined. Likewise, knowledge of node voltages permits determination of responses in the network. A voltage always represents the potential at one point with respect to another point. Node voltage is the voltage of a given principal node with respect to one particular node called the *reference node*. Usually the reference node is the ground of the circuit, though it can be any principal node designated as the reference node for the circuit.

An equation is written at all principal nodes except the reference node. These equations satisfy the KCL. The unknown variables in the equations are node voltages. With the resulting node equations, the unknown node voltages are found by determinants.

Guidelines for writing node equations are as follows:

1 The circuit must contain only current sources. When voltage sources appear in the circuit, they must be converted into equivalent current sources before analysis.

2 The polarity of the node voltage is arbitrary. The text assumes node voltages to be positive with respect to the reference node. Actual polarities are indicated in the sign of the answer.

3 Currents are assumed to flow into a node through resistors attached to the node. Source currents are assumed to flow out of a node. A current source forcing current away from a node is considered positive. A source forcing current into a node is taken as negative. The format for the node equation is

Source current leaving node = component current entering node

By counting the principal nodes in the circuit, the required number of node equations can be found using Eq. (4-2).

Number of node equations = number of principal nodes − 1

$$NE = PN - 1 \tag{4-2}$$

EXAMPLE 4-20 For the circuit in Fig. 4-37, determine the required number of node equations. Indicate node voltages on the circuit. Write the node equations and solve for the node voltages.

SOLUTION

$PN = 4$

$NE = 4 - 1 = 3$ node equations required

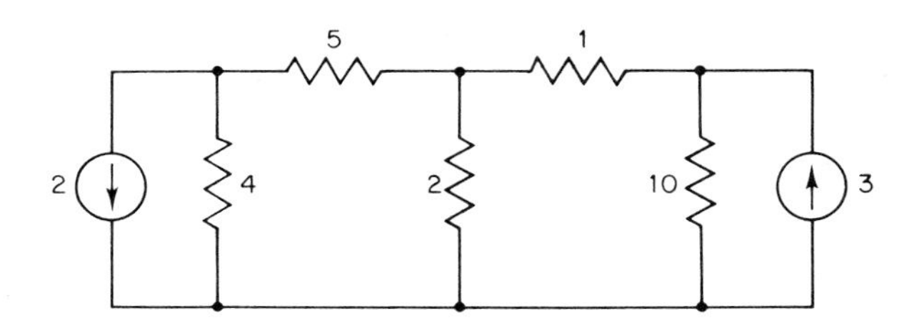

FIGURE 4-37
Circuit for Example 4-20.

Select the bottom node as reference node (see Fig. 4-38).
Using the format

Source current leaving node = component current entering node

we have:

First node equation:

$$2 = \frac{E_1}{4} + \frac{E_1 - E_2}{5}$$

$$= (\tfrac{1}{4} + \tfrac{1}{5})E_1 - \tfrac{1}{5}E_2 \tag{4-3}$$

Second node equation:

$$0 = \frac{E_2 - E_1}{5} + \frac{E_2}{2} + \frac{E_2 - E_3}{1}$$

$$0 = -\tfrac{1}{5}E_1 + (\tfrac{1}{5} + \tfrac{1}{2} + \tfrac{1}{1})E_2 - \tfrac{1}{1}E_3 \tag{4-4}$$

Third node equation:

$$-3 = \frac{E_3 - E_2}{1} + \frac{E_3}{10}$$

$$-3 = -\tfrac{1}{1}E_2 + (\tfrac{1}{1} + \tfrac{1}{10})E_3 \tag{4-5}$$

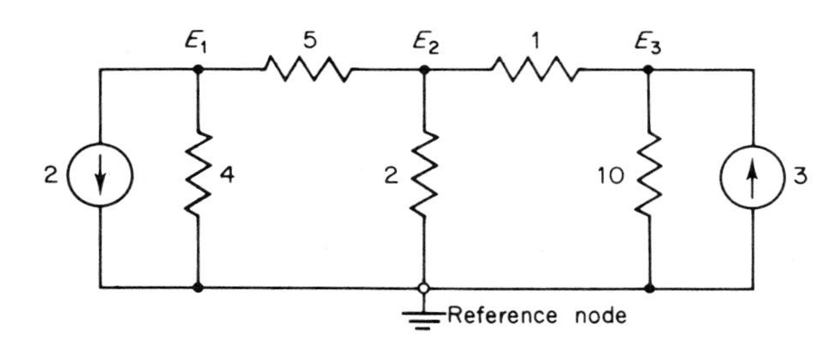

FIGURE 4-38

Grouping equations for a determinant solution gives

$$\begin{aligned} 2 &= 0.45E_1 - 0.2E_2 \\ 0 &= -0.2E_1 + 1.7E_2 - E_3 \\ -3 &= -E_2 + 1.1E_3 \end{aligned}$$

$$E_1 = \frac{\begin{vmatrix} 2 & -0.2 & 0 \\ 0 & 1.7 & -1 \\ -3 & -1 & 1.1 \end{vmatrix}}{\begin{vmatrix} 0.45 & -0.2 & 0 \\ -0.2 & 1.7 & -1 \\ 0 & -1 & 1.1 \end{vmatrix}} = \frac{1.14}{0.347} = 3.29 \text{ V}$$

$$E_2 = \frac{\begin{vmatrix} 0.45 & 2 & 0 \\ -0.2 & 0 & -1 \\ 0 & -3 & 1.1 \end{vmatrix}}{0.347} = \frac{-0.91}{0.347} = -2.62 \text{ V}$$

$$E_3 = \frac{\begin{vmatrix} 0.45 & -0.2 & 2 \\ -0.2 & 1.7 & 0 \\ 0 & -1 & -3 \end{vmatrix}}{0.347} = \frac{-1.78}{0.347} = -5.13 \text{ V}$$

As a check, the node voltages can be substituted into the three node equations.

$$2 = (0.45)(3.29) - (0.2)(-2.62) = 1.48 + 0.52$$

$$2 = 2 \qquad \text{checks}$$

$$0 = (-0.2)(3.29) + (1.7)(-2.62) - (-5.13) = -0.658 - 4.47 + 5.13$$

$$0 = 0 \qquad \text{checks}$$

$$-3 = -(-2.62) + 1.1(-5.13) = 2.62 - 5.62$$

$$-3 = -3 \qquad \text{checks}$$

////

Notice that in the first node equation [Eq. (4-3)], the coefficient for E_1 is equal to the sum of the conductances between node 1 and all other principal nodes. The coefficient for E_2 in the second node equation [Eq. (4-4)] is the sum of the conductances between node 2 and all other principal nodes. Likewise, E_3's coefficient in the third node equation [Eq. (4-5)] is the sum of conductances between the third node and all

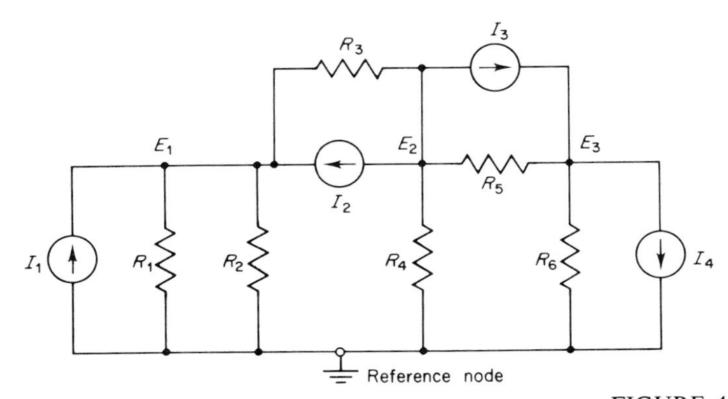

FIGURE 4-39
Circuit for Example 4-21.

other principal nodes. Coefficients for the other voltage terms are simply the negative of the conductance between the node represented by the voltage term and the node representing the node equation. For example, E_2 has a coefficient $-\frac{1}{5}$ in the first node equation [Eq. (4-3)]. The conductance between the first and second node is $\frac{1}{5}$. Zero is the coefficient of E_3 in the first node equation [Eq. (4-3)]. This is valid since there is no conductance directly attached between the first and third nodes. One cannot pass through a node (like the second node) in determining the conductance between nodes (like the first and third nodes). In conclusion:

1 The conductive coefficient for the voltage E_k, when writing the kth node equation, is the sum of the conductances between the kth node and all other principal nodes.

2 The conductive coefficients for the other voltages E_l, E_m, E_n, etc., when writing the kth node equation, is the negative of the conductance between the kth and lth node, kth and mth node, kth and nth node, etc.

EXAMPLE 4-21 Write the node equations for the circuit in Fig. 4-39.

SOLUTION

First node equation: $$-I_1 - I_2 = \left(\frac{1}{R_1} + \frac{1}{R_2} + \frac{1}{R_3}\right)E_1 - \frac{1}{R_3}E_2$$

Second node equation: $$I_2 + I_3 = -\frac{1}{R_3}E_1 + \left(\frac{1}{R_3} + \frac{1}{R_4} + \frac{1}{R_5}\right)E_2 - \frac{1}{R_5}E_3$$

Third node equation: $$I_4 - I_3 = -\frac{1}{R_5}E_2 + \left(\frac{1}{R_5} + \frac{1}{R_6}\right)E_3$$ ////

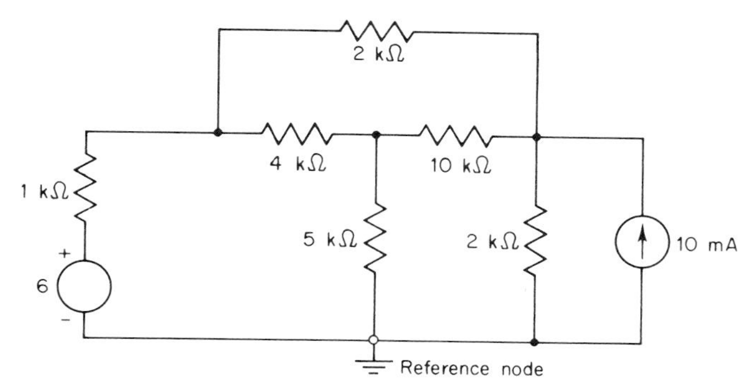

FIGURE 4-40
Circuit for Example 4-22.

EXAMPLE 4-22 Determine the node voltages in the circuit of Fig. 4-40.

SOLUTION The voltage source must be converted into an equivalent current source (see Fig. 4-41).

First node equation: $$6\text{ mA} = \left(\frac{1}{1\text{ k}\Omega} + \frac{1}{4\text{ k}\Omega} + \frac{1}{2\text{ k}\Omega}\right)E_1 - \frac{1}{4\text{ k}\Omega}E_2 - \frac{1}{2\text{ k}\Omega}E_3$$

Second node equation: $$0 = -\frac{1}{4\text{ k}\Omega}E_1 + \left(\frac{1}{4\text{ k}\Omega} + \frac{1}{5\text{ k}\Omega} + \frac{1}{10\text{ k}\Omega}\right)E_2 - \frac{1}{10\text{ k}\Omega}E_3$$

Third node equation:

$$-10\text{ mA} = -\frac{1}{2\text{ k}\Omega}E_1 - \frac{1}{10\text{ k}\Omega}E_2 + \left(\frac{1}{2\text{ k}\Omega} + \frac{1}{10\text{ k}\Omega} + \frac{1}{2\text{ k}\Omega}\right)E_3$$

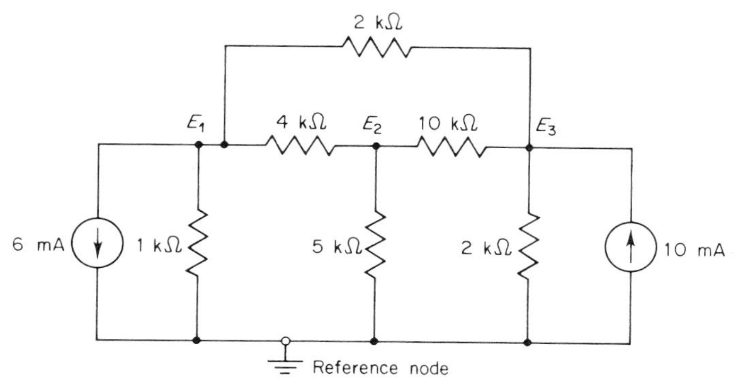

FIGURE 4-41
Circuit after source conversion.

After evaluating coefficients

$$6 \times 10^{-3} = (1.75 \times 10^{-3})E_1 - (0.25 \times 10^{-3})E_2 - (0.5 \times 10^{-3})E_3$$
$$0 = (-0.25 \times 10^{-3})E_1 + (0.55 \times 10^{-3})E_2 - (0.1 \times 10^{-3})E_3$$
$$-10 \times 10^{-3} = (-0.5 \times 10^{-3})E_1 - (0.1 \times 10^{-3})E_2 + (1.1 \times 10^{-3})E_3$$

Both sides of each equation can be multiplied by 1,000 to reduce labor in carrying along a factor of 10^{-3}.

$$6 = 1.75E_1 - 0.25E_2 - 0.5E_3$$
$$0 = -0.25E_1 + 0.55E_2 - 0.1E_3$$
$$-10 = -0.5E_1 - 0.1E + 1.1E_3$$

$$E_1 = \frac{\begin{vmatrix} 6 & -0.25 & -0.5 \\ 0 & 0.55 & -0.1 \\ -10 & -0.1 & 1.1 \end{vmatrix}}{\begin{vmatrix} 1.75 & -0.25 & -0.5 \\ -0.25 & 0.55 & -0.1 \\ -0.5 & -0.1 & 1.1 \end{vmatrix}} = \frac{0.57}{0.809} = 0.705 \text{ V}$$

$$E_2 = \frac{\begin{vmatrix} 1.75 & 6 & -0.5 \\ -0.25 & 0 & -0.1 \\ -0.5 & -10 & 1.1 \end{vmatrix}}{0.809} = \frac{-1.05}{0.809} = -1.3 \text{ V}$$

$$E_3 = \frac{\begin{vmatrix} 1.75 & -0.25 & 6 \\ -0.25 & 0.55 & 0 \\ -0.5 & -0.1 & -10 \end{vmatrix}}{0.809} = \frac{-7.19}{0.809} = -8.88 \text{ V}$$

////

If a voltage source does not have resistance in series with it, the conversion to an equivalent current source cannot be accomplished. Then the voltage remains in the circuit during the analysis. The difference in voltage between the nodes where the source is connected is equal to the value of the voltage source. The following example depicts this situation.

EXAMPLE 4-23 Find the node voltages in the circuit of Fig. 4-42.

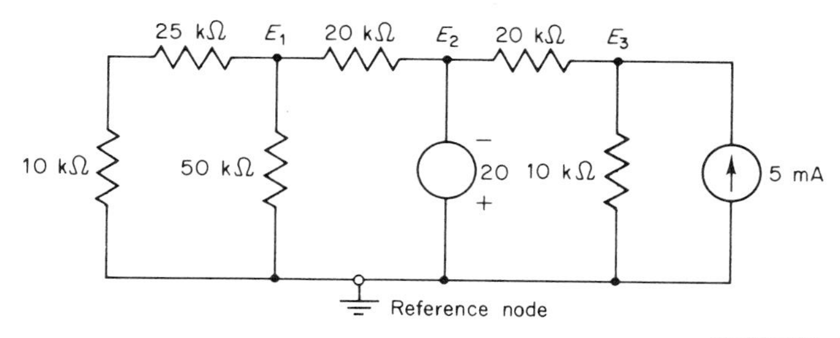

FIGURE 4-42
Circuit for Example 4-23.

SOLUTION The voltage source cannot be converted.

First node equation: $0 = \left(\frac{1}{35\text{ k}\Omega} + \frac{1}{50\text{ k}\Omega} + \frac{1}{20\text{ k}\Omega}\right)E_1 - \frac{1}{20\text{ k}\Omega}E_2$

Second node equation: $E_2 = -20\text{ V}$ by inspection

Third node equation: $-5\text{ mA} = -\frac{1}{20\text{ k}\Omega}E_2 + \left(\frac{1}{20\text{ k}\Omega} + \frac{1}{10\text{ k}\Omega}\right)E_3$

E_1 and E_3 can be found without determinants since E_2 is a known quantity in each equation.

$$0 = (0.129 \times 10^{-3})E_1 - (0.5 \times 10^{-4})(-20)$$

$$E_1 = \frac{-1 \times 10^{-3}}{0.129 \times 10^{-3}} = -7.75\text{ V}$$

$$-5 \times 10^{-3} = -(0.5 \times 10^{-4})(-20) + (0.15 \times 10^{-3})E_3$$
$$-6 \times 10^{-3} = (0.15 \times 10^{-3})E_3$$
$$E_3 = -40\text{ V}$$

////

EXAMPLE 4-24 Write the node equations for the circuit in Fig. 4-43. Solve for the node voltages. Resketch the original circuit and indicate the node voltages, current through each resistor, and voltage drop across each resistor.

SOLUTION

First node equation: $-2\text{ mA} = \left(\frac{1}{2\text{ k}\Omega} + \frac{1}{5\text{ k}\Omega} + \frac{1}{4\text{ k}\Omega} + \frac{1}{1\text{ k}\Omega}\right)E_1 - \frac{1}{1\text{ k}\Omega}E_2$

Second node equation: $-7\text{ mA} = -\frac{1}{1\text{ k}\Omega}E_1 + \left(\frac{1}{1\text{ k}\Omega} + \frac{1}{2\text{ k}\Omega}\right)E_2$

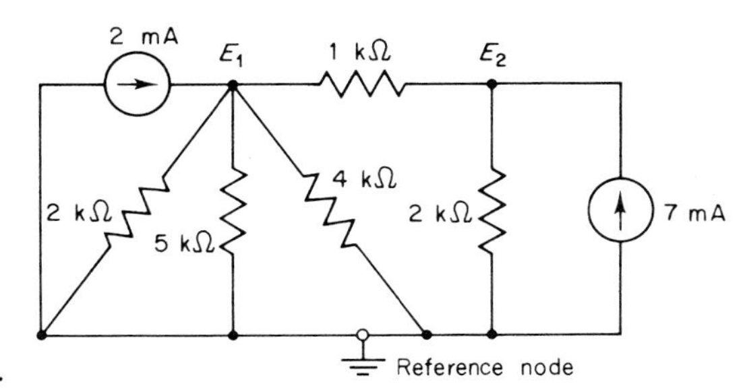

FIGURE 4-43
Circuit for Example 4-24.

Multiplying both equations by 1×10^3,

$$-2 = 1.95E_1 - E_2$$

$$-7 = -E_1 + 1.5E_2$$

$$E_1 = \frac{\begin{vmatrix} -2 & -1 \\ -7 & 1.5 \end{vmatrix}}{\begin{vmatrix} 1.95 & -1 \\ -1 & 1.5 \end{vmatrix}} = \frac{-10}{1.92} = -5.2 \text{ V}$$

$$E_2 = \frac{\begin{vmatrix} 1.95 & -2 \\ -1 & -7 \end{vmatrix}}{1.92} = \frac{-15.65}{1.92} = -8.15 \text{ V}$$

When the node voltages are known, the component currents and voltages can be found (see Fig. 4-44). ////

EXAMPLE 4-25 Provide the same information required in Example 4-24 but change the reference node (see Fig. 4-45).

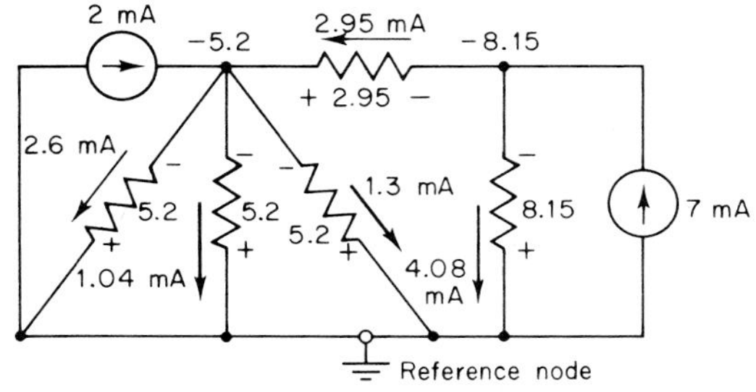

FIGURE 4-44
Circuit indicating component voltages and currents.

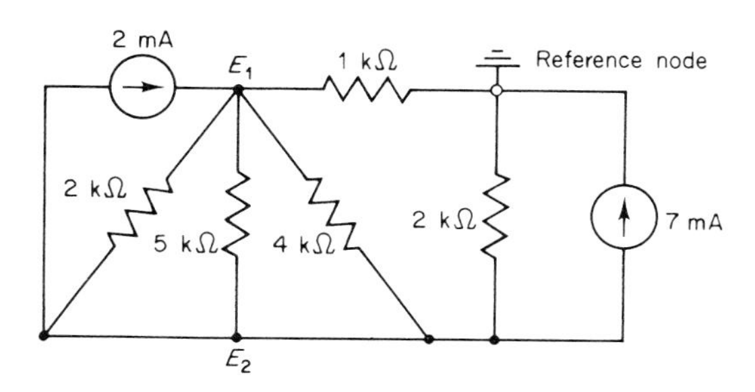

FIGURE 4-45
Circuit for Example 4-25.

SOLUTION

First node equation:

$$-2\ \text{mA} = \left(\frac{1}{2\ \text{k}\Omega} + \frac{1}{5\ \text{k}\Omega} + \frac{1}{4\ \text{k}\Omega} + \frac{1}{1\ \text{k}\Omega}\right)E_1 - \left(\frac{1}{2\ \text{k}\Omega} + \frac{1}{5\ \text{k}\Omega} + \frac{1}{4\ \text{k}\Omega}\right)E_2$$

Second node equation:

$$9\ \text{mA} = -\left(\frac{1}{2\ \text{k}\Omega} + \frac{1}{5\ \text{k}\Omega} + \frac{1}{4\ \text{k}\Omega}\right)E_1 + \left(\frac{1}{2\ \text{k}\Omega} + \frac{1}{5\ \text{k}\Omega} + \frac{1}{4\ \text{k}\Omega} + \frac{1}{2\ \text{k}\Omega}\right)E_2$$

Multiplying both equations by 1×10^3 gives

$$-2 = 1.95E_1 - 0.95E_2$$

$$9 = -0.95E_1 + 1.45E_2$$

$$E_1 = \frac{\begin{vmatrix} -2 & -0.95 \\ 9 & 1.45 \end{vmatrix}}{\begin{vmatrix} 1.95 & -0.95 \\ -0.95 & 1.45 \end{vmatrix}} = \frac{5.65}{1.92} = 2.95\ \text{V}$$

$$E_2 = \frac{\begin{vmatrix} 1.95 & -2 \\ -0.95 & 9 \end{vmatrix}}{1.92} = \frac{15.65}{1.92} = 8.15\ \text{V}$$

Figure 4-46 shows the node voltages and component responses.
Note that the circuit responses are identical in Figs. 4-44 and 4-46. Thus, selection of the reference node is arbitrary. ////

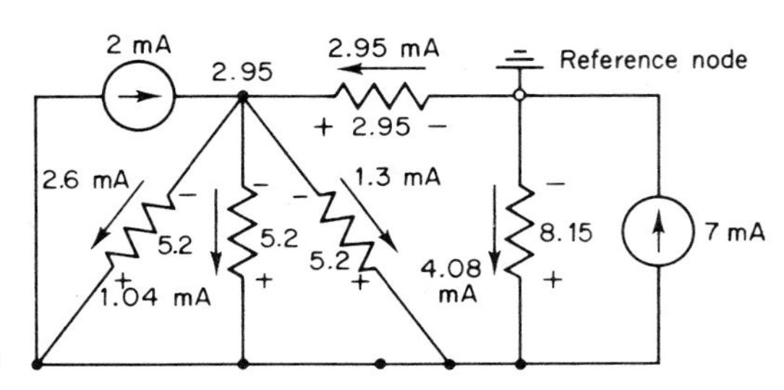

FIGURE 4-46
Circuit indicating component voltages and currents.

4-8 SUPERPOSITION

With either mesh-current analysis or node-voltage analysis, the voltage or current response at any point in a circuit can be found. The response is determined by writing network equations and solving the resulting simultaneous equations. Superposition applied to a circuit removes the requirement of taking all sources into consideration when writing network equations. With superposition, only the response to each source is considered. The principle may be stated as follows:

> **Principle of superposition** Response at any location in a circuit to the excitations is the algebraic sum of the responses to the individual excitations when each is applied one at a time with the others set equal to zero.

The resulting response due to all sources is the algebraic sum of individual responses. Setting an excitation to zero implies zero voltage or zero current. An ideal voltage source set equal to zero is equivalent to replacing the two terminals of the source with a short circuit. An ideal current source is equivalent to an open circuit when set equal to zero. Consider the circuit in Fig. 4-47. With E_1 being the only source not set equal to zero, Fig. 4-48 results. Figure 4-49 represents all sources but I_1 set equal to zero. The result of setting all sources but E_2 equal to zero is shown in Fig. 4-50.

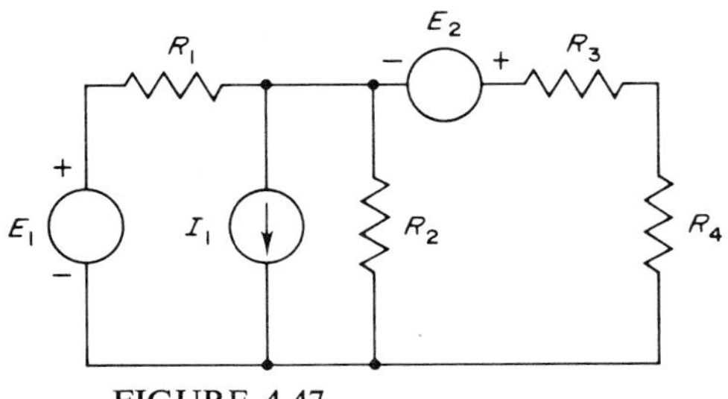

FIGURE 4-47

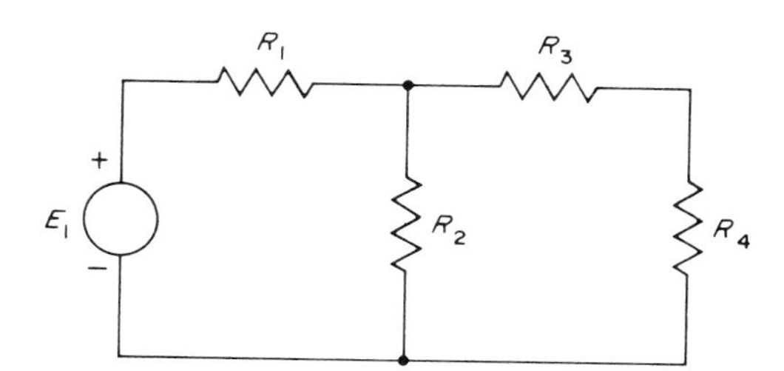

FIGURE 4-48
All sources but E_1 set equal to zero.

An initial current direction or voltage polarity is assumed for the desired response. All component excitation effects are taken in the assumed direction or polarity. Component effects are added, and the result indicates the magnitude and actual direction or polarity for the overall response. The following example should help clarify the procedure.

EXAMPLE 4-26 For the assumed current in the circuit of Fig. 4-51, determine the magnitude and actual direction using superposition

SOLUTION First set the 10-V source equal to zero, and solve for the I component due to the 3-mA current source (see Fig. 4-52). Current entering the 3-kΩ resistor can be found by using the CDR. With the current in the 3-kΩ resistor known, the CDR can be applied again to get the current in the 2-kΩ resistor. This current is the desired current due to the 3-mA source. I_{3mA} can also be found by using mesh or node equations.

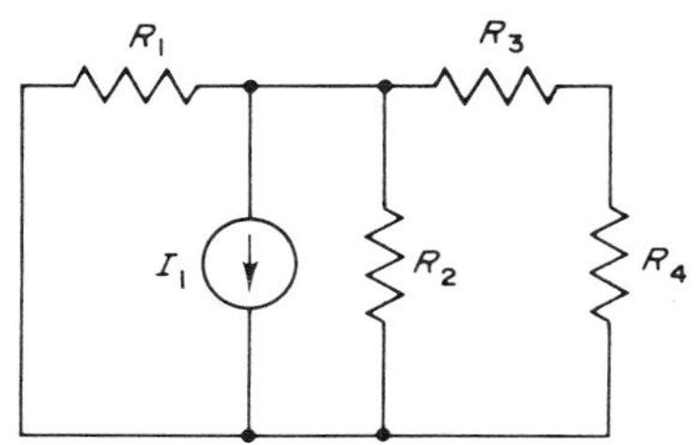

FIGURE 4-49
All sources but I_1 set equal to zero.

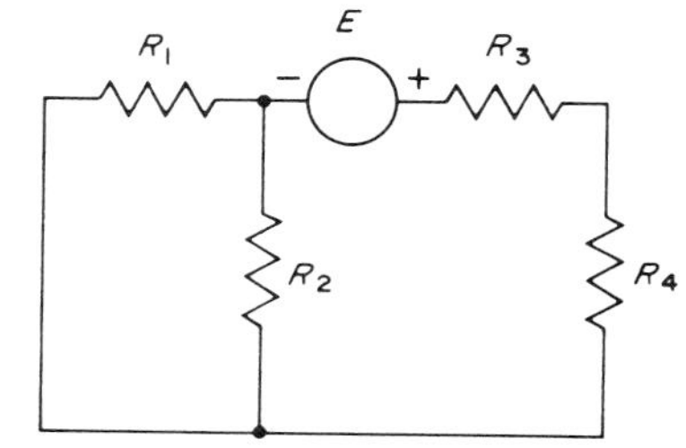

FIGURE 4-50
All sources but E_2 set equal to zero.

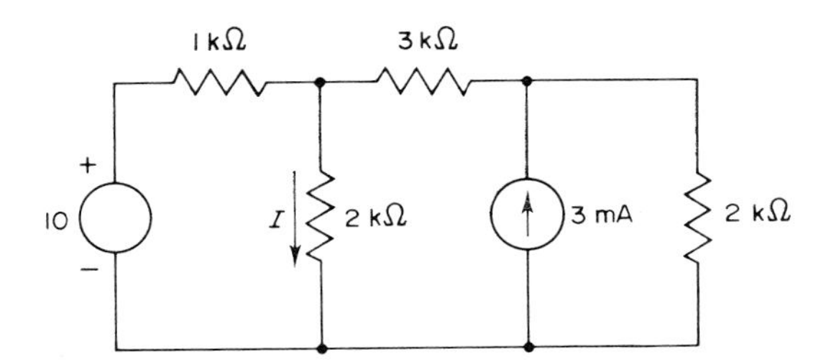

FIGURE 4-51
Circuit for Example 4-26.

Using node equations, let the bottom node be the reference node.

First node equation: $0 = \left(\frac{1}{1\ \text{k}\Omega} + \frac{1}{2\ \text{k}\Omega} + \frac{1}{3\ \text{k}\Omega}\right)E_1 - \frac{1}{3\ \text{k}\Omega}E_2$

Second node equation: $-3\ \text{mA} = -\frac{1}{3\ \text{k}\Omega}E_1 + \left(\frac{1}{3\ \text{k}\Omega} + \frac{1}{2\ \text{k}\Omega}\right)E_2$

$$0 = 1.83E_1 - 0.33E_2$$

$$-3 = -0.33E_1 + 0.83E_2$$

$$E_1 = \frac{\begin{vmatrix} 0 & -0.33 \\ -3 & 0.83 \end{vmatrix}}{\begin{vmatrix} 1.83 & -0.33 \\ -0.33 & 0.83 \end{vmatrix}} = \frac{-1}{1.41} = -0.71\ \text{V}$$

$I_{3\text{mA}}$ is in the correct direction for a negative E_1 voltage. Therefore,

$$I_{3\text{mA}} = \frac{0.71}{2\ \text{k}\Omega} = 0.355\ \text{mA}$$

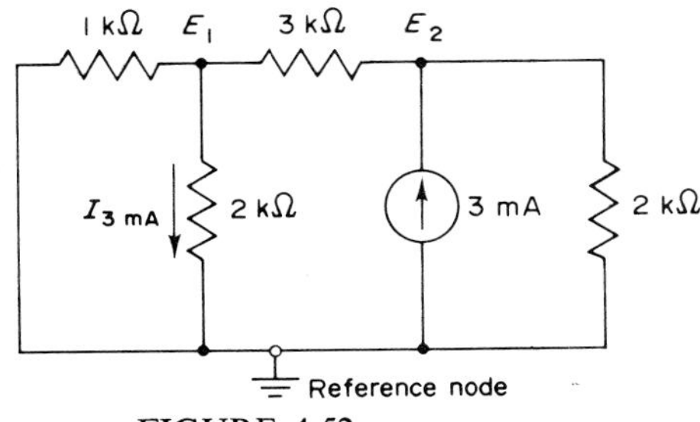

FIGURE 4-52

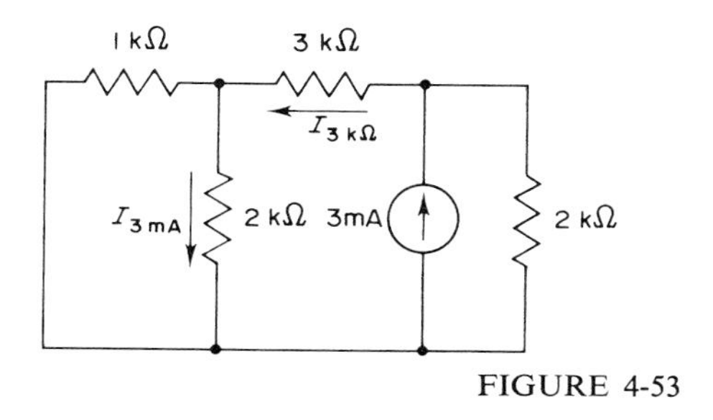

FIGURE 4-53

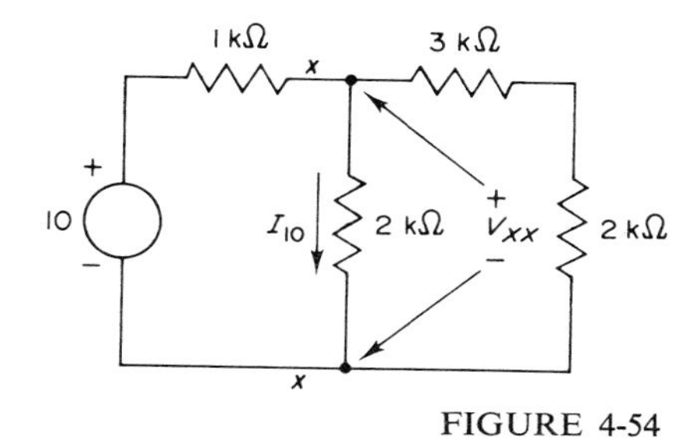

FIGURE 4-54

Using the CDR, the resistance to the left of the current source is 3 kΩ + 1 kΩ || 2 kΩ.

$$\text{Resistance to left} = 3\ \text{k}\Omega + \frac{(1\ \text{k}\Omega)(2\ \text{k}\Omega)}{3\ \text{k}\Omega} = 3.67\ \text{k}\Omega \text{ (see Fig. 4-53).}$$

$$I_{3\text{k}\Omega} = 3\ \text{mA}\,\frac{1/3.67\ \text{k}\Omega}{1/3.67\ \text{k}\Omega + 1/2\ \text{k}\Omega}$$

$$= 3\ \text{mA}\,\frac{2\ \text{k}\Omega}{5.67\ \text{k}\Omega} = 1.06\ \text{mA}$$

$$I_{3\text{mA}} = 1.06\ \text{mA}\,\frac{1/2\ \text{k}\Omega}{1/2\ \text{k}\Omega + 1/1\ \text{k}\Omega}$$

$$= 1.06\ \text{mA}\,\frac{1\ \text{k}\Omega}{3\ \text{k}\Omega} = 0.355\ \text{mA} \qquad \text{same as node analysis}$$

Next set the 3 mA source equal to zero and solve for the I component due to the 10-V source (see Fig. 4-54).

I_{10} can easily be found using mesh equations or the VDR. Using the VDR, the resistance to the right of terminals x-x is 2 kΩ || (3 kΩ + 2 kΩ).

$$\text{Resistance to right} = \frac{(2\ \text{k}\Omega)(5\ \text{k}\Omega)}{7\ \text{k}\Omega} = 1.43\ \text{k}\Omega$$

$$V_{xx} = 10\,\frac{1.43\ \text{k}\Omega}{1\ \text{k}\Omega + 1.43\ \text{k}\Omega} = 5.88\ \text{V}$$

I_{10} is in the opposite direction for polarity of V_{xx}. Therefore,

$$I_{10} = -\frac{5.88}{2\ \text{k}\Omega} = -2.94\ \text{mA}$$

$$I = I_{3\text{mA}} + I_{10}$$

$$= 0.355\ \text{mA} - 2.940\ \text{mA}$$

$$= -2.59\ \text{mA}.$$

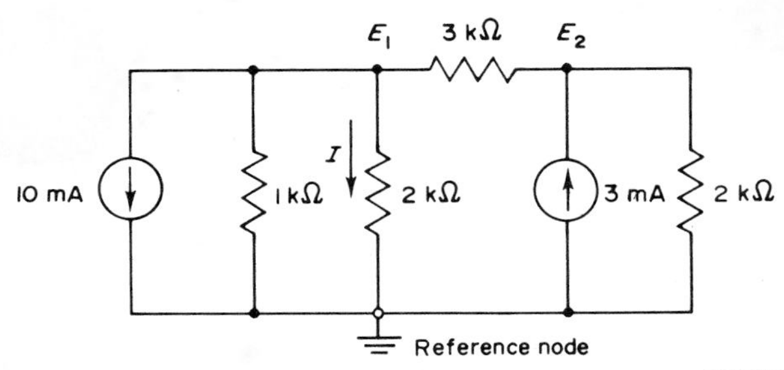

FIGURE 4-55
Analysis of entire circuit.

This result using superposition can be verified by writing node or mesh equations for the entire circuit in Fig. 4-51. The node-equation result is shown in Fig. 4-55.

First node equation: $10 \text{ mA} = \left(\frac{1}{1\text{ k}\Omega} + \frac{1}{2\text{ k}\Omega} + \frac{1}{3\text{ k}\Omega}\right)E_1 - \frac{1}{3\text{ k}\Omega}E_2$

Second node equation: $-3 \text{ mA} = -\frac{1}{3\text{ k}\Omega}E_1 + \left(\frac{1}{3\text{ k}\Omega} + \frac{1}{2\text{ k}\Omega}\right)E_2$

$$10 = 1.83E_1 - 0.33E_2$$
$$-3 = -0.33E_1 + 0.83E_2$$

$$E_1 = \frac{\begin{vmatrix} 10 & -0.33 \\ -3 & 0.83 \end{vmatrix}}{\begin{vmatrix} 1.83 & -0.33 \\ -0.33 & 0.83 \end{vmatrix}} = \frac{7.3}{1.41} = 5.17 \text{ V}$$

I is in the opposite direction for polarity of E_1. Therefore,

$$I = -\frac{5.17}{2\text{ k}\Omega} = -2.59 \text{ mA} \qquad ////$$

Various approaches can be used in solving for the response of a circuit with a single excitation. The approach selected depends upon the circuit configuration. Some of the approaches for a particular circuit are simpler than others. The problem solver must make the final decision, which becomes easier with experience.

EXAMPLE 4-27 Solve for the indicated voltage drop using the principle of superposition (see Fig. 4-56).

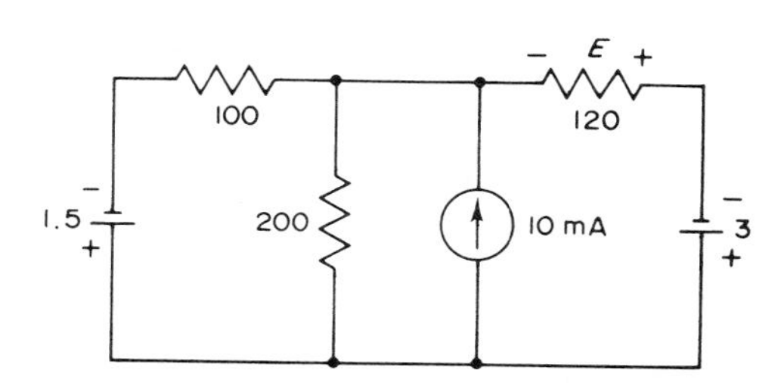

FIGURE 4-56
Circuit for Example 4-27.

SOLUTION The assumed polarity of the drop is shown on the circuit.

CASE 1 The response to the 1.5-V source (Fig. 4-57), using the VDR, is

$$E_{1.5} = 1.5 \frac{200 \| 120}{100 + (200 \| 120)}$$

$$E_{1.5} = 1.5 \tfrac{75}{175} = 0.643 \text{ V}$$

CASE 2 The response to the 10-mA source (Fig. 4-58), using the CDR for current in 120-Ω resistor is,

$$I_{120} = 10 \text{ mA} \frac{\frac{1}{120}}{\frac{1}{100} + \frac{1}{200} + \frac{1}{120}}$$

$$= 10 \text{ mA} \frac{8.33 \times 10^{-3}}{23.33 \times 10^{-3}} = 3.57 \text{ mA}$$

$$E_{10\text{mA}} = (3.57 \text{ mA})(120) = 0.429 \text{ V}$$

CASE 3 The response to the 3-V source (Fig. 4-59), using the VDR, is

$$E_3 = -3 \frac{120}{120 + (100 \| 200)}$$

$$E_3 = -3 \left(\tfrac{120}{187}\right) = -1.925 \text{ V}$$

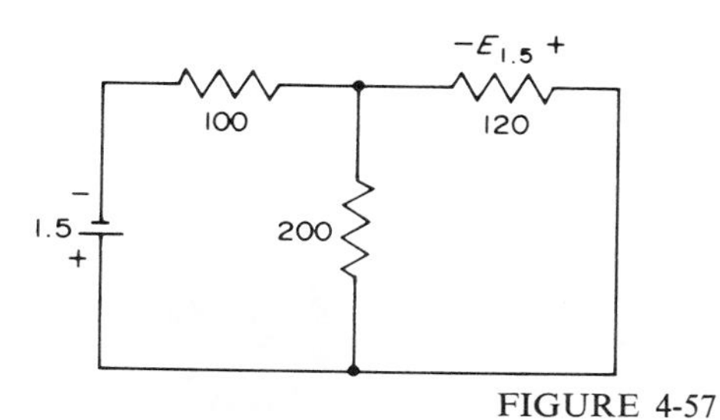

FIGURE 4-57

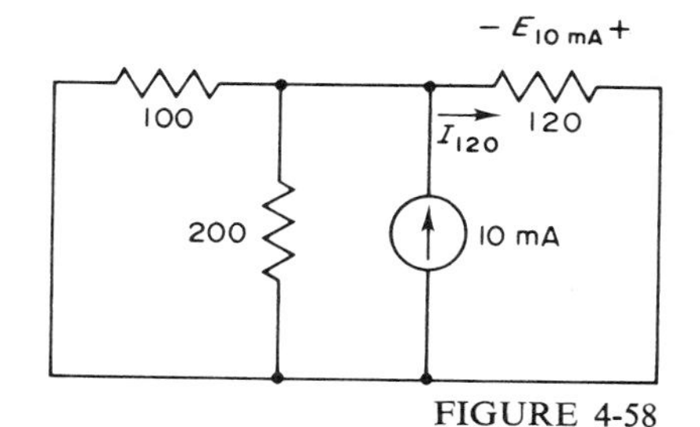

FIGURE 4-58

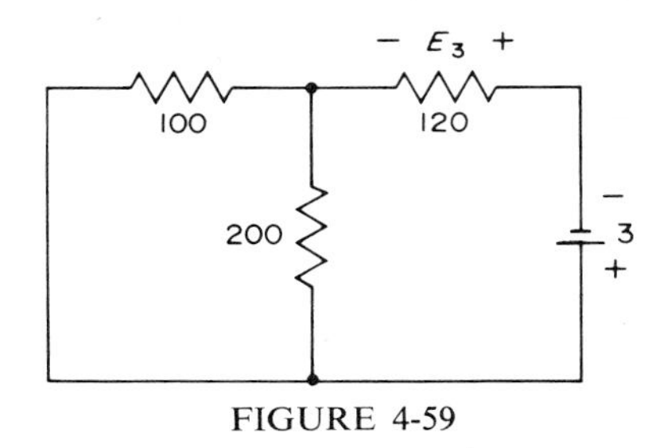

FIGURE 4-59

$$\begin{aligned}\text{Total response } E &= E_{1.5} + E_{10\text{mA}} + E_3 \\ &= 0.643 + 0.429 - 1.925 \\ &= -0.853 \text{ V}\end{aligned}$$

The magnitude of the voltage drop across the 120-Ω resistor is 0.853 V. The actual polarity is opposite the assumed polarity. ////

4-9 THÉVENIN'S THEOREM

Thévenin's theorem is a method of circuit analysis that involves reducing a complex circuit to a simple equivalent circuit. Equivalent-circuit analysis is particularly useful in tube and transistor circuitry (see Fig. 4-60).

> **Thévenin's theorem** Any two-terminal network may be replaced by a simple series circuit consisting of an equivalent emf E_{th} and an equivalent internal resistance R_{th}. The equivalent series circuit will provide the same current through a load as the original circuit if
>
> *1* The equivalent emf E_{th} is the voltage seen between the terminals of the original network with the load resistance removed.

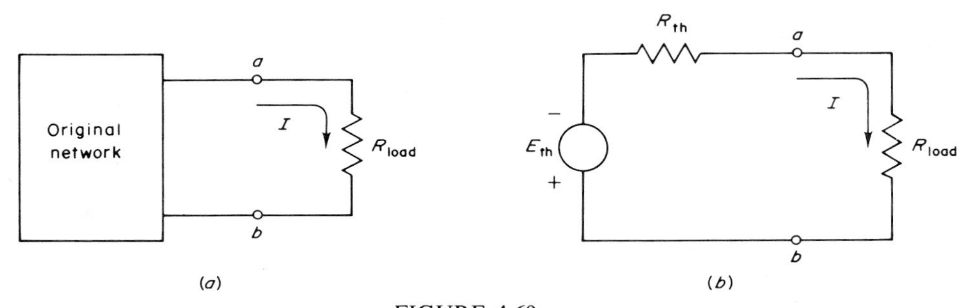

FIGURE 4-60
Original network and resultant Thévenin equivalent circuit.

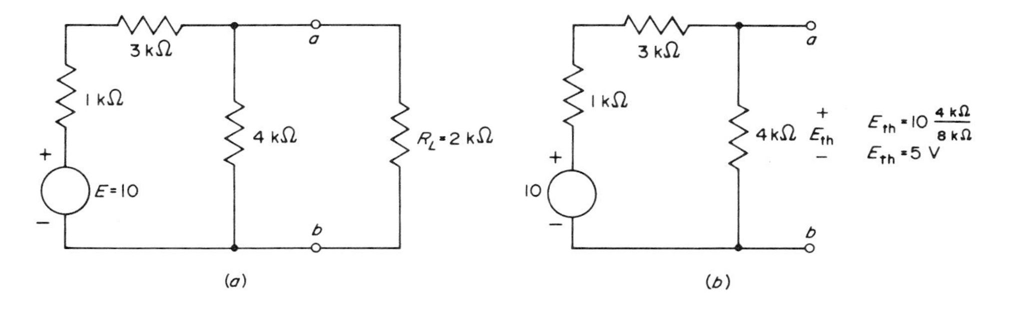

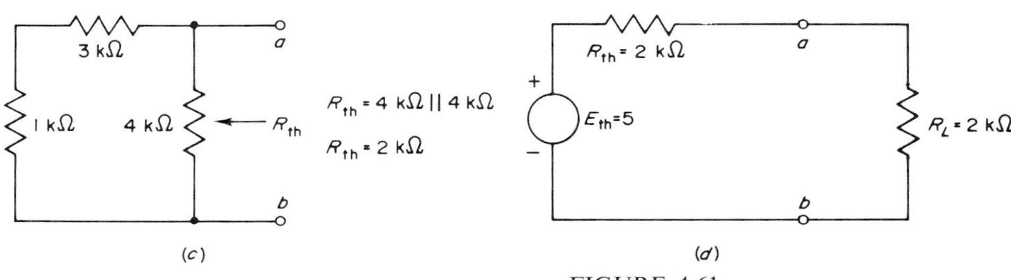

FIGURE 4-61
Conversion of network to Thévenin circuit.

2 The equivalent resistance R_{th} is the resistance seen between the terminals of the original network when the internal sources are set equal to zero.

Figure 4-61 shows the application of Thévenin's theorem to a network. The current through the load resistor R_L will be the same in the original and Thévenin circuits.

It must be pointed out that Thévenin's theorem establishes equivalency for R_L only. The current, voltage, and power of R_L in Fig. 4-61*d* are therefore the same as in Fig. 4-61*a*. The power supplied to the circuit of Fig. 4-61*d* by E_{th} is not the same as the power supplied to the circuit of Fig. 4-61*a* by E.

Thévenin's theorem has the effect of removing a portion of a circuit to make the remaining circuitry easier to analyze. Use of Thévenin's theorem is further demonstrated by the following examples.

EXAMPLE 4-28 Determine the Thévenin equivalent circuit for the 10-kΩ resistor in Fig. 4-62. From the equivalent circuit, find the voltage drop across and current through the resistor. Compare these responses with the results obtained by analyzing the entire circuit.

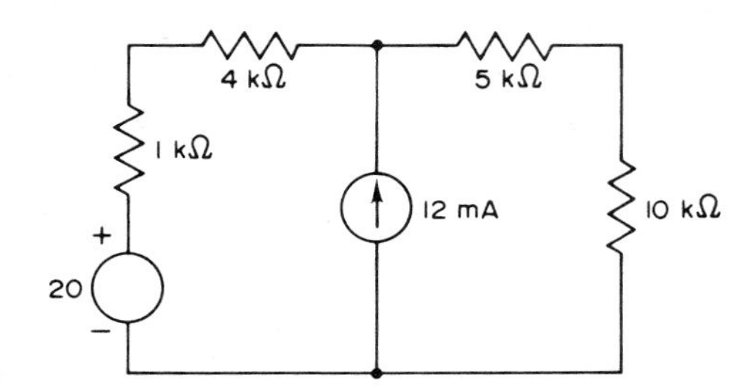

FIGURE 4-62
Circuit for Example 4-28.

SOLUTION The internal sources must be set equal to zero for finding R_{th} (see Fig. 4-63).

By inspection,

$$R_{th} = 10 \text{ k}\Omega$$

The voltage at the terminals where the 10-kΩ resistor is removed becomes E_{th}. The polarity of E_{th} is assumed (see Fig. 4-64). This terminal voltage is equal to the voltage across the 12-mA source since no voltage drop occurs across the 5-kΩ resistor with the 10 kΩ removed. The current source will force a constant 12 mA in the single loop. Figure 4-65 shows the resulting voltage drops:

$$(4 \text{ k}\Omega)(12 \text{ mA} = 48 \text{ V})$$

$$(1 \text{ k}\Omega)(12 \text{ mA} = 12 \text{ V})$$

Kirchhoff's voltage law states that the voltage sources in a loop equal the voltage drops around the loop. Therefore, the drop across the current source is 40 V with the polarity indicated in Fig. 4-65. The resulting Thévenin equivalent

$$20 = 12 + 48 + E_{th}$$

$$E_{th} = -40 \text{ V}$$

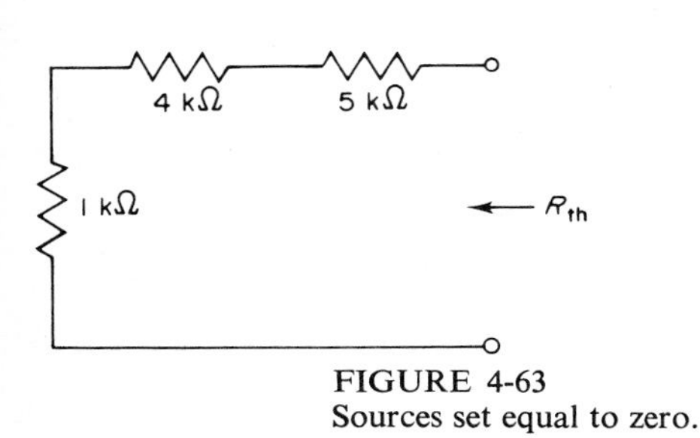

FIGURE 4-63
Sources set equal to zero.

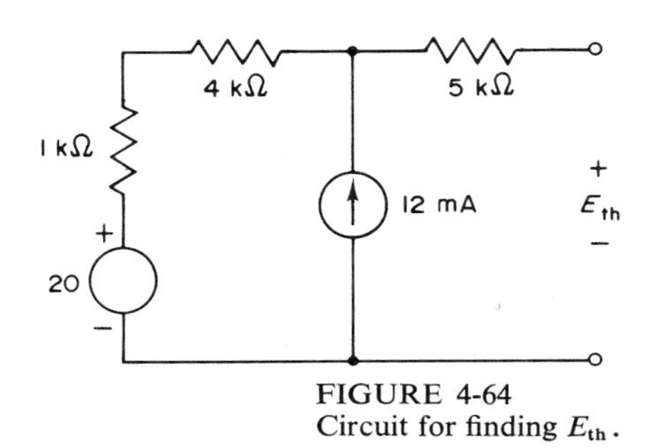

FIGURE 4-64
Circuit for finding E_{th}.

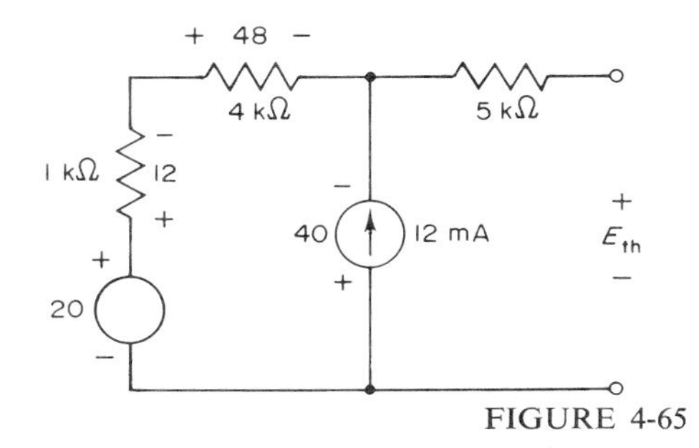

FIGURE 4-65

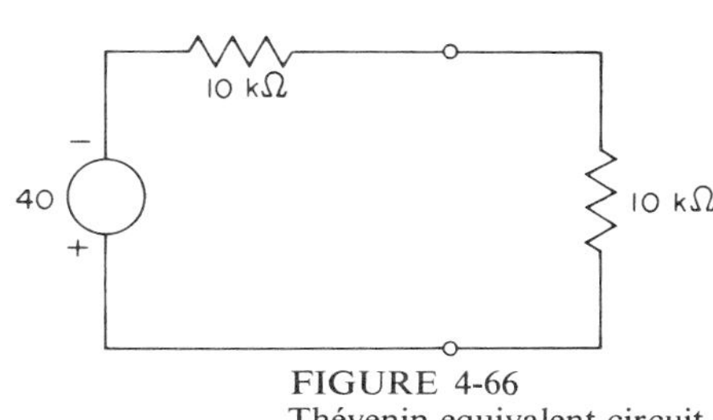

FIGURE 4-66
Thévenin equivalent circuit.

circuit is shown in Fig. 4-66. With the equivalent circuit, it is very easy to determine responses at the 10-kΩ resistor (see Fig. 4-67). To verify the responses at the 10-kΩ resistor, the entire circuit will be analyzed. The circuit of Fig. 4-62 can be analyzed using node equations after the voltage source is converted as shown in Fig. 4-68.

Node equation:
$$-8 \text{ mA} = \left(\frac{1}{5 \text{ k}\Omega} + \frac{1}{15 \text{ k}\Omega}\right)E_1$$
$$-8 = 0.267E_1$$
$$E_1 = -30.0 \text{ V}$$

The current through the 10-kΩ resistor is downward since E_1 is negative.

$$I_{10\text{k}\Omega} = \frac{30.0}{15 \text{ k}\Omega} = 2 \text{ mA} \qquad \text{downward}$$
$$E_{10\text{k}\Omega} = (2 \text{ mA})(10 \text{ k}\Omega) = 20 \text{ V} \qquad \text{negative on top}$$

Responses check using either method. ////

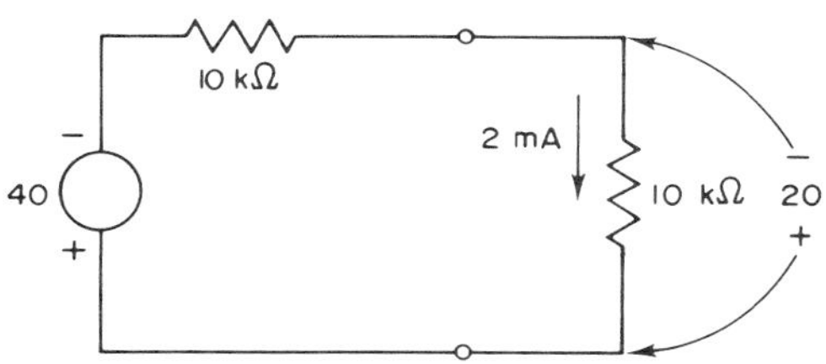

FIGURE 4-67
Responses at 10-kΩ resistor.

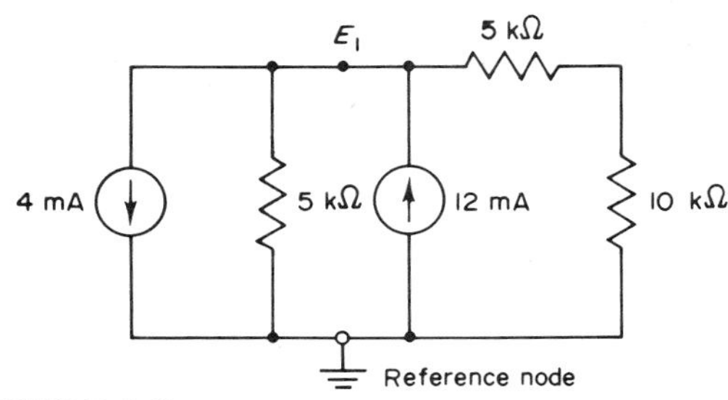

FIGURE 4-68
Circuit for node analysis.

EXAMPLE 4-29 Determine the Thévenin equivalent circuit for the 4-kΩ resistor shown in Fig. 4-69.

SOLUTION Sources are set to zero for R_{th} (see Fig. 4-70).

$$R_{th} = 1\ \text{k}\Omega \| 2\ \text{k}\Omega = 0.67\ \text{k}\Omega$$

The 4-kΩ resistor is removed to find E_{th} (see Fig. 4-71).

$$I = \frac{16 + 9}{3\ \text{k}\Omega} = 8.34\ \text{mA}$$

$$E_{1\text{k}\Omega} = (8.34\ \text{mA})(1\ \text{k}\Omega) = 8.34\ \text{V}$$

$$E_{th} = 8.34 - 9.00 = -0.66\ \text{V}$$

////

The resulting equivalent circuit is shown in Fig. 4-72.

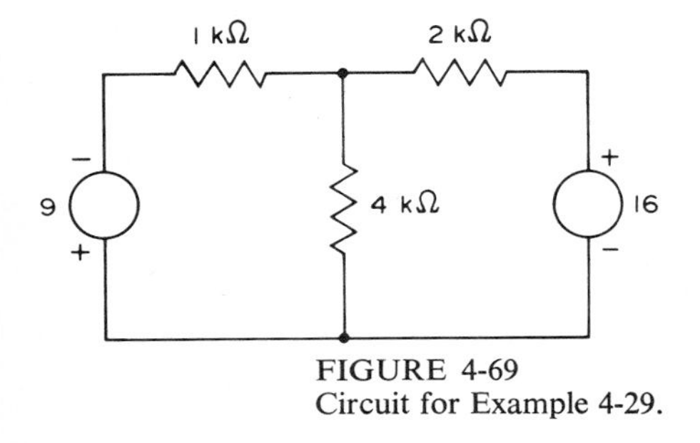

FIGURE 4-69
Circuit for Example 4-29.

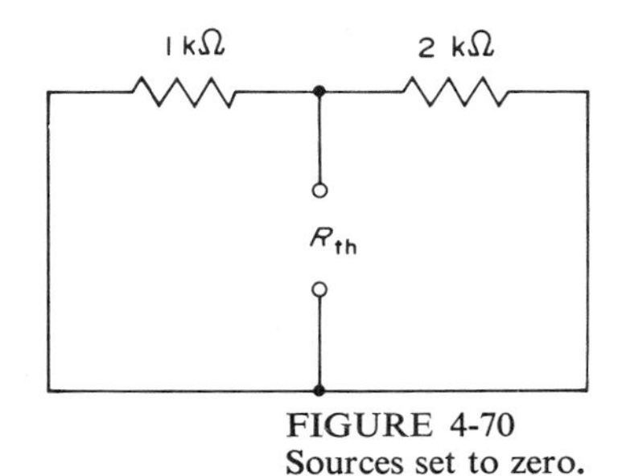

FIGURE 4-70
Sources set to zero.

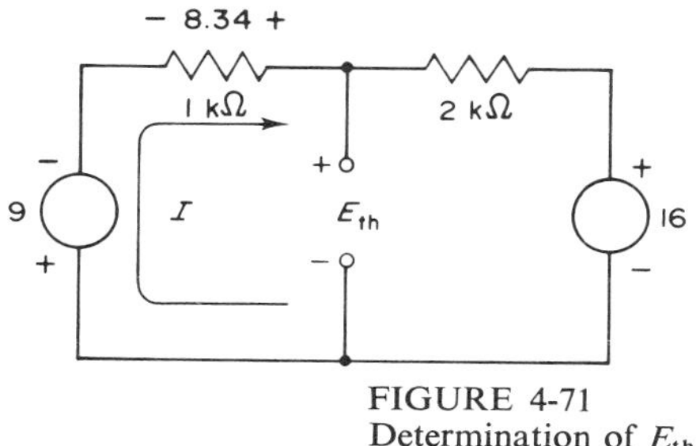

FIGURE 4-71
Determination of E_{th}.

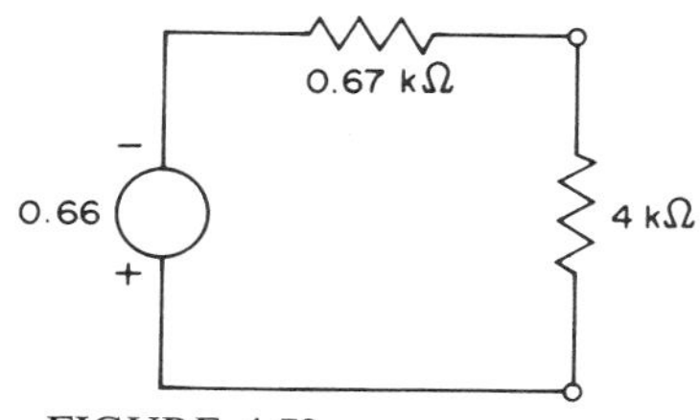

FIGURE 4-72
Thévenin circuit for Example 4-29.

EXAMPLE 4-30 Find the Thévenin circuit to the left of points x-y (see Fig. 4-73).

SOLUTION Setting the 9 mA source to zero yields $R_{th} = 6$ kΩ.
E_{th} is equal to the voltage across the 2-kΩ resistor. The entire 9 mA from the current source flows through the 2-kΩ resistor.

$$E_{th} = 18 \text{ V} \qquad ////$$

EXAMPLE 4-31 Find the Thévenin circuit for an actual current source shown in Fig. 4-76.

SOLUTION Setting $I = 0$ results in

$$R_{th} = R$$
$$E_{th} = IR \qquad ////$$

This validates the conversion between sources shown previously in Fig. 4-24.

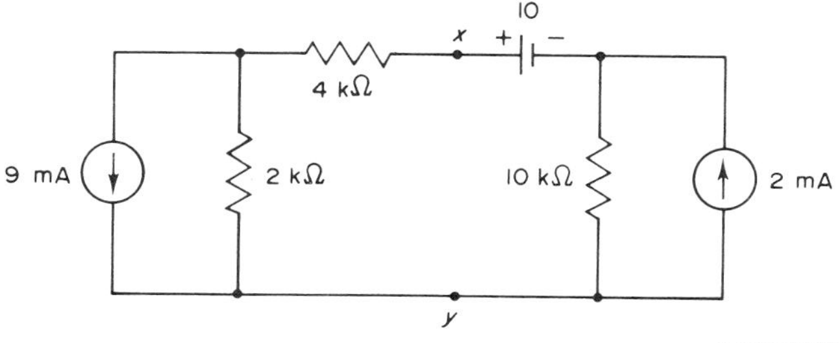

FIGURE 4-73
Circuit for Example 4-30.

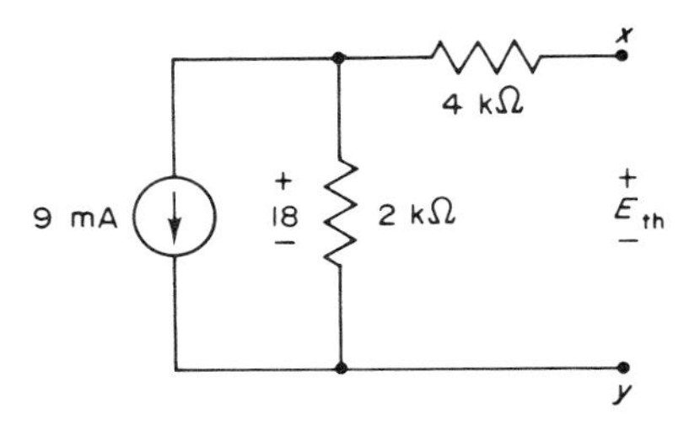

FIGURE 4-74
Circuit for E_{th}.

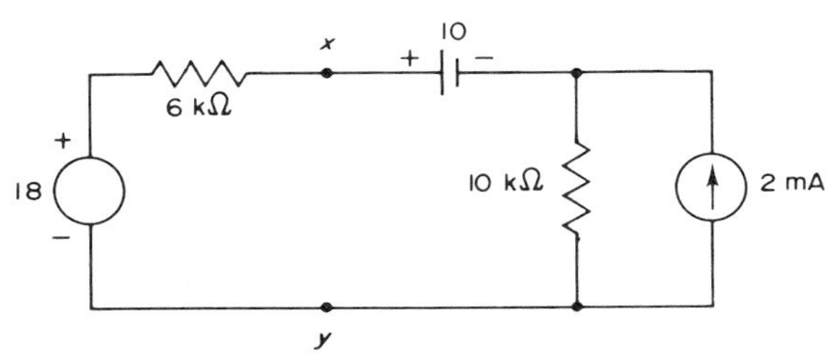

FIGURE 4-75
Thévenin circuit for Example 4-30.

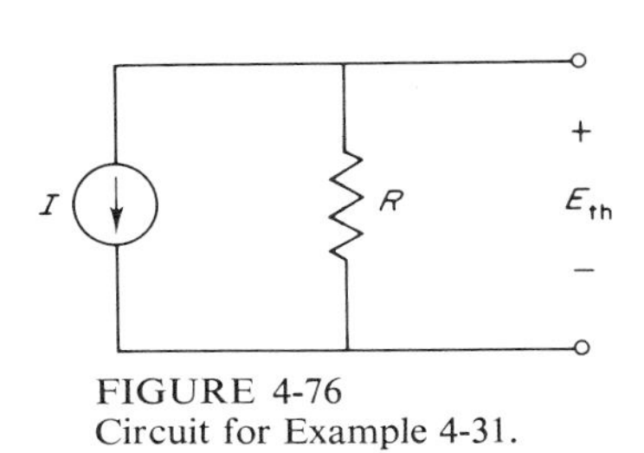

FIGURE 4-76
Circuit for Example 4-31.

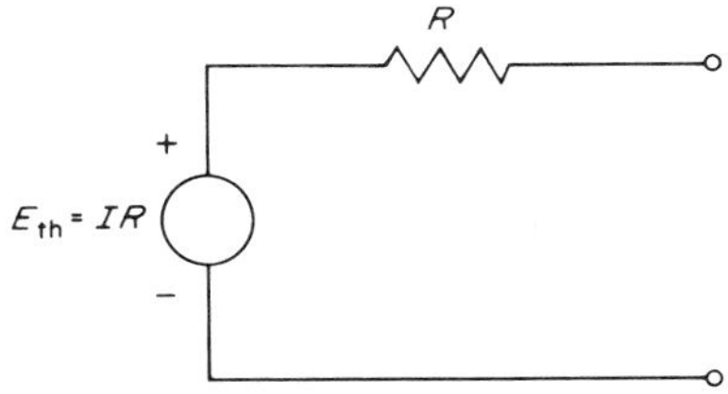

FIGURE 4-77
Thévenin equivalent for Example 4-31.

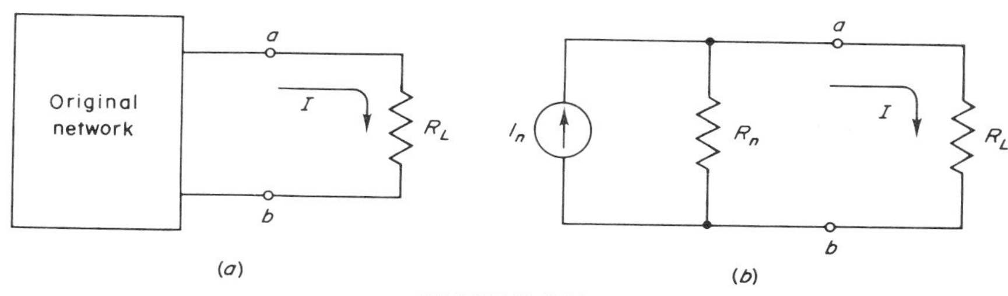

FIGURE 4-78
Original network and resultant Norton equivalent circuit.

4-10 NORTON'S THEOREM

Norton's theorem provides a second method of reducing complex circuitry to a simple equivalent circuit.

Norton's Theorem Any network can be reduced, considering any two terminals of the network, to a simple parallel circuit consisting of an equivalent resistance R_n shunted across a constant current source I_n.

Reduction of a network to a Norton equivalent circuit is shown in Fig. 4-78.

The equivalent circuit will provide the same current through R_L as the original circuit if:

1 The constant current I_n is the current that would flow through a short circuit between the two terminals being considered in the original network.

2 The equivalent resistance R_n is the resistance seen between the two terminals being considered in the original network, as in Thévenin's theorem. Internal sources are set equal to zero.

Figure 4-79 shows the application of Norton's theorem to the same circuit that Thévenin's theorem was applied to in Fig. 4-61.

The direction of I_n in Fig. 4-79*d* is oriented to produce the short-circuit current direction at terminals *a-b* in Fig. 4-79*b*.

In Norton's theorem, as in Thévenin's theorem, the equivalency is established for the chosen load terminals. Therefore, the equivalency applies only at the load.

Application of Norton's theorem is further illustrated in the following examples.

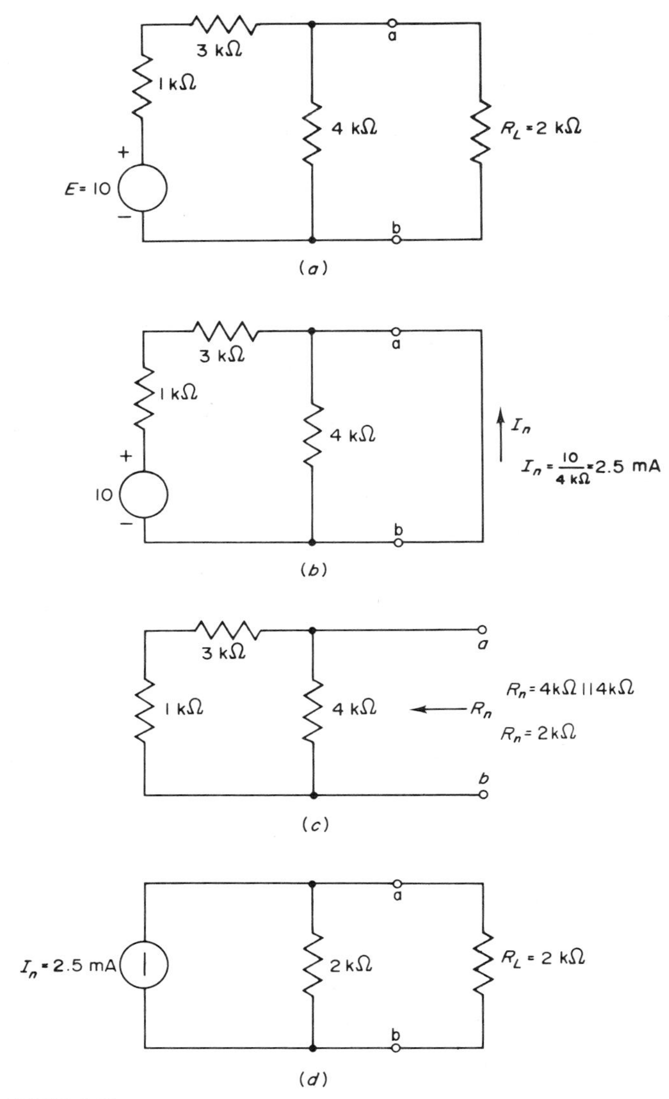

FIGURE 4-79
Conversion to Norton circuit.

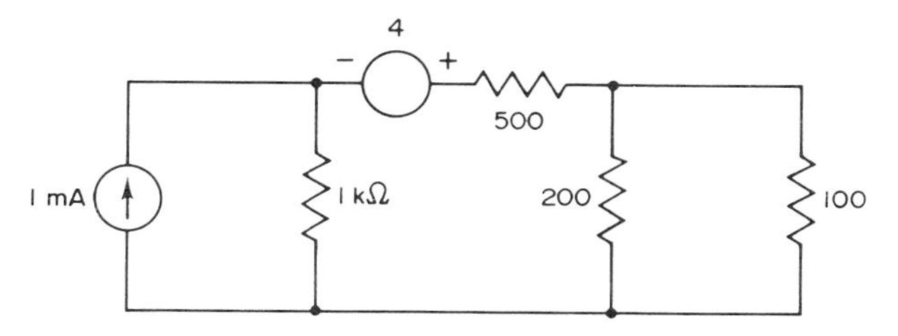

FIGURE 4-80
Circuit for Example 4-32.

EXAMPLE 4-32 Determine the Norton equivalent circuit for the 100-Ω resistor in Fig. 4-80. After obtaining the Norton circuit, find the responses at the 100-Ω resistor. Show that the same responses result through a complete circuit analysis.

SOLUTION Internal sources must be set to zero for R_n (see Fig. 4-81).

$$R_n = 200 \| 1.5\ \text{k}\Omega = \frac{(200)(1{,}500)}{1{,}700}$$

$$= 176\ \Omega$$

A short is placed across the terminals to obtain I_n. The direction of I_n is arbitrarily assumed (see Fig. 4-82). The short circuit shunts the 200-Ω resistor. A single loop results if the current source is converted into a voltage source (see Fig. 4-83).

$$I_n = \frac{1-4}{1.5\ \text{k}\Omega} = -2\ \text{mA}$$

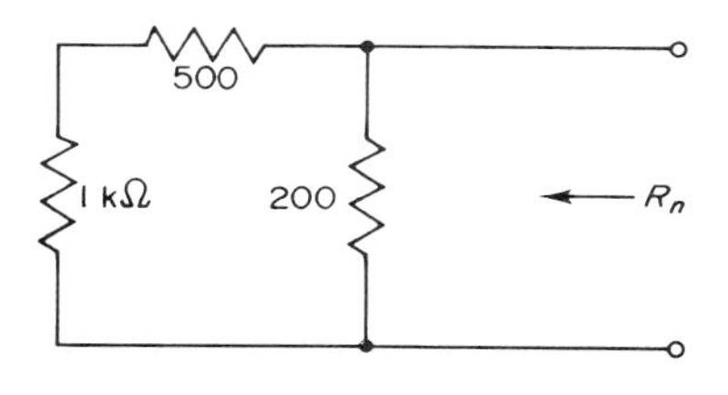

FIGURE 4-81
Circuit for R_n.

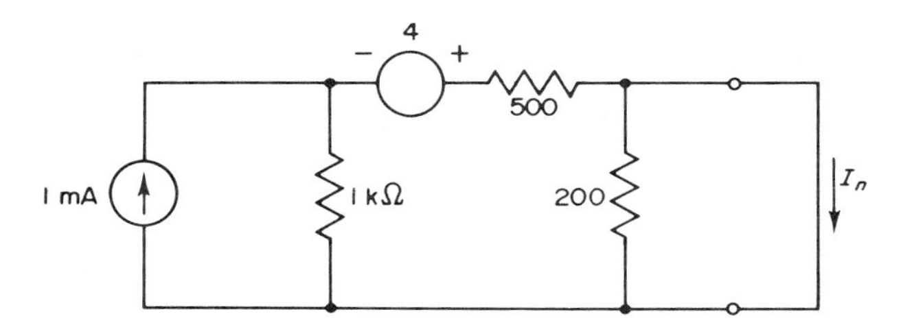

FIGURE 4-82
Circuit for I_n.

Thus, the short-circuit current flows in the upward direction. The resulting Norton circuit is shown in Fig. 4-84. The current in the 100-Ω resistor is in the upward direction. Its magnitude can be found by using the CDR.

$$I = 2\text{ mA}\,\frac{\frac{1}{100}}{\frac{1}{100}+\frac{1}{176}}$$

$$= 2\text{ mA}\,\frac{0.01}{0.0157} = 1.28\text{ mA}$$

$$E = (1.28\text{ mA})(100) = 0.128\text{ V}$$

Responses are shown in Fig. 4-85.

The original circuit in Fig. 4-80 will be analyzed using mesh equations. Figure 4-86 shows the circuit after conversion.

First mesh equation: $-3 = (1.7\text{ k}\Omega)\,I_1 - (0.2\text{ k}\Omega)I_2$

Second mesh equation: $0 = (-0.2\text{ k}\Omega)\,I_1 + (0.3\text{ k}\Omega)I_2$

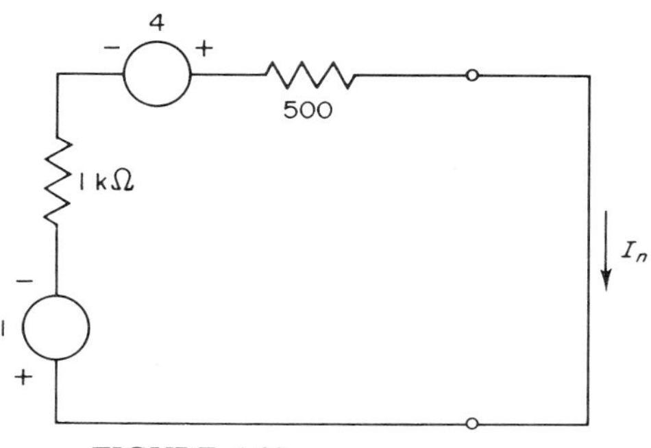

FIGURE 4-83

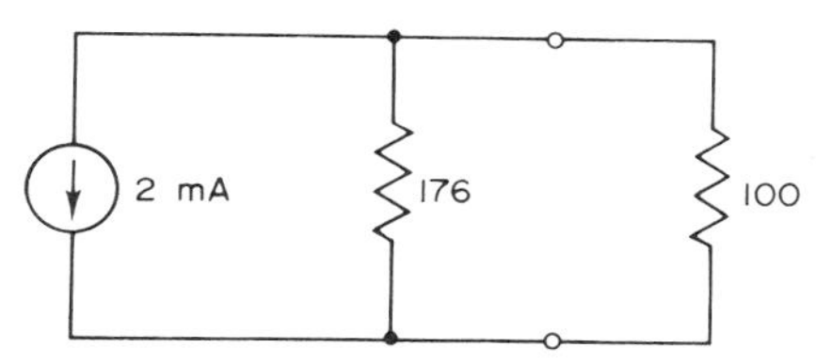

FIGURE 4-84
Norton circuit for Example 4-32.

Solving for I_2,

$$I_2 = \frac{\begin{vmatrix} 1.7\text{ k}\Omega & -3 \\ -0.2\text{ k}\Omega & 0 \end{vmatrix}}{\begin{vmatrix} 1.7\text{ k}\Omega & -0.2\text{ k}\Omega \\ -0.2\text{ k}\Omega & 0.3\text{ k}\Omega \end{vmatrix}} = \frac{-0.6 \times 10^3}{0.47 \times 10^6}$$

$$= -1.28\text{ mA}$$

Current through the 100-Ω resistor is upward at a magnitude of 1.28 mA

$$E = (1.28\text{ mA})\,(100) = 0.128\text{ V} \qquad \text{negative on bottom}$$

Responses check using either method. ////

EXAMPLE 4-33 Determine the Norton equivalent circuit below points x-y in Fig. 4-87.

SOLUTION Sources are set to zero to find R_n (see Fig. 4-88).

$$R_n = 2\text{ k}\Omega \| (4\text{ k}\Omega + 10\text{ k}\Omega + 5\text{ k}\Omega)$$

$$= \frac{(2\text{ k}\Omega)(19\text{ k}\Omega)}{21\text{ k}\Omega} = 1.81\text{ k}\Omega$$

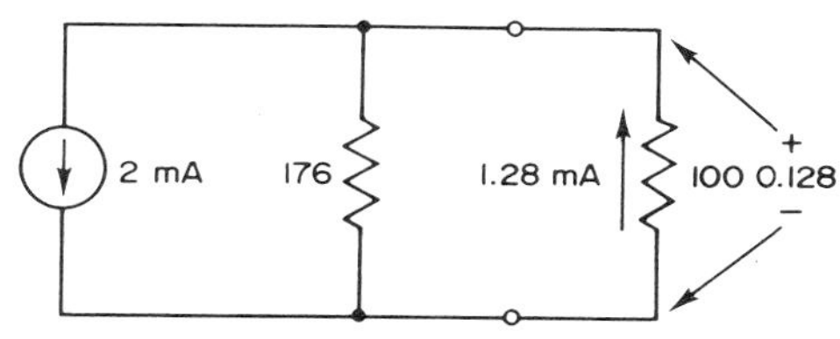

FIGURE 4-85
Responses at 100-Ω resistor.

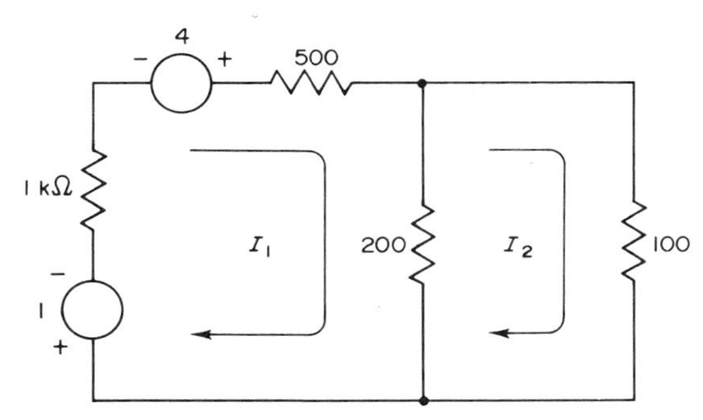

FIGURE 4-86
Circuit for mesh equations.

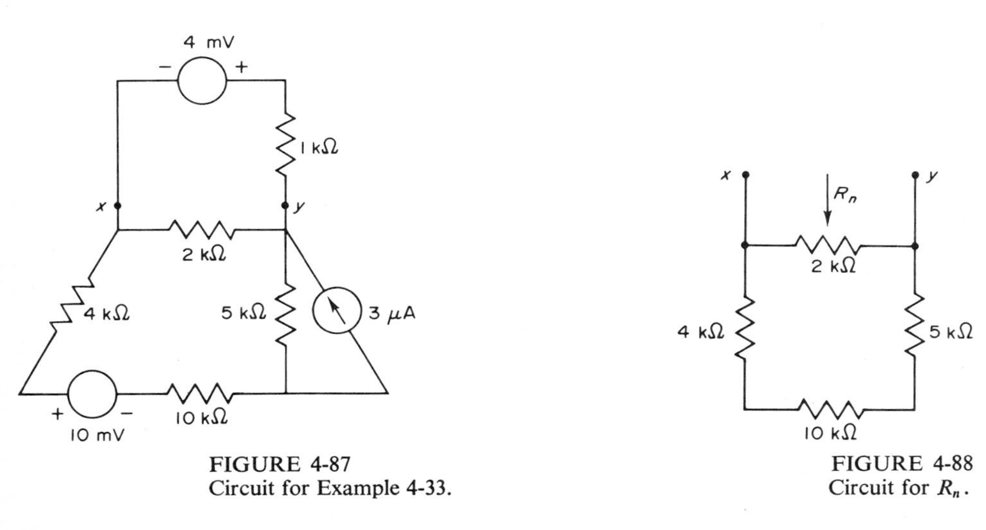

FIGURE 4-87
Circuit for Example 4-33.

FIGURE 4-88
Circuit for R_n.

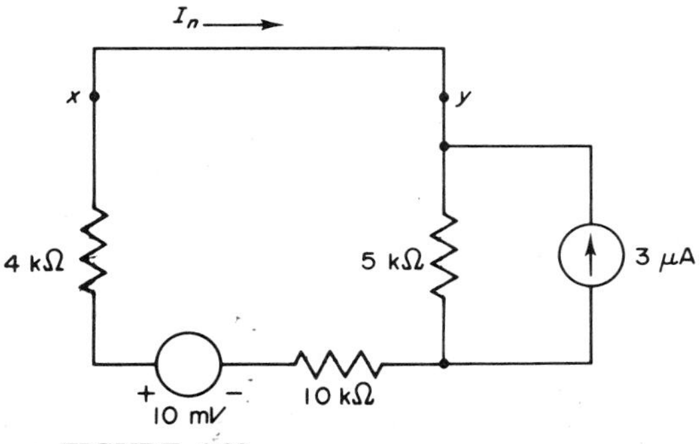

FIGURE 4-89
Circuit for I_n.

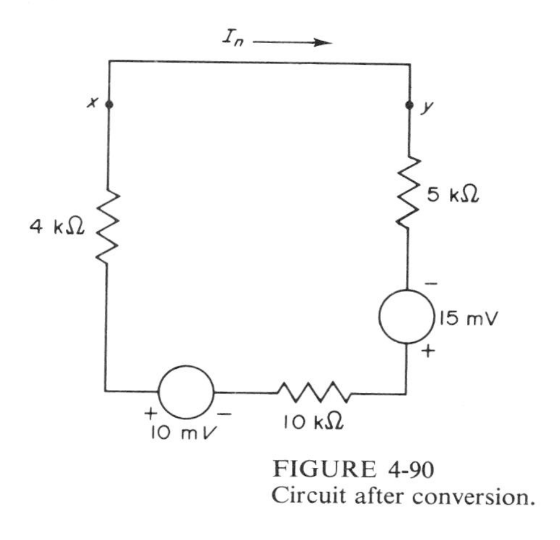

FIGURE 4-90
Circuit after conversion.

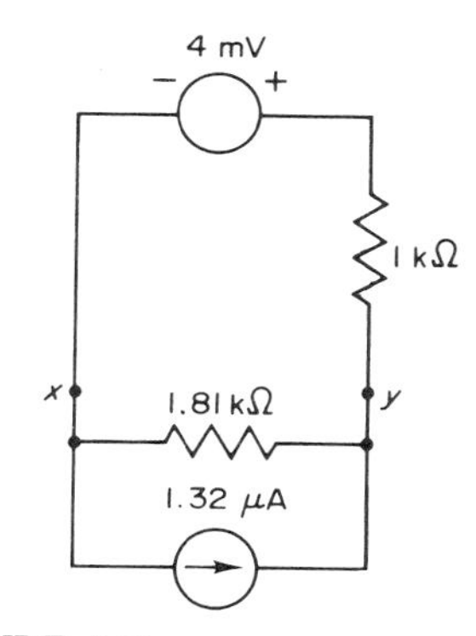

FIGURE 4-91
Norton circuit for Example 4-33.

Terminals x-y are shorted to find I_n. The short will shunt out the 2-kΩ resistor. The direction of I_n is arbitrarily selected (see Fig. 4-89). The current source will be converted to obtain a single loop (see Fig. 4-90).

$$I_n = \frac{-25 \text{ mV}}{19 \text{ k}\Omega} = -1.32\ \mu\text{A}$$

I_n actually flows in the opposite direction. The resulting circuit is shown in Fig. 4-91.

////

Care must be exercised in choosing the terminals at which Norton's theorem is to be applied. The main consideration in selecting the terminals is that the remaining circuitry be as easy as practical to analyze.

4-11 RECIPROCITY THEOREM

This theorem applies to networks with a single excitation. The excitation may be either a voltage source or current source. Reciprocity is valid when the excitation is a voltage source and the response is a current or the excitation is a current source and the response is a voltage.

Reciprocity theorem The ratio of the response to the excitation is the same whether the response is at point m and the excitation at point k or the response is at k and the excitation is at m.

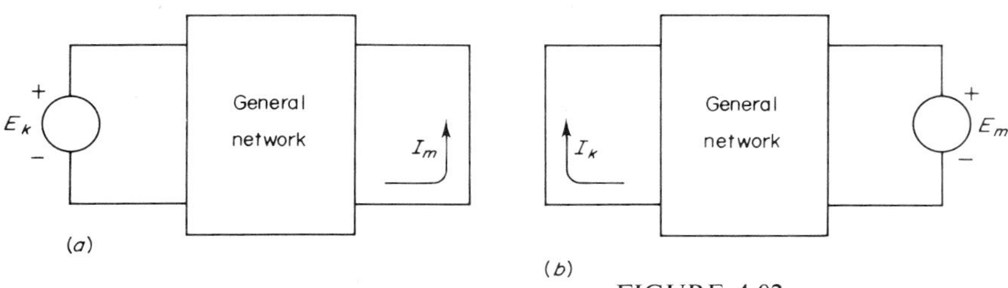

FIGURE 4-92
Voltage excitation and current response.

In Fig. 4-92*a* a voltage is applied at location k, and the current response is at location m. The direction of I_m coincides with the direction that the voltage source forces current into the network. Figure 4-92*b* shows a voltage applied at location m and the current response at location k. By reciprocity, $I_m/E_k = I_k/E_m$. The ratio of current at a location with respect to a voltage source at another location is called the *transfer conductance*. In Fig. 4-93, a similar situation exists with current-source excitation and voltage response. The polarity of the voltage response agrees with the direction of current flow from the current source. By reciprocity, $E_m/I_k = E_k/I_m$. These voltage-current ratios are called *transfer resistances*.

Not only is the concept of source-response exchange useful, but the transfer conductance or transfer resistance between points is constant in a given network. With the constant transfer ratio the response for many values of excitation can be found quite easily.

EXAMPLE 4-34 Show that reciprocity is valid in the circuits of Fig. 4-94. For the following excitation values, determine response current: 2, 4, 8, 10, 17.

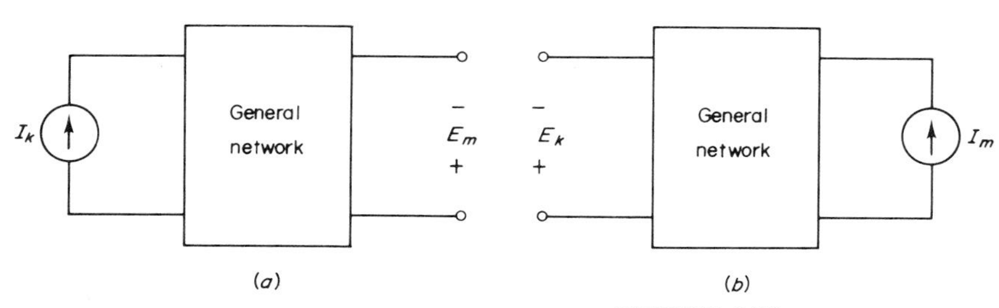

FIGURE 4-93
Current excitation and voltage response.

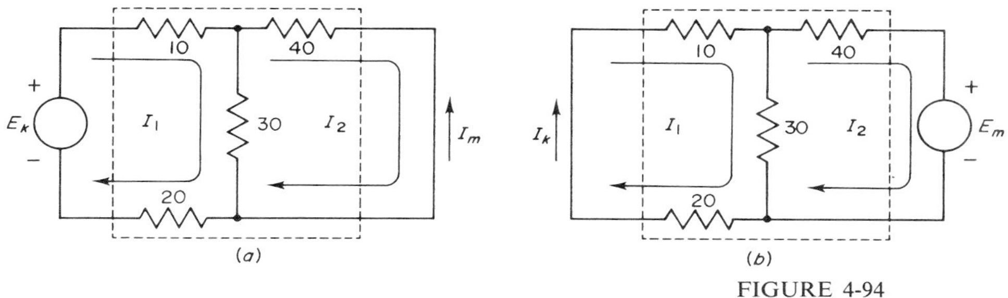

FIGURE 4-94
Circuits for Example 4-34.

SOLUTION Write mesh equations for Fig. 4-94*a*:

First mesh equation: $-E_k = 60I_1 - 30I_2$

Second mesh equation: $0 = -30I_1 + 70I_2$

Solve for I_2.

$$I_2 = \frac{\begin{vmatrix} 60 & -E_k \\ -30 & 0 \end{vmatrix}}{\begin{vmatrix} 60 & -30 \\ -30 & 70 \end{vmatrix}} = \frac{-30E_k}{3.3 \times 10^3} = -9.1 \times 10^{-3}E_k$$

$$I_m = -I_2 = 9.1 \times 10^{-3}E_k$$

Therefore, Transfer conductance: $\dfrac{I_m}{E_k} = 9.1 \times 10^{-3}$ mhos

Writing mesh equations for Fig. 4-94*b*, we have:

First mesh equation: $0 = 60I_1 - 30I_2$

Second mesh equation: $E_m = -30I_1 + 70I_2$

Solving for I_1 gives

$$I_1 = \frac{\begin{vmatrix} 0 & -30 \\ E_m & 70 \end{vmatrix}}{\begin{vmatrix} 60 & -30 \\ -30 & 70 \end{vmatrix}} = \frac{30E_m}{3.3 \times 10^{-3}} = 9.1 \times 10^{-3}E_m$$

$$I_k = I_1 = (9.1 \times 10^{-3})E_m$$

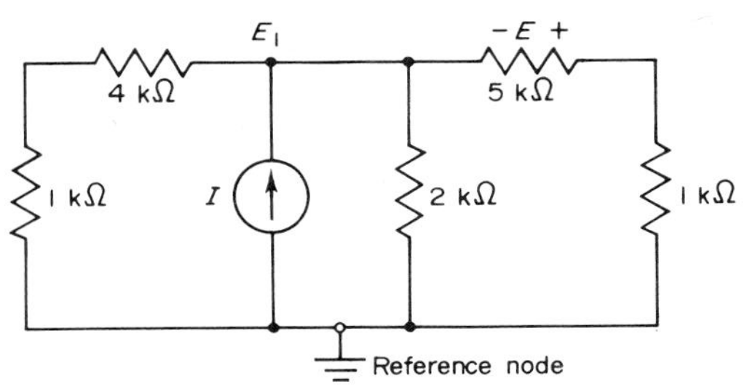

FIGURE 4-95
Circuit for Example 4-35.

Therefore, transfer conductance $= \dfrac{I_k}{E_m} = 9.1 \times 10^{-3}$ mhos

Thus,

$$\frac{I_m}{E_k} = \frac{I_k}{E_m}$$

For the given assortment of excitation values, response currents can easily be obtained from the transfer conductance.

E_k or E_m	I_m or I_k
2	$(2)(9.1 \times 10^{-3}) = 18.2$ mA
4	$(4)(9.1 \times 10^{-3}) = 36.4$ mA
8	$(8)(9.1 \times 10^{-3}) = 72.8$ mA
10	$(10)(9.1 \times 10^{-3}) = 91$ mA
17	$(17)(9.1 \times 10^{-3}) = 155$ mA

EXAMPLE 4-35 Determine the transfer resistance for the location shown in the circuit of Fig. 4-95.

SOLUTION A single node equation will lead to the drop across the 5-kΩ resistor. Node equation:

$$-I = \left(\frac{1}{5\text{ k}\Omega} + \frac{1}{2\text{ k}\Omega} + \frac{1}{6\text{ k}\Omega}\right)E_1$$

$$= (0.867 \times 10^{-3})E_1$$

$$E_1 = (-1.15 \times 10^3)I$$

The node voltage is negative. The polarity of E corresponds to the negative node voltage.

Using the VDR,

$$E = (1.15 \times 10^3)I \frac{5\ \text{k}\Omega}{6\ \text{k}\Omega} = (0.96 \times 10^3)I$$

Therefore, transfer resistance $\dfrac{E}{I} = 0.96 \times 10^3\ \Omega$

The same value of transfer resistance is obtained when the excitation and response are exchanged (see Fig. 4-96).

First node equation: $-I = \left(\dfrac{1}{5\ \text{k}\Omega} + \dfrac{1}{2\ \text{k}\Omega} + \dfrac{1}{5\ \text{k}\Omega}\right)E_1 - \dfrac{1}{5\ \text{k}\Omega} E_2$

Second node equation: $I = -\dfrac{1}{5\,\text{k}\Omega} E_1 + \left(\dfrac{1}{5\,\text{k}\Omega} + \dfrac{1}{1\,\text{k}\Omega}\right) E_2$

$$-I = (0.9 \times 10^{-3})E_1 - (0.2 \times 10^{-3})E_2$$

$$I = (-0.2 \times 10^{-3})E_1 + (1.2 \times 10^{-3})E_2$$

Multiplying by 1×10^3 gives

$$(-1 \times 10^3)I = 0.9E_1 - 0.2E_2$$

$$(1 \times 10^3)I = -0.2E_1 + 1.2E_2$$

Solving for E_1, we have

$$E_1 = \frac{\begin{vmatrix} (-1 \times 10^3)I & -0.2 \\ (1 \times 10^3)I & 1.2 \end{vmatrix}}{\begin{vmatrix} 0.9 & -0.2 \\ -0.2 & 1.2 \end{vmatrix}} = \frac{(-1 \times 10^3)I}{1.04} = (-0.96 \times 10^3)I$$

E_1 is a negative node voltage. E agrees with the negative node voltage.

$$E = 0.96 \times 10^3 I$$

Therefore, Transfer resistance $\dfrac{E}{I} = 0.96 \times 10^3\ \Omega$

This agrees with the previous value of transfer resistance. ////

Reciprocity does not have any relationship for a voltage response due to a voltage source or current response due to a current source.

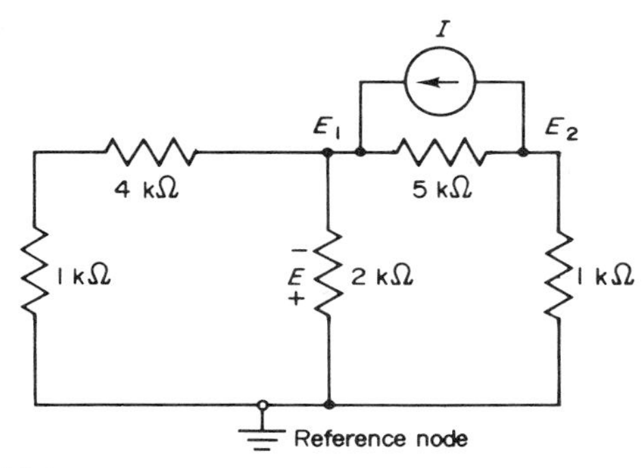

FIGURE 4-96
Exchange of excitation and response.

4-12 DELTA-Y OR PI-T TRANSFORMATIONS

In actual circuit configurations, it is at times impossible to resolve a circuit in terms of series, parallel, or combinations of series and parallel arrangements. In circuits that are this complex, transformation from one type of configuration to another may reduce the resulting circuit to one in which series and parallel circuit methods can be used.

The configuration of components shown in Fig. 4-97 is referred to as a delta connection. The configuration of components shown in Fig. 4-98 is referred to as a Y connection. It will be shown that the delta connection of Fig. 4-97 can be replaced by an equivalent Y connection, as shown in Fig. 4-98. It will likewise be shown that a Y connection can be replaced by an equivalent delta connection.

The resistance between points A-B of Fig. 4-97 is

$$R_{AB} = \frac{R_2(R_1 + R_3)}{R_1 + R_2 + R_3} \tag{4-6}$$

The resistance between points A-B of Fig. 4-98 is

$$R_{AB} = R_a + R_b \tag{4-7}$$

If the delta connection is to be equivalent to the Y connection, the resistance between points A-B of both connections must be equal. Therefore, Eq. (4-6) is set equal to Eq. (4-7).

$$R_a + R_b = \frac{R_1R_2 + R_2R_3}{R_1 + R_2 + R_3} \tag{4-8}$$

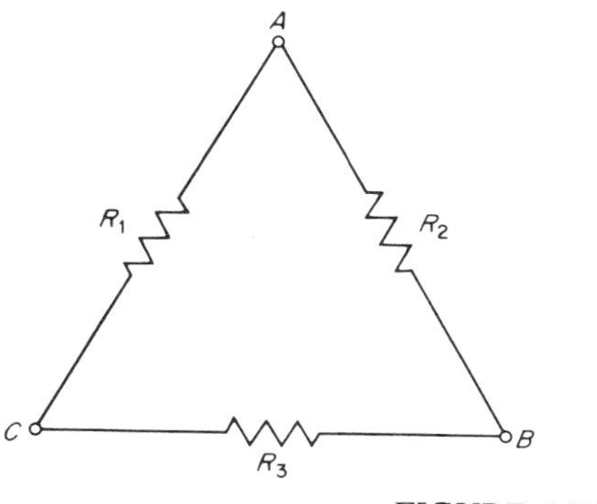

FIGURE 4-97
Delta connection.

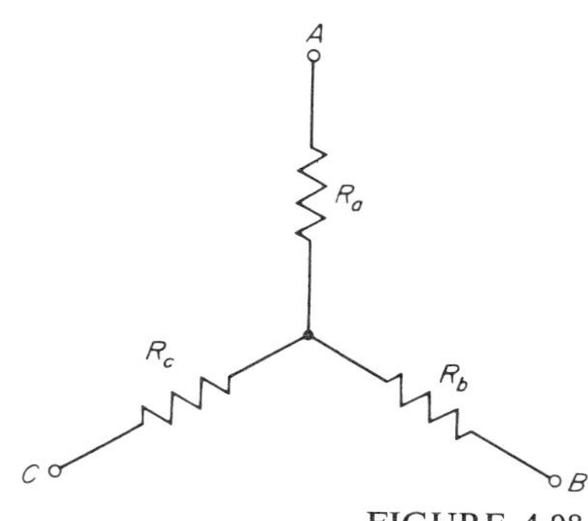

FIGURE 4-98
Y connection.

Similarly for points A-C

$$R_a + R_c = \frac{R_1R_2 + R_1R_3}{R_1 + R_2 + R_3} \tag{4-9}$$

and for points B-C

$$R_b + R_c = \frac{R_1R_3 + R_2R_3}{R_1 + R_2 + R_3} \tag{4-10}$$

Subtracting Eq. (4-10) from Eq. (4-8),

$$R_a - R_c = \frac{R_1R_2 - R_1R_3}{R_1 + R_2 + R_3} \tag{4-11}$$

Adding Eqs. (4-9) and (4-11)

$$2R_a = \frac{2R_1R_2}{R_1 + R_2 + R_3}$$

$$R_a = \frac{R_1R_2}{R_1 + R_2 + R_3} \tag{4-12}$$

By similar algebraic procedure,

$$R_b = \frac{R_2R_3}{R_1 + R_2 + R_3} \tag{4-13}$$

and

$$R_c = \frac{R_1R_3}{R_1 + R_2 + R_3} \tag{4-14}$$

The values of R_a, R_b, and R_c given by Eqs. (4-12) to (4-14) represent the values for a Y-connected circuit that can replace a delta-connected circuit.

From Eqs. (4-12) to (4-14), R_1, R_2, and R_3 can be found. R_1, R_2, and R_3 are found in terms of R_a, R_b, and R_c, the Y-connection values. Thus R_1, R_2, and R_3 yield values for an equivalent delta-connected circuit that can replace the original Y-connected circuit.

Dividing Eq. (4-12) by Eq. (4-13),

$$\frac{R_a}{R_b} = \frac{R_1}{R_3}$$

Therefore,

$$R_1 = \frac{R_3 R_a}{R_b} \tag{4-15}$$

Dividing Eq. (4-12) by Eq. (4-14),

$$\frac{R_a}{R_c} = \frac{R_2}{R_3}$$

Therefore,

$$R_2 = \frac{R_3 R_a}{R_c} \tag{4-16}$$

Substituting the results of Eqs. (4-15) and (4-16) into Eq. (4-14),

$$R_c = \frac{(R_3 R_a/R_b)R_3}{R_3 R_a/R_b + R_3 R_a/R_c + R_3}$$

$$= \frac{R_3 R_a/R_b}{R_a/R_b + R_a/R_c + 1}$$

$$\frac{R_3 R_a}{R_b} = \frac{R_c R_a}{R_b} + R_a + R_c$$

$$R_3 R_a = R_c R_a + R_a R_b + R_c R_b$$

$$R_3 = \frac{R_c R_a + R_a R_b + R_c R_b}{R_a} \tag{4-17}$$

By repeating the algebraic process used to find R_3, R_2 and R_1 can also be found.

$$R_2 = \frac{R_c R_a + R_a R_b + R_c R_b}{R_c} \tag{4-18}$$

and

$$R_1 = \frac{R_c R_a + R_a R_b + R_c R_b}{R_b} \tag{4-19}$$

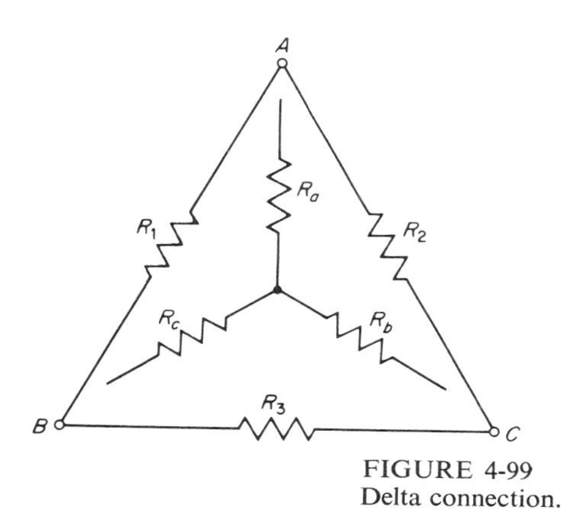

FIGURE 4-99
Delta connection.

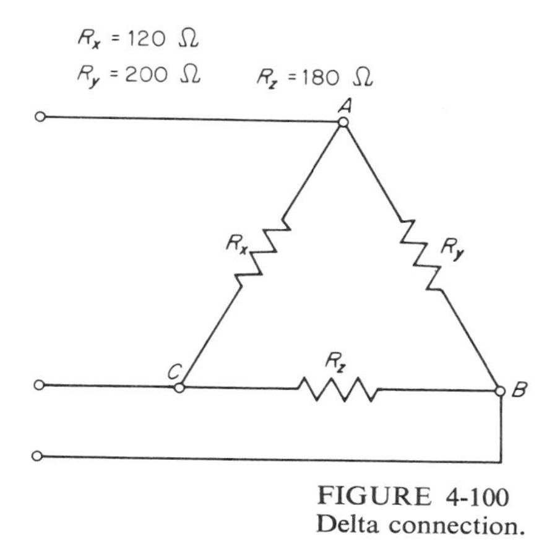

FIGURE 4-100
Delta connection.

The values of R_1, R_2, R_3 given by Eqs. (4-17) to (4-19) represent the values of a delta-connected circuit that can replace a Y-connected circuit.

If the delta-connected circuit and Y-connected circuit are superimposed, as in Fig. 4-99, it can be seen that a desired resistance in the Y connection equals the product of the adjacent resistances divided by the sum of the resistances in the delta connection. A desired resistance in the delta connection is the sum of the products of the resistances in the Y connection, taken two at a time, divided by the resistance opposite in the Y connection.

EXAMPLE 4-36 Convert the delta connection of Fig. 4-100 to an equivalent Y connection.

SOLUTION The equivalent Y connection is shown in Fig. 4-101.

$$R_1 = \frac{R_x R_y}{R_x + R_y + R_z} = \frac{(120)(200)}{120 + 200 + 180} = \frac{(120)(200)}{500} = 48\ \Omega$$

$$R_2 = \frac{R_x R_z}{R_x + R_y + R_z} = \frac{(120)(180)}{500} = 43.2\ \Omega$$

$$R_3 = \frac{R_y R_z}{R_x + R_y + R_z} = \frac{(200)(180)}{500} = 72\ \Omega$$

////

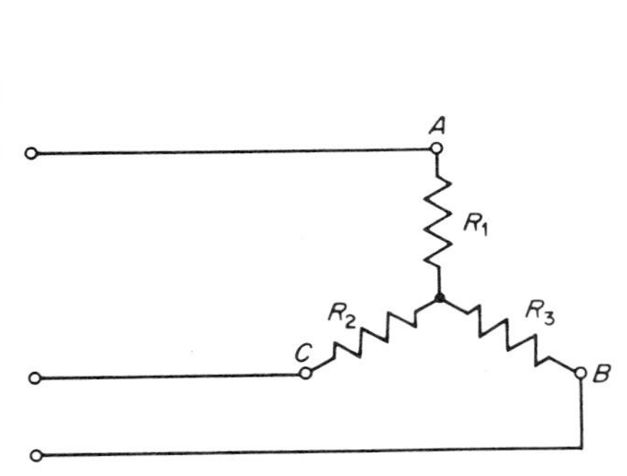

FIGURE 4-101
Equivalent Y connection for Fig. 4-100.

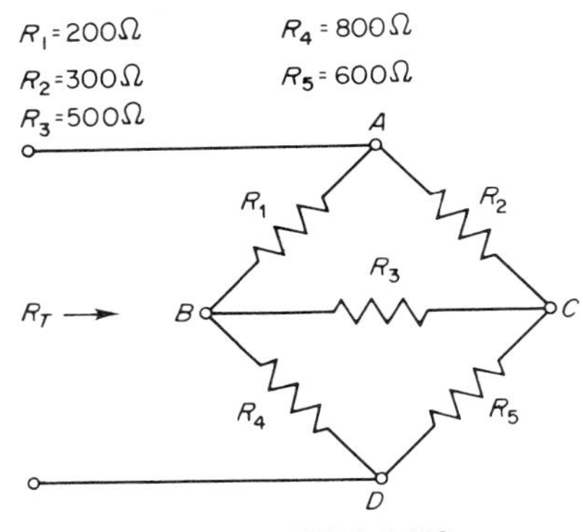

FIGURE 4-102
Circuit for Example 4-37.

EXAMPLE 4-37 Find the total resistance of Fig. 4-102.

SOLUTION

1 If the delta connection consisting of R_1, R_2, and R_3 is converted into an equivalent Y connection, the circuit of Fig. 4-103 results.

$$R_a = \frac{R_1 R_2}{R_1 + R_2 + R_3} = \frac{(200)(300)}{200 + 300 + 500} = \frac{(200)(300)}{1{,}000} = 60\ \Omega$$

$$R_b = \frac{R_1 R_3}{R_1 + R_2 + R_3} = \frac{(200)(500)}{1{,}000} = 100\ \Omega$$

$$R_c = \frac{R_2 R_3}{R_1 + R_2 + R_3} = \frac{(300)(500)}{1{,}000} = 150\ \Omega$$

2 The total resistance can now be found by ordinary series-parallel methods.

$$R_T = R_a + \frac{(R_b + R_4)(R_c + R_5)}{R_b + R_4 + R_c + R_5}$$

$$= 60 + \frac{(100 + 800)(150 + 600)}{1{,}650} = 60 + \frac{(900)(750)}{1{,}650} = 60 + 409 = 469\ \Omega$$

////

ALTERNATIVE SOLUTION

1 If the Y connection consisting of R_2, R_3, and R_5 is converted into an equivalent delta connection, the circuit of Fig. 4-104 results.

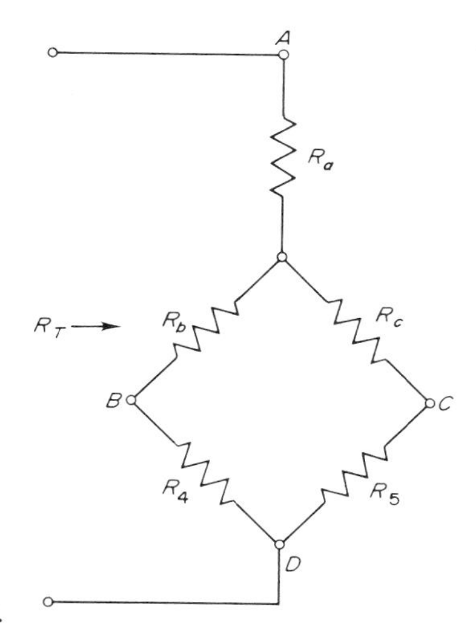

FIGURE 4-103
Equivalent of the circuit of Fig. 4-102.

$$R_x = \frac{R_2 R_3 + R_3 R_5 + R_5 R_2}{R_5}$$

$$= \frac{(300)(500) + (500)(600) + (600)(300)}{600} = \frac{63 \times 10^4}{600} = 1{,}050\ \Omega$$

$$R_y = \frac{R_2 R_3 + R_3 R_5 + R_5 R_2}{R_3} = \frac{63 \times 10^4}{500} = 1{,}260\ \Omega$$

$$R_z = \frac{R_2 R_3 + R_3 R_5 + R_5 R_2}{R_2} = \frac{63 \times 10^4}{300} = 2{,}100\ \Omega$$

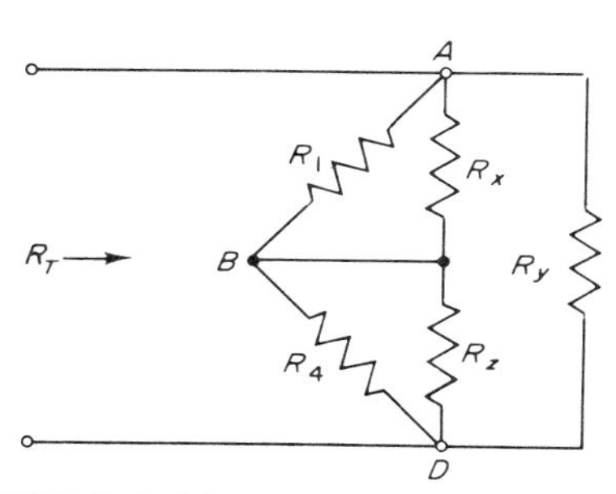

FIGURE 4-104
Alternative equivalent of the circuit of Fig. 4-102.

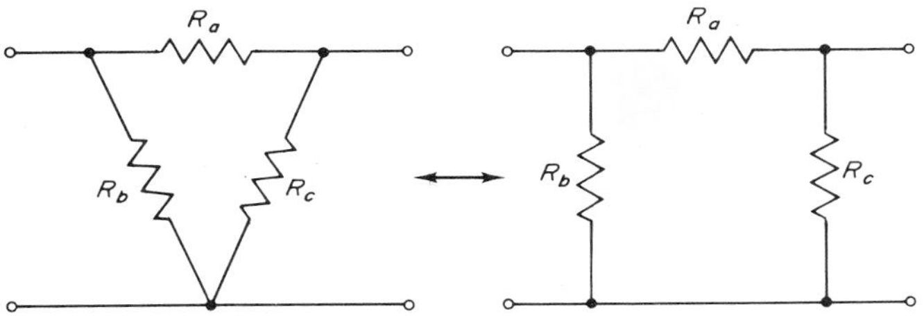

FIGURE 4-105
Delta and pi equivalence.

2 The total resistance can now be found by ordinary series-parallel methods.

$$R_{1x} = \frac{R_1 R_x}{R_1 + R_x} = \frac{(200)(1{,}050)}{1{,}250} = 168\ \Omega$$

$$R_{4z} = \frac{R_4 R_z}{R_4 + R_z} = \frac{(800)(2{,}100)}{2{,}900} = 580\ \Omega$$

$$R_T = \frac{(R_{1x} + R_{4z})R_y}{R_{1x} + R_{4z} + R_y}$$

$$= \frac{(168 + 580)(1{,}260)}{2{,}008} = \frac{(748)(1{,}260)}{2{,}008} = 469\ \Omega \qquad ////$$

Delta and Y circuits are frequently referred to as pi and T circuits. The delta is equivalent to the pi, as shown in Fig. 4-105, and the Y is equivalent to the T, as shown in Fig. 4-106.

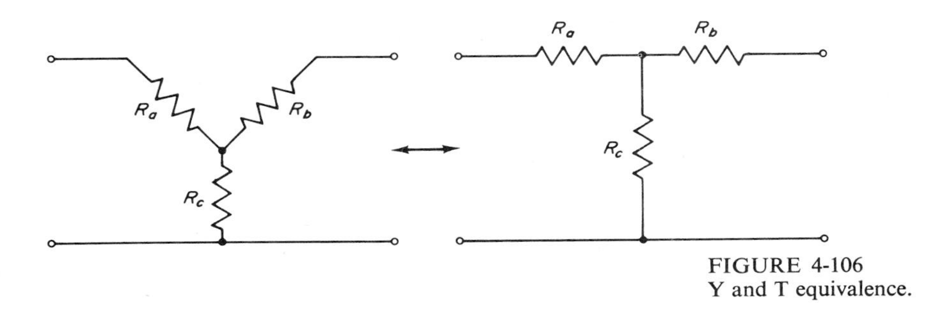

FIGURE 4-106
Y and T equivalence.

PROBLEMS

4-1 Determine the value of E_{xy} in each of the four circuits in Fig. 4-107. Use the VDR and assume the given polarity of E_{xy} in the circuits.

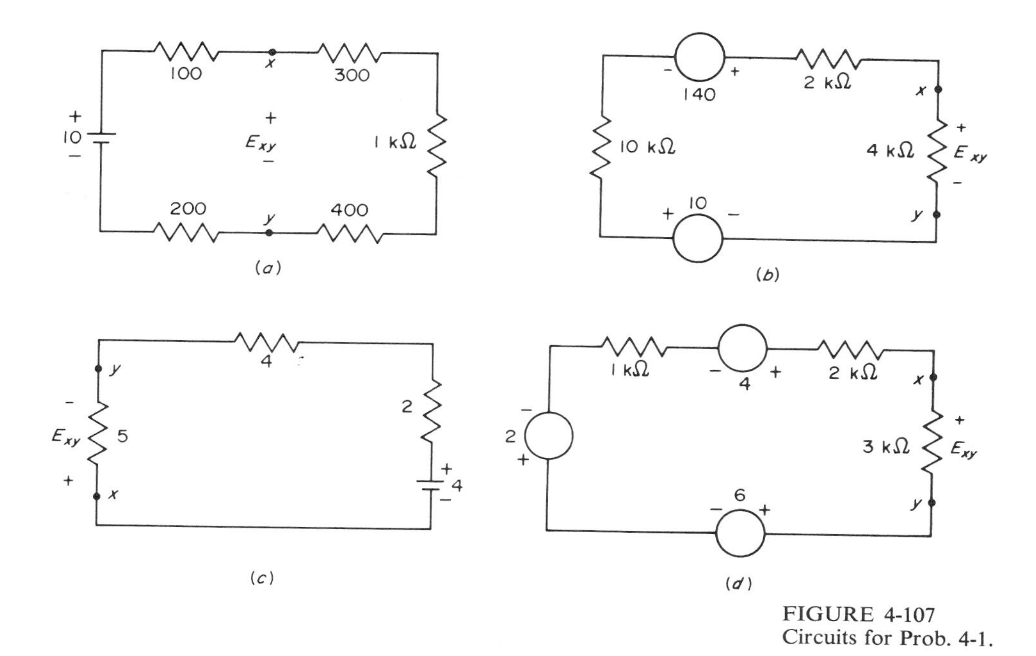

FIGURE 4-107
Circuits for Prob. 4-1.

4-2 Using the VDR, determine the unknown quantity in the circuits of Fig. 4-108.

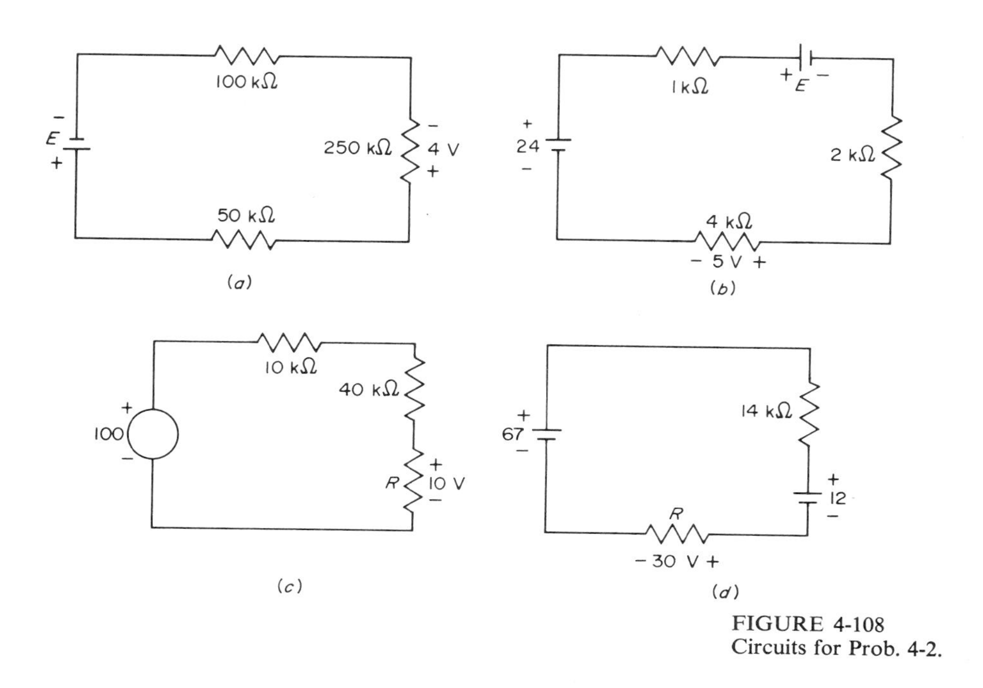

FIGURE 4-108
Circuits for Prob. 4-2.

4-3 Transistor amplifiers commonly use the biasing circuit shown in Fig. 4-109. For good design, the base bias current I_B is much less than the voltage-divider currents I_1 and I_2. Thus I_B is negligible compared to I_1 or I_2. If the voltage drop across R_2 is 1.7 V, find R_2.

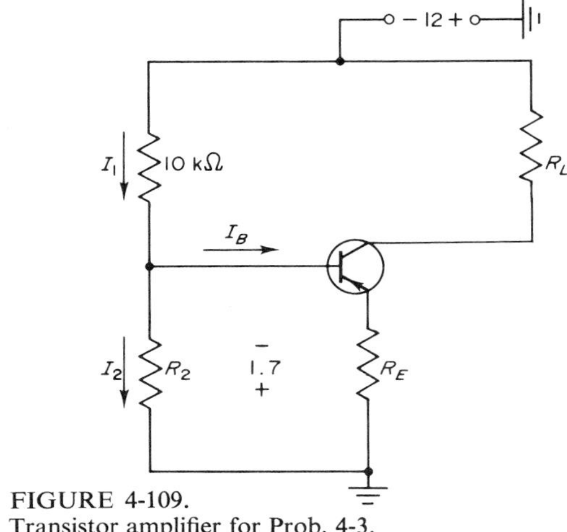

FIGURE 4-109.
Transistor amplifier for Prob. 4-3.

4-4 For each of the circuits in Fig. 4-110, write the KVL equation and solve for the assumed loop current.

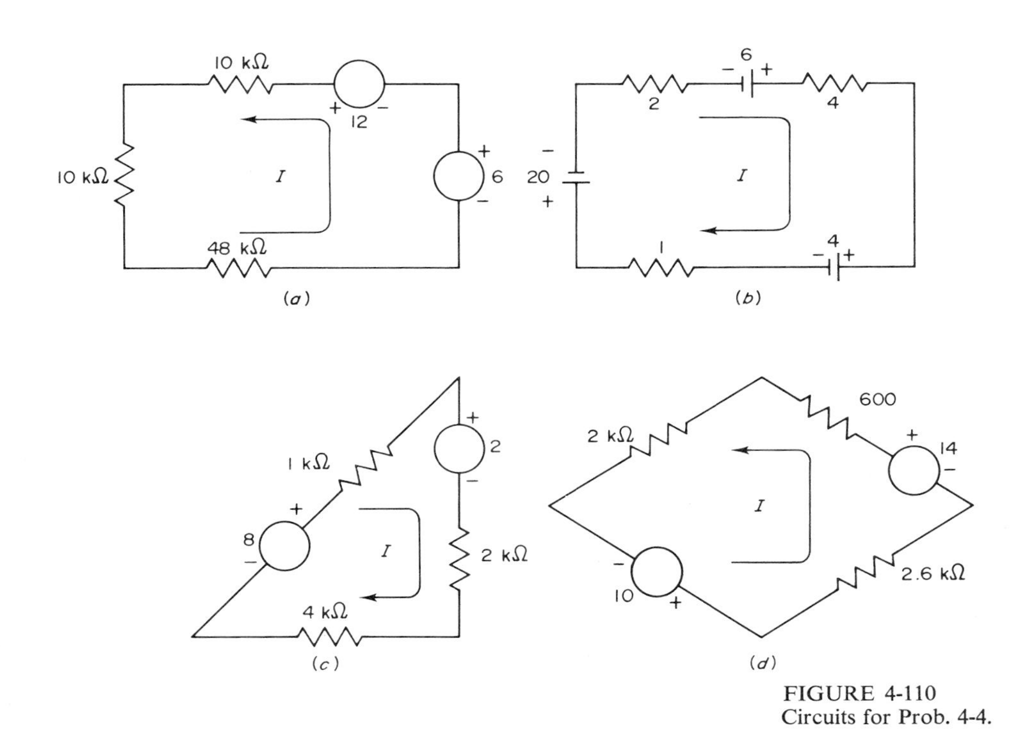

FIGURE 4-110
Circuits for Prob. 4-4.

4-5 Using the CDR, determine the indicated current in the circuits of Fig. 4-111.

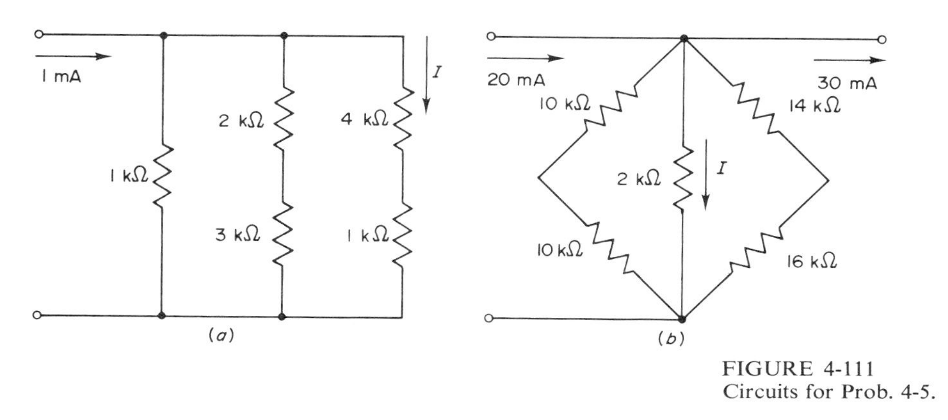

FIGURE 4-111
Circuits for Prob. 4-5.

4-6 Calculate the required input current for the desired branch current shown in the circuits of Fig. 4-112. Use the CDR. Branch current is the actual direction of flow.

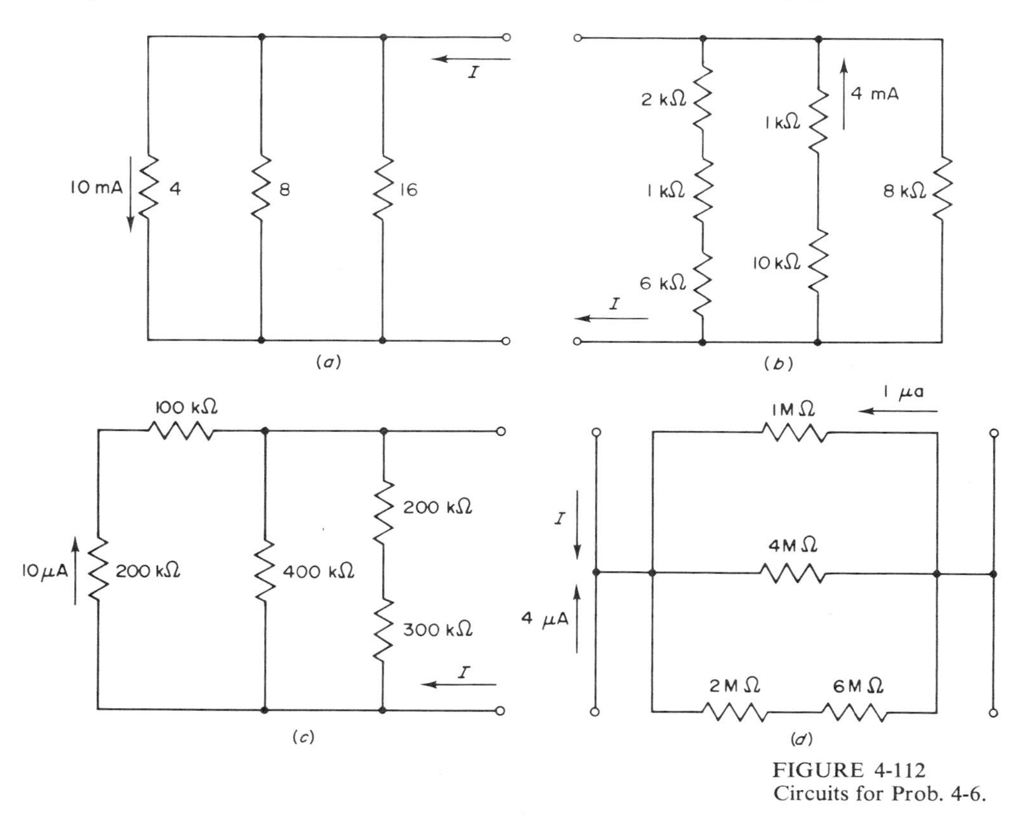

FIGURE 4-112
Circuits for Prob. 4-6.

4-7 An approximate dc equivalent circuit for a transistor amplifier is shown in Fig. 4-113. If the dc base current $I_B = 20\ \mu\text{A}$, calculate the current in R_i.

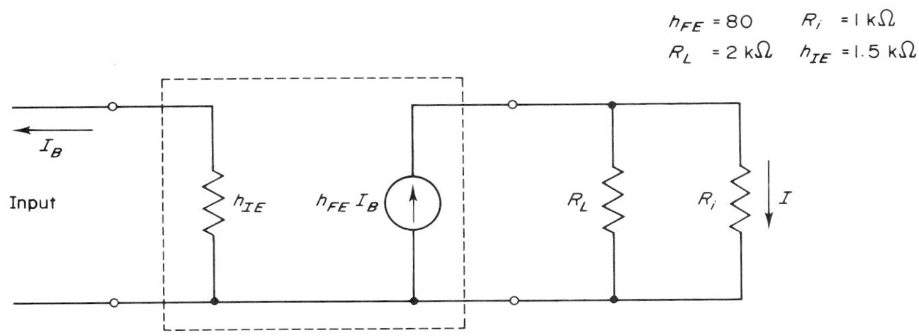

FIGURE 4-113
Circuit for Prob. 4-7.

4-8 Find the value of the unknown currents in the circuits of Fig. 4-114.

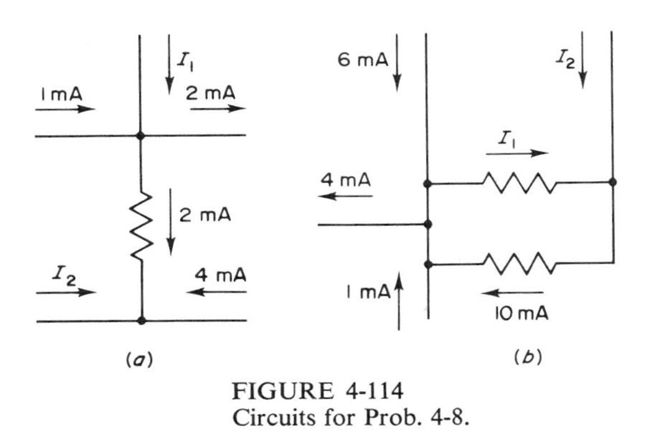

FIGURE 4-114
Circuits for Prob. 4-8.

4-9 For each of the circuits in Fig. 4-115, write the KCL equation and solve for the assumed node-voltage polarity. Consider that the bottom node is a reference node and write the equation for the top node.

4-10 Consider the circuit of Fig. 4-116. Write the three mesh equations. Solve for the unknown mesh currents.

4-11 Calculate the mesh currents in the circuit of Fig. 4-117.

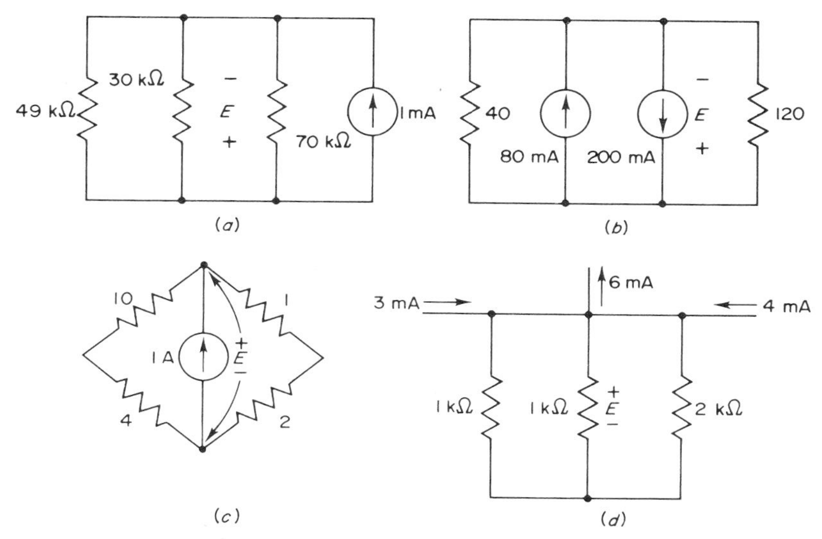

FIGURE 4-115
Circuits for Prob. 4-9.

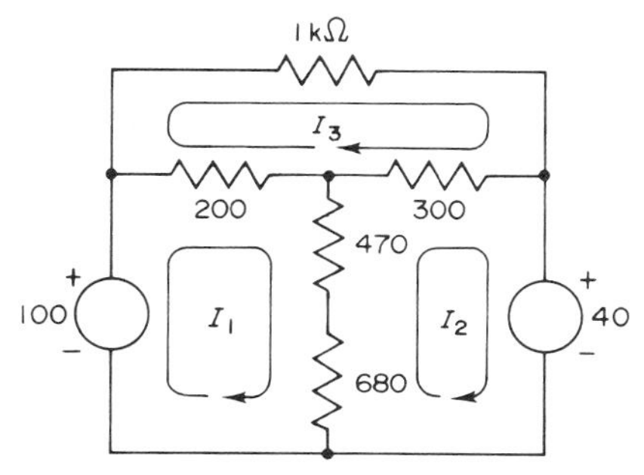

FIGURE 4-116
Circuit for Prob. 4-10.

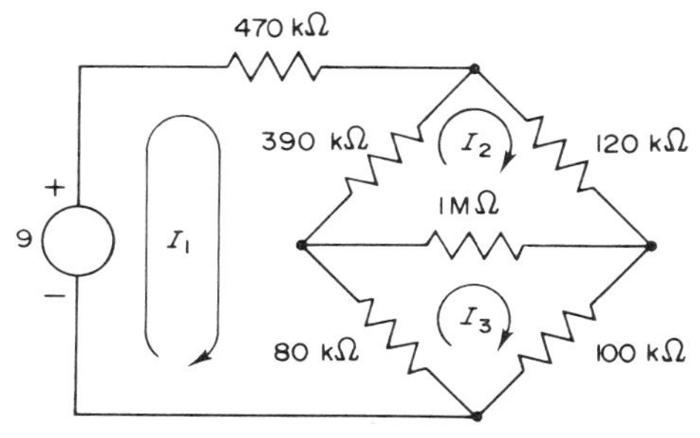

FIGURE 4-117
Circuit for Prob. 4-11.

4-12 Calculate the mesh currents in the circuit of Fig. 4-118.

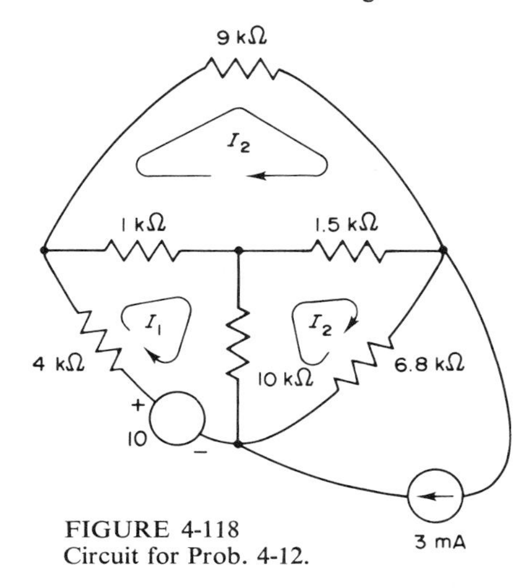

FIGURE 4-118
Circuit for Prob. 4-12.

4-13 Determine the current through the 8-Ω resistor in the circuit of Fig. 4-119. Use mesh-current analysis.

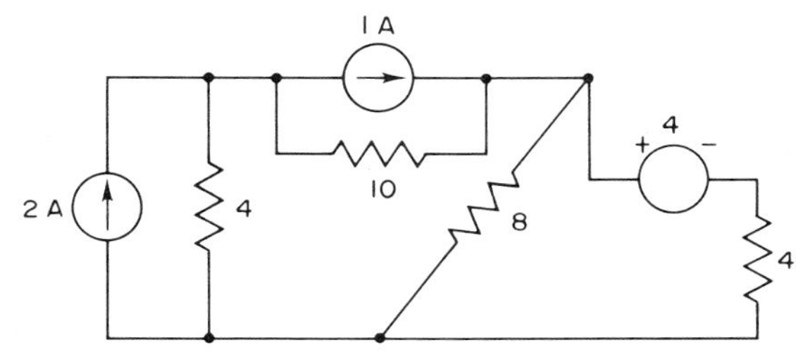

FIGURE 4-119
Circuit for Prob. 4-13.

4-14 Determine the voltage drop across the 2- and 12-kΩ resistors using mesh-current analysis (see Fig. 4-120).

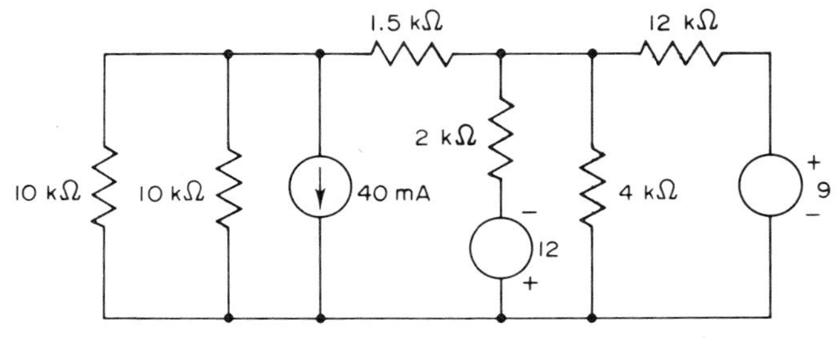

FIGURE 4-120
Circuit for Prob. 4-14.

4-15 Apply mesh-current analysis to the circuit of Fig. 4-121. Determine the voltage at terminals *a-b*.

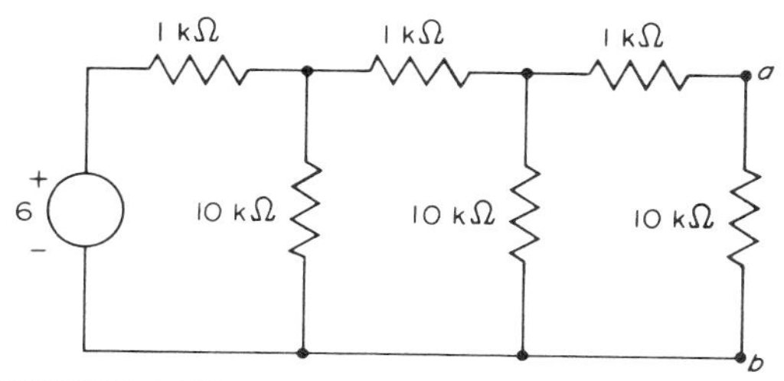

FIGURE 4-121
Circuit for Prob. 4-15.

4-16 Solve for the mesh currents in Fig. 4-122.

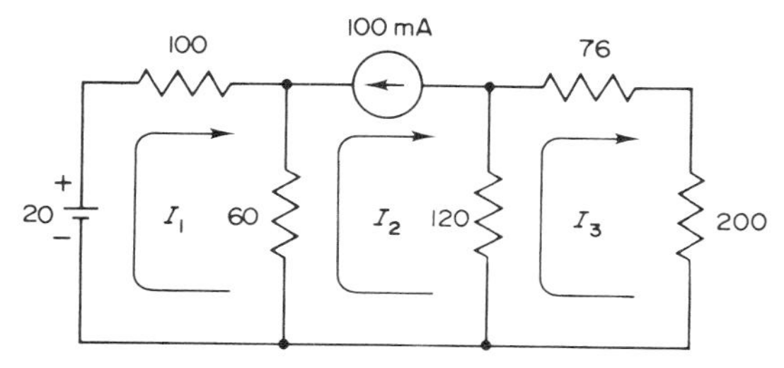

FIGURE 4-122
Circuit for Prob. 4-16.

4-17 Using mesh analysis, determine the voltage across the 10-kΩ resistor at terminals *a-b* of the circuit shown in Fig. 4-123. *Hint*: Rearrange the circuit to facilitate solution.

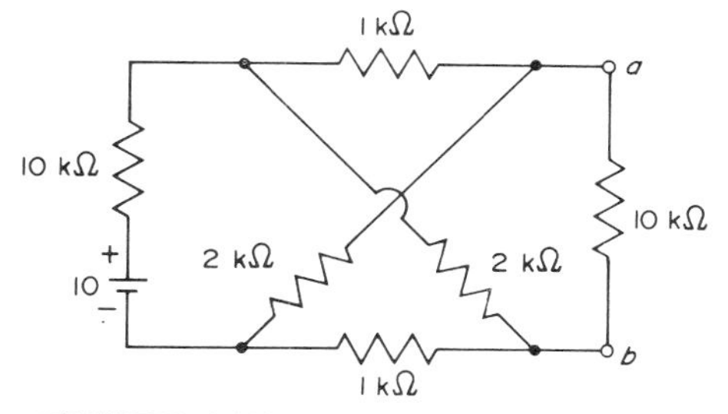

FIGURE 4-123
Circuit for Prob. 4-17.

4-18 Specify the number of mesh equations required for the circuits in Fig. 4-124.

4-19 Specify the number of node equations required for the circuits in Fig. 4-124.

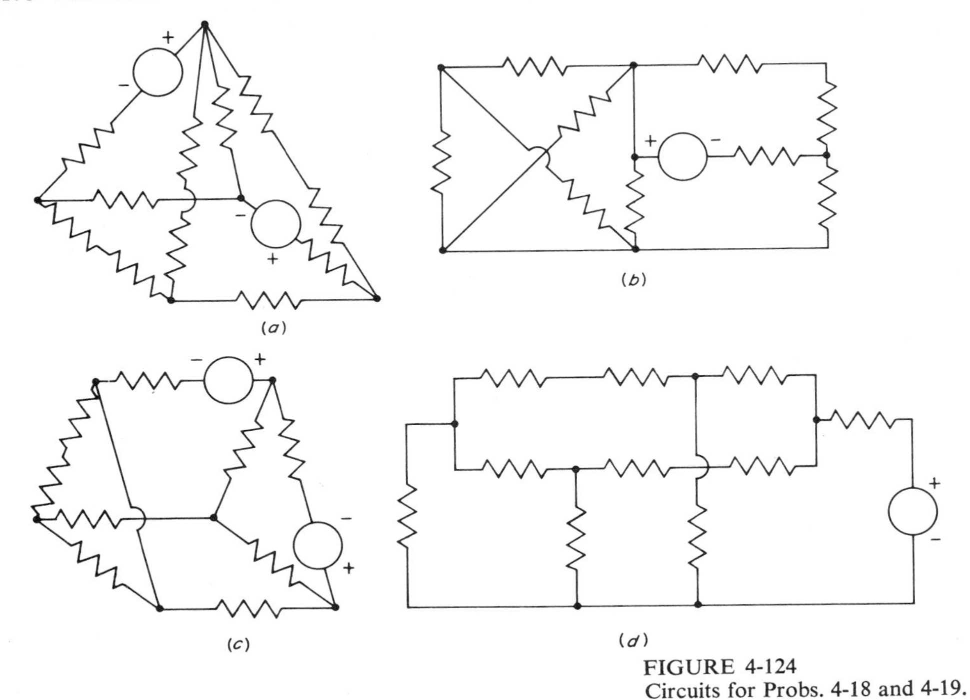

FIGURE 4-124
Circuits for Probs. 4-18 and 4-19.

4-20 Determine the node voltages for the circuit in Fig. 4-125.

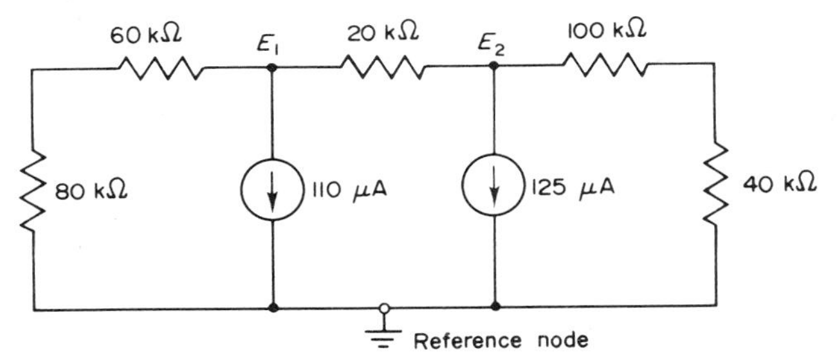

FIGURE 4-125
Circuit for Prob. 4-20.

4-21 Determine the node voltages for the circuit in Fig. 4-126.

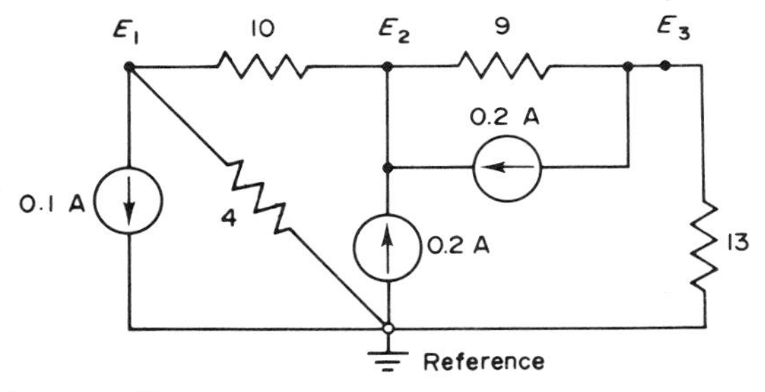

FIGURE 4-126
Circuit for Prob. 4-21.

4-22 Determine the node voltages for the circuit in Fig. 4-127.

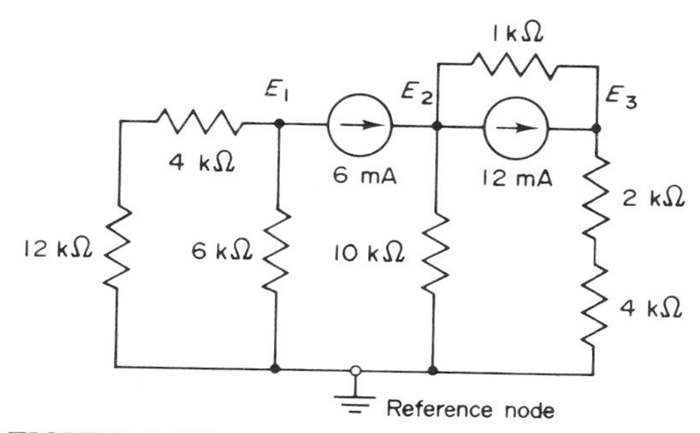

FIGURE 4-127
Circuit for Prob. 4-22.

4-23 Determine the node voltages for the circuit in Fig. 4-128.

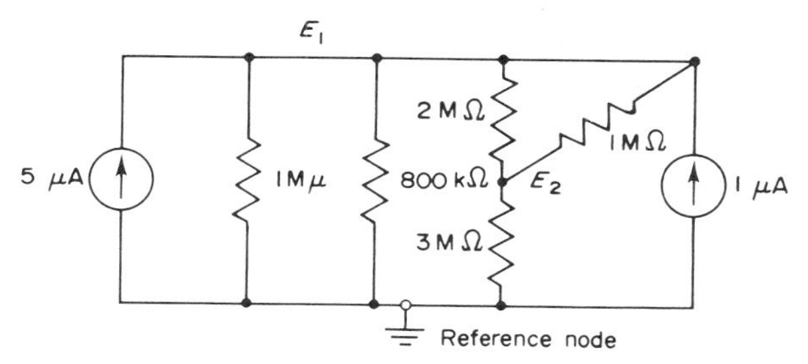

FIGURE 4-128
Circuit for Prob. 4-23.

4-24 Determine the node voltages for the circuit in Fig. 4-129.
4-25 Determine the node voltages for the circuit in Fig. 4-130.

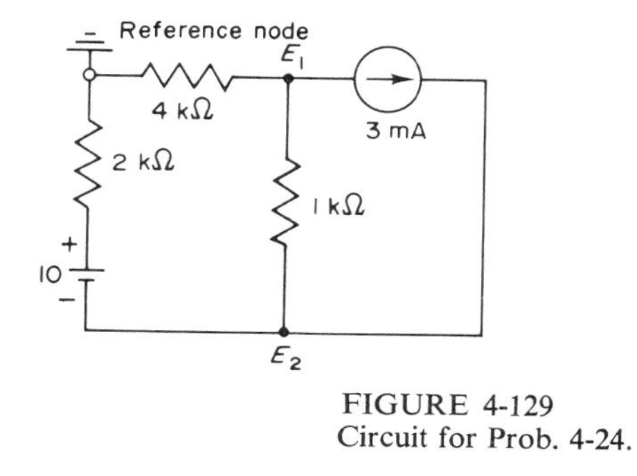

FIGURE 4-129
Circuit for Prob. 4-24.

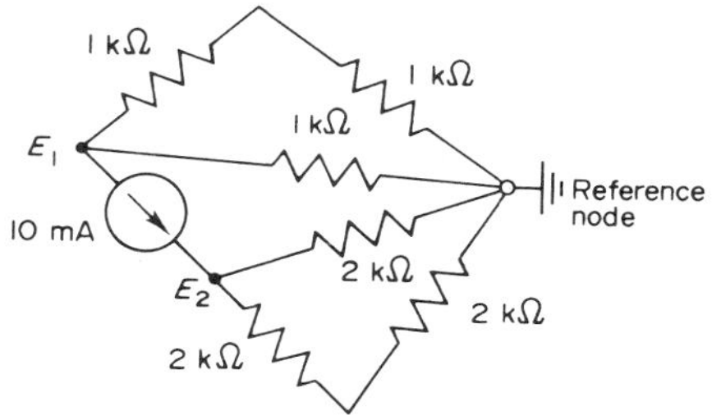

FIGURE 4-130
Circuit for Prob. 4-25.

4-26 Determine the node voltages for the circuit in Fig. 4-131.

4-27 Determine the node voltages for the circuit in Fig. 4-132.

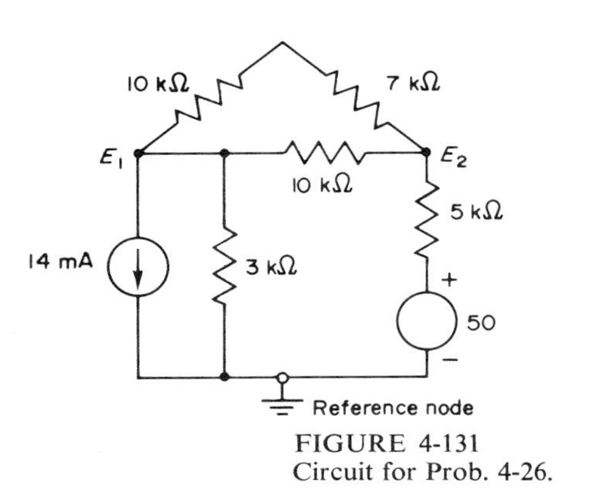

FIGURE 4-131
Circuit for Prob. 4-26.

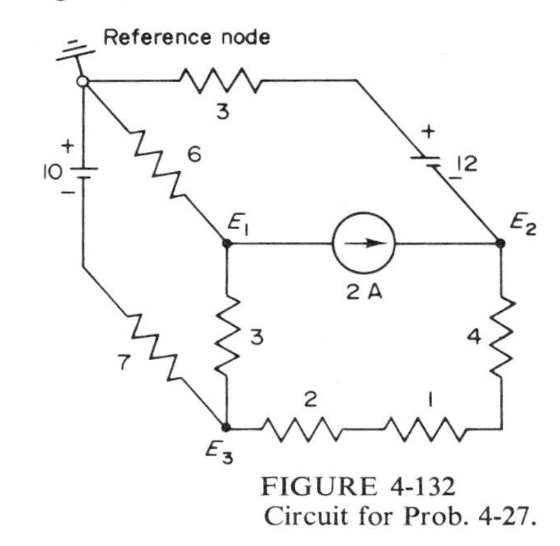

FIGURE 4-132
Circuit for Prob. 4-27.

4-28 Using node-voltage analysis, find the magnitude and direction of the current through the 10-kΩ resistor (see Fig. 4-133).

4-29 Using node-voltage analysis, find the magnitude and polarity of the voltage drop across the 120-kΩ resistor (see Fig. 4-134).

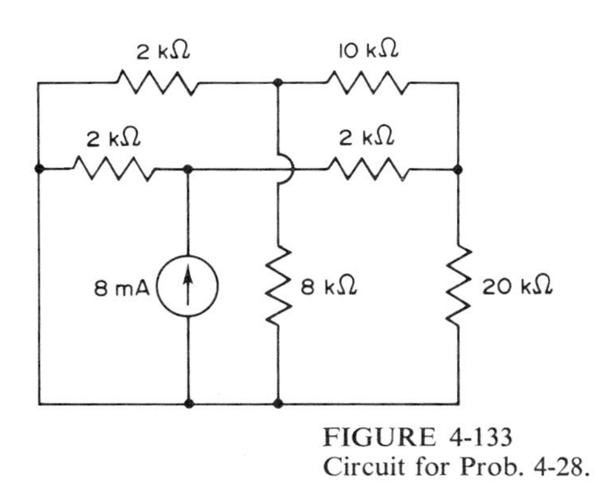

FIGURE 4-133
Circuit for Prob. 4-28.

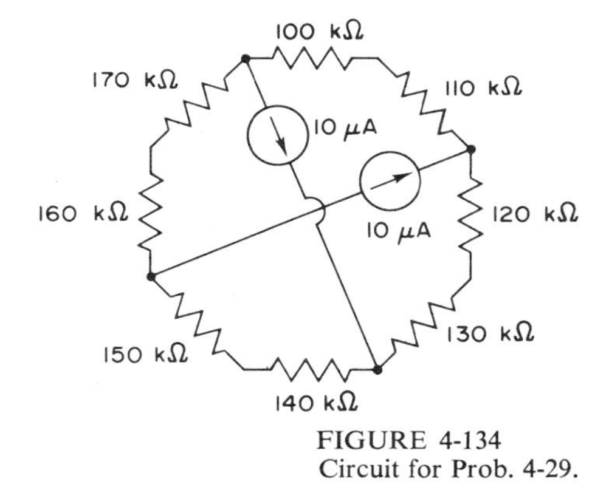

FIGURE 4-134
Circuit for Prob. 4-29.

4-30 Calculate the magnitude of current for the source if E_2 is to be 4 V. Also find the other node voltages (see Fig. 4-135).

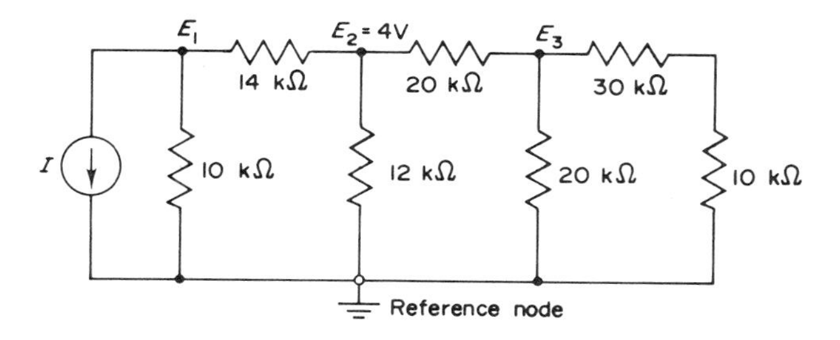

FIGURE 4-135
Circuit for Prob. 4-30.

4-31 Find the current through the 1-kΩ resistor in each of the circuits. Use superposition (see Fig. 4-136). Assume given current direction.

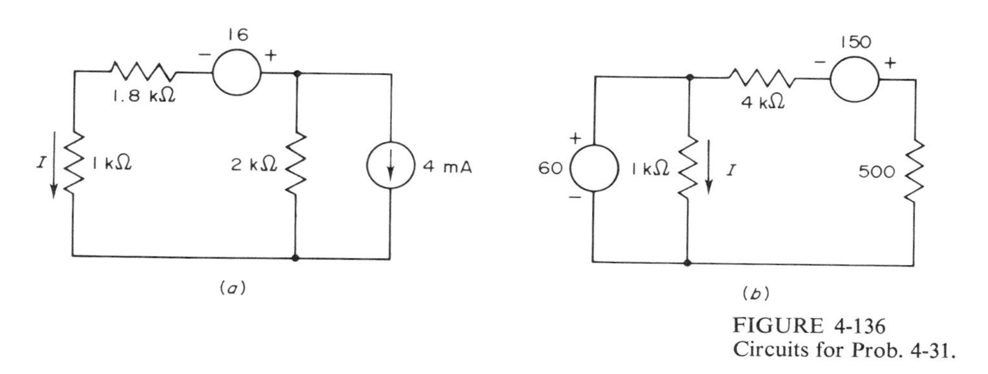

FIGURE 4-136
Circuits for Prob. 4-31.

4-32 Using superposition, determine the current through each of the indicated resistors in Fig. 4-137. Assume the given current direction.

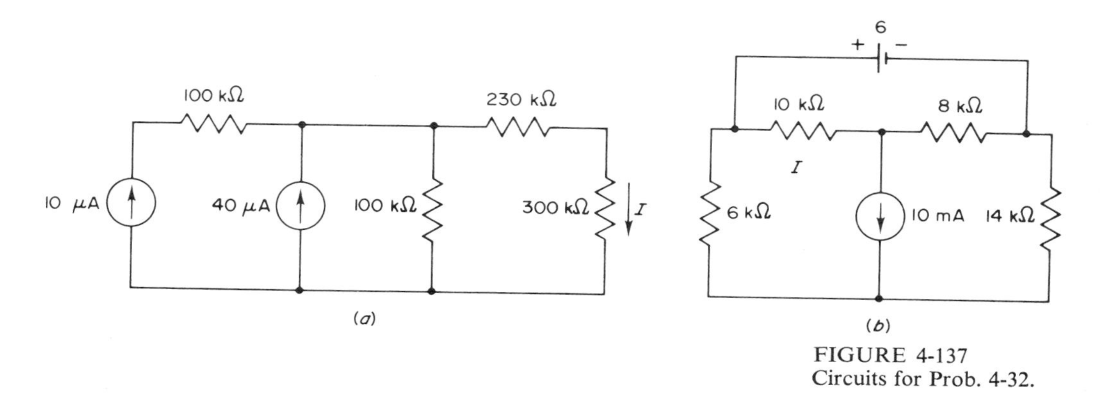

FIGURE 4-137
Circuits for Prob. 4-32.

4-33 Determine the value of the assumed voltage drop in the circuits of Fig. 4-138 using superposition.

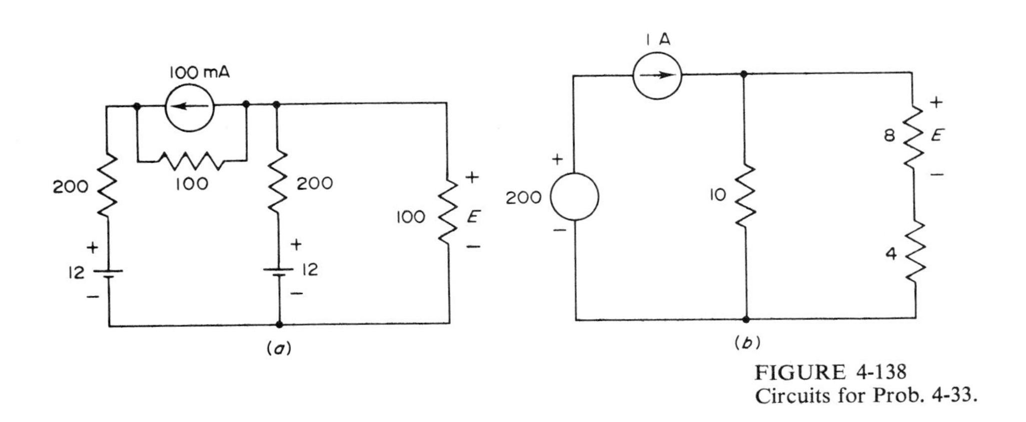

FIGURE 4-138
Circuits for Prob. 4-33.

4-34 Determine the value of the assumed voltage drop and current flow in the circuit of Fig. 4-139 using superposition.

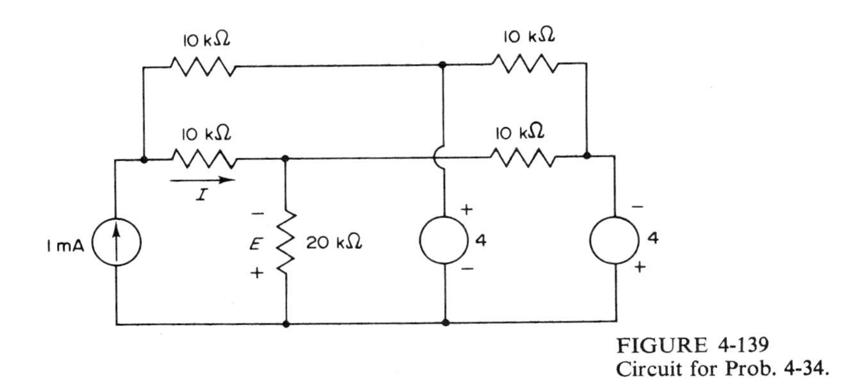

FIGURE 4-139
Circuit for Prob. 4-34.

4-35 An equivalent circuit for a transistor amplifier is shown in Fig. 4-140. Find the Thévenin equivalent for R_L in the output portion.

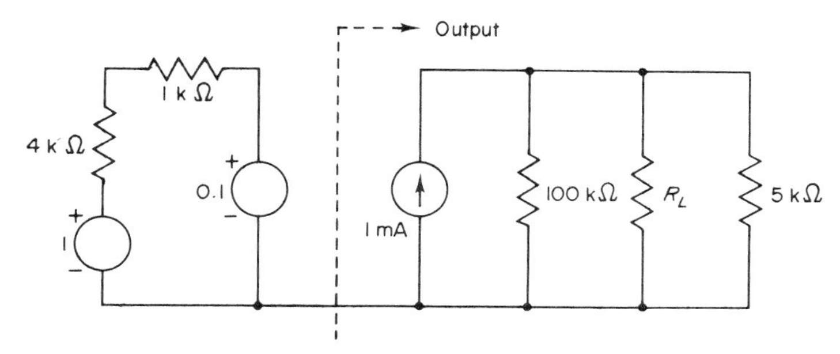

FIGURE 4-140
Circuit for Prob. 4-35.

4-36 A common emitter amplifier is shown in Fig. 4-141. Determine the Thévenin equivalent of the bias circuit to the left of points *x*-*y*.

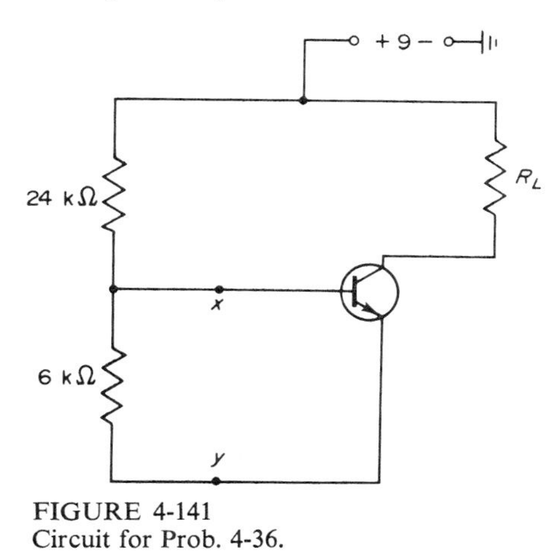

FIGURE 4-141
Circuit for Prob. 4-36.

4-37 Determine the Thévenin equivalent for the 40-kΩ resistor in the circuits of Fig. 4-142.

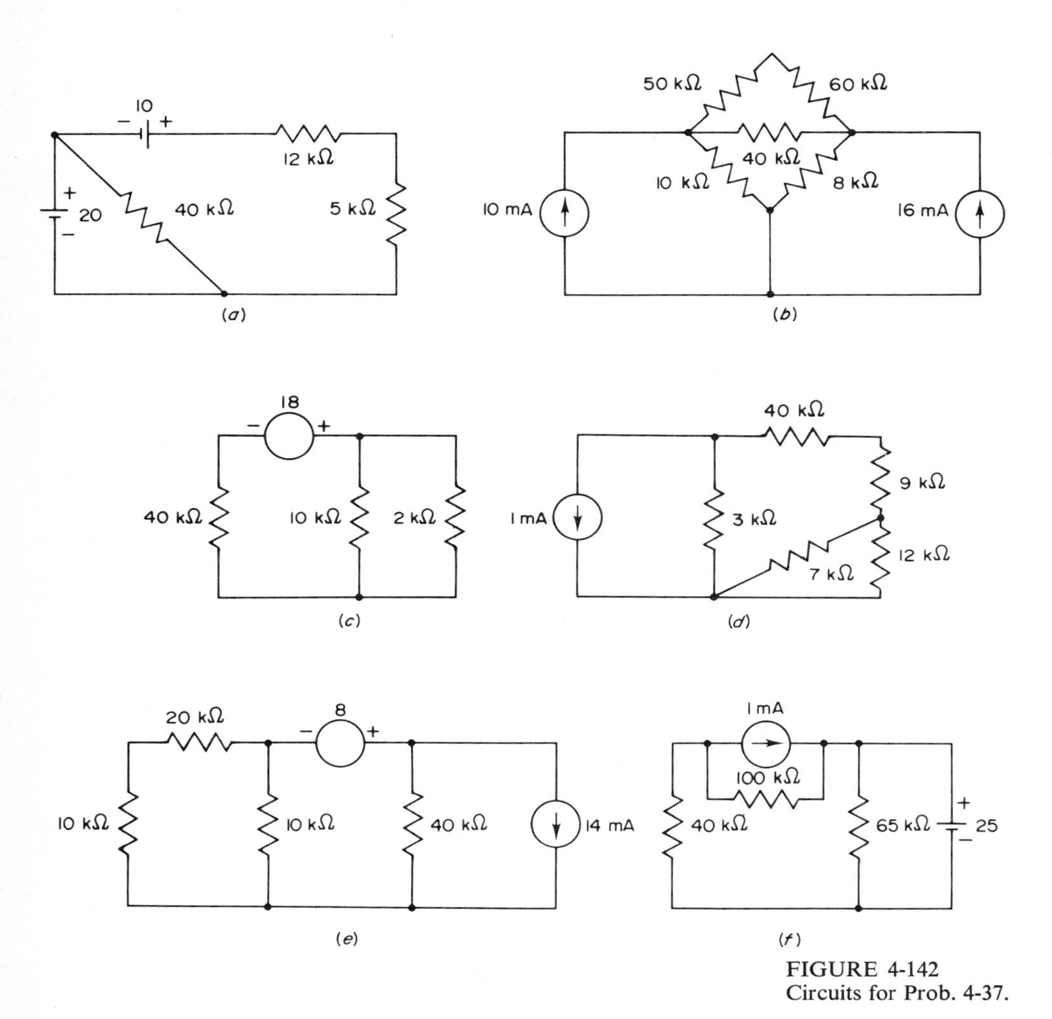

FIGURE 4-142
Circuits for Prob. 4-37.

4-38 Determine the Thévenin equivalent at the specified points in the circuits of Fig. 4-143.

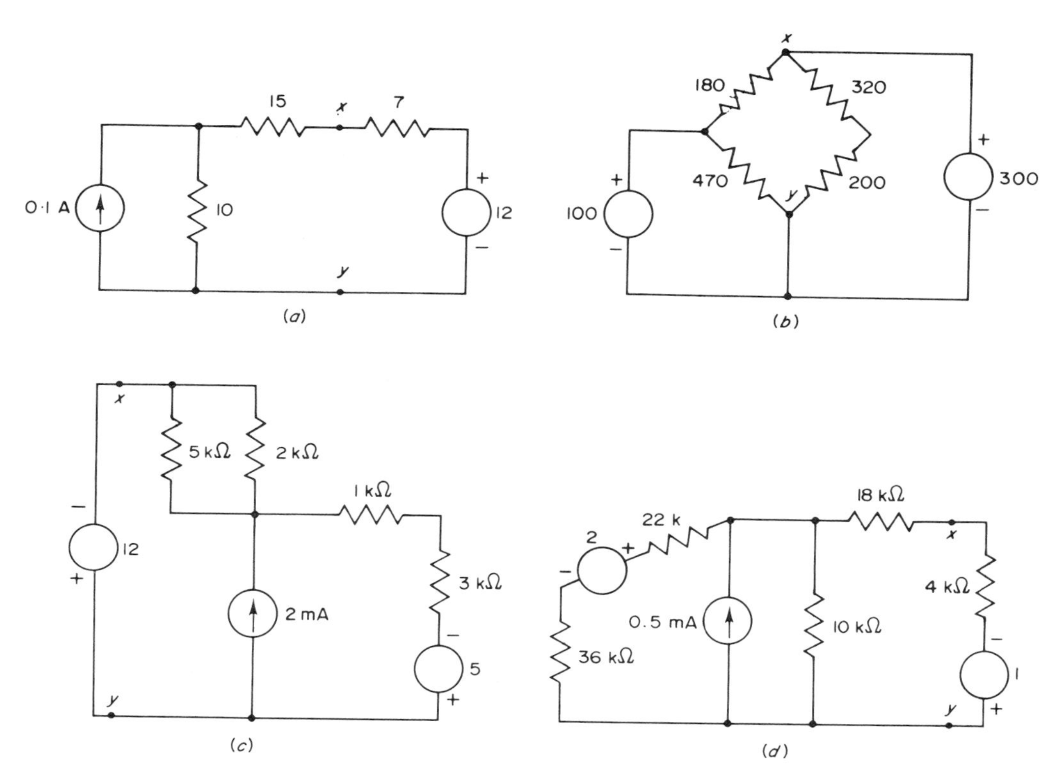

FIGURE 4-143
Circuits for Prob. 4-38: (*a*) to the left of *x*-*y*, (*b*) at points *x*-*y*, (*c*) to the right of *x*-*y*, (*d*) to the left of *x*-*y*.

4-39 Determine the Norton equivalent circuit for the 1-kΩ resistor in the circuits of Fig. 4-144.

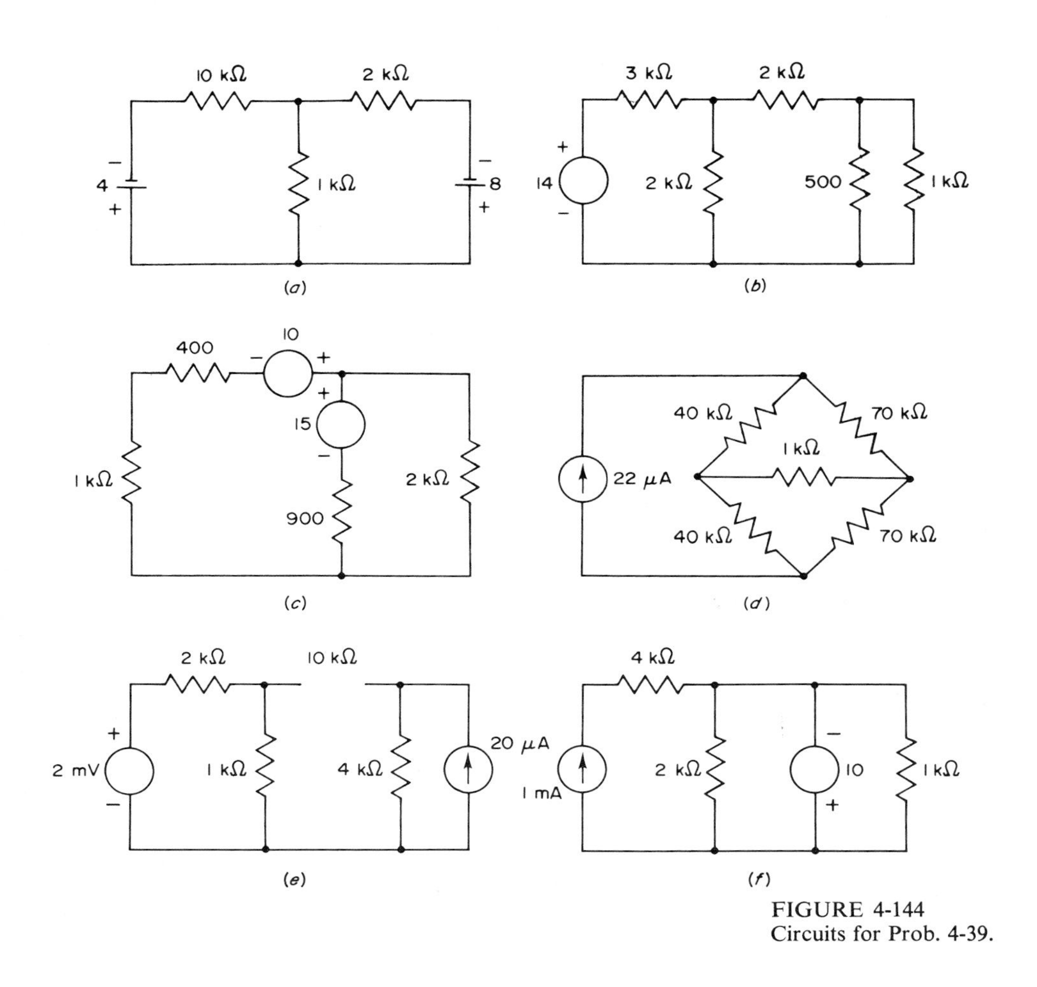

FIGURE 4-144
Circuits for Prob. 4-39.

4-40 Find the Norton equivalent circuit at the points indicated in the circuits of Fig. 4-145.

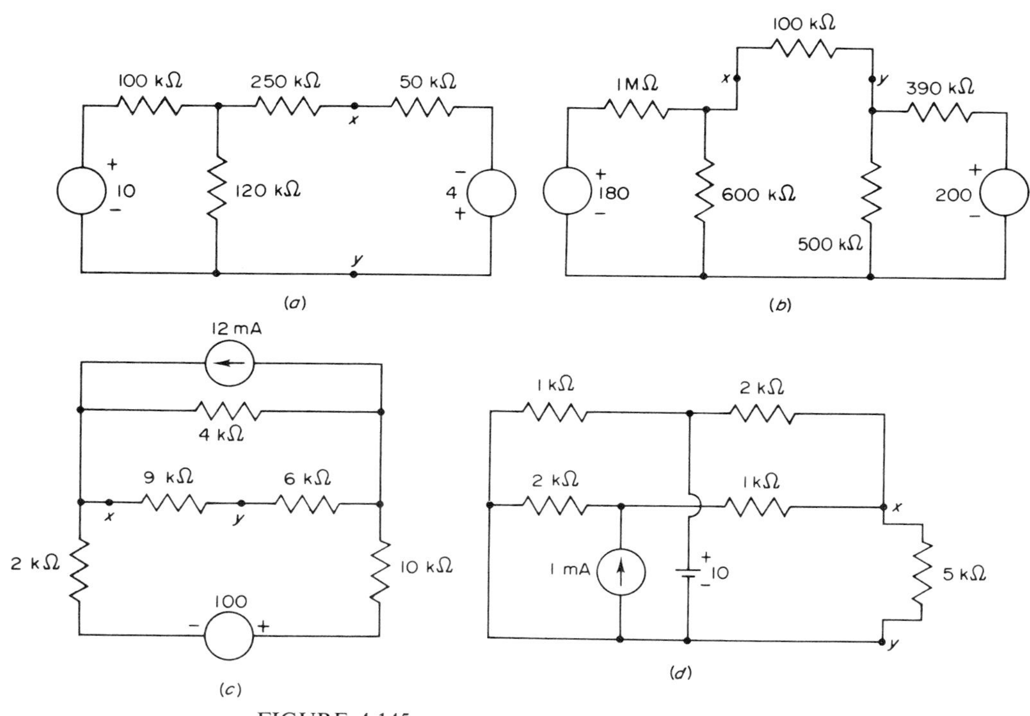

FIGURE 4-145
Circuits for Prob. 4-40: (*a*) to the left of *x*-*y*, (*b*) below *x*-*y*, (*c*) at *x*-*y*, (*d*) to the left of *x*-*y*.

4-41 The following data were obtained at two terminals of a general resistive network:

Terminal voltage	10 V	0
Terminal current	0	8 mA

Find the Thévenin and Norton equivalents.

4-42 Find the Thévenin and Norton equivalents of a network which produces the following experimental data.

Terminal voltage	5 V	0.455 V
Terminal current	0	45.5 mA

4-43 Find the Thévenin and Norton equivalents of a network which produces the following experimental data.

Terminal voltage	0	20 mV
Terminal current	12 μA	10 μA

4-44 Use the circuit in Fig. 4-146 to demonstrate the validity of the reciprocity theorem.

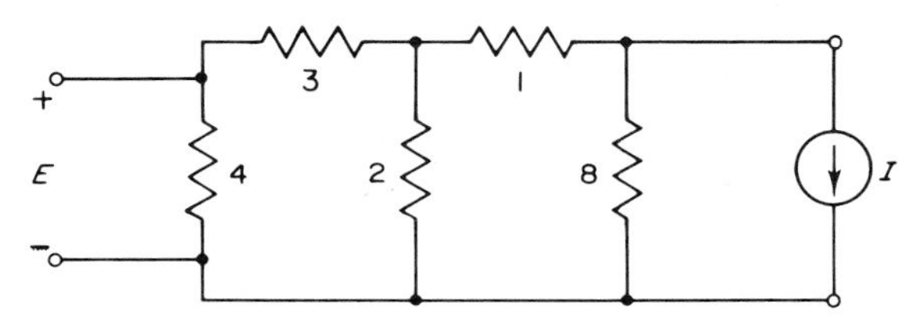

FIGURE 4-146
Circuit for Prob. 4-44.

4-45 Find the transfer resistance at the locations shown in Fig. 4-147.

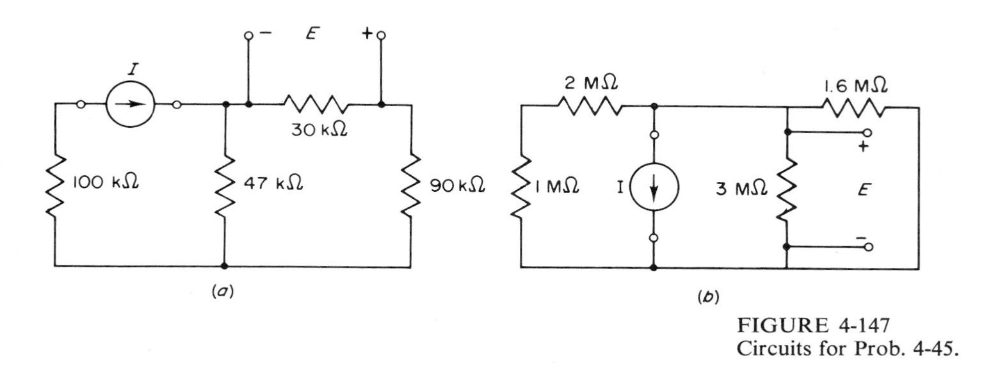

FIGURE 4-147
Circuits for Prob. 4-45.

4-46 Find the transfer conductance at the locations shown in Fig. 4-148.

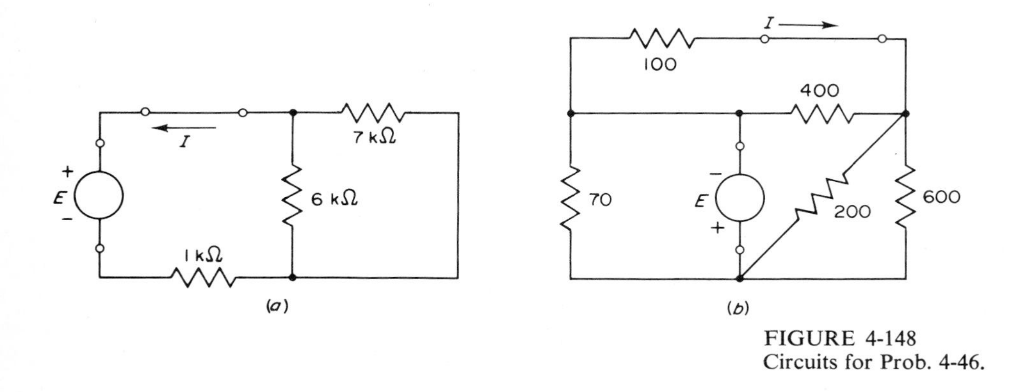

FIGURE 4-148
Circuits for Prob. 4-46.

4-47 Convert the T circuits in Fig. 4-149 to equivalent pi circuits.

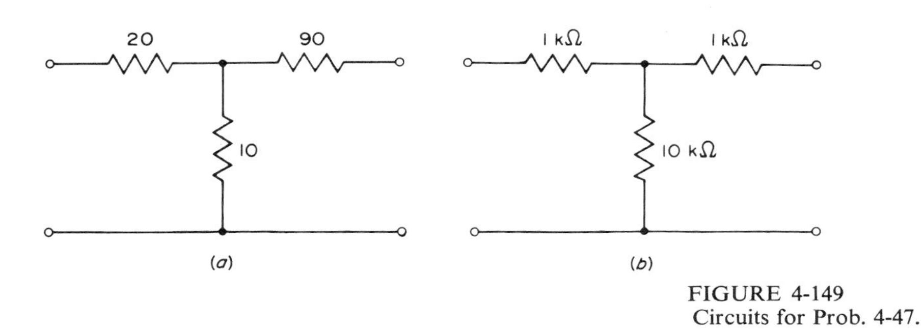

FIGURE 4-149
Circuits for Prob. 4-47.

4-48 Convert the delta circuits in Fig. 4-150 to equivalent Y circuits.

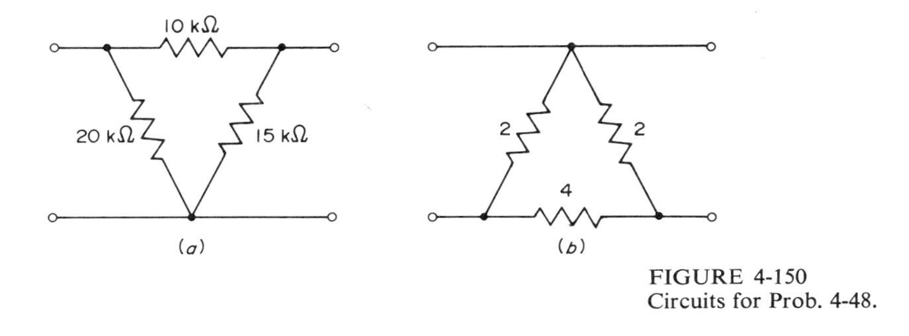

FIGURE 4-150
Circuits for Prob. 4-48.

5

CAPACITANCE

5-1 INTRODUCTION

Capacitance is the ability of a circuit to store energy in the form of an electrostatic field. Before discussing capacitance and the relationship of current and voltage in capacitive circuits, the nature of static electricity and the electrostatic field must be examined.

5-2 STATIC ELECTRICITY

Preceding chapters have dealt with electrons in motion. It was found that directed motion of electrons is due to electromotive force or voltage. Static electricity is electricity at rest and therefore deals with the properties of charge at rest.

5-3 THE ELECTROSTATIC FIELD

The electron is defined as having a negative charge. If a number of electrons are isolated at a point in space, that point has a net negative charge proportional to the number of electrons. Q represents the charge in coulombs that has been isolated at

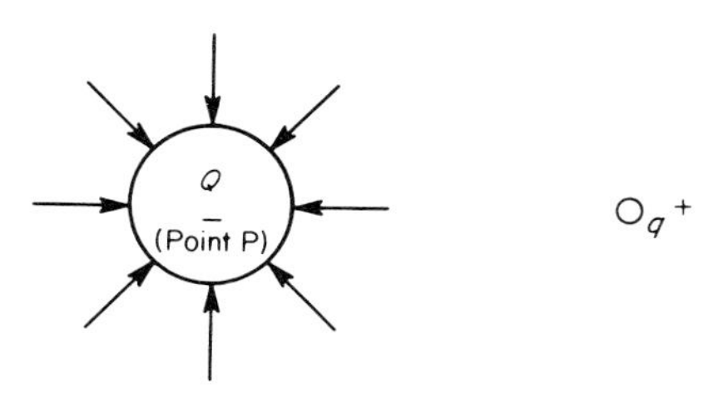

FIGURE 5-1
Electrostatic field in the vicinity of a negative charge.

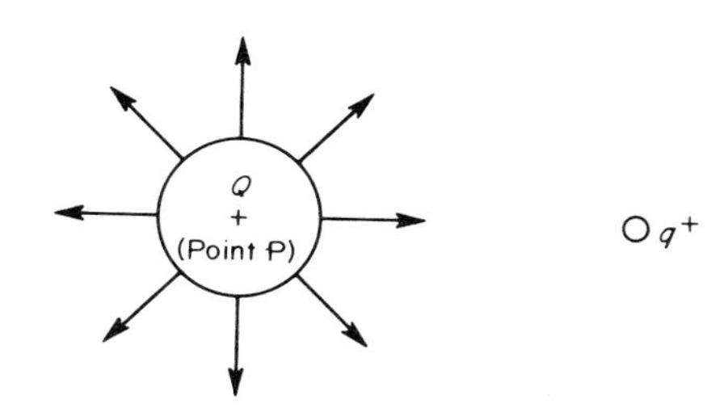

FIGURE 5-2
Electrostatic field in the vicinity of a positive charge.

point P in Fig. 5-1. If a tiny positive test charge q is placed in the environment of the isolated charge Q, there is a force of attraction between the charge Q and the test charge. Regardless of the position of the test charge, the force of attraction is such that the positive test charge tends to move toward the isolated charge. This force of attraction is symbolized by the radial lines in Fig. 5-1. The lines represent the *electrostatic force field*, and the arrowheads indicate the direction a small positive test charge would move in when placed in the electrostatic field. Whenever charge exists, an electrostatic field is associated with it. A positive charge Q isolated at point P in space has an electrostatic field as shown in Fig. 5-2. The radial lines representing the electrostatic field now point away from charge Q. It has been stated that these lines indicate the direction a small positive test charge q would move in in the field of the isolated charge. The arrows therefore indicate repulsion of the test charge in these circumstances.

In Fig. 5-3, two isolated charges Q_1 and Q_2 are shown with their associated electrostatic field. Q_1 is a negative charge, and Q_2 is a positive charge. The direction of the field again indicates the direction a positive test charge would move due to this charge arrangement. The field lines may also be thought of as leaving a positive charge and ending on a negative charge.

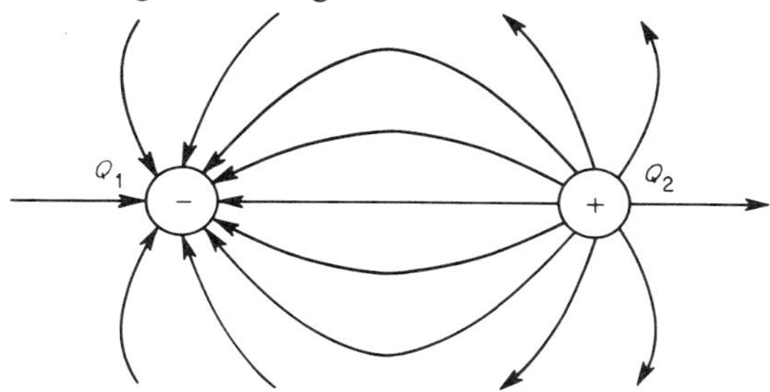

FIGURE 5-3
Electrostatic field between positive and negative charges.

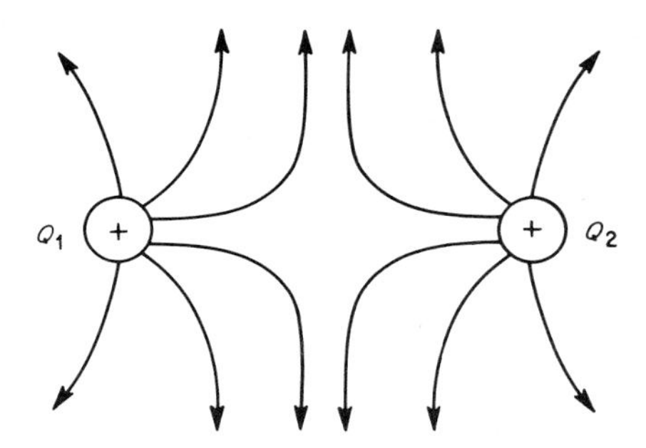

FIGURE 5-4
Electrostatic field between two positive charges.

Figure 5-4 shows the field associated with two isolated charges when both are positive.

The force of attraction and repulsion of point charges is best demonstrated by small metal spheres suspended on strings. When charge is placed on these spheres, repulsion results if the charges are of like polarity and attraction results if the charges are unlike. These conditions are shown in Fig. 5-5. The angular displacement of the strings from the vertical is due to electrostatic forces of attraction or repulsion. The magnitude of the displacement, which is related to the magnitude of the force, depends upon Q_1 and Q_2, the charges on the spheres, the distance between the spheres, and the medium surrounding the spheres. This relationship is given by Eq. (5-1), which is the mathematical expression of Coulomb's law.

$$F = k\,\frac{Q_1 Q_2}{d^2} \tag{5-1}$$

Equation (5-1) yields the force F in newtons when Q_1 and Q_2 are in coulombs and the distance d between the charges is in meters. In the mks system, k is approximately equal to 9×10^9 in free space.

EXAMPLE 5-1 A charge of 1 C is located 0.04 m from a charge of 0.5 C. (1) Find the force of attraction in newtons if the charges are unlike charges. (2) Find the force of attraction in pounds. (1 N = 0.2248 lb.)

SOLUTION

1
$$F = k\,\frac{Q_1 Q_2}{d^2} = \frac{(9 \times 10^9)(1)(0.5)}{(0.04)^2} = 2.81 \times 10^{12}\ \text{N}$$

2
$$F\,(\text{lb}) = F\,(\text{N}) \times 0.2248 = 0.632 \times 10^{12}\ \text{lb}$$

////

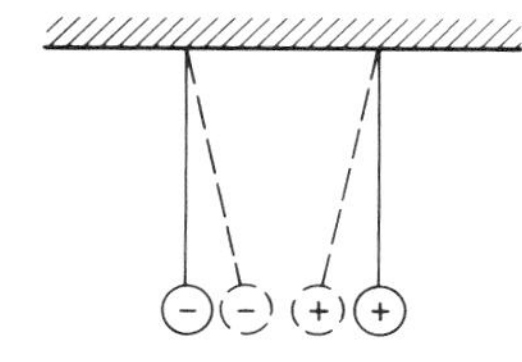

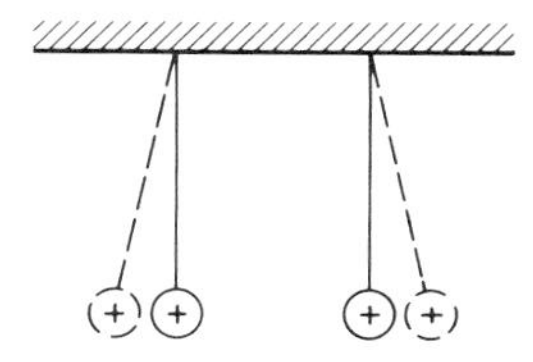

FIGURE 5-5
Repulsion and attraction of electrostatic charges.

Up to this point, charges isolated in space have been discussed. Consider now two conductive plates separated by an insulating material. This condition is shown in Fig. 5-6. A negative charge is placed on plate 1 and an equal positive charge is placed on plate 2. Between these two plates an electrostatic force field exists that will force a negative charge placed there to move toward plate 2. Likewise, a positive charge placed between the plates would be forced toward plate 1. The charge accumulated on the plates establishes a difference of potential or voltage. The field strength between the plates depends on the voltage between the plates and the distance between the plates. This relationship is given by Eq. (5-2). In Eq. (5-2), the field strength is in volts per meter when the potential difference is in volts and d is in meters.

$$\mathrm{E} = \text{Field strength} = |\vec{\mathrm{E}}| = \frac{E}{d}\ \mathrm{V/m} \tag{5-2}$$

Field strength is a measure of the force on a charge located in the medium between the plates. Since field strength is a vector quantity represented by $\vec{\mathrm{E}}$, $\mathrm{E} = |\vec{\mathrm{E}}|$ represents the magnitude of the vector quantity. Field strength is often referred to as the electric field intensity.

5-4 CAPACITANCE

If a battery is connected to a pair of parallel conductive plates, as shown in Fig. 5-7, the applied voltage exerts a force on the free electrons in the conductor, forcing electrons to accumulate on the upper plate and moving electrons off the lower plate.

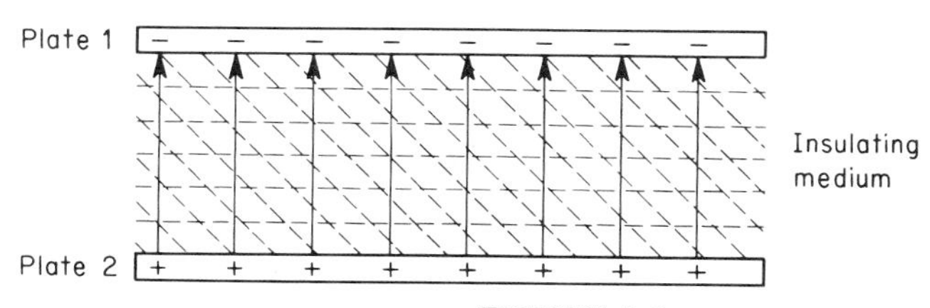

FIGURE 5-6
Electrostatic field between two conductive plates.

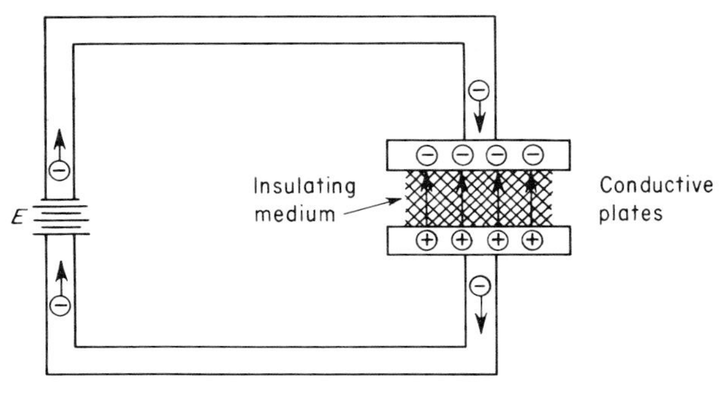

FIGURE 5-7
Charging capacitor.

This gives the upper plate a negative charge and the lower plate a positive charge. The battery moves electrons onto the upper plate until the force of the electrons on the upper plate is equal and opposite to the force of the battery that tries to move more electrons onto the upper plate. When charge moves in the circuit, it is continuous throughout the circuit even though there is theoretically no charge moving through the insulating medium between the plates. Rather, the electrostatic force is exerted across the medium.

When as much charge has accumulated as a given battery can force onto the plates, the potential difference across the plates equals the battery voltage. *The ratio between the charge accumulated and the voltage across the plates is defined as capacitance.* This relationship is shown in Eq. (5-3). In this equation, C is the symbol for capacitance in farads when charge Q is in coulombs and potential difference E is in volts.

$$C = \frac{Q}{E} \tag{5-3}$$

A pair of parallel plates separated by an insulating medium has one farad (F) of capacitance when a potential difference of one volt accumulates a charge of one coulomb on the plates. The farad is an impractically large unit; the microfarad (μF, 10^{-6} F) and the picofarad (pF, 10^{-12} F) are the common units of capacitance.

A mechanical analog will illustrate further the concept of capacitance. A pump is inserted in a waterline as shown in Fig. 5-8. A flexible diaphragm is placed across the pipe. As the pump applies pressure to the system, water moves in the direction indicated in Fig. 5-8. The water moved in the upper portion exerts pressure

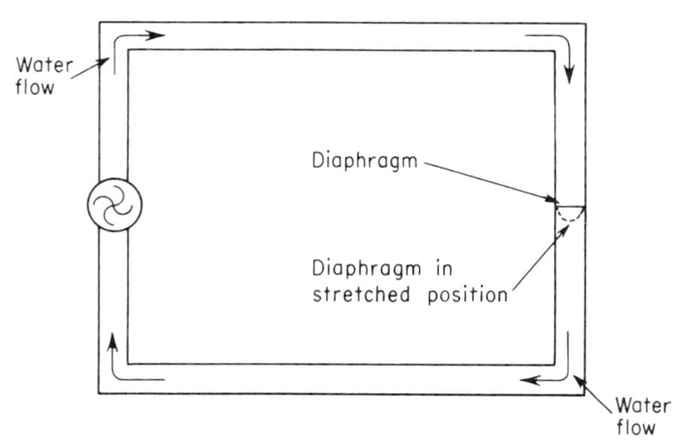

FIGURE 5-8
Hydraulic analog of charging capacitor.

on the diaphragm, causing it to stretch as shown. This causes water to move on the other side of the diaphragm without water actually passing through the diaphragm. This condition is similar to that of a capacitor, where current flows on both sides of the capacitor, but none actually through the capacitor. The diaphragm in Fig. 5-8 is stretched an amount that depends upon the pressure exerted by the pump. When the diaphragm is stretched to a point where the force of the pump just balances the resisting force of the stretched diaphragm, water ceases to flow. When the pump is removed, the diaphragm returns to its original position, forcing the water to move in the opposite direction. This action is shown in Fig. 5-9. In the process of stretching the diaphragm, energy is stored, and as the diaphragm returns to its original position energy is returned, causing the water to move.

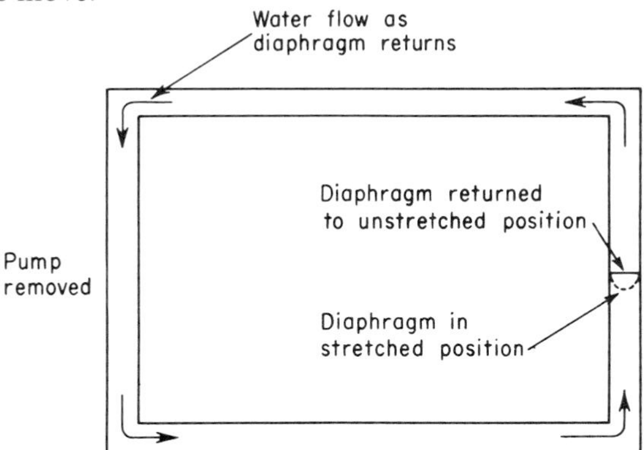

FIGURE 5-9
Hydraulic analog of discharging capacitor.

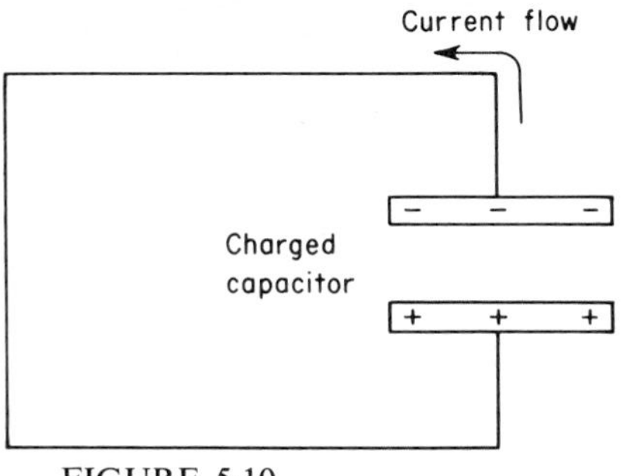

FIGURE 5-10
Discharging capacitor.

In the circuit of Fig. 5-7, if the battery is removed once the charge has accumulated on the plates of the capacitor and the plates of the capacitor are connected together through a conductor as shown in Fig. 5-10, the charge moves to equalize the difference of potential across the capacitor. As the charge moves through the circuit in Fig. 5-10, the energy stored in the electrostatic field is returned in the form of current. Electrons leave the negatively charged plate and enter the positively charged plate until the plates are in an uncharged condition and the capacitor is no longer capable of developing a current. If the capacitor is no longer capable of developing current, it is no longer storing energy.

5-5 FACTORS AFFECTING CAPACITANCE

As defined by Eq. (5-3), capacitance is the ratio of charge to potential difference.

The *electrostatic flux lines* shown in Sec. 5-3 represent the electrostatic force field due to the charge. In the mks system of units flux ψ is equated to the charge in coulombs setting up the force field.

$$\psi = Q \tag{5-4}$$

The flux density that results from charge distribution across the surface of a parallel-plate capacitor is

$$D = \frac{\psi}{A} = \frac{Q}{A} \tag{5-5}$$

where A is the area of the plate in square meters and Q is the charge in coulombs. Equation (5-5) yields the flux density D in coulombs per square meter.

The flux density for a given parallel-plate capacitor depends on the field strength and the permittivity ε of the insulating medium between the plates. The permittivity of the insulating medium is a measure of the ease with which a flux field is established.

$$D = \varepsilon \mathrm{E} \tag{5-6}$$

E represents the field strength in volts per meter. Substituting Eqs. (5-2) and (5-5) into Eq. (5-6) yields

$$\frac{Q}{A} = \varepsilon \frac{E}{d} \tag{5-7}$$

Rearranging Eq. (5-7),

$$\frac{Q}{E} = \varepsilon \frac{A}{d}$$

But the ratio of charge to potential difference is defined as capacitance. Therefore,

$$C = \varepsilon \frac{A}{d} \tag{5-8}$$

The permittivity of a vacuum is $\varepsilon_0 = 8.85 \times 10^{-12}$ F/m. The permittivity of any insulating medium divided by the permittivity of vacuum yields relative permittivity.

$$\varepsilon_r = \frac{\varepsilon}{\varepsilon_0} \tag{5-9}$$

Therefore,

$$\varepsilon = \varepsilon_0 \varepsilon_r = (8.85 \times 10^{-12})\varepsilon_r \text{ F/m} \tag{5-10}$$

Substituting Eq. (5-10) into Eq. (5-8),

$$C = \frac{(8.85 \times 10^{-12})\varepsilon_r A}{d} \text{ F} \tag{5-11}$$

Equation (5-11) yields the capacitance in farads when A is in square meters and d is in meters. ε_r, the relative permittivity, is tabulated in Table 5-1.

Table 5-1 RELATIVE PERMITTIVITY OF VARIOUS DIELECTRICS

Dielectric	Relative permittivity
Air	1.0006
Barium-strontium titanate (ceramic)	7,500.0
Porcelain	6.0
Transformer oil	4.0
Bakelite	7.0
Paper	2.5
Rubber	3.0
Teflon	2.0
Mica	5.0

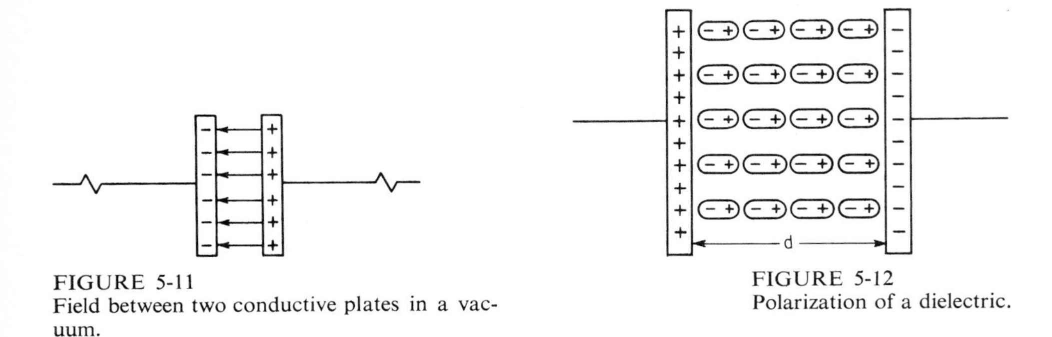

FIGURE 5-11
Field between two conductive plates in a vacuum.

FIGURE 5-12
Polarization of a dielectric.

The capacitance of the parallel-plate capacitor is shown by Eq. (5-11) to be directly proportional to the area of the plates and inversely proportional to the distance separating the plates.

The effect of the insulating material given by ε_r is shown by investigation of the electrostatic field first with a vacuum between the plates and then with an insulating material between the plates.

If a parallel-plate capacitor is situated in a vacuum, no particles of matter exist between these plates. The electrostatic field with a vacuum separating the plates is shown in Fig. 5-11. If in place of the vacuum an insulating material is placed between the plates of the capacitor, the molecular structure of the material will be polarized. In an atom of the insulating material, the electrons are bound tightly to the nucleus, but it must be remembered that an atom is constructed of negative electrons and positive protons. When a negative charge is placed on one plate of a capacitor and a positive charge on the other, the atoms in the insulating material are polarized as shown in Fig. 5-12. Polarization of these atoms is the result of the electrons of the atoms being attracted toward the positive plate while the protons are attracted toward the negative plate. It must be noted that the electrons and protons are still bound in their respective atoms but the atoms are distorted somewhat to net the effect shown in Fig. 5-12. When the insulating material is polarized as shown, the flux lines are set up with less opposition, and therefore a larger capacitance is obtained for a given field strength.

5-6 DIELECTRIC STRENGTH

The amount of voltage that can be placed across the plates without breakdown of the insulator between the plates depends upon the dielectric strength of the insulator. If a large enough electrostatic field exists between the plates of a capacitor, the force on the electrons in the insulating material causes the electrons to be pulled free of their nuclei. This results in a breakdown of the insulator, commonly referred to as the *dielectric*. When the dielectric breaks down and conduction occurs between the plates, the capacitor is no longer effective. The dielectric strengths of various insulators are listed in Table 5-2. Dielectric strength is given in volts per mil (1 mil = 0.001 in.). The voltage rating of a capacitor depends on the type of insulating material and its thickness.

EXAMPLE 5-2 What is the capacitance of a parallel-plate capacitor when the area of the plate is 0.5 m^2 and the plates are separated by 0.0001 m? The dielectric used is paper.

SOLUTION

$$C = \frac{(8.85 \times 10^{-12})\varepsilon_r A}{d} = \frac{(8.85 \times 10^{-12})(2.5)(0.5)}{1 \times 10^{-4}} = 11.05 \times 10^{-8} = 0.1105\ \mu\text{F}$$

////

EXAMPLE 5-3 Calculate the capacitance if the dimensions are the same as in Example 5-2 and the dielectric is air.

SOLUTION

$$C = \frac{(8.85 \times 10^{-12})(1.0006)(0.5)}{1 \times 10^{-4}} = 4.42 \times 10^{-8}\ \text{F} = 0.0442\ \mu\text{F}$$ ////

EXAMPLE 5-4 Find the voltage rating for the capacitors of Examples 5-2 and 5-3.

Table 5-2 DIELECTRIC STRENGTH OF VARIOUS MATERIALS

Dielectric	Dielectric strength, V/mil
Air	75
Barium-strontium titanate	75
Porcelain	200
Transformer oil	400
Bakelite	400
Paper	500
Rubber	700
Teflon	1,500
Glass	3,000
Mica	5,000

SOLUTION

1 The distance between the plates must be converted from meters to inches (1 m = 39.37 in.):

$$d\ (\text{in.}) = (0.0001)(39.37) = 3.937 \times 10^{-3}\ \text{in.}$$

2 In Example 5-2 the dielectric was paper, for which the dielectric strength is 500 V/mil.

$$E\ (\text{V}) = (500\ \text{V/mil})(3.937) = 1{,}970\ \text{V}$$

3 In Example 5-3 the dielectric was air, for which the dielectric strength is 75 V/mil.

$$E\ (\text{V}) = (75\ \text{V/mil})(3.937) = 295\ \text{V}$$

The value of voltage found in steps 2 and 3 represents the maximum voltage that can be applied to the capacitors without breaking down their respective dielectrics. ////

EXAMPLE 5-5 Calculate the capacitance of a parallel-plate capacitor having a plate area of 0.5 m^2 and a paper dielectric with a thickness of 0.0004 m.

SOLUTION

$$C = \frac{(8.85 \times 10^{-12})(2.5)(0.5)}{4 \times 10^{-4}} = 2.76 \times 10^{-8}\ \text{F} = 0.0276\ \mu\text{F} \qquad ////$$

EXAMPLE 5-6 Calculate the voltage rating for the capacitor of Example 5-5.

SOLUTION

1 Convert d in meters to d in inches.

$$d(\text{in.}) = 4 \times 10^{-4} \times 39.37 = 15.73 \times 10^{-3}\ \text{in.}$$

2

$$E\ (\text{V}) = 500 \times 15.73 = 7{,}865\ \text{V} \qquad ////$$

Comparison of Examples 5-6 and 5-4 shows that by increasing the distance between the plates the breakdown voltage is increased but at the cost of decreased capacitance.

5-7 LEAKAGE RESISTANCE

It must be pointed out that there is no perfect insulator. That is, every material has some free electrons within its structure. These free electrons move in response

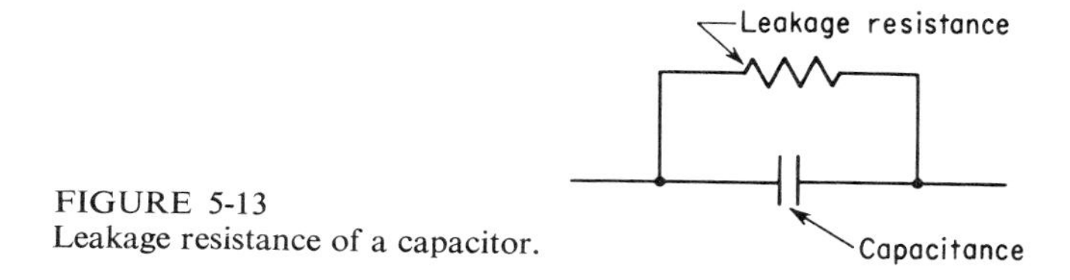

FIGURE 5-13
Leakage resistance of a capacitor.

to the potential between the plates of a capacitor. The current created by the movement of these few electrons through a dielectric is referred to as *leakage current*. The effect of leakage current may be represented by a large resistance in parallel with the capacitor, as shown in Fig. 5-13. The larger the value of the leakage resistance, the more nearly correct it is to neglect this effect when working with capacitive circuits.

Leakage resistance of most capacitors is of the order of $10^9\ \Omega$.

5-8 ELECTROLYTIC CAPACITORS

In some capacitor applications, the capacitance needed for proper operation of a circuit is too great to be achieved with ordinary capacitor construction. Examination of Eq. (5-11) indicates several means of increasing capacitance. The area of the plates can be increased, but obviously there is a practical limit to plate size. Otherwise, a capacitor would be larger than the equipment in which it is to be used. The dielectric constant could be improved by using a different dielectric, but practical dielectrics do not have constants in the thousands. The only remaining factor is the distance separating the plates. If this distance could be reduced to one thousandth of what it is in the usual capacitor of similar plate area, then the capacitance would be 1,000 times as great. This is achieved in *electrolytic capacitors*. See Fig. 5-14.

Electrolytic capacitors are constructed with one plate of aluminum and the other plate of a suitable electrolyte such as aluminum hydroxide. When voltage of the correct polarity is applied, a thin insulating film forms between the aluminum plate and the hydroxide. This microscopically thin film now becomes the dielectric. Thus it is possible to have large capacitance in a space no larger than that occupied by a paper capacitor twenty to thirty times smaller in capacitance. To further increase the capacitance, the aluminum plate is etched chemically to roughen its surface and thereby increase its area. The electrolyte may be a liquid or a paste, and the plate is usually a spiral of aluminum foil. The entire assembly is then mounted in either a can or a cardboard container.

Ordinary electrolytic capacitors must be used only in dc circuits, since operation with the polarity reversed removes the dielectric film. Elimination of the dielectric

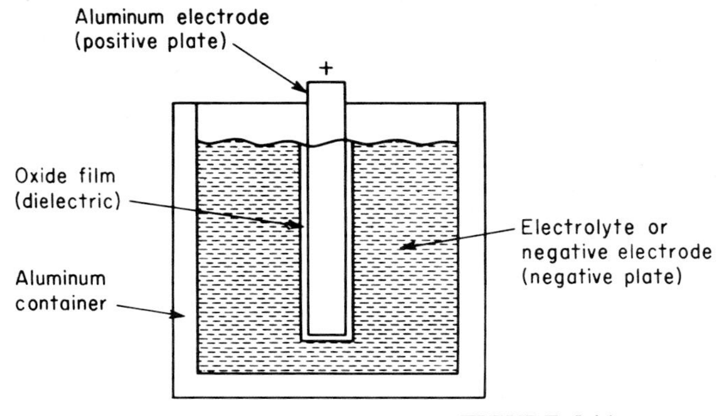

FIGURE 5-14
Construction of an electrolytic capacitor.

results in large leakage current, which damages both the circuit and the electrolytic capacitor.

Electrolytic capacitors are characterized by high leakage currents and relatively low values of breakdown voltage.

5-9 PAPER CAPACITORS

Paper capacitors are constructed of two long strips of aluminum foil separated by a waxed-paper dielectric (see Fig. 5-15). The plates of foil and the dielectric are spiral-wound to form a cylinder. The entire assembly is then sealed with wax or plastic. Leakage losses are usually low and vary with the rated voltage of the capacitor and the size of the capacitor. The larger the capacitance, the greater its leakage losses. The higher the voltage rating of the capacitor, the lower its leakage losses.

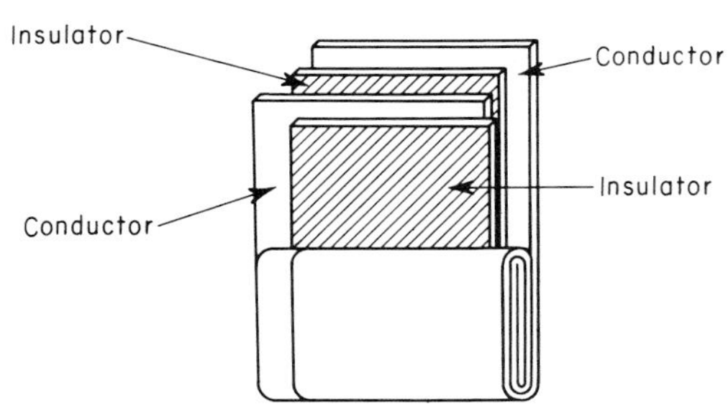

FIGURE 5-15
Construction of a paper capacitor.

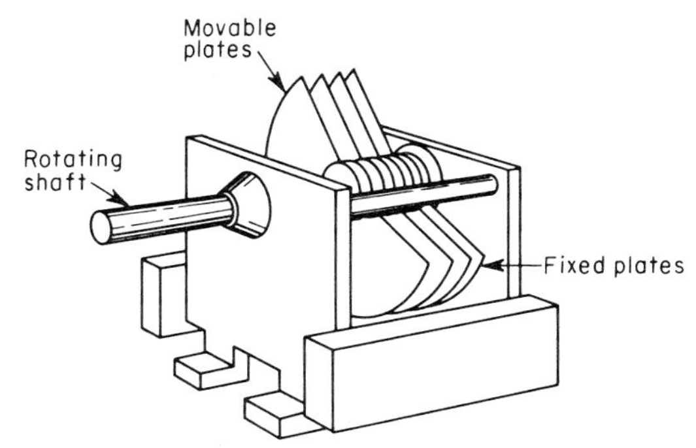

FIGURE 5-16
Construction of a variable capacitor.

5-10 VARIABLE CAPACITORS

There are many applications for capacitors that can be adjusted over a range of capacitance values. The most common variable capacitor consists of a set of fixed plates and a set of movable plates using air as the dielectric. The movable plates are attached to a shaft which, when rotated, intermeshes them with the fixed plate to varying degrees. The effective area between the plates thus varies with the degree to which the plates are meshed. As capacitance is proportional to the effective area between the plates, the capacitance is thus varied. See Fig. 5-16.

In addition to the types of capacitors already discussed, there are numerous other types for special and varied applications.

5-11 CAPACITORS IN SERIES AND PARALLEL

Consider a number of capacitors connected in parallel, as shown in Fig. 5-17. Equation (5-3) states that capacitance is the ratio of the charge stored to the voltage across the capacitor. The total capacitance can be calculated.

SOLUTION

1 By applying Eq. (5-3) to each capacitor and to the total effective capacitance.

$$C_1 = \frac{Q_1}{E} \qquad C_2 = \frac{Q_2}{E} \qquad C_3 = \frac{Q_3}{E} \qquad C_t = \frac{Q_t}{E}$$

2 The total charge Q_t is the sum of the charges on each capacitor.

$$Q_t = Q_1 + Q_2 + Q_3 \tag{5-12}$$

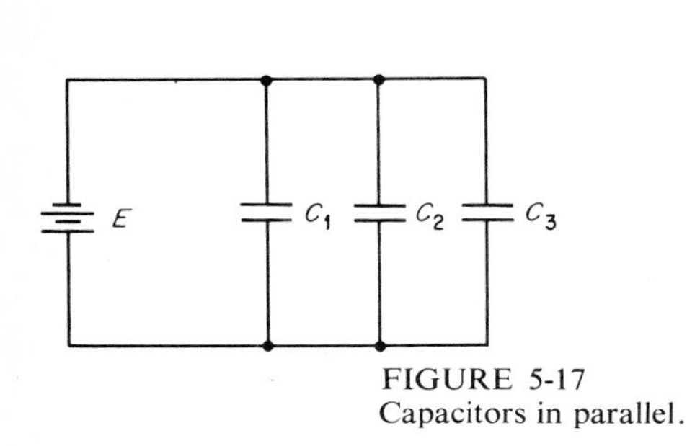

FIGURE 5-17
Capacitors in parallel.

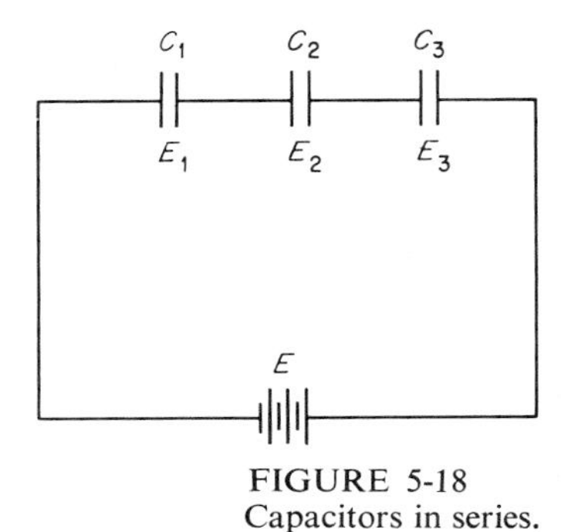

FIGURE 5-18
Capacitors in series.

3 From Eq. (5-3), it follows that $Q = CE$, and if the charge on each of the parallel capacitors is written in this form and substituted into Eq. (5-12), Eq. (5-13) results.

$$C_t E = C_1 E + C_2 E + C_3 E \tag{5-13}$$

4 Dividing both sides of Eq. (5-13) by the common factor E,

$$C_t = C_1 + C_2 + C_3 \qquad //// \tag{5-14}$$

Equation (5-14) states mathematically that the total capacitance of a number of capacitors in parallel is the sum of the branch capacitances. The result of combining capacitors in parallel may be thought of as simply increasing the total effective plate area by the plate area of each capacitor added in parallel.

In Fig. 5-18, three capacitors are connected in series, and the following computations resolve the effective capacitance of this configuration.

SOLUTION

1 The voltages shown in Fig. 5-18 are related as follows:

$$E = E_1 + E_2 + E_3 \tag{5-15}$$

2 The charges on all capacitors must be the same, since the capacitors are connected in series and any charge movement in one part of the circuit must take place in all parts of the series circuit. Solving Eq. (5-3) for voltage in terms of capacitance and charge, the following results are obtained for each of the series capacitors and the total effective capacitance:

$$E = \frac{Q}{C_t} \qquad E_1 = \frac{Q}{C_1} \qquad E_2 = \frac{Q}{C_2} \qquad E_3 = \frac{Q}{C_3}$$

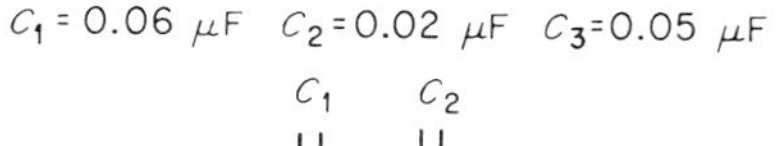

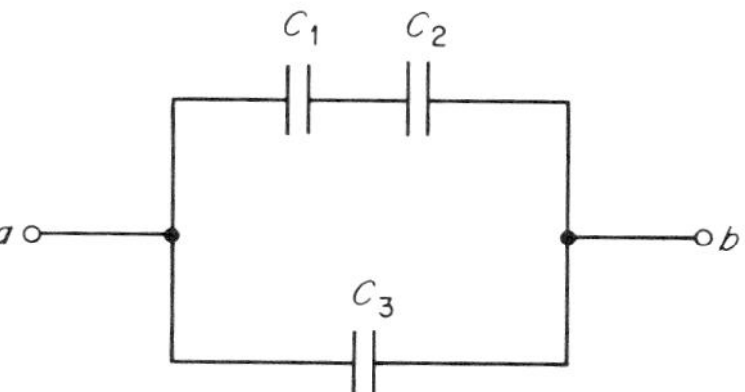

FIGURE 5-19
Series-parallel combination of capacitors in Example 5-7.

3 Substituting these results into Eq. (5-15),

$$\frac{Q}{C_t} = \frac{Q}{C_1} + \frac{Q}{C_2} + \frac{Q}{C_3} \tag{5-16}$$

4 Dividing both sides of Eq. (5-16) by the common factor Q,

$$\frac{1}{C_t} = \frac{1}{C_1} + \frac{1}{C_2} + \frac{1}{C_3}$$

and

$$C_t = \frac{1}{1/C_1 + 1/C_2 + 1/C_3} \qquad //// \tag{5-17}$$

Equation (5-17) states that the total capacitance of a number of capacitors connected in series equals the reciprocal of the sum of the reciprocals of the series capacitors.

Note that Eq. (5-14) for capacitors in parallel is similar to the equation for resistors in series and Eq. (5-17) is similar to the equation for resistors in parallel.

EXAMPLE 5-7 In the circuit of Fig. 5-19 calculate the total capacitance between points a and b,

SOLUTION

1 C_1 and C_2 are in series. Thus,

$$C_{1\text{-}2} = \frac{C_1 C_2}{C_1 + C_2} = \frac{(0.06 \times 10^{-6})(0.02 \times 10^{-6})}{0.08 \times 10^{-6}} = 0.015\ \mu\text{F}$$

2 $C_{1\text{-}2}$ is in parallel with C_3. Thus,

$$C_{a\text{-}b} = C_{1\text{-}2} + C_3 = 0.015 + 0.05 = 0.065\ \mu\text{F} \qquad ////$$

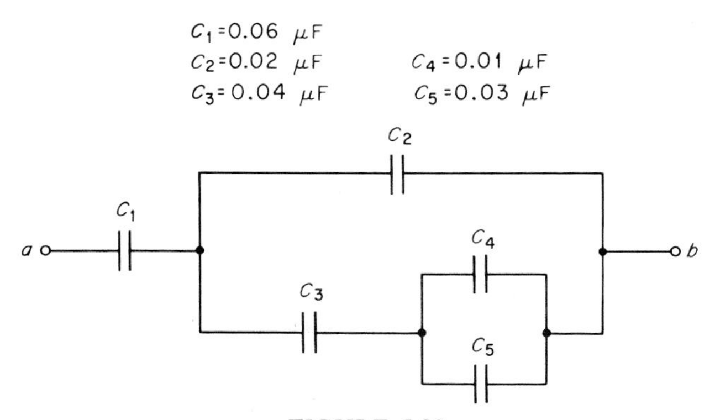

FIGURE 5-20
Series-parallel combination of capacitors in Example 5-8.

EXAMPLE 5-8 In the circuit of Fig. 5-20, calculate the total capacitance between points *a* and *b*.

SOLUTION

1 C_4 and C_5 are in parallel. Simply add them.

$$C_{4\text{-}5} = C_4 + C_5 = 0.01 + 0.03 = 0.04\ \mu\text{F}$$

2 C_3 and $C_{4\text{-}5}$ are in series.

$$C_{3\text{-}4\text{-}5} = \frac{C_{4\text{-}5}\, C_3}{C_{4\text{-}5} + C_3} = \frac{(0.04)(0.04)}{0.08} = 0.02\ \mu\text{F}$$

3 $C_{3\text{-}4\text{-}5}$ and C_2 are in parallel. Therefore add them.

$$C_{2\text{-}3\text{-}4\text{-}5} = C_{3\text{-}4\text{-}5} + C_2 = 0.02 + 0.02 = 0.04\ \mu\text{F}$$

4 C_1 and $C_{2\text{-}3\text{-}4\text{-}5}$ are in series.

$$C_{a\text{-}b} = \frac{C_1 C_{2\text{-}3\text{-}4\text{-}5}}{C_1 + C_{2\text{-}3\text{-}4\text{-}5}} = \frac{(0.06)(0.04)}{0.1} = 0.024\ \mu\text{F} \qquad ////$$

EXAMPLE 5-9 If 75 V is applied to the circuit in Fig. 5-19 at terminals *a-b*, find the voltage across each capacitor.

SOLUTION

1 The voltage applied at *a-b* is directly across $C_{1\text{-}2}$ and C_3.

$$E_3 = 75\ \text{V}$$

$$E_{1\text{-}2} = 75\ \text{V}$$

2 The charge on the equivalent capacitor $C_{1\text{-}2}$ is

$$Q_{1\text{-}2} = E_{1\text{-}2}\,C_{1\text{-}2} = (75)(0.015 \times 10^{-6}) = 1.125 \times 10^{-6}\ \text{C}$$

3 The charge on $C_{1\text{-}2}$ is the charge on C_1 and C_2.

$$E_1 = \frac{Q_{1\text{-}2}}{C_1} = \frac{1.125 \times 10^{-6}}{0.06 \times 10^{-6}} = 18.75\ \text{V}$$

$$E_2 = \frac{Q_{1\text{-}2}}{C_2} = \frac{1.125 \times 10^{-6}}{0.02 \times 10^{-6}} = 56.25\ \text{V}$$

Note:

$$E_1 + E_2 = E_{1\text{-}2} = 18.75 + 56.25 = 75\ \text{V}$$ ////

5-12 CURRENT IN A CAPACITIVE CIRCUIT

From the definition of capacitance, $C = Q/E$. Therefore,

$$Q = CE \tag{5-18}$$

Capacitance C in Eq. (5-18) is a constant, and if the voltage across the capacitor is constant, then the charge Q on the capacitor remains constant. When the charge on a capacitor is constant, there is no movement of charge in the circuit and therefore no current. If the voltage across a capacitor is changing, the charge must change accordingly. When the voltage across a capacitor is increasing, the charge on the capacitor must increase, and this movement of charge constitutes a current. If charge increasing on the capacitor is defined as a *positive current*, then an increase of voltage gives a positive current. When the voltage across a capacitor is decreasing, the charge must decrease accordingly, and as a result a current in the opposite direction, or a *negative current*, occurs. The more rapidly the voltage across the capacitor changes, the greater must be the current, since current is the rate of charge movement. This relationship is shown in Eq. (5-19).

$$i = C\frac{de_c}{dt} \tag{5-19}$$

Equation (5-19) simply states that the current i at any instant of time equals the capacitance in farads times the rate of change of the voltage in volts per second. The quantity de_c/dt is the mathematical notation for instantaneous rate of change of voltage.

If, in place of the instantaneous rate of change, the average rate of change $\Delta e_c/\Delta t$ is used, the average current is obtained. Equation (5-20) shows this relationship.

$$i_{av} = C\frac{\Delta e_c}{\Delta t} \tag{5-20}$$

In Eq. (5-20), Δe_c represents the change in the voltage across the capacitor during change in time Δt. Δe_c can be expressed as $e_{c-2} - e_{c-1}$, where e_{c-2} is the voltage across the capacitor at time t_2, and e_{c-1} is the voltage across the capacitor at time t_1. Thus

$$\Delta e_c = e_{c-2} - e_{c-1} \tag{5-21}$$

and

$$\Delta t = t_2 - t_1 \tag{5-22}$$

Substituting the results of Eqs. (5-21) and (5-22) into Eq. (5-20),

$$i_{av} = C\left(\frac{e_{c-2} - e_{c-1}}{t_2 - t_1}\right) \tag{5-23}$$

EXAMPLE 5-10 The voltage across an 8-μF capacitor is 15 V, and 2 s later the voltage across the capacitor is 45 V. What is the average value of current during this change of voltage?

SOLUTION

$$i_{av} = C\frac{e_2 - e_1}{t_2 - t_1}$$

$$e_2 = 45 \text{ V} \qquad t_2 = 2 \text{ s}$$

$$e_1 = 15 \text{ V} \qquad t_1 = 0 \text{ s}$$

$$i_{av} = 8 \times 10^{-6}\left(\frac{45 - 15}{2 - 0}\right) = (8 \times 10^{-6})\left(\frac{30}{2}\right) = 120\ \mu\text{A}$$

EXAMPLE 5-11 In Fig. 5-21, the voltage across a capacitor is plotted versus time. Find the average current in a circuit containing a 0.5-μF capacitor (1) from time $t = 0$ ms to $t = 5$ ms, (2) from time $t = 5$ ms to $t = 15$ ms, and (3) from time $t = 15$ ms to $t = 17$ ms.

SOLUTION

1
$$i_{av} = 0.5 \times 10^{-6}\frac{350 - 0}{5 \times 10^{-3} - 0} = 0.035 \text{ A}$$

2
$$i_{av} = 0.5 \times 10^{-6}\frac{350 - 350}{(15 \times 10^{-3}) - (5 \times 10^{-3})} = 0$$

3
$$i_{av} = 0.5 \times 10^{-6}\frac{300 - 350}{(17 \times 10^{-3}) - (15 \times 10^{-3})} = 0.5 \times 10^{-6}\frac{-50}{2 \times 10^{-3}}$$

$$= -0.0125 \text{ A}$$

////

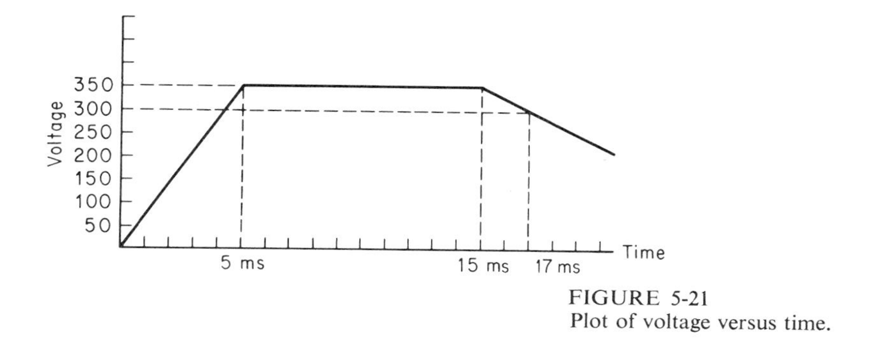

FIGURE 5-21
Plot of voltage versus time.

5-13 DC SOURCES IN RC CIRCUITS

Figure 5-22 shows a resistor and a capacitor connected in series. When the switch is closed, current begins to flow. At the instant the switch is closed there is no charge on the capacitor, and therefore there is no voltage across the capacitor. The applied voltage is across the resistor at the instant the switch is closed, and the current in the circuit at this instant is $i = E/R$. As current flows, however, a charge accumulates on

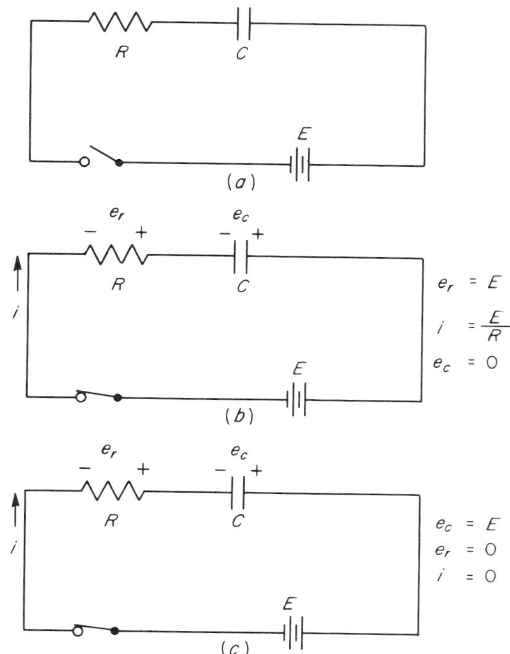

FIGURE 5-22
RC circuit: (*a*) Switch open, (*b*) switch just closed, (*c*) switch closed for a long time.

the capacitor and develops a voltage across the capacitor. As the voltage across the capacitor increases, the drop across the resistor decreases. The current in the circuit, at any instant, equals

$$i = \frac{e_r}{R} \tag{5-24}$$

As shown by Eq. (5-24), the current in the circuit is directly proportional to the voltage across the resistor. As the capacitor continues to accumulate charge, the current in the circuit decreases. The gradual reduction of current in the circuit is accounted for further if one considers that as the capacitor takes on charge, that charge opposes further addition of charge to the capacitor, thus gradually reducing current.

The current continues to decrease gradually until the capacitor is charged to the value of the applied voltage E. At this time the current is zero. The time relation of i, e_r, and e_c is plotted in Fig. 5-23.

It can be seen that at any time, the sum of e_r and e_c equals the applied voltage E.

$$E = e_r + e_c \tag{5-25}$$

At time t_0, e_r equals E, and e_c equals zero.

$$E = E + 0$$

At time t_1, $e_r = 0.368E$, and $e_c = 0.632E$.

$$E = 0.368E + 0.632E$$

At time t_2, $e_r = 0.1355E$, and $e_c = 0.8645E$.

$$E = 0.1355E + 0.8645E$$

Using Eq. (5-25), and substituting for e_r,

$$E = iR + e_c \tag{5-26}$$

The current in a capacitive circuit is given by Eq. (5-19). Substituting this value of current into Eq. (5-26),

$$E = RC\frac{de_c}{dt} + e_c \tag{5-27}$$

Equation (5-27) can be solved only by application of differential equation theory. The solution is

$$e_c = E(1 - \varepsilon^{-t/RC}) \tag{5-28}$$

where E = applied voltage
R = resistance through which capacitor is charging
t = time after the capacitor begins to charge,
ε = naperian or natural base = 2.718

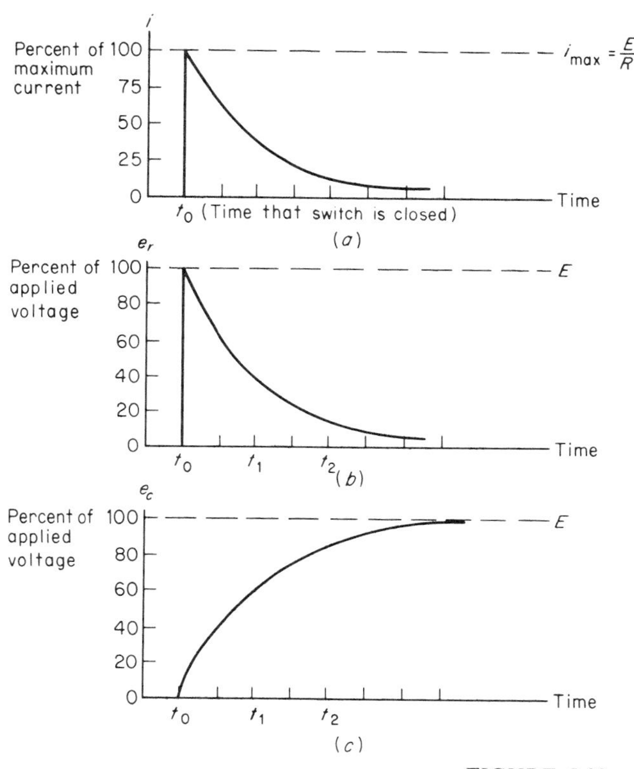

FIGURE 5-23
RC circuit functions versus time.

Evaluating Eq. (5-19) with the value of e_c given in Eq. (5-28) yields the equation for the current in a dc resistance-capacitance circuit.

$$i = \frac{E}{R} \varepsilon^{-t/RC} \tag{5-29}$$

Equation (5-29) is found by differentiating Eq. (5-28) to obtain de_c/dt, the instantaneous rate of change of e_c.

The voltage across the resistance can now be calculated.

$$\begin{aligned} e_r &= iR \\ &= \frac{E}{R} \varepsilon^{-t/RC} R \\ &= E\varepsilon^{-t/RC} \end{aligned} \tag{5-30}$$

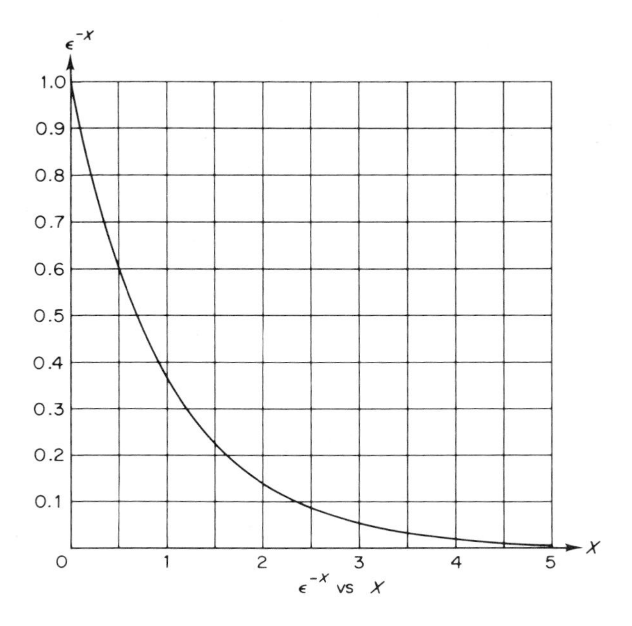

FIGURE 5-24
ϵ^{-x} vs. x.

If the results of Eqs. (5-28) and (5-30) are substituted into Eq. (5-25), the consistency of the equations is shown

$$
\begin{aligned}
E &= e_r + e_c \\
&= E\varepsilon^{-t/RC} + E(1 - \varepsilon^{-t/RC}) \\
&= E\varepsilon^{-t/RC} + E - E\varepsilon^{-t/RC} \\
&= E
\end{aligned}
$$

Figure 5-24 shows the relationship between ε^{-x} and x. Appendix Table A-9 gives a more precise tabular listing. Let x represent t/RC. The time required for the capacitor to obtain a charge depends on the magnitudes of resistance and capacitance. If Eq. (5-28) is evaluated when time equals RC s, the following relation results:

$$
\begin{aligned}
e_c &= E(1 - \varepsilon^{-t/RC}) \\
&= E(1 - \varepsilon^{-RC/RC}) \\
&= E(1 - \varepsilon^{-1}) \\
&= E(1 - 0.368) = 0.632E \qquad \text{for} \quad t = RC \text{ s}
\end{aligned}
$$

This result indicates that the capacitor has charged to 63.2 percent of the applied voltage E in RC s. *The product RC is referred to as the time constant of the resistance-*

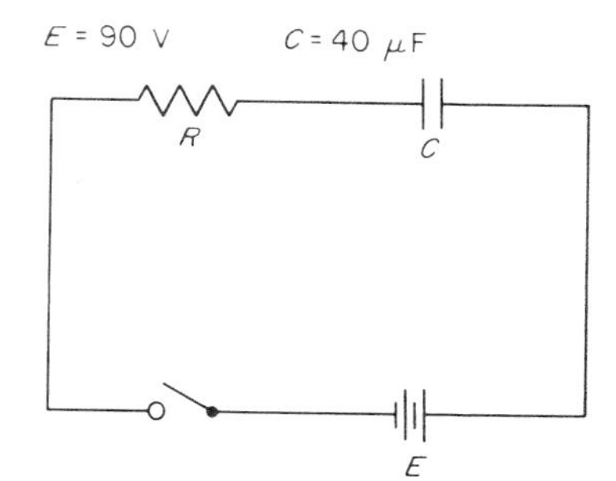

FIGURE 5-25
RC series circuit for Example 5-12.

capacitance circuit. As time increases, the quantity $\varepsilon^{-t/RC}$ becomes smaller and smaller. As $\varepsilon^{-t/RC}$ becomes smaller, the voltage e_c approaches the applied voltage E. Mathematically the quantity $\varepsilon^{-t/RC}$ could equal zero, and thus e_c equal E, only when time becomes infinite. In five time constants, however, the value of e_c becomes so close to the applied voltage that for most practical purposes e_c is considered equal to the applied voltage.

$$\begin{aligned} e_c &= E(1 - \varepsilon^{-t/RC}) \\ &= E(1 - \varepsilon^{-5RC/RC}) \\ &= E(1 - 0.0067) \\ &= 0.9933E \qquad \text{for} \quad t = 5RC \text{ s} \end{aligned}$$

EXAMPLE 5-12 In Fig. 5-25, calculate e_c 1 s after closing the switch when (1) $R = 50$ kΩ, (2) $R = 25$ kΩ, and (3) $R = 100$ kΩ.

SOLUTION

1 $$\begin{aligned} e_c &= E(1 - \varepsilon^{-t/RC}) \\ &= 90(1 - \varepsilon^{-1/(50 \times 10^3 \times 40 \times 10^{-6})}) \\ &= 90(1 - \varepsilon^{-1/2}) \qquad \varepsilon^{-1/2} = 0.606 \\ &= 90(1 - 0.606) \\ &= 90 \times 0.394 = 35.4 \text{ V} \end{aligned}$$

2 $$\begin{aligned} e_c &= 90(1 - \varepsilon^{-1/(25 \times 10^3 \times 40 \times 10^{-6})}) = 90(1 - \varepsilon^{-1}) \qquad \varepsilon^{-1} = 0.368 \\ &= 90(1 - 0.368) = 90 \times 0.632 = 56.8 \text{ V} \end{aligned}$$

3 $$\begin{aligned} e_c &= 90(1 - \varepsilon^{-1/(100 \times 10^3 \times 40 \times 10^{-6})}) = 90(1 - \varepsilon^{-1/4}) \qquad \varepsilon^{-1/4} = 0.779 \\ &= 90(1 - 0.779) = 90 \times 0.221 = 19.9 \text{ V} \end{aligned}$$

////

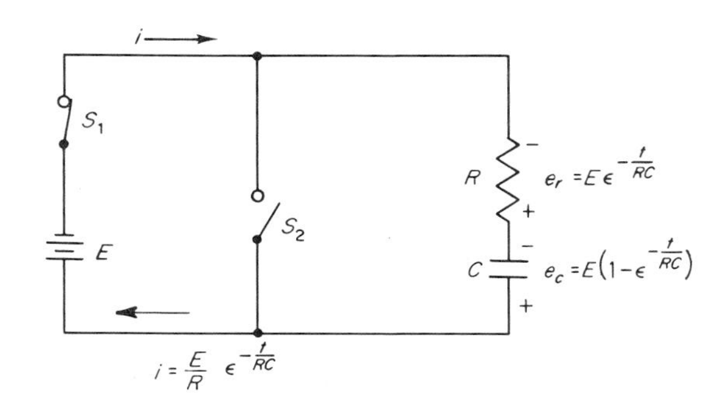

FIGURE 5-26
Charging capacitor C.

5-14 DISCHARGE IN AN RC CIRCUIT

Figure 5-26 shows a circuit arrangement. When switch S_1 is closed, the capacitor charges to applied voltage E, as previously discussed. Once the capacitor is fully charged, S_1 is opened and S_2 is closed. The circuit conditions are shown in Fig. 5-26 for charging the capacitor. Figure 5-27a shows circuit values at the instant S_1 is opened and S_2 is closed, while Fig. 5-27b shows circuit conditions after S_2 has been closed for a relatively long period of time. When S_2 is closed, there is present on the capacitor an accumulated charge which represents a difference of potential E_c. E_c in Fig. 5-27a is the initial value of voltage across the capacitor at the instant discharge of the capacitor begins. The voltage across the capacitor causes current to

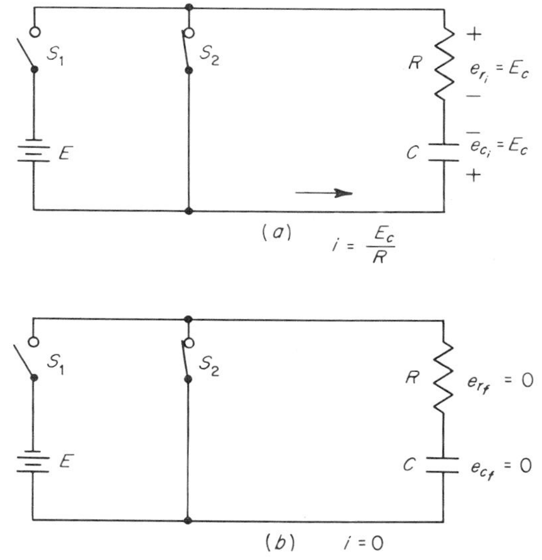

FIGURE 5-27
Discharging capacitor C.

flow in the direction shown by Fig. 5-27*a*. The direction of current is, of course, opposite to the direction during charging of the capacitor. While the capacitor is discharging, the potential difference across the capacitor is being reduced, and therefore the current is diminishing. This diminishing process continues until the capacitor no longer stores any charge. At all times, the value of voltage drop across the resistor e_r must equal the value of e_c.

$$e_r = e_c$$

$$Ri = e_c$$

$$RC\frac{de_c}{dt} = e_c \tag{5-31}$$

The value of e_c that satisfies Eq. (5-31) is

$$e_c = E_c \varepsilon^{-t/RC} \tag{5-32}$$

where E_c = voltage across the capacitor when discharge is initiated, V
t = time after initiation of discharge, s
R = resistance through which the discharge takes place, Ω
C = capacitance being discharged, F

It follows that the value of e_r during discharge must be

$$e_r = E_c \varepsilon^{-t/RC} \tag{5-33}$$

and

$$i = \frac{e_r}{R} = \frac{E_c}{R}\varepsilon^{-t/RC} \tag{5-34}$$

It must be noted that the current flowing during discharge of the capacitor, as given by Eq. (5-34), is opposite in direction to the current that charges the capacitor, as given by Eq. (5-29). If the charging current is considered positive, the discharge current then must be considered negative. This current polarity relation is demonstrated in Fig. 5-28*a*. Furthermore, if the current reverses direction, the polarity of the voltage drop across the resistor during discharge, as given by Eq. (5-33), must be opposite that of the voltage drop across the resistor during charging, as given by Eq. (5-30). Fig. 5-28*b* demonstrates this. Fig. 5-28*c* shows the voltage relationship across the capacitor.

EXAMPLE 5-13 In the circuit of Fig. 5-29, calculate the values of e_c, e_{r1}, and i (1) 0.005 s after closing S_1, (2) 0.01 s after closing S_1, and (3) 0.06 s after closing S_1.

SOLUTION

1 By Eq. 5-28

$$e_c = E(1 - \varepsilon^{-t/RC}) = 80(1 - \varepsilon^{-0.005/(6\times10^3)(2\times10^{-6})}) = 80(1 - \varepsilon^{-0.005/0.012})$$

$$= 80(1 - 0.659) = (80)(0.341) = 27.28 \text{ V}$$

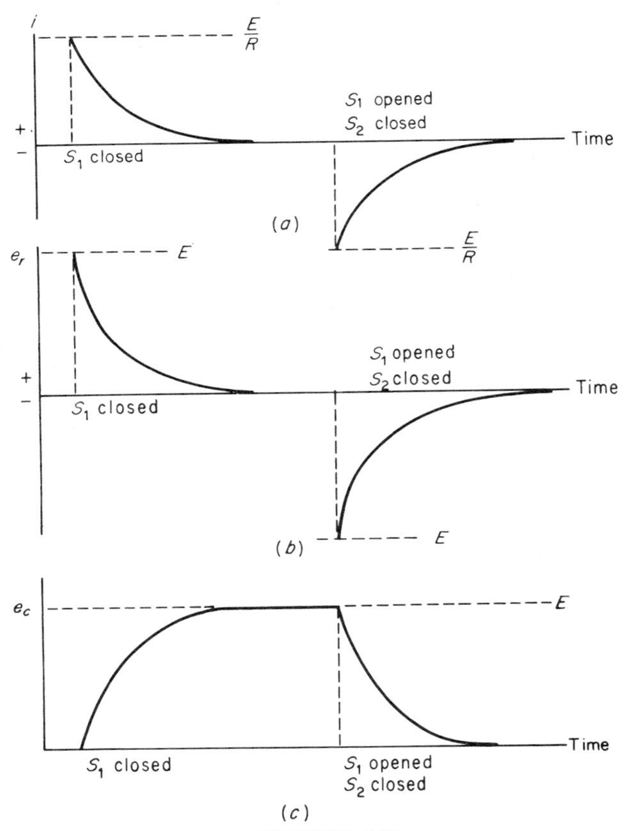

FIGURE 5-28
Charge and discharge currents and voltages versus time.

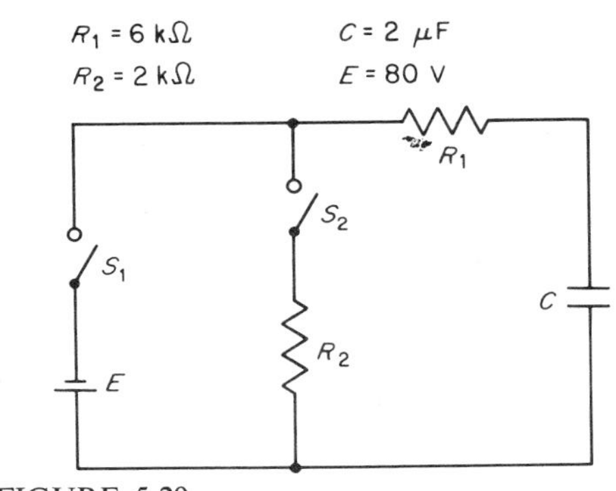

FIGURE 5-29
RC circuit for Example 5-13.

By Eq. (5-29),

$$i = \frac{E}{R} \varepsilon^{-t/RC} = \frac{80}{6 \times 10^3} 0.659 = (0.01333)(0.659) = 8.78 \text{ mA}$$

By Eq. (5-30),

$$e_{r_1} = E\varepsilon^{-t/RC} = (80)(0.659) = 52.72 \text{ V}$$

or, by Eq. (5-25),

$$e_{r_1} = E - e_c = 80 - 27.28 = 52.72 \text{ V}$$

2 $e_c = 80(1 - \varepsilon^{-0.01/(6 \times 10^3)(2 \times 10^{-6})}) = (80)(1 - 0.4346) = (80)(0.56) = 45.2 \text{ V}$

$$i = \frac{(80)(0.434)}{6 \times 10^3} = 5.78 \text{ mA}$$

$$e_{r_1} = E - e_c = 80 - 45.2 = 34.8 \text{ V}$$

3 The time constant of the charging path for the capacitor is

$$R_1C = (6 \times 10^3)(2 \times 10^{-6}) = 12 \text{ ms}$$

Five times R_1C is

$$5R_1C = (5)(12) \text{ ms} = 60 \text{ ms}$$

Therefore, when $t = 60 \text{ ms} = 0.06 \text{ s}$

$$e_c \approx 80 \text{ V}$$

Since the capacitor is almost fully charged, the current $i \cong 0$.
Since the current is zero, the voltage drop across the resistor $e_{r_1} \cong 0$. ////

EXAMPLE 5.14 If the capacitor in Fig. 5-29 is allowed to charge to 80 V and then S_1 is opened and S_2 closed, calculate the value of e_c and i, 0.03 s after the described switching action has occurred.

SOLUTION

1 Use Eq. (5-32).

$$e_c = E_c\, \varepsilon^{-t/RC} = 80\, \varepsilon^{-0.03/(8 \times 10^3)(2 \times 10^{-6})}$$

$$= 80\, \varepsilon^{-0.03/0.016} = (80)(0.1535) = 12.28 \text{ V}$$

$$i = \frac{E_c}{R} \varepsilon^{-t/RC} = \frac{80}{8 \times 10^3} \varepsilon^{-0.03/(8 \times 10^3)(2 \times 10^{-6})}$$

$$= (0.010)(0.1535) = 1.535 \text{ mA}$$

Note: The resistance through which the capacitor discharges is the sum of R_1 and R_2. The current flows in the opposite direction with respect to the current in Example 5-13. ////

EXAMPLE 5-15 If the capacitor in Fig. 5-29 is allowed to charge to 60 V and then S_1 is opened and S_2 closed, calculate (1) the value of e_c and i, 0.03 s after the described switching action has occurred, (2) the value of e_c and i, 0.01 s after the described switching action has occurred, and (3) the time necessary for the discharge current to go to zero after the described switching action has occurred.

SOLUTION

1 Use Eq. (5-32).

$$e_c = E_c\,\varepsilon^{-t/RC} = 60\,\varepsilon^{-0.03/(8\times10^3)(2\times10^{-6})} = (60)(0.1535) = 9.2\text{ V}$$

Use Eq. (5-34).

$$i = \frac{E_c}{R}\,\varepsilon^{-t/RC} = \frac{60}{8\times10^3}\,0.1535 = 1.15\text{ mA}$$

2

$$e_c = 60\,\varepsilon^{-0.01/(8\times10^3)(2\times10^{-6})} = (60)(0.535) = 32.1\text{ V}$$

$$i = \frac{60}{8\times10^3}\,\varepsilon^{-0.01/(8\times10^3)(2\times10^{-6})} = \frac{60}{8\times10^3}\,0.535 = 4\text{ mA}$$

3 The time constant of the discharge path is

$$RC = (R_1 + R_2)C = (8\times10^3)(2\times10^{-6}) = 16\text{ ms}$$

and

$$5RC = 5\times16\text{ ms} = 80\text{ ms}$$

Therefore, 80 ms after S_1 is opened and S_2 is closed

$$i \approx 0$$

////

EXAMPLE 5-16 In the circuit of Fig. 5-29, calculate the time that elapses after closing S_1 until 35 V is seen across the capacitor.

SOLUTION

1 Use Eq. (5-28)

$$e_c = E(1 - \varepsilon^{-t/RC})$$

$$35 = 80(1 - \varepsilon^{-t/(6\times10^3)2\times10^{-6}})$$

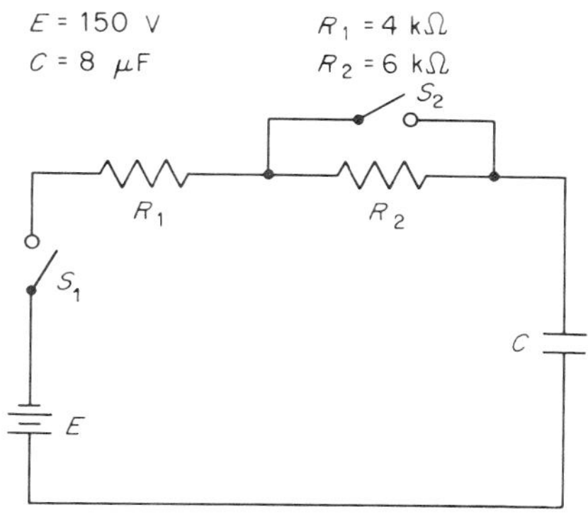

FIGURE 5-30
RC circuit for Example 5-17.

$$\tfrac{35}{80} = 1 - \varepsilon^{-t/0.012}$$

$$0.437 - 1 = -\varepsilon^{-t/0.012}$$

$$0.563 = \varepsilon^{-t/0.012}$$

2 From the tables or slide rule

$$\varepsilon^{-0.575} = 0.563$$

Therefore,

$$\frac{-t}{0.012} = -0.575$$

$$t = (0.575)(0.012) = 6.9 \text{ ms} \qquad ////$$

In the preceding examples demonstrating the charge or discharge of a capacitor through resistance, the resistance of the current path was not changed. The resistance of the charge path was held constant, and then the resistance of the discharge path was held constant. In Example 5-17 the resistance of the charge path changes during charging of the capacitor.

EXAMPLE 5-17 In the circuit of Fig. 5-30, S_1 is closed, and 0.04 s later S_2 is also closed. Find e_c 0.04 s after S_2 is closed.

SOLUTION

1 The resistance of the charging path during the 0.04 s after closing S_1 is

$$R = R_1 + R_2$$

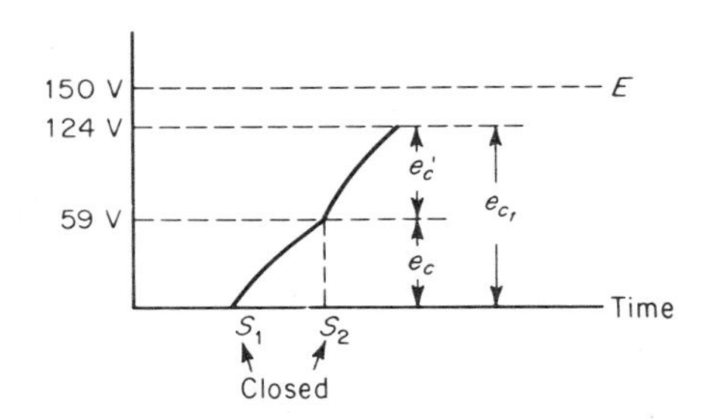

FIGURE 5-31
Plot of capacitor voltage for Example 5-17.

2 The voltage across the capacitor can be found 0.04 s after closing S_1 by Eq. (5-28).

$$e_c = E(1 - \varepsilon^{-t/RC}) = (150)(1 - \varepsilon^{-0.04/0.08})$$
$$= 150(1 - \varepsilon^{-0.5}) = (150)(1 - 0.6065) = (150)(0.3935) = 59 \text{ V}$$

3 When S_2 is closed, the resistance of the charging path is reduced by the magnitude of R_2. Therefore,

$$R = R_1$$

4 If the time when S_2 is closed is treated as a reference time, then the capacitor has yet to charge to E, but the capacitor has already charged to 59 V. The total remaining charge to take place in e_c is

$$E_c' = E - 59 = 150 - 59 = 91 \text{ V}$$

5 The change in e_c that takes place in the time after closing S_2 can now be calculated.

$$e_c' = E_c'(1 - \varepsilon^{-t/RC}) = (91)(1 - \varepsilon^{-0.04/(4\times 10^3)(8\times 10^{-6})})$$
$$= 91(1 - \varepsilon^{-1.25}) = (91)(1 - 0.286) = (91)(0.714) = 65 \text{ V}$$

6 e_c' represents the change in voltage across the capacitor after S_2 is closed. The total voltage across the capacitor is

$$e_{ct} = e_c + e_c' = 59 + 65 = 124 \text{ V}$$

7 The above conditions are shown graphically in Fig. 5-31. ////

5-15 POWER AND ENERGY IN THE RC CIRCUIT

While a capacitor is being charged, energy is being stored in the form of an electrostatic field. The instantaneous rate at which energy is being stored is the *instantaneous power*, the product of e_c and i, as given by Eq. (5-35)

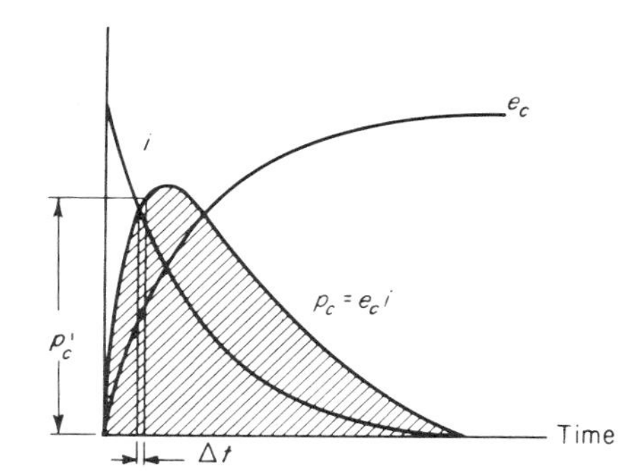

FIGURE 5-32
Power versus time.

$$p_c = e_c i \tag{5-35}$$

Since the current and the potential difference across the capacitor are constantly changing, the power varies. This relationship is plotted in Fig. 5-32.

The area beneath the power curve represents the energy stored in the capacitor. This relation can be seen if a narrow vertical strip of area is considered. The width of this strip is Δt s, and the height of the strip is p'_c. The product of p'_c and Δt represents the area of the strip, as shown in Eq. (5-36).

$$\text{Area} = p'_c \, \Delta t \tag{5-36}$$

In Chap. 3, it was shown that power in watts times time in seconds yields energy in joules. Therefore, the area in Eq. (5-36) represents the amount of energy stored in Δt s. If the entire shaded area of Fig. 5-32 were divided into narrow strips of Δt width and the areas of all strips determined and then summed, the total area beneath the curve would be known. The total area therefore represents the total energy stored by the capacitor.

The total energy stored is found to be

$$W_c = \frac{CE_c^2}{2} \text{ J} \tag{5-37}$$

where W_c = energy stored, J
C = capacitance, F
E_c = potential difference across capacitor, V

When the capacitor is discharged, the energy stored during charging is delivered to the circuit through which the capacitor discharges.

EXAMPLE 5-18 Find the energy stored in the capacitor of Example 5-17 when there is a potential difference of 150 V across the capacitor.

SOLUTION Using Eq. (5-37),

$$W_c = \frac{CE_c^2}{2} = \frac{(8 \times 10^{-6})(150)^2}{2} = 9 \times 10^{-2} \text{ J} \qquad ////$$

PROBLEMS

5-1 When 75 V is applied to a capacitor, a charge of 5×10^{-6} C is accumulated. What is the capacitance of the capacitor in farads and microfarads?
5-2 If 150 V is applied to a capacitor and 5×10^{-6} C of charge is accumulated, what is the capacitance of the capacitor in farads and microfarads?
5-3 If 200 V is applied to a 0.005-μF capacitor, what is the charge in coulombs on the capacitor?
5-4 A 0.02-μF capacitor has a charge of 0.05×10^{-6} C. What is the potential difference across the capacitor?
5-5 What is the capacitance of a parallel-plate capacitor having a plate area of 0.25 m^2 and a distance separating the plates of 0.0008 m? Use mica as the dielectric.
5-6 Repeat Prob. 5-5 with glass as the dielectric.
5-7 Find the voltage ratings of the capacitors in Probs. 5-5 and 5-6.
5-8 What is the total capacitance of a 15-μF capacitor and a 20-μF capacitor connected in series?
5-9 What is the total capacitance of three 6-μF capacitors connected in series?
5-10 What is the total capacitance of a 0.002-μF capacitor connected in parallel with a 0.008-μF capacitor?
5-11 Find the total capacitance of Fig. 5-33.
5-12 If 250 V is applied to terminals *a-b* of Fig. 5-33, find the charge on each capacitor.
5-13 Find the total capacitance of Fig. 5-34.
5-14 The voltage across C_3 in Fig. 5-34 is 250 V. Find the voltage applied to terminals *a-b*.
5-15 The voltage across a 0.05-μF capacitor is 35 V, and 3 ms later the voltage across the capacitor is 82 V. What is the average current?
5-16 The voltage across a 0.003-μF capacitor is 25 V, and 0.015 ms later the voltage across the capacitor is 105 V. What is the average current?
5-17 The voltage across a given capacitor changes from 40 V to 60 V in 0.002 ms. The average current is 0.9 A. Find the capacitance.
5-18 In the circuit of Fig. 5-35, find (*a*) e_r and e_c 0.5 sec after closing S_1, (*b*) the maximum current, (*c*) the time constant, and (*d*) the time after closing S_1 until 160 V is seen across the capacitor.
5-19 In the circuit of Fig. 5-35, the capacitance is increased to 35 μF. Repeat the calculations of Prob. 5-18.

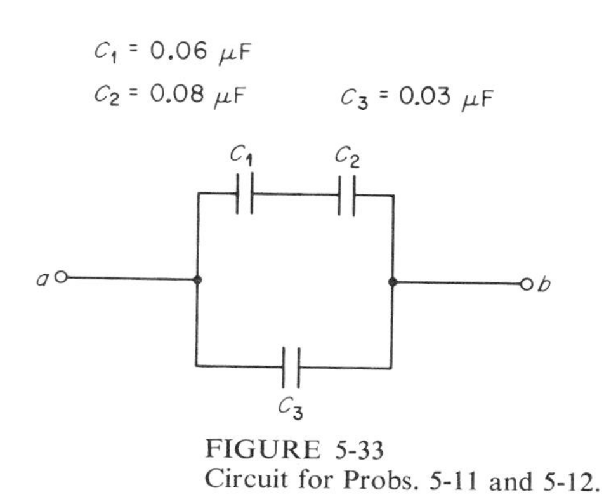

FIGURE 5-33
Circuit for Probs. 5-11 and 5-12.

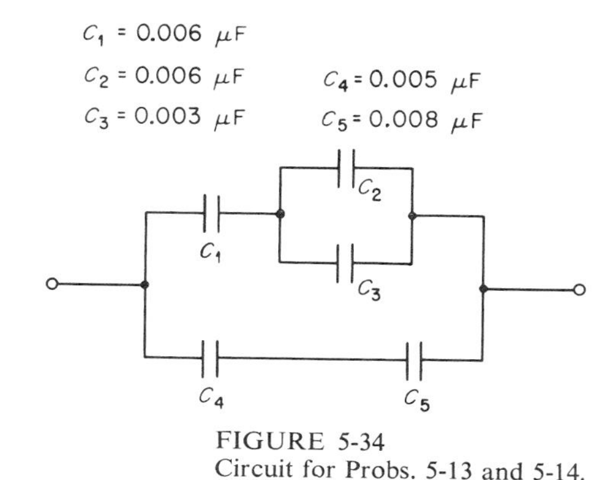

FIGURE 5-34
Circuit for Probs. 5-13 and 5-14.

5-20 In the circuit of Fig. 5-35, the capacitance is decreased to 5 μF. Repeat the calculations of Prob. 5-18.

5-21 In the circuit of Fig. 5-35, the resistance is increased to 75 kΩ. Repeat the calculations of Prob. 5-18.

5-22 In the circuit of Fig. 5-35, the resistance is decreased to 20 kΩ. Repeat the calculations of Prob. 5-18.

5-23 In the circuit of Fig. 5-36, calculate (*a*) e_c 0.2 s after closing S_1, (*b*) the time required for e_c to equal 60 V, and (*c*) the time after closing S_1 at which $e_{R_1} = 25$ V.

5-24 In the circuit of Fig. 5-36, S_1 is closed until the capacitor has charged to 60 V. S_1 is then opened, and S_2 is closed. Calculate (*a*) the voltage across R_2 0.3 s after closing S_2, and (*b*) the time required after closing S_2 for the voltage across the capacitor to drop to 20 V.

5-25 In the circuit of Fig. 5-36, S_1 is closed until the capacitor has charged to 35 V. S_1 is then opened, and S_2 is closed. Calculate (*a*) the current flowing 0.6 s after closing S_2,

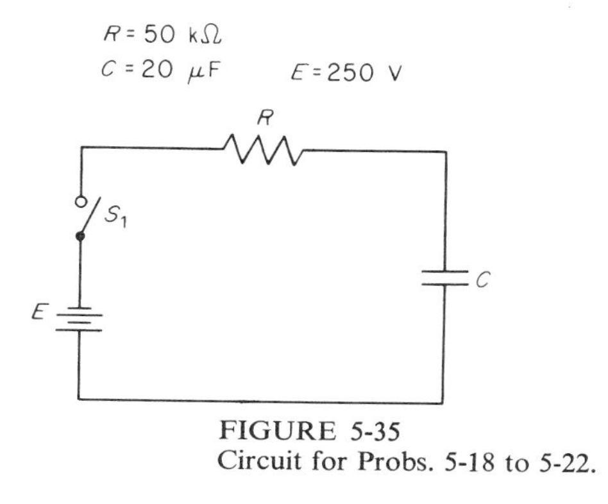

FIGURE 5-35
Circuit for Probs. 5-18 to 5-22.

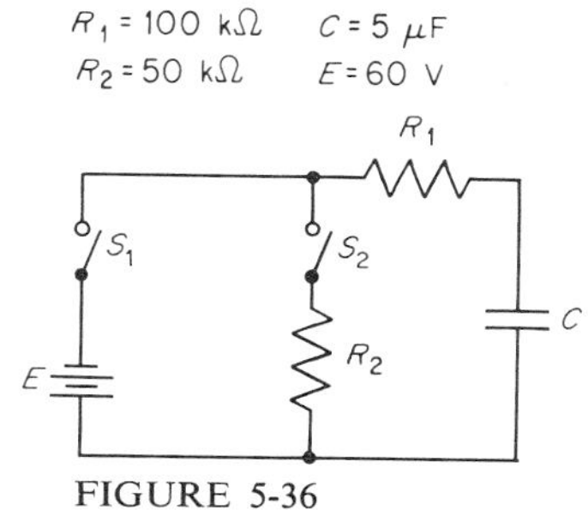

FIGURE 5-36
Circuit for Probs. 5-23 to 5-25.

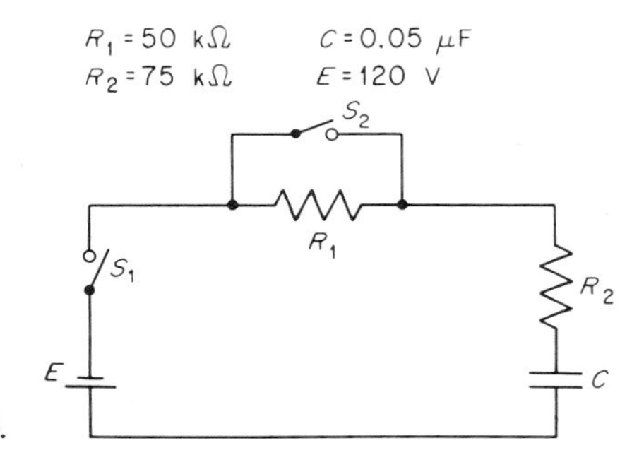

FIGURE 5-37
Circuit for Probs. 5-26 to 5-28.

(*b*) the voltage across R_1 1.8 s after closing S_2, and (*c*) the time required after closing S_2 for the voltage across the capacitor to be 5 V.

5-26 In the circuit of Fig. 5-37, S_1 is closed, and 0.002 s later S_2 is closed. Calculate the voltage across the capacitor 0.001 s after closing S_2.

5-27 In the circuit of Fig. 5-37, S_1 is closed, and when the voltage across the capacitor rises to 70 V, S_2 is closed. (*a*) How many seconds after closing S_1 is S_2 closed? (*b*) How many seconds after closing S_2 does the voltage across the capacitor reach 110 V?

5-28 In the circuit of Fig. 5-37 S_1 is closed, and when the voltage across the capacitor rises to 50 V, S_2 is closed. (*a*) How many seconds after closing S_1 is S_2 closed? (*b*) How many seconds after closing S_2 does the voltage across the capacitor reach 100 V?

6

ELECTROMAGNETISM AND MAGNETIC CIRCUITS

6-1 INTRODUCTION

The phenomenon of magnetic attraction was known to the ancients as early as the first century of the Christian era. The only magnetic material at that time was the natural magnet, lodestone (magnetite ore). The Chinese were probably the first to use a form of bar magnet as a crude compass. In all the centuries between the first and nineteenth, magnetism and electricity were thought to be unrelated phenomena. However, in 1819 the Danish physicist Oersted discovered that a compass needle is deflected when brought near a wire carrying current and thereby demonstrated that electric current is a source of magnetism. In the following decade, other investigators, notably Ampère and Faraday, made further progress toward understanding the relationship between electricity and magnetism. It is remarkable that in a short span of little more than 10 years the major portion of the body of knowledge about magnetism and electromagnetism was developed.

Magnetism and electromagnetism play vital roles in nearly all electric and electronic devices. Transformers, motors, generators, relays, computers, television receivers, loudspeakers, radios, home appliances, and almost any other electronic devices you might mention depend on magnets or electromagnetism for their successful operation.

It is possible to magnetize certain materials by stroking them with lodestone, but today all useful magnets take advantage of the magnetic properties of electric currents. Many devices, such as relays and transformers, must act as magnets only for short intervals, and this behavior is brought about electrically. Therefore, in this chapter, we shall begin with the study of electromagnetism, and then turn to the study of magnetic circuits and magnetic materials.

Probably the most difficult part of working magnetic circuit problems is handling the units. Three different systems of units are commonly used in magnetic circuits. The text will adopt the rationalized mks system of units. The cgs and English systems are discussed at the end of this chapter. Using a single system of units makes concepts and formulas much easier to understand.

6-2 THE MAGNETIC FIELD

A magnetic field is any region in which a magnetic force is exerted. From careful observation of magnetic forces, it is possible to describe a magnetic field accurately without actually seeing the field itself. For example, if a bar magnet is placed under a sheet of glass and iron filings are sprinkled over the glass, the filings distribute themselves over the glass in a definite pattern. This pattern shows that there is a force exerted between the ends of the bar magnet in such a manner that the filings lie along groups of lines between the ends of the bar magnet. Figure 6-1 illustrates the magnetic field around a bar magnet.

It can be said, therefore, that a field of force exists between the ends of the magnet and the field can be represented by lines. *These lines do not indicate a flow of any substance*. They are merely a representation of the location and magnitude of magnetic forces.

The *poles* of a magnet are those regions at the ends of a magnet where the magnetic field is concentrated. The polarity of the poles is defined in terms of the compass. The compass needle is a small bar magnet. When brought near one of the poles of a bar magnet, the needle points directly at the pole. When the end of the compass needle that ordinarily points north on the earth points to a pole of a magnet, that pole is regarded as the *south pole of the magnet*. When the compass is brought near the other pole of the magnet, the other end of the compass needle points to this pole, indicating that this pole is a *north pole*. The behavior of the compass needle is due to a fundamental rule of magnetism: *Unlike poles attract each other, while like poles repel each other*. Figure 6-2 shows the application of a compass to establish the polarity of a bar magnet.

The lines of force in a magnetic field are referred to as *lines of induction*, or simply *magnetic flux* Φ. Figure 6-3 shows the magnetic field (the flux) of a horseshoe magnet.

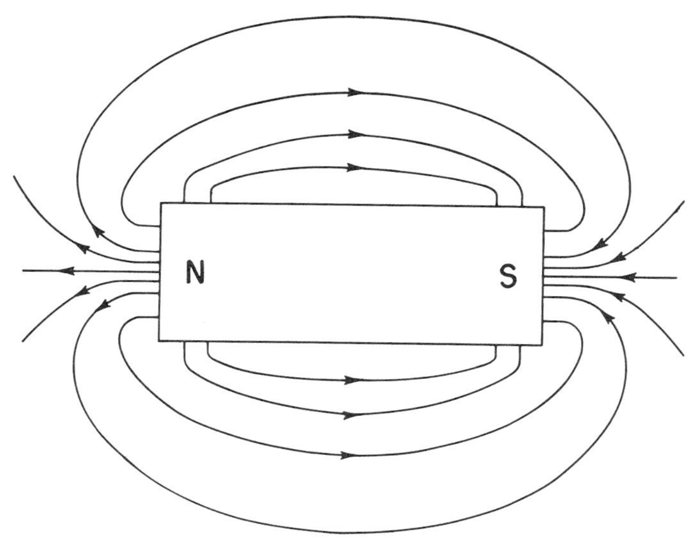

FIGURE 6-1
Magnetic field of a bar magnet.

Magnetic fields have several properties which are summarized in the following statements:

1 The lines of force are always complete loops.
2 The assumed direction of the lines is such that they are defined as leaving the north pole of the magnet, returning to the south pole, and then from the south to the north pole within the magnet (see Fig. 6-4).
3 Lines of force represent energy along their length, and they always tend to shorten, in a manner similar to stretched rubber bands.

An interesting corollary to statement 3 is that magnetic flux always follows the path of least opposition. This is not necessarily the shortest path in terms of distance, but it is the "shortest" path in terms of least opposition to magnetic lines of force.

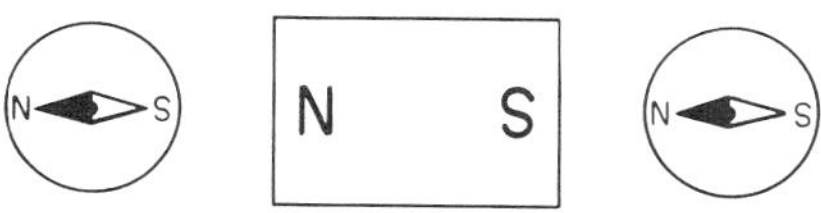

FIGURE 6-2
Unlike poles attract; like poles repel.

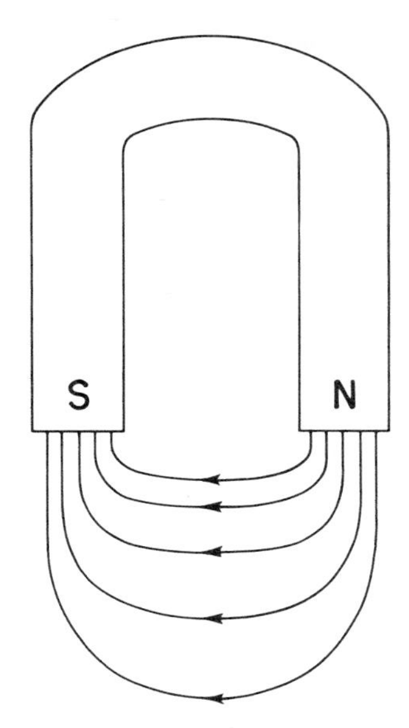

FIGURE 6-3
Field of a horseshoe magnet.

This may be stated: *The magnetic field always arranges itself so that the maximum number of lines of flux is established.*

4 Lines of force exert a sidewise spreading effect upon each other, which causes the field to spread. This spread is due to mutual repulsion between lines.

Whenever electric current flows in a conductor, a magnetic field surrounds the conductor. This field is circular and concentric with the conductor. Each line of flux is a complete circle; neither north nor south pole is present. The greatest concentration of flux is at the surface of the conductor, and the *flux density* decreases with distance from the surface of the wire. The direction of flux depends on the direction of the current. If the current is reversed, the direction of the flux is reversed.

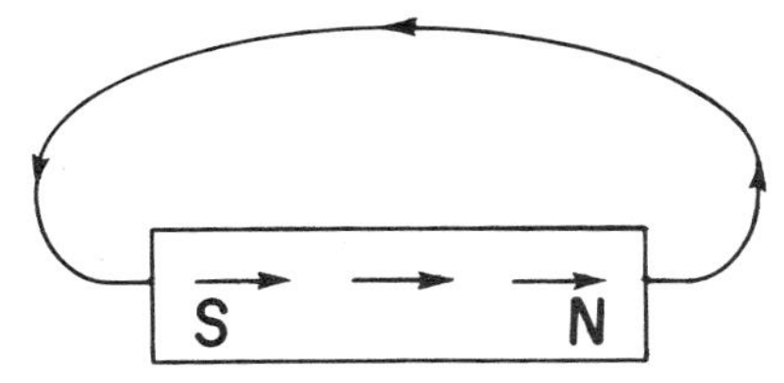

FIGURE 6-4
Flux leaves the north pole and enters the south pole of a magnet.

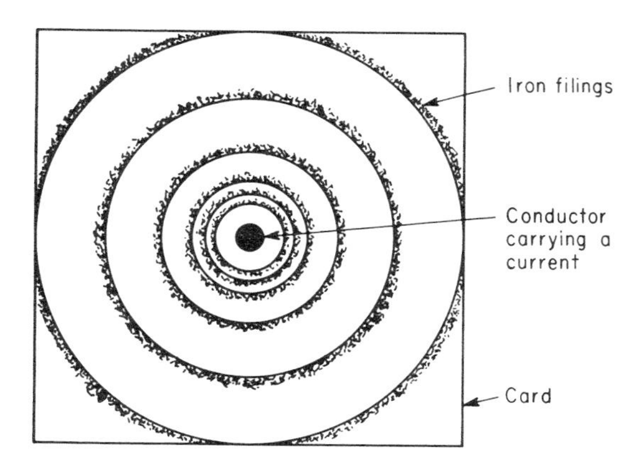

FIGURE 6-5
Field around a conductor.

The field around a conductor carrying a current can be observed by the use of a small compass. When the compass is brought near the wire, the needle aligns itself tangent to lines of flux, with the north pole of the compass pointing in the direction of the flux. As the compass is moved around the wire, the needle turns to maintain the tangential relationship between the flux lines and itself. If the direction of current is reversed, the needle reverses its direction, but it remains always tangent to the flux.

The flux around a current-carrying conductor can also be demonstrated by the use of iron filings. The conductor is passed vertically through a horizontal piece of cardboard. A large current (60 to 80 A) is passed through the conductor, and iron filings are sprinkled on the cardboard. The filings align themselves along the lines of flux and clearly demonstrate the pattern of the magnetic field. Figure 6-5 illustrates this experiment.

The direction of flux around a conductor can be determined by a simple rule: *Grasp the conductor with the left hand so that the thumb points in the direction of the current. The fingers then point in the direction of the lines of flux.*

If wire is bent to form a single loop and current sent through it, the magnetic flux around it is additive in the center of the loop. Figure 6-6 shows the magnetic flux about such a loop of wire.

If a single conductor is coiled into many loops close together, like a helix, some of the flux from each turn of the helix adds to similar flux produced by each of the other turns. Flux links the entire coil and is especially concentrated along the axis of the helix. This field is much stronger because each turn of the coil is adjacent to another turn whose field has the same direction. The individual fields add up to create a much stronger field. A cross section of such a coil is shown in Fig. 6-7, and it can readily be seen how the magnetic flux of each turn adds to the flux of adjacent turns to develop a pronounced common magnetic field. The coil has magnetic

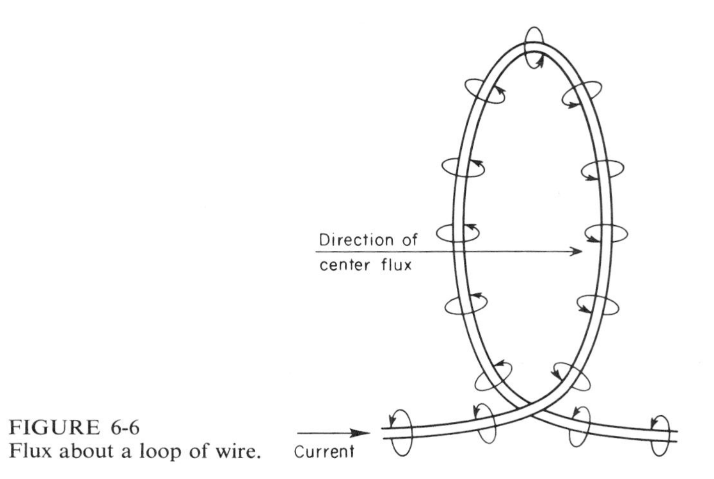

FIGURE 6-6
Flux about a loop of wire.

poles and acts as a bar magnet, but it exhibits these properties only as long as current flows in the coil.

The flux density within a coil can be greatly increased by winding it around a core of magnetic material, such as soft iron. Such a coil is an *electromagnet.* The explanation for the great increase in flux will appear later.

The polarity of the field set up by a coil can be found by use of the left-hand rule: *Grasp the coil so that the fingers point in the direction of the current; the thumb then points to the north pole of the coil.*

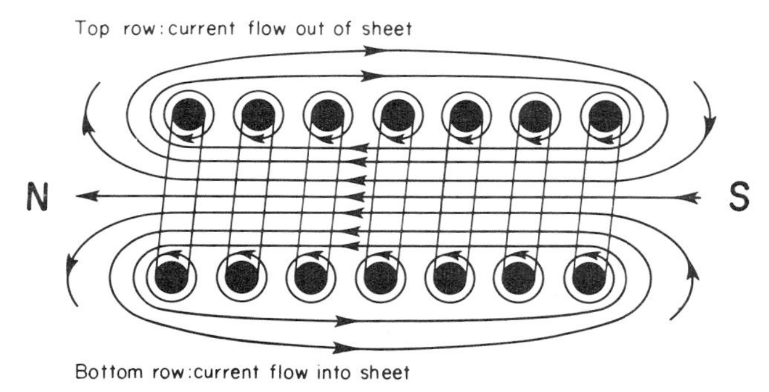

FIGURE 6-7
Cross section of a coil, showing the flux.

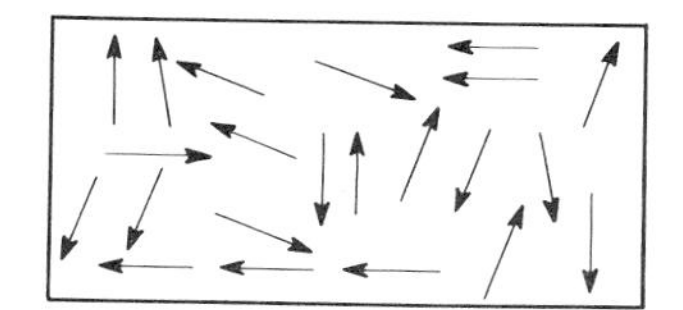

FIGURE 6-8
Random orientation in a nonmagnetic material.

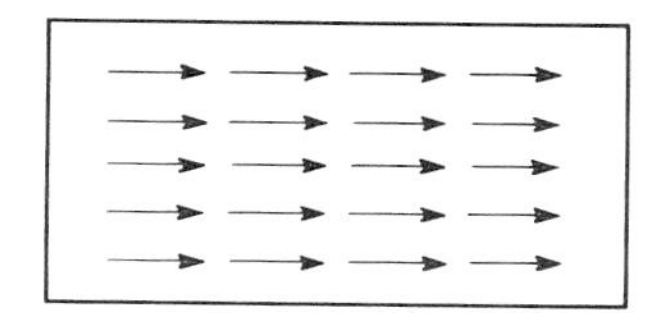

FIGURE 6-9
Orientation in a magnetic material.

6-3 THE MOLECULAR THEORY OF MAGNETISM

In Chap. 1, it was pointed out that all matter consists of molecules and atoms. The molecular theory of magnetism is based on the theory that all atoms and molecules have magnetic properties and that individual molecules behave like small bar magnets, referred to as *domains*. In magnetic materials, according to this theory, the molecular domains align themselves so that their magnetic properties add to one another, and *macroscopic* pieces of these materials exhibit magnetic properties. In nonmagnetic materials, the domains do not become oriented in the presence of a magnetic field. Nonmagnetic materials may be regular, orderly structures, such as carbon, quartz, and zinc, but the molecular theory of magnetism explains that they are nonmagnetic because their molecular domains are fixed in random relationships to one another. Figure 6-8 illustrates the random magnetic properties of a nonmagnetic material.

Magnetic materials may be *permanent* or *temporary*. In both types of material, the molecules are oriented by a magnetic field. In the temporary magnetic material, the molecules return to random orientation as soon as a magnetizing force is removed. Permanent magnetic materials are those which retain magnetic orientation for considerable periods of time after a magnetizing force has been removed. Figure 6-9 illustrates the molecular domain arrangement of a magnetic material under the influence of a magnetic field. Saturation occurs when all the domains become aligned.

Lodestone, as already mentioned, is a natural magnetic material. Other matter can be given magnetic properties if brought under the influence of a magnetic field. For example, if a nail is picked up by a magnet, the nail itself acts as a magnet for a short period of time.

What makes the nail behave like a magnet? When the magnet picks up the nail, the forces exerted by the magnet orient the molecules of the nail. When the nail is removed from the magnet, it acts as a magnet because its molecules are so arranged that their magnetic forces add to each other.

Why does the nail act as a magnet for only a short period of time? Again the answer is provided by the molecular theory of magnetism. When the magnetic field of the magnet is removed, the molecules of the nail gradually drift back to their original positions, and the nail loses its magnetism.

Magnetic materials are classified in accordance with the length of time they retain magnetic properties after a magnetizing force is removed. Matter that retains magnetic properties for considerable periods of time, often as much as 75 years, is referred to as *permanent magnetic material.* Other matter, such as soft iron, exhibits magnetic properties only when a magnetizing force is present; when a magnetizing force is removed, the matter no longer acts as a magnet. It is classed as a *temporary magnetic material.* Both types of material have important applications in electronics.

The ability to retain magnetic properties after a magnetizing force has been removed is called *retentivity.*

6-4 FLUX DENSITY

Flux density B is a measure of the flux passing through a cross-sectional area per unit of area. B is expressed in Eq. (6-1), where area A is the area at right angles to the direction of the flux.

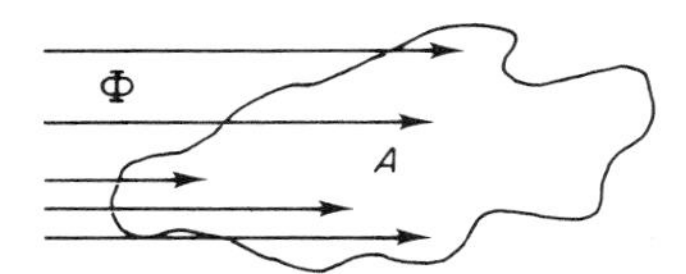

$$B = \frac{\Phi}{A} \tag{6-1}$$

The unit of flux is the weber (Wb), and area is in square meters. Thus, the unit of flux density is webers per square meter, which is also called the tesla (T).

EXAMPLE 6-1 Determine the flux density for 4×10^{-5} Wb of flux passing through $0.01\ \text{m}^2$.

SOLUTION

$$B = \frac{\Phi}{A}$$

$$= \frac{4 \times 10^{-5}}{0.01} = 4 \times 10^{-3}\ \text{Wb/m}^2 \qquad ////$$

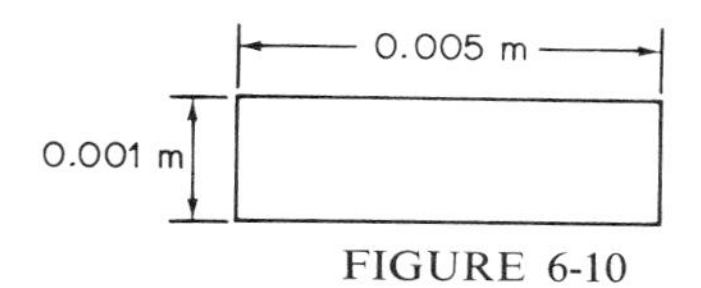

FIGURE 6-10

EXAMPLE 6-2 Find the amount of flux to provide a flux density of 3×10^{-2} Wb/m^2 in the cross-sectional area of Fig. 6-10.

SOLUTION

$$A = (5 \times 10^{-3})(1 \times 10^{-3}) = 5 \times 10^{-6} \text{ m}^2$$

$$\Phi = BA$$

$$= (3 \times 10^{-2})(5 \times 10^{-6}) = 1.5 \times 10^{-7} \text{ Wb}$$

////

The weber is a large unit of flux since it represents 10^8, or 100 million, lines of magnetic induction.

6-5 PERMEABILITY AND RELUCTANCE

Permeability μ is a measure of the ability of a substance to conduct magnetic flux. Permeability in magnetic circuits is analogous to conductivity in electric circuits. The permeability of free space, μ_0, is $4\pi \times 10^{-7}$. Materials with permeabilities slightly less than that of free space are *diamagnetic*, and those slightly greater are *paramagnetic*. Materials such as iron, steel, nickel, etc., have high permeabilities and are referred to as *ferromagnetic*.

Relative permeability μ_r is the permeability of a substance compared to that of free space (vacuum). Because relative permeability is a ratio, it is dimensionless and has the same magnitude in all systems of units. The relative permeability of free space, by definition, equals 1.

$$\mu_r = \frac{\mu}{\mu_0} \tag{6-2}$$

where μ is the permeability of some substance and μ_0 is the permeability of vacuum.

Air, wood, paper, and nonferrous materials have relative permeabilities approximately equal to the relative permeability of free space, and in all computations the relative permeability of these materials is considered to be 1. On the other hand, nickel, iron, cobalt, and various alloys of these metals have high permeabilities which vary from 100 to thousands of times the permeability of air. Table 6-1 indicates relative permeabilities for various materials.

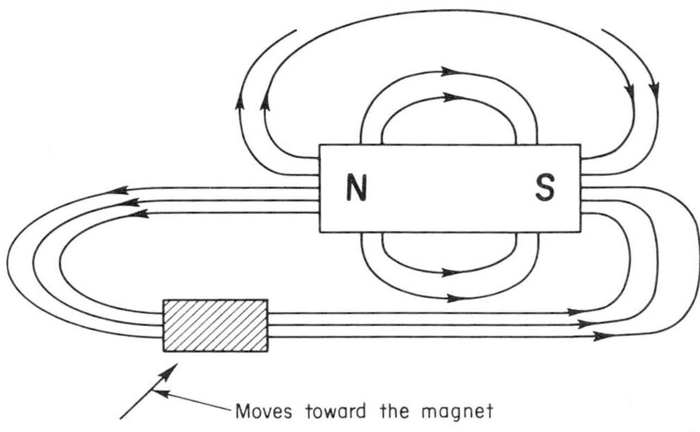

FIGURE 6-11
Magnetic attraction.

Permeability is not constant for a given magnetic substance but varies with the previous magnetic history and the degree of flux density of the material. In Sec. 6-2, magnetic lines of force were said to encounter opposition in traveling from the north to the south pole of a magnet. This opposition to lines of force is called *reluctance* $\mathcal{R}$.

Reluctance varies for different materials. The reluctance of air is several thousand times the reluctance of soft iron. This helps to explain why iron and other magnetic materials are attracted by magnets. In statement 3 about the properties of magnetic fields (Sec. 6-2), it was said that lines of force follow the path of least opposition. If a piece of iron is placed within the field of a magnet, the lines of force pass through the iron rather than through the air around it. Since lines of flux always try to shorten, the iron is attracted to the magnet. Figure 6-11 illustrates this behavior of flux.

Electric resistance varies with type of material and is directly proportional to the length of a conductor and inversely proportional to its cross-sectional area. The reluctance of magnetic materials varies in the same manner. Reluctance depends

Table 6-1 RELATIVE PERMEABILITIES

Classification	Material	μ_r
Diamagnetic	Bismuth	0.99
Paramagnetic	Platinum	1.00002
Ferromagnetic	Nickel	50
	Transformer iron	5,500
	Permalloy	30,000–80,000

upon the magnetic properties of the material and is inversely proportional to the cross-sectional area of the magnetic path and directly proportional to the length of the magnetic path. Reluctance is expressed as

$$\mathscr{R} = \frac{l}{\mu A} \tag{6-3}$$

where l = length of magnetic path, m
μ = permeability of material
A = cross-sectional area of magnetic path, m^2

EXAMPLE 6.3 Determine the permeabilities of bismuth, nickel, and transformer iron using Table 6-1.

SOLUTION

$$\mu = \mu_r \mu_0$$

Bismuth:

$$\mu = (0.99)(4\pi \times 10^{-7}) = 12.45 \times 10^{-7}$$

Nickel:

$$\mu = (50)(4\pi \times 10^{-7}) = 6.28 \times 10^{-5}$$

Transformer iron:

$$\mu = (5.5 \times 10^3)(4\pi \times 10^{-7}) = 6.9 \times 10^{-3}$$

////

EXAMPLE 6-4 Find the reluctance for each material in the previous example if the material is cylindrical with a radius of 0.001 m and a length of 0.02 m (see Fig. 6-12).

SOLUTION

$$\mathscr{R} = \frac{l}{\mu A}$$

$$A = \pi r^2 = \pi(1 \times 10^{-6}) = \pi \times 10^{-6} m^2$$

Bismuth:

$$\mathscr{R} = \frac{2 \times 10^{-2}}{(1.245 \times 10^{-6})(\pi \times 10^{-6})} = 5.12 \times 10^9$$

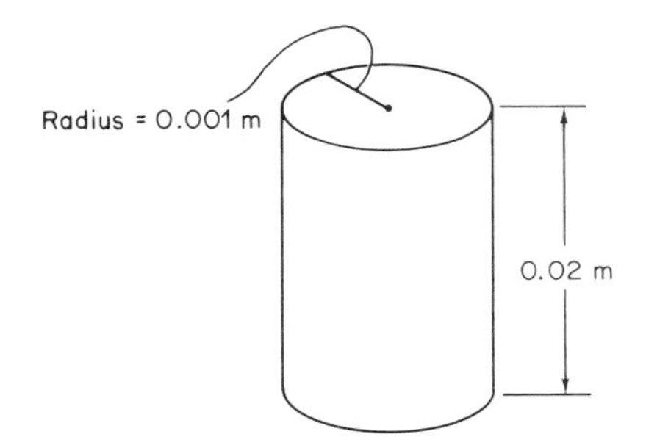

FIGURE 6-12
Circuit for Example 6-4.

Nickel:
$$\mathscr{R} = \frac{2 \times 10^{-2}}{(6.28 \times 10^{-5})(\pi \times 10^{-6})} = 1.01 \times 10^{8}$$

Transformer iron:
$$\mathscr{R} = \frac{2 \times 10^{-2}}{(6.9 \times 10^{-3})(\pi \times 10^{-6})} = 9.23 \times 10^{5}$$
////

Materials with larger permeabilities have smaller reluctances.

6-6 OHM'S LAW FOR THE MAGNETIC CIRCUIT

Magnetomotive force is that force which causes magnetic flux. It is analogous to electromotive force in electric circuits.

The only practical source of magnetomotive force (mmf) is electromagnetism. Lodestone is a natural source of mmf, but almost all modern applications of magnetism depend on magnetic fields that are established electrically.

Magnetomotive force is the product of the current flowing through a coil and the number of turns of the coil:

$$F = NI \qquad \text{AT} \tag{6-4}$$

where AT is the abbreviation for ampere-turns. A magnetic circuit is defined as any region where closed lines of magnetic flux exist.

Magnetic circuits may be compared with electric circuits, but one important exception must be noted. In electric circuits there are both insulators and conductors. There is a wide range of resistivities, with the resistance of some insulators approaching infinity. Therefore, switches can be inserted in electric circuits, and current can

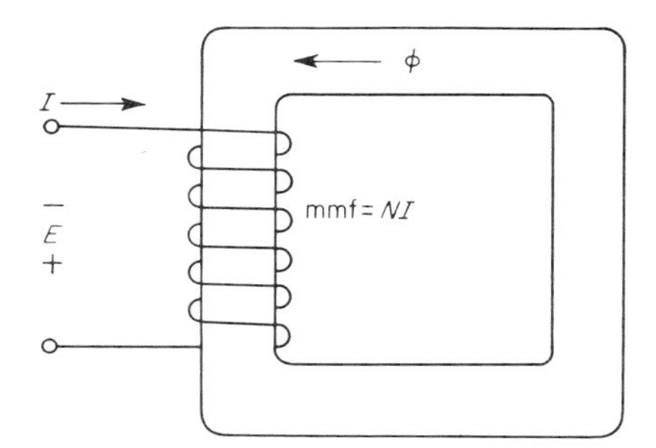

FIGURE 6-13
Basic magnetic circuit.

be turned on and off. *There are no magnetic insulators.* If a source of magnetomotive force is present, it will set up magnetic flux. As long as there is magnetomotive force, there is no way to "switch" flux off or on. The permeability of the best magnetic materials is only 10^4 times the permeability of air, whereas copper may have 10^{18} times the conductivity of an insulator. Taking this exception into account, comparison of a magnetic circuit with an electric circuit is quite valid. Figure 6-13 shows a simple magnetic circuit. In a magnetic circuit, flux is analogous to the current, reluctance is analogous to the resistance, and magnetomotive force is analogous to the emf of an electric circuit.

The flux in a magnetic circuit is directly proportional to the applied magnetomotive force and inversely proportional to the reluctance of the magnetic path. This statement is identical in form to Ohm's law for electric circuits and is therefore referred to as "Ohm's law for magnetic circuits."

$$\Phi = \frac{F}{\mathscr{R}} \text{ Wb} \tag{6-5}$$

Component parameters may be substituted for F and $\mathscr{R}$ in Eq. (6-5). The result is

$$\Phi = \frac{\mu A N I}{l} \text{ Wb} \tag{6-6}$$

EXAMPLE 6-5 Given the magnetic circuit of Fig. 6-14, calculate the reluctance, flux, and flux density. The average length of the path is 1×10^{-2} m, and the cross-sectional area of the path is 1×10^{-5} m^2. The permeability of the material is 0.22.

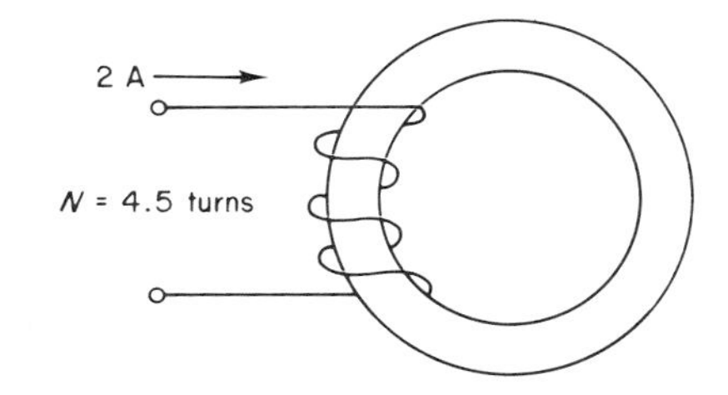

FIGURE 6-14
Circuit for Example 6-5.

SOLUTION

$$\mathscr{R} = \frac{l}{\mu A} = \frac{1 \times 10^{-2}}{(0.22)(1 \times 10^{-5})}$$

$$\mathscr{R} = 4.55 \times 10^{3}$$

$$F = NI = (4.5)(2) = 9 \text{ AT}$$

$$\Phi = \frac{F}{\mathscr{R}} = \frac{9}{4.55 \times 10^{3}}$$

$$= 1.98 \times 10^{-3} \text{ Wb}$$

$$B = \frac{\Phi}{A} = \frac{1.98 \times 10^{-3}}{1 \times 10^{-5}}$$

$$= 1.98 \times 10^{2} \text{ Wb/m}^2$$

////

EXAMPLE 6-6 Rework the previous example assuming that the number of turns is doubled.

SOLUTION

$$\mathscr{R} = 4.55 \times 10^{3}$$

$$F = NI = (9)(2) = 18 \text{ AT}$$

$$\Phi = \frac{F}{\mathscr{R}} = \frac{18}{4.55 \times 10^{3}}$$

$$= 3.96 \times 10^{-3} \text{ Wb}$$

$$B = \frac{\Phi}{A} = 3.96 \times 10^{2} \text{ Wb/m}^2$$

////

6-7 MAGNETIC FIELD INTENSITY

Magnetic field intensity is a measure of the strength of a magnetic field. The magnetic drop is found when the intensity is multiplied by the length of the magnetic path in the circuit. Magnetic drop is completely analogous to voltage drop in electric circuits. In computing magnetic circuits, the reader will find that magnetic field intensity can be used as easily as voltage drop is used in electric circuits.

Magnetic field intensity is the magnetomotive force per unit of length, measured along the path of magnetic flux. Magnetic field intensity is frequently referred to as magnetizing force, or *magnetic gradient*, and is represented by the symbol H.

$$H = \frac{NI}{l} \text{ AT/m} \tag{6-7}$$

Field intensity and field density are related through the permeability of the material.

$$H = \frac{NI}{l} = \frac{F}{l} = \frac{\Phi \mathcal{R}}{l} = \frac{\Phi l/\mu A}{l}$$

$$= \frac{\Phi}{\mu A} \quad \text{and} \quad B = \frac{\Phi}{A}$$

Therefore,

$$H = \frac{B}{\mu} \tag{6-8}$$

EXAMPLE 6-7 A current of 1 A is passed through a coil consisting of 200 turns. The coil is a toroid and has an average length of 6 cm. Calculate the magnetomotive force, magnetic field intensity, and flux density; $\mu = 0.02$.

SOLUTION

$$F = NI = 200 \text{ AT}$$

$$H = \frac{F}{l} = \frac{200}{6 \times 10^{-2}} = 33.3 \times 10^2 \text{ AT/m}$$

$$B = \mu H = (2 \times 10^{-2})(33.3 \times 10^2)$$

$$= 66.6 \text{ Wb/m}^2$$

////

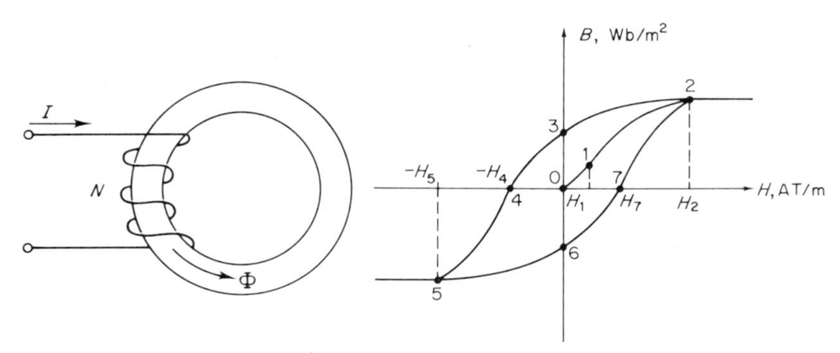

FIGURE 6-15

6-8 HYSTERESIS AND MAGNETIZATION CURVES

The relationship between the magnetic field intensity and flux density is not linear in ferromagnetic materials. This implies that B will not double if H is doubled. Though Eq. (6-8) has the appearance of a linear equation between B and H, it is not, since the permeability is not constant for various values of H.

The test circuit and curve in Fig. 6-15 show the relationship between B and H. With the core initially unmagnetized, the magnetic field intensity is slightly increased from zero to H_1. This is accomplished by an increase in current I. The curve follows the path from 0 to 1. An additional increase in H to H_2 places the core in saturation. Basically, this means that all domains in the material are aligned and that further increase in H will produce no further increase in flux density B. Decreasing H from H_2 to zero results in the path between points 2 and 3. Notice that at point 3 the material maintains a flux density without an applied magnetic field intensity. This value of B is referred to as the *residual flux density*. Residual flux density is the result of many domains remaining in alignment after the field intensity is removed. It is because of residual flux density that permanent magnets exist.

A negative field intensity must be applied from point 3 to point 4 to cause the flux density to return to zero. The value of field intensity H_4 is called the *coercive force*. Decreasing field intensity to H_5 causes the core to saturate in a direction opposite to the first saturation. Returning H to zero results in a residual flux density at point 6. The magnitude is the same at 6 as it was at 3, but the flux direction is reversed. Then increasing H from zero to H_7 produces a flux density of zero.

After the initial magnetization from 0 to 2, the path at points 234567 will be followed for applied field intensities from H_2 to $-H_5$. The curve described by the path 234567 is called a *hysteresis curve*. The word hysteresis is derived from a Greek word meaning "lag behind." As the test circuit indicates, B lags behind changes in H.

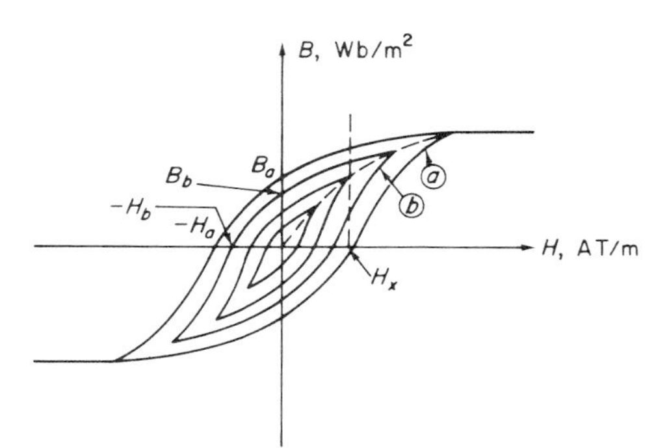

FIGURE 6-16
Set of hysteresis curves.

The size of the hysteresis curve for a particular material is dependent upon the peak values of applied field intensity. For example, Fig. 6-16 shows a set of hysteresis curves for some ferromagnetic material. Curve a is obtained by saturating the core with the required field intensity. The inside curves result from smaller peak field intensities. None of the inside curves reached saturation. As mentioned previously, B_a and B_b are residual flux densities for two of the curves a and b. The residual flux density caused by core saturation is called the *retentivity* of a magnetic material. Thus B_a is the retentivity. H_a and H_b are the coercive forces for curves a and b. The coercive force corresponding to core saturation is the *coercivity* of a magnetic material. Thus H_a is the coercivity.

The dashed line connecting the tips of the hysteresis curves in Fig. 6-16 is called the *magnetization curve*. It is used to correlate a value of H to a value for B. For example, an applied H_x along curve b would result in two values for flux density. In order to have a one-to-one correspondence between B and H, the dashed curve is used to obtain a compromise between the two points on the curve. Figure 6-17 shows a set of magnetization curves for various materials.

EXAMPLE 6-8 A 100-turn coil is wound on a core of cast iron. The length of the core is 0.05 m. Determine the required input current to produce a flux density of 0.6 Wb/m^2.

SOLUTION Figure 6-17 indicates that a magnetic field intensity of 1,900 AT/m is needed.

$$H = 1{,}900 = \frac{NI}{l}$$

$$I = \frac{1{,}900l}{N} = \frac{(1.9 \times 10^3)(5 \times 10^{-2})}{100}$$

$$= 0.95 \text{ A}$$

////

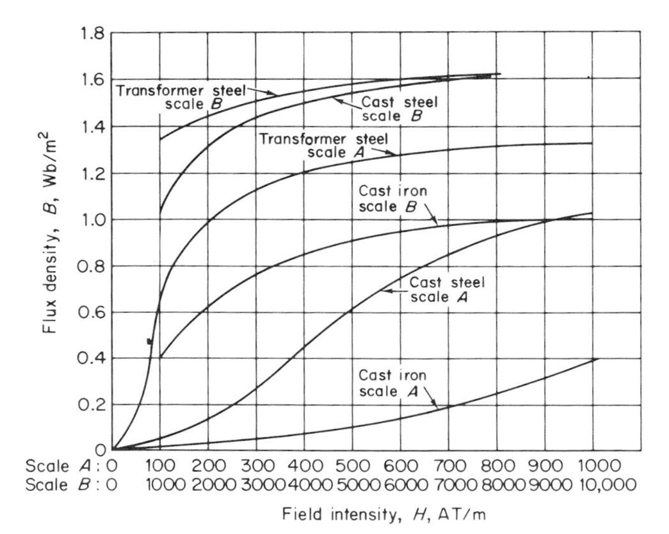

FIGURE 6-17
Magnetization curves.

6-9 THE SERIES MAGNETIC CIRCUIT

The series magnetic circuit is similar to the series electric circuit. There is only one flux path through a series magnetic circuit, just as there is only one current path through a series electric circuit. The rules of series magnetic circuits resemble the rules of series electric circuits.

A law for magnetic circuits that is similar to Kirchhoff's voltage law is Ampère's circuital law.

Ampère's Circuital Law The algebraic sum of the mmf sources producing the assumed direction of flux around any closed loop is equal to the sum of the mmf drops.

The mmf drop is found by taking the product of the magnetic field intensity and the length of the magnetic path:

$$F = Hl \tag{6-9}$$

A series magnetic circuit may take several forms. The circuit could consist of one material or several materials, or it could contain an air gap.

Figure 6-18 shows a magnetic circuit containing two mmf sources and three types of magnetic material. Φ is the assumed direction of flux. Everywhere the flux

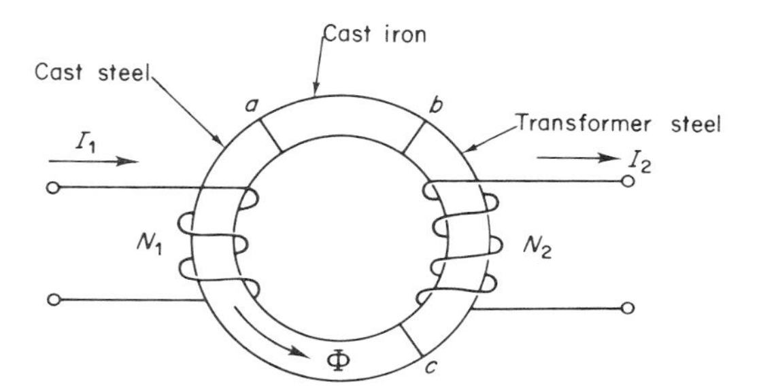

FIGURE 6-18

is the same in the series circuit. The mmf source due to I_1 would produce the assumed direction of flux. Algebraically, source 1 is considered positive. Source 2 is negative since it would produce a flux direction opposite to the assumed direction. If the desired flux density is known, the values of field intensity H_{ab}, H_{bc}, and H_{ac} can be found from the magnetization curves in Fig. 6-17. Ampère's circuital-law equation is

$$F_1 - F_2 = H_{ab} l_{ab} + H_{bc} l_{bc} + H_{ac} l_{ac}$$
$$N_1 I_1 - N_2 I_2 = H_{ab} l_{ab} + H_{bc} l_{bc} + H_{ac} l_{ac}$$

The terms l_{ab}, l_{bc}, and l_{ac} represent the average lengths of the magnetic paths.

EXAMPLE 6-9 A toroid of cast steel is to have a flux density of 0.4 Wb/m^2. The average diameter is 0.1 m. A coil of 1,000 turns is wound around the toroid. Determine the required input current and the value of μ for the material.

SOLUTION The magnetic path length is equivalent to the average circumference.

$$l = \pi d = \pi(0.1) = 0.314 \text{ m}$$

Figure 6-17 yields the field intensity for $B = 0.4 \text{ Wb/m}^2$

$$H = 370 \text{ AT/m}$$

Writing Ampère's circuital-law equation, we have

$$NI = Hl$$
$$I = \frac{Hl}{N} = \frac{(370)(0.314)}{1{,}000}$$
$$= 0.116 \text{ A}$$

Permeability is given by

$$\mu = \frac{B}{H} = \frac{0.4}{0.37 \times 10^3} = 1.08 \times 10^{-3}$$

The relative permeability is

$$\mu_r = \frac{\mu}{\mu_0} = \frac{1.08 \times 10^{-3}}{4\pi \times 10^{-7}}$$

$$= 860 \qquad ////$$

EXAMPLE 6-10 Rework the previous example for a flux density of 1.4 Wb/m^2.

SOLUTION

$$l = 0.314 \text{ m}$$

From Fig. 6-17, $H = 2{,}600$ AT/m.

$$NI = Hl$$

$$I = \frac{Hl}{N} = \frac{(2.6 \times 10^3)(0.314)}{1 \times 10^3}$$

$$= 0.816 \text{ A}$$

$$\mu = \frac{B}{H} = \frac{1.4}{2.6 \times 10^3} = 5.38 \times 10^{-4}$$

$$\mu_r = \frac{\mu}{\mu_0} = \frac{5.38 \times 10^{-4}}{4\pi \times 10^{-7}} = 428 \qquad ////$$

As can be seen from the two examples, μ was not constant.

EXAMPLE 6-11 For the magnetic circuit shown in Fig. 6-19, find the required current to produce a flux of 1×10^{-4} Wb in the direction indicated.

SOLUTION

$$B = \frac{\Phi}{A} = \frac{1 \times 10^{-4}}{1 \times 10^{-4}} = 1 \text{ Wb/m}^2$$

From Fig. 6-17:

Transformer steel: $H_{ab} = 210$ AT/m

Cast steel: $H_{bcda} = 920$ AT/m

$$l_{ab} = 0.1 \text{ m}$$

$$l_{bcda} = 0.2 \text{ m}$$

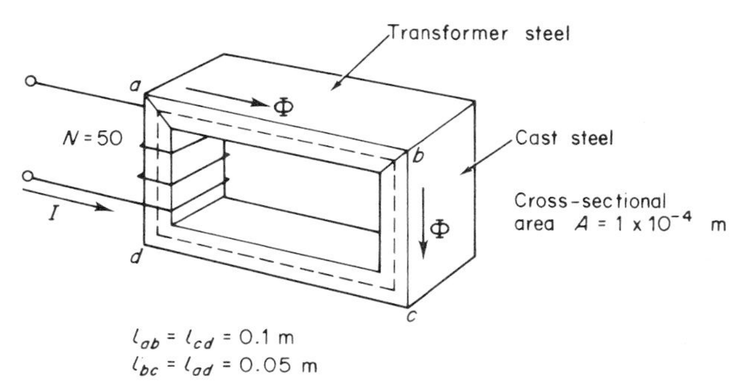

FIGURE 6-19
Circuit for Example 6-11.

Applying Ampère's circuital law gives

$$-NI = H_{ab}l_{ab} + H_{bcda}l_{bcda}$$

$$I = -\frac{(210)(0.1) + (920)(0.2)}{50}$$

$$= -\tfrac{205}{50}$$

$$= -4.1 \text{ A}$$

////

The negative sign indicates that the current must be reversed to provide the required direction of flux.

When an air gap occurs in the magnetic circuit, the lines of flux tend to spread outside the area of the gap. This effect is called *fringing* (see Fig. 6-20). Fringing will be neglected, and the flux is assumed to be contained in the area of the air gap (see Fig. 6-21). Since the permeability of air is nearly equal to that of free space, the permeability in the air gap is taken as μ_0.

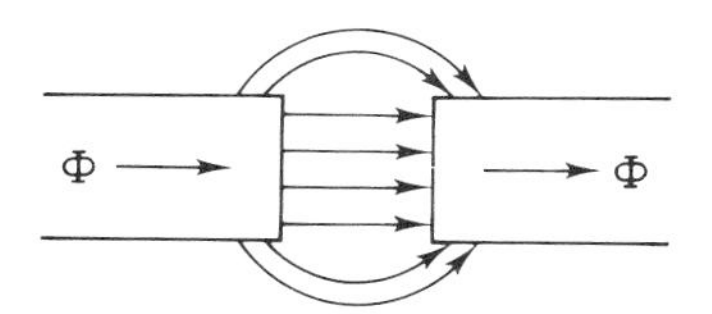

FIGURE 6-20
Fringing.

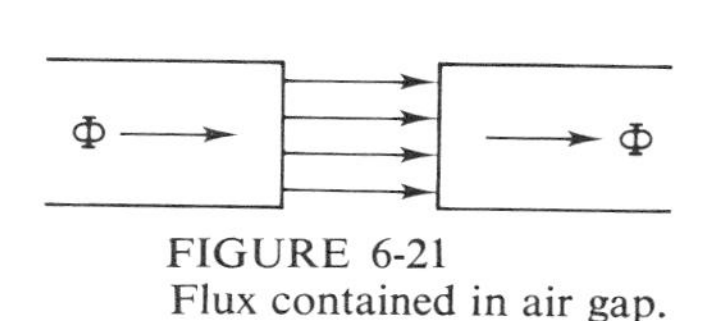

FIGURE 6-21
Flux contained in air gap.

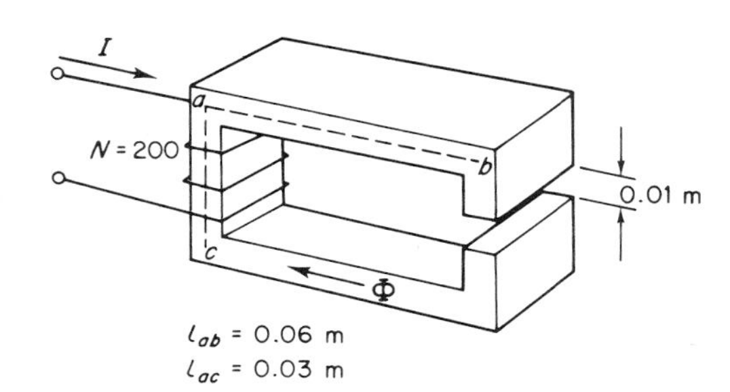

FIGURE 6-22
Circuit for Example 6-12.

EXAMPLE 6-12 Determine the current for a flux density of 0.3 Wb/m^2. The magnetic material is cast iron (see Fig. 6-22).

SOLUTION From Fig. 6-17,

$$H = 880 \text{ AT/m}$$

For the air gap

$$H = \frac{B}{\mu_0} = \frac{3 \times 10^{-1}}{4\pi \times 10^{-7}} = 2.38 \times 10^5 \text{ AT/m}$$

Applying Ampère's circuital law gives

$$NI = (8.8 \times 10^2)(0.17) + (2.38 \times 10^5)(1 \times 10^{-2})$$

$$= 150 + 2{,}380 = 2{,}430 \text{ AT}$$

$$I = \frac{2{,}430}{200} = 12.15 \text{ A}$$

////

EXAMPLE 6-13 Calculate the number of turns in coil 2 for a flux density of 0.6 Wb/m^2. The magnetic circuit is shown in Fig. 6-23.

SOLUTION Referring to the magnetization curves in Fig. 6-17,

Transformer steel: $H_{ab} = 95$ AT/m

Cast steel: $H_{ac} = 490$ AT/m

Cast iron: $H_{de} = 1{,}900$ AT/m

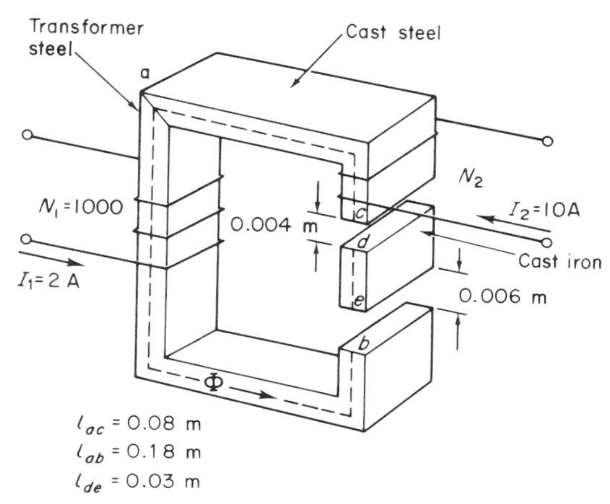

FIGURE 6-23
Circuit for Example 6-13.

For air gaps:

$$H_{cd} = H_{eb} = \frac{B}{\mu_0} = \frac{6 \times 10^{-1}}{4\pi \times 10^{-7}} = 4.78 \times 10^5 \text{ AT/m}$$

Ampère's circuital law gives

$$\begin{aligned} N_1 I_1 + N_2 I_2 &= H_{ab} l_{ab} + H_{ac} l_{ac} + H_{de} l_{de} + H_{cd} l_{cd} + H_{eb} l_{eb} \\ &= (95)(0.18) + (490)(0.08) + (1{,}900)(0.03) + (4.78 \times 10^5)(4 \times 10^{-3}) \\ &\quad + (4.78 \times 10^5)(6 \times 10^{-3}) \\ &= 17.1 + 39.2 + 57 + 1{,}910 + 2{,}870 \\ &= 4{,}893.3 \\ N_2 &= \frac{4{,}893.3 - 2000}{10} \\ &= 289.33 \text{ turns} \end{aligned}$$

////

Notice in the last two examples that the air gap produced by far the largest mmf drop. This is simply because air is nonmagnetic and offers a large opposition to flux flow.

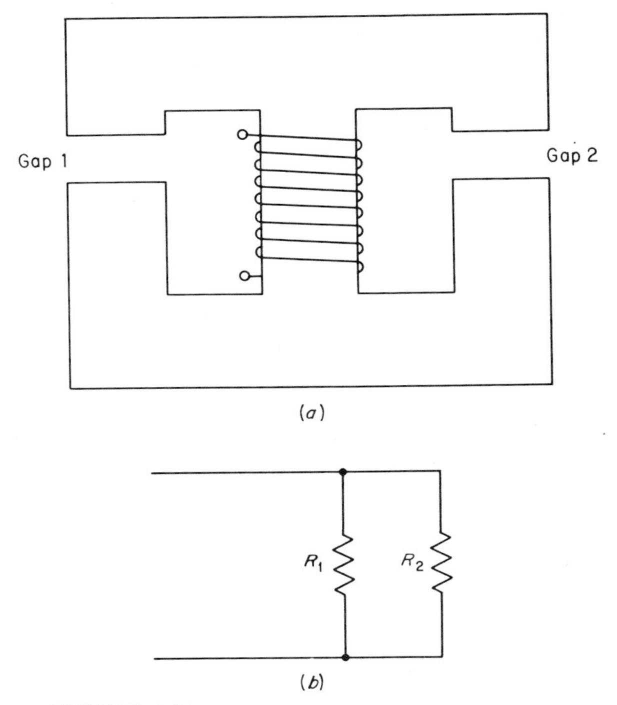

FIGURE 6-24
(*a*) Parallel magnetic circuit. (*b*) Electrical analog of parallel magnetic circuit.

6-10 THE PARALLEL MAGNETIC CIRCUIT

Parallel magnetic circuits are handled in a manner similar to parallel electric circuits. Just as the total current in a parallel electric circuit is the sum of the branch currents, the total flux of a parallel magnetic circuit is the sum of the fluxes of the branches of the circuit. Similarly, the magnetomotive forces across all branches are alike, just as the voltages across all branches of a parallel circuit are alike.

A parallel magnetic circuit is shown in Fig. 6-24. The air gaps are analogous to resistances in an electric circuit, while the magnetic materials may be considered analogous to the connecting wires of a parallel electric circuit.

It is important that the reader understand that *there is no truly parallel magnetic circuit*. The parallel magnetic circuit is an approximation; it is based on the assumption that the reluctance of air gaps is so much greater than the reluctance of the magnetic materials in the circuit that the total reluctance of each branch, for all practical purposes, equals the reluctance of the gap itself.

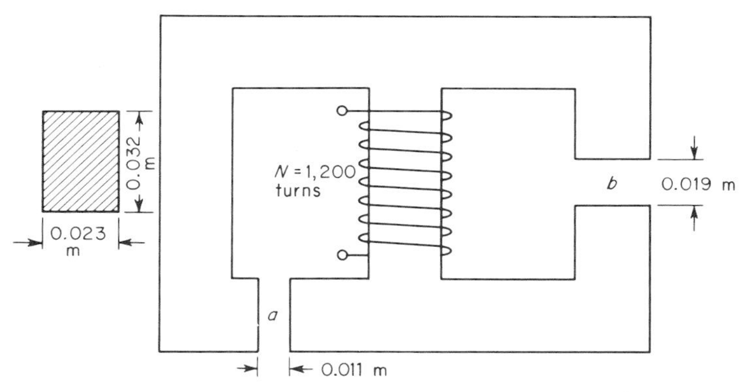

FIGURE 6-25
Circuit for Example 6-14.

EXAMPLE 6-14 Given the magnetic circuit of Fig. 6-25, calculate (1) the flux density within each gap when the current is 3.41 A and (2) the current necessary to create a flux density of 1.6 Wb/m^2 in air gap *b*. When this current flows, what is the flux across gap *a*?

SOLUTION Knowing the type of magnetic material is not essential with an assumption that the air-gap reluctance is much larger than the reluctance of the material. The applied mmf source is in parallel with gaps *a* and *b*.

1

$$F = NI = (1.2 \times 10^3)(3.41)$$
$$= 4.08 \times 10^3 \text{ AT}$$
$$NI = H_b l_b = H_a l_a$$
$$H_b = \frac{NI}{l_b} = \frac{4.08 \times 10^3}{1.9 \times 10^{-2}} = 2.14 \times 10^5 \text{ AT/m}$$
$$H_a = \frac{NI}{l_a} = \frac{4.08 \times 10^3}{1.1 \times 10^{-2}} = 3.71 \times 10^5 \text{ AT/m}$$
$$B_b = \mu_0 H_b = (4\pi \times 10^{-7})(2.14 \times 10^5)$$
$$= 0.269 \text{ Wb/m}^2$$
$$B_a = \mu_0 H_a = (4\pi \times 10^{-7})(3.71 \times 10^5)$$
$$= 0.466 \text{ Wb/m}^2$$

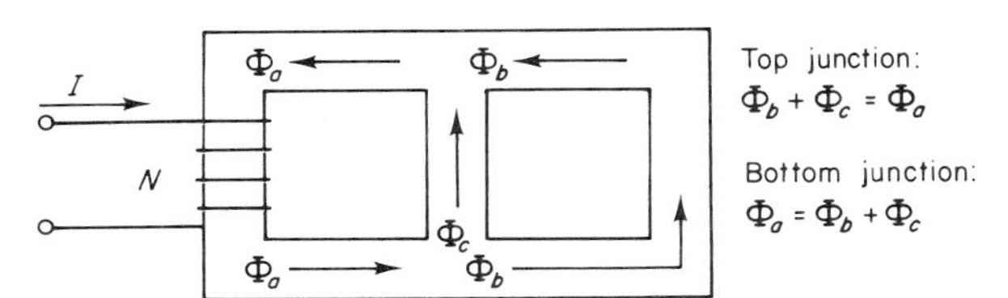

FIGURE 6-26
A series-parallel circuit.

2 When we know that the flux density at air gap b is $B_b = 1.6$ Wb/m², we can find the magnetic field intensity.

$$H_b = \frac{B_b}{\mu_0} = \frac{1.6}{4\pi \times 10^{-7}}$$
$$= 1.27 \times 10^6 \text{ AT/m}$$
$$NI = H_b l_b$$
$$I = \frac{H_b l_b}{N} = \frac{(1.27 \times 10^6)(1.9 \times 10^{-2})}{1.2 \times 10^3}$$
$$= 20.1 \text{ A}$$

The field intensity H_a is obtained with

$$NI = H_a l_a$$
$$H_a = \frac{NI}{l_a} = \frac{2.72 \times 10^4}{1.1 \times 10^{-2}}$$
$$= 2.2 \times 10^6 \text{ AT/m}$$
$$B_a = \mu_0 H_a = (4\pi \times 10^{-7})(2.2 \times 10^6)$$
$$= 2.76 \text{ Wb/m}^2$$
$$\Phi_a = B_a A = (2.76)(2.3 \times 10^{-2})(3.2 \times 10^{-2})$$
$$= 2.04 \times 10^{-3} \text{ Wb}$$

////

6-11 THE SERIES-PARALLEL MAGNETIC CIRCUIT

In the previous section the parallel magnetic circuit was discussed. Such circuits are unlikely to be encountered in actual practice. In such devices as transformers, motors, and generators, the magnetic circuit is usually a series-parallel circuit. The rules for flux and mmf in the series-parallel magnetic circuit are the same as the rules for current and voltage in the series-parallel electric circuit. Figure 6-26 shows the

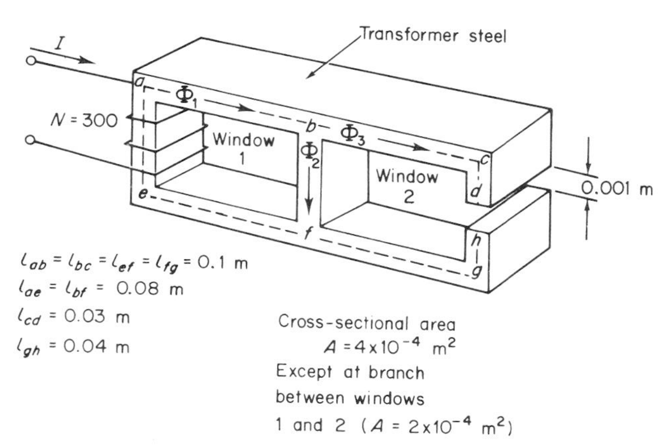

FIGURE 6-27
Circuit for Example 6-15.

flux distribution in a series-parallel magnetic circuit. The flux conditions at a junction are similar to Kirchhoff's current law. The sum of the fluxes into a junction is equal to the sum of the fluxes out of a junction.

EXAMPLE 6-15 Calculate the coil current required to establish a flux of 3×10^{-4} Wb in the air gap (see Fig. 6-27).

SOLUTION

1 Investigate the sections containing flux Φ_3.

Flux density at air gap: $$B_{dh} = \frac{\Phi_3}{A_{dh}} = \frac{3 \times 10^{-4}}{4 \times 10^{-4}} = 0.75 \text{ Wb/m}^2$$

Thus $$B_{dh} = B_{bc} = B_{cd} = B_{fg} = B_{gh} = 0.75 \text{ Wb/m}^2$$

Field intensity at gap: $$H_{dh} = \frac{B_{dh}}{\mu_0} = \frac{7.5 \times 10^{-1}}{4\pi \times 10^{-7}} = 5.97 \times 10^5 \text{ AT/m}$$

The field intensity from Fig. 6-17, for $B = 0.75$ Wb/m^2, is

$$H_{bc} = H_{cd} = H_{fg} = H_{gh} = 115 \text{ AT/m}$$

mmf drop at gap: $$H_{dh} l_{dh} = (5.97 \times 10^5)(1 \times 10^{-3}) = 597 \text{ AT}$$

mmf drop at sections: $H_{bc}l_{bc} = H_{fg}l_{fg} = (115)(0.1) = 11.5 \text{ AT}$

$$H_{cd}l_{cd} = (115)(0.03) = 3.45 \text{ AT}$$

$$H_{gh}l_{gh} = (115)(0.04) = 4.6 \text{ AT}$$

2 Investigate the section containing flux Φ_2. The mmf drop across this section is not known since Φ_2 is not known. However, Ampère's circuital law can be applied to window 2 to find this mmf drop. Notice that no mmf sources are present in the paths around window 2. Applying Ampère's circuital law gives

$$0 = H_{bc}l_{bc} + H_{cd}l_{cd} + H_{dh}l_{dh} + H_{fg}l_{fg} + H_{gh}l_{gh} - H_{bf}l_{bf}$$

The term $H_{bf}l_{bf}$ is negative since the direction of Φ_2 is opposite Φ_3 around the window.

$$0 = 628 - H_{bf}l_{bf}$$

Therefore $H_{bf}l_{bf} = 628 \text{ AT}$

$$H_{bf} = \frac{628}{8 \times 10^{-2}} = 7{,}850 \text{ AT/m}$$

From Fig. 6-17,

$$B_{bf} = 1.62 \text{ Wb/m}^2$$

$$\Phi_2 = B_{bf}A_2 = (1.62)(2 \times 10^{-4})$$

$$= 3.24 \times 10^{-4} \text{ Wb}$$

3 Investigate the sections containing flux Φ_1.

$$\Phi_1 = \Phi_2 + \Phi_3$$

$$= 6.24 \times 10^{-4} \text{ Wb}$$

The flux density along paths *ab*, *ae*, and *ef* is

$$B_{ab} = B_{ae} = B_{ef} = \frac{\Phi_1}{A_1} = \frac{6.24 \times 10^{-4}}{4 \times 10^{-4}} = 1.56 \text{ Wb/m}^2$$

The field intensity from Fig. 6-17 is

$$H_{ab} = H_{ae} = H_{ef} = 4.3 \times 10^3 \text{ AT/m}$$

The mmf drop along paths *ab*, *ae*, and *ef* is

$$H_{ab}l_{ab} = (4.3 \times 10^3)(0.1) = 430 \text{ AT}$$

$$H_{ae}l_{ae} = (4.3 \times 10^3)(0.08) = 344 \text{ AT}$$

$$H_{ef}l_{ef} = (4.3 \times 10^3)(0.1) = 430 \text{ AT}$$

Applying Ampère's circuital law gives

$$NI = H_{ab}l_{ab} + H_{bf}l_{bf} + H_{ae}l_{ae} + H_{ef}l_{ef}$$

$$300I = 1{,}832$$

$$I = 6.11 \text{ A}$$

////

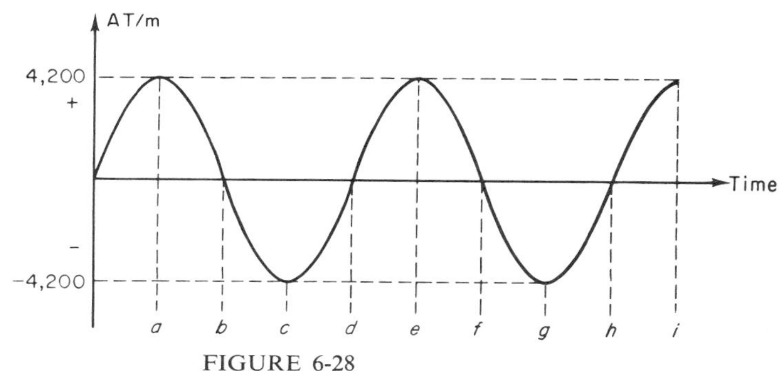

FIGURE 6-28
Magnetizing force when alternating current is applied to a solenoid.

6-12 POWER LOSSES IN MAGNETIC MATERIALS

Magnetic materials energized by alternating currents develop two power losses that are not present with direct currents. These losses arise from *hysteresis* and *eddy currents.*

Hysteresis losses occur because all magnetic materials have some degree of residual flux density. That is, all materials retain some magnetization after all magnetizing force has been removed. When an attempt is made to reverse the polarization of the magnetic material by reversing the direction of current through a coil, electric energy is required to overcome the residual magnetism of the material. For example, consider an alternating current with a peak amplitude of 3.5 A applied to a 1,200-turn solenoid that has a core of magnetic material. Since magnetomotive force is a function of current, the magnetomotive forces set up by the coil vary in both amplitude and direction as the current changes. Figure 6-28 is a graph of ampere-turns per meter over several cycles of alternating current.

If it is assumed that the magnetic core has had no previous magnetic experience, a *B-H* curve can be drawn for the core over several cycles of ac input to the coil. From zero to time interval *a*, the magnetization follows the virginal *B-H* curve for the core material. From *a* to *b*, the magnetizing force is decreasing to zero, but the residual flux density of the core material prevents the flux density from becoming zero at the same instant that the magnetizing force reaches zero. To make the flux density of the core material zero, it is necessary to reverse the magnetizing force. The amount of magnetizing force necessary to remove the residual magnetism of the core material is the *coercive force.* Some portion of the changing magnetizing force from *b* to *c* must be used to overcome the residual magnetism of the core. When the magnetomotive force reaches *c*, the flux density of the core is maximum. From *c* to *d* the magnetizing force is again decreasing to zero, and once again a residual flux is present in the core when the magnetizing force reaches zero. The residual flux must be overcome by magnetizing force in a positive direction, and this is accomplished by a

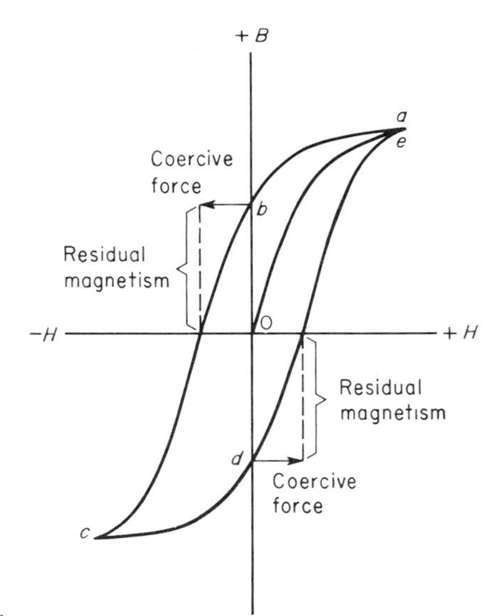

FIGURE 6-29
A typical hysteresis loop.

portion of the magnetizing force from *d* to *e*. At *e* the flux density is again maximum. From this point on, all future cycles of input will follow the curve traced by points *a* to *b*, *b* to *c*, *c* to *d*, and *d* to *e*. Figure 6-29 illustrates the hysteresis curve that the core followed as a result of ac input to the coil. The area within the loop represents electric energy per cycle used in overcoming the residual magnetism of the core material.

Hysteresis loss is encountered in transformers, motors, and generators. In magnetic computing elements, hysteresis is put to useful purposes. Here a core of magnetic material must have a high retentivity to store information but must be capable of switching polarity when a current pulse is passed through a coil wound upon the core. Hysteresis is valuable here because a large current is needed to switch the core and the core is insensitive to stray magnetic forces (see Fig. 6-30).

Eddy currents circulate within a magnetic material when the magnetic field is varying, as in the case of operation with alternating currents. These circulating currents cause I^2R losses that reduce the efficiency of the device. The name *eddy currents* is derived from the manner in which these currents flow, much like eddies in a pool of water.

6-13 APPLICATIONS OF MAGNETIC MATERIALS

It was pointed out in Sec. 6-3 that there are two types of magnetic materials, temporary and permanent. Both types find wide applications in electronics.

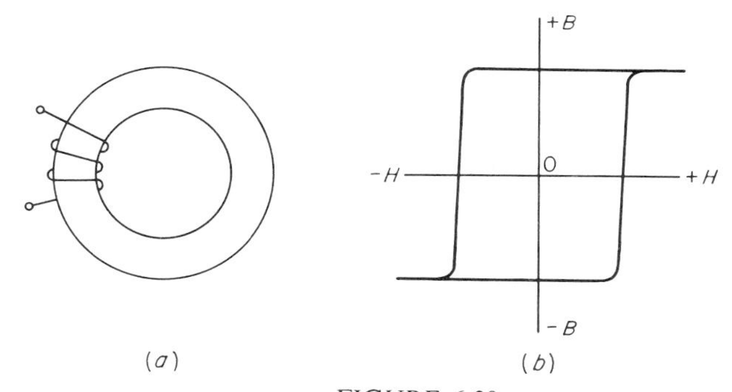

FIGURE 6-30
A magnetic computer element and its hysteresis loop.

Temporary magnetic materials are used wherever magnetizing force is changing. Typical uses of temporary magnetic material are for pole pieces in motors and generators and for cores in choke coils, saturable reactors, and transformers. An interesting application of temporary magnetic material is in shielding circuit components from magnetic fields. The component to be shielded is surrounded by a shell of magnetic material such as soft iron. Since magnetic lines of flux follow the path of least reluctance, external magnetic fields remain in the material of the shield and do not affect the component inside the shield. Shielding also prevents devices like transformers from spreading magnetic fields that could disturb other circuits. Figure 6-31 shows the use of temporary magnetic material as a shield. The ideal temporary magnetic material would have a very narrow hysteresis loop.

Permanent magnets are used wherever a magnetic field of constant value is needed. Examples are loudspeakers, earphones, some types of microphones, magnetron oscillators in radar, and magnetic recording devices.

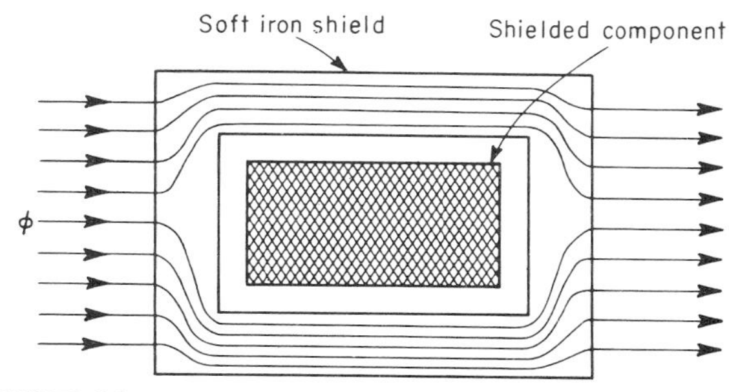

FIGURE 6-31
Permeability shielding.

6-14 SYSTEMS OF MAGNETIC UNITS

In this chapter, we have used one system of magnetic units, the rationalized mks system. The rationalized cgs (gaussian) system and the English system are also in common use. In the cgs system of magnetic units, the unit of flux is the maxwell (Mx), the unit of flux density is the gauss (G), the unit of mmf is the gilbert (Gb), and the unit of magnetic field intensity is the gilbert per centimeter, or oersted (Oe). In the English system of magnetic units, the flux is in lines, flux density is in lines per square inch, mmf is in ampere-turns, and field intensity is in ampere-turns per inch.

Accurate results can be achieved in any of the systems, but all quantities for a particular problem must be expressed in the same system of units. Table 6-2 lists the magnetic units of all three systems.

Conversions from one system to another are simple arithmetic transformations. Much confusion can be avoided by using Tables 6-3 to 6-5.

Problems expressed in cgs or English units may be converted to the rationalized mks units then solved. The result then undergoes conversion to the original system of units. A more direct approach is to remain in the same system until final solution. Since the magnetization curves in Fig. 6-17 are expressed in mks units, some conversion must take place for other unit systems

The value of the free-space permeability is different in each system. As stated previously, the value of μ_0 in the mks system is $4\pi \times 10^{-7}$. The other values can be obtained through unit conversions. From Eq. (6-8),

$$H = \frac{B}{\mu}$$

Table 6-2 MAGNETIC UNITS IN MKS, CGS, AND ENGLISH SYSTEMS

Magnetic term and symbol	mks rationalized	cgs rationalized	English
Flux Φ	Wb	Mx	line
Flux density B	Wb/m^2 (T)	G	line/in.^2
Magnetomotive force F	AT	Gb	AT
Field intensity H	AT/m	Gb/cm (oe)	AT/in.

Table 6-3 CONVERSION OF CGS TO MKS MAGNETIC UNITS

cgs unit	mks unit	
1 Mx (Φ)	10^{-8}	Wb
1 G (B)	10^{-4}	Wb/m^2 (T)
1 Gb (F)	0.7958	AT
1 Gb/cm (H) (oe)	79.6	AT/m

Therefore
$$\mu = \frac{B}{H}\frac{\text{Wb/m}^2}{AT/\text{m}} = \frac{B}{H}\frac{\text{Wb}}{\text{AT}\cdot\text{m}}$$

Thus the units for permeability in the mks system are henrys per meter or webers per ampere turn-meter. The unit henry (H) is used very often in describing a coil. The henry is equivalent to webers per ampere-turn. We find μ_0 for the cgs system as follows:

$$\mu_0 = 4\pi \times 10^{-7}\,\frac{\text{Wb}}{\text{AT}\cdot\text{m}}\,\frac{10^8\ \text{Mx}}{1\ \text{Wb}}\,\frac{1\ \text{AT}}{1.257\ \text{Gb}}\,\frac{1\ \text{m}}{100\ \text{cm}}$$
$$= 4\pi \times 10^{-7}\,\frac{10^8\ \text{Mx}}{1.257 \times 10^2\ \text{Gb}\cdot\text{cm}}$$
$$= 1\ \text{Mx/Gb}\cdot\text{cm} \qquad \text{cgs system}$$

We find μ_0 for the English system in a similar way:

$$\mu_0 = 4\pi \times 10^{-7}\,\frac{\text{Wb}}{\text{AT}\cdot\text{m}}\,\frac{10^8\ \text{lines}}{\text{Wb}}\,\frac{1\ \text{AT}}{1\ \text{A}}\,\frac{1\ \text{m}}{39.37\ \text{in.}}$$
$$= 4\pi \times 10^{-7}\,\frac{10^8\ \text{lines}}{39.37\ \text{AT}\cdot\text{in.}}$$
$$= 3.2\ \text{lines/AT}\cdot\text{in.} \qquad \text{English system.}$$

Table 6-6 summarizes the values of μ_0 in the three systems.

Table 6-4 CONVERSION OF ENGLISH TO MKS MAGNETIC UNITS

English unit	mks unit	
1 line (Φ)	10^{-8}	Wb
1 lines/in.^{2} (B)	1.552×10^{-5}	Wb/m^2 (T)
1 AT (F)	1	AT
1 AT/in. (H)	39.4	AT/m

Table 6-5 CONVERSION OF MKS TO CGS AND ENGLISH MAGNETIC UNITS

mks unit	cgs unit	English unit
1 Wb (Φ)	10^8 Mx	10^8 lines
1 Wb/m^2 (B) (T)	10^4 G	64.52×10^3 lines/in.^{2}
1 AT (F)	1.257 Gb	1 AT
1 AT/m (H)	1.257×10^{-2} Oe	2.54×10^{-2} AT/in.

EXAMPLE 6-16 In the magnetic circuit of Fig. 6-32 what current must flow through the coil to establish a total flux of 8,000 Mx?

SOLUTION Solve for area in square centimeters.

$$A = 1.25^2 = 1.56 \text{ cm}^2$$

Solve for flux density in gauss.

$$B = \frac{\Phi}{A} = \frac{8{,}000}{1.56} = 5{,}130 \text{ G}$$

Field intensity can be found from Fig. 6-17 once gauss are converted to webers per square meter using Table 6-3.

$$B = 5{,}130 \text{ G} \frac{10^{-4} \text{ Wb/m}^2}{1 \text{ G}} = 0.513 \text{ Wb/m}^2$$

$$H = 1{,}460 \text{ AT/m}$$

Use Table 6-5 to convert H to gilberts per centimeter.

$$H = 1{,}460 \frac{\text{AT}}{\text{m}} \frac{1.257 \text{ Gb}}{1 \text{ AT}} \frac{1 \text{ m}}{100 \text{ cm}}$$

$$= 18.3 \text{ Gb/cm} = 18.3 \text{ Oe}$$

The average length of the magnetic path is

$$l = (2)(2.75 \text{ cm}) + (2)(6.25 \text{ cm}) = 18 \text{ cm}$$

Applying Ampère's circuital law gives

$$NI = Hl = (18.3)(18) = 329 \text{ Gb.}$$

Use Table 6-3 to convert gilberts to ampere-turns.

$$NI = 329 \text{ Gb} \frac{0.7958 \text{ AT}}{1 \text{ Gb}}$$

$$= 262 \text{ AT}$$

$$I = \tfrac{262}{300} = 0.874 \text{ A}$$ ////

Table 6-6 PERMEABILITY OF FREE SPACE

	mks	cgs	English
μ_0	$4\pi \times 10^{-7}$	1	3.2

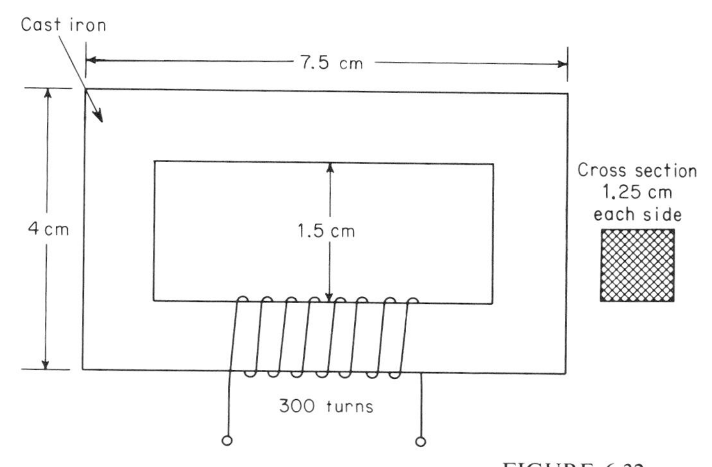

FIGURE 6-32
Magnetic circuit for Example 6-16.

PROBLEMS

6-1 A solenoid with an air core is 0.15 m long and has an inside diameter of 0.06 m. What current is needed to establish a magnetic field intensity of 4,200 AT/m if the coil consists of 630 turns?

6-2 A certain coil with an air core has a flux density of 0.17 wb/m^2. If the coil consists of 1,100 turns and is 0.3 m long, what is the current through the coil?

6-3 A coil consisting of fine wire is wound upon a wooden ring. The turns of the coil completely cover the ring. A current of 73 mA flows through the coil and establishes a flux density of 100 G. Figure 6-33 shows the dimensions of the ring. Calculate the number of turns of wire.

6-4 An air gap is 2.1 cm long, 5.21 cm high, and 3.25 cm wide. It is desired to establish a flux of 0.32 wb across this air gap. What magnetizing force is required in order to obtain the desired flux? Give in ampere-turns per meter.

6-5 In the solenoid of Fig. 6-34, what flux density in gauss is set up by a current of 230 mA?

6-6 A magnetic circuit has a flux of 0.86 Wb. The mmf applied to the circuit is 459 AT. What is the reluctance of the circuit?

6-7 A magnetic circuit with a reluctance of 2,300 units has a flux of 561 mWb. What is the magnetomotive force applied to this circuit?

6-8 A magnetic core has a cross-sectional area of 0.25 in.^{2}. The flux threading the core is 1×10^{-4} Wb. What is the flux density in gauss, webers per square meter, and kilolines per square inch?

6-9 A solenoid has a flux of 0.076 Wb and a flux density of 0.25 Wb/m^2. What is the inside diameter of the solenoid?

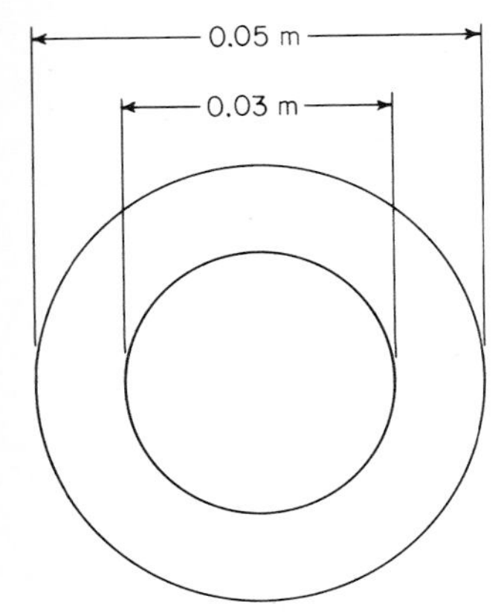

FIGURE 6-33
Ring for Prob. 6-3.

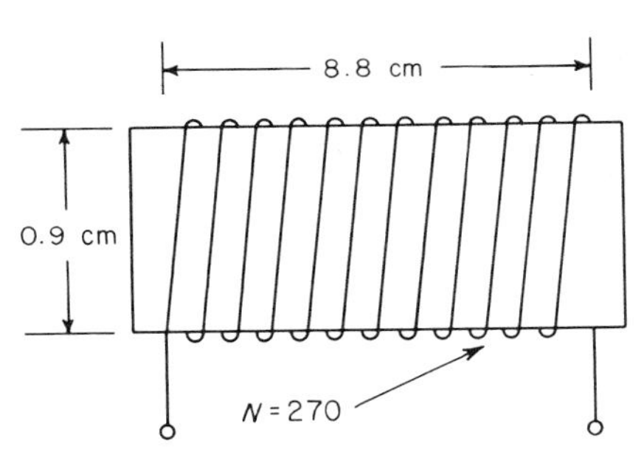

FIGURE 6-34
Coil for Prob. 6-5.

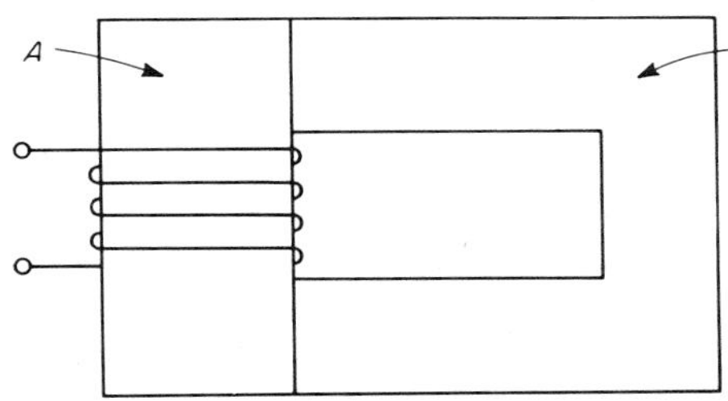

FIGURE 6-35
Circuit for Prob. 6-10.

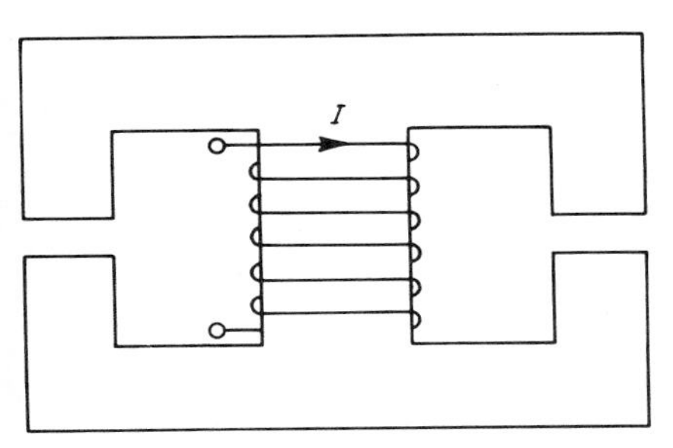

FIGURE 6-36
Circuit for Prob. 6-11.

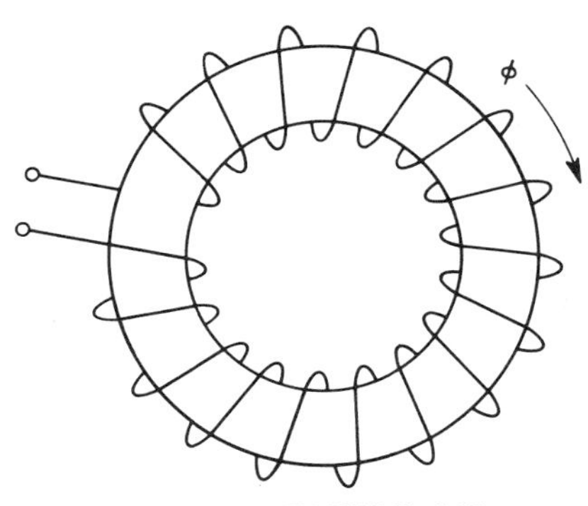

FIGURE 6-37
Toroid for Prob. 6-12.

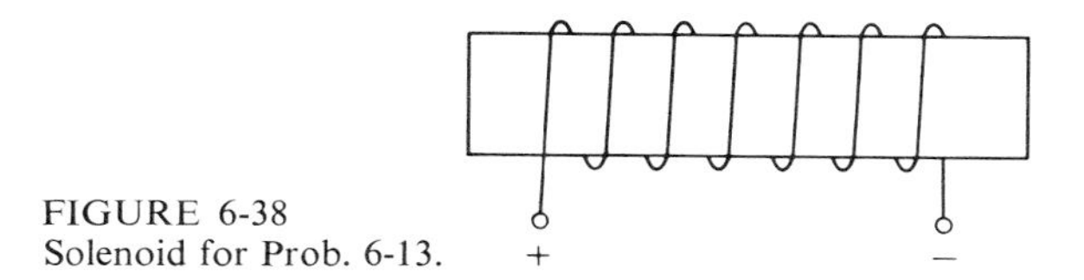

FIGURE 6-38
Solenoid for Prob. 6-13.

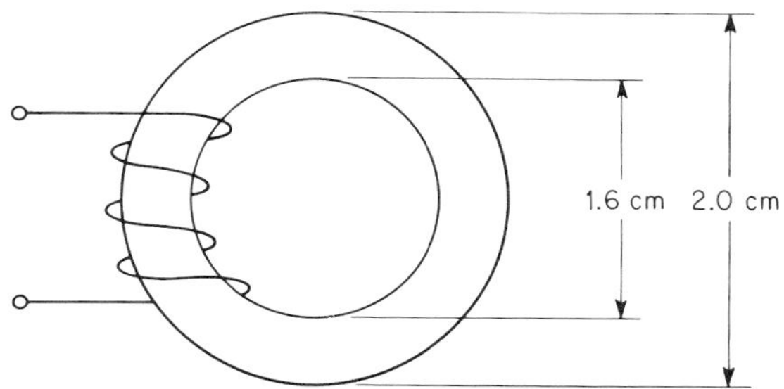

FIGURE 6-39
Magnetic circuit for Prob. 6-14.

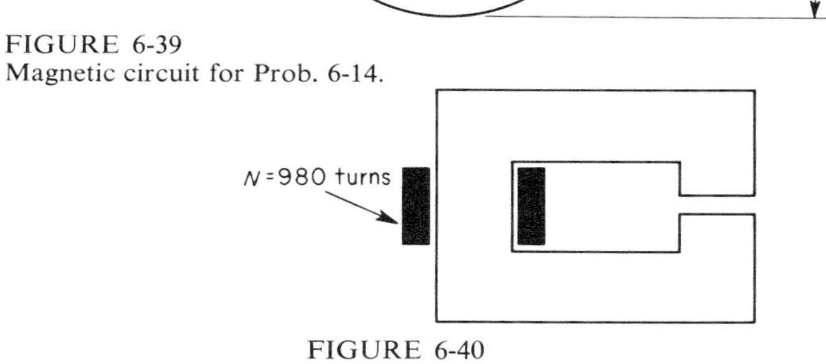

FIGURE 6-40
Coil for Prob. 6-15.

6-10 Given the magnetic circuit of Fig. 6-35 when the flux density in piece A is 5,400 G, what is the flux density of piece B in gauss?

6-11 Given the magnetic circuit of Fig. 6-36, show the direction of the flux through each section of the circuit.

6-12 The flux in the toroid of Fig. 6-37 is established in a clockwise direction. What is the direction of current through the coil?

6-13 Indicate the north and south poles of the solenoid of Fig. 6-38.

6-14 In the magnetic circuit of Fig. 6-39, the flux density of the cast-steel ring is 9,200 G and the current through the coil is 1.1 A. Determine the number of turns wound upon the ring.

6-15 Consider the choke coil of Fig. 6-40. The spacing of the air gap is 3 mm. The cross-sectional area of the core is 1 cm^2; mean length of the magnetic path is 5.8 cm. The core material is transformer steel, and the current through the coil is 240 mA. What is the total flux within the air gap? Specify the flux in maxwells. Assume the gap reluctance is much larger than the reluctance.

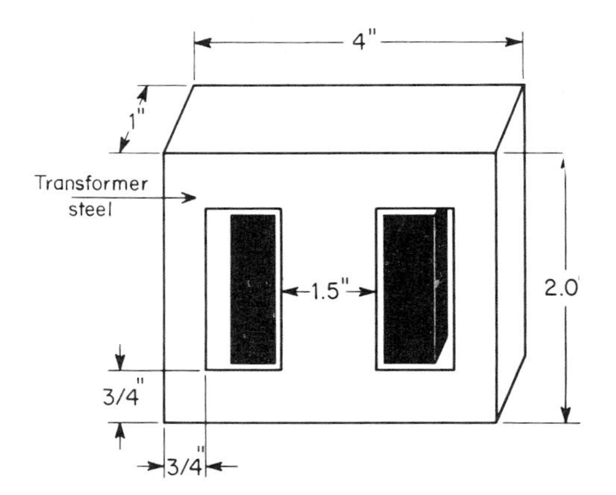

FIGURE 6-41
Transformer for Prob. 6-16.

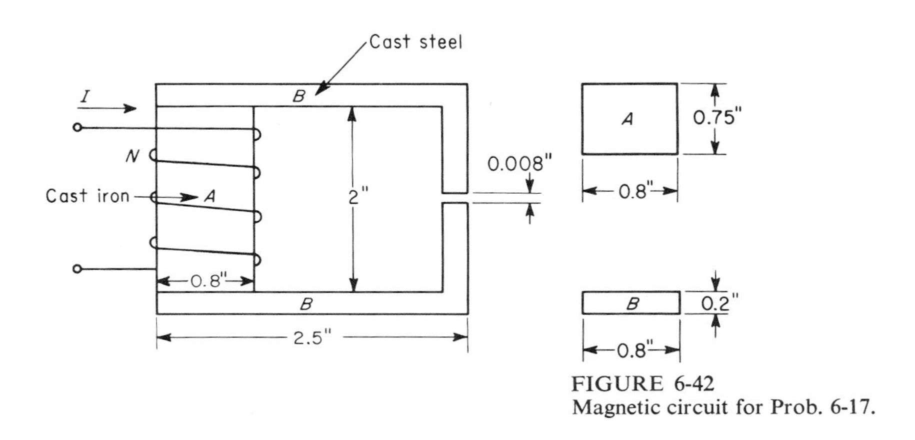

FIGURE 6-42
Magnetic circuit for Prob. 6-17.

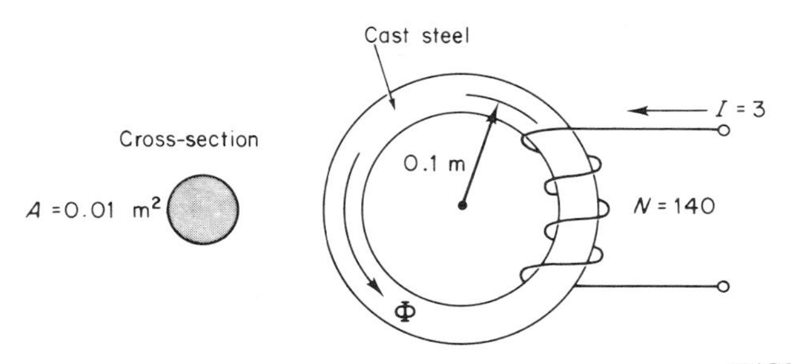

FIGURE 6-43
Circuit for Prob. 6-18.

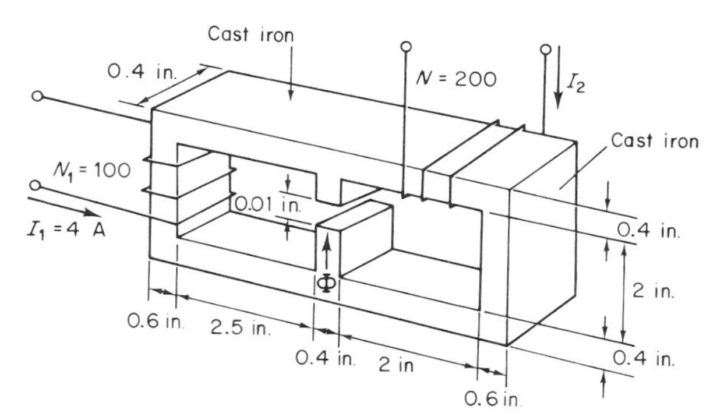

FIGURE 6-44
Circuit for Prob. 6-19.

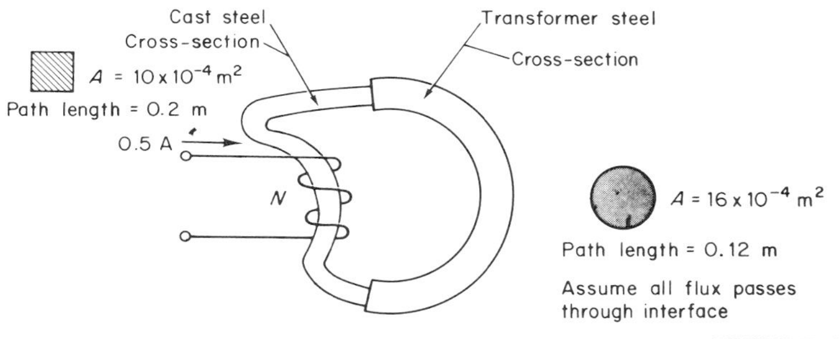

FIGURE 6-45
Circuit for Prob. 6-20.

6-16 A transformer has the dimensions shown in Fig. 6-41. There are 760 turns wound upon the center core. If the current is 120 mA, what is the flux density of the center-core material?

6-17 Given the series magnetic circuit of Fig. 6-42, where flux across the air gap is 4.2×10^5 lines, determine the magnetomotive force needed to establish this flux.

6-18 Determine the magnetic flux in the circuit of Fig. 6-43. Specify the flux in webers.

6-19 Calculate the current I_2 required to provide a flux of 8,000 lines in the air gap of Fig. 6-44.

6-20 Find the number of turns to provide a flux of 6.4×10^{-4} Wb in the transformer steel (see Fig. 6-45).

7

ELECTROMAGNETIC INDUCTION

7-1 ELECTROMAGNETIC INDUCTION

The generation of electricity is always a process of changing energy from some other form into electric energy. The most practical and most common process converts mechanical energy into electric energy. Generators, electric motors, transformers, and many other devices depend upon electromagnetic induction.

In 1831, Michael Faraday in England and Joseph Henry in the United States discovered that voltage is induced across the terminals of a conductor when the conductor is moved across a magnetic field or when the magnetic field is moved across the conductor. Figure 7-1 shows a stationary magnetic field and a moving conductor, and Fig. 7-2 illustrates the induction of voltage in a stationary conductor by a moving magnetic field.

Note that the motion between the conductor and the field must be in such a manner that lines of force are "cut" by the conductor. If the relative motion between conductor and field is such that the lines of force are parallel to the motion, there is no induced emf (see Fig. 7-3).

The discussion in the preceding paragraphs might imply that mechanical displacement must take place between a magnetic field and a conductor in order to induce

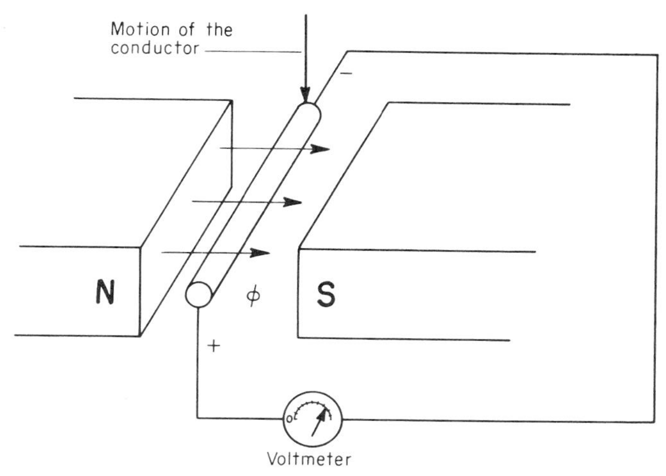

FIGURE 7-1
Moving a conductor through a magnetic field generates a voltage.

voltage. However, voltages are induced in still another manner. Whenever the flux surrounding a conductor changes in magnitude, direction, or both, voltage is induced in the conductor. In this case, no mechanical action of any kind occurs; the induced voltage is the result of a change in the relationship between the flux and the conductor. A magnetic field represents stored energy (Sec. 7-10), and any change in the magnetic field around a conductor must produce a reaction in the conductor. This reaction appears as an induced voltage.

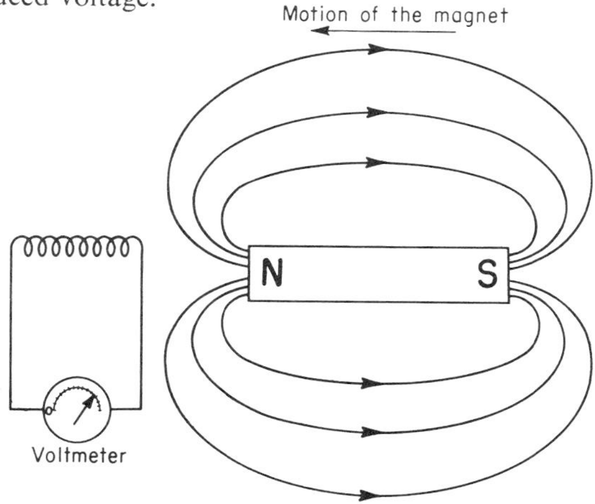

FIGURE 7-2
Changing the field around a conductor induces a voltage.

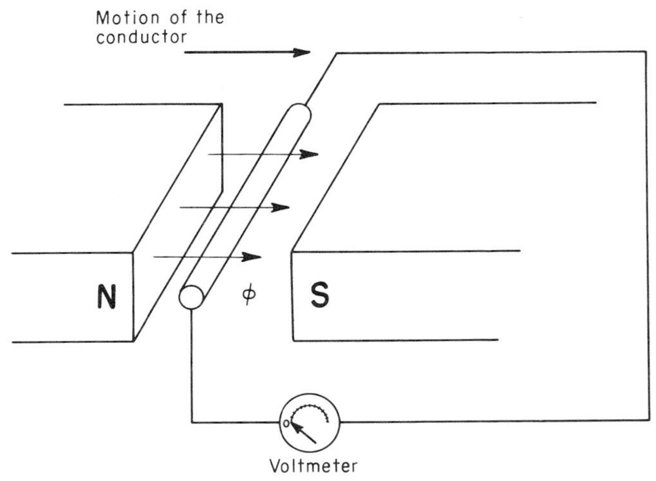

FIGURE 7-3
When a conductor moves parallel to lines of force, no emf is generated.

It has been stated that a magnetic field is always present around a conductor that is carrying a current (Chap. 6). This statement is one part of a fundamental physical law:

> A current always has associated with it a magnetic field that is at right angles to the current. Conversely, a changing magnetic field always has associated with it an emf that is at right angles to the direction of the lines of flux.

The emf that is always associated with a changing magnetic field is an induced emf.

Induced voltages can be developed by mechanical motion between conductors and magnetic fields, or they can be produced by changes in the amplitude or direction of a magnetic field around a conductor. It is common practice to refer to voltages developed as a result of mechanical motion as *generated voltages* and to refer to voltages developed by a change in flux about a conductor as *induced voltages*. In either case, the voltage is the result of electromagnetic induction.

7-2 FLUX LINKAGES

The flux encircling a conductor or a coil is said to be linking the conductor or coil. Flux linkages are the product of the number of turns of a coil and the number of lines of flux encircling the coil. Figure 7-4 illustrates the flux linkages that are established

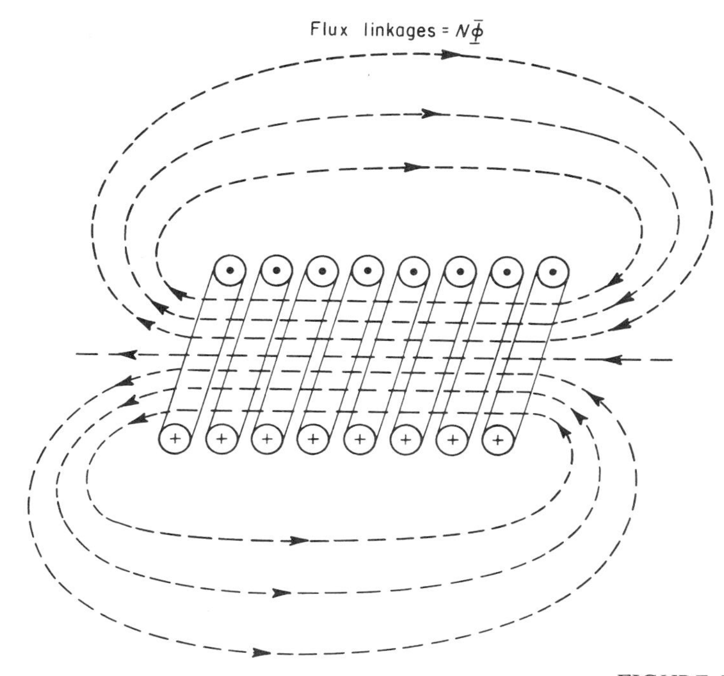

FIGURE 7-4
Flux around a solenoid.

when current flows through a solenoid. For the sake of simplicity, the turns of the solenoid are drawn with no space between turns.

$$\text{Flux linkages} = N\Phi \tag{7-1}$$

7-3 FARADAY'S LAW

Faraday discovered that the magnitude of an induced voltage varies directly with the rate at which lines of force are cut by a conductor. This discovery is one of the foundations of modern electricity and electronics and is the basis for Faraday's law.

> **Faraday's Law** The voltage induced in a conductor by a reaction between the conductor and a magnetic field is directly proportional to the time rate of change of flux linkages, or flux cuttings.

Note that in the statement of Faraday's law, flux cuttings and change in flux linkages have the same meaning. In most discussions of electromagnetic induction, change in flux linkages is the preferred expression.

A flux change of 10^8 lines induces voltage of 1 V per turn in a solenoid.

Where the unit of flux is the weber (10^8 lines), the instantaneous voltage induced in a coil by a flux change is

$$e = N\frac{d\Phi}{dt} \qquad \text{V} \tag{7-2}$$

(The notation $d\Phi/dt$ designates the rate of change of flux during an infinitesimally small time interval. Hence, Eq. (7-2) gives the instantaneous voltage induced in a coil or a winding.)

If flux is given in maxwells, instantaneous induced voltage is found by multiplying the right side of Eq. (7-2) by a factor of 10^{-8}.

$$e = N\frac{d\Phi}{dt} \times 10^{-8} \qquad \text{V}$$

The average voltage induced in a coil or a conductor can be found by taking the total flux change over the time interval required for the flux change:

$$E = N\frac{\Delta\Phi}{\Delta t} \qquad \text{V} \tag{7-3}$$

where Φ is the flux in webers and t is time in seconds.

When a conductor is moved through a uniform field at a constant velocity, the induced voltage can be computed as

$$E = Blv \sin\theta \qquad \text{V} \tag{7-4}$$

where B = flux density, Wb/m^2
l = length (within field) of conductor, m
v = velocity of conductor, m/s
θ = angle between direction of motion of conductor and direction of field

EXAMPLE 7-1 A 120-turn solenoid establishes a flux of 550 mWb. What is the average voltage induced in the coil if the flux changes to 675 mWb in 2.3 ms?

SOLUTION The change in flux is 0.125 Wb over a time interval of 2.3×10^{-3} s. Solve by Eq. (7-3)

$$E = 120\frac{0.125}{2.3 \times 10^{-3}} = 6.52 \times 10^3 \text{ V}$$

////

EXAMPLE 7-2 A certain solenoid develops an induced voltage of 12.8 V when flux changes 0.016 Wb in a time interval of 500 ms. What is the number of turns of the solenoid?

SOLUTION Rearranging Eq. (7-3),

$$N = E\frac{\Delta t}{\Delta\Phi} = 12.8\,\frac{500 \times 10^{-3}}{16 \times 10^{-3}} = 400 \text{ turns} \qquad ////$$

7-4 LENZ'S LAW

In the discussion of electromagnetic induction (See 7-1), it was stated that a change in flux linkages induces voltage in a conductor because a magnetic field represents stored energy. When an attempt is made to reduce or increase flux linkages, and hence the energy stored in the field, opposition to the change is encountered. This opposition is the induced voltage itself and is always of such polarity as to oppose the change in the magnetic field. This is summarized in the rule known as Lenz's law.

Lenz's Law The polarity of an induced voltage is always in such a direction as to establish a current that opposes any change in the magnetic field surrounding a conductor or a coil.

This means that the induced voltage of a coil is in *series-opposition* to the source voltage whenever there is an attempt to increase the flux linkages of the coil. Conversely, the induced voltage will be in *series-aiding* with the source voltage when an attempt is made to decrease the flux linkages of a coil. Hence, a coil acts as a load when current through the coil is increasing, but it acts as a source of voltage when current is decreasing.

Figure 7-5*a* shows the polarity of the voltage induced in a coil as current is increasing, while Fig. 7-5*b* shows the polarity of the voltage induced in a coil as current is decreasing.

7-5 INDUCTANCE

Inductance is the property of a circuit that opposes *changes* in current. It has been stated (Sec. 7-4, Lenz's law) that opposition to a change in flux is seen as an induced voltage across the terminals of a conductor or a coil. Inductance is a proportionality factor that permits a mathematical statement of the relationship involving induced voltage, rate of change of flux, and rate of change of current.

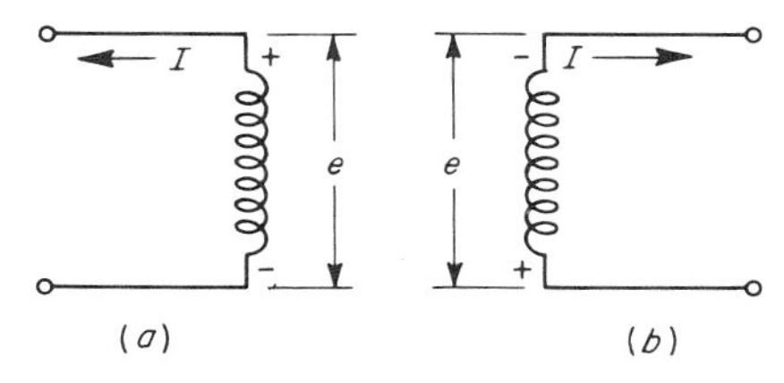

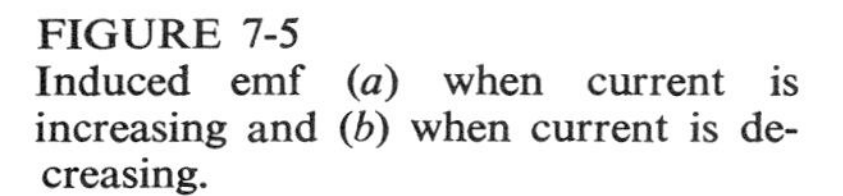

FIGURE 7-5
Induced emf (*a*) when current is increasing and (*b*) when current is decreasing.

The unit of inductance is the *henry* (H), and a circuit is said to have an inductance of one henry when a current change of one ampere per second induces an emf of one volt.

It follows from the definition of inductance that this is a circuit property which is effective only when current is changing; when current through a circuit is constant, inductance has no effect.

7-6 SELF-INDUCTANCE

Since a current in a conductor establishes flux linkages, any change in current changes flux linkages. A change in flux linkages develops an emf of self-induction in accordance with Lenz's law. This *counter emf* may be expressed by Eq. (7-2) (Faraday's law), or it may be expressed in terms of inductance and the rate of change of current. (It is common practice to use the term inductance in place of self-inductance.)

$$e = N\frac{d\Phi}{dt} \qquad \text{V} \tag{7-2}$$

$$e = L\frac{di}{dt} \qquad \text{V} \tag{7-5}$$

where L is the inductance in henrys and di/dt is the instantaneous rate of change of current.

Setting Eqs. (7-2) and (7-5) equal to each other and solving for L,

$$L = N\frac{d\Phi}{di} \qquad \text{H} \tag{7-6}$$

where N is the number of turns of the coil and $d\Phi/di$ is the instantaneous rate of change of the flux, in webers per unit current.

Average inductance is found by the equation

$$L = N\frac{\Delta\Phi}{\Delta i} \qquad \text{H} \tag{7-7}$$

where flux is in webers and current is in amperes.

When the length of a solenoid is much greater than its diameter and the permeability of the magnetic path is constant, the inductance of the coil can be found without determining the flux change and the current change in the coil. From Eq. (6-5),

$$\Phi = \frac{F}{\mathcal{R}}$$

$$\Phi = \frac{\mu NIA}{l} \qquad \text{Wb}$$

Therefore,

$$\Delta\Phi = \frac{\mu N\,\Delta i\,A}{l} \qquad \text{Wb}$$

If this value of $\Delta\Phi$ is substituted into Eq. (7-7), then

$$L = \frac{\mu N^2 A}{l} \tag{7-8}$$

where μ = permeability of magnetic path, mks units
N = total turns of coil
A = area, m^2
l = length of coil, m

Equation (7-8) indicates a very important characteristic of inductance: *Inductance varies as the square of the number of turns of the coil.*

EXAMPLE 7-3 A coil wound upon a wooden cylinder is 12 cm long and has a diameter of 2 cm. If the coil consists of 150 turns, what is its inductance?

SOLUTION

1 The permeability of the wooden cylinder is $4\pi \times 10^{-7}$ H/m.

2 The area of the coil is found.

$$A = \frac{\pi d^2}{4} = \frac{(3.14)(2 \times 10^{-2})^2}{4} = 3.14 \times 10^{-4}\ \text{m}^2$$

3 The remaining factors of Eq. (7-8) are known, and a result is readily obtained.

$$L = \frac{\mu N^2 A}{l} = \frac{(12.56 \times 10^{-7})(150)^2(3.14 \times 10^{-4})}{12 \times 10^{-2}} = 73.9\ \mu\text{H} \qquad ////$$

The unit of inductance is the henry; however, smaller units are millihenrys (mH) and microhenrys (μH). Practical values of inductance range from a few microhenrys to hundreds of henrys.

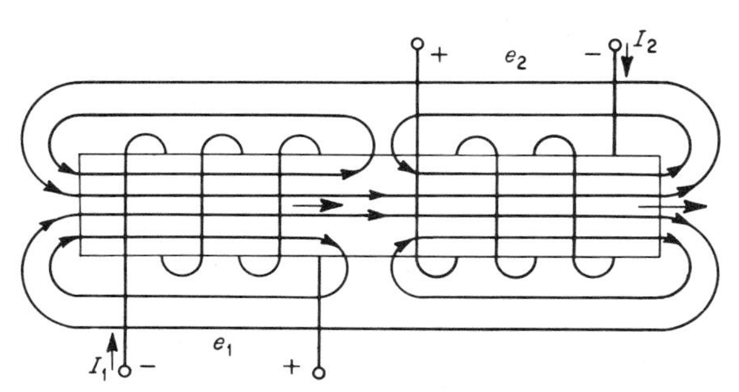

FIGURE 7-6
Coils mutually coupled by magnetic field.

EXAMPLE 7-4 Find the inductance of the coil of Example 7-3 if a magnetic material with a permeability of 5×10^{-2} H/m is used in place of the wooden coil form.

SOLUTION Use Eq. (7-8) and substitute the proper values.

$$L = \frac{(5 \times 10^{-2})(150)^2(3.14 \times 10^{-4})}{12 \times 10^{-2}} = 2.94 \text{ H} \qquad ////$$

7-7 MUTUAL INDUCTANCE

When the position of one coil with respect to another is such that the flux from each coil links the turns of the other, the two coils are mutually coupled, and *mutual inductance* exists between them.

As a result of mutual inductance, a flux change in one coil will induce voltage in the other. This induced voltage is proportional to the rate of change of flux linkages, exactly like an induced voltage that results from self-inductance. It is important to note that the flux linkages are the result of flux from one coil linking the turns of another coil. Figure 7-6 illustrates mutual inductance between two coils.

In accordance with Lenz's law, the voltage that is induced in coil 2 by a change of current in coil 1 is

$$e_2 = M\frac{di_1}{dt} \qquad \text{V} \tag{7-9}$$

The voltage induced in coil 1 by a change of current in coil 2 is

$$e_1 = M\frac{di_2}{dt} \qquad \text{V} \tag{7-10}$$

where di_1/dt = instantaneous rate of change of current in coil 1
di_2/dt = instantaneous rate of change of current in coil 2
e_1 = instantaneous voltage induced in coil 1
e_2 = instantaneous voltage induced in coil 2
M = mutual inductance, H

Note the similarities between Eq. (7-5), which gives the voltage of self-induction, and Eqs. (7-9) and (7-10).

The mutual inductance that exists between coils depends upon the self-inductance of the coils and the degree of coupling between coils, which is referred to as the *coefficient of coupling.* The coefficient of coupling is the factor that states the proportion between the mutual flux linkages that actually occur in a given situation and the maximum possible mutual flux linkages. Maximum possible flux linkages occur when every line of flux from coil 1 links every turn of coil 2 and every line of flux from coil 2 links every turn of coil 1. Under these conditions, the coefficient of coupling is 1. When the coefficient of coupling is less than 1, some of the flux from each coil does not link turns of the other coil.

The voltages of mutual induction may be stated in terms of Faraday's law.

$$e_1 = kN_1 \frac{d\Phi_2}{dt} \qquad (7\text{-}11)$$

$$e_2 = kN_2 \frac{d\Phi_1}{dt} \qquad (7\text{-}12)$$

where e_1 = instantaneous voltage induced in coil 1 by a change in flux of coil 2
e_2 = instantaneous voltage induced in coil 2 by a change in flux of coil 1
N_1 = total turns of coil 1
N_2 = total turns of coil 2
k = coefficient of coupling between coils 1 and 2

The average voltage induced in coil 1 can be found by

$$E_1 = kN_1 \frac{\Delta\Phi_2}{\Delta t} \quad \text{V} \qquad (7\text{-}13)$$

where t is in seconds and Φ is in webers.

The average voltage induced in coil 2 can be found in a similar manner.

$$E_2 = kN_2 \frac{\Delta\Phi_1}{\Delta t} \quad \text{V} \qquad (7\text{-}14)$$

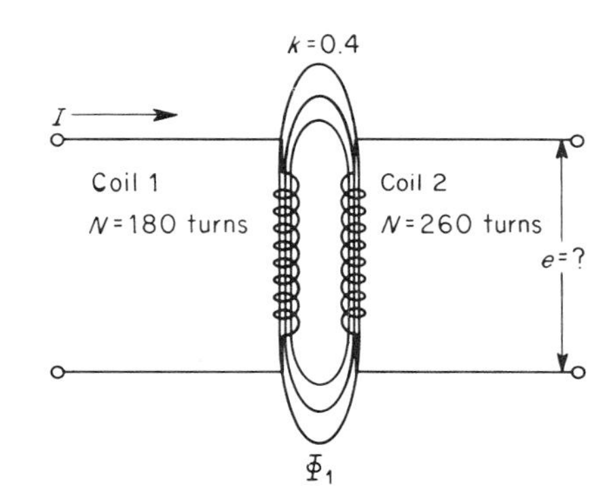

FIGURE 7-7
Circuit for Example 7-5.

EXAMPLE 7-5 In the circuit of Fig. 7-7, flux Φ_1 changes from 600 to 160 mWb in a time interval of 0.125 s. The coefficient of coupling is 0.4. What is the average voltage developed across coil 2?

SOLUTION Equation (7-14) yields the average voltage developed in coil 2 as a result of a flux change in coil 1.

$$E_2 = (0.4)(260)\frac{440 \times 10^{-3}}{125 \times 10^{-3}} = 366 \text{ V} \qquad ////$$

Equations (7-9) to (7-12) may be used to solve for mutual inductance. If Eqs. (7-9) and (7-12) are solved for M, then

$$M = kN_2 \frac{d\Phi_1}{di_1} \tag{7-15}$$

If Eqs. (7-10) and (7-11) are solved for M, then

$$M = kN_1 \frac{d\Phi_2}{di_2} \tag{7-16}$$

The mutual inductance between coils is the same going from coil 1 to coil 2 as it is going from coil 2 to coil 1 (providing the permeability of both magnetic paths is constant). Therefore Eqs. (7-15) and (7-16) result in equal quantities. If Eq. (7-15) is multiplied by Eq. (7-16), the result is

$$M^2 = k^2 N_1 \frac{d\Phi_1}{di_1} N_2 \frac{d\Phi_2}{di_2}$$

However, $L_1 = N_1\, d\Phi_1/di_1$, and $L_2 = N_2\, d\Phi_2/di_2$ [Eq. (7.6)]. Therefore,

$$M^2 = k^2 L_1 L_2$$

$$M = k\sqrt{L_1 L_2} \qquad \text{H} \tag{7-17}$$

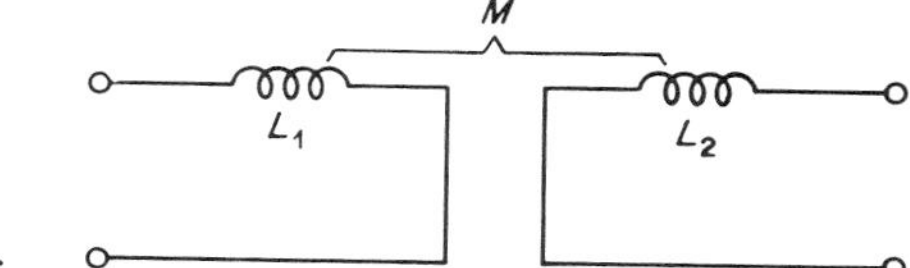

FIGURE 7-8
Circuit for Example 7-6.

EXAMPLE 7-6 Given the circuit of Fig. 7-8, calculate the mutual inductance when the coefficient of coupling is 0.2. Assume that $L_1 = 250\ \mu\text{H}$ and $L_2 = 400\ \mu\text{H}$.

SOLUTION

$$M = k\sqrt{L_1 L_2}$$
$$= 0.2\sqrt{(250 \times 10^{-6})(400 \times 10^{-6})} = (0.2)(316 \times 10^{-6}) = 63.2\ \mu\text{H} \qquad ////$$

The coefficient of coupling is derived from Eq. (7-17)

$$k = \frac{M}{\sqrt{L_1 L_2}} \tag{7-18}$$

EXAMPLE 7-7 In the circuit of Fig. 7-8, what is the coefficient of coupling if the mutual inductance is 120 μH?

SOLUTION

$$k = \frac{M}{\sqrt{L_1 L_2}}$$
$$= \frac{120 \times 10^{-6}}{\sqrt{(250 \times 10^{-6})(400 \times 10^{-6})}} = \frac{120 \times 10^{-6}}{316 \times 10^{-6}} = 0.38 \qquad ////$$

Mutually coupled coils may be connected to each other, or they may be insulated from one another. In most transformers, the coils are insulated from one another, while in many electronic circuits mutual inductance exists between coils that are connected to one another.

7-8 COILS IN SERIES AND PARALLEL

Figure 7-9 shows a series circuit consisting of two coils that are *not* mutually coupled. The total inductance of the circuit is readily found.

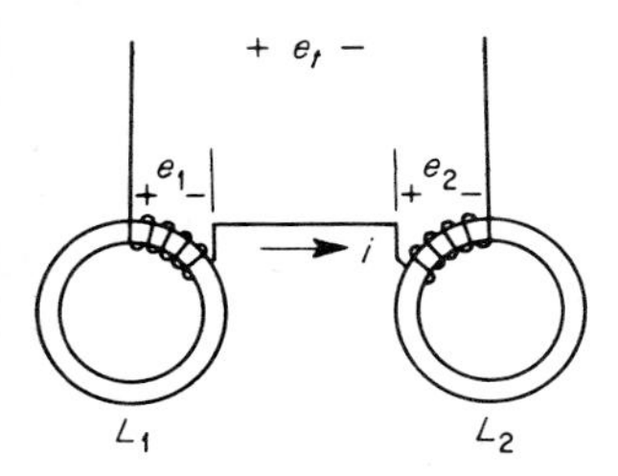

FIGURE 7-9
Coils in series with no mutual inductance between them.

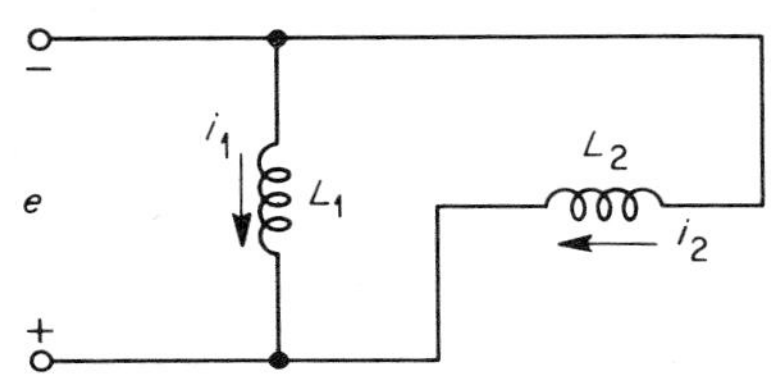

FIGURE 7-10
Coils in parallel with no mutual inductance between them.

1 $e_t = e_1 + e_2$ where

$$e_1 = L_1 \frac{di}{dt} \qquad e_2 = L_2 \frac{di}{dt} \qquad e_t = L_t \frac{di}{dt}$$

2 Substitute into step 1.

$$L_t \frac{di}{dt} = L_1 \frac{di}{dt} + L_2 \frac{di}{dt}$$

3 Divide step 2 by di/dt.

$$L_t = L_1 + L_2$$

A general equation for the total inductance of coils in series is

$$L_t = L_1 + L_2 + \cdots + L_n \tag{7-19}$$

Figure 7-10 is a circuit consisting of coils in parallel that are not mutually coupled because the flux from coil 1 does not cut the turns of coil 2, and the flux from coil 2 does not cut the turns of coil 1. Whenever a magnetic field is at right angles to the axis of a coil, mutual flux linkages are reduced to negligible amounts. Basic circuit theory permits the total inductance to be readily computed.

1 The voltages across all branches of a parallel circuit are alike. Therefore,

$$e = L_1 \frac{di_1}{dt}$$

$$e = L_2 \frac{di_2}{dt}$$

$$e = L_t \left(\frac{di_1}{dt} + \frac{di_2}{dt}\right)$$

2 Solve for e/L_1.

$$\frac{e}{L_1} = \frac{di_1}{dt}$$

3 Solve for e/L_2.

$$\frac{e}{L_2} = \frac{di_2}{dt}$$

4 Solve for e/L_t.

$$\frac{e}{L_t} = \frac{di_1}{dt} + \frac{di_2}{dt}$$

5 Substitute from steps 2 and 3 into step 4.

$$\frac{e}{L_t} = \frac{e}{L_1} + \frac{e}{L_2}$$

6 Divide both sides by e.

$$\frac{1}{L_t} = \frac{1}{L_1} + \frac{1}{L_2}$$

7 Take the reciprocal of both sides.

$$L_t = \frac{1}{1/L_1 + 1/L_2}$$

The total inductance of two coils in parallel is computed most easily by

$$L_t = \frac{L_1 L_2}{L_1 + L_2} \tag{7-20}$$

Where the parallel circuit consists of more than two coils in parallel, the general solution is

$$L_t = \frac{1}{\dfrac{1}{L_1} + \dfrac{1}{L_2} + \dfrac{1}{L_3} + \cdots + \dfrac{1}{L_n}} \tag{7-21}$$

Note that inductance of coils in series and parallel circuits is calculated in precisely the same way as resistance of series and parallel circuits.

When mutual inductance exists between coils in series, it may add to the total inductance or it may subtract from the total inductance. When coils are connected so that the polarities of their magnetic fields aid each other, the coils are in *series-aiding*. Figure 7-11 is a sketch of coils connected in series-aiding.

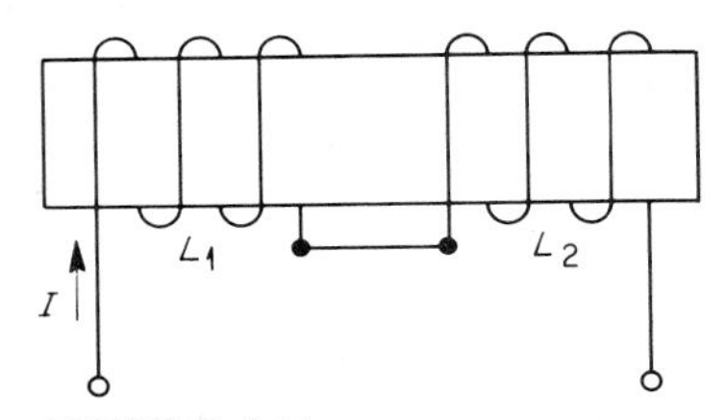

FIGURE 7-11
Coils connected in series-aiding, with mutual inductance between them.

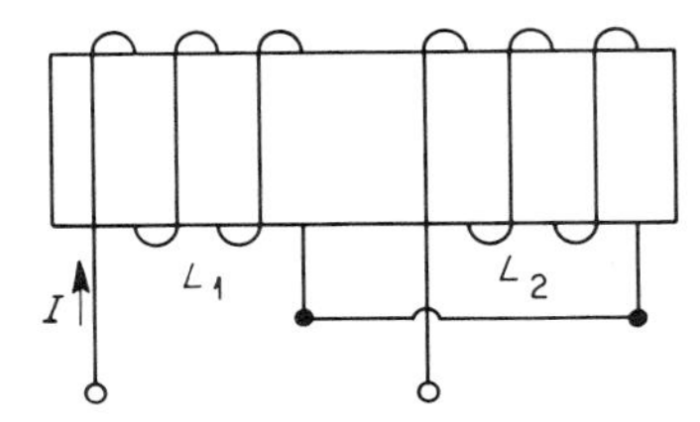

FIGURE 7-12
Coils connected in series-opposing, with mutual inductance between them.

When the magnetic fields of series-connected coils oppose each other, the coils are in *series-opposing*. Figure 7-12 shows a series circuit with the coils connected in series-opposing.

The total inductance of the circuit of Fig. 7-11 is

$$L_t = L_1 + L_2 + 2M$$

The total inductance of the circuit of Fig. 7-12 is

$$L_t = L_1 + L_2 - 2M$$

A general equation for the total inductance of coils in series is, therefore,

$$L_t = L_1 + L_2 \pm 2M \tag{7-22}$$

EXAMPLE 7-8 Given the circuit of Fig. 7-11, where $L_1 = 55$ mH, $L_2 = 80$ mH, and $k = 0.8$, find the total inductance of the circuit.

SOLUTION

1 $$L_t = L_1 + L_2 + 2M$$

2 $$M = k\sqrt{L_1 L_2} = 0.8\sqrt{(55 \times 10^{-3})(80 \times 10^{-3})} = 53 \text{ mH}$$

3 $$L_t = 80 \text{ mH} + 55 \text{ mH} + 106 \text{ mH} = 241 \text{ mH}$$

EXAMPLE 7-9 Given the circuit of Fig. 7-12, where $L_1 = 4$ H and $L_2 = 7$ H, find the coefficient of coupling if the total inductance is 1 H.

SOLUTION

1 $$L_t = L_1 + L_2 - 2M$$

Therefore,

$$M = \frac{L_1 + L_2 - L_t}{2} = 5 \text{ H}$$

2

$$k = \frac{M}{\sqrt{L_1 L_2}} = \frac{5}{\sqrt{28}} = 0.945 \qquad ////$$

EXAMPLE 7-10 Two coils are connected in series-aiding. The total inductance is 800 mH. The coils are then connected in series-opposing, and the total inductance is found to be 500 mH. If the inductance of one coil is 350 mH, what is the inductance of the other?

SOLUTION

1
$$L_t = L_1 + L_2 + 2M \qquad \text{aiding}$$

2
$$L_t' = L_1 + L_2 - 2M \qquad \text{opposing}$$

3 Add steps 1 and 2.

$$L_t + L_t' = 2L_1 + 2L_2$$

Substituting,

$$1{,}300 \text{ mH} = 700 \text{ mH} + 2L_2$$

$$L_2 = 300 \text{ mH} \qquad ////$$

The total inductance for two coils in parallel when there is mutual inductance between them is found with Eq. (7-23). The negative sign is used

$$L_t = \frac{L_1 L_2 - M^2}{L_1 + L_2 \pm 2M} \tag{7-23}$$

when the mutual inductance is aiding and positive sign for the opposite situation.

EXAMPLE 7-11 Two parallel-connected coils have a mutual inductance of 78 mH. The inductance of one coil is 150 mH, and the inductance of the other coil is 235 mH. What is the total inductance of the parallel circuit? The mutual inductance is aiding.

SOLUTION

$$L_t = \frac{L_1 L_2 - M^2}{L_1 + L_2 - 2M}$$

$$= \frac{(150 \times 10^{-3})(235 \times 10^{-3}) - (78 \times 10^{-3})^2}{(150 \times 10^{-3}) + (235 \times 10^{-3}) - (156 \times 10^{-3})} = 154 \text{ mH} \qquad ////$$

7-9 TIME CONSTANTS IN INDUCTIVE CIRCUITS

Inductance is a circuit property that is analogous to inertia in mechanical systems. In mechanical systems, inertia causes backlash and time delay. Inductance has similar effects in electric circuits.

All circuits have inductance, whether the circuits contain coils or not. Inductance is defined as that property of a circuit which opposes changes in current. This opposition to current change appears as an induced voltage. When the current through a conductor changes, the flux surrounding the conductor also changes. An emf that opposes the change of current develops along the conductor. When the rate of change of current is small, the effect of self-inductance is ignored. For example, in the discussion of Ohm's law and electric circuits, inductance was not considered because the self-inductance of connecting wires is negligible compared to the resistance of the circuit. However, when circuits are used under conditions where the rate of change of current is very high, as in radar, even a short length of wire may possess considerable inductance.

In order to observe the effect of inductance upon the rate of current rise, the circuit of Fig. 7-13 will be analyzed. In this circuit consider that S_1 and S_2 operate simultaneously but in opposite directions. That is, when S_1 closes, S_2 opens; when S_1 opens, S_2 closes.

Assume that at time t_0 switch S_1 is closed. Current begins to flow. However, the expanding flux, as a result of the current, develops a voltage across the coil. The polarity of the induced voltage is such that it opposes the battery voltage, and its magnitude depends upon the rate of change of current. At the instant the switch is closed, the current is zero. The current then rises gradually to a value determined by the applied voltage and the resistance of the circuit. The final value of the current is completely independent of the inductance of the circuit. Figure 7-14 shows the rate of rise of current for the circuit of Fig. 7-13.

A mathematical analysis of the rise of current in an inductive circuit may be made with the circuit in Fig. 7-13. The sum of the voltages around this circuit at any instant must equal the applied voltage.

$$E = iR + L\frac{di}{dt} \qquad (7\text{-}24)$$

where E = applied voltage

L = inductance of circuit, H

R = resistance of circuit, Ω

i = current flow at any instant in time, A

di/dt = instantaneous rate of change of current

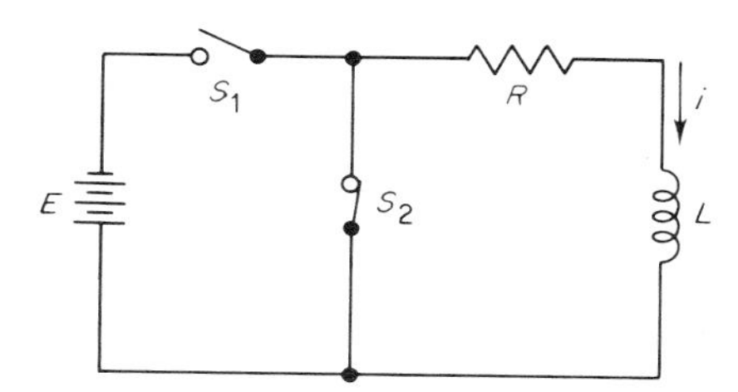

FIGURE 7-13
Circuit that illustrates transients in inductive circuits.

Equation (7-24) is a differential equation. Solving this equation for the current i gives

$$i = \frac{E}{R}(1 - \varepsilon^{-(R/L)t}) \tag{7-25}$$

where $\varepsilon = 2.718$.

Equations (7-24) and (7-25) provide a simple explanation for the graph of Fig. 7-14. At time zero, the current is zero in accordance with Eq. (7-25). Also, at this instant all the applied voltage appears across the coil. At a later time, the exponential term in Eq. (7-25) becomes so small that the current in the circuit approaches E/R, and all of the drop in the circuit is across the resistor.

This same analysis can be made with Eq. (7-24). At time t_0 the rate of current change is nearly infinite, and the induced voltage equals the applied voltage. At time t_1 the rate of current change is practically zero, and the induced voltage is negligible. Then all of the input voltage is dropped across the resistor.

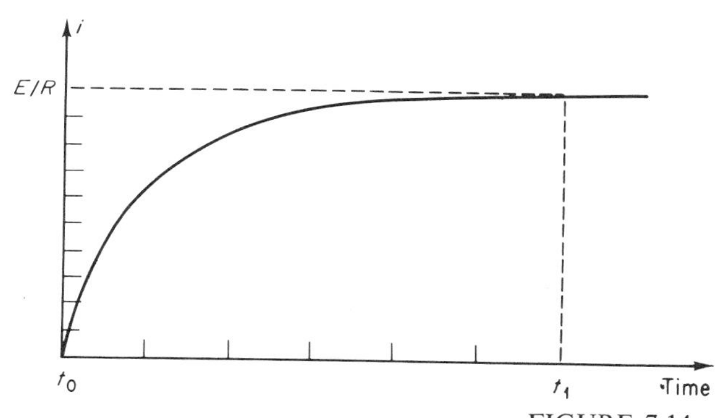

FIGURE 7-14
Current rise in an inductive circuit.

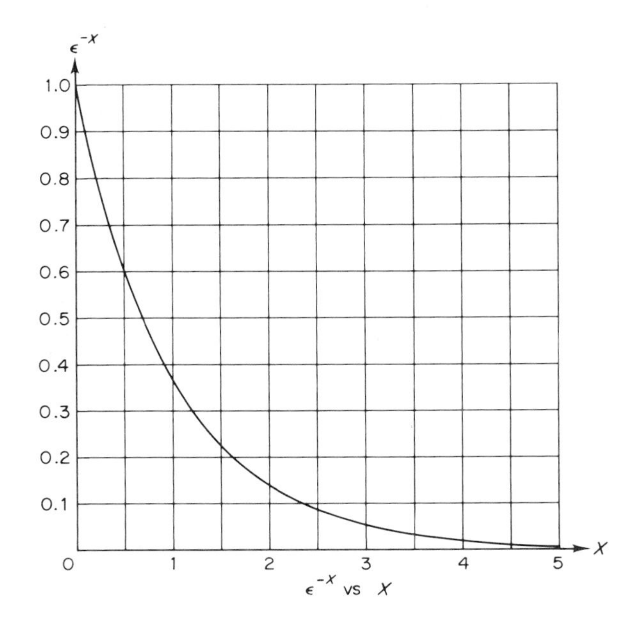

FIGURE 7-15
ε^{-x} vs. x.

Equation (7-25) permits calculation of current in an inductive circuit at any instant between the time that the current is started and the time that it reaches its final value of E/R.

Figure 7-15 shows the relationship between ε^{-x} and x. Appendix Table A-9 gives a more precise tabular listing. Let x represent $(R/L)t$.

EXAMPLE 7-12 In the circuit of Fig. 7-13, the resistance of the circuit is 22 kΩ, the inductance is 780 mH, and the applied emf is 45 V. What is the current 10 μs after switch S_1 is closed?

SOLUTION

$$i = \frac{E}{R}(1 - \varepsilon^{-(R/L)t})$$

$$= \frac{45}{22 \times 10^3}(1 - \varepsilon^{-(22 \times 10^3/780 \times 10^{-3})(10 \times 10^{-6})})$$

$$= (2.04 \times 10^{-3})(1 - \varepsilon^{-0.282})$$

$$= (2.04 \times 10^{-3})(1 - 0.754)$$

$$= (2.04 \times 10^{-3})(0.246) = 0.504 \text{ mA}$$

////

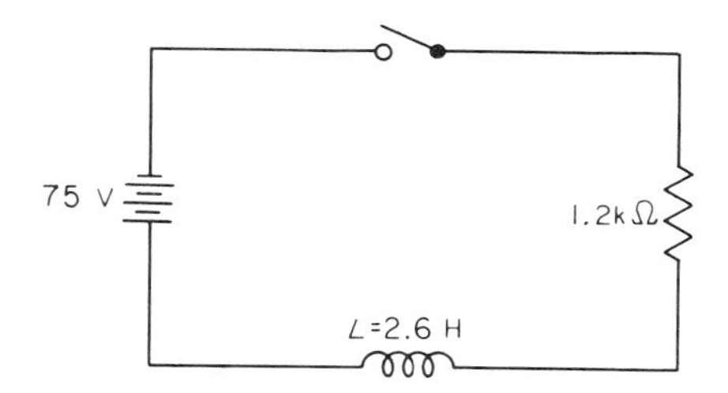

FIGURE 7-16
Circuit for Example 7-14.

EXAMPLE 7-13 Given the same conditions as in Example 7-12, find the current 10 ms after S_1 is closed.

SOLUTION

$$i = \frac{E}{R}(1 - \varepsilon^{-(R/L)t})$$

$$= (2.04 \times 10^{-3})(1 - \varepsilon^{-(22 \times 10^3/780 \times 10^{-3})(10^{-2})})$$

$$= (2.04 \times 10^{-3})(1 - \varepsilon^{-282})$$

$$= (2.04 \times 10^{-3})(1 - 0) = 2.04 \text{ mA} \qquad ////$$

Note: In this solution ε^{-282} has been assumed equal to zero. This is an approximation. The error is so infinitesimal, however, that the assumption is justified.

EXAMPLE 7-14 Given the circuit of Fig. 7-16, determine at what time after the switch is closed the current reaches 25 mA.

SOLUTION

1 Let the exponent $(R/L)t = x$.

2 Solve for ε^{-x}.

$$i = \frac{E}{R}(1 - \varepsilon^{-x})$$

$$\frac{iR}{E} = 1 - \varepsilon^{-x}$$

$$\varepsilon^{-x} = 1 - \frac{iR}{E} = 1 - \frac{(25 \times 10^{-3})(1.2 \times 10^3)}{75}$$

3 Substitute values.

$$\varepsilon^{-x} = 0.6$$

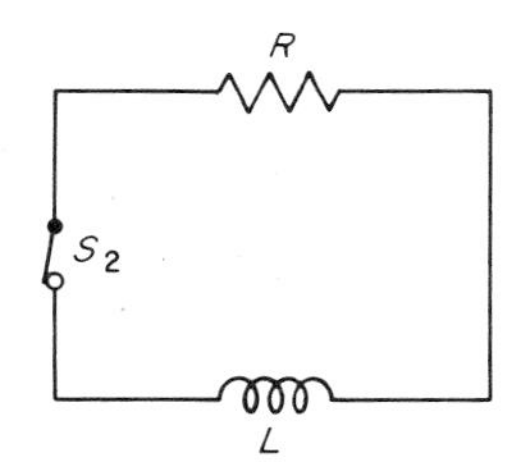

FIGURE 7-17
Equivalent circuit when inductance is discharging.

4 The most direct solution for x is obtained from Table A-9.

$$x = 0.51$$

5 Solve for t.

$$\frac{R}{L}t = 0.51$$

$$t = \frac{0.51 \times 2.6}{1.2 \times 10^3} = 1.105 \times 10^{-3} = 1.1 \text{ ms} \qquad ////$$

When an attempt is made to stop current in an inductive circuit, inductance causes the decay of current to be gradual. This, of course, is due to changes in flux linkages that occur as the current decreases. In this case, the induced voltage resulting from the flux change is so directed as to keep current flowing.

Figure 7-13 permits mathematical analysis of this situation. Assume that switch S_1 has been closed for a long time and the current through the circuit has reached its E/R value. Now let switch S_1 be opened. Switch S_2 closes at the same instant. The equivalent circuit at this instant is shown in Fig. 7-17.

The total voltage of the circuit is zero, and the equation for the sum of the voltages around the circuit is

$$0 = iR + L\frac{di}{dt} \qquad (7\text{-}26)$$

Equation (7-26) is a differential equation; when solved for current i it yields

$$i = \frac{E}{R}\varepsilon^{-(R/L)t} \qquad (7\text{-}27)$$

where E/R is the initial current.

It is important to note that Eq. (7-27) can be written in the general form

$$i = I\varepsilon^{-(R/L)t} \qquad (7\text{-}28)$$

where I is the current flowing at the instant current decay starts.

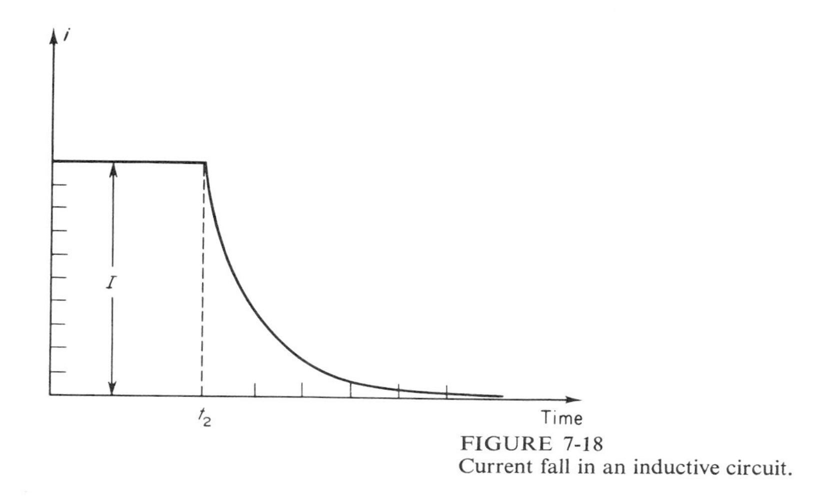

FIGURE 7-18
Current fall in an inductive circuit.

Figure 7-18 illustrates the decay of current through the coil and resistor in the circuit of Fig. 7-13 when switch S_1 is opened.

EXAMPLE 7-15 Given the circuit of Fig. 7-19, find the current through the coil 0.05 s after the switch is closed.

SOLUTION

$$i = I\varepsilon^{-(R/L)t} = 0.5\varepsilon^{-(5/8)(0.05)}$$

$$= 0.5\,\varepsilon^{-0.0313} = (0.5)(0.9691) = 0.485 \text{ A} \qquad ////$$

A quantity called *time constant* is often used in solving inductive circuits. The time constant of an inductive circuit is the time that would be required for the current to reach maximum value if the average rate of current rise were equal to the initial rate of rise. Time constant is also the time that would be required for the current to reach zero if its initial rate of decrease remained constant, rather than decreasing exponentially.

In either case, the rate of change would be constant. This would require that the exponent of epsilon be -1. This can be achieved only by multiplying R/L by L/R.

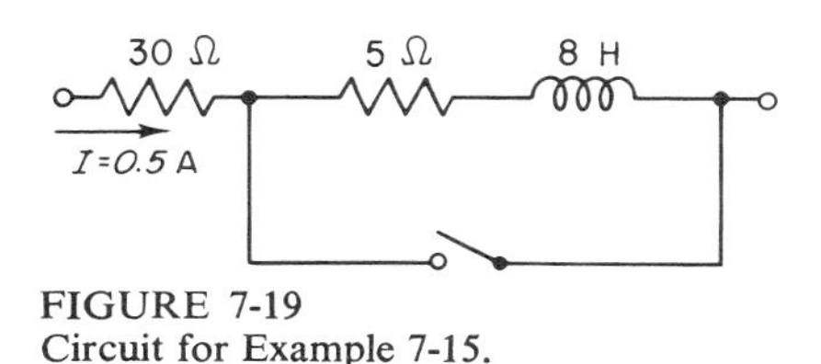

FIGURE 7-19
Circuit for Example 7-15.

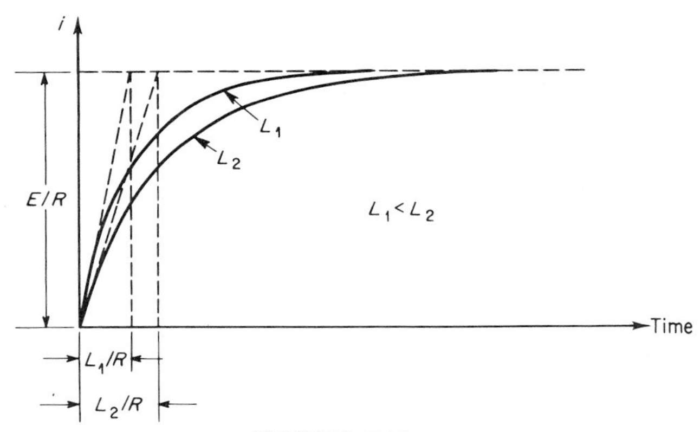

FIGURE 7-20
Current rise in inductive circuit where L_2 is greater than L_1.

The fraction L/R is the time constant of an inductive circuit. The constant is of importance because it permits rapid determination of the rate of rise or fall of current in inductive circuits. When Eq. (7-25) is solved for a time equal to the L/R time constant, the current is 0.632 times the E/R value of the current.

$$i = \frac{E}{R}(1 - \varepsilon^{-1}) = \frac{E}{R}(1 - 0.368) = 0.632\,\frac{E}{R}$$

When Eq. (7-28) is solved for a time equal to the L/R time constant, the current is 0.368 times the initial current.

$$i = I\varepsilon^{-1} = 0.368I$$

Resistance in an inductive circuit has the following effects upon the current:

1 The maximum current equals the applied voltage divided by the resistance.
2 The rate of current rise is proportional to the resistance.
3 The time constant of the circuit is inversely proportional to the resistance.

Inductance in a circuit has the following effects upon current:

1 The rate of current rise is inversely proportional to the inductance.
2 The time constant of the circuit is directly proportional to the inductance.
3 The inductance has no effect at all upon the final value of the current.

Figure 7-20 illustrates the rate of current rise in a circuit for two different values of inductance. The resistance of the circuit is constant.

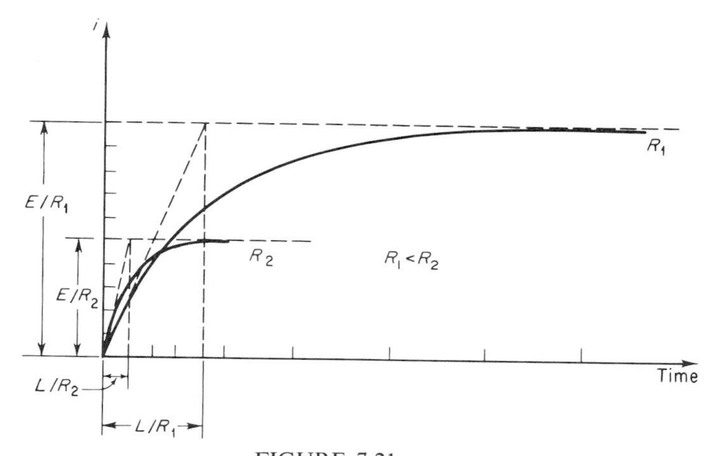

FIGURE 7-21
Current rise in inductive circuits where R_2 is greater than R_1.

Figure 7-21 illustrates the rate of current rise in a circuit for two different values of resistance. The inductance of the circuit is constant.

Thus far, in all discussion of the effect of inductance in a circuit, much attention has been given to current in the circuit, but little thought has been given to the voltages in the circuit. Figure 7-22 illustrates the current and voltages across the components

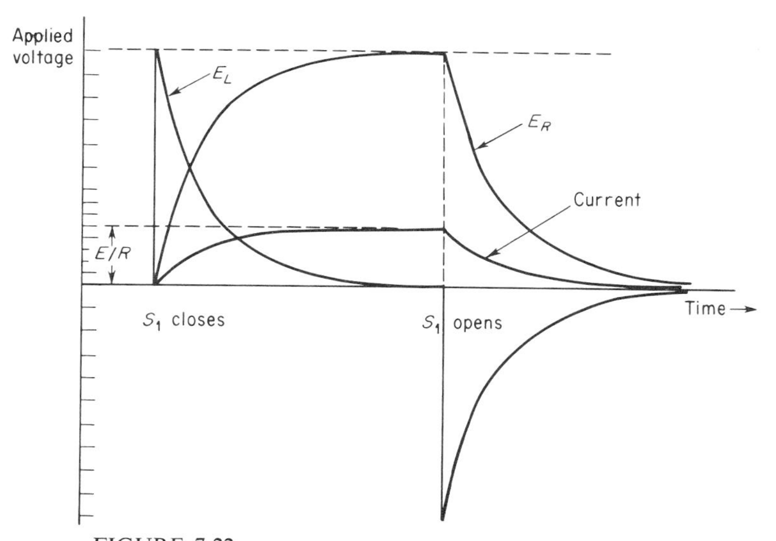

FIGURE 7-22
Voltage and current variations during a complete cycle for the circuit of Figure 7-13.

of the circuit of Fig. 7-13 during a complete cycle of operation, from the time switch S_1 is closed to the time S_1 is opened.

7-10 ENERGY STORAGE IN A MAGNETIC FIELD

In Chap. 3, energy was defined as the ability to do work. The magnetic field must therefore have energy to account for the fact that it exerts forces upon magnetic materials. The source of energy for a magnetic field is electromagnetism. When current is passed through a coil, a large portion of the electric energy is diverted to establish the magnetic field. The magnetic field is established by work done in overcoming the opposition offered by the inductance of the coil.

Computation of the energy contained in a magnetic field requires the use of calculus. In this case, it is used to sum all the minute increases in current during the time current is rising. The resultant equation for the energy stored in a magnetic field is

$$W = \frac{LI^2}{2} \quad \text{J} \tag{7-29}$$

where L = inductance, H
I = current, A
W = energy, J

[In Eq. (7-29) the current is actually the total amount of change of the current. Since the current usually starts from zero, the total current is normally used in the equation. However, in a case where current changes and it is necessary to calculate the change in energy stored, replace I^2 by $I_1{}^2 - I_2{}^2$, where I_1 is the original current and I_2 is the new current.]

EXAMPLE 7-16 Calculate the energy stored in its magnetic field when the inductance of a coil is 3.3 H and the current is 2 A.

SOLUTION

$$W = \frac{LI^2}{2} = \frac{(3.3)(2^2)}{2} = 6.6 \text{ J} \qquad ////$$

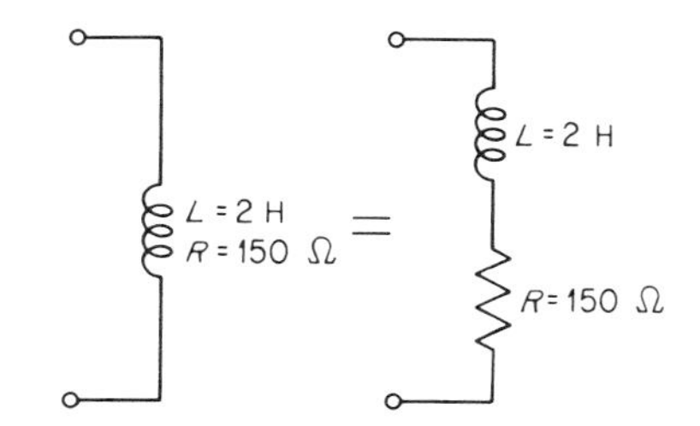

FIGURE 7-23
A coil and its series equivalent.

EXAMPLE 7-17 What is the energy returned to the circuit if the current in the circuit of Example 7-16 is reduced to 0.5 A?

SOLUTION

$$W = \frac{L(I_1^2 - I_2^2)}{2} = \frac{3.3(4 - 0.25)}{2} = 6.1875 \text{ J} \qquad ////$$

It has already been pointed out that all circuits have the properties of resistance and inductance. Very often these properties are lumped into a single circuit component, such as a coil. When the resistance of a coil is very small compared to its inductance, it is common practice to ignore the resistance. However, when the resistance of a coil cannot be ignored, analysis of the problem is greatly simplified if the inductance and the resistance of the coil are treated as if they were in separate parts of the circuit. That is, the resistance of the coil is considered as a resistance in series with the coil. In later chapters, we shall see that an equivalent resistance can be shown in parallel with the inductance of the coil. Figure 7-23 shows a coil with both resistance and inductance, and the series equivalent of this coil.

PROBLEMS

7-1 A 150-turn solenoid develops an average induced voltage of 12 V as a result of a change in flux. If the total time for the flux change is 250 ms, what is the total flux change?

7-2 Over what time interval must a flux change of 750 mWb occur if a 220-turn solenoid develops an average induced voltage of 81 V?

7-3 Given the circuit of Fig. 7-24, show the polarity of the induced voltage across the coil when the switch is opened.

7-4 What is the inductance of a 75-turn coil if a flux change of 12 Mx occurs with a current change of 10^{-1} A?

7-5 A coil wound on a wooden core has a length of 3 in. and a diameter of 0.275 in. How many turns must be wound for the inductance to be 125 μH? Be sure to convert inches into meters.

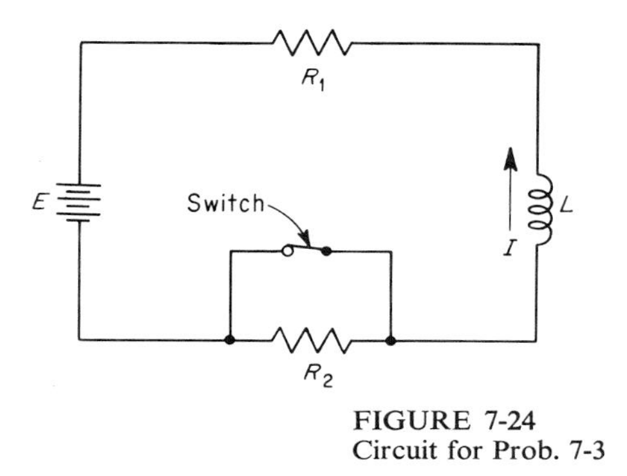

FIGURE 7-24
Circuit for Prob. 7-3

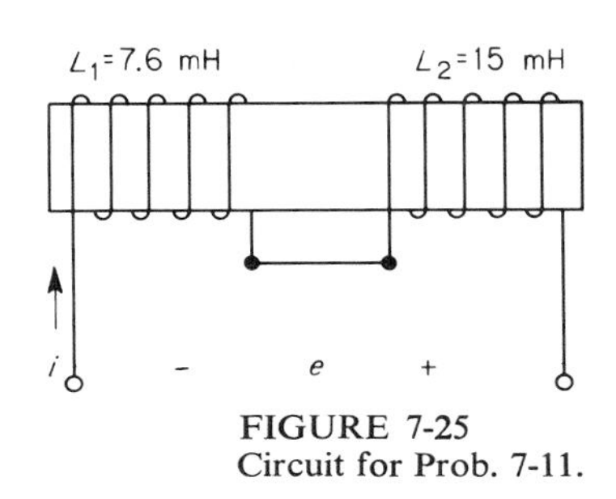

FIGURE 7-25
Circuit for Prob. 7-11.

7-6 What is the inductance of a coil wound upon a ferromagnetic core if the diameter of the coil is 22 mm, the length of the coil is 15 cm, the permeability of the magnetic path is 17×10^{-4}, and the number of turns is 625?

7-7 The coefficient of coupling between two coils is 0.46. The flux changes from 680 to 236 mWb in the first coil in a time interval of 450 ms. If the second coil consists of 195 turns, what is the induced voltage in the second coil?

7-8 What is the coefficient of coupling between two coils if a flux change of 65,000 lines in one coil induces a voltage of 53.2 V in the other coil, which has 340 turns? The time interval is 3.73 ms.

7-9 What is the mutual inductance between two coils if the coefficient of coupling is 0.8, the inductance of one coil is 75 mH, and the inductance of the other coil is 180 mH?

7-10 What is the coefficient of coupling between the coils of Fig. 7-8 if the mutual inductance between them is 150 μH?

7-11 Given the circuit of Fig. 7-25, find the total inductance of the circuit if the coefficient of coupling between the coils is 0.78.

7-12 Two coils are connected in series. The inductance of one coil is 15 H, and the inductance of the other coil is unknown. The total inductance of the circuit, including mutual inductance, is measured and is found to be 31 H's. The coils are then connected in series-opposing and the total inductance is found to be 1.9 H's. Solve for the unknown inductance.

7-13 In the circuit of Fig. 7-26, the total inductance when the switch is at position *A* is 208 mH. When the switch is at position *B*, what is the total inductance? What is the coefficient of coupling between the coils?

7-14 In the circuit of Fig. 7-27, coils 2 and 3 are connected in series-opposing. What is the total inductance of the circuit? The coefficient of coupling between coils 2 and 3 is 0.86. Coil 1 is not mutually coupled to coils 2 and 3.

7-15 What is the total inductance of the circuit of Fig. 7-28? The mutual inductance between coils is 150 mH. This mutual inductance is opposing.

7-16 For the circuit of Fig. 7-29, at what time after switch S_1 is closed does the current reach 780 mA?

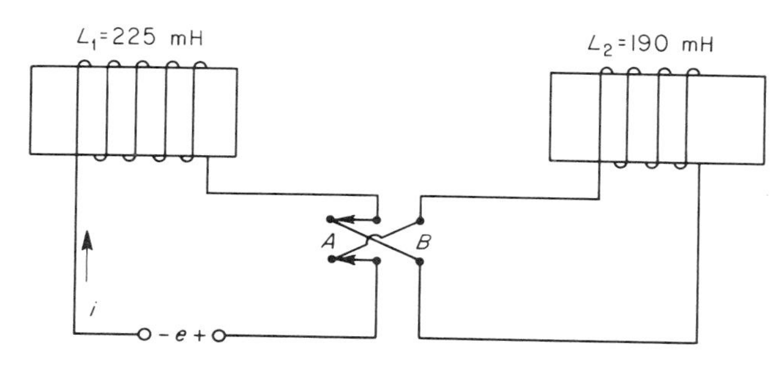

FIGURE 7-26
Circuit for Prob. 7-13.

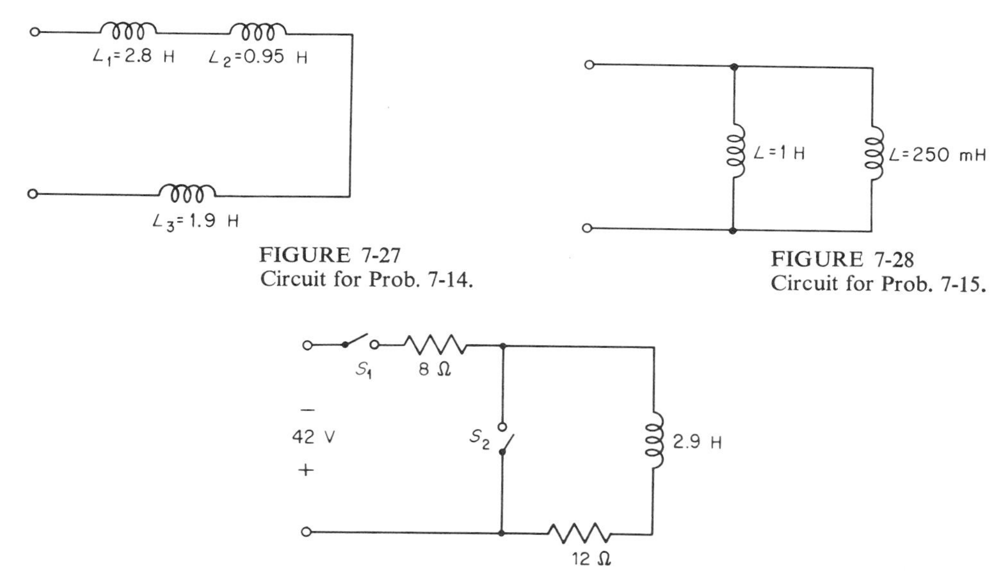

FIGURE 7-27
Circuit for Prob. 7-14.

FIGURE 7-28
Circuit for Prob. 7-15.

FIGURE 7-29
Circuit for Probs. 7-16 and 7-17.

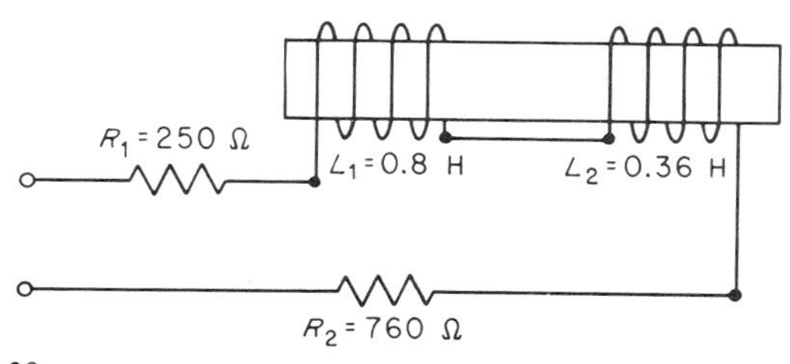

FIGURE 7-30
Circuit for Probs. 7-18 and 7-20.

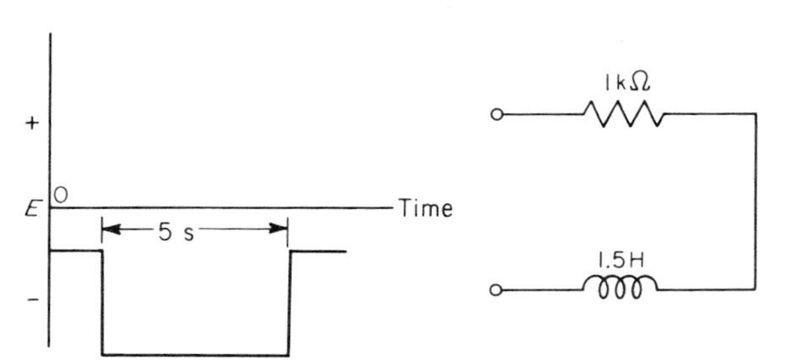

FIGURE 7-31
Circuit for Prob. 7-19.

7-17 In the circuit of Fig. 7-29, assume that switch S_1 has been closed so long that the current through the circuit is limited only by resistance. How soon after switch S_2 is closed does the current through the coil drop to 200 mA?

7-18 For the circuit of Fig. 7-30, where the coefficient of coupling is 0.77, find the time constant.

7-19 Sketch the waveforms of voltage across the resistor and the coil in the circuit of Fig. 7-31 when the voltage shown is applied to the circuit.

7-20 What is the time constant of the circuit of Fig. 7-30 if the coils are connected in series-opposing?

7-21 A coil has an inductance of 0.5 H. What resistance must be placed in series with the coil to give the combination a time constant of 2.1×10^{-3} s?

7-22 A long solenoid consists of 450 turns of wire. The diameter of the solenoid is 0.75 in., the length of the solenoid is 8 in., and the permeability of the magnetic path is 55×10^{-5}. If the current through the solenoid is 5.1 A, what is the energy stored in the magnetic field surrounding the solenoid?

7-23 If the current in the solenoid of Prob. 7-22 is decreased from 5.1 to 4 A, how much energy is returned to the circuit?

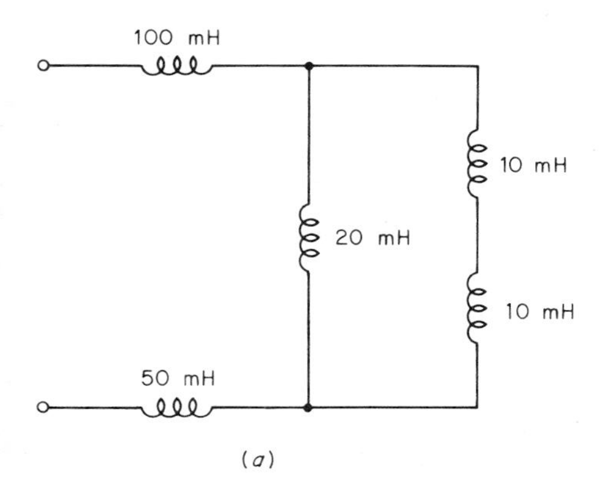

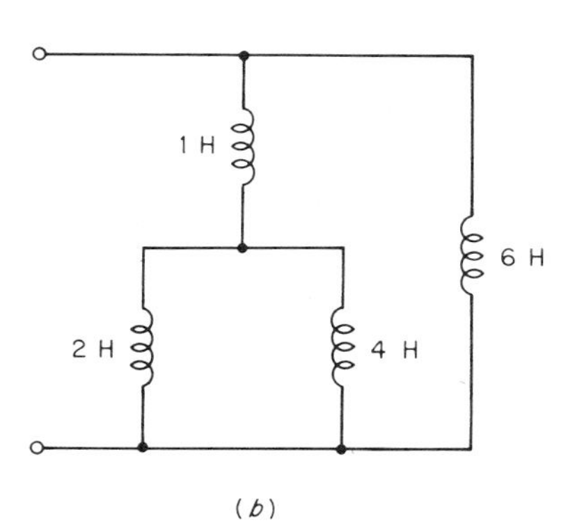

FIGURE 7-32
Circuits for Prob. 7-25.

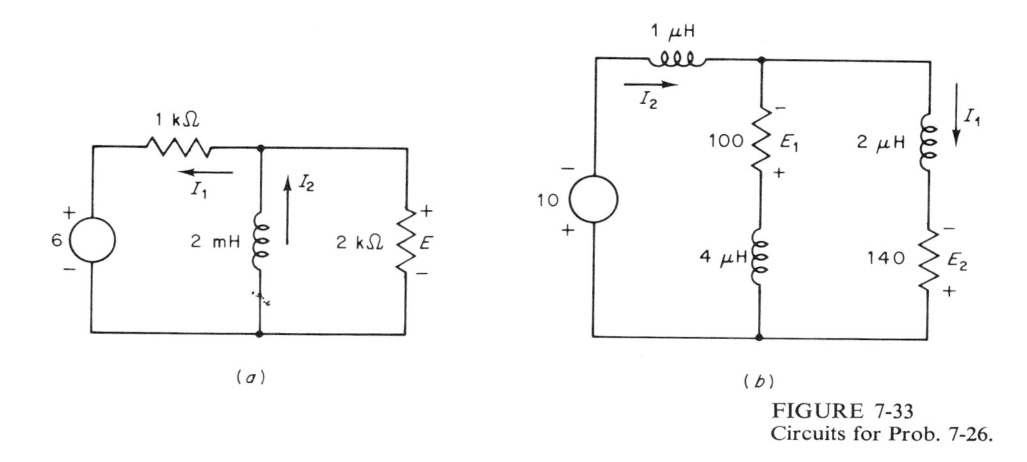

FIGURE 7-33
Circuits for Prob. 7-26.

7-24 A relay coil has a resistance of 450 Ω and an inductance of 6.8 H. How soon after 22 V is applied to the coil does the relay operate? The relay will operate when 40 mA flows through the coil.

7-25 Find the total inductance for the circuits in Fig. 7-32. Assume no mutual coupling.

7-26 Determine the indicated voltages and currents for the circuits in Fig. 7-33. Assume source applied for a long time.

8

ALTERNATING VOLTAGES AND CURRENTS

8-1 INTRODUCTION

An alternating voltage is *any* voltage that varies in both amplitude and polarity with respect to time. The voltage may vary in a regular, predictable manner or in an irregular, nonrepetitive manner with respect to time. In either case, the voltage is considered to be an alternating voltage. Figure 8-1 shows an alternating voltage that varies in a regular manner with respect to time.

The alternating voltage illustrated in Fig. 8-2 is a video signal of the type used in television broadcasting. The amplitude of the voltage at any instant depends on the light from the scene being televised and is obviously not a predictable quantity.

An alternating current is any current that varies in both amplitude and direction. As with alternating voltages, there are no limitations on rate of change or waveshape. An alternating current is simply a current that changes amplitude and direction with time.

Alternating currents and voltages are widely used to distribute electric power. However, the uses of alternating voltages and currents extend far beyond this application. All electronic communication systems, electronic computers, and electronic instrumentation systems require alternating currents and voltages as well as direct

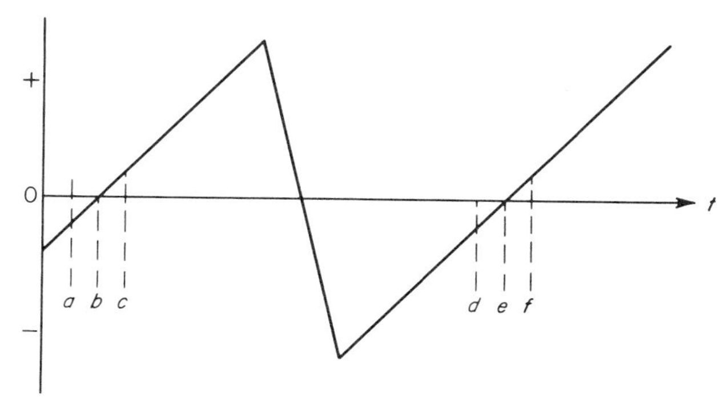

FIGURE 8-1
A sawtooth waveform of voltage.

currents and voltages. When alternating voltages and currents supply electric power to operate other devices, the ac (the notation ac is common usage to denote either an alternating voltage or current or both) is usually produced by huge alternators (ac generators) operated by power companies.

Electronic devices may also be used to develop ac voltages and currents. In this case, the source of the alternating voltages and currents is a circuit called an *oscillator*. An oscillator is an electronic circuit that converts direct current into alternating current. The actual power that is supplied by an oscillator depends on the size of the components of the circuit, the efficiency of the oscillator, and the magnitude of the dc input power. In practice, output may range from the few milliwatts supplied by the oscillator of a superheterodyne radio receiver to the millions of watts of pulse power supplied by the oscillator of a missile-warning radar.

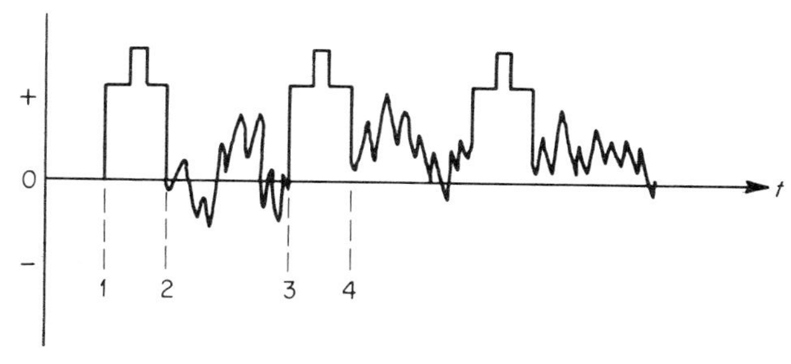

FIGURE 8-2
A typical video signal.

8-2 FREQUENCY AND PERIODICITY

DC voltages and currents are easily defined in magnitude. Alternating voltages and currents, however, cannot be exactly defined in terms of magnitude only. All alternating voltages and currents have three characteristics: *amplitude*, *frequency*, and *phase*. This section is concerned with the frequency characteristic of ac voltages and currents.

In Sec. 8-1, it was stated that any voltage or current that changes polarity or direction is considered to be ac. However, the great majority of all ac voltages and currents change in magnitude and direction at predetermined rates. That is, an ac voltage rises to a maximum value, decreases from maximum to zero, then rises to maximum value of the opposite polarity, and again decreases to zero. It repeats this process continuously.

A cycle of alternating voltage or current consists of one complete transition from some point on an ac waveform to the same point on the following ac waveform. For example, one cycle of the ac waveform of Fig. 8-1 may be measured between points a and d, b and e, or c and f. In Fig. 8-2, one cycle may be measured between points 1 and 3 or between points 2 and 4.

The number of cycles per second is defined as the frequency of an ac voltage or current. The unit of frequency is the hertz (Hz). One hertz is equivalent to one cycle per second. The unit hertz was selected to honor a pioneer in electromagnetic wave research, Heinrich Rudolf Hertz. Common power-line frequency in the United States is 60 Hz, while the frequency of a radio broadcasting station may be 1M Hz. Television stations operate at frequencies on the order of 10^8 Hz.

Some fundamental mathematical relationships can now be written relating frequency to the time of one cycle. Since frequency equals cycles per second, it follows that

$$T = \frac{1}{f} \tag{8-1}$$

where T is the time of one cycle, in seconds, and f is the frequency in hertz. When the time of one cycle is known, frequency is found by

$$f = \frac{1}{T} \tag{8-2}$$

EXAMPLE 8-1 A radio station operates at a frequency of 1,420 kHz. What is the time of one cycle?

SOLUTION

$$T = \frac{1}{1.42 \times 10^6} = 0.705 \times 10^{-6} \text{ s} = 0.705\ \mu\text{s} \qquad ////$$

EXAMPLE 8-2 A certain signal has a period (the time of one cycle) of 250 μs. What is the frequency of this signal?

SOLUTION

$$f = \frac{1}{250 \times 10^{-6}} = 4 \times 10^3 \text{ Hz} = 4 \text{ kHz} \qquad ////$$

8-3 SINUSOIDAL VOLTAGES AND CURRENTS

A unique waveform for ac voltage or current is the sine wave. In preceding sections, it has been stated that an alternating voltage or current may have any waveshape. This is indeed true, but this very fact could make mathematical analysis of ac circuits very laborious. However, it can be shown mathematically and demonstrated graphically that *any waveshape, no matter how irregular, consists of various combinations of sinusoidal waveshapes*. Therefore, the unique feature of the sine wave is that it is basic to all ac voltages and currents!

In the sine wave, one complete cycle is represented by 360° or 2π radians (rad). Hence, if the period of a sine wave is 0.2 s, then each degree of the cycle represents 0.556 ms. At any instant, the amplitude of the sine wave equals the product of the maximum value of the sine wave and the sine of the angle corresponding to time. The equation for a sine wave of voltage is

$$e = E_{\max} \sin \theta \tag{8-3}$$

The equation for a sine wave of current is written in a similar manner.

$$i = I_{\max} \sin \theta \tag{8-4}$$

In Eqs. (8-3) and (8-4), θ is any angle.

Figure 8-3 is a sine wave of voltage, showing the substitution of angular measure in degrees and radians for time.

EXAMPLE 8-3 Given a sine wave of current with a maximum amplitude of 45 mA, find the amplitude of current at (1) 25°, (2) 70°, and (3) 210°.

SOLUTION

1 $\quad i = (45 \times 10^{-3}) \sin 25° = (45 \times 10^{-3})(0.422) = 19 \text{ mA}$

2 $\quad i = (45 \times 10^{-3}) \sin 70° = (45 \times 10^{-3})(0.939) = 42.3 \text{ mA}$

3 $\quad i = (45 \times 10^{-3}) \sin 210° = (45 \times 10^{-3})(-0.5) = -22.5 \text{ mA} \qquad ////$

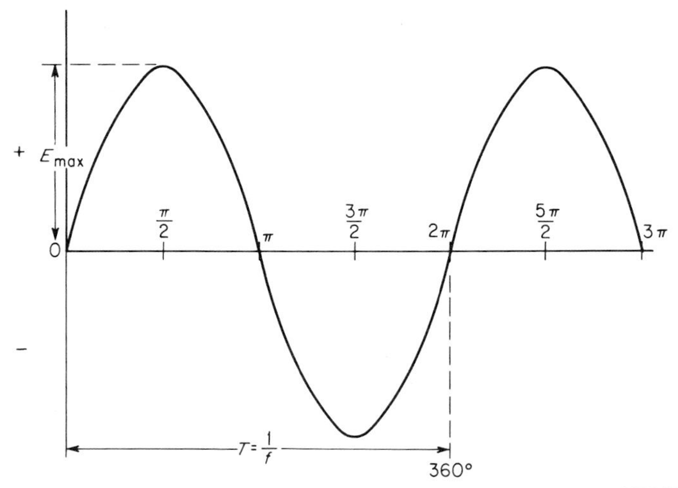

FIGURE 8-3
Sine wave of voltage.

EXAMPLE 8-4 A 1,000-Hz sine wave signal has a maximum amplitude of 24 V. What is the instantaneous amplitude of the signal 50 μs before it reaches its peak positive value?

SOLUTION

1 Determine the period of the signal.

$$T = \frac{1}{f} = 1{,}000\ \mu\text{s}$$

2 Determine the number of degrees per microsecond.

$$1{,}000\ \mu\text{s} = 360°$$
$$1\ \mu\text{s} = 0.36°$$

3 Therefore 50 μs corresponds to 18°(50 × 0.36).

4 The peak maximum positive value of a sine wave occurs at 90°. Therefore, 18° before peak positive value is 72°(90° − 18°).

5 The solution to the problem is found with Eq. (8-3).

$$e = E_{\text{max}} \sin 72° = 22.8\ \text{V}$$ ////

In addition to the graphic representation of a sine wave, like that in Fig. 8-3, a sine wave may be represented by a radius vector, or *phasor*. The phasor has a constant amplitude equal to the maximum amplitude of the sine wave, and the instantaneous

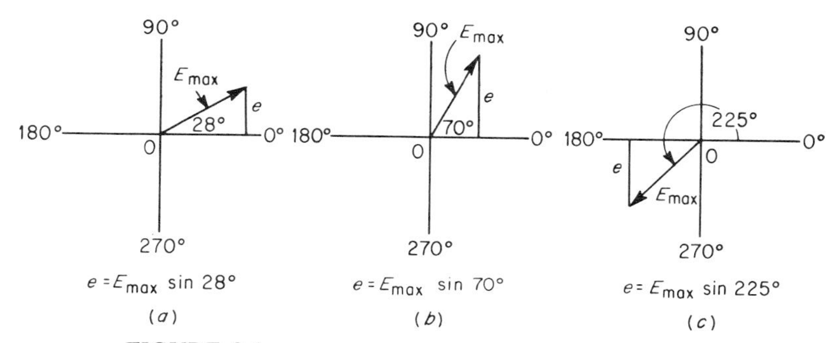

FIGURE 8-4
Phasor representation of a sine wave of voltage showing instantaneous values at (*a*) 28°, (*b*) 70°, and (*c*) 225°.

value of the sine wave is the product of the phasor and the sine of the angle between the phasor and the origin. The phasor representation is extremely useful in adding and subtracting alternating voltages and currents. Figure 8-4 illustrates phasor representation for the sine wave of Fig. 8-3.

Discussion of phasor representation of sine waves leads logically to another useful concept. *Angular velocity* is normally associated with rotating machinery. However, a phasor representing a sine wave may be visualized as a rotating vector, and as such it too has angular velocity. As indicated in Fig. 8-4, the positive direction of rotation is counterclockwise (ccw).

Velocity is the ratio of distance to time. The angular velocity of a sine wave is the "distance" of one cycle, in radians, divided by the period of the sine wave. Angular velocity is represented by lowercase omega (ω).

$$\omega = \frac{2\pi}{T} \tag{8-5}$$

However, $T = 1/f$ [Eq. (8-2)]. If this value of T is substituted into Eq. (8-5), then

$$\omega = 2\pi f \qquad \text{rad/s} \tag{8-6}$$

EXAMPLE 8-5 What is the angular velocity of a sine wave of voltage if the frequency of the signal is 5.8 MHz?

SOLUTION From Eq. (8-6),

$$\omega = (2\pi)(5.8 \times 10^6) = 36.4 \times 10^6 \text{ rad/s} \qquad ////$$

EXAMPLE 8-6 A certain signal has an angular velocity of $1{,}200 \times 10^3$ rad/s. What is the frequency of this signal?

SOLUTION

$$\omega = 2\pi f$$

$$f = \frac{\omega}{2\pi} = \frac{1{,}200 \times 10^3}{2\pi} = 191 \text{ kHz} \qquad ////$$

Equations (8-3) and (8-4) may be rewritten in terms of Eq. (8-6). The angular velocity of a sine wave is a constant, and the particular angle of a sine wave at any instant is a direct function of time. Hence, if the angular velocity is multiplied by time in seconds, the product is an angle in radians.

$$\text{Radians} = \text{angular velocity} \times \text{time}$$

Equations (8-3) and (8-4), when radian measure is used, are written

$$e = E_{max} \sin \omega t \tag{8-7}$$

$$i = I_{max} \sin \omega t \tag{8-8}$$

Equations (8-3) and (8-4) are used where direct computations are required. Equations (8-7) and (8-8) are used where derivations or concepts are to be handled. As a matter of fact, a numerical solution with Eqs. (8-7) and (8-8) requires that radians be converted into degrees (1 rad = 57.3°).

EXAMPLE 8-7 A certain signal has a frequency of 440 Hz. If the maximum amplitude of the signal is 500 V, what is the amplitude of the signal 10^{-4} s after the start of a cycle?

SOLUTION

1

$$e = E_{max} \sin (2\pi \times 440 \times 10^{-4} \text{ rad})$$

$$e = E_{max} \sin (27.6 \times 10^{-2} \text{ rad})$$

2 Convert radians to degrees.

3

$$57.3 \times 0.276 = 15.8°$$

$$e = E_{max} \sin 15.8° = 500 \times 0.272 = 136 \text{ V} \qquad ////$$

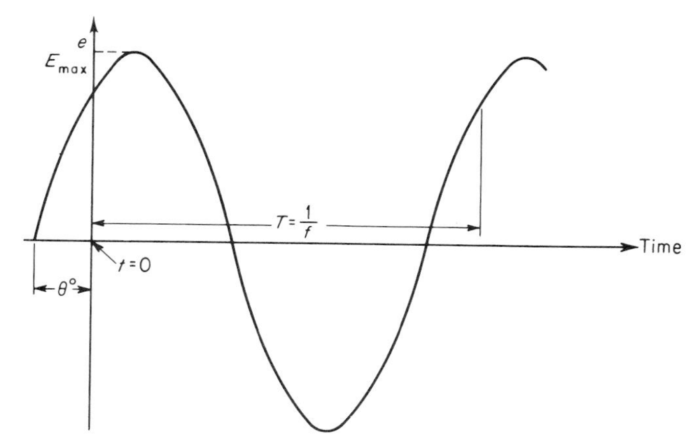

FIGURE 8-5
Voltage waveform.

8-4 PHASE ANGLE AND PHASE DIFFERENCE

In Sec. 8-3, it was noted that all ac voltages and currents have three characteristics, frequency, amplitude, and phase. In this section, the phase characteristics of a sine wave will be discussed.

In the equation of a sine wave, the independent variable is time. In both representations of the sine wave, by graph or by phasor, an angular notation has been substituted for time. It should be apparent from the equation of the sine wave that all sinusoids have zero amplitude at the time that the angular equivalent of time is zero. It is conventional to represent a sine wave as starting at 0°. However, it is equally permissible to consider a sine wave as starting at any other point on its cycle. Figure 8-5 illustrates a sine wave of voltage that is not zero at the start of its cycle.

When a sine wave is considered to start at some magnitude other than zero, the fact must be indicated in the equation of the wave. The angular displacement of a wave from 0° to the point on its cycle where the wave is considered to begin is its *phase angle*. For example, in Fig. 8-5 θ is the phase angle of the wave.

The equation for the voltage waveform of Fig. 8-5 is written

$$e = E_{max} \sin (\omega t + \theta°) \tag{8-9}$$

Figure 8-6 illustrates a sine wave of current described by the equation

$$i = I_{max} \sin (\omega t - \phi°) \tag{8-10}$$

In ac circuits that contain capacitance, inductance, or both, the phase angles of current and voltage can differ from one another. That is, the current in the circuit

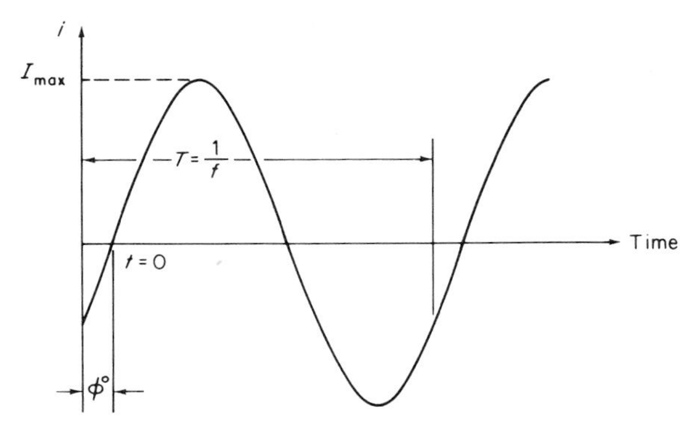

FIGURE 8-6
Current waveform.

may reach maximum or minimum at different times than the voltage. This time between alternating quantities is called *phase difference* and is expressed in degrees. Phase difference may also express the time displacement between waves of different frequencies that are present in the same circuit.

It should be apparent that the phase difference between sine waves of different frequencies is constantly changing. However, it is often convenient to express the phase difference between signals of different frequencies at some particular instant in time. When alternating quantities of the same frequency reach positive maxima (or

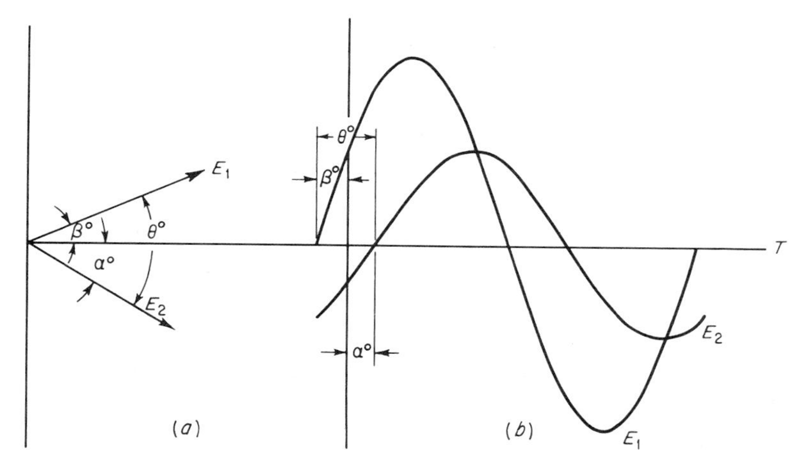

FIGURE 8-7
Two phasor quantities of the same frequency.

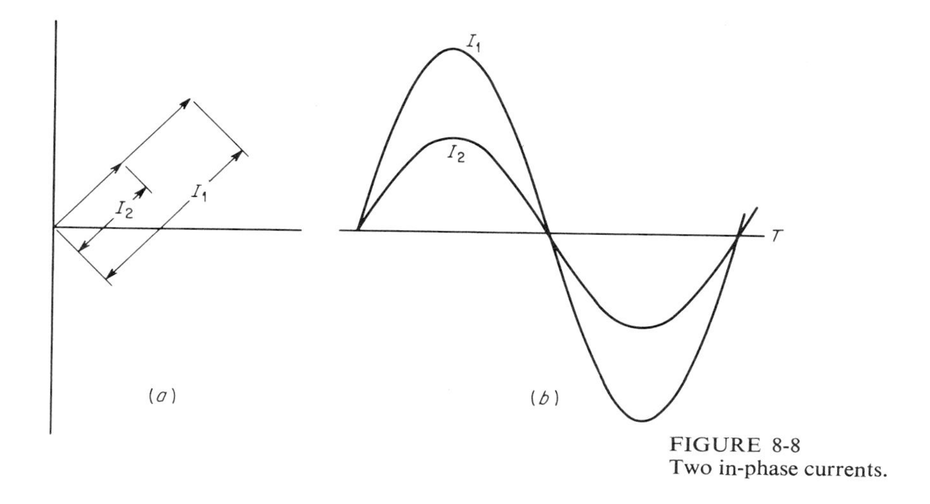

FIGURE 8-8
Two in-phase currents.

any other convenient reference point on the cycle) at the same instant, the quantities are said to be *in phase:* the phase difference between them is 0°.

Figure 8-7*a* shows two phasors of the same frequency displaced from each other by $\theta°$. E_1 is said to be leading E_2 by $\theta°$ (counterclockwise rotation of phasors, as noted earlier, is the positive direction).

The equations for voltages E_1 and E_2 in Fig. 8-7 are

$$e_1 = E_1 \sin(\omega t + \beta°)$$
$$e_2 = E_2 \sin(\omega t - \alpha°)$$

where $\theta = \beta + \alpha$.

Note that the phase difference between E_1 and E_2 is the difference between angles β and α. It may be stated that E_1 leads the reference axis by $\beta°$ and E_2 lags this same reference by $\alpha°$. Figure 8-8 illustrates two currents that are in phase with each other. Part *a* is the phasor representation of these currents, and part *b* shows the currents as sinusoids.

8-5 THE AVERAGE VALUE OF A SINE WAVE

The average value of *any* current or voltage is the value that would be indicated by a *dc meter*. This concept is of particular value in electronics, since many voltages and currents are combinations of direct current and sinusoids. The concept of average values is of particular importance in rectifier circuits.

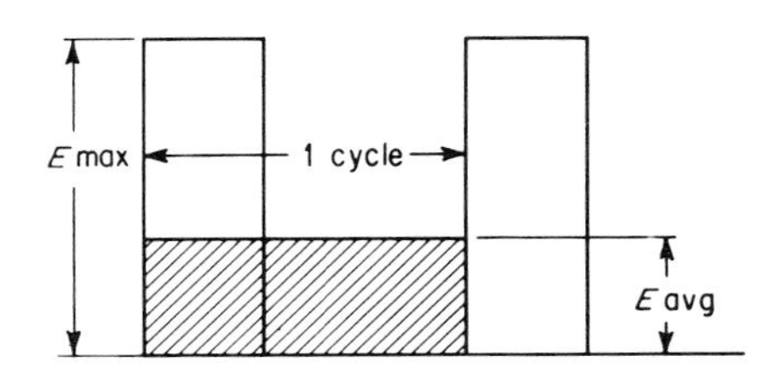

FIGURE 8-9
Average value of a pulse.

The average value of any curve is the area enclosed by the curve divided by the base of the curve. Figure 8-9 shows one cycle of a rectangular pulse of voltage, and it illustrates the average value of this pulse *over one cycle.*

It is apparent that the average value of a sine wave over a complete cycle is zero, since the average of one half of the cycle is exactly equal but opposite in polarity to the average of the other half. The average value of a sine wave is obtained by assuming that it has been rectified. That is, both halves of the waveform are assumed to be positive. A rectified sine wave is shown in Fig. 8-10.

Calculation of the average value of a rectified sine wave is accomplished with integral calculus. This process yields the average value of the curve from 0 to π rad. This average value is also that of the rectified sine wave over a full cycle. The average value of voltage is

$$E_{\mathrm{av}} = 0.636E_{\mathrm{max}} \tag{8-11}$$

The average value of a sine wave of current is

$$I_{\mathrm{av}} = 0.636I_{\mathrm{max}} \tag{8-12}$$

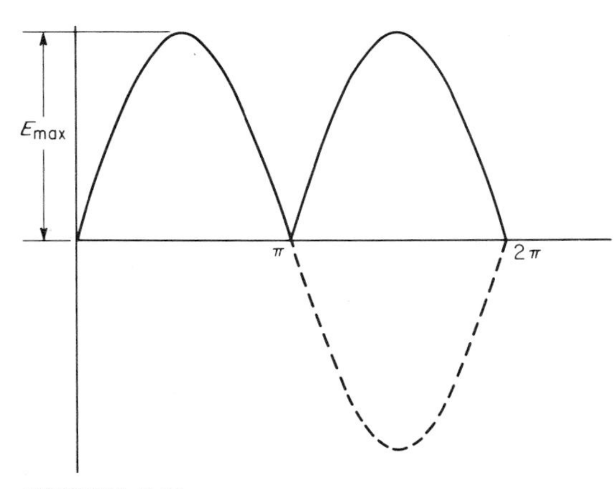

FIGURE 8-10
Rectified sine wave.

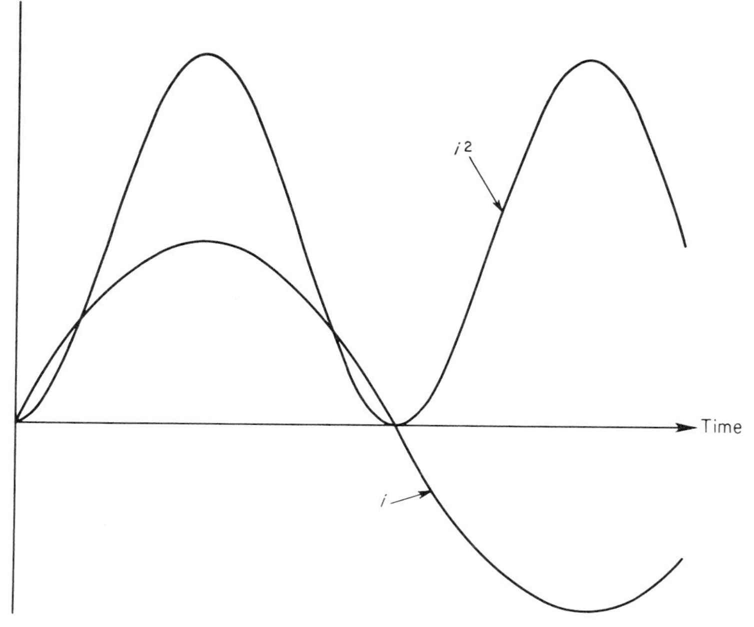

FIGURE 8-11
Plot of a sine wave of current and the square of this current.

8-6 THE EFFECTIVE VALUE OF A SINE WAVE

The effective value of a current or voltage waveform is that value which will dissipate the same power as a *numerically equal* direct current or voltage. For example, an alternating current of 2 A effective value dissipates exactly the same power as 2 A direct current. Note that no consideration is given to the waveshape of the alternating current; we simply state that an effective current of 2 A ac develops the same power as 2 A dc. In short, the effective value is defined in terms of power dissipation. It follows, then, that the effective value of a sine wave of current or voltage is that value which will dissipate the same power as a numerically equal direct current or voltage.

The power dissipated in a resistance is I^2R or E^2/R. To solve for the effective value of a sine wave of current, the following relationship can be set up:

$$(I_{\text{eff}})^2 R = (I_{\text{dc}})^2 R \tag{8-11}$$

Figure 8-11 illustrates a sine wave of current and the square of the current. Note that the ordinates of the square of the current are always positive and that the frequency is twice the frequency of the sine wave of current. The effective value of the sine wave of current must be found with calculus, and it is the square root of the average value of the current-squared waveform. The effective value is frequently referred to as the root-mean-square (rms) value. The effective values of a sine wave are

$$I_{eff} = 0.707 I_{max} \tag{8-13}$$

$$E_{eff} = 0.707 E_{max} \tag{8-14}$$

EXAMPLE 8-8 What is the peak value of a sine wave of voltage that dissipates the same power as 115 V dc?

SOLUTION The effective value of the voltage is 115 V. From Eq. (8-14), the peak value is

$$E_{max} = \frac{115}{0.707} = 1.414 \times 115 = 163 \text{ V} \qquad ////$$

EXAMPLE 8-9 A certain current has a peak value of 235 mA. What is the effective value of this current?

SOLUTION The rms value is found with Eq. (8-13).

$$I_{eff} = (0.707)(235 \times 10^{-3}) = 166 \times 10^{-3} \text{ A} = 166 \text{ mA} \qquad ////$$

Note that lowercase letters are used to indicate instantaneous values of current or voltage whenever the current or voltage is variable with time. Definite values of current or voltage are indicated with capital letters (E_{max}, I_{max}, etc.). Average or effective values are always indicated by capital letters with the notation E_{av}, I_{av}, E_{eff}, or I_{eff}.

PROBLEMS

8-1 A sine wave of voltage has an instantaneous amplitude of 12 V at an angle of 120°. Find (*a*) the peak amplitude of the wave, (*b*) the effective value of the wave, and (*c*) the average value of the rectified wave.

8-2 Given the voltage equation $e = 196 \sin 377t$, find (*a*) the frequency of the voltage, (*b*) the effective value of the voltage, and (*c*) the amplitude of the voltage at the instant $t = 100$ ms.

8-3 What is the period of the voltage in the equation of Prob. 8-2?

8-4 A sine wave of voltage has a period of 31 μs and a peak amplitude of 45 mV. Write the general equation for this voltage.

8-5 Given the current equation $i = 14 \sin (12.5 \times 10^3 t)$mA, find (*a*) the frequency of the current and (*b*) the instantaneous amplitude of the current 225° after the start of a cycle.

8-6 Ten microseconds after the start of a cycle a certain sine wave of voltage reaches one-half its maximum amplitude. Find (*a*) the frequency of the voltage and (*b*) the period of the sine wave.

8-7 A 250-Hz sine wave of voltage has a peak amplitude of 48 μV. What is the instantaneous amplitude of this voltage 1 ms after the signal has gone through zero in a negative direction?

8-8 A sine wave of current has a peak amplitude of 15 mA. (*a*) What is the average value of this current? (*b*) What is the effective value of this current? (*c*) What is the instantaneous amplitude of this current at 205°?

8-9 The period of a certain sine wave is 10 ms. What is its angular velocity?

8-10 A certain voltage has the equation $e = E_{max} \sin (\omega t + 45°)$. (*a*) Draw a graph representing the voltage. (*b*) How many degrees are needed for the voltage to reach its maximum negative value?

8-11 A certain current is represented by the equation $i = I_{max} \sin (\omega t - 210°)$. (*a*) Plot a graph of this current. (*b*) What is the amplitude of this current at the reference point?

8-12 Plot a graph of the voltage $e = E_{max} \sin (\omega t + 180°)$.

8-13 Two currents enter a junction in a circuit. One is represented by the equation $i = I_{max} \sin (\omega t + 85°)$. The phase difference between the two currents is 61°. Write the two equations that can represent the other current.

8-14 What is the phase difference between two voltages when one is represented by the equation $e = E_{max} \sin (\omega t - 18°)$ and the other is represented by $e = E_{max} \sin (\omega t - 240°)$?

8-15 A sinusoidal current has a peak amplitude of 1.2 A. What is the effective value of this current?

8-16 If the frequency of the current of Prob. 8-15 is 60 Hz, what is the instantaneous amplitude of the current at 150° of the cycle?

8-17 A sinusoidal current develops a power of 150 W in a 440-Ω resistor. What is the effective value of this current?

8-18 A peak voltage of 25 V develops 50 mW of power in a certain resistor. What is the value of the resistance?

8-19 A load draws 12 A (rms) from a power line. If the power dissipated in the load is 2,400 W, what is the peak amplitude of the line voltage?

8-20 What rms current flows through a load resistance of 2,500 Ω if the power developed in the load is 140 μW?

9

RESISTANCE, INDUCTANCE, AND CAPACITANCE IN AC CIRCUITS

9-1 INTRODUCTION

Generally, the waveforms for the alternating voltages and currents of a number of electronic devices do not differ appreciably from a sinusoid. Further, the nonsinusoidal waveforms can be separated into sinusoidals for purposes of analysis. It will be demonstrated later that all alternating waveforms are composed of sinusoids. Therefore, in all discussion of voltages and currents in this chapter, the sinusoid will be used as the basis of mathematical analysis.

In following sections, the effects of the properties of the electric circuit will be examined. These circuit properties, or circuit parameters, are resistance, inductance, and capacitance. It must be pointed out that it is impossible to construct a purely resistive, purely inductive, or purely capacitive circuit. In discussing pure resistance, pure inductance, or pure capacitance, we are considering hypothetical circuits.

9-2 THE PURE RESISTANCE CIRCUIT

A sinusoidal voltage is applied to a resistance as shown in Fig. 9-1. The resulting time variations in the current and voltage waveforms are shown in Fig. 9-2.

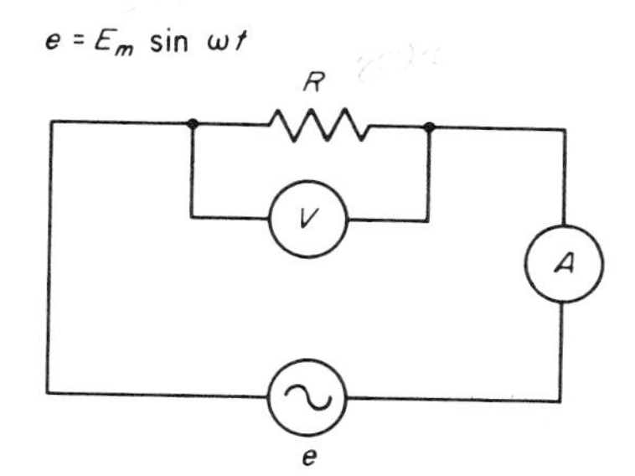

FIGURE 9-1
Pure resistive circuit.

From the waveforms in Fig. 9-2, the following can be concluded:

1 Both waves are sinusoidal and have the same frequency.
2 The waves are in phase.

The waveform relations of Fig. 9-2 can be expressed mathematically. The voltage applied to the resistance of Fig. 9-1 is

$$e = E_m \sin \omega t$$

Applying Ohm's law at any instant,

$$i = \frac{e}{R} = \frac{E_m \sin \omega t}{R} = I_m \sin \omega t$$

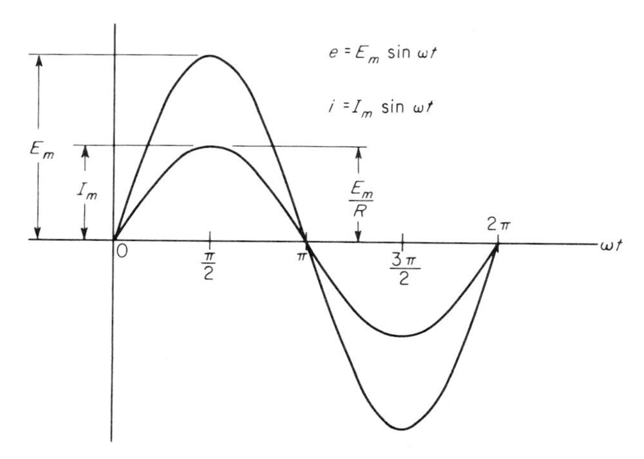

FIGURE 9-2
In-phase current and voltage.

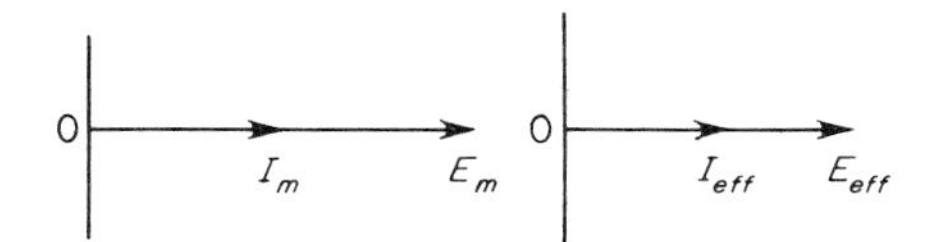

FIGURE 9-3
Phasor diagram of maximum and effective values.

where

$$I_m = \frac{E_m}{R}$$

Figure 9-3 shows two phasor diagrams. In one the current phasor and voltage phasor are shown as maximum values. In the other, they are shown as effective values. Since these revolving time phasors represent sinusoids, it would be more proper to use maximum values. In practice, it is more convenient to use effective values, and furthermore, the effective values are those which would be indicated by meter readings. Effective values will be used on all phasor diagrams.

Ohm's law can thus be applied to the effective magnitudes of the phasor current and phasor voltage. Conductance G can be used in the Ohm's law equations for the phasor current and voltage just as it is in dc circuits.

To demonstrate the Ohm's-law relations of alternating current and voltage to a resistance the following example is given.

EXAMPLE 9-1 The voltage applied to the resistance of Fig. 9-1 is $e = 150 \sin 377t$. The resistance is 800 Ω. Calculate (1) the maximum value of current, (2) the instantaneous current at $t = 0.002$ s, and (3) effective value of current.

SOLUTION

1

$$I_m = \frac{E_m}{R} = \frac{150}{800} = 0.188 \text{ A}$$

2 Calculating the instantaneous voltage when $t = 0.002$ s,

$$e = 150 \sin[(377)(0.002)] = 150 \sin 0.754 \qquad 0.754 \text{ rad} = 43.2°$$
$$= 150 \sin 43.2° = (150)(0.682) = 102.5 \text{ V}$$

The instantaneous current is

$$i = \frac{e}{R} = \frac{102.5}{800} = 0.128 \text{ A}$$

The instantaneous current can also be calculated directly.

$$i = I_m \sin \omega t = 0.188 \sin 377t$$
$$= 0.188 \sin[(377)(0.002)] = 0.188 \sin 0.754$$
$$= 0.188 \sin 43.2^\circ = (0.188)(0.682) = 0.128 \text{ A}$$

3

$$I_{eff} = \frac{E_{eff}}{R} = \frac{(0.707)(150)}{800} = 0.133 \text{ A}$$

Also $I_{eff} = 0.707 I_m = (0.707)(0.188) = 0.133$ A. ////

The average power dissipated in a resistance is given by the equation

$$P = E_{eff} I_{eff} \cos \theta \tag{9-1}$$

In Eq. (9-1), voltage and current are given as effective values: $\cos \theta$ is the power factor, and θ is the phase angle between current and voltage. Since in a pure resistance circuit the current and voltage are in phase, the angle $\theta = 0$ and $\cos \theta = 1$. Equation (9-1) therefore reduces to Eq. (9-2) for a purely resistive circuit.

$$P = E_{eff} I_{eff} \tag{9-2}$$

Since the effective values of current and voltage were defined in Chap. 8 in terms of equivalent dc quantities, Eq. (9-2) is consistent.

9-3 THE PURE INDUCTANCE CIRCUIT

Inductance was defined in Chap. 7 as the property of a circuit to oppose *change* in current. This opposition results in induced emf. The induced emf is proportional to the rate at which current is changing as well as the magnitude of the inductance. This relationship was given by the equation

$$e_L = L \frac{di}{dt} \tag{7-5}$$

If a sinusoidal current is flowing in the inductance shown in Fig. 9-4, the induced voltage across the inductance can be plotted versus time.

The plot of current in Fig. 9-5 increases from time t_0 to t_1. The current is increasing at a decreasing rate, and at time t_1 the instantaneous rate of change of current is zero. Therefore, e_L is zero at time t_1. From time t_1 to t_2 the current is decreasing; the rate of change of current is negative and e_L is negative. At time t_2, the rate of change of current is maximum, and thus e_L is maximum in the negative direction. At time t_3, the rate of change of current is zero, and thus e_L is again zero.

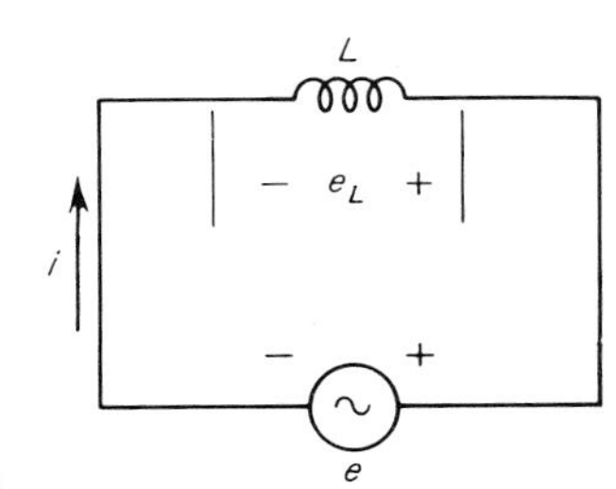

FIGURE 9-4
Pure inductive circuit.

From t_3 to t_4, the current is increasing, the rate of change is positive, and e_L is positive, reaching maximum at t_4, where the rate of change is maximum.

The maximum positive value of voltage occurs 90° ahead of the maximum positive value of current. The current is said to be lagging the voltage by 90 electrical degrees. This phase relationship can be derived mathematically by applying calculus.

From Eq. (7-5), which is

$$e_L = L\frac{di}{dt}$$

where

$$i = I_m \sin \omega t$$

by differentiation, di/dt is found.

$$\frac{di}{dt} = \omega I_m \cos \omega t$$

but

$$\cos \omega t = \sin(\omega t + 90°) \qquad \text{trigonometric identity}$$

Therefore,

$$e_L = \omega L I_m \sin(\omega t + 90°)$$

By the general form of a periodic function,

$$E_m = \omega L I_m$$

and

$$\omega L = \frac{E_m}{I_m} \tag{9-3}$$

Since the ratio of volts to amperes is defined as opposition to current in ohms the quantity ωL is measured in ohms. The quantity ωL is called the *inductive reactance* and is symbolized X_L.

$$X_L = \omega L$$
$$= 2\pi f L \qquad \Omega \tag{9-4}$$

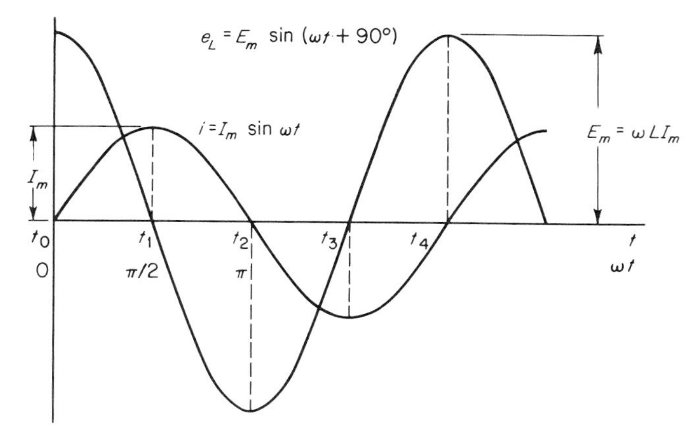

FIGURE 9-5
Voltage leading current by 90°; the voltage-current relationship in a pure inductive circuit.

If $E_m = 1.414E_{\text{eff}}$ and $I_m = 1.414I_{\text{eff}}$ are substituted into Eq. (9-3), it is seen that the ratio of the effective values of voltage and current also equals the inductive reactance

$$X_L = \omega L = \frac{E_m}{I_m} = \frac{E_{\text{eff}}}{I_{\text{eff}}} \tag{9-5}$$

The reciprocal of inductive reactance is called *inductive susceptance* and is given the symbol B_L. The unit of inductive susceptance is the mho when the frequency is in hertz and the inductance is in henrys.

$$B_L = \frac{1}{X_L}$$

$$= \frac{1}{2\pi f L} \quad \text{mhos} \tag{9-6}$$

EXAMPLE 9-2 An alternating current with a frequency of 2k Hz and a maximum value of 0.15 A flows in a coil having 175-mH inductance. (1) Find the maximum voltage developed across the inductance. (2) Find the effective value of the voltage across the inductance. (3) Write the periodic functions representing the voltage and current.

SOLUTION

1 $E_m = I_m X_L = (0.15)(2\pi f L) = (0.15)(2\pi)(2 \times 10^3)(0.175) = 330 \text{ V}$

2 $E_{\text{eff}} = 0.707E_m = (0.707)(330) = 233 \text{ V}$

The effective value of the voltage can also be found by first calculating the effective value of the current.

$I_{\text{eff}} = (0.707)(0.15) = 0.106$ A

Then, $E_{\text{eff}} = I_{\text{eff}} X_L = (0.106)(2.2 \times 10^3) = 233$ V

3 If the current is taken as the reference,

$$i = I_m \sin \omega t = 0.15 \sin 12{,}570t \text{ A}$$

The voltage is leading the current in an inductive circuit.

$$e = E_m \sin(\omega t + 90°) = 330 \sin(12{,}570t + 90°) \text{ V} \qquad ////$$

EXAMPLE 9-3 A voltage across an inductance is 40 V when the current is 120 mA. The frequency of current and voltage is 400 Hz. Find the inductance. The given voltage and current are effective values.

SOLUTION The magnitude of the inductive reactance can be found.

$$X_L = \frac{E_{\text{eff}}}{I_{\text{eff}}} = \frac{40}{120 \times 10^{-3}} = 333\ \Omega$$

The inductance can now be calculated.

$$X_L = 2\pi f L$$

$$L = \frac{X_L}{2\pi f} = \frac{333}{(6.28)(400)} = 0.133 \text{ H} \qquad ////$$

EXAMPLE 9-4 If 150 V effective is measured across a 75-mH coil and the frequency of the voltage is 3.5 kHz, calculate the magnitude of the effective current.

SOLUTION

$$X_L = 2\pi f L = (6.28)(3.5 \times 10^3)(7.5 \times 10^{-2}) = 1{,}650\ \Omega$$

$$I_{\text{eff}} = \frac{E_{\text{eff}}}{X_L} = \frac{150}{1{,}650} = 91 \text{ mA} \qquad ////$$

The power relationship in an inductive circuit can be analyzed by writing Eq. (3-15) with instantaneous values.

$$p = ei \tag{9-7}$$

Applying Eq. (9-7) to Fig. 9-5, we see that the instantaneous power is positive from t_0 to t_1, negative from t_1 to t_2, positive from t_2 to t_3, and negative from t_3 to t_4.

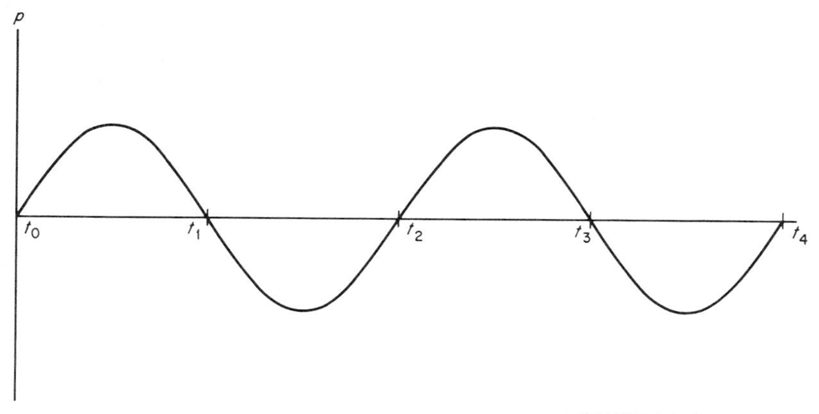

FIGURE 9-6
Power in a pure inductive circuit.

The instantaneous power is plotted in Fig. 9-6. *Positive power* indicates that energy is taken from the source, and *negative power* that energy is returned to the source. Since over one complete cycle, from t_0 to t_4, as much energy is returned as is taken from the source, the net energy taken from the source is zero. Power over a complete cycle is therefore zero. This supports the definition that inductance is the property of a circuit to store energy in the form of a magnetic field. Thus, when current is increasing in magnitude, the magnetic field is building up and storing energy from the source. When current is decreasing in magnitude, the magnetic field is collapsing and returning energy to the source.

The average power is also given by Eq. (9-1).

$$P = E_{\text{eff}} I_{\text{eff}} \cos \theta = E_{\text{eff}} I_{\text{eff}} \cos 90^\circ = 0$$

9-4 THE PURE CAPACITANCE CIRCUIT

A change in voltage across a capacitor results in a current that is proportional to both the rate of voltage change and the capacitance. The relations are given by the equation

$$i_c = C \frac{de}{dt} \tag{5-19}$$

A sinusoidal voltage is applied to a capacitor in the circuit of Fig. 9-7. The current is plotted versus time with reference to voltage e in Fig. 9-8.

The plot of voltage in Fig. 9-8 increases from t_0 to t_1. The voltage is increasing at a decreasing rate, and at t_1 the rate of change of voltage is zero. At time t_1, the

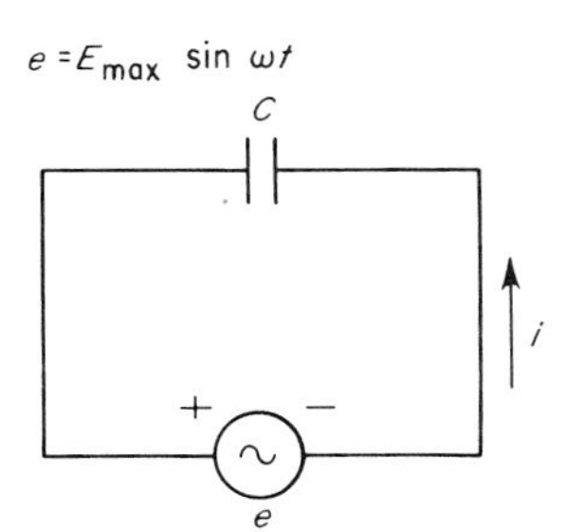

FIGURE 9-7
Pure capacitive circuit.

current must then be zero. From time t_1 to t_2, the voltage is decreasing, and at t_2 the voltage is changing at a maximum rate. The current is negative from t_1 to t_3 and maximum negative at t_2. At t_3, the rate of change of voltage is instantaneously zero, and therefore the current is zero. From t_3 to t_4, the voltage is increasing at an increasing rate, and maximum instantaneous rate of change occurs at t_4. Therefore i is maximum at t_4.

From the waveforms of Fig. 9-8, it is seen that the maximum positive current occurs 90 electrical degrees ahead of the maximum positive voltage. The current is said to be leading the voltage by 90° in a purely capacitive circuit. This phase relationship is derived mathematically by applying calculus.

Using Eq. (5-19),

$$i_c = C\frac{de}{dt}$$

and

$$e = E_m \sin \omega t \quad \text{V}$$

By differentiation, de/dt is found.

$$\frac{de}{dt} = \omega E_m \cos \omega t$$

Therefore,

$$i_c = \omega C E_m \cos \omega t = \frac{E_m}{1/\omega C} \sin(\omega t + 90°)$$

By the general form of a periodic function,

$$I_m = \frac{E_m}{1/\omega C}$$

and

$$\frac{1}{\omega c} = \frac{E_m}{I_m} \tag{9-8}$$

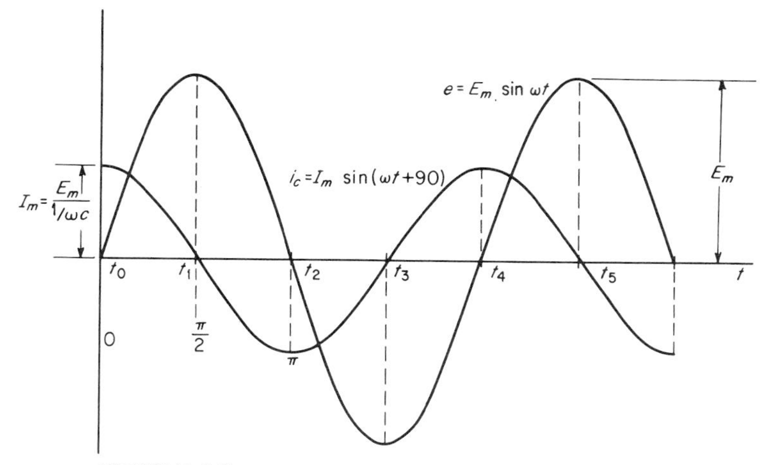

FIGURE 9-8
Current leading voltage by 90°; the current-voltage relationship in a pure capacitive circuit.

Since the ratio of volts to amperes is defined as opposition to current in ohms, the quantity $1/\omega C$ is in ohms. The quantity $1/\omega C$ is called *capacitive reactance* and is symbolized X_C.

$$X_C = \frac{1}{\omega C} = \frac{1}{2\pi f C} \qquad \Omega \tag{9-9}$$

X_C can be shown to be the ratio of effective values of current and voltage in the same manner as was shown for X_L in the previous section.

$$X_C = \frac{E_{\text{eff}}}{I_{\text{eff}}} \qquad \Omega \tag{9-10}$$

The reciprocal of capacitive reactance is called *capacitive susceptance* and is symbolized B_C. The unit of capacitive susceptance is the mho when the frequency is in hertz and the capacitance is in farads.

$$B_C = \frac{1}{X_C} = 2\pi f C \quad \text{mhos} \tag{9-11}$$

EXAMPLE 9-5 A voltage having an effective value of 220 V and a frequency of 20 kHz is applied to an 0.08-μF capacitor. (1) Calculate the effective value of the current. (2) Calculate the maximum value of the current. (3) Write the periodic functions representing current and voltage.

SOLUTION

1

$$I_{\text{eff}} = \frac{E_{\text{eff}}}{X_C} \qquad X_C = \frac{1}{\omega_c} = \frac{1}{(2\pi)(2 \times 10^4)(0.8 \times 10^{-7})} = 99.5\ \Omega$$

$$= \frac{220}{99.5} = 2.21\ \text{A}$$

2 $I_m = 1.414 I_{\text{eff}} = 3.12$ A

3 If the voltage is taken as a reference,

$$e = E_m \sin \omega t = (1.414)(220)\sin 125{,}700t = 311 \sin 125{,}700t$$

The current leads the voltage by 90° in a capacitive circuit.

$$i = I_m \sin (\omega t + 90°) = 3.12 \sin(125{,}700t + 90°)$$

These periodic functions can also be written with reference to the current.

$$i = I_m \sin \omega t \quad \text{A}$$

With reference to the current, the voltage lags by 90°.

$$e = E_m \sin(\omega t - 90°) \quad \text{V}$$

////

EXAMPLE 9-6 When 250 V is applied to a 0.05-μF capacitor, a current of 0.6 A is measured. Find the frequency. Voltage and current are effective values.

SOLUTION The magnitude of X_C can be found.

$$X_C = \frac{E_{\text{eff}}}{I_{\text{eff}}} = \frac{250}{0.6} = 416\ \Omega$$

The frequency can now be calculated from

$$X_C = \frac{1}{2\pi f C}$$

$$f = \frac{1}{2\pi X_C C} = \frac{1}{(6.28)(416)(5 \times 10^{-8})} = 7{,}650\ \text{Hz}$$

////

The power relation in a capacitive circuit can be analyzed with Eq. (3-15) on the basis of instantaneous values. The product of e and i of Fig. 9-8 is plotted in Fig. 9-9. This is the same as was done in the preceding section for instantaneous power in a pure inductive circuit. The average power in a purely capacitive circuit is zero, and it was in the purely inductive circuit.

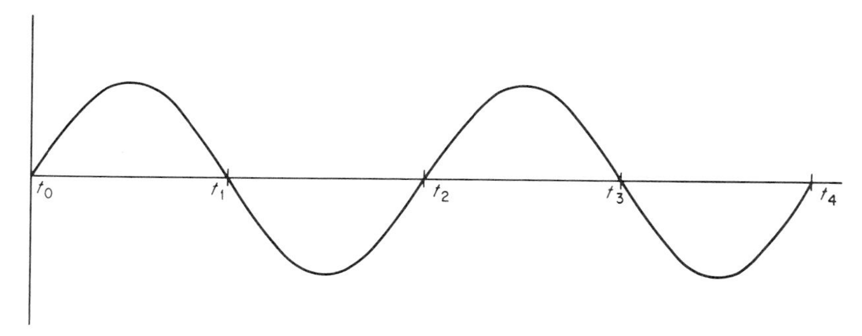

FIGURE 9-9
Power in a pure capacitive circuit.

This relationship is justified by considering that from t_0 to t_1 the voltage is increasing, charge is being stored, and energy is being stored by the capacitor. From t_1 to t_2, the voltage is decreasing, charge is returned by the capacitor, and energy is returned to the source.

The average power is again given by Eq. (9-1).

$$P = E_{\text{eff}} I_{\text{eff}} \cos \theta = E_{\text{eff}} I_{\text{eff}} \cos 90^\circ = 0$$

9-5 THE j OPERATOR

The *j operator* rotates a quantity through an angle of 90° ccw. The j operator, also called the *complex operator*, is the square root of -1. In some texts i is used to denote $\sqrt{-1}$, but this notation is avoided in electronics because of possible confusion with instantaneous current.

If a magnitude A is located on a set of coordinate axes as shown in Fig. 9-10, jA is 90° ccw, j^2A is 180° ccw, j^3A is 270° ccw, and j^4A is 360° ccw. This exponential progression is shown in Table 9-1.

Table 9-1

$j = \sqrt{-1}$	90° ccw
$j^2 = -1$	180° ccw
$j^3 = -j$	270° ccw
$j^4 = +1$	360° ccw
$j^5 = j$	450° ccw
$j^6 = -1$	540° ccw
$j^7 = -j$	630° ccw
$j^8 = +1$	720° ccw

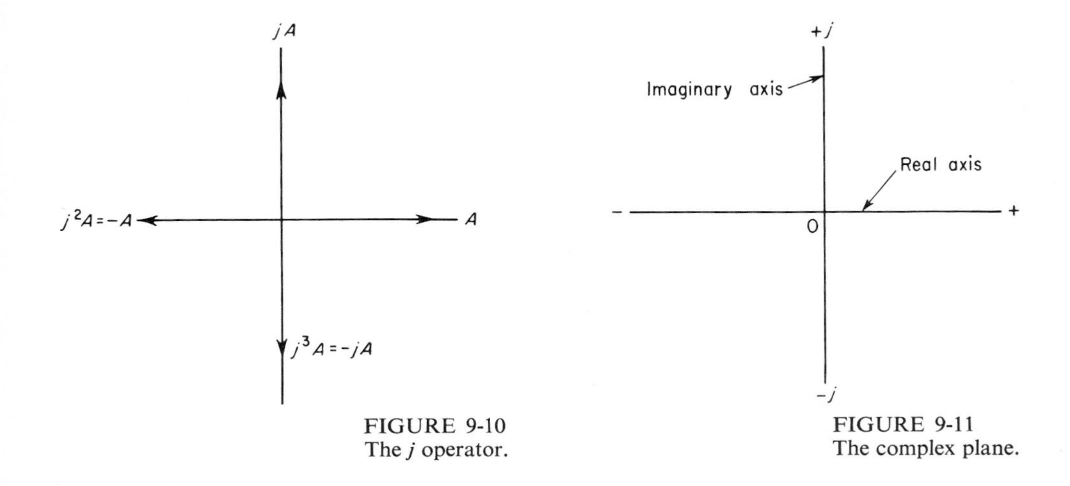

FIGURE 9-10
The j operator.

FIGURE 9-11
The complex plane.

With the j operator any quantity can be located on a complex plane with a plus and minus j axis. The complex plane is shown in Fig. 9-11.

The horizontal axis is referred to as the *real axis*, and the vertical axis is referred to as the *imaginary axis*. A j appearing before a value indicates an imaginary number. Imaginary is a term derived from an old idea that all numbers must be real to exist.

Quantity $j3$ would be located 3 units from the origin in the j, or vertical, direction. Quantity $a + jb$ is a general complex number, where a represents a magnitude on the horizontal axis, or the real portion of the complex number, and b represents a magnitude on the vertical axis, or the imaginary portion of the complex number. A quantity represented by $4 + j3$ would then be plotted as shown in Fig. 9-12. This complex number can also be defined by giving the magnitude of the radial distance r from the origin of the complex plane to the point $4 + j3$ and the angle between this radius and the positive real axis. The complex number $4 + j3$ can then be given by

$$\theta = \tan^{-1}(\tfrac{3}{4})$$

and

$$r = \frac{3}{\sin \theta}$$

Thus,

$$4 + j3 = 5 \angle 36.8^\circ$$

Consider the general complex number $a + jb$. The reference angle θ is

$$\theta = \tan^{-1} \frac{b}{a}$$

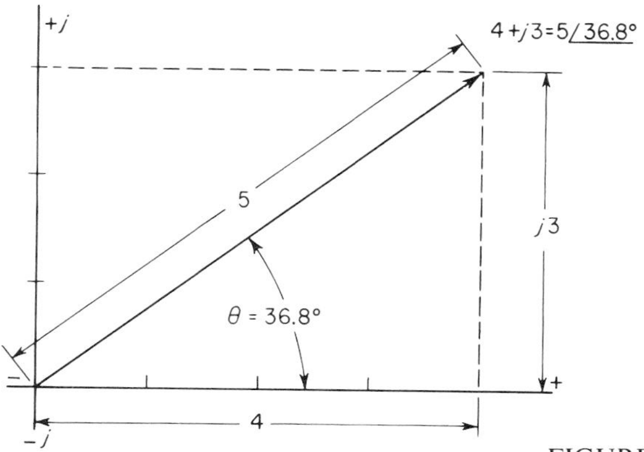

FIGURE 9-12
Polar and rectangular forms.

and $$r = \frac{b}{\sin \theta} = \frac{a}{\cos \theta} = \sqrt{a^2 + b^2}$$

Thus, $$a + jb = r \angle \theta$$

For this complex number, $a + jb$ is called the *rectangular form*, while $r \angle \theta$ is called the *polar form*. If a complex number is given in polar form, its rectangular form can be found.

$$a = r \cos \theta$$

and $$b = r \sin \theta$$

The general equation for converting from polar to rectangular form is

$$\mathbf{A} = A \angle \pm\theta = A \cos \theta \pm jA \sin \theta \tag{9-12}$$

Boldface **A** indicates a phasor quantity having both magnitude and direction. The general equations for converting from rectangular to polar form are

$$\mathbf{A} = \frac{a}{\cos \theta} \angle \tan^{-1} \frac{b}{a} \tag{9-13}$$

or

$$\mathbf{A} = \frac{b}{\sin \theta} \angle \tan^{-1} \frac{b}{a} \tag{9-14}$$

These general relations are shown in Fig. 9-13.

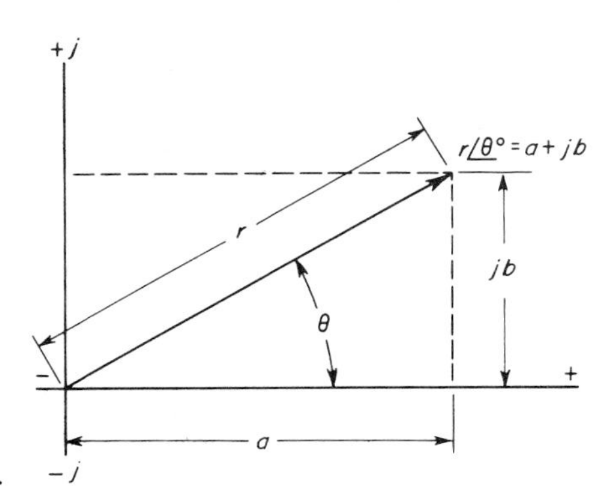

FIGURE 9-13
General polar and rectangular forms.

EXAMPLE 9-7 Given the following complex numbers, convert those in polar form to rectangular form and those in rectangular form to polar form. (1) $300 - j175$, (2) $-40 + j60$, (3) $85 - j90$, (4) $40 \angle -45°$, (5) $200 \angle 150°$, (6) $0.5 \angle 215°$. Refer to trigonometric tables in the Appendix.

SOLUTION

1

$$\theta = \tan^{-1} \frac{-175}{300} = -30.3° = 329.7°$$

$$r = \frac{300}{\cos(-30.3°)} = \frac{-175}{\sin(-30.3°)}$$

$$r = \frac{300}{0.863} = \frac{-175}{-0.505} = 347$$

$$300 - j175 = 347 \angle 329.7° \text{ or } 347 \angle -30.3°$$

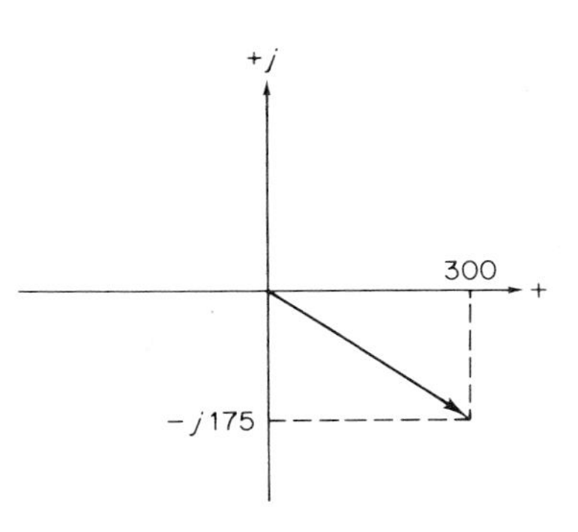

2

$$\theta = \tan^{-1} \frac{60}{-40} = 123.6°$$

$$r = \frac{-40}{\cos 123.6°} = \frac{60}{\sin 123.6°}$$

The angle may be converted to an acute value for the table.

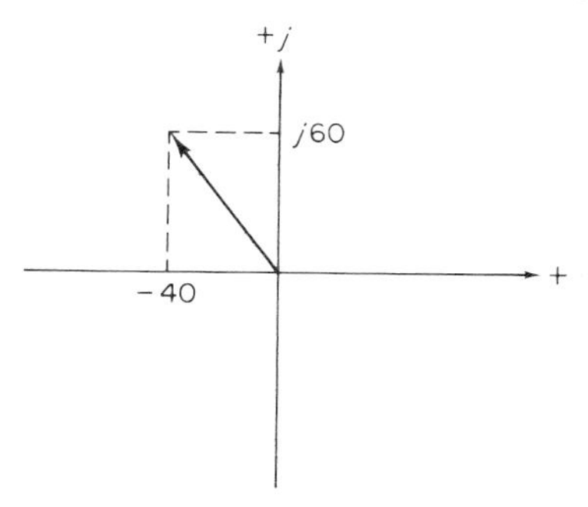

$$\cos 123.6° = -\cos 56.4°$$

$$\sin 123.6° = \sin 56.4°$$

$$r = 72.2$$

$$-40 + j60 = 72.2 \angle 123.6°$$

3

$$\theta = \tan^{-1} \frac{-90}{85} = -46.6° = 313.4°$$

$$r = \frac{85}{\cos(-46.6°)} = \frac{-90}{\sin(-46.6°)} = 124$$

$$85 - j90 = 124 \angle -46.6°$$

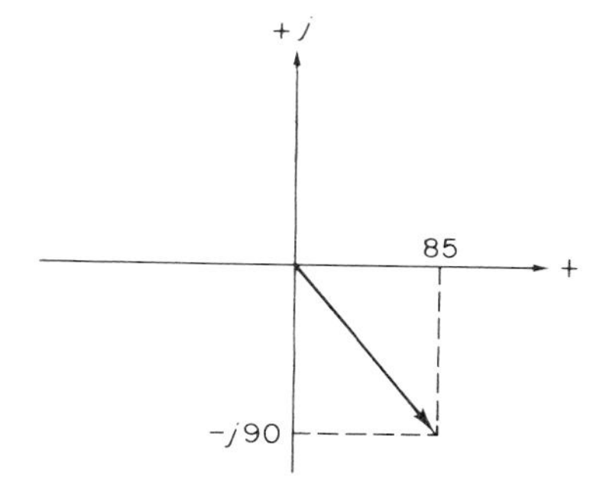

4

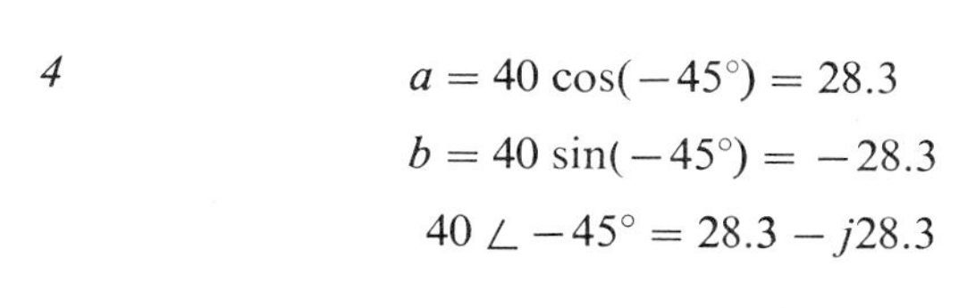

$$a = 40 \cos(-45°) = 28.3$$

$$b = 40 \sin(-45°) = -28.3$$

$$40 \angle -45° = 28.3 - j28.3$$

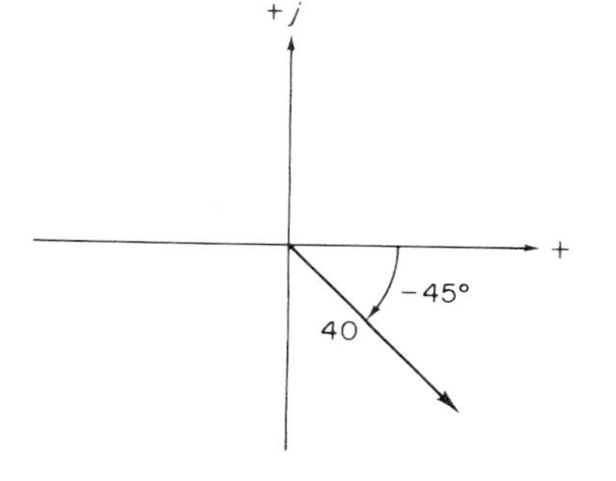

5 $a = 200 \cos 150° = 200(-\cos 30°) = -173$

$b = 200 \sin 150° = 200 \sin 30° = 100$

$$200 \angle 150° = -173 + j100$$

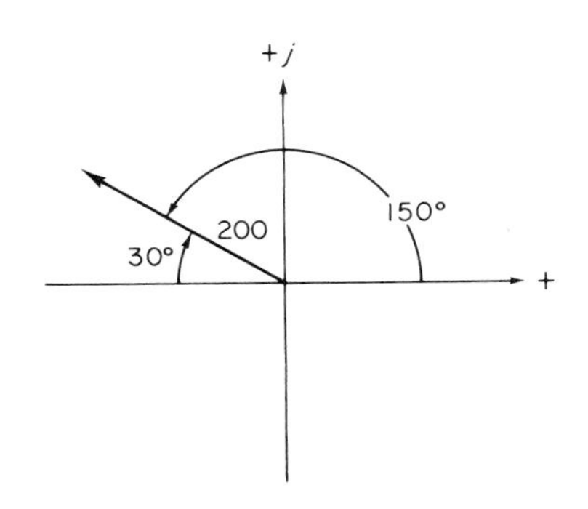

6 $a = 0.5 \cos 215° = 0.5(-\cos 35°) = -0.41$

$b = 0.5 \sin 215° = 0.5(-\sin 35°) = -0.286$

$$0.5 \angle 215° = -0.41 - j0.286 \qquad ////$$

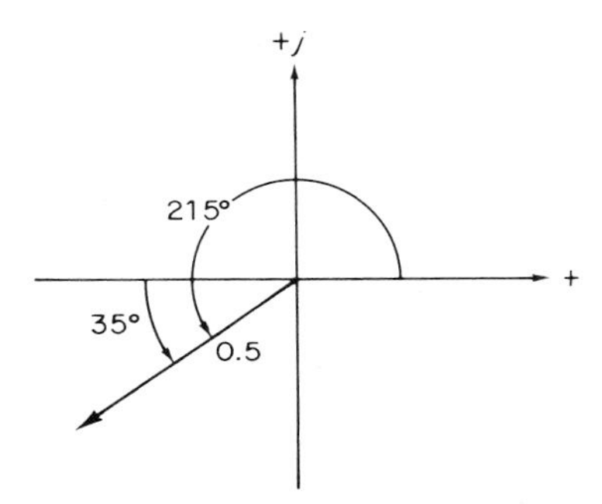

Complex numbers may be added, subtracted, multiplied, or divided. Two or more complex numbers must be added or subtracted in rectangular form.

$$(4 + j3) + (7 - j5) = 4 + 7 + j(3 - 5) = 11 - j2$$

and $$(7 - j4) - (5 - j2) = 7 - 5 + j(-4 + 2) = 2 - j2$$

Complex numbers may be multiplied or divided in either polar or rectangular form. However, it is most conveniently done in polar form.

Multiplying,

$$(250 \angle 70°)(50 \angle 30°) = (250)(50) \angle (70° + 30°) = 12.5 \times 10^3 \angle 100°$$

and $$(12 \angle -60°)(15 \angle 20°) = (12)(15) \angle (-60° + 20°) = 180 \angle -40°$$

Dividing,

$$\frac{0.12 \angle 120°}{0.40 \angle 70°} = \frac{0.120}{0.40} \angle (120° - 70°) = 0.3 \angle 50°$$

$$\frac{8 \angle 16°}{2 \angle -40°} = \frac{8}{2} \angle [16° - (-40)] = 4 \angle 56°$$

It must be noted that a complex number has the following special forms:

$$\mathbf{A} = r \angle 0^\circ = a + j0$$
$$= r \angle 90^\circ = 0 + jb$$
$$= r \angle -90^\circ = 0 - jb$$
$$= r \angle 180^\circ = -a + j0$$

9-6 USE OF COMPLEX NUMBERS

If complex numbers are used to describe sinusoidal currents and voltages, the mathematics of complex numbers can be applied to alternating currents and voltages. Table 9-2 shows the relationship between sinusoidal functions and phasor representations. The magnitude of the phasor quantity represents the effective value of the sinusoidal function.

If an ac voltage is given by $\mathbf{E} = E_{\text{eff}} \angle \theta^\circ$ (resistance does not cause a phase shift) the current phasor through a resistance would be

$$\mathbf{I} = \frac{E_{\text{eff}} \angle \theta^\circ}{R \angle 0^\circ}$$

$$\mathbf{I} = \frac{E_{\text{eff}}}{R} \angle \theta^\circ = I_{\text{eff}} \angle \theta^\circ$$

where the magnitude of $\mathbf{I}$ is

$$I_{\text{eff}} = \frac{E_{\text{eff}}}{R} \qquad \text{A}$$

and the phase angle between current and voltage is zero.

This approach can be applied to the purely inductive circuit of Sec. 9-3.

The voltage applied to a purely inductive circuit is

$$\mathbf{E} = E_{\text{eff}} \angle \gamma^\circ$$

Table 9-2

Sinusoidal function	Phasor
$\sqrt{2}\,(10) \sin \omega t$	$10 \angle 0^\circ$
$41 \sin(\omega t - 30^\circ)$	$(0.707)(41) \angle -30^\circ = 29 \angle -30^\circ$
$0.04 \sin(\omega t + \theta^\circ)$	$(0.707)(0.04) \angle \theta^\circ = 0.0283 \angle \theta^\circ$
$141.4 \cos(\omega t + 105^\circ)$	$100 \angle 195^\circ$

where γ is any angle. From Sec. 9-3, it was found that the current lags the applied voltage by 90°.

$$\mathbf{I} = I_{\text{eff}} \angle \gamma^\circ - 90^\circ$$

$$= \frac{E_{\text{eff}}}{X_L} \angle \gamma^\circ - 90^\circ = \frac{E_{\text{eff}} \angle \gamma^\circ}{X_L \angle 90^\circ}$$

converting $X_L \angle 90^\circ$ to rectangular form gives

$$\mathbf{I} = \frac{E_{\text{eff}} \angle \gamma^\circ}{jX_L} = \frac{\mathbf{E}}{jX_L}$$

Therefore the effect of inductance on magnitude and phase is given by the complex number

$$jX_L = j\omega L = j2\pi f L \qquad \Omega \tag{9-15}$$

To repeat this process for the purely capacitive circuit of Sec. 9-4, apply the following voltage to a purely capacitive circuit.

$$\mathbf{E} = E_{\text{eff}} \angle \gamma^\circ$$

where γ is any angle. The current leads the applied voltage by 90°.

$$\mathbf{I} = I_{\text{eff}} \angle \gamma^\circ + 90^\circ$$

$$= \frac{E_{\text{eff}}}{X_C} \angle \gamma^\circ + 90^\circ = \frac{E_{\text{eff}} \angle \gamma^\circ}{X_C \angle -90^\circ}$$

convert $X_C \angle -90^\circ$ to rectangular form

$$\mathbf{I} = \frac{E_{\text{eff}} \angle \gamma^\circ}{-jX_C} = \frac{\mathbf{E}}{-jX_C}$$

Therefore the effect of capacitance on magnitude and phase is given by the complex number

$$-jX_C = -j\frac{1}{\omega C} = -j\frac{1}{2\pi f C} \qquad \Omega \tag{9-16}$$

Plotting resistance, capacitive reactance, and inductive reactance on a complex plane yields the results shown in Fig. 9-14.

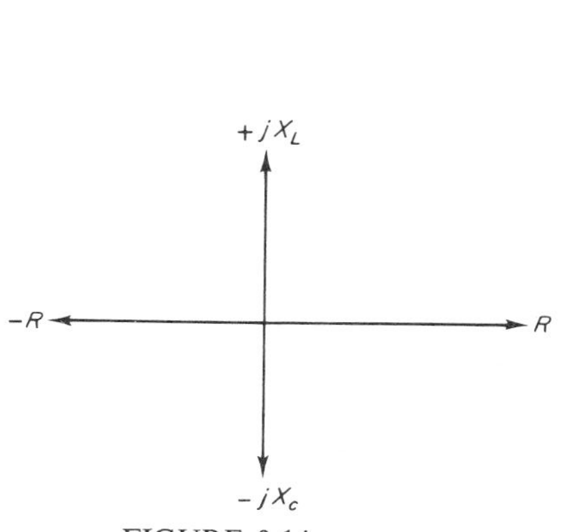

FIGURE 9-14
Resistance and reactance on the complex plane.

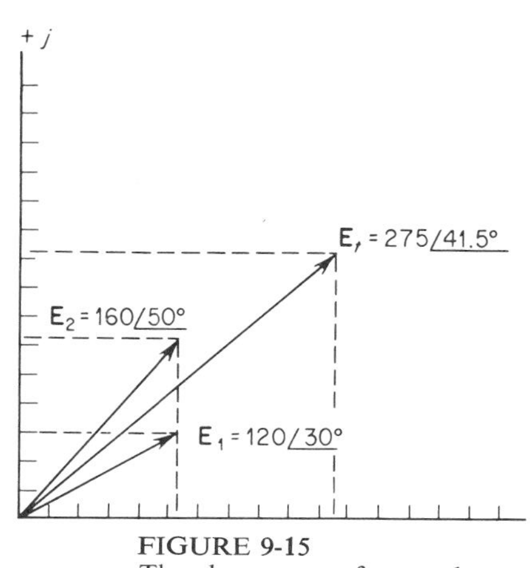

FIGURE 9-15
The phasor sum of two voltages.

The complex notation is useful in adding and subtracting sinusoidal currents and voltages in phasor form.

To calculate the sum of two voltages given in phasor form by

$$\mathbf{E}_1 = 120 \angle 30^\circ \text{ V}$$

$$\mathbf{E}_2 = 160 \angle 50^\circ \text{ V}$$

the voltages must be converted into rectangular form.
Therefore,

$$\mathbf{E}_1 = 104 + j60 \text{ V}$$

$$\mathbf{E}_2 = 103 + j122.5 \text{ V}$$

Adding,

$$\mathbf{E}_t = \mathbf{E}_1 + \mathbf{E}_2 = (104 + j60) + (103 + j122.5) = 207 + j182.5 \text{ V}$$

The sum may then be stated in phasor form.

$$\mathbf{E} = 275 \angle 41.5^\circ \text{ V}$$

This relationship is shown graphically in Fig. 9-15.

EXAMPLE 9-8 Voltage $\mathbf{E} = 150 \angle 40^\circ$ V is applied to a purely capacitive circuit. The frequency of the applied voltage is 15 kHz and the capacitance is 0.06 μF. (1) Calculate the current phasor. (2) Sketch the current and voltage phasors on a complex plane. (3) Write the periodic functions for current and voltage.

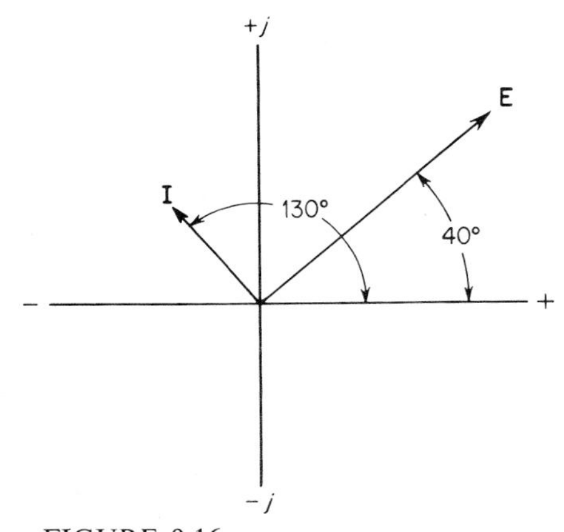

FIGURE 9-16
Voltage and current phasors for Example 9-8.

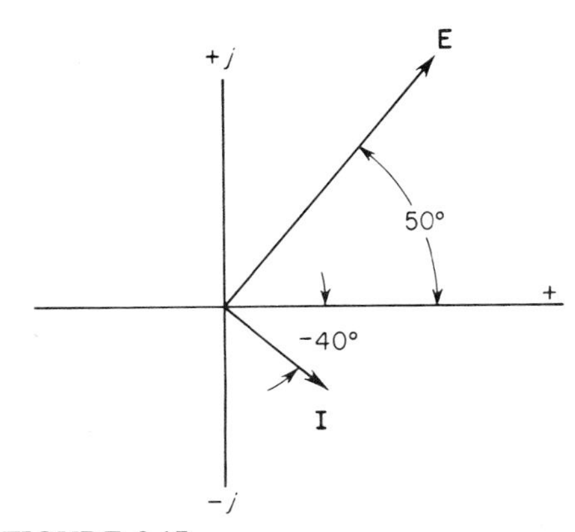

FIGURE 9-17
Current and voltage phasors for Example 9-9.

SOLUTION

1 $$-jX_C = -j\frac{1}{\omega C} \tag{9-17}$$

$$= -j\frac{1}{(6.28)(15 \times 10^3)(0.06 \times 10^{-6})} = -j177\ \Omega = 177 \angle -90^\circ\ \Omega$$

$$\mathbf{I} = \frac{\mathbf{E}}{-jX_C} = \frac{150 \angle 40^\circ}{177 \angle -90^\circ} = 0.848 \angle [40^\circ - (-90^\circ)] = 0.848 \angle 130^\circ \text{ A}$$

Note that regardless of the phase angle of the voltage, the current leads the voltage by 90 electrical degrees.

2 See Fig. 9-16.

$$\mathbf{E} = 150 \angle 40^\circ \text{ V} \qquad \sqrt{2}(150) = 212$$

$$e = 212 \sin(\omega t + 40^\circ)$$

$$= 212 \sin[(94.3 \times 10^3)t + 40^\circ] \qquad \text{V}$$

$$\mathbf{I} = 0.848 \angle 130^\circ \text{ A} \qquad \sqrt{2}(0.848) = 1.2$$

$$i = 1.2 \sin(\omega t + 130^\circ)$$

$$= 1.2 \sin[(94.3 \times 10^3)t + 130^\circ] \qquad \text{A}$$

////

EXAMPLE 9-9 Current $\mathbf{I} = 0.45 \angle -40^\circ$ flows in a purely inductive circuit having 780 Ω inductive reactance. Calculate the phasor voltage and sketch the current and voltage phasors on the complex plane (see Fig. 9-17).

SOLUTION

$$\mathbf{E} = jX_L\mathbf{I} = (780 \angle 90^\circ)(0.45 \angle -40^\circ) = 351 \angle 50^\circ \text{ V}$$

////

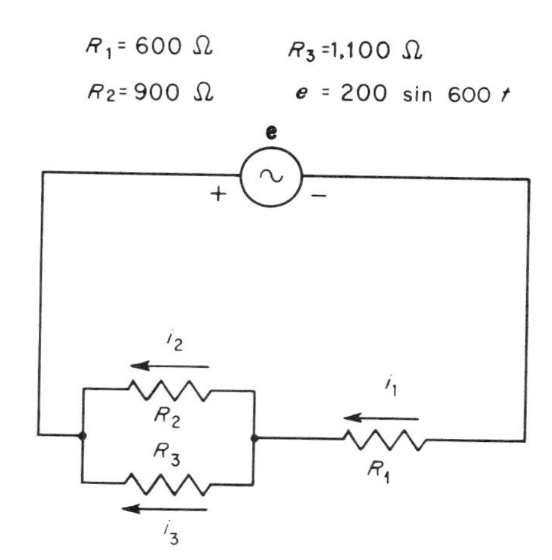

FIGURE 9-18
Circuit for Probs. 9-3 to 9-6.

PROBLEMS

9-1 Calculate the maximum current through a 2-kΩ resistor if the effective value of the applied voltage is 70 V.

9-2 The current in a resistive circuit is given by the periodic function $i = 0.4 \sin 875t$. The effective value of the applied voltage is 150 V. (*a*) Calculate the resistance. (*b*) Write the periodic function of the voltage. (*c*) Calculate the current and voltage at time $t = 0.001$ s. (*d*) Show that the ratio of voltage to current as calculated in (*c*) also yields the resistance.

9-3 Given the circuit of Fig. 9-18, calculate the effective voltage across each of the resistors. Polarity on voltage source and current directions indicate circuit conditions at some instant in time. Using the VDR would be appropriate.

9-4 Write the periodic function for each of the currents of Fig. 9-18.

9-5 Calculate i_1, i_2, and i_3 at time $t = 0.004$ s and show that $i_2 + i_3 = i_1$ (Fig. 9-18).

9-6 Calculate the power dissipated in each of the resistors of Fig. 9-18.

9-7 Calculate E_4 in the circuit of Fig. 9-19.

9-8 Write the periodic function for e in Fig. 9-19.

9-9 Calculate p_2 at time $t = 0.04$ ms (Fig. 9-19).

9-10 Write the periodic function for i_1 (Fig. 9-19).

9-11 Calculate i_4 at $t = 0.7$ ms (Fig. 9-19).

9-12 Calculate p_3 at time $t = 0.9$ ms (Fig. 9-19).

9-13 Determine the inductive reactance X_L of a 10 mH inductor operating at the following sinusoidal frequencies: (*a*) 100 Hz, (*b*) 1.5 kHz, (*c*) 12 kHz, (*d*) 80 kHz, (*e*) 1 MHz, (*f*) dc.

9-14 Calculate the inductance for the various inductive reactances at $f = 20$ kHz: (*a*) $X_L = 1$ Ω, (*b*) $X_L = 8$ Ω, (*c*) $X_L = 900$ Ω, (*d*) $X_L = 4$ kΩ, (*e*) $X_L = 20$ kΩ, (*f*) $X_L = 100$ k

9-15 Find the value of inductance for each inductive reactance and frequency: (*a*) $X_L =$ 420 Ω at $f = 1$ kHz, (*b*) $X_L = 49$ kΩ at $f = 400$ kHz, (*c*) $X_L = 9$ Ω at $f = 60$ Hz, (*d*) $X_L =$ 1 kΩ at $f = 1$ MHz.

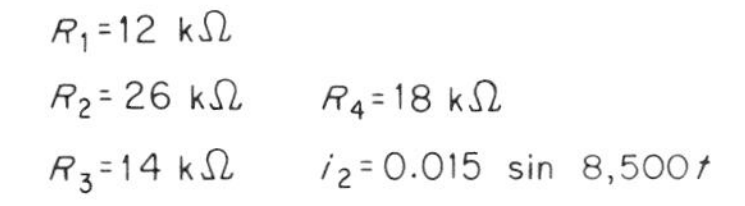

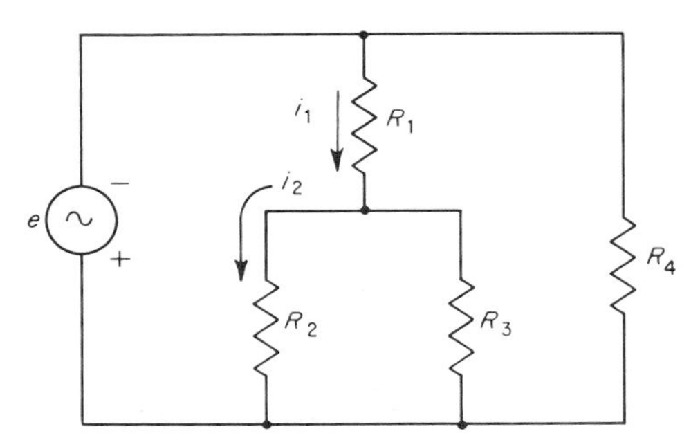

FIGURE 9-19
Circuit for Probs. 9-7 to 9-12.

9-16 Find the inductive susceptance B_L for the following inductance and frequency combinations: (*a*) $L = 1$ mH at $f = 800$ Hz, (*b*) $L = 80$ μH at $f = 1$ MHz, (*c*) $L = 12$ μH at $f = 31$ MHZ, (*d*) $L = 4$ H at $f = 60$ Hz.

9-17 Determine the capacitive reactance X_c of a 0.005-μF capacitor operating at the following sinusoidal frequencies: (*a*) 100 Hz, (*b*) 1.5 Hz, (*c*) 12 kHz, (*d*) 80 kHz, (*e*) 1 MHz, (*f*) dc.

9-18 Calculate the capacitance for the various capacitive reactances at $f = 20$ kHz: (*a*) $X_C = 1$ Ω, (*b*) $X_C = 8$ Ω, (*c*) $X_C = 900$ Ω, (*d*) $X_C = 4$ kΩ, (*e*) $X_C = 20$ kΩ, (*f*) $X_C = 100$ kΩ.

9-19 Find the value of capacitance for each capacitive reactance and frequency: (*a*) $X_C = 420$ Ω at $f = 1$ kHz, (*b*) $X_C = 49$ kΩ, at $f = 400$ kHz, (*c*) $X_C = 9$ Ω at $f = 60$ Hz, (*d*) $X_C = 1$ kΩ at $f = 1$ MHz.

9-20 Find the capacitive susceptance B_C for the following capacitance and frequency combinations: (*a*) $C = 0.01$ μF at $f = 10$ kHz, (*b*) $C = 0.24$ μF at $f = 2$ kHz, (*c*) $C = 0.58$ μF at $f = 40$ kHz, (*d*) $C = 19$ μF at $f = 60$ Hz.

9-21 The following voltages are applied across a 0.005-μF capacitor. (1) Calculate the capacitive reactance. (2) Calculate the maximum value of current. (3) Write the periodic function for the current. (*a*) $e = 400 \sin 377t$, (*b*) $e = 130 \sin (4{,}500t + 40°)$, (*c*) $e = 18 \sin 6{,}500t$, (*d*) $e = \sin (941{,}000t + 90°)$.

9-22 The effective current in a purely capacitive circuit is 0.8 A. The voltage applied is $e = 140 \sin (1.4 \times 10^6)t$. Calculate the capacitance.

9-23 The following voltages are applied to a 0.002 μF capacitor. (1) Calculate the maximum currents. (2) Calculate the effective currents. (3) Write the periodic functions for current and voltage.

(*a*)	$\mathbf{E} = 100 \angle 0°$ V	at $f = 1$ kHz
(*b*)	$\mathbf{E} = 40 \angle 100°$ V	at $f = 14$ kHz
(*c*)	$\mathbf{E} = 250 \angle -60°$	at $f = 7$ kHz
(*d*)	$\mathbf{E} = 12 \angle -18°$ V	at $f = 350$ kHz.

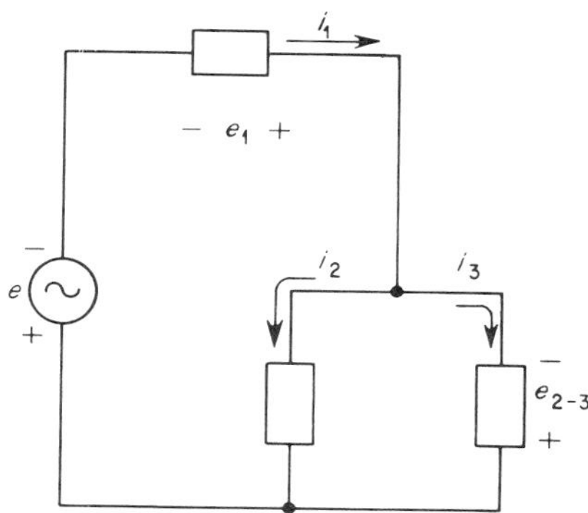

FIGURE 9-20
Circuit for Probs. 9-28 to 9-32.

9-24 Current $\mathbf{I} = 0.05 \angle 0°$ A flows in a purely capacitive circuit when a sinusoidal voltage with a maximum value of 20 V is applied. $C = 0.3\ \mu$F. Calculate the frequency.

9-25 A 200-V effective, 60-Hz source is connected to a 15-mH coil. Calculate the effective current. Write the periodic functions for current and voltage.

9-26 Current $i = 0.08 \sin 3{,}500t$ flows in an inductance of 0.5 H. Calculate the effective voltage and write the periodic function for the voltage.

9-27 For the following voltages and reactances, (1) calculate the maximum current, (2) calculate L, (3) write the periodic function for the current, and (4) show current and voltage phasors on a complex plane.

(*a*)	$e = 350 \sin (1{,}000t + 20°)$	$X_L = 60\ \Omega$
(*b*)	$e = 25 \sin (1{,}500t + 50°)$	$X_L = 60\ \Omega$
(*c*)	$e = 110 \sin (377t - 70°)$	$X_L = 60\ \Omega$
(*d*)	$e = 600 \sin (900t - 60°)$	$X_L = 80\ \Omega$
(*e*)	$e = 90 \sin (754t + 15°)$	$X_L = 140\ \Omega$
(*f*)	$e = 12 \sin (48{,}000t - 80°)$	$X_L = 9\ \text{k}\Omega$

9-28 In the circuit of Fig. 9-20, $\mathbf{I}_1 = 0.4 \angle -30°$ and $\mathbf{I}_3 = 0.4 \angle 30°$. Calculate $\mathbf{I}_2$.

9-29 In the circuit of Fig. 9-20, $\mathbf{E}_1 = 150 \angle 50°$ and $\mathbf{E}_{2-3} = 60 \angle 20°$. Calculate $\mathbf{E}$. Show the voltages on a complex plane.

9-30 In the circuit of Fig. 9-20, $i_2 = 0.05 \sin (500t + 20°)$ and $i_3 = 0.09 \sin (500t - 40°)$. Calculate $\mathbf{I}_1$ and write the periodic function for i_1. Show the current phasors on a complex plane.

9-31 In the circuit of Fig. 9-20, $e = 156 \sin 377t$ and $e_{2-3} = 60 \sin (377t - 40°)$. Calculate $\mathbf{E}_1$ and write the periodic function for e_1. Show the voltages on the complex plane.

9-32 In the circuit of Fig. 9-20, $i_3 = 0.7 \sin (400t + 130°)$ and $i_2 = 0.5 \sin (400t + 190°)$. Calculate $\mathbf{I}_1$, and write the periodic function for i_1. Show the currents on a complex plane.

9-33 For the following phasor voltages and inductive reactances: (1) calculate $\mathbf{I}$, (2) show $\mathbf{E}$ and $\mathbf{I}$ on the complex plane, and (3) calculate the frequency if $L = 0.5$ mH.

(*a*)	$\mathbf{E} = 400 \angle 40°$	$jX_L = j600$
(*b*)	$\mathbf{E} = 220 \angle -120°$	$jX_L = j320$
(*c*)	$\mathbf{E} = 60 \angle 150°$	$jX_L = j400$
(*d*)	$\mathbf{E} = 12 \angle 30°$	$jX_L = j8{,}700$

9-34 For the following phasor currents and capacitive reactances: (1) calculate **E**, (2) show **E** and **I** on the complex plane, and (3) calculate the frequency if $C = 0.0004\ \mu\text{F}$.

(*a*)	$\mathbf{I} = 4 \angle -20°$ mA,	$-jX_C = -j20\ \text{k}\Omega$
(*b*)	$\mathbf{I} = 18 \angle 40°\ \mu\text{A}$	$-jX_C = -j1.2\ \text{M}\Omega$
(*c*)	$\mathbf{I} = 2 \angle 10°$ A	$-jX_C = -j10\ \Omega$
(*d*)	$\mathbf{I} = 26 \angle 170°$ mA	$-jX_C = -j2\ \text{k}\Omega$

9-35 $i = 0.270 \sin (350t + 40°)$, and $e = 420 \sin (350t - 50°)$. Calculate the reactance in the circuit.

9-36 $e = 175 \sin (4{,}500t + 20°)$, and $i = 0.5 \sin (4{,}500t - 70°)$. Calculate the reactance in the circuit.

9-37 For each pair of voltage and current, indicate whether the circuit component is a resistor, inductor, or capacitor. Also find the value of the component.

(*a*)	$e = 10 \sin (1{,}000t + 70°)$ V	$i = 2 \sin (1{,}000t - 20°)$ mA
(*b*)	$e = 1.8 \sin (72{,}000t - 10°)$ mV	$i = 12 \sin (72{,}000t + 80°)\ \mu\text{A}$
(*c*)	$e = 22 \sin (377t + 12°)$ V	$i = 14 \sin (377t + 12°)$ mA
(*d*)	$e = 142 \sin (7{,}100t + 100°)$ mV	$i = 2 \sin (7{,}100t - 170°)$ mA.

9-38 Determine the power factor and average power in the circuit responses of Prob. 9-37.

10

SERIES AND PARALLEL AC CIRCUITS

10-1 INTRODUCTION

In Chap. 9 the concepts of pure reactive and resistive circuits were discussed. With these pure circuit elements, the concept of the complex plane was developed. Practical circuits, however, contain a combination of these elements.

10-2 IMPEDANCE AND ADMITTANCE

Impedance is the general expression for opposition to current in circuits containing L, C, or R with sinusoidal excitations. Impedance may be pure resistance or pure reactance, but usually it is a combination of resistance and reactance. The symbol $\mathbf{Z}$ is used for impedance, which is expressed in ohms. Impedance takes the general phasor form

$$\mathbf{Z} = R \pm jX \tag{10-1}$$

Impedance is shown on the complex plane in Fig. 10-1.

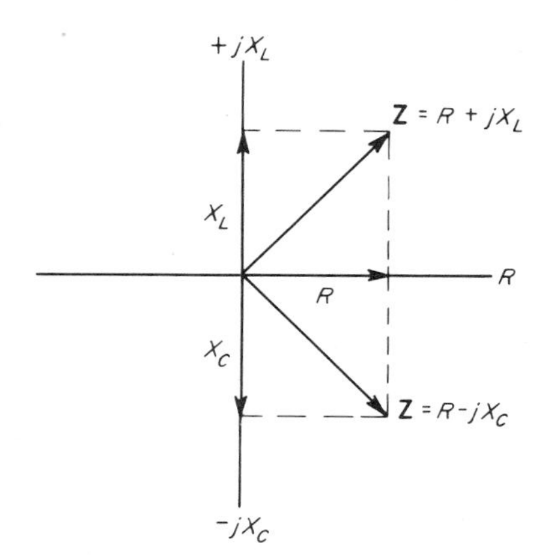

FIGURE 10-1
Impedance on the complex plane.

Impedance is the ratio of phasor voltage to phasor current in an ac circuit, as shown by Eq. (10-2).

$$\mathbf{Z} = \frac{\mathbf{E}}{\mathbf{I}} \tag{10-2}$$

Equation (10-2) is the general Ohm's law relationship for ac circuits, and the equation can be handled algebraically in the usual way. It must be remembered that computations involving phasors must be done as outlined in Chap. 9.

The impedance of the circuit of Fig. 10-2 is

$$\mathbf{Z} = R + jX_L = 400 + j300 = 500 \angle 36.8^\circ$$

The impedance of the circuit of Fig. 10-3 is

$$\mathbf{Z} = R - jX_C = 100 - j150 = 180 \angle -56.4^\circ$$

Admittance has the symbol Y and is defined as the reciprocal of impedance. Admittance is expressed in mhos.

$$\mathbf{Y} = \frac{1}{\mathbf{Z}}$$

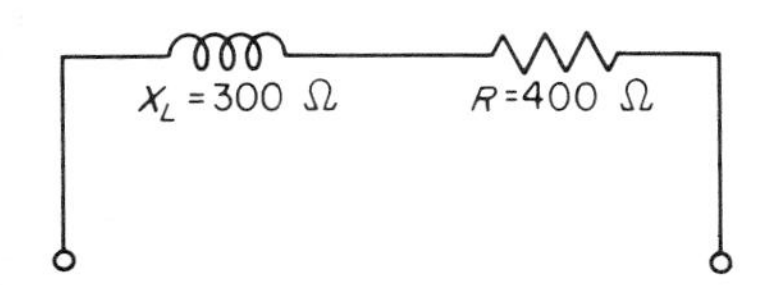

FIGURE 10-2
Resistance and inductive reactance in series.

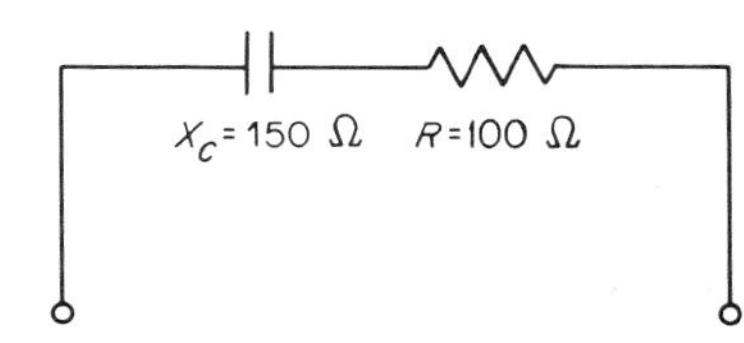

FIGURE 10-3
Resistance and capacitive reactance in series.

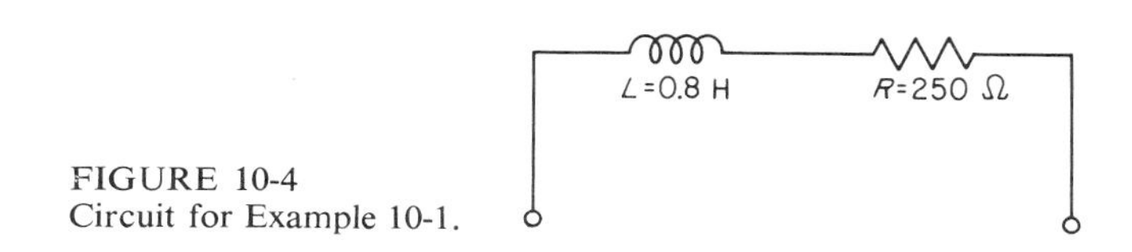

FIGURE 10-4
Circuit for Example 10-1.

EXAMPLE 10-1 Calculate the impedance and admittance of the circuit of Fig. 10-4 at a frequency of 150 Hz.

SOLUTION

$$jX_L = j\omega L = j(6.28)(150)(0.8) = j754\Omega$$

$$\mathbf{Z} = R + jX_L = 250 + j754 = 791 \angle 71.6^\circ\ \Omega$$

$$\mathbf{Y} = \frac{1}{\mathbf{Z}} = \frac{1}{791 \angle 71.6^\circ} = 0.00126 \angle -71.6^\circ \text{ mho} \qquad ////$$

10-3 THE SERIES AC CIRCUIT

On the basis of the discussion of impedance and the Ohm's law equation (10-2), series ac circuits can now be investigated.

Consider the circuit of Fig. 10-5. The impedance **Z** of the circuit is the phasor sum of the inductive reactance and the resistance, since these components are in series. As mentioned previously, the polarities and current direction represent the responses and excitation at some instant in time.

$$\mathbf{Z} = R + jX_L = 900 + j650$$

Applying Eq. (10-2),

$$\mathbf{I} = \frac{\mathbf{E}}{\mathbf{Z}} = \frac{120 \angle 0^\circ}{900 + j650}$$

Converting $900 + j650$ to polar form,

$$\mathbf{I} = \frac{120 \angle 0^\circ}{1{,}110 \angle 35.8^\circ} = 0.108 \angle -35.8^\circ$$

Since the circuit of Fig. 10-5 contains inductive reactance, it is to be expected that the current will lag the voltage. However, since the circuit is not pure inductance, the current will not lag by a full 90°. The current phasor was calculated and found to be lagging by 35.8°. The phase relationship between applied voltage **E** and current **I** is shown on the complex plane of Fig. 10-6.

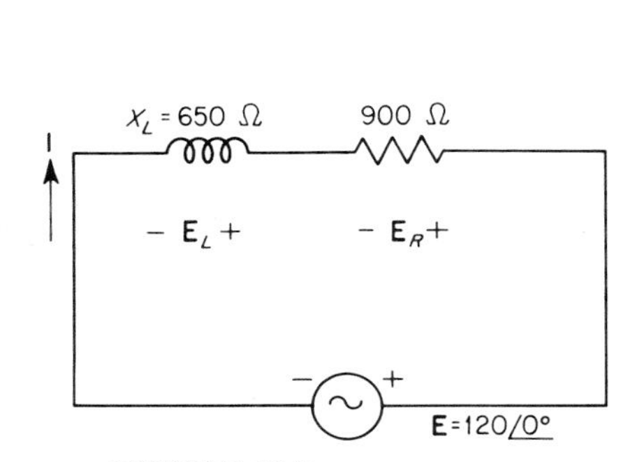

FIGURE 10-5
Resistance and inductive reactance, showing voltage phasors.

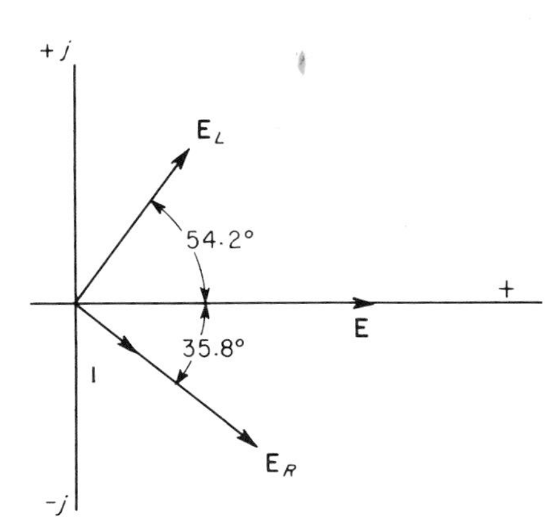

FIGURE 10-6
Phasors of Fig. 10-5.

In a series circuit, the sum of the phasors of the voltage drops must equal the applied voltage phasor.

For the circuit of Fig. 10-5,

$$\mathbf{E} = \mathbf{E}_R + \mathbf{E}_L$$
$$= R\mathbf{I} + jX_L\mathbf{I} = (R + jX_L)\mathbf{I}$$

but

$$\mathbf{Z} = R + jX_L \tag{10-3}$$

Therefore,

$$\mathbf{E} = \mathbf{ZI}$$

This is simply the algebraic transformation of Eq. (10-2). The value of $\mathbf{E}_R$ and $\mathbf{E}_L$ can now be calculated.

$$\mathbf{E}_R = R\mathbf{I} = (900 \angle 0^\circ)(0.108 \angle -35.8^\circ) = 97.2 \angle -35.8^\circ$$

and

$$\mathbf{E}_L = (jX_L)\mathbf{I} = (650 \angle 90^\circ)(0.108 \angle -35.8^\circ) = 70.2 \angle 54.2^\circ$$

Plotting $\mathbf{E}_R$ and $\mathbf{E}_L$ on the complex plane of Fig. 10-6 shows that the voltage across the resistance is in phase with the current. The voltage across the inductance leads the current by 90°.

The phasor sum of $\mathbf{E}_R$ and $\mathbf{E}_L$ is given by Eq. (10-3).

$$\mathbf{E} = \mathbf{E}_R + \mathbf{E}_L = (97.2 \angle -35.8^\circ) + (70.2 \angle 54.2^\circ)$$
$$= (79 - j57) + (41 + j57) = 120 + j0 = 120 \angle 0^\circ$$

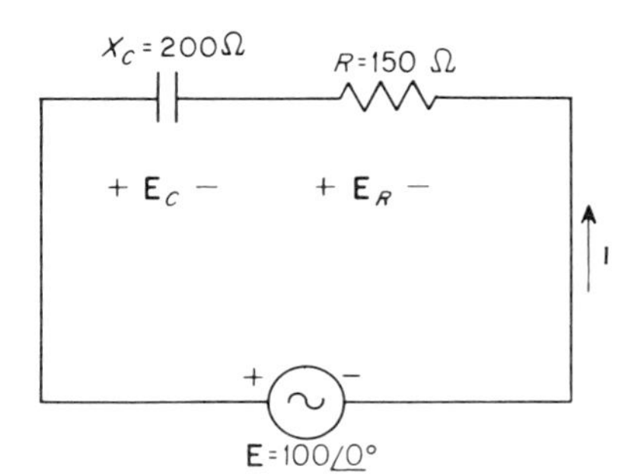

FIGURE 10-7
Resistance and capacitive reactance, showing voltage drops.

In the circuit of Fig. 10-7, a capacitor and a resistor are shown in series. The current is calculated

$$\mathbf{I} = \frac{\mathbf{E}}{\mathbf{Z}} = \frac{100\ \angle 0^\circ}{150 - j200} = \frac{100\ \angle 0^\circ}{250\ \angle -53.1^\circ} = 0.4\ \angle 53.1^\circ$$

The reactive component of the circuit of Fig. 10-7 is capacitive, and so the current can be expected to lead the applied voltage by some angle. As shown by the phasor calculation, the angle of lead is 53.1°. The phase relation of applied voltage **E** and the current is illustrated in Fig. 10-8.

Calculating $\mathbf{E}_C$ and $\mathbf{E}_R$,

$$\mathbf{E}_C = (-jX_C)\mathbf{I} = (200\ \angle -90^\circ)(0.4\ \angle 53.1^\circ) = 80\ \angle -36.9^\circ$$

$$\mathbf{E}_R = R\mathbf{I} = (150\ \angle 0^\circ)(0.4\ \angle 53.1^\circ) = 60\ \angle 53.1^\circ$$

Applied voltage **E** must be the phasor sum of $\mathbf{E}_R$ and $\mathbf{E}_C$.

$$\mathbf{E} = \mathbf{E}_R + \mathbf{E}_C = 60\ \angle 53.1^\circ + 80\ \angle -36.9^\circ = (36 + j48) + (64 - j48)$$
$$= 100 + j0 = 100\ \angle 0^\circ$$

Plotting $\mathbf{E}_R$ and $\mathbf{E}_C$ on the complex plane of Fig. 10-8 shows that $\mathbf{E}_R$ and **I** are in phase, and that **I** leads $\mathbf{E}_C$ by 90°.

In the circuit of Fig. 10-9, inductance, capacitance, and resistance are shown in series. The impedance can be expressed

$$\mathbf{Z}_t = R + jX_L - jX_C = 400 + j400 - j150 = 400 + j250 = 471\ \angle 32^\circ$$

The current can now be calculated.

$$\mathbf{I} = \frac{\mathbf{E}}{\mathbf{Z}} = \frac{50\ \angle 0^\circ}{471\ \angle 32^\circ} = 0.106\ \angle -32^\circ$$

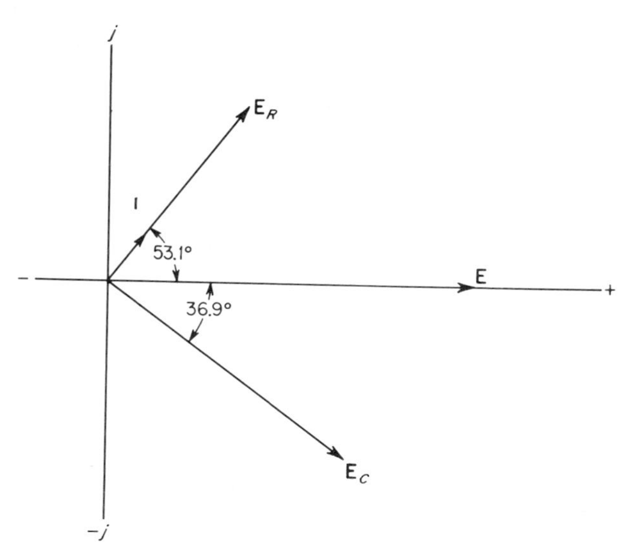

FIGURE 10-8
Phasors of Fig. 10-7.

The current therefore lags the applied voltage. The net reactive component appears as inductive reactance. The net reactance has a value of 250 Ω, as shown by the rectangular form of $\mathbf{Z}_t$. The circuit of Fig. 10-9 could be replaced with a resistance of 400 Ω and an inductance whose reactance is 250 Ω. Under these conditions the current that would result would be the same as the current of Fig. 10-9. It must be remembered that this equivalent relationship would be true only at the frequency for which X_L and X_C are given in Fig. 10-9. If the frequency were increased, X_L would

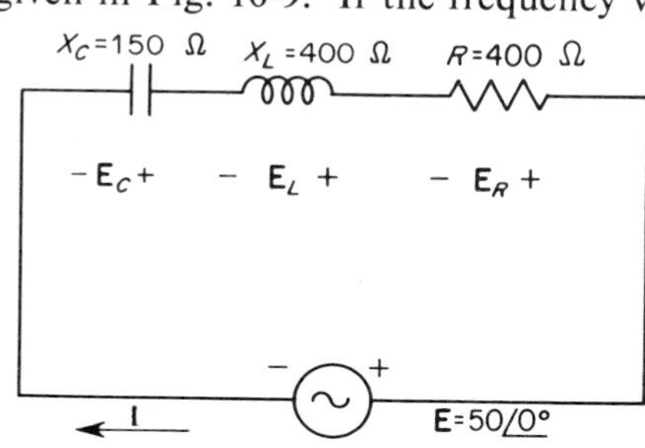

FIGURE 10-9
Resistance, inductive reactance, and capacitive reactance in series.

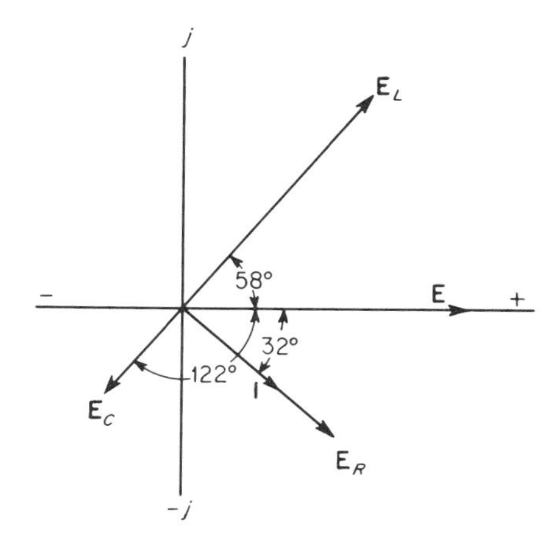

FIGURE 10-10
Phasors of Fig. 10-9.

increase, and X_C would decrease. The resulting net inductive reactance would be increased, and a new value of inductance would be required in an equivalent circuit. When frequency is decreased, X_C is increased, and X_L is decreased. If X_C becomes larger than X_L, the net reactive component becomes capacitive reactance. The equivalent circuit at such a frequency would consist of resistance and an appropriate value of capacitance.

The voltages across the components of Fig. 10-9 are

$$\mathbf{E}_L = jX_L\mathbf{I} = (400 \angle 90^\circ)(0.106 \angle -32^\circ) = 42.4 \angle 58^\circ$$

$$\mathbf{E}_C = (-jX_C)\mathbf{I} = (150 \angle -90^\circ)(0.106 \angle -32^\circ) = 15.9 \angle -122^\circ$$

$$\mathbf{E}_R = R\mathbf{I} = (400 \angle 0^\circ)(0.106 \angle -32^\circ) = 42.4 \angle -32^\circ$$

The current and voltage phasors for the circuit of Fig. 10-9 are shown on the complex plane of Fig. 10-10.

In Fig. 10-10, the current is in phase with $\mathbf{E}_R$, leading $\mathbf{E}_C$ by 90°, and lagging $\mathbf{E}_L$ by 90°. The phasor sum of the voltage drops equals applied voltage $\mathbf{E}$.

$$\begin{aligned}\mathbf{E} &= \mathbf{E}_R + \mathbf{E}_L + \mathbf{E}_C \\ &= (42.4 \angle -32^\circ) + (42.4 \angle 58^\circ) + (15.9 \angle -122^\circ) \\ &= (36 - j22.5) + (22.5 + j36) + (-8.5 - j13.5) \\ &= (36 + 22.5 - 8.5) + j(36 - 22.5 - 13.5) = 50 + j0 = 50 \angle 0^\circ\end{aligned}$$

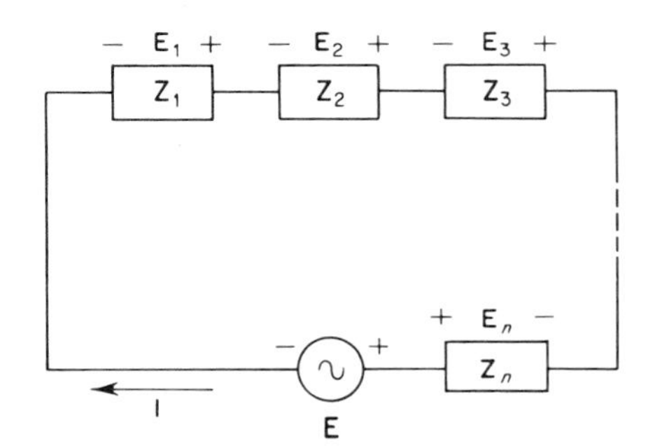

FIGURE 10-11
General series-impedance circuit.

The general relations of the series ac circuit are the same as those of a series dc circuit. The general relations are here given with reference to Fig. 10-11.

1. The current **I** is the common factor in a series circuit.
2. $\mathbf{E} = \mathbf{E}_1 + \mathbf{E}_2 + \mathbf{E}_3 + \cdots + \mathbf{E}_n$.
3. $\mathbf{Z}_t = \mathbf{E}/\mathbf{I} = \mathbf{Z}_1 + \mathbf{Z}_2 + \mathbf{Z}_3 + \cdots + \mathbf{Z}_n$.
4. Equation (10-2) is applicable to the total circuit or to any part of the circuit.

Example 10-2 further demonstrates the effect of frequency on a series *LCR* circuit.

EXAMPLE 10-2 In the circuit of Fig. 10-12, the voltage across the capacitor is $24 \angle 0°$. The circuit is operating at a frequency of 500 kHz. (1) Calculate the applied voltage. (2) Calculate the components that could replace the circuit at 500 kHz. (3) Sketch the current and voltage phasors on a complex plane. (4) Write the periodic functions for the currents and the voltages.

SOLUTION

1

$$-jX_C = -j\frac{1}{\omega C} = -j\frac{1}{(6.28)(5 \times 10^5)(5 \times 10^{-10})} = -j636\ \Omega$$

$$jX_L = j\omega L = j(6.28)(5 \times 10^5)(7 \times 10^{-5}) = j220\ \Omega$$

$$\mathbf{Z} = R + jX_L - jX_C = 250 + j220 - j636 = 250 - j416$$

The current phasor can be calculated with the voltage across the capacitor as the reference.

$$\mathbf{I} = \frac{\mathbf{E}_C}{-jX_C} = \frac{24 \angle 0°}{636 \angle -90°} = 0.0377 \angle 90°$$

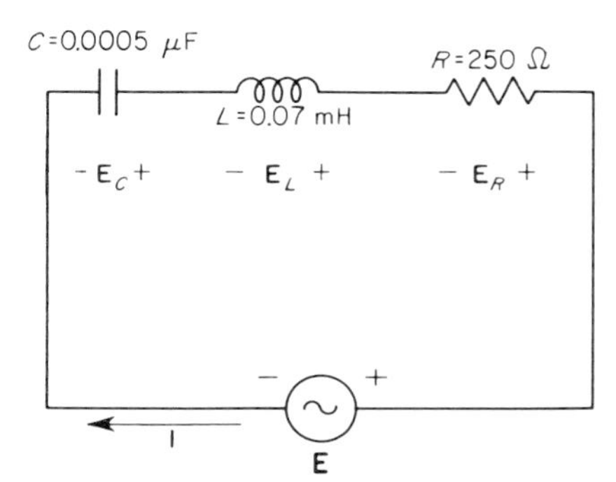

FIGURE 10-12
Circuit for Example 10-2.

The applied voltage can now be calculated.

$$\mathbf{E} = \mathbf{ZI} = (250 - j416)(0.0377 \angle 90^\circ) = (486 \angle -59^\circ)(0.0377 \angle 90^\circ) = 18.32 \angle 31^\circ$$

2 From the calculation of the impedance, it can be seen that the net reactance is $-j416$. Therefore, the net reactance is capacitive.

$$X'_C = \frac{1}{2\pi f C'}$$

$$C' = \frac{1}{2\pi f X'_C} = \frac{1}{(6.28)(5 \times 10^5)(4.16 \times 10^2)} = 0.00765 \times 10^{-7} = 765 \text{ pF}$$

The circuit of Fig. 10-12 can be replaced at 500 kHz with the circuit of Fig. 10-13, as far as the source is concerned.

3 First calculating $\mathbf{E}_R$ and $\mathbf{E}_L$,

$$\mathbf{E}_R = R\mathbf{I} = (250 \angle 0^\circ)(0.0377 \angle 90^\circ) = 9.42 \angle 90^\circ$$

$$\mathbf{E}_L = jX_L\mathbf{I} = (220 \angle 90^\circ)(0.0377 \angle 90^\circ) = 8.3 \angle 180^\circ$$

The current and voltage phasors are shown in Fig. 10-14. Had the applied voltage been used as the reference axis, there would have been no change in the magnitude of the current and voltages. The current and voltage phasors are shown in Fig. 10-15 with the applied voltage as the reference.

4 Writing the periodic functions, using the phase relationships shown in Fig. 10-14

$$i = (1.414 \times 0.0377) \sin(\omega t + 90^\circ) = 0.0534 \sin(\omega t + 90^\circ)$$

$$e = (1.414 \times 18.32) \sin(\omega t + 31^\circ) = 25.9 \sin(\omega t + 31^\circ)$$

$$e_C = (1.414 \times 24) \sin \omega t = 33.9 \sin \omega t$$

$$e_R = (1.414 \times 9.42) \sin(\omega t + 90^\circ) = 13.32 \sin(\omega t + 90^\circ)$$

$$e_L = (1.414 \times 8.3) \sin(\omega t + 180^\circ) = 11.75 \sin(\omega t + 180^\circ)$$

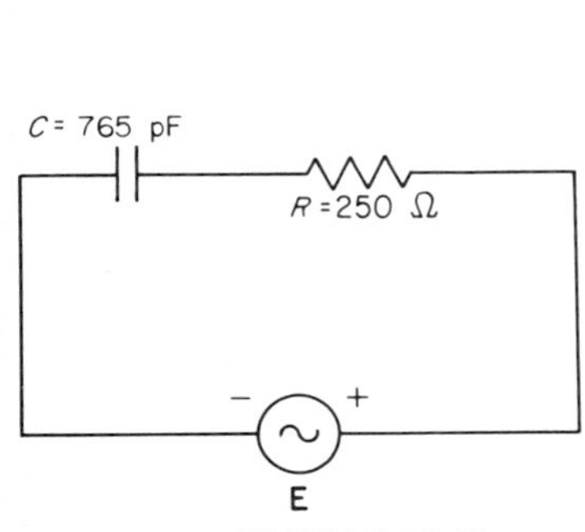

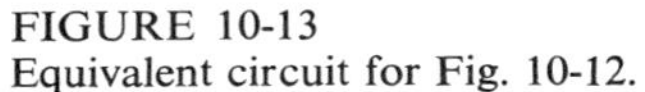
FIGURE 10-13
Equivalent circuit for Fig. 10-12.

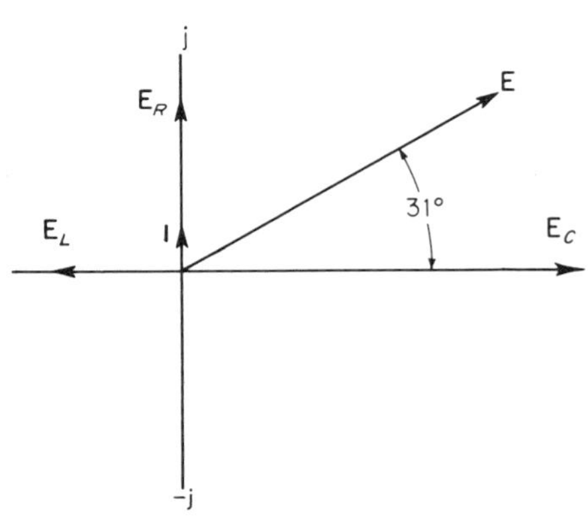

FIGURE 10-14
Phasors for Fig. 10-12.

Writing the periodic functions using the phase relationships shown in Fig. 10-15,

$$i = 0.0534 \sin (\omega t + 59^\circ)$$

$$e = 25.9 \sin \omega t$$

$$e_C = 33.9 \sin (\omega t - 31^\circ)$$

$$e_R = 13.32 \sin (\omega t + 59^\circ)$$

$$e_L = 11.75 \sin (\omega t + 149^\circ)$$

////

It must be noted that in Example 10-2 it makes no difference in the magnitude of the current and voltage phasors which phasor is used as the reference. Further, the phase angles between **I** and $\mathbf{E}_R$, $\mathbf{E}_L$, and $\mathbf{E}_C$ are unchanged.

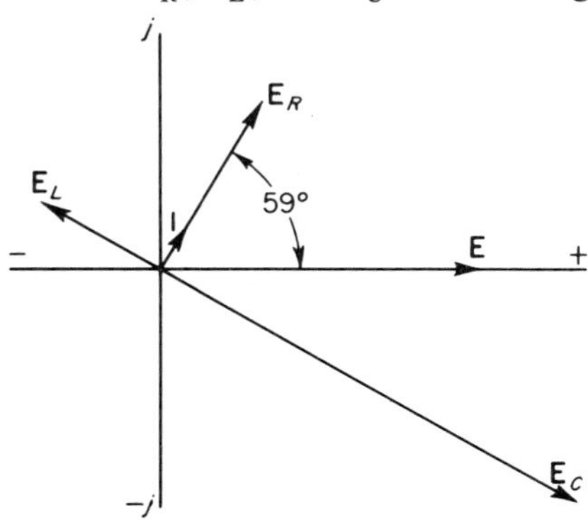

FIGURE 10-15
Phasors for Fig. 10-12 with applied voltage as reference.

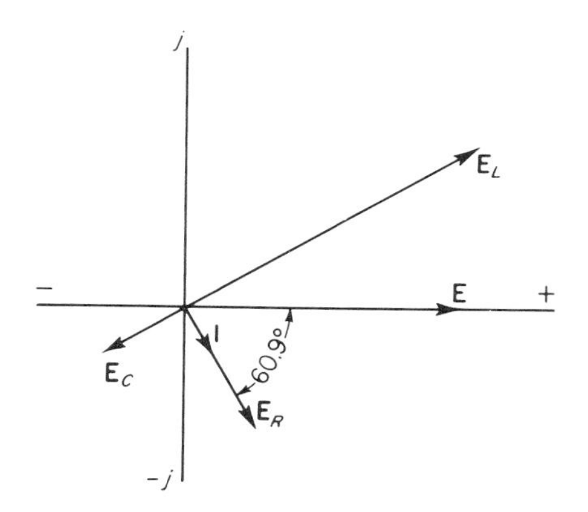

FIGURE 10-16
Phasors for Example 10-3.

EXAMPLE 10-3 Applied voltage **E** of Fig. 10-12 is held constant at 18.32 $\angle 0°$ while the frequency is increased to 1.5 MHz. (1) Calculate the current and voltage across each component. (2) Calculate the components that could replace the circuit of Fig. 10-12 at 1.5 MHz. (3) Sketch the current and voltage phasors on a complex plane.

SOLUTION

1

$$-jX_C = -j\frac{1}{\omega C} = -j\frac{1}{(6.28)(1.5 \times 10^6)(5 \times 10^{-10})} = -j212\ \Omega$$

$$jX_L = j\omega L = j(6.28)(1.5 \times 10^6)(7 \times 10^{-5}) = j660\ \Omega$$

$$\mathbf{Z} = R + jX_L - jX_C = 250 + j660 - j212 = 250 + j448 = 514 \angle 60.9°$$

The current can now be calculated using the applied voltage as the reference.

$$\mathbf{I} = \frac{\mathbf{E}}{\mathbf{Z}} = \frac{18.32 \angle 0°}{514 \angle 60.9°} = 0.0356 \angle -60.9°$$

$$\mathbf{E}_C = (-jX_C)\mathbf{I} = (212 \angle -90°)(0.0356 \angle -60.9°) = 7.55 \angle -150.9°$$

$$\mathbf{E}_L = (jX_L)\mathbf{I} = (660 \angle 90°)(0.0356 \angle -60.9°) = 23.5 \angle 29.1°$$

$$\mathbf{E}_R = R\mathbf{I} = (250 \angle 0°)(0.0356 \angle -60.9°) = 8.9 \angle -60.9°$$

2 The total impedance of the circuit of Fig. 10-12 was calculated in (1) to be $250 + j448$. Therefore, the net reactive component is inductive reactance, equal to 448 Ω.

$$X_L' = 448\ \Omega = 2\pi f L'$$

$$L' = \frac{X_L'}{2\pi f} = \frac{448}{(6.28)(1.5 \times 10^6)} = 47.5 \times 10^{-6}\ \text{H}$$

At a frequency of 1.5 MHz, the circuit of Fig. 10-12 could be replaced by an inductance of 47.5μH in series with a 250-Ω resistor.

3 See Fig. 10-16.

////

10-4 THE PARALLEL AC CIRCUIT

Figure 10-17 shows a parallel circuit. The current in each branch can be calculated.

$$\mathbf{I}_1 = \frac{\mathbf{E}}{\mathbf{Z}_1} = \frac{100 \angle 0^\circ}{400 + j500} = \frac{100 \angle 0^\circ}{640 \angle 51.35^\circ} = 0.156 \angle -51.35^\circ$$

$$\mathbf{I}_2 = \frac{\mathbf{E}}{\mathbf{Z}_2} = \frac{100 \angle 0^\circ}{600 + j400} = \frac{100 \angle 0^\circ}{721 \angle 33.7^\circ} = 0.138 \angle -33.7^\circ$$

The total current in the circuit of Fig. 10-17 is

$$\mathbf{I}_t = \mathbf{I}_1 + \mathbf{I}_2 = 0.156 \angle -51.35^\circ + 0.138 \angle -33.7^\circ$$
$$= (0.0975 - j0.122) + (0.1148 - j0.0765) = 0.2123 - j0.1985 = 0.291 \angle -43.1^\circ$$

The series-equivalent impedance of Fig. 10-17 can now be calculated.

$$\mathbf{Z}_t = \frac{\mathbf{E}}{\mathbf{I}_t} = \frac{100 \angle 0}{0.291 \angle -43.1^\circ} = 344 \angle 43.1^\circ\ \Omega$$

The series-equivalent impedance can also be calculated using,

$$\mathbf{Z}_t = \frac{\mathbf{Z}_1\mathbf{Z}_2}{\mathbf{Z}_1 + \mathbf{Z}_2} = \frac{(640 \angle 51.35^\circ)(721 \angle 33.7^\circ)}{(400 + j500) + (600 + j400)} = \frac{(640 \angle 51.35^\circ)(721 \angle 33.7^\circ)}{1{,}000 + j900}$$
$$= \frac{(640 \angle 51.35^\circ)(721 \angle 33.7^\circ)}{1{,}347 \angle 42^\circ} = 344 \angle 43.1^\circ\ \Omega$$

The total admittance of Fig. 10-17 is

$$\mathbf{Y}_t = \mathbf{Y}_1 + \mathbf{Y}_2 = \frac{1}{\mathbf{Z}_1} + \frac{1}{\mathbf{Z}_2}$$
$$= \frac{1}{640 \angle 51.35^\circ} + \frac{1}{721 \angle 33.7^\circ}$$
$$= 0.00156 \angle -51.35^\circ + 0.00138 \angle 33.7^\circ$$
$$= (0.000975 - j0.00122) + (0.001148 - j0.000765)$$
$$= 0.002123 - j0.001985 = 0.00291 \angle -43.1^\circ \text{ mho}$$

Likewise,

$$\mathbf{Y}_t = \frac{1}{\mathbf{Z}_t} = \frac{1}{344 \angle 43.1^\circ} = 0.00291 \angle -43.1^\circ \text{ mho}$$

The general relations of a parallel ac circuit are the same as those of a parallel dc circuit, except that phasors are used in ac circuit relations. The general relations are here given with reference to Fig. 10-18.

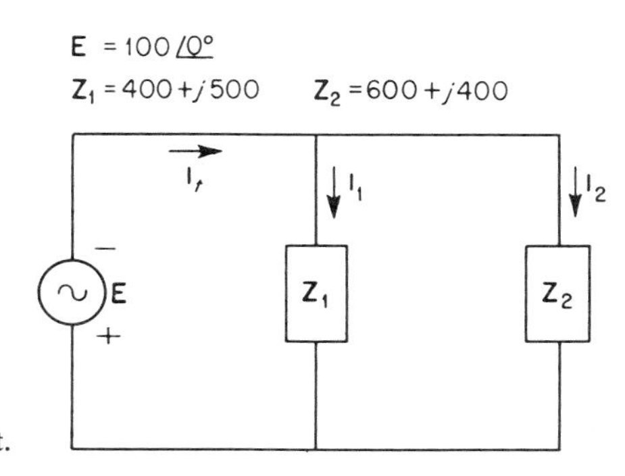

FIGURE 10-17
Parallel-impedance circuit.

1 The voltage **E** is the common factor in a parallel circuit.
2 $\mathbf{I}_t = \mathbf{I}_1 + \mathbf{I}_2 + \cdots + \mathbf{I}_n$.
3 $\mathbf{Y}_t = \mathbf{Y}_1 + \mathbf{Y}_2 + \cdots + \mathbf{Y}_n$ and $\mathbf{Z}_t = 1/\mathbf{Y}_t$.
4 Equation 10-2 is applicable to an entire circuit or to any part of it.

EXAMPLE 10-4 Given the circuit of Fig. 10-19, (1) calculate the current in each branch, (2) calculate the equivalent impedance of the circuit, and (3) show the components that would make up an equivalent impedance for the circuit at 150 kHz.

SOLUTION

1

$$\mathbf{Z}_1 = R_1 + jX_L$$
$$= 60 + j(6.28)(1.5 \times 10^5)(3 \times 10^{-4}) = 60 + j282 = 289 \angle 78^\circ\ \Omega$$
$$\mathbf{I}_1 = \frac{\mathbf{E}}{\mathbf{Z}_1}$$
$$= \frac{60^\circ \angle 0^\circ}{289 \angle 78^\circ} = 0.2075 \angle -78^\circ = 0.0431 - j0.203$$
$$\mathbf{Z}_2 = R_2 - jX_C$$
$$= 80 - j\frac{1}{(6.28)(1.5 \times 10^5)(1.06 \times 10^{-8})}$$
$$= 80 - j100 = 128 \angle -51.35^\circ\ \Omega$$
$$\mathbf{I}_2 = \frac{\mathbf{E}}{\mathbf{Z}_2}$$
$$= \frac{60 \angle 0^\circ}{128 \angle -51.35^\circ} = 0.468 \angle 51.35^\circ = 0.293 + j0.366$$

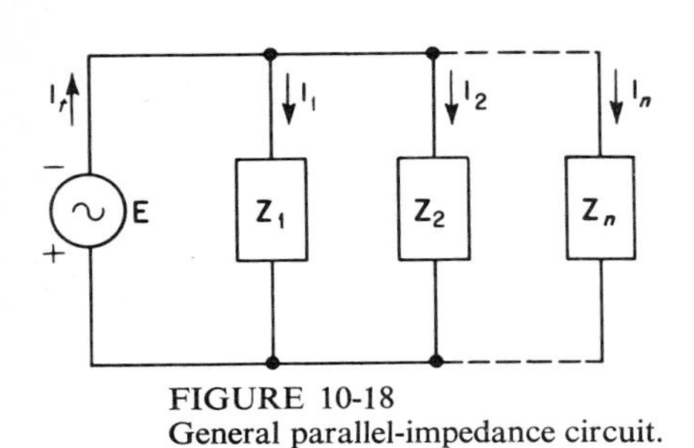

FIGURE 10-18
General parallel-impedance circuit.

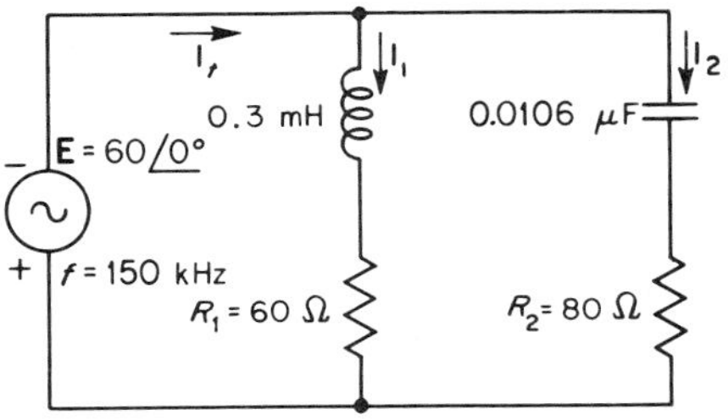

FIGURE 10-19
Circuit for Example 10-4.

2 The series-equivalent impedance can be calculated by any of the several methods. Calculating $\mathbf{I}_t$,

$$\begin{aligned}\mathbf{I}_t &= \mathbf{I}_1 + \mathbf{I}_2 \\ &= (0.0431 - j0.203) + (0.293 + j0.366) \\ &= 0.3361 + j0.163 = 0.374 \angle 25.9^\circ\end{aligned}$$

$\mathbf{Z}_t$ can now be calculated.

$$\mathbf{Z}_t = \frac{\mathbf{E}}{\mathbf{I}_t} = \frac{60 \angle 0^\circ}{0.374 \angle 25.9^\circ} = 160.5 \angle -25.9^\circ \ \Omega$$

$\mathbf{Z}_t$ can be calculated in another way.

$$\begin{aligned}\mathbf{Z}_t &= \frac{\mathbf{Z}_1\mathbf{Z}_2}{\mathbf{Z}_1 + \mathbf{Z}_2} \\ &= \frac{(289 \angle 78^\circ)(128 \angle -51.35^\circ)}{(60 + j282) + (80 - j100)} = \frac{(289 \angle 78^\circ)(128 \angle -51.35^\circ)}{140 + j182} \\ &= \frac{(289 \angle 78^\circ)(128 \angle -51.35^\circ)}{229 \angle 52.4^\circ} = 160.5 \angle -25.9^\circ \ \Omega\end{aligned}$$

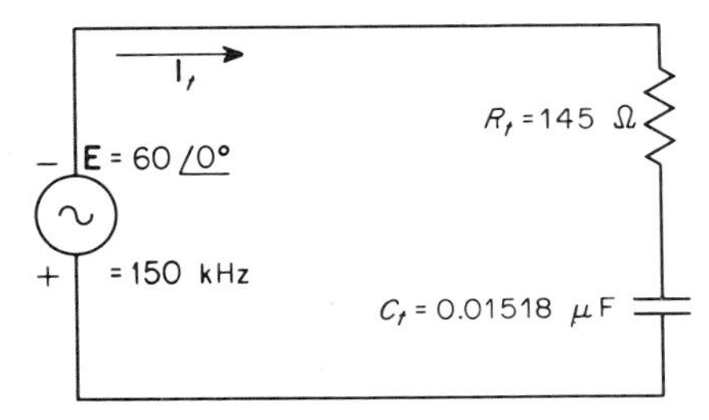

FIGURE 10-20
Series equivalent for Fig. 10-19.

3 Converting $\mathbf{Z}_t$ to rectangular form makes it easier to see what components make up the series-equivalent impedance of the circuit.

$$\mathbf{Z}_t = 160.5 \angle -25.9^\circ = 145 - j70$$

From this it can be seen that a resistor of 145 Ω in series with a capacitor having a reactance of 70 Ω are the components that make up a circuit equivalent to that of Fig. 10-19 at a frequency of 150 kHz.
Further,

$$C_t = \frac{1}{2\pi f X_{Ct}}$$

$$= \frac{1}{(6.28)(1.5 \times 10^5)(7 \times 10^1)} = 0.01518 \times 10^{-6}\text{ F} = 0.01518\ \mu\text{F}$$

The series-equivalent circuit is shown in Fig. 10-20. ////

It must be remembered that the circuit of Fig. 10-20 is the equivalent circuit for the circuit of Fig. 10-19 *only at a frequency of* 150 kHz. If the frequency of the applied voltage is changed, the equivalent of the circuit also changes. This is demonstrated in Example 10-5.

EXAMPLE 10-5 If the magnitude of the applied voltage in the circuit of Fig. 10-19 is held constant but its frequency is reduced to 25 kHz, (1) calculate the current in each branch and the total current, (2) calculate the equivalent impedance, and (3) show the equivalent circuit and calculate the values of its components.

SOLUTION

1 $\mathbf{Z}_1 = R_1 + jX_L$

$$= 60 + j(6.28)(2.5 \times 10^4)(3 \times 10^{-4}) = 60 + j47.1 = 76.1 \angle 38.2^\circ\ \Omega$$

$$\mathbf{I}_1 = \frac{\mathbf{E}}{\mathbf{Z}_1}$$

$$= \frac{60 \angle 0^\circ}{76.1 \angle 38.2^\circ} = 0.787 \angle -38.2^\circ = 0.619 - j0.485$$

$$\mathbf{Z}_2 = R_2 - jX_C$$

$$= 80 - j\frac{1}{(6.28)(2.5 \times 10^4)(1.06 \times 10^{-8})} = 80 - j600 = 605 \angle -82.4^\circ\ \Omega$$

$$\mathbf{I}_2 = \frac{60 \angle 0^\circ}{605 \angle -82.4^\circ} = 0.0992 \angle 82.4^\circ = 0.0131 + j0.0984$$

$$\mathbf{I}_t = \mathbf{I}_1 + \mathbf{I}_2$$

$$= (0.619 - j0.485) + (0.0131 + j0.0984) = 0.6321 - j0.3866$$

$$= 0.742 \angle -31.4^\circ$$

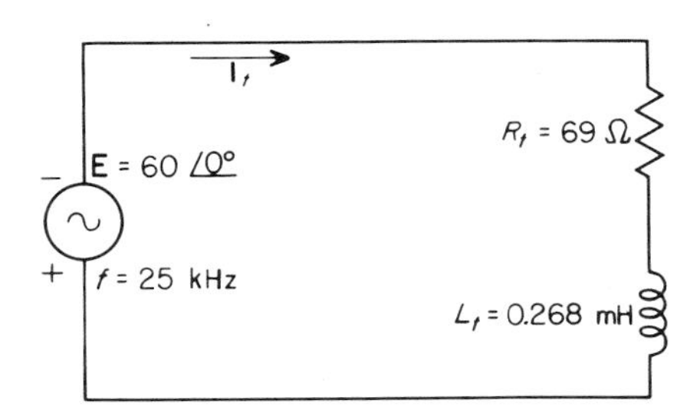

FIGURE 10-21
Series-equivalent circuit for Example 10-5.

2
$$\mathbf{Z}_t = \frac{\mathbf{E}}{\mathbf{I}}$$

$$= \frac{60 \angle 0^\circ}{0.742 \angle -31.4^\circ} = 80.9 \angle 31.4^\circ = 69 + j42.1$$

Check,

$$\mathbf{Z}_t = \frac{\mathbf{Z}_1\mathbf{Z}_2}{\mathbf{Z}_1 + \mathbf{Z}_2}$$

$$= \frac{(76.1 \angle 38.2^\circ)(605 \angle -82.4^\circ)}{(60 + j47.1) + (80 - j600)} = \frac{(76.1 \angle 38.2^\circ)(605 \angle -82.4^\circ)}{140 - j553.9}$$

$$= \frac{(76.1 \angle 38.2^\circ)(605 \angle -82.4^\circ)}{570 \angle -75.8^\circ} = 80.9 \angle 31.4^\circ = 69 + j42.1$$

$$R_t = 69\ \Omega \qquad X_{Lt} = 42.1\ \Omega$$

3 $$L_t = \frac{X_{Lt}}{2\pi f}$$

$$= \frac{42.1}{(6.28)(2.5 \times 10^4)} = 2.68 \times 10^{-4}\ \text{H} = 0.268\ \text{mH} \qquad ////$$

The resulting equivalent circuit is shown in Fig. 10-21.

Examples 10-4 and 10-5 demonstrate how the characteristics of ac circuits are affected by the operating frequency of the source. When the frequency changes, the characteristics of the circuit may change from capacitive to inductive. The equivalent resistance can also be changed by frequency variation. The dependency of circuit impedance on frequency has many practical applications, which will be discussed later.

10-5 SERIES-PARALLEL AC CIRCUITS

The general relations of series and parallel ac circuits have been given and demonstrated in Secs. 10-3 and 10-4. A great number of circuit configurations consist of series and parallel connections that can be handled individually, reducing the overall circuit to a simple series equivalent.

The exact methods of solving series-parallel circuits vary widely with particular circuit configurations, information given, and solutions required. No general rules can be given, but the following examples demonstrate several of these methods.

EXAMPLE 10-6 In the circuit of Fig. 10-22, (1) calculate **E.** (2) If the impedances are given for a frequency of 250 kHz, calculate the series-equivalent components. (3) Show on a complex plane all current and voltage phasors. (4) Write the periodic function for all the currents and voltages. (5) Show that Kirchhoff's laws hold for the instantaneous values of current and voltage at time $t = 0.3\ \mu$s.

SOLUTION

1

$$\mathbf{Z}_{23} = \mathbf{Z}_2 + \mathbf{Z}_3$$

$$= (30 + j150) + (50 - j240) = 80 - j90$$

$$\mathbf{E}_{23} = \mathbf{Z}_{23}\mathbf{I}_{23}$$

$$= (80 - j90)(0.4\ \angle 0°) = (120\ \angle -48°)(0.4\ \angle 0°) = 48\ \angle -48° = \mathbf{E}_4$$

$$\mathbf{I}_4 = \frac{\mathbf{E}_4}{\mathbf{Z}_4}$$

$$= \frac{48\ \angle -48°}{90 + j400} = \frac{48\ \angle -48°}{410\ \angle 77.3°} = 0.117\ \angle -125.3°$$

$$\mathbf{I}_1 = \mathbf{I}_{23} + \mathbf{I}_4$$

$$= 0.4\ \angle 0° + 0.117\ \angle -125.3° = (0.4 + j0) + (-0.067 - j0.096)$$

$$= 0.333 - j0.096 = 0.346\ \angle -16.05°$$

$$\mathbf{E}_1 = \mathbf{Z}_1\mathbf{I}_1$$

$$= (40 + j280)(0.346\ \angle -16.05°) = (282\ \angle 81.86°)(0.346\ \angle -16.05°)$$

$$= 97.7\ \angle 65.81°$$

$$\mathbf{E} = \mathbf{E}_1 + \mathbf{E}_{23}$$

$$= 97.7\ \angle 65.81° + 48\ \angle -48° = (40 + j89) + (31.85 - j35.8)$$

$$= 71.85 + j53.2 = 89.4\ \angle 36.55°$$

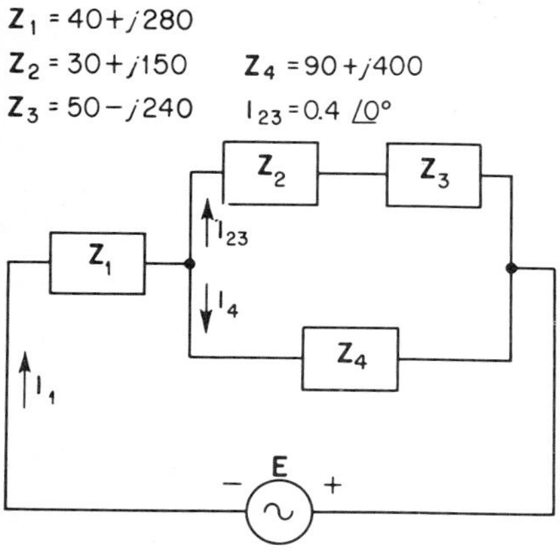

$\mathbf{Z}_1 = 40 + j280$
$\mathbf{Z}_2 = 30 + j150$ $\mathbf{Z}_4 = 90 + j400$
$\mathbf{Z}_3 = 50 - j240$ $\mathbf{I}_{23} = 0.4 \angle 0°$

FIGURE 10-22
Circuit for Example 10-6.

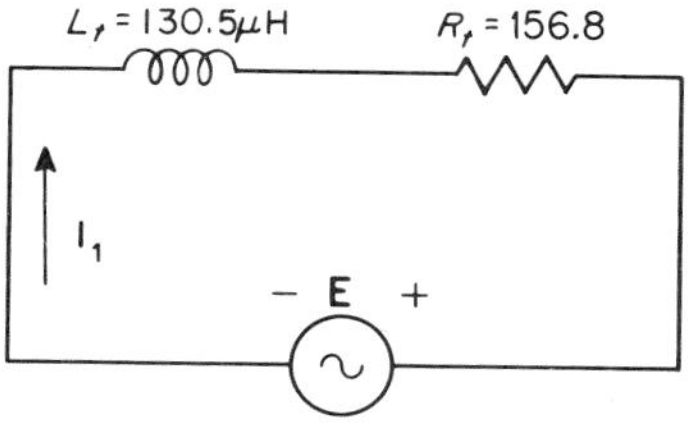

$L_t = 130.5\,\mu H$ $R_t = 156.8$

FIGURE 10-23
Series-equivalent components for Fig. 10-22.

2 $\mathbf{Z}_t = \dfrac{\mathbf{E}}{\mathbf{I}_1}$

$$= \frac{89.4 \angle 36.55°}{0.346 \angle -16.05°} = 258 \angle 52.6° = 156.8 + j205$$

Therefore,

$R_t = 156.8\ \Omega$

$X_{Lt} = 205\ \Omega$

$$L_t = \frac{X_{Lt}}{2\pi f}$$

$$= \frac{205}{(6.28)(2.5 \times 10^5)} = 13.05 \times 10^{-5}\ \text{H} = 130.5\ \mu\text{H}$$

See Fig. 10-23.

3 $\mathbf{E}_2 = \mathbf{Z}_2 \mathbf{I}_{23}$

$$= (30 + j150)(0.4 \angle 0°) = (153 \angle 78.7°)(0.4 \angle 0°) = 61.3 \angle 78.7°$$

$\mathbf{E}_3 = \mathbf{Z}_3 \mathbf{I}_{23}$

$$= (50 - j240)(0.4 \angle 0°) = (245 \angle -78.2°)(0.4 \angle 0°) = 98 \angle -78.2°$$

See Fig. 10-24 for the phasor plot.

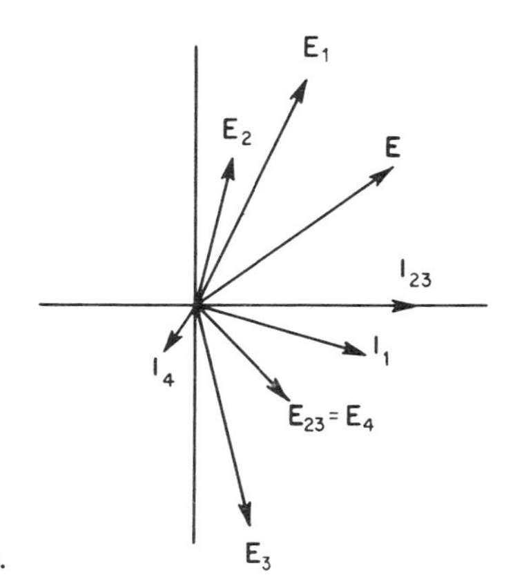

FIGURE 10-24
Phasors for Fig. 10-22.

4 $e = (1.414)(89.4) \sin(\omega t + 36.55°) = 126.3 \sin(\omega t + 36.55°)$

$e_1 = (1.414)(97.7) \sin(\omega t + 65.81°) = 138.2 \sin(\omega t + 65.81°)$

$e_2 = (1.414)(61.3) \sin(\omega t + 78.7°) = 86.6 \sin(\omega t + 78.7°)$

$e_3 = (1.414)(98) \sin(\omega t - 78.2°) = 139 \sin(\omega t - 78.2°)$

$e_4 = (1.414)(48) \sin(\omega t - 48°) = 68 \sin(\omega t - 48°)$

$i_1 = (1.414)(0.346) \sin(\omega t - 16.05°) = 0.49 \sin(\omega t - 16.05°)$

$i_{23} = (1.414)(0.4) \sin \omega t = 0.566 \sin \omega t$

$i_4 = (1.414)(0.117) \sin(\omega t - 125.3°) = 0.1658 \sin(\omega t - 125.3°)$

5 $e = e_1 + e_{23}$

$= 138.2 \sin [(2\pi)(2.5 \times 10^5)(0.3 \times 10^{-6}) + 65.81°]$

$+ 68 \sin [(2\pi)(2.5 \times 10^5)(.3 \times 10^{-6}) - 48°]$

$= 138.2 \sin(27° + 65.81°) + 68 \sin(27° - 48°)$

$= 138.2 \sin 92.81° + 68 \sin(-21°) = 138 - 24.8 = 113.2$

$e = 126.3 \sin(27° + 36.55°) = 126.3 \sin 63.55° = 113.2$

$i_1 = i_{23} + i_4$

$= 0.566 \sin 27° + 0.1658 \sin(27° - 125.3°)$

$= 0.566 \sin 27° + 0.1658 \sin(-98.3°) = 0.257 - 0.164 = 0.093$

$i_1 = 0.49 \sin(27° - 16.05°) = 0.49 \sin(10.95°) = 0.093$ ////

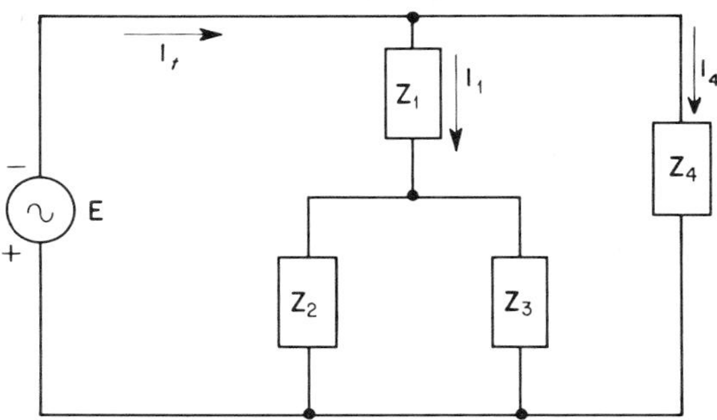

FIGURE 10-25
Circuit for Example 10-7.

EXAMPLE 10-7 Calculate $\mathbf{Z}_4$ of Fig. 10-25.

SOLUTION

$$\mathbf{I}_4 = \mathbf{I}_t - \mathbf{I}_1$$
$$= (0.32 \angle -36.7^\circ) - (0.122 \angle -33.3^\circ) = (0.256 - j0.191) - (0.102 - j0.067)$$
$$= 0.154 - j0.124 = 0.198 \angle -38.8^\circ$$

$$\mathbf{E}_4 = \mathbf{E}_1 + \mathbf{E}_{23} = \mathbf{Z}_1\mathbf{I}_1 + \mathbf{Z}_{23}\mathbf{I}_1 = \mathbf{Z}_1\mathbf{I}_1 + \frac{\mathbf{Z}_2\mathbf{Z}_3}{\mathbf{Z}_2 + \mathbf{Z}_3}\mathbf{I}_1$$

$$= (400 \angle 50^\circ)(0.122 \angle -33.3^\circ) + \frac{(200 \angle -38^\circ)(370 \angle -58^\circ)}{(157.5 - j123) + (196 - j314)}(.122 \angle -33.3^\circ)$$

$$= 48.8 \angle 16.7^\circ + \frac{(200 \angle -38^\circ)(370 \angle -58^\circ)}{353.5 - j437}(0.122 \angle -33.3^\circ)$$

$$= 48.8 \angle 16.7^\circ + \frac{(200 \angle -38^\circ)(370 \angle -58^\circ)}{562 \angle -51^\circ}(0.122 \angle -33.3^\circ)$$

$$= 48.8 \angle 16.7^\circ + (131.8 \angle -45^\circ)(0.122 \angle -33.3^\circ)$$

$$= 48.8 \angle 16.7^\circ + 16.08 \angle -78.3^\circ = (46.8 + j14.05) + (3.26 - j15.7)$$

$$= 50.06 \angle -2^\circ$$

$$\mathbf{Z}_4 = \frac{\mathbf{E}_4}{\mathbf{I}_4} = \frac{50.06 \angle -2^\circ}{0.198 \angle -38.8^\circ} = 253 \angle 36.8^\circ \ \Omega$$

////

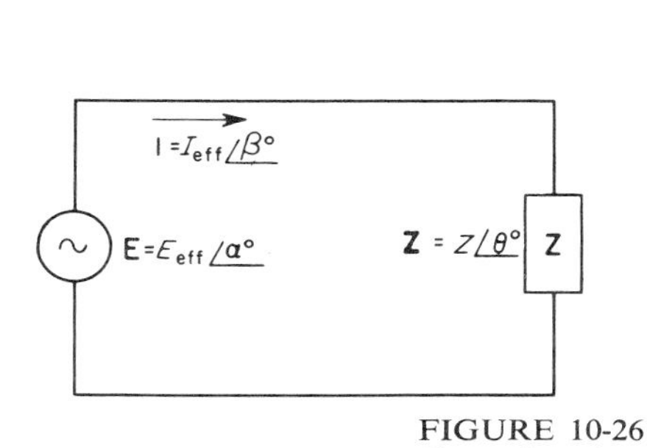

FIGURE 10-26
General ac circuit.

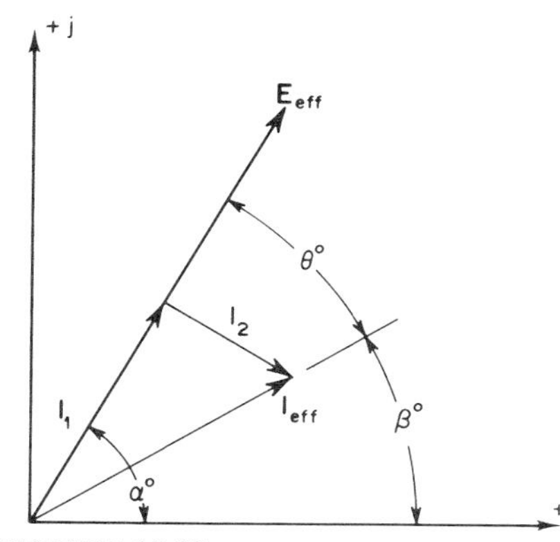

FIGURE 10-27
Current and voltage phasors for Fig. 10-26.

10-6 POWER IN AC CIRCUITS

In Chap. 9, it was shown that no power is dissipated in a purely inductive or capacitive circuit. It was also shown that the power dissipated in a purely resistive circuit is the product of effective current and voltage. The ac power relations in a circuit will now be discussed.

A general ac circuit is shown in Fig. 10-26. The current and voltage phasors are shown in Fig. 10-27. The angle between the current phasor and the voltage phasor is θ, the angle of the impedance **Z**. The current phasor can be broken into two components, one coinciding with the voltage phasor, and one perpendicular to the voltage phasor. The coincident component of current is

$$I_1 = I_{eff} \cos \theta$$

The power dissipated in the circuit of Fig. 10-26 is

$$P_{true} = E_{eff} I_{eff} \cos \theta \qquad (10\text{-}4)$$

The power given by Eq. (10-4) is called *true power* because it is power actually dissipated in the circuit. It has the customary power unit, the watt.

The normal (perpendicular) component of the current does not dissipate power because 90° difference between the voltage phasor and the normal current phasor indicates that energy is alternately stored and released to the circuit.

$$I_2 = I_{eff} \sin \theta$$

The product of voltage and the normal component of current is called *reactive power*. Reactive power is given by Eq. (10-5).

$$P_{reac} = E_{eff} I_{eff} \sin \theta \tag{10-5}$$

The unit of reactive power is the var, which stands for volt-ampere reactive. It is a measure of the energy that is not dissipated.

The product of the *magnitudes* of the current and voltage phasors is called *apparent power*. Apparent power is given by Eq. (10-6).

$$P_{appar} = E_{eff} I_{eff} \tag{10-6}$$

The volt-ampere (VA) is the unit of apparent power.

The ratio of true power to apparent power is called the *power factor* of a circuit.

$$PF = \frac{P_{true}}{P_{appar}} = \frac{E_{eff} I_{eff} \cos \theta}{E_{eff} I_{eff}} = \cos \theta \tag{10-7}$$

As shown by Eq. (10-7), the power factor is the $\cos \theta$. θ is the angle between current and voltage, given with reference to the voltage. If the current lags the voltage, the power factor is said to be a lagging one, and if the current leads the voltage, the power factor is said to be a leading one. The angle of lead or lag is determined by the angle of the impedance.

The range of the power factor is from $\cos -90°$ to $\cos 90°$.

EXAMPLE 10-8 In the circuit of Fig. 10-28, calculate (1) true power, (2) apparent power, and (3) power factor. The phasor diagram is shown in Fig. 10-29.

SOLUTION

1

$$\mathbf{I} = \frac{\mathbf{E}}{\mathbf{Z}} = \frac{80 \angle 30°}{720 \angle 56.3°} = 0.111 \angle -26.3°$$

$$P_{true} = E_{eff} I_{eff} \cos \theta = (80)(0.111) \cos 56.3° = 4.94 \text{ W}$$

2

$$P_{appar} = E_{eff} I_{eff} = (80)(0.111) = 8.9 \text{ VA}$$

3

$$\text{PF} = \cos \theta = \cos 56.3° = 0.555 \qquad \text{lagging}$$

Check:

$$\text{PF} = \frac{P_{true}}{P_{appar}} = \frac{4.94}{8.9} = 0.555 \qquad \text{lagging}$$

////

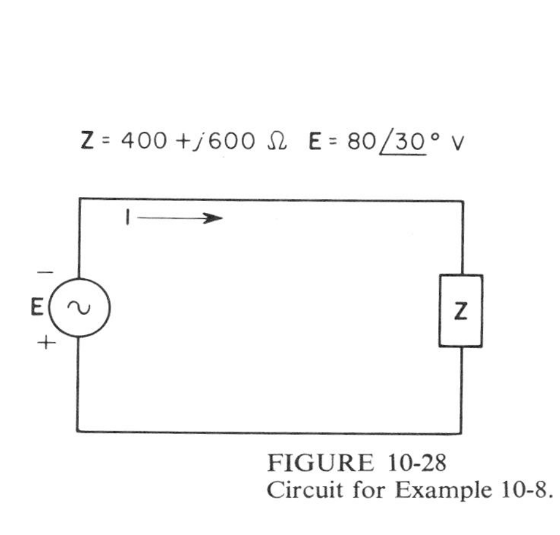

FIGURE 10-28
Circuit for Example 10-8.

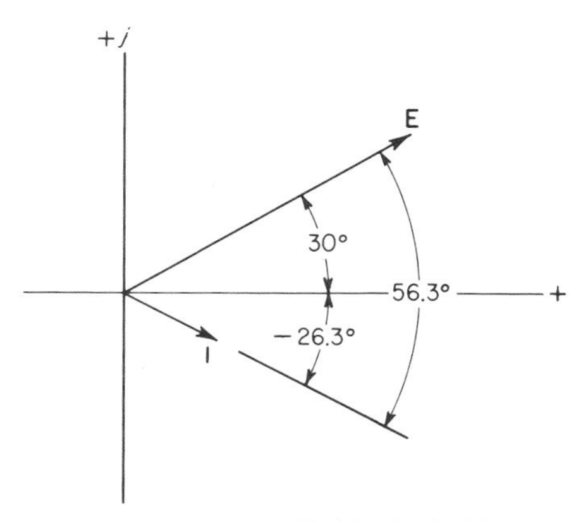

FIGURE 10-29
Phasors for Fig. 10-28.

EXAMPLE 10-9 In the circuit of Fig. 10-30, calculate (1) true power, (2) apparent power, and (3) power factor. The phasor diagram is shown in Fig. 10-31.

SOLUTION

1

$$\mathbf{E} = \mathbf{ZI} = (300 \angle -60°)(0.4 \angle 40°) = 120 \angle -20°$$

$$P_{\text{true}} = E_{\text{eff}} I_{\text{eff}} \cos \theta = (120)(0.4) \cos 60° = 24 \text{ W}$$

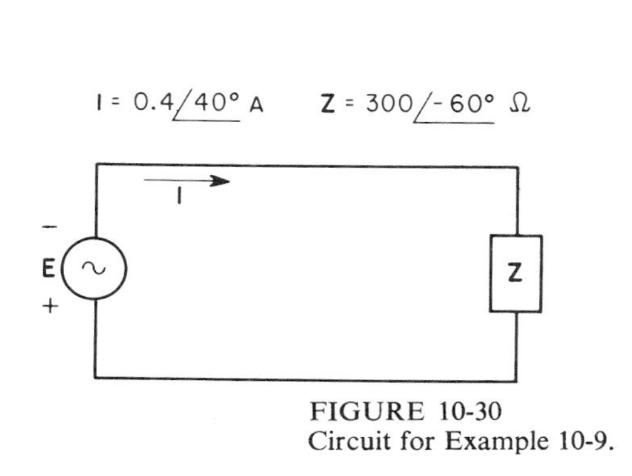

FIGURE 10-30
Circuit for Example 10-9.

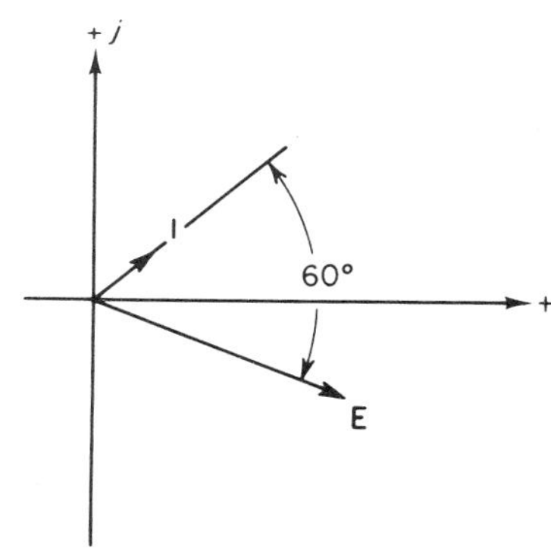

FIGURE 10-31
Phasors for Fig. 10-30.

2 $$P_{\text{appar}} = E_{\text{eff}} I_{\text{eff}} = (120)(0.4) = 48 \text{ VA}$$

3 $$\text{PF} = \cos 60° = 0.5 \qquad \text{leading}$$

Check

$$\text{PF} = \frac{P_{\text{true}}}{P_{\text{appar}}} = \frac{24}{48} = 0.5 \qquad \text{leading}$$ ////

PROBLEMS

10-1 A resistor of 300 Ω and a 0.005-μF capacitor are in series. Calculate the impedance in both polar and rectangular forms at frequencies of (*a*) 500 kHz, (*b*) 250 kHz, (*c*) 200 kHz, (*d*) 180 kHz, (*e*) 100 kHz, and (*f*) 50 kHz.

10-2 A resistor of 500 Ω and a 5-mH coil are connected in series. Calculate the impedance in polar and rectangular forms at frequencies of (*a*) 500 kHz, (*b*) 100 kHz, (*c*) 50 kHz, (*d*) 20 kHz, (*e*) 10 kHz, (*f*) 3 kHz, and (*g*) 800 Hz.

10-3 An impedance is given as $\mathbf{Z} = 400 \angle 40° \ \Omega$. Calculate the resistances and inductances at frequencies of (*a*) 60 kHz, (*b*) 3,000 Hz, (*c*) 200 kHz, (*d*) 15 kHz, and (*e*) 35 kHz.

10-4 An impedance is given as $\mathbf{Z} = 250 \angle -70° \ \Omega$. Calculate the resistances and capacitances at frequencies of (*a*) 2.5 MHz, (*b*) 750 kHz, (*c*) 420 kHz, (*d*) 120 kHz, and (*e*) 12 kHz.

10-5 An impedance is given as $\mathbf{Z} = 15 \times 10^3 \angle 56° \ \Omega$. Calculate the resistances and inductances at frequencies of (*a*) 2.4 MHz, (*b*) 2 MHz, (*c*) 1.2 MHz, (*d*) 700 kHz, and (*e*) 400 kHz.

10-6 An impedance is given as $\mathbf{Z} = 20 \times 10^3 \angle -25° \ \Omega$. Calculate the resistances and capacitances at frequencies of (*a*) 50 kHz, (*b*) 120 kHz, (*c*) 150 kHz, (*d*) 210 kHz, and (*e*) 340 kHz.

10-7 A resistance of 800 Ω is in series with an 8-mF capacitor. Calculate the impedance in polar and rectangular forms at frequencies of (*a*) 10 Hz, (*b*) 30 Hz, (*c*) 60 Hz, (*d*) 100 Hz, (*e*) 200 Hz, and (*f*) 1 kHz.

10-8 In the circuit of Fig. 10-32, $\mathbf{I} = 0.35 \angle 40°$ A and $\mathbf{E} = 150 \angle 70°$ V. Calculate the impedance.

10-9 In the circuit of Fig. 10-32, $\mathbf{I} = 0.08 \angle -40°$ A and $\mathbf{E} = 90 \angle 20°$ V. Calculate the impedance.

10-10 In the circuit of Fig. 10-32, $\mathbf{I} = 0.075 \angle 65°$ A and $\mathbf{Z} = 200 + j350 \ \Omega$. Calculate the voltage $\mathbf{E}$.

10-11 In the circuit of Fig. 10-32, $\mathbf{E} = 400 \angle 0°$ V and $\mathbf{Z} = 3{,}500 - j2{,}700 \ \Omega$. Calculate the current $\mathbf{I}$.

10-12 In the circuit of Fig. 10-32, $\mathbf{I} = 0.005 \angle -60°$ A and $\mathbf{Z} = 600 + j400 \ \Omega$. Calculate the voltage $\mathbf{E}$.

10-13 In the circuit of Fig. 10-33, $\mathbf{E} = 20 \angle 0°$ at a frequency of 110 kHz. (*a*) Calculate the total impedance of the circuit. (*b*) Calculate the voltage drop across each component of the circuit. (*c*) Show that the phasor sum of the voltages across the components is equal to the applied voltage.

FIGURE 10-32
Circuit for Probs. 10-8 to 10-12.

10-14 In the circuit of Fig. 10-33, $\mathbf{E} = 35 \angle 0°$ V at a frequency of 135 kHz. Repeat the calculations of Prob. 10-13.

10-15 In the circuit of Fig. 10-33, $\mathbf{E} = 25 \angle 0°$ V at a frequency of 146 kHz. Repeat the calculations of Prob. 10-13.

10-16 In the circuit of Fig. 10-33, $\mathbf{E} = 40 \angle 30°$ V at a frequency of 90 kHz. Repeat the calculations of Prob. 10-13.

10-17 Calculate the inductance necessary to replace L and C in Fig. 10-33 for the conditions of Prob. 10-14.

10-18 Calculate the capacitance necessary to replace L and C in Fig. 10-33 for the conditions of Prob. 10-13.

10-19 Calculate the capacitance necessary to replace L and C in Fig. 10-33 for the conditions of Prob. 10-16.

10-20 Calculate the inductance necessary to replace L and C in Fig. 10-33 for the conditions of Prob. 10-15.

10-21 In the circuit of Fig. 10-34, $R = 150\ \Omega$, $\mathbf{E}_R = 70$ V, $L = 0.5$ mH, $X_L = 300\ \Omega$, and $C = 0.002\ \mu$F. Find (*a*) the frequency , (*b*) X_C, (*c*) $\mathbf{Z}_t$, (*d*) $\mathbf{I}$, and (*e*) $\mathbf{E}_C$.

10-22 In the circuit of Fig. 10-34, $R = 600\ \Omega$, $\mathbf{E}_L = 25 \angle 90°$ V, $\mathbf{E}_C = 40 \angle -90°$ V, freq. = 4.2 MHz, and $\mathbf{E}_R = 42$ V. Find (*a*) C, (*b*) L, (*c*) $\mathbf{E}$, and (*d*) $\mathbf{Z}_t$.

10-23 Calculate the equivalent impedance of Fig. 10-35 at the following frequencies: (*a*) 90 kHz, (*b*) 50 kHz, (*c*) 140 kHz, (*d*) 200 kHz, and (*e*) 170 kHz.

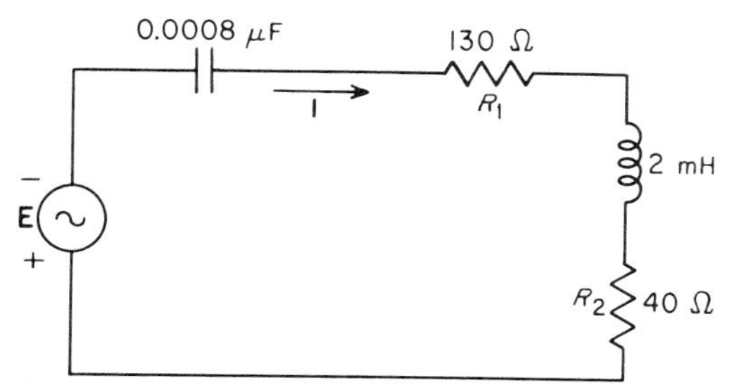

FIGURE 10-33
Circuit for Probs. 10-13 to 10-20.

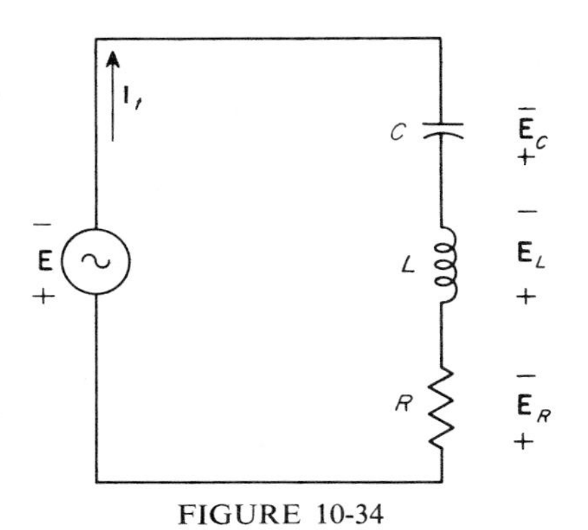

FIGURE 10-34
Circuit for Probs. 10-21 and 10-22.

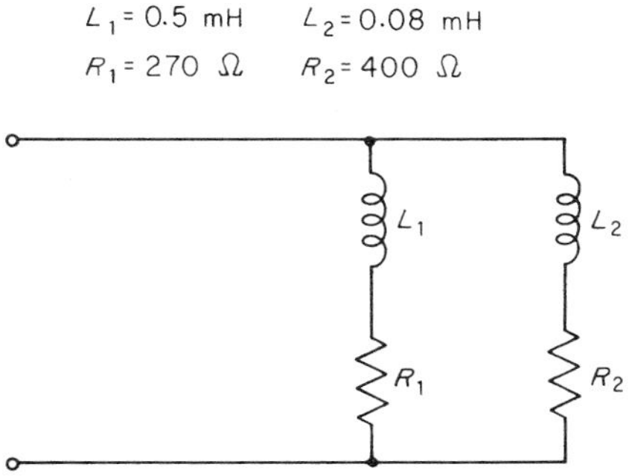

$L_1 = 0.5$ mH $L_2 = 0.08$ mH
$R_1 = 270\ \Omega$ $R_2 = 400\ \Omega$

L_1 L_2 R_1 R_2

FIGURE 10-35
Circuit for Prob. 10-23.

10-24 Calculate the series-equivalent impedance of Fig. 10-36 at the following frequencies: (*a*) 40 kHz, (*b*) 4 kHz, (*c*) 120 kHz, (*d*) 160 kHz, and (*e*) 200 kHz.

10-25 Calculate the admittance of each branch, the total admittance, and the impedance of the circuit of Fig. 10-37 at the following frequencies: (*a*) 300 kHz, (*b*) 350 kHz, (*c*) 200 kHz, (*d*) 270 kHz, (*e*) 120 kHz, and (*f*) 140 kHz.

10-26 In the circuit of Fig. 10-38, $\mathbf{Z}_1 = 150 + j1{,}000\ \Omega$, $\mathbf{Z}_2 = 40 - j300\ \Omega$, $\mathbf{Z}_3 = 30 + j650\ \Omega$, and $\mathbf{E} = 120\ \angle 40°$. Find (*a*) $\mathbf{I}_1$, (*b*) $\mathbf{I}_2$, (*c*) $\mathbf{I}_3$, (*d*) $\mathbf{I}_t$, (*e*) $\mathbf{Y}_1$, (*f*) $\mathbf{Y}_t$, and (*g*) $\mathbf{Z}_t$.

10-27 In the circuit of Fig. 10-38, the impedances are as given in Prob. 10-26, and $\mathbf{I}_1 = 0.005\ \angle 20°$ A. Find (*a*) $\mathbf{E}$, (*b*) $\mathbf{I}_2$, (*c*) $\mathbf{I}_3$, and (*d*) $\mathbf{I}_t$.

10-28 In the circuit of Fig. 10-38, $\mathbf{Y}_1 = 0.003\ \angle -75°$ mho, $\mathbf{Y}_2 = 0.008\ \angle 68°$mho, $\mathbf{Y}_3 = 0.0016\angle -62°$ mho, and $\mathbf{I}_2 = 0.004 + j0.003$ A. Find (*a*) $\mathbf{E}$, (*b*) $\mathbf{I}_1$, (*c*) $\mathbf{I}_3$, (*d*) $\mathbf{Y}_t$, (*e*) $\mathbf{Z}_t$, and (*f*) $\mathbf{I}_t$.

10-29 In the circuit of Fig. 10-38, $\mathbf{I}_1 = 0.0005\ \angle 40°$ A, $\mathbf{I}_2 = 0.0009\ \angle 70°$ A, $\mathbf{I}_3 = 0.0003\ \angle 65°$ A, and $\mathbf{Z}_2 = 6{,}500\ \angle -56°\ \Omega$. Find (*a*) $\mathbf{Z}_1$, (*b*) $\mathbf{Z}_3$, and (*c*) $\mathbf{Y}_t$.

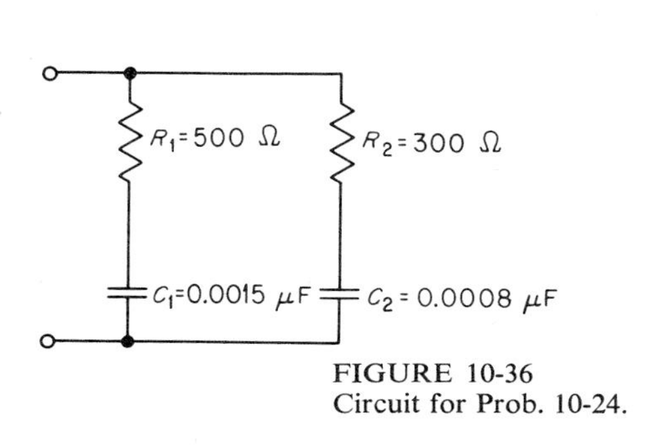

FIGURE 10-36
Circuit for Prob. 10-24.

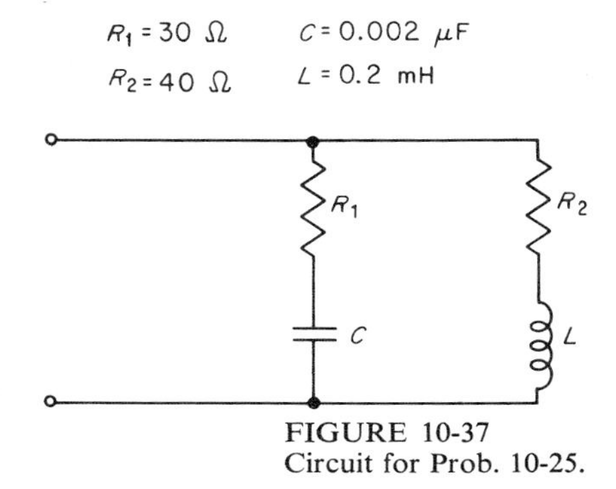

FIGURE 10-37
Circuit for Prob. 10-25.

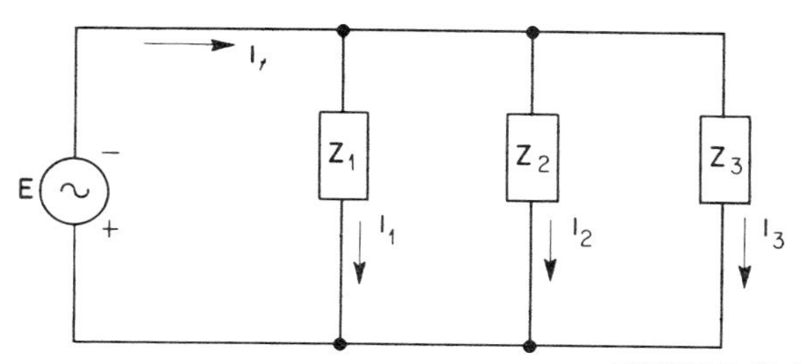

FIGURE 10-38
Circuit for Probs. 10-26 to 10-30.

10-30 In the circuit of Fig. 10-38, $\mathbf{I}_t = 0.0083 \angle 38°$, $\mathbf{I}_t = 0.0087 \angle 70°$ A, $\mathbf{I}_2 = 0.0003 \angle 65°$ A, and $\mathbf{Z}_2 = 2{,}600 \angle -65°$. Find (*a*) $\mathbf{I}_3$, (*b*) $\mathbf{Z}_3$, (*c*) $\mathbf{Z}_t$, and (*d*) $\mathbf{Y}_t$.

10-31 Calculate the total impedance of the circuit of Fig. 10-39.

10-32 If $\mathbf{E}$ of Fig. 10-39 is $50 \angle 20°$ V, calculate the branch currents and the total current.

10-33 Calculate the branch admittances and the total admittance of Fig. 10-39.

10-34 If $\mathbf{E}$ of Fig. 10-39 is $60 \angle 0°$ V, calculate the branch and total currents.

10-35 Calculate the series-equivalent components of Fig. 10-39 at frequencies of (*a*) 25 kHz, (*b*) 40 kHz, and (*c*) 100 kHz.

10-36 Calculate $\mathbf{Z}_t$ of Fig. 10-40.

10-37 If applied voltage $\mathbf{E}$ of Fig. 10-40 is $140 \angle 0°$ V, calculate (*a*) $\mathbf{I}_2$, (*b*) $\mathbf{I}_3$, and (*c*) $\mathbf{E}_1$.

10-38 If $\mathbf{I}_3$ of Fig. 10-40 is $0.002 \angle 40°$ A, calculate (*a*) $\mathbf{I}_2$, (*b*) $\mathbf{I}_1$, and (*c*) $\mathbf{E}$.

10-39 If $\mathbf{E}_1$ of Fig. 10-40 is $100 \angle 30°$ V, calculate (*a*) $\mathbf{I}_2$, (*b*) $\mathbf{I}_3$, and (*c*) $\mathbf{E}$.

10-40 Calculate the total impedance of the circuit of Fig. 10-41

10-41 If $\mathbf{E}_4$ of Fig. 10-41 is $106 \angle 40°$ V, calculate (*a*) $\mathbf{I}_2$ and (*b*) $\mathbf{E}$.

10-42 If $\mathbf{I}_2$ of Fig. 10-41 is $0.03 \angle 63°$ A, calculate (*a*) $\mathbf{E}_3$, (*b*) $\mathbf{E}_4$, and (*c*) $\mathbf{E}_1$.

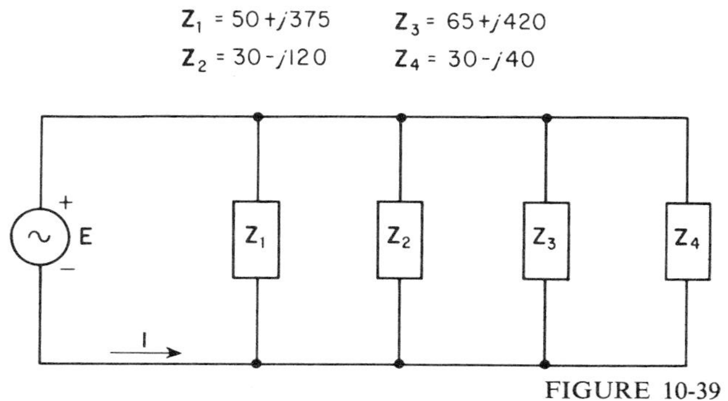

FIGURE 10-39
Circuit for Probs. 10-31 to 10-35.

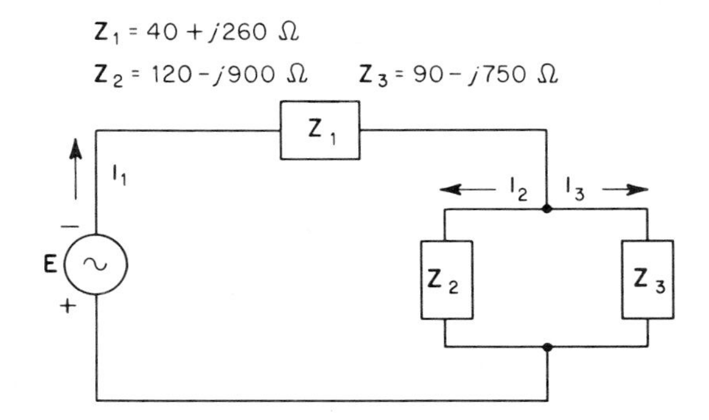

FIGURE 10-40
Circuit for Probs. 10-36 to 10-39.

10-43 If $\mathbf{I}_t$ of Fig. 10-41 is 0.09 $\angle 10°$ A, calculate (*a*) $\mathbf{E}$, (*b*) $\mathbf{I}_2$, and (*c*) $\mathbf{E}_4$.

10-44 If $\mathbf{I}_{3-4}$ of Fig. 10-41 is 0.6 $\angle 30°$ A, calculate (*a*) $\mathbf{E}_4$, (*b*) $\mathbf{E}_2$, and (*c*) $\mathbf{I}_t$.

10-45 If $\mathbf{E}$ of Fig. 10-41 is 500 $\angle 0°$ V, calculate (*a*) $\mathbf{I}_{3-4}$, (*b*) $\mathbf{E}_4$, and (*c*) $\mathbf{E}_1$.

10-46 If $\mathbf{E}$ of Fig. 10-41 is 500 $\angle 30°$ V, calculate parts *a*, *b*, and *c* of Prob. 10-45.

10-47 If $\mathbf{E}$ of Fig. 10-41 is 500 $\angle -60°$ V, calculate parts *a*, *b*, and *c* of Prob. 10-45.

10-48 If $\mathbf{E}$ of Fig. 10-41 is 120 $\angle 140°$ V, calculate (*a*) $\mathbf{E}_2$, (*b*) $\mathbf{E}_4$, and (*c*) $\mathbf{I}_t$.

10-49 If $\mathbf{E}_3$ of Fig. 10-41 is 140 $\angle 160°$ V, calculate (*a*) $\mathbf{E}_2$, (*b*) $\mathbf{E}_1$, and (*c*) $\mathbf{E}$.

10-50 If $\mathbf{I}_t$ of Fig. 10-41 is 0.7 $\angle 130°$ A, calculate $\mathbf{E}_4$.

10-51 Calculate the equivalent admittance of $\mathbf{Z}_6$ and $\mathbf{Z}_7$ of Fig. 10-42 if $\mathbf{Z}_1 = 350 \angle 30°$ Ω, $\mathbf{Z}_2 = 600 \angle -70°$ Ω, $\mathbf{Z}_3 = 450 \angle 60°$ Ω, $\mathbf{Z}_4 = 50 \angle 50°$ Ω, $\mathbf{Z}_5 = 30 \angle -38°$ Ω, $\mathbf{Z}_6 = 150 \angle 40°$ Ω, and $\mathbf{Z}_7 = 560 \angle -80°$ Ω.

10-52 Calculate the total impedance of the circuit of Fig. 10-42.

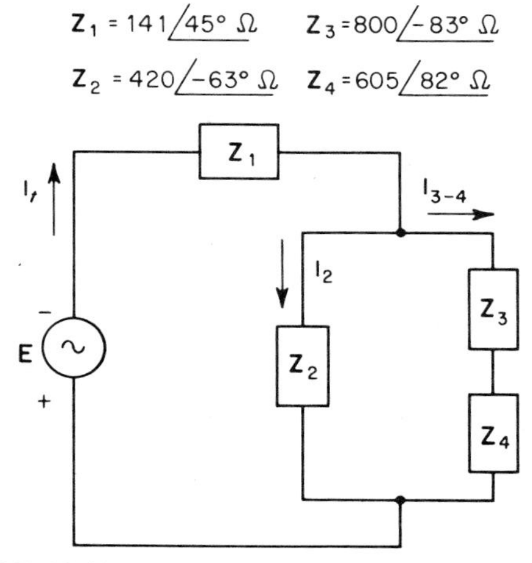

FIGURE 10-41
Circuit for Probs. 10-40 to 10-50.

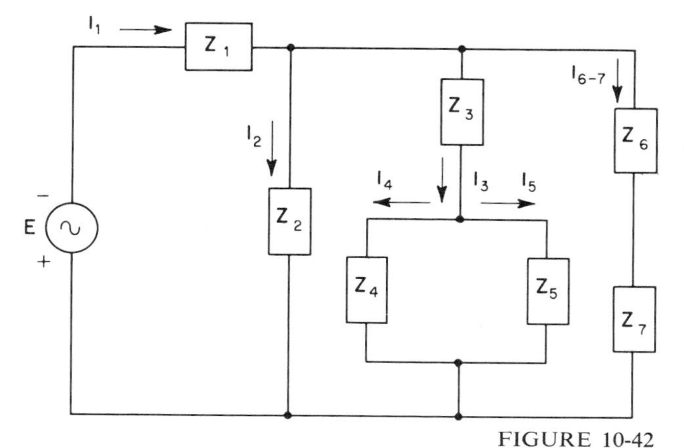

FIGURE 10-42
Circuit for Probs. 10-51 to 10-60.

10-53 If $\mathbf{I}_5$ of Fig. 10-42 is 0.09 $\angle 30°$ A, calculate $\mathbf{E}_7$.
10-54 If $\mathbf{I}_1$ of Fig. 10-42 is 0.3 $\angle 60°$ A, calculate $\mathbf{I}_4$.
10-55 If $\mathbf{E}$ of Fig. 10-42 is 160 $\angle 120°$ V, calculate $\mathbf{E}_6$.
10-56 If $\mathbf{E}_7$ of Fig. 10-42 is 60 $\angle -48°$ V, calculate $\mathbf{E}_3$.
10-57 If $\mathbf{I}_3$ of Fig. 10-42 is 0.036 $\angle -20°$ A, calculate $\mathbf{I}_5$.
10-58 If $\mathbf{I}_2$ of Fig. 10-42 is 0.05 $\angle 36°$ A, calculate $\mathbf{E}$.
10-59 If $\mathbf{E}_1$ of Fig. 10-42 is 200 $\angle -100°$ V, calculate $\mathbf{E}_7$.
10-60 If $\mathbf{I}_{6-7}$ of Fig. 10-42 is 0.093 $\angle -150°$ A, calculate $\mathbf{I}_2$.
10-61 Calculate the true power, reactive power, and apparent power for the information of Prob. 10-6 if $E = 12 \angle 20°$ V.
10-62 Repeat Prob. 10-61 using the information of Prob. 10-7 if $E = 40 \angle -30°$ V.
10-63 Repeat Prob. 10-61 using the information of Prob. 10-8.
10-64 Repeat Prob. 10-61 using the information of Prob. 10-10.
10-65 Repeat Prob. 10-61 using the information of Prob. 10-11.
10-66 Repeat Prob. 10-61 using the information of Prob. 10-12.
10-67 Repeat Prob. 10-61 using the information of Prob. 10-21.
10-68 Repeat Prob. 10-61 using the information of Prob. 10-22.
10-69 Calculate true power in each impedance and total true power using the information given in Prob. 10-26.
10-70 Repeat Prob. 10-69 using the information given in Prob. 10-27.
10-71 Repeat Prob. 10-69 using the information given in Prob. 10-28.
10-72 Repeat Prob. 10-69 using the information given in Prob. 10-37.
10-73 Repeat Prob. 10-69 using the information given in Prob. 10-41.
10-74 Repeat Prob. 10-69 using the information of Prob. 10-48.
10-75 Repeat Prob. 10-69 using the information of Prob. 10-49.
10-76 Repeat Prob. 10-69 using the information of Prob. 10-50.

11

NETWORK ANALYSIS OF CIRCUITS WITH AC EXCITATIONS

11-1 INTRODUCTION

In Chap. 4, a group of rules, laws, and theorems was developed and used in solving circuit problems containing dc excitations. These same techniques, restated, can be applied to circuits with ac excitations. In this chapter, the rules, laws, and theorems of Chap. 4 are restated for ac circuits and demonstrated by examples. Also included is a section on ac bridge circuits, The network-analysis treatment is limited to circuits where all sources operate at the same frequency.

11-2 SOURCE CONVERSIONS

Conversion between a voltage source and current source is accomplished in a manner similar to that for dc circuits. Figure 11-1 illustrates the conversion process.

EXAMPLE 11-1 Convert the voltage source in Fig. 11-2 into an equivalent current source.

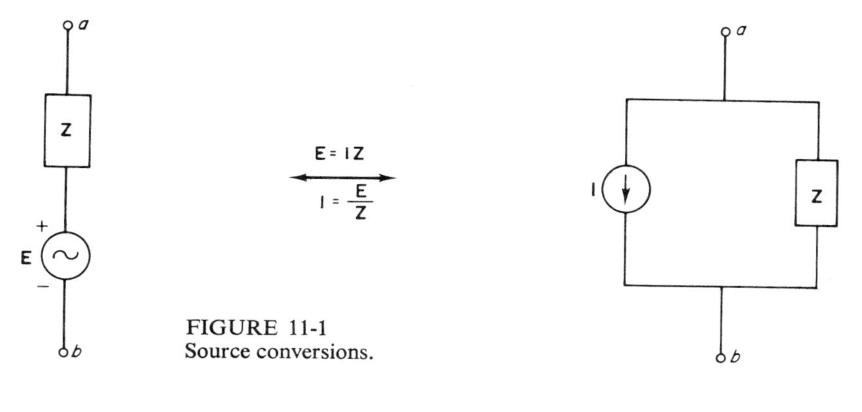

FIGURE 11-1
Source conversions.

SOLUTION

$$\mathbf{I} = \frac{\mathbf{E}}{\mathbf{Z}}$$

$$\mathbf{Z} = 1{,}000 - j500 = 1{,}116 \angle -26.6^\circ$$

$$\mathbf{I} = \frac{32 \angle 70^\circ}{11.2 \times 10^2 \angle -26.6^\circ} = 28.6 \angle 96.6^\circ \text{ mA.} \qquad ////$$

The resulting equivalent current source is shown in Fig. 11-3.

11-3 VOLTAGE DIVISION AND CURRENT DIVISION RULES

An exchange of impedance for resistance in the VDR and admittance for conductance in the CDR revises the two rules for circuits with ac excitations.

Voltage Division Rule The voltage drop across an impedance in a series circuit is equal to the applied voltage multiplied by the ratio of the impedance where the voltage drop occurs to the sum of the impedances in the series circuit.

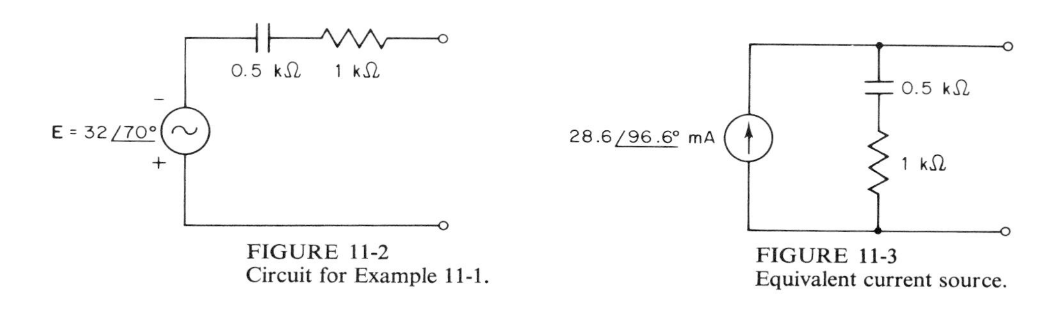

FIGURE 11-2
Circuit for Example 11-1.

FIGURE 11-3
Equivalent current source.

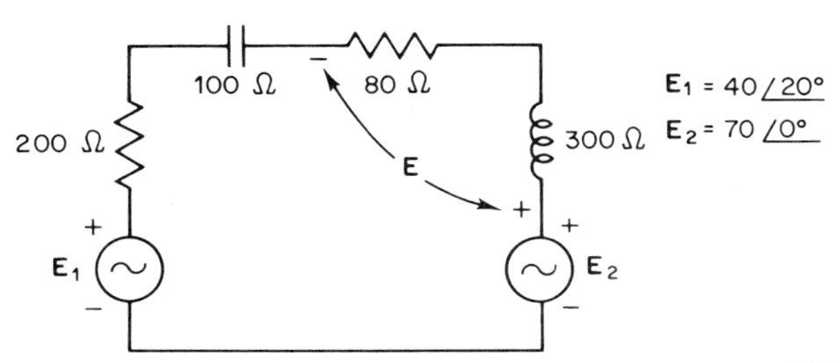

FIGURE 11-4
Circuit for Example 11-2.

Current Division Rule The current in a branch of a parallel circuit is equal to the applied current multiplied by the ratio of the admittance where the current flow occurs to the sum of the admittances in the circuit.

EXAMPLE 11-2 Determine the voltage drop indicated in the circuit of Fig. 11-4.

SOLUTION Using the VDR,

$$\mathbf{E} = (\mathbf{E}_2 - \mathbf{E}_1)\,\frac{80 + jX_L}{280 + jX_L - jX_C}$$

$$\mathbf{E}_1 = 40\ \angle 20^\circ = 37.6 + j13.7\ \text{V}$$

$$\mathbf{E} = (70 - 37.6 - j13.7)\,\frac{80 + j300}{280 + j200}$$

$$= 35\ \angle -23^\circ\ \frac{321\ \angle 75^\circ}{345\ \angle 35.5^\circ}$$

$$= 32.6\ \angle 16.5^\circ\ \text{V}$$

////

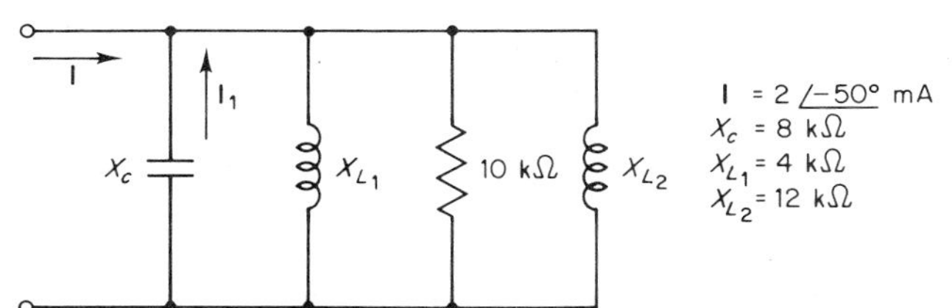

FIGURE 11-5
Circuit for Example 11-3.

EXAMPLE 11-3 Calculate the current through the capacitor in the circuit of Fig. 11-5.

SOLUTION Using the CDR,

$$\mathbf{I}_1 = -\mathbf{I}\frac{j\dfrac{1}{X_C}}{\dfrac{1}{10\text{ k}\Omega} + j\dfrac{1}{X_C} - j\dfrac{1}{X_{L_1}} - j\dfrac{1}{X_{L_2}}}$$

$$\mathbf{I}_1 = 2\angle 130^\circ \frac{1.25\times 10^{-4}\angle 90^\circ}{2.32\times 10^{-4}\angle -64.5^\circ}\text{ mA}$$

$$= 1.08\angle -75.5^\circ\text{ mA}$$

////

11-4 KIRCHHOFF'S LAWS

Kirchhoff's laws for circuits with ac excitations are as follows.

Kirchhoff's Voltage Law (KVL) The phasor sum of the voltages around any closed loop is zero.

Kirchhoff's Current Law (KCL) The phasor sum of the currents at a node is zero.

The current and voltage equations are derived in the same fashion as those for dc circuits. The algebraic manipulation of phasor quantities is no different from that of dc quantities until numerical quantities are introduced.

When more than one ac voltage source is part of the circuit to be analyzed, the relative polarities of the sources must be given. The relative polarity, or sense, of a source is given with respect to its phasor values.

11-5 MESH-CURRENT ANALYSIS

The current in each mesh is determined by writing the KVL equations for the meshes in the circuit and solving for the unknowns with determinants. Circuits with ac sources require phasor manipulation, whereas circuits with dc sources require algebraic manipulation. The list of guidelines in Chap. 4 is repeated in shortened form for reference.

1. The circuit must only contain voltage sources. Use source conversion if necessary.
2. Assume clockwise direction for mesh currents.
3. Voltage sources are positive if they produce the assumed mesh-current direction.
4. Use determinants to solve for unknown mesh currents.

EXAMPLE 11-4 Determine the mesh currents in the circuit of Fig. 11-6.

SOLUTION

First mesh equation: $$-\mathbf{E}_1 = (\mathbf{Z}_1 + \mathbf{Z}_2)\mathbf{I}_1 - \mathbf{Z}_2\mathbf{I}_2 \tag{11-1}$$

Second mesh equation: $$\mathbf{E}_2 = -\mathbf{Z}_2\mathbf{I}_1 + (\mathbf{Z}_2 + \mathbf{Z}_3)\mathbf{I}_2 \tag{11-2}$$

$$\mathbf{Z}_1 + \mathbf{Z}_2 = 60 + j104 + 110 + j102 = 170 + j206 = 245 \angle 57^\circ$$

$$\mathbf{Z}_2 + \mathbf{Z}_3 = 110 + j102 + 123 + j158 = 233 + j260 = 350 \angle 48^\circ$$

Substituting values into Eq. (11-1) gives

$$-95 \angle 0^\circ = 245 \angle 57^\circ \mathbf{I}_1 - 150 \angle 43^\circ \mathbf{I}_2 \tag{11-3}$$

Substituting values into Eq. (11-2),

$$105 \angle 40^\circ = -150 \angle 43^\circ \mathbf{I}_1 + 350 \angle 48^\circ \mathbf{I}_2 \tag{11-4}$$

Solving Eqs. (11-3) and (11-4) using determinants, we have

$$\mathbf{I}_1 = \frac{\begin{vmatrix} -95 \angle 0^\circ & -150 \angle 43^\circ \\ 105 \angle 40^\circ & 350 \angle 48^\circ \end{vmatrix}}{\begin{vmatrix} 245 \angle 57^\circ & -150 \angle 43^\circ \\ -150 \angle 43^\circ & 350 \angle 48^\circ \end{vmatrix}} = \frac{-3.32 \times 10^4 \angle 48^\circ + 1.57 \times 10^4 \angle 83^\circ}{8.85 \times 10^4 \angle 105^\circ - 2.25 \times 10^4 \angle 86^\circ}$$

$$= \frac{2.24 \angle -156^\circ}{6.5 \angle 111^\circ} = 0.345 \angle 93^\circ \text{ A}$$

$$\mathbf{I}_2 = \frac{\begin{vmatrix} 245 \angle 57^\circ & -95 \angle 0^\circ \\ -150 \angle 43^\circ & 105 \angle 40^\circ \end{vmatrix}}{6.5 \times 10^4 \angle 111^\circ} = \frac{2.58 \times 10^4 \angle 97^\circ - 1.43 \times 10^4 \angle 43^\circ}{6.5 \times 10^4 \angle 111^\circ}$$

$$= \frac{2.06 \angle 130^\circ}{6.5 \angle 111^\circ} = 0.317 \angle 19^\circ \text{ A}$$

////

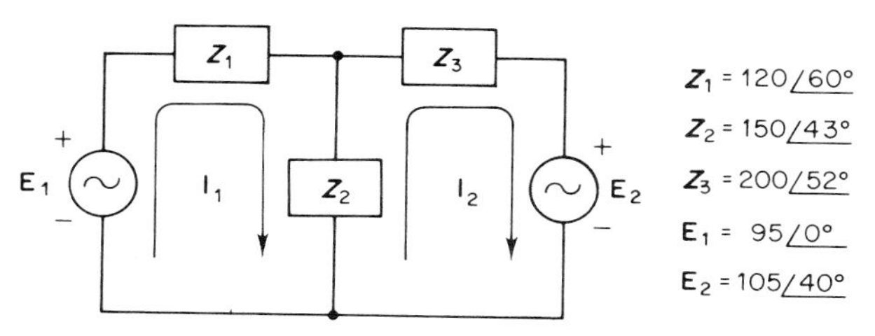

FIGURE 11-6
Circuit for Example 11-4.

11-6 NODE-VOLTAGE ANALYSIS

Principal node voltages with respect to the reference node are determined in much the same manner that dc node voltages were found in Chap. 4. The significant difference is the use of phasors and complex numbers in ac circuits. As in Chap. 4, the node equations are written to satisfy the KCL. Guidelines for node analysis are repeated from Chap. 4.

1 The circuit must contain only current sources. Use source conversion if necessary.
2 Assume node voltage positive with respect to the reference node.
3 A current source forcing current away from a node is considered positive.

EXAMPLE 11-5 Determine the node voltages for the circuit shown in Fig. 11-7.

SOLUTION The voltage source must be converted to an equivalent current source.

$$\mathbf{I} = \frac{\mathbf{E}}{\mathbf{Z}} = \frac{12 \angle 60^\circ}{20\ \text{k}\Omega \angle -90^\circ} = 0.6 \angle 150^\circ \text{ mA}$$

The resulting circuit is shown in Fig. 11-8.

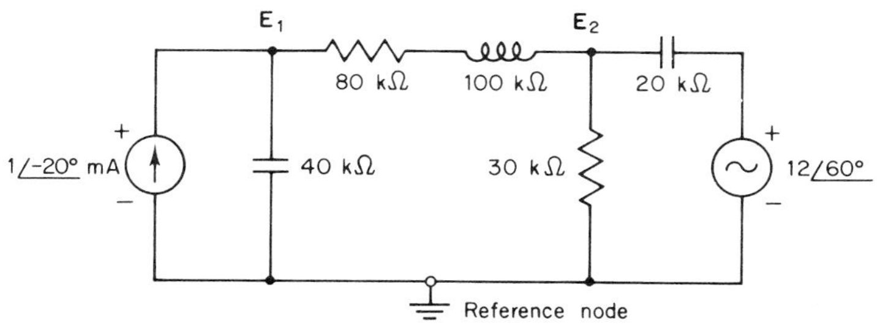

FIGURE 11-7
Circuit for Example 11-5.

First node equation:

$$-1 \times 10^{-3} \angle -20^\circ = \left(j \frac{1}{40\ \text{k}\Omega} + \frac{1}{80\ \text{k}\Omega + j100\ \text{k}\Omega}\right)\mathbf{E}_1 - \left(\frac{1}{80\ \text{k}\Omega + j100\ \text{k}\Omega}\right)\mathbf{E}_2 \tag{11-5}$$

Second node equation:

$$0.6 \times 10^{-3} \angle 150^\circ = -\left(\frac{1}{80\ \text{k}\Omega + j100\ \text{k}\Omega}\right)\mathbf{E}_1 + \left(j\frac{1}{20\ \text{k}\Omega} - j\frac{1}{30\ \text{k}\Omega} + \frac{1}{80\ \text{k}\Omega + j100\ \text{k}\Omega}\right)\mathbf{E}_2 \tag{11-6}$$

Multiplying Eq. (11-5) by 1×10^3 and evaluating coefficients gives

$$-1 \angle -20^\circ = 1.95 \times 10^{-2} \angle 76^\circ \mathbf{E}_1 - 0.785 \times 10^{-2} \angle -51^\circ \mathbf{E}_2 \tag{11-7}$$

Multiplying Eq. (11-6) by 1×10^3 and evaluating coefficients gives

$$0.6 \angle 150^\circ = -0.785 \times 10^{-2} \angle -51^\circ \mathbf{E}_1 + 1.16 \times 10^{-2} \angle 65^\circ \mathbf{E}_2 \tag{11-8}$$

Multiplying Eqs. (11-7) and (11-8) by 1×10^2 to simplify for determinant evaluation,

$$-100 \angle -20^\circ = 1.95 \angle 76^\circ \mathbf{E}_1 - 0.785 \angle -51^\circ \mathbf{E}_2 \tag{11-9}$$

$$60 \angle 150^\circ = -0.785 \angle -51^\circ \mathbf{E}_1 + 1.16 \angle 65^\circ \mathbf{E}_2 \tag{11-10}$$

Solve Eqs. (11-9) and (11-10) using determinants.

$$\mathbf{E}_1 = \frac{\begin{vmatrix} -100 \angle -20^\circ & -0.785 \angle -51^\circ \\ 60 \angle 150^\circ & 1.16 \angle 65^\circ \end{vmatrix}}{\begin{vmatrix} 1.95 \angle 76^\circ & -0.785 \angle -51^\circ \\ -0.785 \angle -51^\circ & 1.16 \angle 65^\circ \end{vmatrix}} = \frac{-116 \angle 45^\circ + 45 \angle 99^\circ}{2.26 \angle 141^\circ - 0.616 \angle -102^\circ}$$

$$= \frac{95 \angle -158^\circ}{2.6 \angle 129^\circ} = 36.5 \angle 73^\circ\ \text{V}$$

$$\mathbf{E}_2 = \frac{\begin{vmatrix} 1.95 \angle 76^\circ & -100 \angle -20^\circ \\ -0.785 \angle -51^\circ & 60 \angle 150^\circ \end{vmatrix}}{2.6 \angle 129^\circ} = \frac{117 \angle -134^\circ - 78.5 \angle -71^\circ}{2.6 \angle 129^\circ}$$

$$= \frac{108 \angle -175^\circ}{2.6 \angle 129^\circ} = 41.5 \angle 56^\circ\ \text{V}$$

////

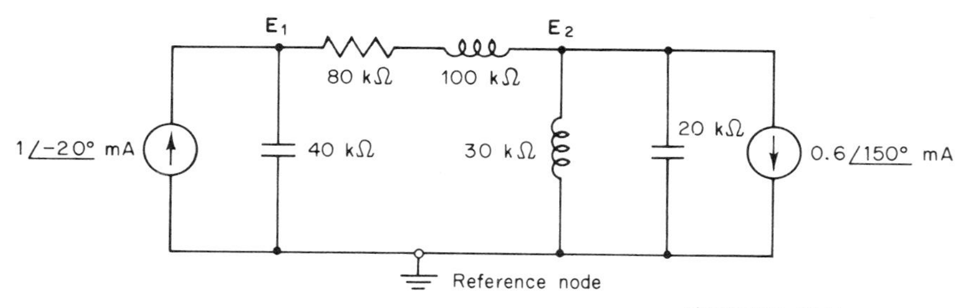

FIGURE 11-8
Circuit after source conversion.

11-7 SUPERPOSITION

Each of a network's ac sources can be taken into account individually by using the principle of superposition, which can be restated for circuits with ac sources.

> **Principle of Superposition** Response at any location in a circuit to the excitations is the phasor sum of the responses to the individual excitations when each is applied one at a time with the others set equal to zero.

The response may represent a voltage between two points or a current at some point.

EXAMPLE 11-6 Calculate the voltage drop $e(t)$ across the inductor in the circuit of Fig. 11-9. The two excitations are $e_1(t) = 17 \sin(4.4 \times 10^6 t - 30°)$ V and $i_2(t) = 212 \sin(4.4 \times 10^6 t + 25°)\,\mu$ A

SOLUTION The sinusoidal excitations must be converted to phasor form.

$$\mathbf{E}_1 = \frac{17}{\sqrt{2}} \angle -30° = 12 \angle -30°$$

$$\mathbf{I}_2 = \frac{212 \times 10^{-6}}{\sqrt{2}} \angle 25° = 1.5 \times 10^{-4} \angle 25°$$

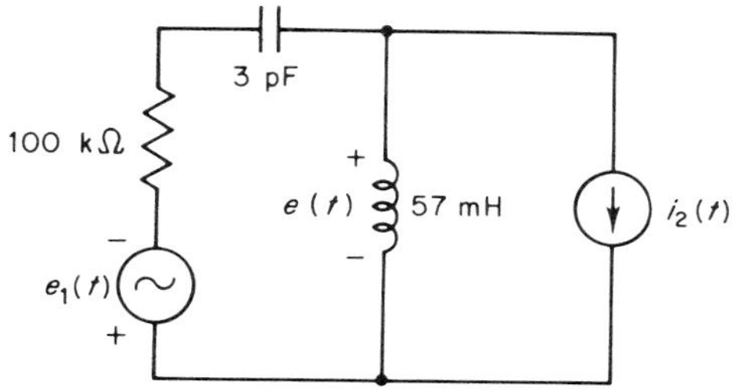

FIGURE 11-9
Circuit for Example 11-6.

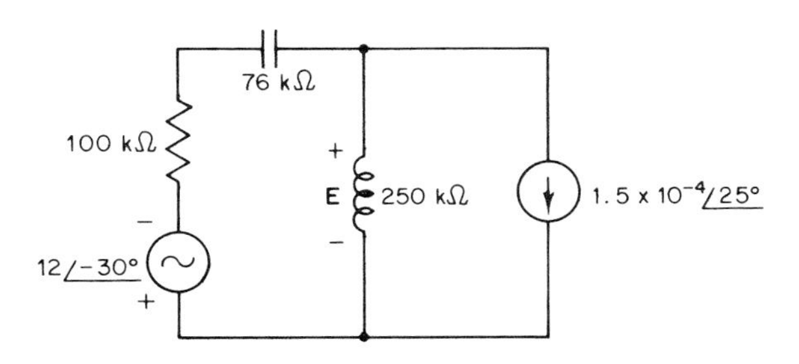

FIGURE 11-10

The frequency of the excitations is given by

$$\omega = 2\pi f = 4.4 \times 10^6 \text{ rad/s}$$

$$f = 700 \text{ kHz}$$

The reactances of the capacitor and inductor are

$$X_C = \frac{1}{\omega C} = \frac{1}{(4.4 \times 10^6)(3 \times 10^{-12})} = 76 \text{ k}\Omega$$

$$X_L = \omega L = (4.4 \times 10^6)(0.57 \times 10^{-1}) = 250 \text{ k}\Omega$$

Figure 11-10 shows the circuit with phasor notation and reactances. The response due to the voltage source is found by setting the current source equal to zero (see Fig. 11-11). The VDR can easily be applied.

$$\mathbf{E}_{E_1} = -12 \angle -30^\circ \frac{2.5 \times 10^5 \angle 90^\circ}{1 \times 10^5 - j0.76 \times 10^5 + j2.5 \times 10^5}$$

$$= -12 \angle -30^\circ \frac{2.5 \times 10^5 \angle 90^\circ}{2.01 \times 10^5 \angle 60^\circ} = -14.9 \angle 0^\circ \text{ V}$$

The response due to the current source is found by setting the voltage source equal to zero (see Fig. 11-12). Product of the parallel equivalent impedance and the current source will yield the desired voltage drop.

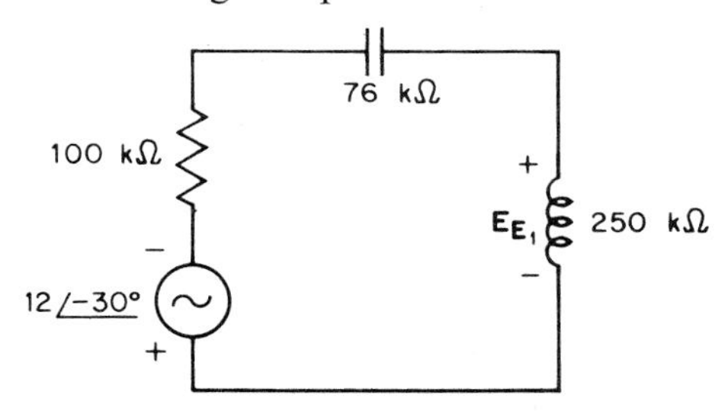

FIGURE 11-11
Current source set to zero.

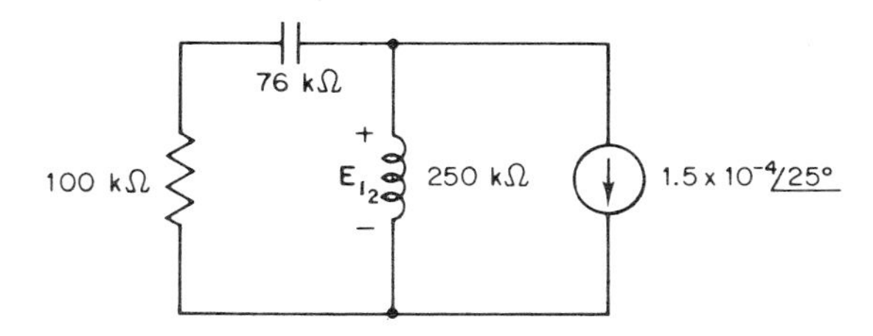

FIGURE 11-12
Voltage source set to zero.

$$100\ \text{k}\Omega - j76\ \text{k}\Omega = 126\ \text{k}\Omega \angle -37^\circ$$

$$\mathbf{Z}_{\text{parallel}} = \frac{(126\ \text{k}\Omega \angle -37^\circ)\ (250\ \text{k}\Omega \angle 90^\circ)}{201\ \text{k}\Omega \angle 60^\circ} = 157\ \text{k}\Omega \angle -7^\circ$$

$$\mathbf{E}_{I_2} = (157\ \text{k}\Omega \angle -7^\circ)(1.5 \times 10^{-4} \angle 25^\circ) = 23.6 \angle 18^\circ\ \text{V}$$

The total phasor response is the sum of the individual responses.

$$\mathbf{E} = \mathbf{E}_{E_1} + \mathbf{E}_{I_2} = -14.9 \angle 0^\circ + 23.6 \angle 18^\circ$$

$$= 10.5 \angle 44^\circ\ \text{V}$$

$$e(t) = 14.8 \sin (4.4 \times 10^6 t + 44^\circ)\ \text{V}$$

////

11-8 THÉVENIN'S THEOREM

When the circuit contains reactive elements and ac sources, the theorem is modified by a simple exchange of impedance for resistance. A modified version of the theorem in Chap. 4 is given for reference.

Thévenin's Theorem Any two-terminal network may be replaced by a simple series circuit consisting of an equivalent emf $\mathbf{E}_{\text{th}}$ and an equivalent internal impedance $\mathbf{Z}_{\text{th}}$. The equivalent series circuit will provide the same current through a load as the original circuit would if:

1 The equivalent emf $\mathbf{E}_{\text{th}}$ is the voltage seen between the terminals of the original network with the load impedance removed.

2 The equivalent impedance $\mathbf{Z}_{\text{th}}$ is the impedance seen between the terminals of the original network when the internal sources are set equal to zero.

Figure 11-13 shows the conversion of a network to its Thévenin equivalent circuit.

EXAMPLE 11-7 Find the Thévenin equivalent circuit at points X-Y in the circuit of Fig. 11-14.

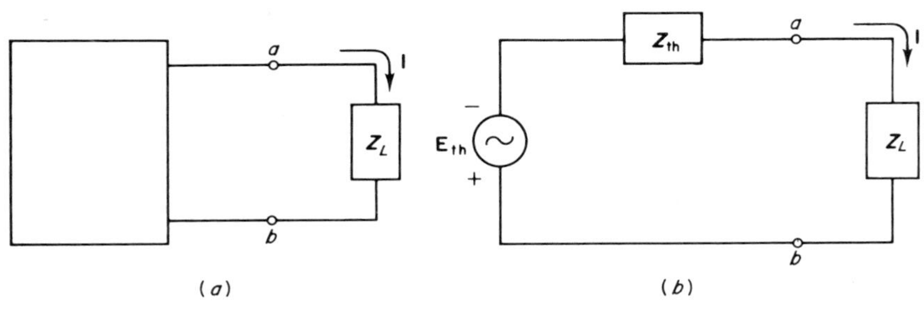

FIGURE 11-13
(*a*) Original network and (*b*) Thévenin equivalent.

SOLUTION

1 $\mathbf{E}_{th}$ is found by removing the 1-Ω inductive load (see Fig. 11-15).
The voltage drop across the 4-Ω inductance is equal to $\mathbf{E}_{th}$ since no drop occurs across the 6-Ω resistance. The VDR can be used to find the voltage across the 4-Ω inductance.

$$\mathbf{E}_{th} = 10 \angle 60^\circ \frac{4 \angle 90^\circ}{6 \angle -90^\circ} = 6.66 \angle -120^\circ \text{ V}$$

2 The voltage source is set equal to zero to find $\mathbf{Z}_{th}$ (see Fig. 11-16).

$$\begin{aligned}\mathbf{Z}_{th} &= \frac{(4 \angle 90^\circ)(10 \angle -90^\circ)}{6 \angle -90^\circ} + \frac{(6)(2 \angle -90^\circ)}{6.3 \angle -18.5^\circ} \\ &= 6.66 \angle 90^\circ + 1.9 \angle -71.5^\circ \\ &= 0.6 + j4.86\end{aligned}$$

The resulting Thévenin equivalent circuit is shown in Fig.11-17. ////

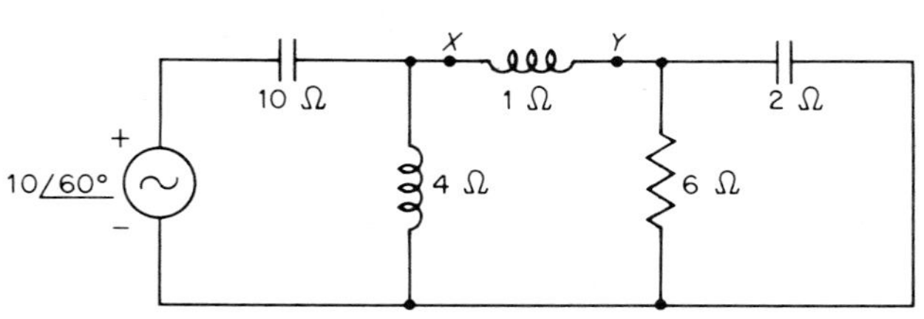

FIGURE 11-14
Circuit for Example 11-7.

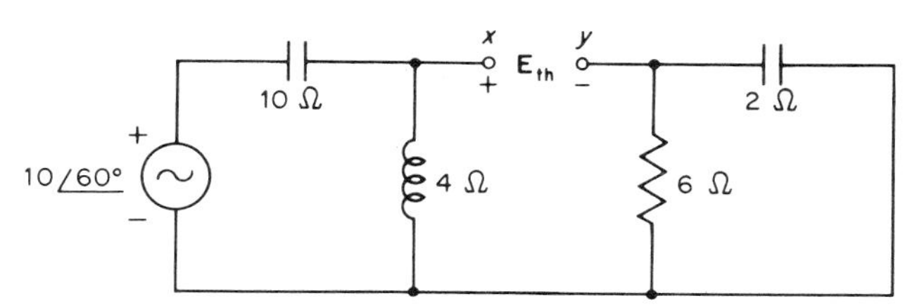

FIGURE 11-15

EXAMPLE 11-8 Determine the Thévenin equivalent circuit for the 2-kΩ resistor shown in Fig. 11-18.

SOLUTION

1 The 2-kΩ resistor is removed to find $\mathbf{E}_{th}$ (see Fig. 11-19). A simple series circuit results after the current source is converted to a voltage source as shown in Fig. 11-20.

$$\mathbf{E}_{th} = 20 \angle 0° - (5 \times 10^3 \angle 90°)\mathbf{I}$$

$$\mathbf{I} = \frac{20 \angle 0° + 10 \angle 50°}{1 \text{ k}\Omega - j7 \text{ k}\Omega + j5 \text{ k}\Omega} = 12.4 \angle 79.5° \text{ mA}$$

$$\mathbf{E}_{th} = 20 - 62 \angle 169.5°$$

$$= 82 \angle -7.7° \text{ V}$$

2 The two sources in Fig. 11-19 are set equal to zero to find $\mathbf{Z}_{th}$ in Fig. 11-21. $\mathbf{Z}_{th}$ is the parallel combination of the 5-kΩ inductor and the series connection of the 7-kΩ capacitor and 1-kΩ resistor.

$$\mathbf{Z}_{th} = \frac{(5 \text{ k}\Omega \angle 90°)(1 \text{ k}\Omega - j7 \text{ k}\Omega)}{1 \text{ k}\Omega + j5 \text{ k}\Omega - j7 \text{ k}\Omega} = 15.8 \text{ k}\Omega \angle 71.5° \ \Omega$$

$$\mathbf{Z}_{th} = 5 \text{ k}\Omega + j15 \text{ k}\Omega$$

The resulting Thévenin equivalent circuit is shown in Fig. 11-22. ////

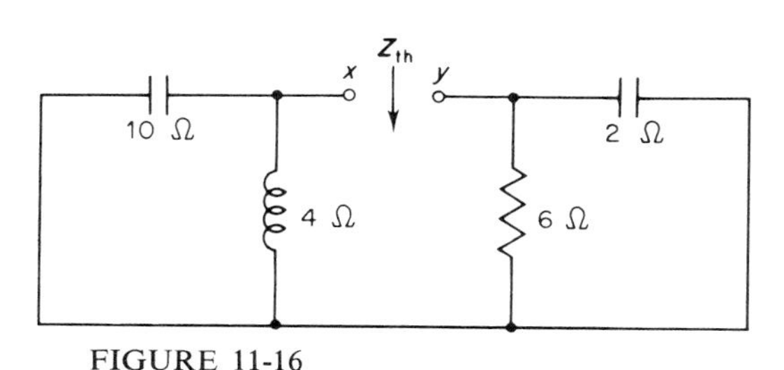

FIGURE 11-16

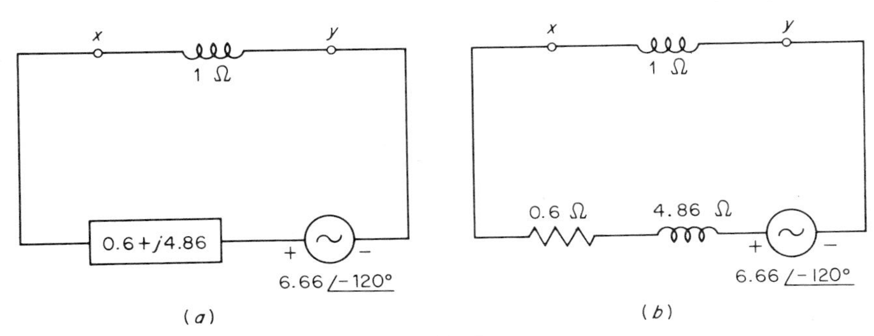

FIGURE 11-17
$\mathbf{Z}_{th}$ as (*a*) block impedance and (*b*) component representation.

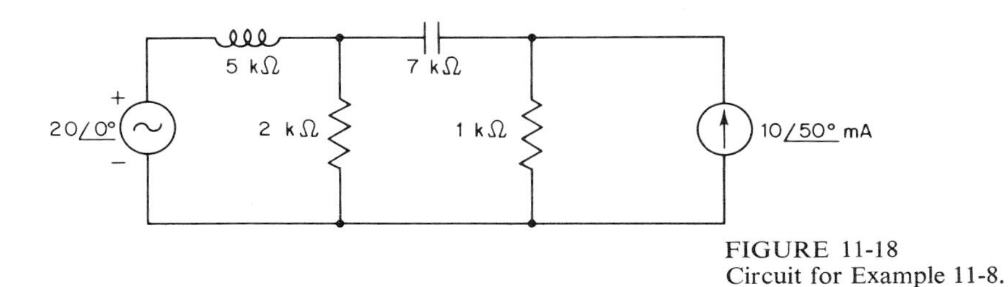

FIGURE 11-18
Circuit for Example 11-8.

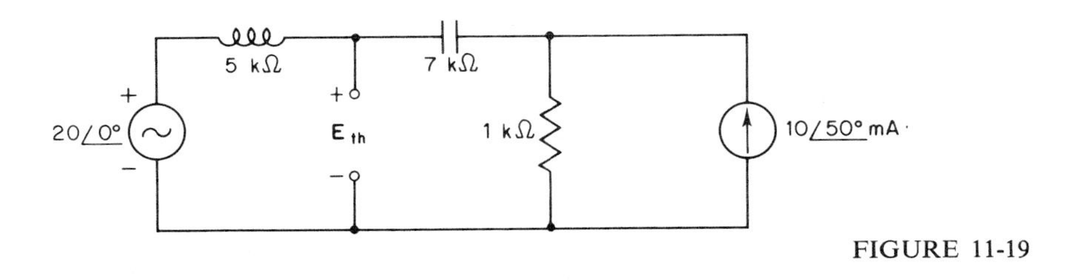

FIGURE 11-19

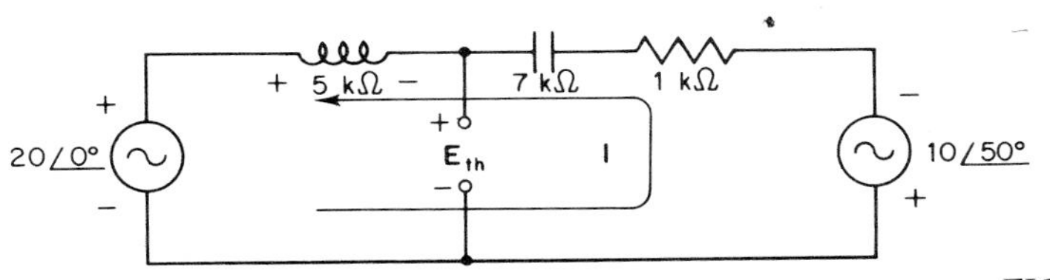

FIGURE 11-20

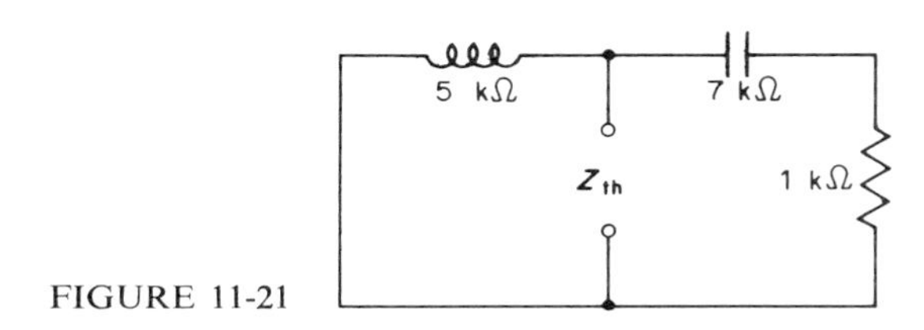

FIGURE 11-21

11-9 NORTON'S THEOREM

As in the case of Thévenin's theorem, Norton's theorem for circuits with reactive elements and ac sources is modified by exchanging impedance for resistance. The theorem stated in Chap. 4 is modified and listed for reference.

Norton's Theorem Any two-terminal network may be replaced by a simple parallel circuit consisting of an equivalent current source $\mathbf{I}_n$ and an equivalent internal impedance $\mathbf{Z}_n$. The equivalent parallel circuit will provide the same current through a load as the original circuit would if:

1 The constant current $\mathbf{I}_n$ is the current that would flow through a short circuit between the two terminals being considered in the original network.

2 The equivalent impedance $\mathbf{Z}_n$ is the impedance seen between the two terminals being considered with the load removed and internal sources set equal to zero. Same as $\mathbf{Z}_{\text{th}}$.

Figure 11-23 shows the conversion of a network to its Norton equivalent circuit.

EXAMPLE 11-9 Determine the Norton equivalent circuit for the 1-kΩ resistor in the circuit of Fig. 11-24.

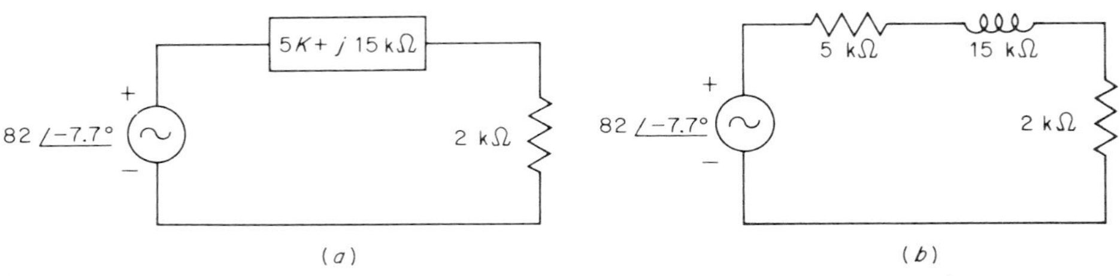

FIGURE 11-22
$\mathbf{Z}_{\text{th}}$ as (*a*) block impedance and (*b*) component representation.

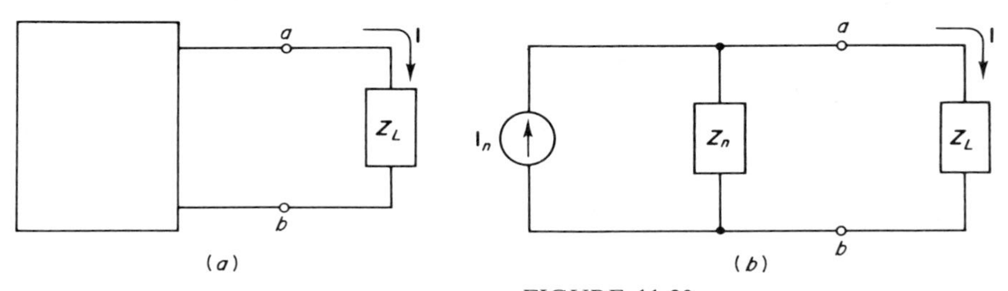

FIGURE 11-23
(*a*) Original network and (*b*) Norton equivalent.

SOLUTION

1 The 1-kΩ resistor is removed and replaced by a short to find $\mathbf{I}_n$ (see Fig. 11-25). The CDR may be used to calculate $\mathbf{I}_n$.

$$\mathbf{I}_n = 4 \times 10^{-3} \angle -70^\circ \frac{1 \times 10^{-4}}{1 \times 10^{-4} - j1 \times 10^{-3} + j0.33 \times 10^{-3}}$$

$$= 0.593 \angle 11.5^\circ \text{ mA}$$

2 To find $\mathbf{Z}_n$, the 1-kΩ resistor is removed and the current source is set equal to zero in Fig. 11-26.

$$\mathbf{Z}_n = 10 \text{ k}\Omega + \frac{(j1 \text{ k}\Omega)(-j3 \text{ k}\Omega)}{-j2 \text{ k}\Omega} = 10 \text{ k}\Omega + j1.5 \text{ k}\Omega$$

The resulting Norton equivalent circuit is shown in Fig. 11-27. ////

EXAMPLE 11-10 Find the Norton equivalent at points *x-y* in the circuit of Fig. 11-28.

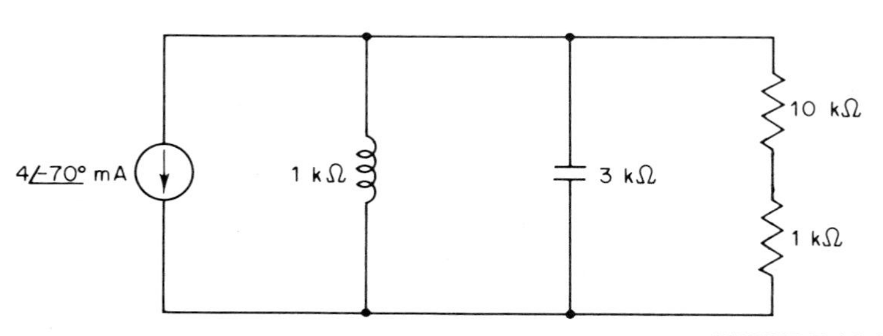

FIGURE 11-24
Circuit for Example 11-9.

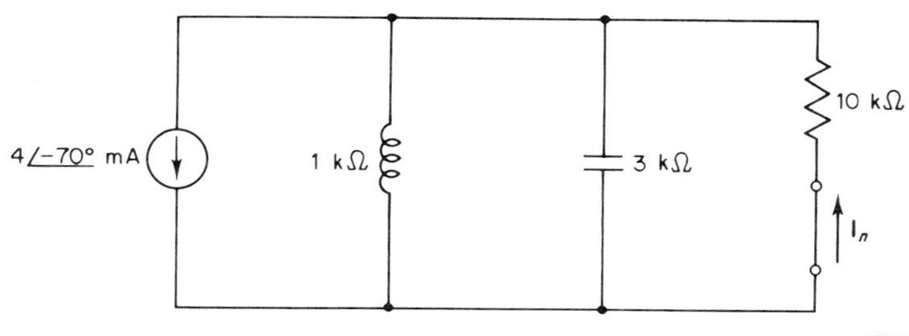

FIGURE 11-25

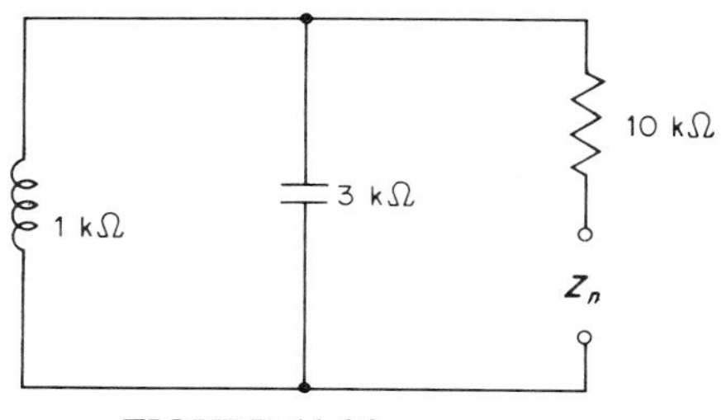

FIGURE 11-26

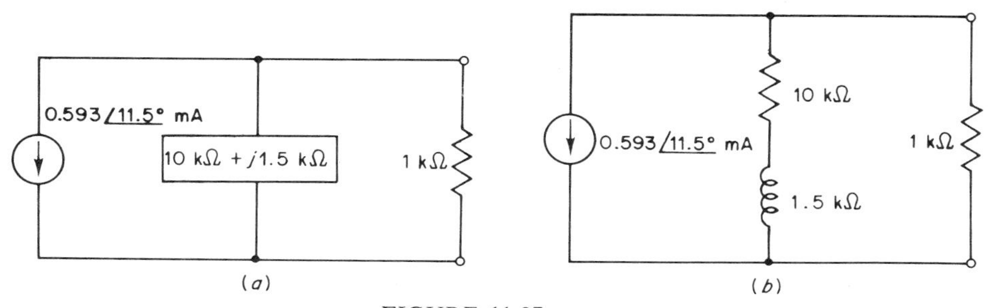

FIGURE 11-27
$\mathbf{Z}_n$ as (*a*) block impedance and (*b*) component representation.

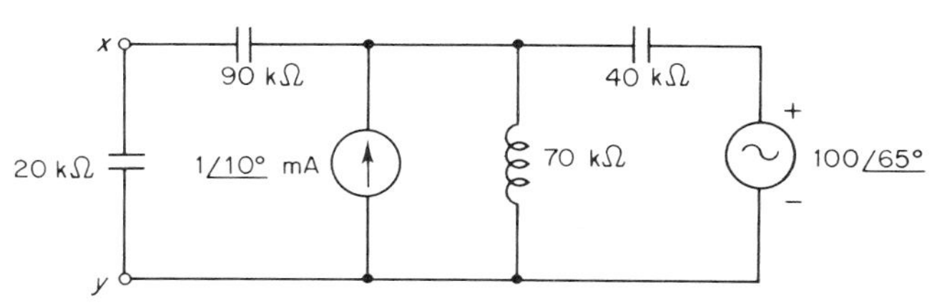

FIGURE 11-28
Circuit for Example 11-10.

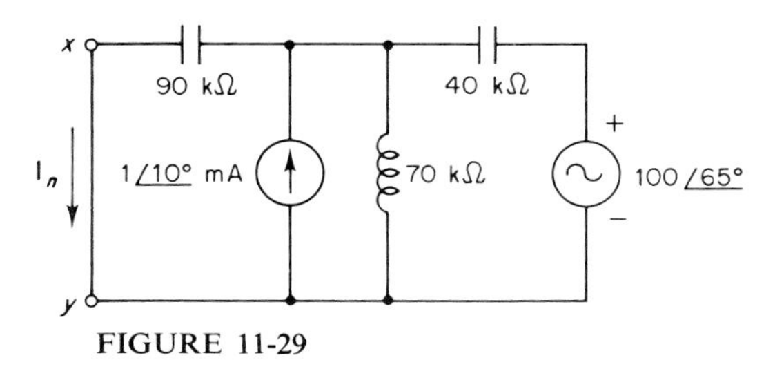

FIGURE 11-29

SOLUTION

1 The capacitor at *x-y* is removed and replaced by a short to find $\mathbf{I}_n$, as shown in Fig. 11-29. The voltage source can be converted to a current source, thus permitting $\mathbf{I}_n$ to be found using the CDR (see Fig. 11-30).

$$\mathbf{I}_n = (1 \angle 10^\circ \text{ mA} - 2.5 \angle 155^\circ \text{ mA}) \frac{1.11 \times 10^{-5} \angle 90^\circ}{2.18 \times 10^{-5} \angle 90^\circ}$$

$$= 1.7 \angle -15.4^\circ \text{ mA}$$

2 To find $\mathbf{Z}_n$, the capacitor at *x-y* is removed and sources are set equal to zero in Fig. 11-31.

$$\mathbf{Z}_n = -j90 \text{ k}\Omega + \frac{(j70 \text{ k}\Omega)(-j40 \text{ k}\Omega)}{j30 \text{ k}\Omega}$$

$$= -j183.5 \text{ k}\Omega$$

The resulting Norton equivalent circuit is shown in Fig. 11-32. ////

11-10 RECIPROCITY THEOREM

The Reciprocity theorem is valid for both resistive circuits with dc sources and resistive circuits or reactive circuits with ac sources. The theorem is restated from Chap. 4.

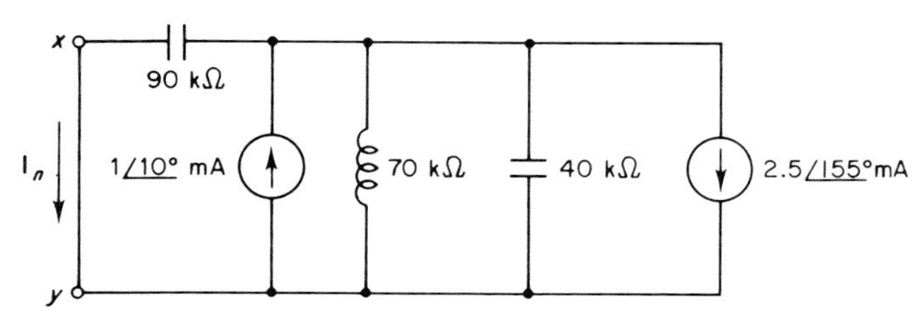

FIGURE 11-30

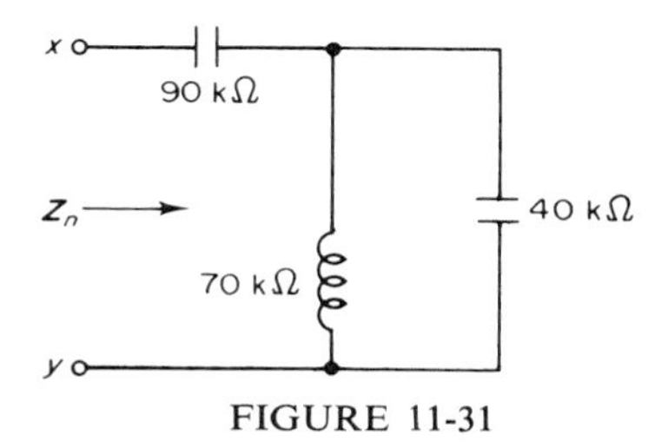

FIGURE 11-31

Reciprocity Theorem The ratio of the response to the excitation is the same whether the response is at point m and the excitation at point k or the response is at k and the excitation is at m. If the excitation is a voltage, then the response must be a current and vice versa.

EXAMPLE 11-11 Demonstrate that the reciprocity theorem is valid in the circuits of Fig. 11-33.

SOLUTION The objective is to prove that $\mathbf{I}_m/\mathbf{E}_k = \mathbf{I}_k/\mathbf{E}_m$.

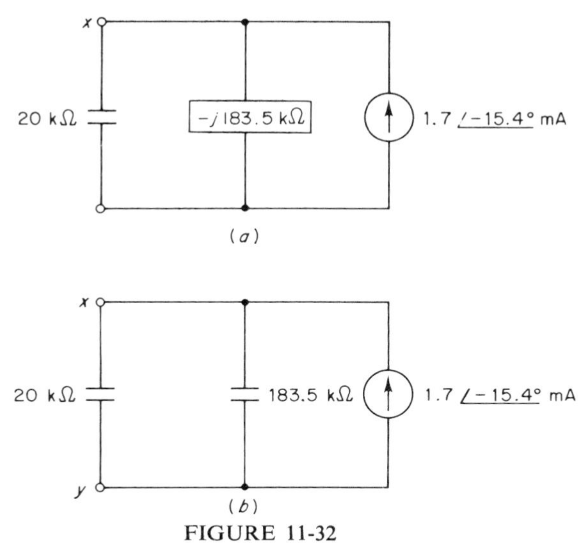

FIGURE 11-32
$\mathbf{Z}_n$ as (*a*) block impedance and (*b*) component representation.

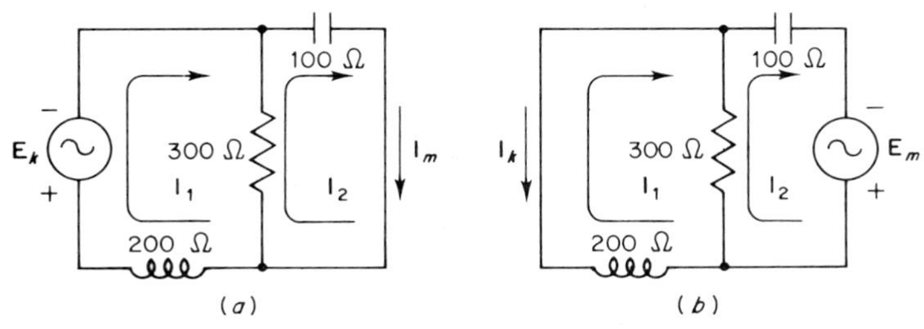

FIGURE 11-33
Circuits for Example 11-11.

Write mesh equations for Fig. 11-33*a*:

First mesh equation: $\mathbf{E}_k = (300 + j200)\ \mathbf{I}_1 - (300)\ \mathbf{I}_2$

Second mesh equation: $0 = -(300)\mathbf{I}_1 + (300 - j100)\mathbf{I}_2$

Rewriting the mesh equations in polar form gives

$$\mathbf{E}_k = 360\ \angle 33.7^\circ \mathbf{I}_1 - 300\mathbf{I}_2$$
$$0 = -300\mathbf{I}_1 + 316\ \angle -18.4^\circ \mathbf{I}_2$$

Since $\mathbf{I}_m = \mathbf{I}_2$, determinants are used to find $\mathbf{I}_2$.

$$\mathbf{I}_2 = \frac{\begin{vmatrix} 360\ \angle 33.7^\circ & \mathbf{E}_k \\ -300 & 0 \end{vmatrix}}{\begin{vmatrix} 360\ \angle 33.7^\circ & -300 \\ -300 & 316\ \angle -18.4^\circ \end{vmatrix}} = \frac{300\mathbf{E}_k}{11.4 \times 10^4\ \angle 15.3^\circ - 9 \times 10^4}$$

$$= 0.835 \times 10^{-2}\ \angle -56^\circ\quad \mathbf{E}_k$$

Transfer conductance: $\mathbf{I}_m/\mathbf{E}_k = 0.835 \times 10^{-2}\ \angle -56^\circ$ mho

Write mesh equations for Fig. 11-33*b*:

First mesh equation: $0 = (300 + j200)\mathbf{I}_1 - (300)\mathbf{I}_2$

Second mesh equation: $-\mathbf{E}_m = -(300)\mathbf{I}_1 + (300 - j100)\mathbf{I}_2$

Rewriting the mesh equations in polar form gives

$$0 = 360\ \angle 33.7^\circ \mathbf{I}_1 - 300\mathbf{I}_2$$
$$-\mathbf{E}_m = -300\mathbf{I}_1 + 316\ \angle -18.4^\circ \mathbf{I}_2$$

Since $\mathbf{I}_k = -\mathbf{I}_1$, determinants are used to find $\mathbf{I}_1$.

$$\mathbf{I}_1 = \frac{\begin{vmatrix} 0 & -300 \\ -\mathbf{E}_m & 316\ \angle -18.4^\circ \end{vmatrix}}{\begin{vmatrix} 360\ \angle 33.7^\circ & -300 \\ -300 & 316\ \angle -18.4 \end{vmatrix}} = \frac{-300\mathbf{E}_m}{359 \times 10^2\ \angle 56^\circ}$$

$$= -\ 0.835 \times 10^{-2}\ \angle -56^\circ \mathbf{E}_m$$

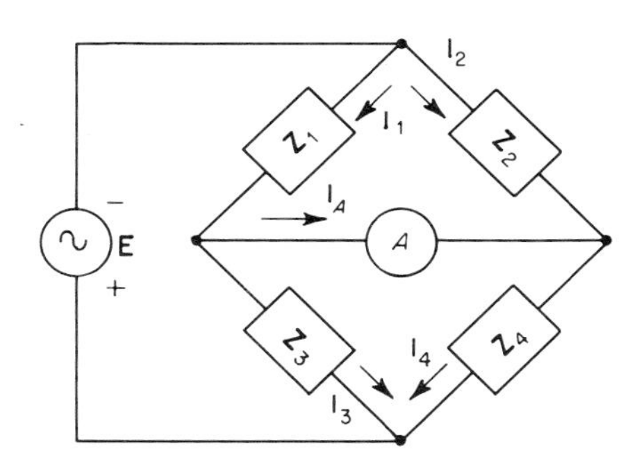

FIGURE 11-34
Wheatstone bridge with an ac source and impedances.

Transfer conductance: $\mathbf{I}_k/\mathbf{E}_m = 0.835 \times 10^{-2} \angle -56°$ mho
Thus,

$$\frac{\mathbf{I}_m}{\mathbf{E}_k} = \frac{\mathbf{I}_k}{\mathbf{E}_m} \qquad ////$$

11-11 AC BRIDGES

The Wheatstone bridge can be used to measure an unknown impedance by balancing the bridge. The bridge circuit is shown in Fig. 11-34.

The bridge is balanced when $\mathbf{I}_A = 0$. Therefore,

$$\mathbf{I}_1 = \mathbf{I}_3 \qquad \text{and} \qquad \mathbf{I}_4 = \mathbf{I}_2$$

Further

$$\mathbf{E}_1 = \mathbf{E}_2$$
$$\mathbf{Z}_1\mathbf{I}_1 = \mathbf{Z}_2\mathbf{I}_2 \tag{11-11}$$

and

$$\mathbf{E}_3 = \mathbf{E}_4$$
$$\mathbf{Z}_3\mathbf{I}_1 = \mathbf{Z}_4\mathbf{I}_2 \tag{11-12}$$

Dividing Eq. (11-11) by Eq. (11-12),

$$\frac{\mathbf{Z}_1\mathbf{I}_1}{\mathbf{Z}_3\mathbf{I}_1} = \frac{\mathbf{Z}_2\mathbf{I}_2}{\mathbf{Z}_4\mathbf{I}_2}$$

$$\frac{\mathbf{Z}_1}{\mathbf{Z}_3} = \frac{\mathbf{Z}_2}{\mathbf{Z}_4}$$

$$\mathbf{Z}_1\mathbf{Z}_4 = \mathbf{Z}_2\mathbf{Z}_3 \tag{11-13}$$

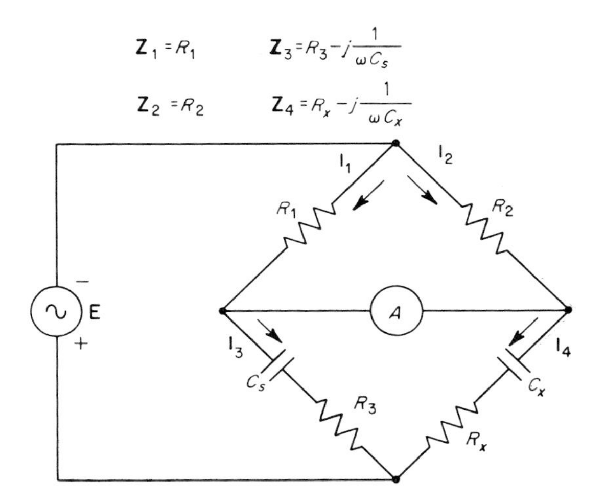

FIGURE 11-35
Capacitance bridge.

It should be noted that to satisfy Eq. (11-13), two relationships must be true:

$$|\mathbf{Z}_1\mathbf{Z}_4| = |\mathbf{Z}_2\mathbf{Z}_3|$$

and

$$\theta_1 + \theta_4 = \theta_2 + \theta_3$$

Equation (11-13) shows the impedance relationship that must exist when the bridge is balanced.

If the impedance bridge is arranged as shown in Fig. 11-35, a capacitance-comparison bridge is formed.

Substituting the impedance values of Fig. 11-35 into Eq. (11-13),

$$R_1\left(R_x - j\frac{1}{\omega C_x}\right) = R_2\left(R_3 - j\frac{1}{\omega C_s}\right)$$

Solving for C_x and R_x,

$$R_1 R_x - j\frac{R_1}{\omega C_x} = R_2 R_3 - j\frac{R_2}{\omega C_s}$$

Equality of two complex numbers requires that their real components as well as their imaginary components be equal. Therefore,

$$R_1 R_x = R_2 R_3$$

$$R_x = \frac{R_2 R_3}{R_1} \tag{11-14}$$

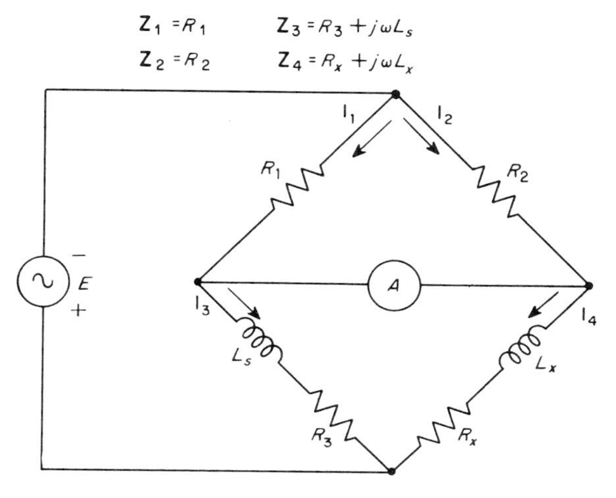

FIGURE 11-36
Inductance bridge.

and

$$-j\frac{R_1}{\omega C_x} = -j\frac{R_2}{\omega C_s}$$

$$\frac{R_1}{C_x} = \frac{R_2}{C_s}$$

$$C_x = \frac{C_s R_1}{R_2} \tag{11-15}$$

An inductance-comparison bridge can be set up to solve for an unknown inductance in the same fashion as for an unknown capacitance. The inductance-comparison bridge is shown in Fig. 11-36.

$$\mathbf{Z}_1 = R_1$$

$$\mathbf{Z}_2 = R_2$$

$$\mathbf{Z}_3 = R_3 + j\omega L_s$$

$$\mathbf{Z}_4 = R_x + j\omega L_x$$

Solving for L_x and R_x by Eq. (11-13),

$$R_1(R_x + j\omega L_x) = R_2(R_3 + j\omega L_s)$$

$$R_1 R_x + j\omega R_1 L_x = R_2 R_3 + j\omega R_2 L_s$$

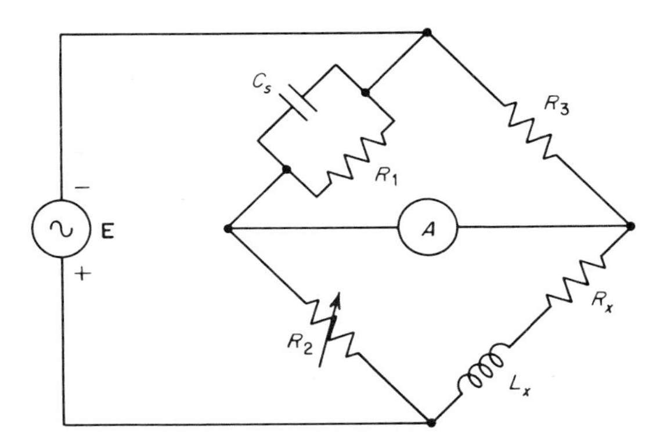

FIGURE 11-37
Maxwell bridge.

Therefore, for balance,

$$R_1 R_x = R_2 R_3$$

$$R_x = \frac{R_2 R_3}{R_1} \tag{11-16}$$

and

$$j\omega R_1 L_x = j\omega R_2 L_s$$

$$L_x = \frac{R_2 L_s}{R_1} \tag{11-17}$$

It is possible to determine an unknown inductance with capacitance standards. This is accomplished in a circuit known as a Maxwell bridge circuit. Using capacitance as a standard has several advantages. Capacitance is influenced to a lesser degree by external fields. Capacitors set up little external field. Further, they are small and inexpensive. The Maxwell bridge circuit is shown in Fig. 11-37.

To obtain the general balance condition of Eq. (11-13), the admittance of $\mathbf{Z}_1$ is written

$$\mathbf{Y}_1 = \frac{1}{R_1} + j\omega C_s$$

since

$$\mathbf{Z}_1 = \frac{1}{\mathbf{Y}_1}$$

Equation (11-13) may now be used.

$$\mathbf{Z}_1 \mathbf{Z}_4 = \mathbf{Z}_2 \mathbf{Z}_3$$

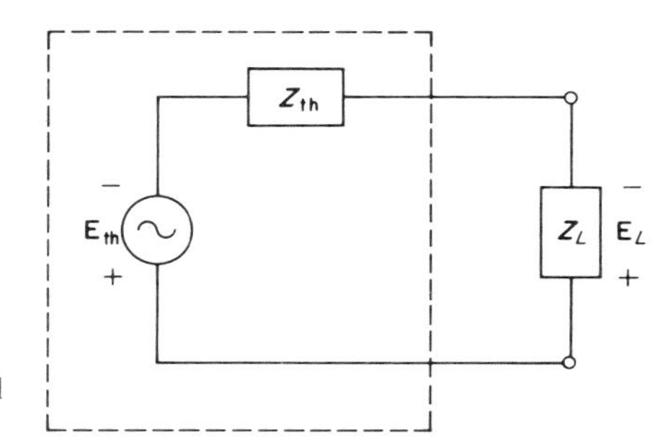

FIGURE 11-38
Thévenin representation of a general network.

where

$$\mathbf{Z}_4 = R_x + j\omega L_x$$
$$\mathbf{Z}_3 = R_3$$
$$\mathbf{Z}_2 = R_2$$

Therefore,

$$\frac{\mathbf{Z}_4}{\mathbf{Y}_1} = \mathbf{Z}_2 \mathbf{Z}_3$$
$$\mathbf{Z}_4 = \mathbf{Z}_2 \mathbf{Z}_3 \mathbf{Y}_1$$
$$R_x + j\omega L_x = R_2 R_3 \left(\frac{1}{R_1} + j\omega C_s\right)$$
$$R_x + j\omega L_x = \frac{R_2 R_3}{R_1} + j\omega C_s R_2 R_3$$

For balance,

$$R_x = \frac{R_2 R_3}{R_1} \tag{11-18}$$

and

$$j\omega L_x = j\omega C_s R_2 R_3$$
$$L_x = C_s R_2 R_3 \tag{11-19}$$

11-12 MAXIMUM POWER THEOREM

The impedance designated as the load in a network will receive maximum power from the source or sources in a network if the impedance of the load is the complex conjugate of the Thévenin impedance seen by the load. Thus, determination of the value of load impedance for maximum power transfer requires the calculation of $\mathbf{Z}_{th}$ for the network. A network with Thévenin values is shown in Fig. 11-38.

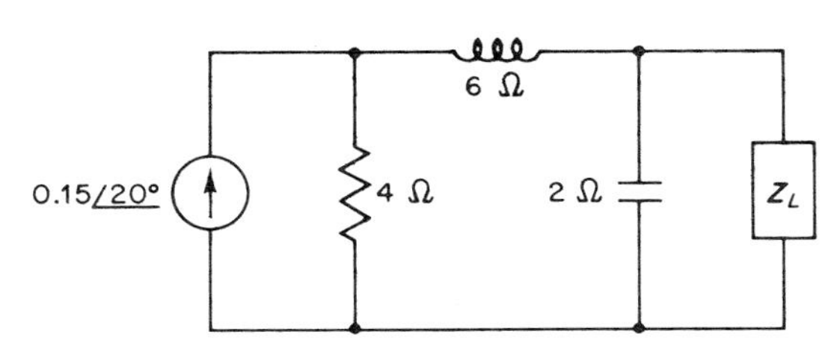

FIGURE 11-39
Circuit for Example 11-12.

If the Thévenin impedance is equal to

$$\mathbf{Z}_{\text{th}} = R \pm jX = A \angle \theta^\circ \tag{11-20}$$

then the load impedance for maximum power transfer is given by

$$\mathbf{Z}_L = R \mp jX = A \angle -\theta^\circ \tag{11-21}$$

The current that flows for maximum power transfer is equal to $\mathbf{E}_{\text{th}}$ divided by the sum of the series impedance

$$\mathbf{I} = \frac{\mathbf{E}_{\text{th}}}{\mathbf{Z}_{\text{th}} + \mathbf{Z}_L} = \frac{\mathbf{E}_{\text{th}}}{2R} \tag{11-22}$$

True power delivered to the load is

$$P_L = E_L I_L \cos \theta \tag{11-23}$$

where

$$E_L = |\mathbf{E}_L| \qquad \text{and} \qquad I_L = |\mathbf{I}_L|$$

In Eq. (11-23), $\theta = 0^\circ$ since the series impedance is resistive, $2R$. Thus, power delivered to the load under conditions of maximum power transfer is given by

$$P_{L,\text{max}} = E_L I_L = \frac{E_{\text{th}}}{2}\frac{E_{\text{th}}}{2R} = \frac{E_{\text{th}}}{4R} \tag{11-24}$$

where

$$E_{\text{th}} = |\mathbf{E}_{\text{th}}|$$

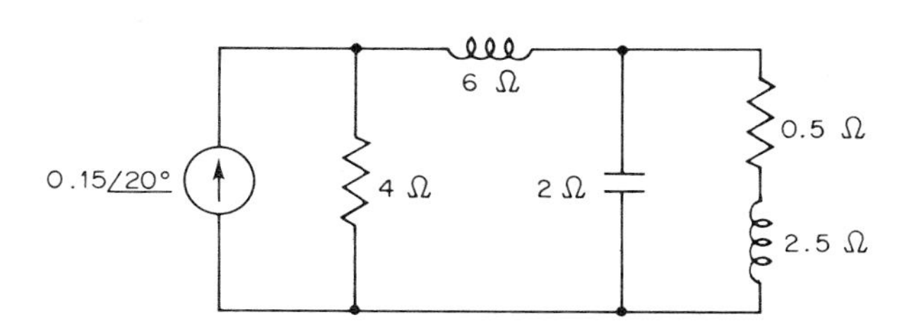

FIGURE 11-40

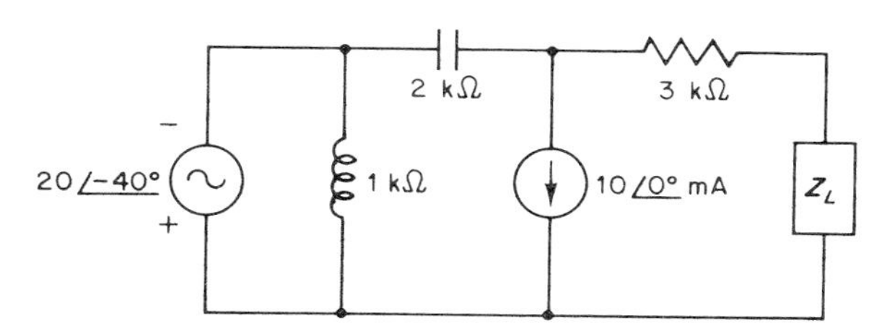

FIGURE 11-41
Circuit for Example 11-13.

EXAMPLE 11-12 Calculate the value of load impedance for maximum load power in the circuit of Fig. 11-39.

SOLUTION

$$\mathbf{Z}_{\text{th}} = \frac{(2\angle -90^\circ)(4+j6)}{4+j6-j2} = \frac{(2\angle -90^\circ)(7.2\angle 56.4^\circ)}{5.65\angle 45^\circ}$$
$$= 2.55\angle -78.6^\circ = 0.5 - j2.5\ \Omega$$

Therefore, the complex conjugate

$$\mathbf{Z}_L = 2.55\angle 78.6^\circ = 0.5 + j2.5\ \Omega$$

Component representation of $\mathbf{Z}_L$ is shown in Fig. 11-40. ////

EXAMPLE 11-13 Find $\mathbf{Z}_L$ and the power delivered to the load under conditions of maximum power transfer (see Fig. 11-41).

SOLUTION The circuit for finding $\mathbf{Z}_{\text{th}}$ is shown in Fig. 11-42.

$$\mathbf{Z}_{\text{th}} = 3\text{ k}\Omega - j2\text{ k}\Omega$$
$$\mathbf{Z}_L = 3\text{ k}\Omega + j2\text{ k}\Omega$$

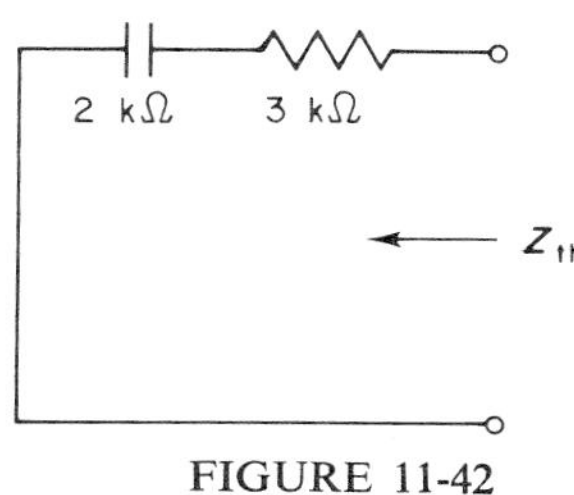

FIGURE 11-42

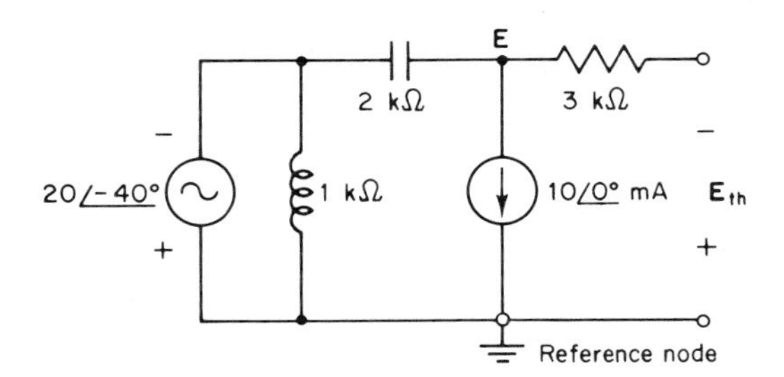

FIGURE 11-43

$\mathbf{E}_{th}$ is equal to the voltage drop across the current source in Fig. 11-43. A single node equation can be written for $\mathbf{E}$

$$10 \angle 0^\circ \text{ mA} = +(0.5 \times 10^{-3} \angle 90^\circ)\,\mathbf{E} - (0.5 \times 10^{-3} \angle 90^\circ)(-20 \angle -40^\circ)$$

$$\mathbf{E} = \frac{8.45 \times 10^{-3} \angle -115^\circ}{0.5 \times 10^{-3} \angle 90^\circ} = 16.9 \angle 155^\circ$$

$$\mathbf{E}_{th} = -\mathbf{E} = 16.9 \angle -25^\circ \text{ V}$$

The term $20 \angle -40^\circ$ appears as $-20 \angle -40^\circ$ in the node equation since node voltages are assumed positive when writing the equation.

$$P_{L,\max} = \frac{E_{th}^2}{4R} = \frac{286}{12 \text{ k}\Omega} = 23.8 \text{ mW} \qquad ////$$

PROBLEMS

11-1 Convert the voltage sources in Fig. 11-44 to equivalent current sources.

11-2 Convert the current sources in Fig. 11-45 to equivalent voltage sources.

11-3 Solve for the indicated voltage drops in Fig. 11-46 using the VDR. Assume the given voltage-drop polarities.

11-4 Solve for the indicated currents in Fig. 11-47 using the CDR. Assume the given current directions.

11-5 Calculate the mesh currents in the circuits of Fig. 11-48.

11-6 Apply mesh-current analysis to the circuits in Fig. 11-49, and find the voltages at terminals *a-b*.

11-7 Calculate the node voltages in the circuits of Fig. 11-50.

11-8 Apply node-voltage analysis to the circuits in Fig. 11-51, and find the indicated currents.

11-9 Use superposition to find the assumed currents in the circuits of Fig. 11-52.

11-10 Use superposition to find the assumed voltage drops in the circuits of Fig. 11-53.

11-11 Determine the Thévenin equivalent circuit for the terminal pair *a-b*. Refer to circuits in Fig. 11-54.

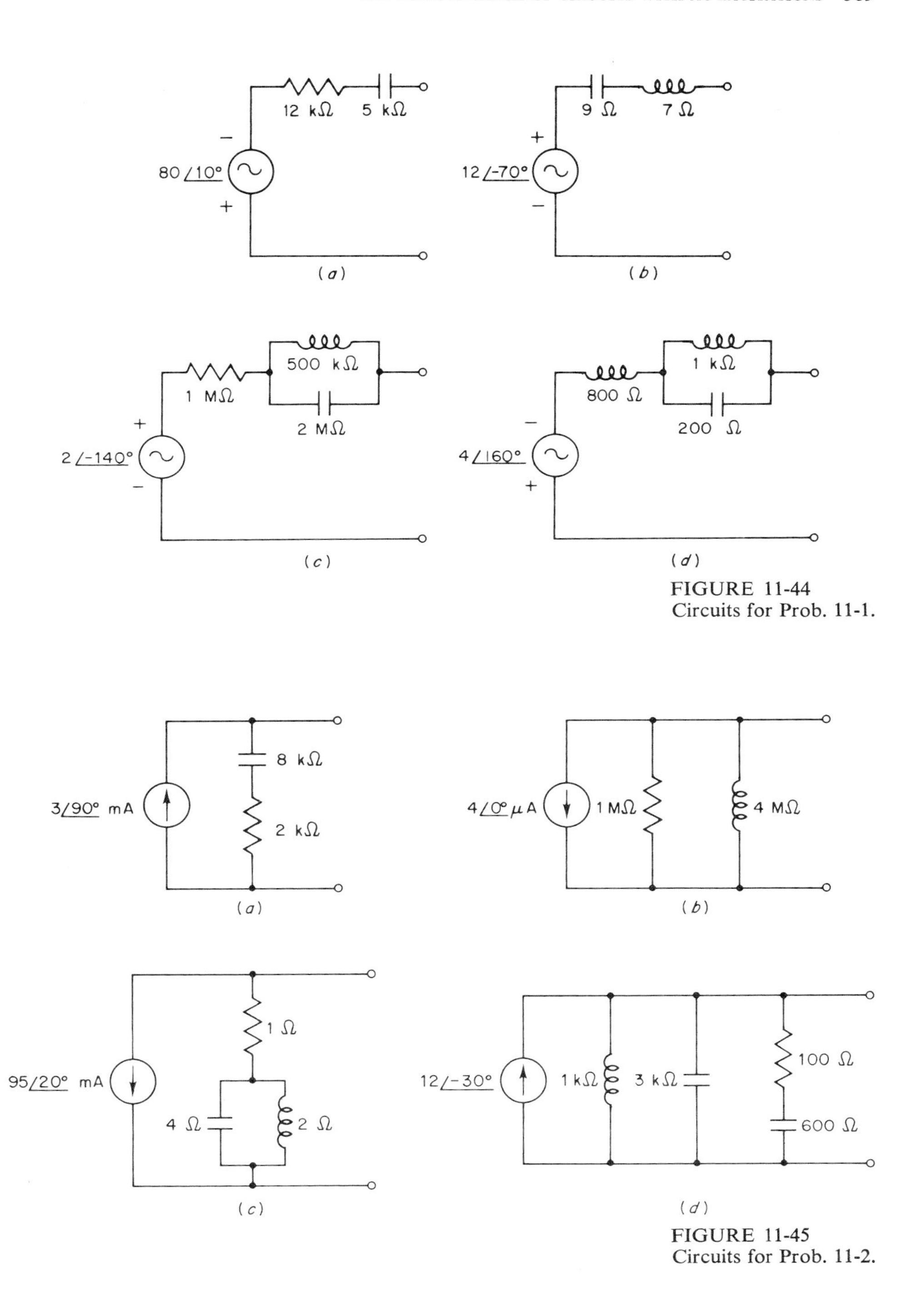

FIGURE 11-44
Circuits for Prob. 11-1.

FIGURE 11-45
Circuits for Prob. 11-2.

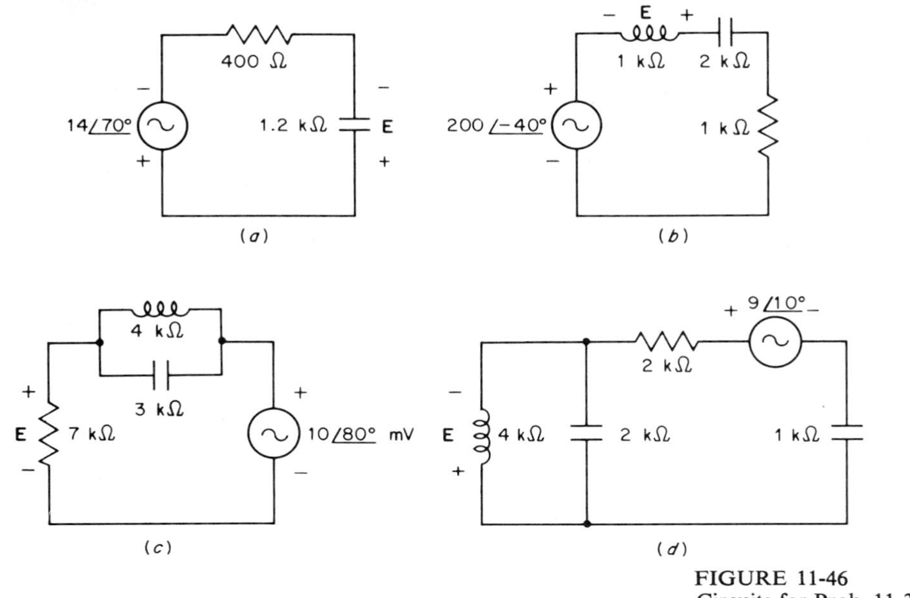

FIGURE 11-46
Circuits for Prob. 11-3.

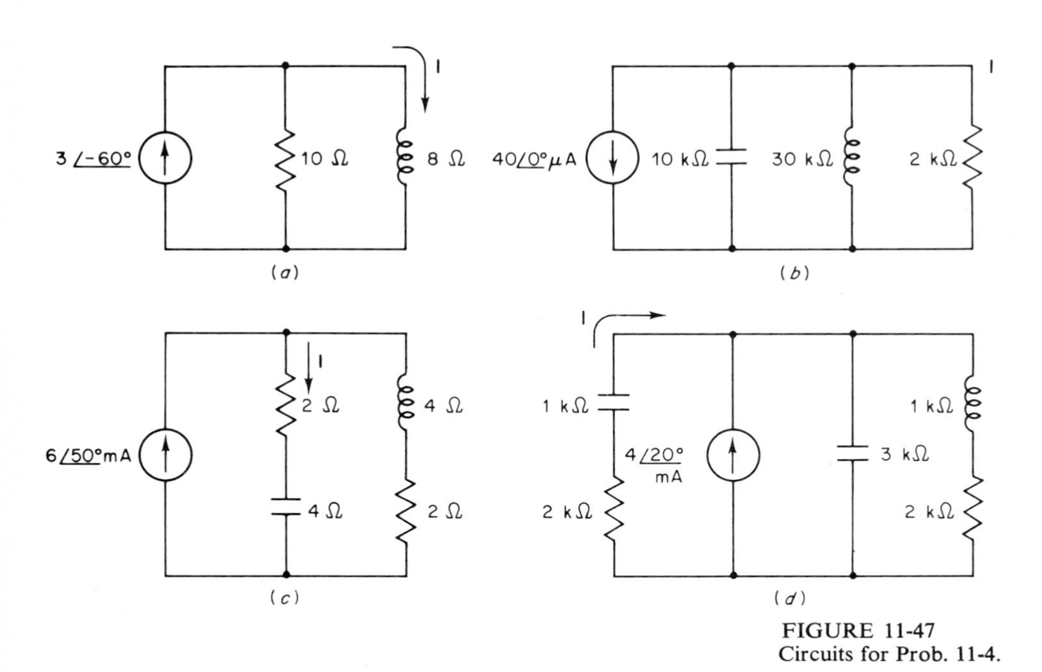

FIGURE 11-47
Circuits for Prob. 11-4.

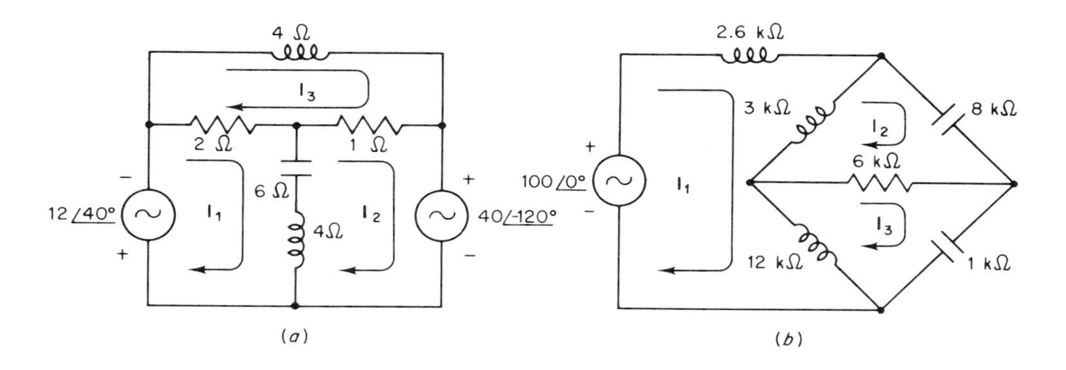

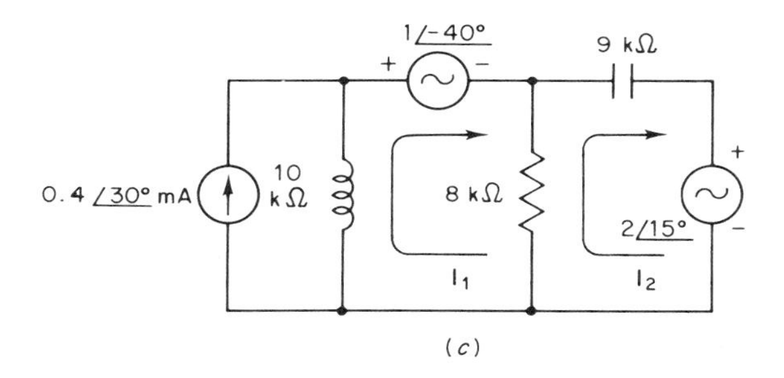

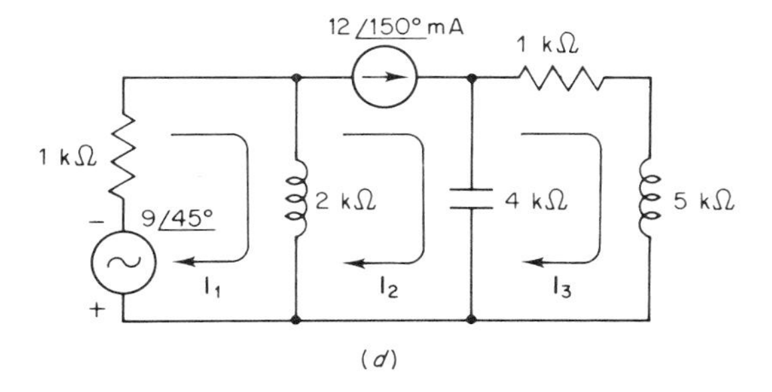

FIGURE 11-48
Circuits for Prob. 11-5.

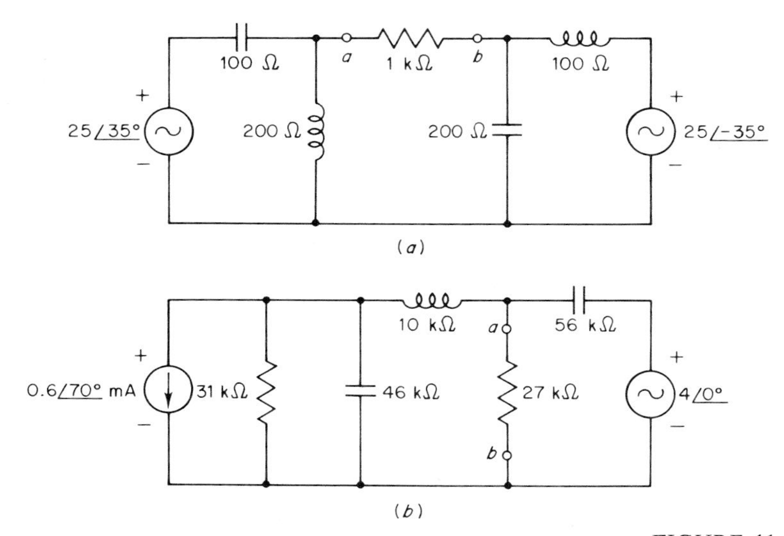

FIGURE 11-49
Circuits for Prob. 11-6.

11-12 Determine the Norton equivalent circuit for the terminal pair *a-b*. Refer to circuits in Fig. 11-55.

11-13 In the circuit shown in Fig. 11-56, determine the current **I**. Exchange the voltage source and current response. Through additional calculations verify the reciprocity theorem.

11-14 In the circuit shown in Fig. 11-57, determine the voltage **E**. Exchange the current source and voltage response. Through additional calculations verify the reciprocity theorem.

11-15 Determine the unknown component values required to balance the bridges in Fig. 11-58.

11-16 Find the load impedance necessary to provide maximum power transfer in each circuit of Fig. 11-59. Also calculate the maximum power in the load.

11-17 A crossover network for the woofer in a hi-fi speaker system is shown in Fig. 11-60. The woofer has an impedance of 8 Ω. This impedance is approximated as a resistance to simplify calculations. Analyze the performance of the crossover at two frequencies.

$$e_1(t) = 10 \sin(6.28 \times 10^3 t)$$

$$e_2(t) = 10 \sin(6.28 \times 10^4 t)$$

For each of the input voltages: (*a*) find the phasor output voltage, (*b*) express the output voltage as a function of time, and (*c*) calculate the true power delivered to the woofer.

11-18 The equivalent circuit of a transistor amplifier in Fig. 11-61 represents an approximation of the circuit under small-signal operation. Determine the output voltage, $e_o(t)$. Input signal is $i_i(t) = 0.6 \sin(7 \times 10^4 t + 35°)$ mA.

$$R_o = 10\ \text{k}\Omega, \quad R_B = 6\ \text{k}\Omega,\ h_{ie} = 1.4\ \text{k}\Omega, \quad h_{fe} = 100, \quad R_L = 2\ \text{k}\Omega,$$

$$R_i = 1\ \text{k}\Omega, \quad C_{C1} = 0.008\ \mu\text{F}, \quad C_{C2} = 0.02\ \mu\text{F}$$

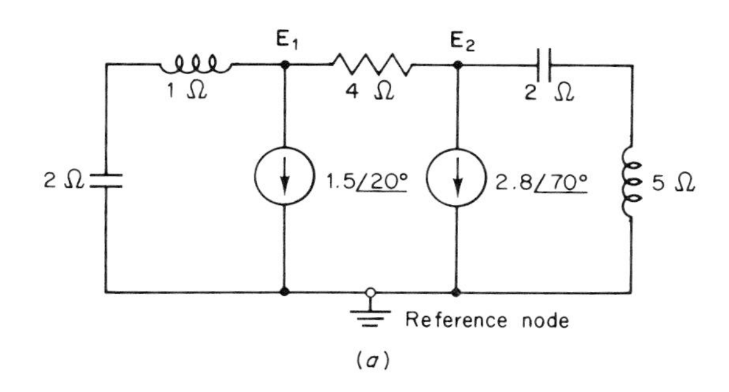

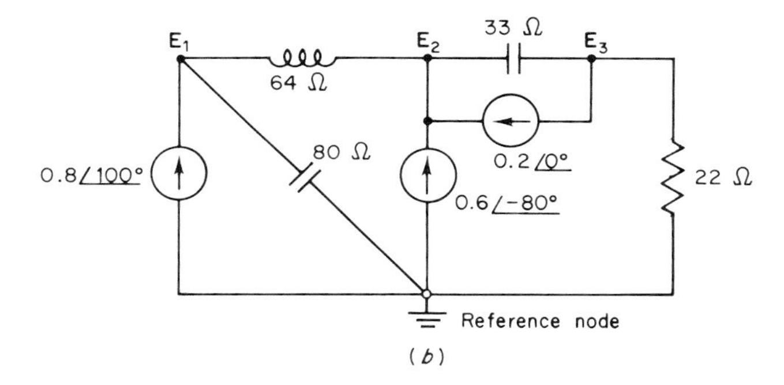

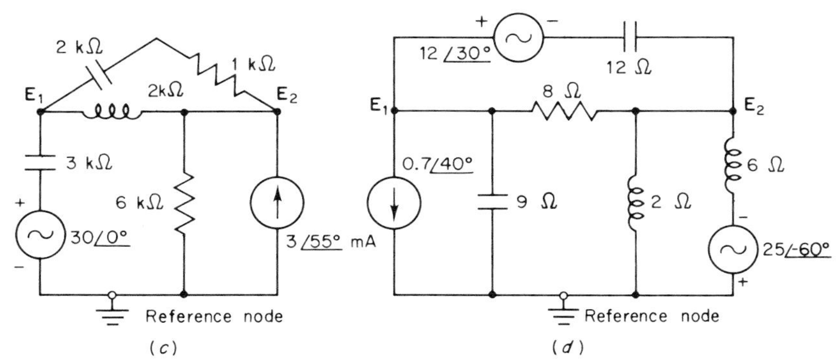

FIGURE 11-50
Circuits for Prob. 11-7.

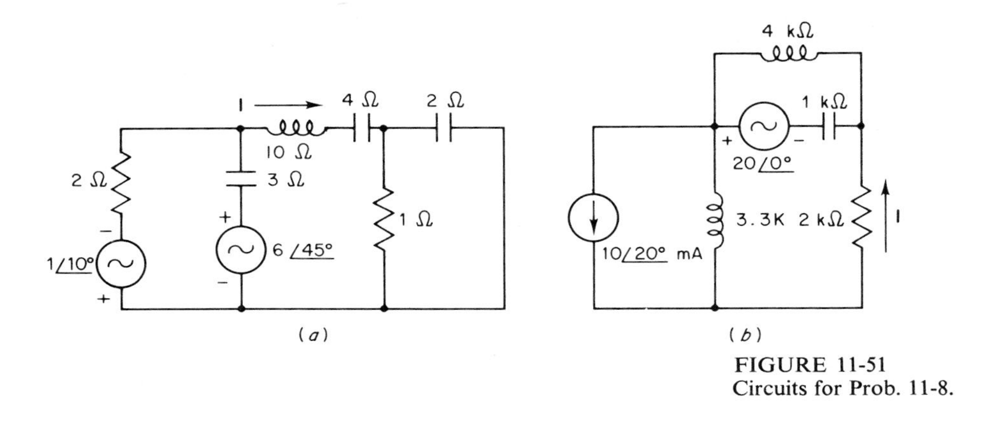

FIGURE 11-51
Circuits for Prob. 11-8.

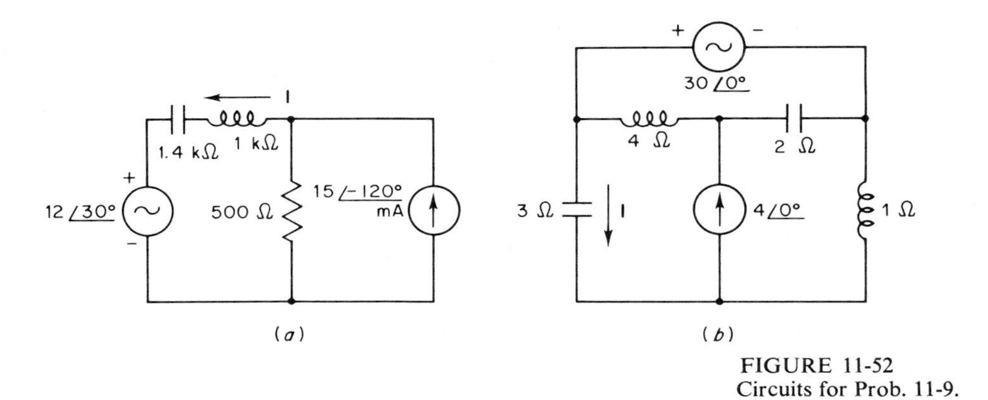

FIGURE 11-52
Circuits for Prob. 11-9.

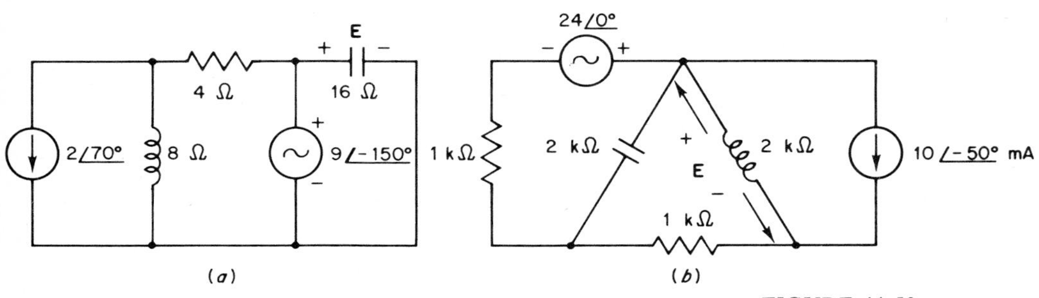

FIGURE 11-53
Circuits for Prob. 11-10.

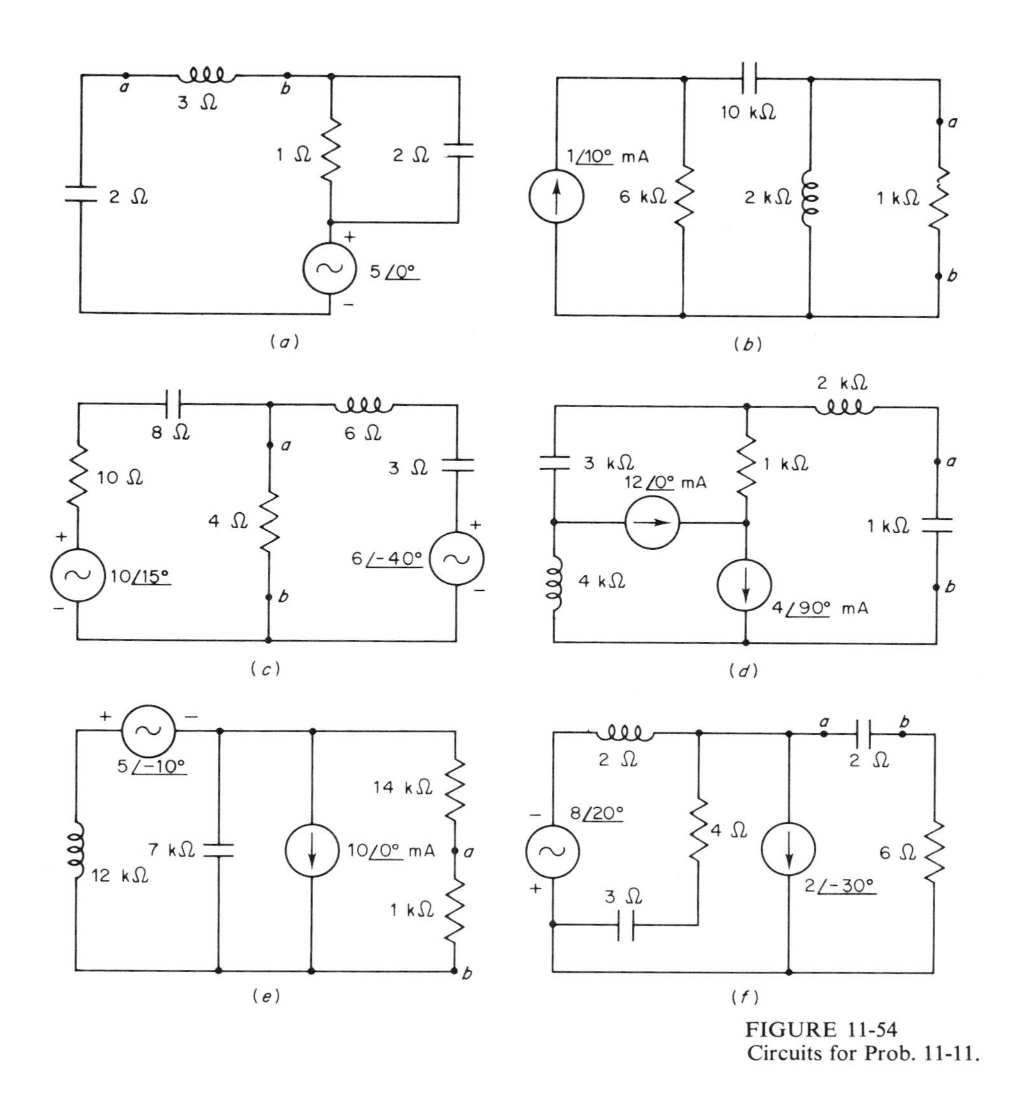

FIGURE 11-54
Circuits for Prob. 11-11.

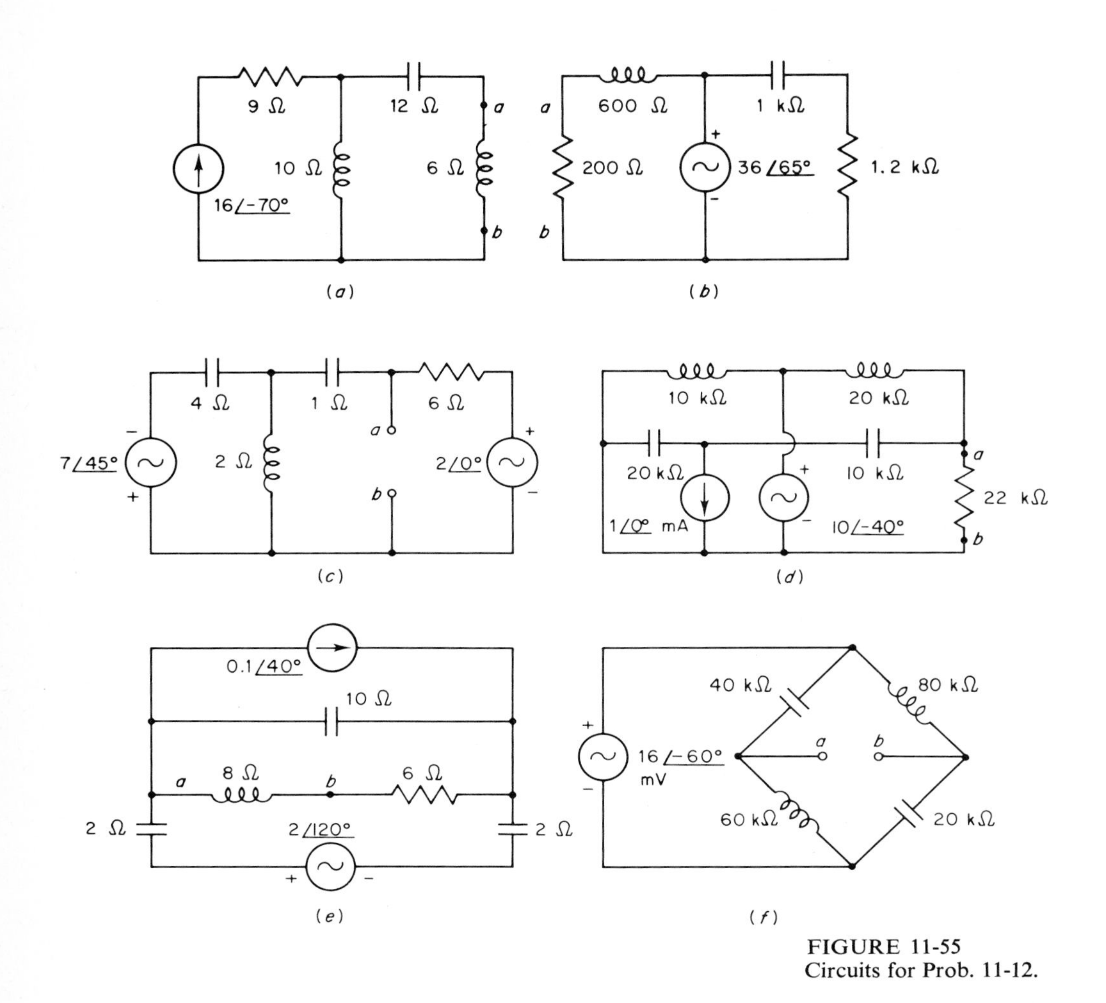

FIGURE 11-55
Circuits for Prob. 11-12.

FIGURE 11-56
Circuit for Prob. 11-13.

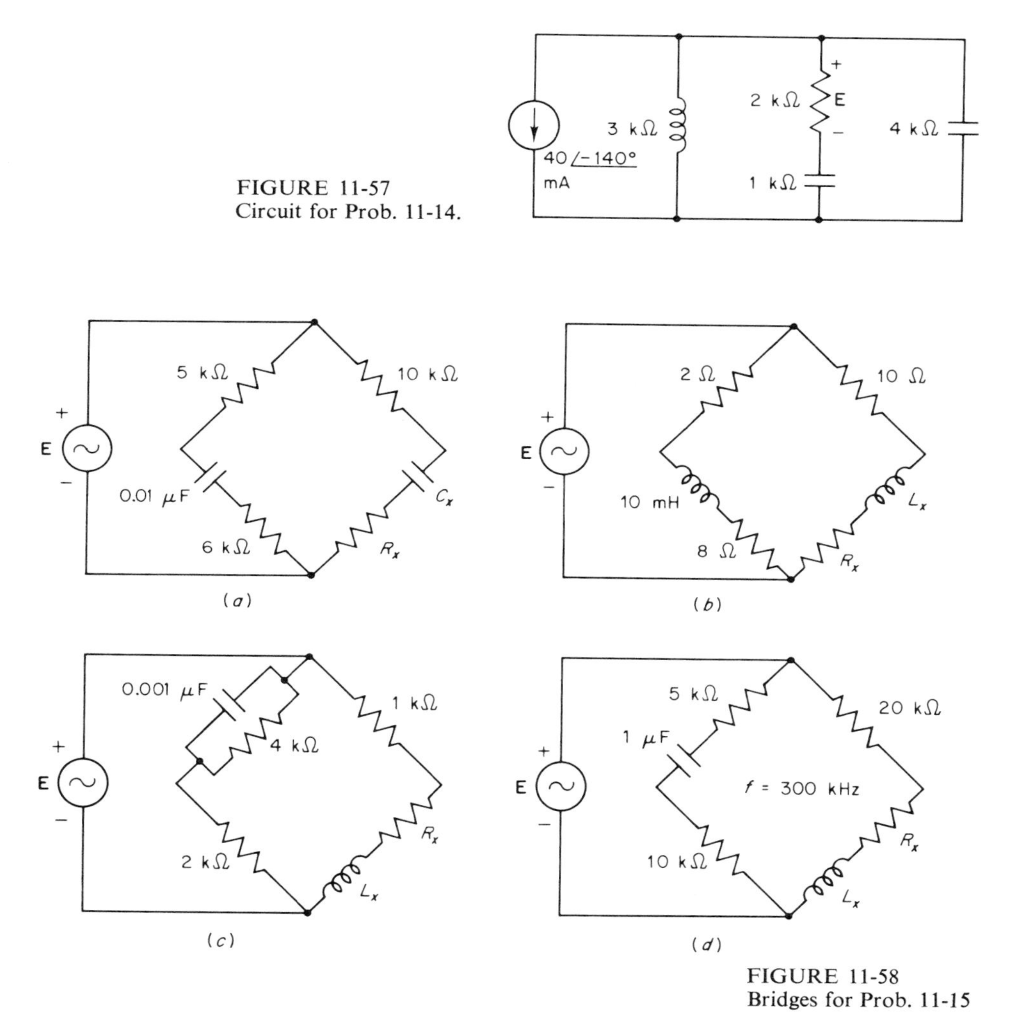

FIGURE 11-57
Circuit for Prob. 11-14.

FIGURE 11-58
Bridges for Prob. 11-15

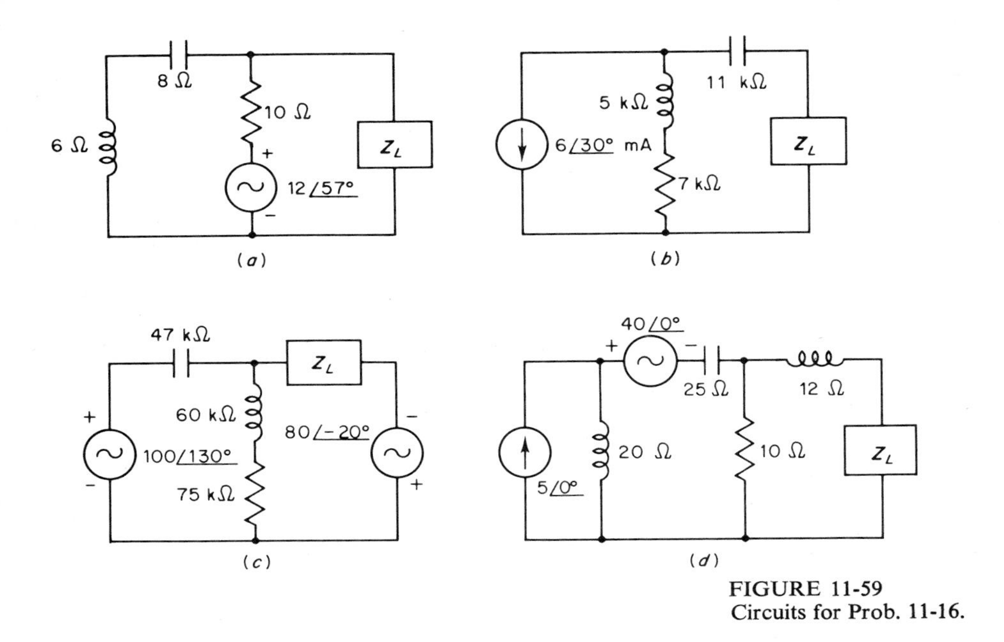

FIGURE 11-59
Circuits for Prob. 11-16.

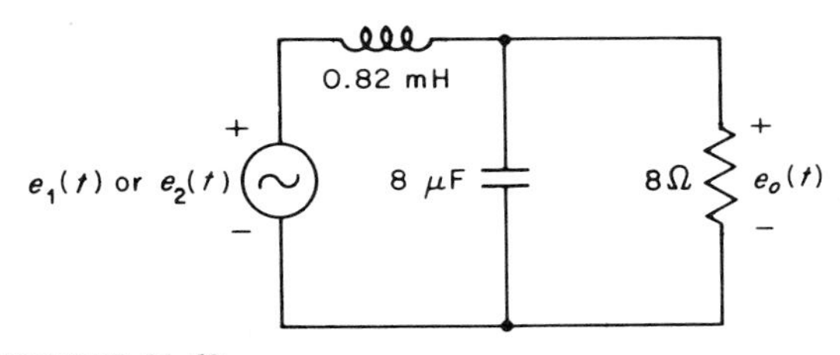

FIGURE 11-60
Circuit for Prob. 11-17.

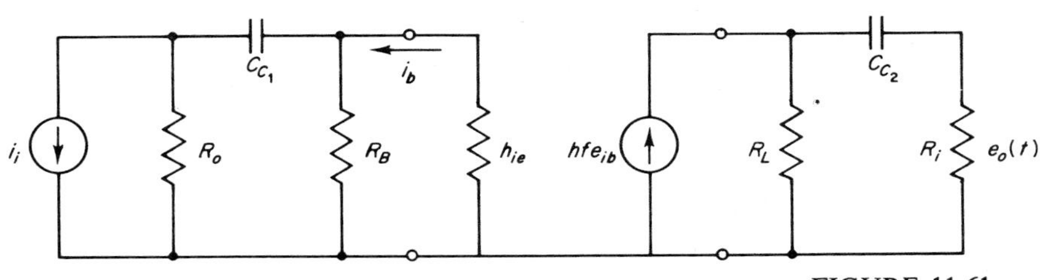

FIGURE 11-61
Circuit for Prob. 11-18.

12

SERIES AND PARALLEL RESONANT CIRCUITS

12-1 INTRODUCTION

In Chap. 9, the concepts of inductive and capacitive reactance were developed. It was shown that inductive reactance is directly proportional to frequency, while capacitive reactance is inversely proportional to frequency. Since inductive reactance and capacitive reactance are in opposition to each other, it can be seen that a circuit containing both inductance and capacitance will have an impedance property that varies with frequency. The circuit containing inductance and capacitance appears capacitive over one range of frequencies and inductive at other frequencies. At some frequency, the reactances will exactly balance one another and the circuit will appear purely resistive. The frequency at which an *LCR* circuit appears as a net resistance is referred to as the *resonant frequency*. The frequency-dependent properties of the *LCR* circuit can be used to select wanted signals or reject unwanted signals on the basis of frequency. Tuning circuits, some filter circuits, and a number of other types of circuits employ the properties of resonant circuits.

Figure 12-1 shows that the effect of frequency on inductive and capacitive reactance.

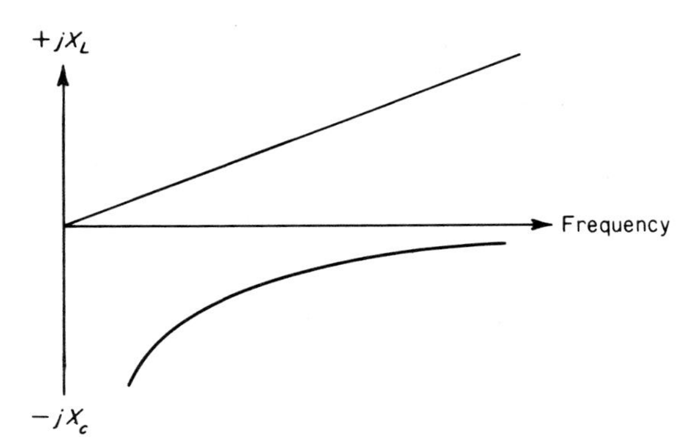

FIGURE 12-1
Inductive and capacitive reactance versus frequency.

12-2 SERIES RESONANCE

R, L, and C are connected in series as shown in Fig. 12-2. The Kirchhoff voltage-loop equation is

$$\mathbf{E} = \mathbf{E}_C + \mathbf{E}_L + \mathbf{E}_R$$

Since current $\mathbf{I}$ is the common factor,

$$\mathbf{E} = -jX_C\mathbf{I} + jX_L\mathbf{I} + R\mathbf{I} \tag{12-1}$$

In order for the voltage phasor in Eq. (12-1) to be at 0°, the sum of the reactive drops must equal zero.

$$-jX_C\mathbf{I} + jX_L\mathbf{I} = 0$$

and

$$-jX_C\mathbf{I} = -jX_L\mathbf{I}$$

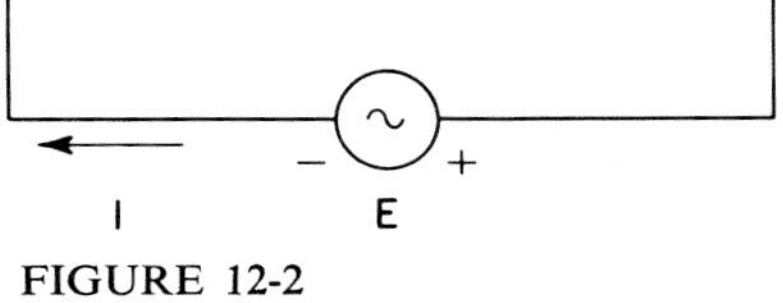

FIGURE 12-2
Series R, L, and C circuit.

Therefore,

$$X_C = X_L \tag{12-2}$$

Equation (12-2) states that when the inductive and capacitive reactance of the series-*LCR* circuit (Fig. 12-2) are equal, the circuit appears purely resistive. When the series-*LCR* circuit appears purely resistive, the circuit is said to be in *series resonance.*

The frequency at which series resonance occurs can be determined from Eq. (12-2) as a function of the reactive components L and C. The *series-resonant frequency* f_o is derived in the following manner.

$$X_C = X_L \tag{12-2}$$

$$\frac{1}{2\pi f_o C} = 2\pi f_o L$$

$$f_o^2 = \frac{1}{4\pi^2 LC}$$

$$f_o = \frac{1}{2\pi\sqrt{LC}} \tag{12-3}$$

Equation (12-3) can be used to determine the resonant frequency if the elements of the series circuit are known. This is demonstrated in Example 12-1.

EXAMPLE 12-1 The elements in the circuit of Fig. 12-2 are $L = 8$ mH, $C = 0.0005$ μF, $R = 500\ \Omega$, and $\mathbf{E} = 30 \angle 0°$ V. Find (1) the resonant frequency, (2) the current, and (3) the voltage across each component.

SOLUTION

1 Using Eq. (12-3),

$$f_o = \frac{1}{2\pi\sqrt{LC}} = \frac{1}{6.28\sqrt{(8 \times 10^{-3})(0.5 \times 10^{-9})}} = 0.0796 \times 10^6 = 79.6 \text{ kHz}$$

2 From Eq. (12-1), it is seen that the sum of the reactive voltages is zero at resonance. Therefore,

$$\mathbf{E} = \mathbf{I}R$$

$$\mathbf{I} = \frac{\mathbf{E}}{R} = \frac{30 \angle 0°}{500 \angle 0°} = 0.06 \angle 0° \text{ A}$$

3 The voltage across each component can now be calculated using the common current of step 2.

$$\mathbf{E}_L = jX_L\mathbf{I}$$
$$= [(6.28)(79.6 \times 10^3)(8 \times 10^{-3}) \angle 90°](0.06 \angle 0°)$$
$$= (4{,}000 \angle 90°)(0.06 \angle 0°) = 240 \angle 90° \text{ V}$$
$$\mathbf{E}_C = -jX_C\mathbf{I}$$
$$= \frac{1 \angle -90°}{(6.28)(79.6 \times 10^3)(5 \times 10^{-10})}(0.06 \angle 0°)$$
$$= (4{,}000 \angle -90°)(0.06 \angle 0) = 240 \angle -90° \text{ V}$$
$$\mathbf{E}_R = R\mathbf{I}$$
$$= (0.06 \angle 0°)(500 \angle 0°) = 30 \angle 0° \text{ V}$$

////

It can be seen from step 3 of Example 12-1 that the voltage drop across the inductance and the capacitance is larger than the applied voltage **E**. The energy supplied by the source is dissipated by the resistance, while the capacitance and the inductance simply exchange energy.

At the resonant frequency, the impedance of a series-*LCR* circuit is purely resistive and at a minimum. The current is therefore maximum at resonance. As frequency is reduced below resonance, the capacitive reactance increases and the inductive reactance decreases. The net reactance at frequencies below resonant frequency is capacitive, and it increases in magnitude as the frequency is further reduced below the resonant frequency.

The impedance of a circuit below resonance is the phasor sum of the resistance and the net capacitive reactance. Since the reactive component of the circuit is capacitive, the current leads the applied voltage by an increasing angle as the frequency is decreased. The magnitude of the current decreases as the frequency is decreased below resonance.

If frequency is increased above the resonant frequency, the inductive reactance increases and the capacitive reactance decreases. The net reactance above the resonant frequency is inductive, and it increases in magnitude as the frequency increases. Impedance increases and current decreases as the frequency increases above resonance. The current lags the applied voltage by an increasing angle as the frequency is increased above the resonant frequency.

Impedance magnitude, current magnitude, and current phase angle are plotted in Fig. 12-3, and Example 12-2 demonstrates the conditions above and below resonance.

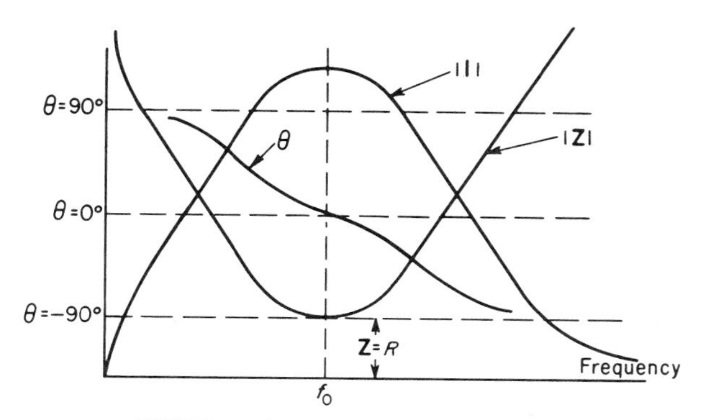

FIGURE 12-3
$|\mathbf{I}|$, $|\mathbf{Z}|$, and θ plotted versus frequency (θ is phase angle of current).

EXAMPLE 12-2 Using the circuit-element values given in Example 12-1 calculate the impedance and current of the circuit of Fig. 12-2 at the following frequencies: (1) 70 kHz, (2) 50 kHz, (3) 90 kHz, and (4) 110 kHz.

SOLUTION

1 At 70 kHz

$$X_L = (6.28)(8 \times 10^{-3})(70 \times 10^3) = 3{,}520\ \Omega$$

$$X_C = \frac{1}{(6.28)(70 \times 10^3)(5 \times 10^{-10})} = 4{,}550\ \Omega$$

$$\mathbf{Z}_t = 500 + j3{,}520 - j4{,}550 = 500 - j1{,}030 = 1{,}143 \angle -64.1^\circ\ \Omega$$

$$\mathbf{I} = \frac{30 \angle 0^\circ}{1{,}143 \angle -64.1^\circ} = 0.0262 \angle 64.1^\circ\ \text{A}$$

2 At 50 kHz,

$$X_L = (6.28)(50 \times 10^3)(8 \times 10^{-3}) = 2{,}510\ \Omega$$

$$X_C = \frac{1}{(6.28)(50 \times 10^3)(5 \times 10^{-10})} = 6{,}360\ \Omega$$

$$\mathbf{Z}_t = 500 + j2{,}510 - j6{,}360 = 500 - j3{,}850 = 3{,}880 \angle -82.6^\circ\ \Omega$$

$$\mathbf{I} = \frac{30 \angle 0^\circ}{3{,}880 \angle -82.6^\circ} = 0.00755 \angle 82.6^\circ\ \text{A}$$

3 At 90 kHz,

$$X_L = (6.28)(90 \times 10^3)(8 \times 10^{-3}) = 4{,}520\ \Omega$$

$$X_C = \frac{1}{(6.28)(90 \times 10^3)(5 \times 10^{-10})} = 3{,}540\ \Omega$$

$$\mathbf{Z}_t = 500 + j4{,}520 - j3{,}540 = 500 + j980 = 1{,}100 \angle 63^\circ\ \Omega$$

$$\mathbf{I} = \frac{30 \angle 0^\circ}{1{,}100 \angle 63^\circ} = 0.0273 \angle -63^\circ\ \text{A}$$

4 At 110 kHz,

$$X_L = (6.28)(110 \times 10^3)(8 \times 10^{-3}) = 5{,}530\ \Omega$$

$$X_C = \frac{1}{(6.28)(110 \times 10^3)(5 \times 10^{-10})} = 2{,}895\ \Omega$$

$$\mathbf{Z}_t = 500 + j5{,}530 - j2{,}895 = 500 + j2{,}635 = 2{,}680 \angle 79.25^\circ\ \Omega$$

$$\mathbf{I} = \frac{30 \angle 0^\circ}{2{,}680 \angle 79.25^\circ} = 0.0112 \angle -79.25^\circ\ \text{A}$$

////

If the resistance of Fig. 12-2 is changed, the frequency response of the impedance will be affected. If the resistance is increased while L and C are held constant, the response curve will be widened. As resistance increases, the net reactance must become larger to have a measurable effect on the impedance. Thus a larger frequency variation is necessary to alter the impedance by a given amount. The effect of increasing resistance while holding L and C constant is shown in Example 12-3.

EXAMPLE 12-3 Calculate the impedance of the circuit of Fig. 12-2 at the frequencies of Example 12-2 if the resistance is changed to (1) 1 kΩ, (2) 2 kΩ.

SOLUTION

1 The net reactance calculated in Example 12-2 is unchanged.

At 70 kHz: $\mathbf{Z}_t = 1{,}000 - j1{,}030 = 1{,}435 \angle -45.8^\circ\ \Omega$

At 50 kHz: $\mathbf{Z}_t = 1{,}000 - j3{,}850 = 3{,}980 \angle -75.4^\circ\ \Omega$

At 90 kHz: $\mathbf{Z}_t = 1{,}000 + j980 = 1{,}400 \angle 44.4^\circ\ \Omega$

At 110 kHz: $\mathbf{Z}_t = 1{,}000 + j2{,}635 = 2{,}820 \angle 69.2^\circ\ \Omega$

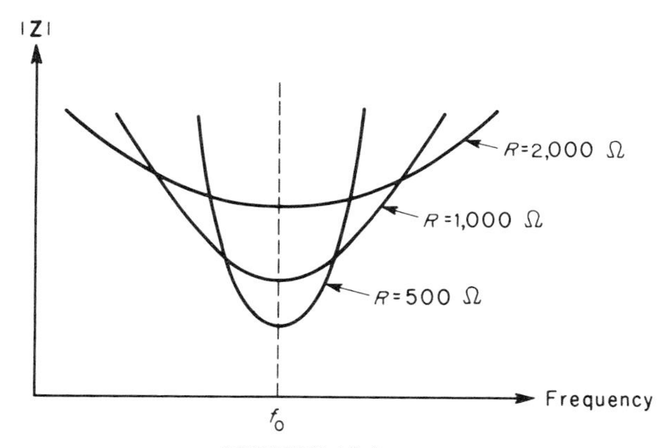

FIGURE 12-4
$|\mathbf{Z}|$ plotted versus frequency for several values of resistance.

It must be noted that as the resistance is increased the current at resonance decreases.

For $R = 1{,}000\ \Omega$, $\mathbf{I}$ at resonance is

$$\mathbf{I} = \frac{30 \angle 0^\circ}{1{,}000 \angle 0^\circ} = 0.03 \angle 0^\circ \text{ A}$$

2 At 70 kHz: $\mathbf{Z}_t = 2{,}000 - j1{,}030 = 2{,}250 \angle -27.25\ \Omega$

At 50 kHz: $\mathbf{Z}_t = 2{,}000 - j3{,}850 = 4{,}340 \angle -62.5\ \Omega$

At 90 kHz: $\mathbf{Z}_t = 2{,}000 + j980 = 2{,}230 \angle 26.1^\circ\ \Omega$

At 110 kHz: $\mathbf{Z}_t = 2{,}000 + j2{,}635 = 3{,}310 \angle 52.7^\circ\ \Omega$ ////

The effect on impedance of increasing resistance while holding L and C constant is shown in Fig. 12-4.

The opposite effect is observed in the frequency response of the impedance when resistance is decreased. As resistance is made smaller, the net reactance necessary to have a measurable effect on the impedance becomes less. A smaller change in frequency yields a comparatively larger change in impedance. Example 12-4 demonstrates the effect of reducing resistance while holding L and C constant.

EXAMPLE 12-4 Calculate the impedance of the circuit of Fig. 12-2 at the frequencies of Example 12-2 if the resistance is changed to (1) 250 Ω and (2) 125 Ω.

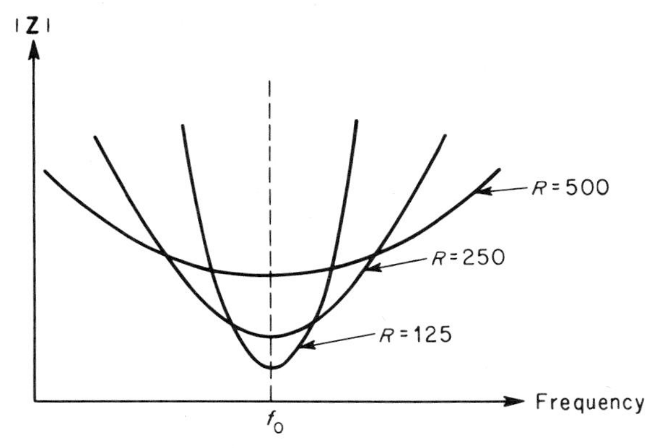

FIGURE 12-5
|**Z**| plotted versus frequency for several values of resistance.

SOLUTIONS

1 The net reactance calculated in Example 12-2 remains unchanged.

At 70 kHz: $\mathbf{Z}_t = 250 - j1{,}030 = 1{,}060 \angle -76.35° \ \Omega$

At 50 kHz: $\mathbf{Z}_t = 250 - j3{,}850 = 3{,}850 \angle -86.28° \ \Omega$

At 90 kHz: $\mathbf{Z}_t = 250 + j980 = 1{,}012 \angle 75.7° \ \Omega$

At 110 kHz: $\mathbf{Z}_t = 250 + j2{,}635 = 2{,}635 \angle 84.55° \ \Omega$

2 At 70 kHz: $\mathbf{Z}_t = 125 - j1{,}030 = 1{,}038 \angle -83.07° \ \Omega$

At 50 kHz: $\mathbf{Z}_t = 125 - j3{,}850 = 3{,}850 \angle -88.14° \ \Omega$

At 90 kHz: $\mathbf{Z}_t = 125 + j980 = 988 \angle 82.73° \ \Omega$

At 110 kHz: $\mathbf{Z}_t = 125 + j2{,}635 = 2{,}635 \angle 87.28° \ \Omega$ ////

The effect on impedance of decreasing resistance while holding L and C constant is shown in Fig. 12-5.

Variation in the frequency response of the series-LCR impedance can be obtained by varying capacitance and inductance. If inductance is increased and capacitance decreased to maintain the same resonant frequency, the change in frequency necessary to cause a significant change in circuit impedance is reduced. Reducing inductance and increasing capacitance has the opposite effect on the frequency response of the impedance.

When the reactive components are varied to alter the frequency response of the impedance, the current at resonance remains unchanged if the resistance remains unchanged.

The effect of altering reactive components to vary the frequency response of the impedance is demonstrated in Example 12-5.

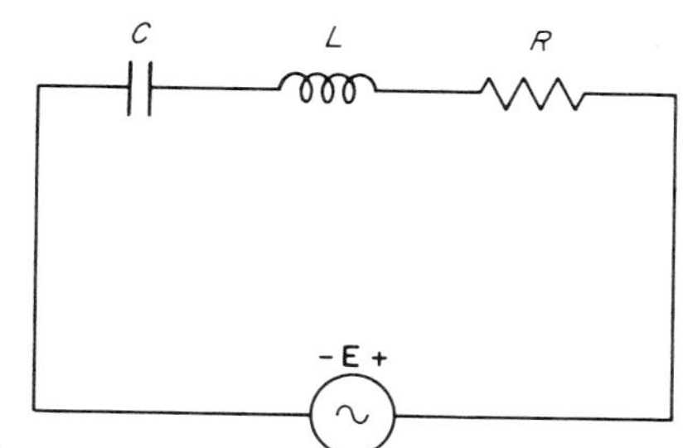

FIGURE 12-6
Circuit for Example 12-5.

EXAMPLE 12-5 For the circuit of Fig. 12-6, calculate the impedance at 150 kHz, 175 kHz, 225 kHz, and 250 kHz under the following conditions:

1 $L = 1.91$ mH $C = 332$ PF $R = 40\ \Omega$

2 $L = 0.957$ mH $C = 663$ PF $R = 40\ \Omega$

3 $L = 0.319$ mH $C = 1{,}990$ PF $R = 40\ \Omega$

SOLUTION The resonant frequency of Fig. 12-6 is the same for all values given because the LC product in each step is the same.

$$f_o = \frac{1}{2\pi\sqrt{LC}} = \frac{1}{6.28\sqrt{(1.91 \times 10^{-3})(33.2 \times 10^{-11})}} = 200 \text{ kHz}$$

1 At 150 kHz: $X_L = (6.28)(150 \times 10^3)(1.91 \times 10^{-3}) = 1{,}800\ \Omega$

$$X_C = \frac{1}{(6.28)(150 \times 10^3)(3.32 \times 10^{-10})} = 3{,}200\ \Omega$$

$$\mathbf{Z}_t = 40 - j1{,}400 = 1{,}400 \angle -88.36^\circ$$

At 175 kHz: $X_L = (6.28)(175 \times 10^3)(1.91 \times 10^{-3}) = 2{,}100\ \Omega$

$$X_C = \frac{1}{(6.28)(175 \times 10^3)(3.32 \times 10^{-10})} = 2{,}740\ \Omega$$

$$\mathbf{Z}_t = 40 - j640 = 640 \angle -86.42^\circ$$

At 225 kHz: $X_L = (6.28)(225 \times 10^3)(1.91 \times 10^{-3}) = 2{,}700\ \Omega$

$$X_C = \frac{1}{(6.28)(225 \times 10^3)(3.32 \times 10^{-10})} = 2{,}130\ \Omega$$

$$\mathbf{Z}_t = 40 + j570 = 570 \angle 85.98^\circ$$

At 250 kHz: $X_L = (6.28)(250 \times 10^3)(1.91 \times 10^{-3}) = 3{,}000\ \Omega$

$$X_C = \frac{1}{(6.28)(250 \times 10^3)(3.32 \times 10^{-10})} = 1{,}920\ \Omega$$

$$\mathbf{Z}_t = 40 + j1{,}080 = 1{,}080 \angle 87.88^\circ$$

2 At 150 kHz: $X_L = (6.28)(150 \times 10^3)(0.957 \times 10^{-3}) = 902\ \Omega$

$$X_C = \frac{1}{(6.28)(150 \times 10^3)(6.63 \times 10^{-10})} = 1{,}600\ \Omega$$

$$\mathbf{Z}_t = 40 - j698 = 698 \angle -86.72^\circ$$

At 175 kHz: $X_L = (6.28)(175 \times 10^3)(0.957 \times 10^{-3}) = 1{,}050\ \Omega$

$$X_C = \frac{1}{(6.28)(175 \times 10^3)(6.63 \times 10^{-10})} = 1{,}370\ \Omega$$

$$\mathbf{Z}_t = 40 - j320 = 323 \angle -82.88^\circ$$

At 225 kHz: $X_L = (6.28)(225 \times 10^3)(0.957 \times 10^{-3}) = 1{,}350\ \Omega$

$$X_C = \frac{1}{(6.28)(225 \times 10^3)(6.63 \times 10^{-10})} = 1{,}065\ \Omega$$

$$\mathbf{Z}_t = 40 + j285 = 288 \angle 82^\circ$$

At 250 kHz: $X_L = (6.28)(250 \times 10^3)(0.957 \times 10^{-3}) = 1{,}500\ \Omega$

$$X_L = \frac{1}{(6.28)(250 \times 10^3)(6.63 \times 10^{-10})} = 960\ \Omega$$

$$\mathbf{Z}_t = 40 + j540 = 540 \angle 85.74^\circ$$

3 At 150 kHz: $X_L = (6.28)(0.319 \times 10^{-3})(150 \times 10^3) = 300\ \Omega$

$$X_C = \frac{1}{(6.28)(150 \times 10^3)(1.99 \times 10^{-9})} = 534\ \Omega$$

$$\mathbf{Z}_t = 40 - j234 = 238 \angle -80.3^\circ$$

At 175 kHz: $X_L = (6.28)(175 \times 10^3)(0.319 \times 10^{-3}) = 350\ \Omega$

$$X_C = \frac{1}{(6.28)(175 \times 10^3)(1.99 \times 10^{-9})} = 456\ \Omega$$

$$\mathbf{Z}_t = 40 - j106 = 113 \angle -69.3^\circ$$

At 225 kHz: $X_L = (6.28)(225 \times 10^3)(0.319 \times 10^{-3}) = 450\ \Omega$

$$X_C = \frac{1}{(6.28)(225 \times 10^3)(1.99 \times 10^{-9})} = 356\ \Omega$$

$$\mathbf{Z}_t = 40 + j94 = 102 \angle 66.9^\circ$$

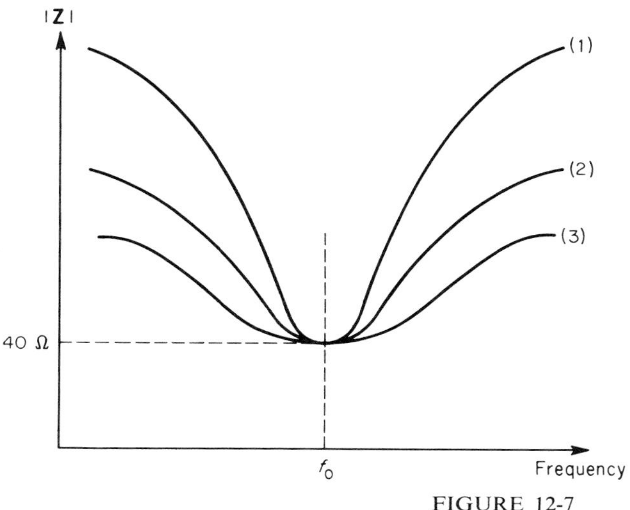

FIGURE 12-7
Comparative results of Example 12-5.

At 250 kHz: $X_L = (6.28)(250 \times 10^3)(0.319 \times 10^{-3}) = 500\ \Omega$

$$X_C = \frac{1}{(6.28)(250 \times 10^3)(1.99 \times 10^{-9})} = 320\ \Omega$$

$$\mathbf{Z}_t = 40 + j180 = 185\ \angle 77.5^\circ$$

Figure 12-7 shows the comparative results of Example 12-5. ////

12-3 ENERGY STORAGE IN A RESONANT CIRCUIT

Previous chapters have shown that inductance and capacitance store energy. In a series-resonant circuit, both inductance and capacitance store energy during a sine wave of current. From 0 to 90° of the current waveform, the inductance stores energy, while the capacitance returns energy to the circuit. From 90 to 180°, the capacitance stores energy and the inductance returns energy to the circuit. During the next quarter cycle, from 180 to 270°, the inductance once again stores energy while the capacitance returns energy to the circuit. From 270 to 360°, it is the capacitor that stores energy while the inductance returns energy to the circuit.

The reader will note that during one complete cycle, the inductance stores energy for one half of the cycle and the capacitance stores energy for the other half of the cycle. In a resonant circuit, the quantities of energy stored by both reactances are equal. It thus appears that the energy is simply transferred back and forth between the inductance and the capacitance. This is indeed true, and if the circuit in question contained no resistance, this interchange of energy between inductance and capacitance would continue indefinitely, even if the initial source of energy were removed.

One important characteristic of any resonant circuit is a measure of the circuit's ability to store energy. Mathematically, this characteristic is defined as the ratio of the energy stored per half cycle to the energy dissipated per half cycle. Since energy storage is the same for both inductance and capacitance at resonance and the resistance of a resonant circuit is mainly the ohmic resistance of the coil, this measure of a resonant circuit's ability to store energy is specified in terms of the inductance.

Let the energy stored be the reactive power, and let the energy dissipated be the power dissipated in the circuit. The ratio of the reactive power to the dissipated power at resonance is the quality factor Q_o.

$$Q_o = \frac{\omega_0 L I^2}{I^2 R} \quad \text{where} \quad I = |\mathbf{I}| \tag{12-4}$$

$$= \frac{\omega_0 L}{R} \tag{12-5}$$

Q_o also determines the *frequency selectivity* of a resonant circuit. As Q_o is increased by decreasing the resistance or increasing the inductance, the impedance increases more rapidly as the frequency is varied above or below resonant frequency. As Q_o is decreased by increasing the resistance or decreasing the inductance, the impedance increases less rapidly as the frequency is varied above or below resonant frequency.

Current is maximum at resonance and decreases as impedance increases. The voltages developed across the components of a series-*LCR* circuit vary with the current. A signal voltage applied to a series-*LCR* circuit at resonant frequency develops maximum voltage across the capacitor. Signals applied to the series-*LCR* circuit at other frequencies develop less voltage across the capacitor. The higher Q_o, the more rapidly the voltage across the capacitor decreases. The lower Q_o, the slower the drop-off of voltage across the capacitor. The larger Q_o, the narrower the band of frequencies that cause significant voltage across the capacitor. The smaller Q_o, the wider the band of frequencies that cause significant voltage across the capacitor. On this basis, a series-*LCR* circuit selects a band of frequencies, and Q_o is indicative of the width of the frequency band that will be selected.

Figure 12-8 shows comparative frequency-response curves for three values of Q_o.

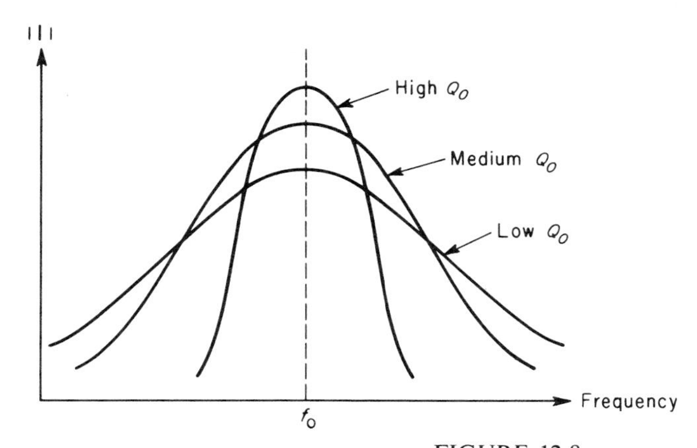

FIGURE 12-8
Frequency reponse for three values of Q_0.

EXAMPLE 12-6 Calculate Q_o for the conditions of Example 12-5.

SOLUTION The resonant frequency was calculated in Example 12-5 to be 200 kHz.

1 $\omega_0 L = 2\pi f_o L = (6.28)(200 \times 10^3)(1.91 \times 10^{-3}) = 2{,}400\ \Omega$

$$Q_o = \frac{\omega_0 L}{R} = \frac{2{,}400}{40} = 60$$

2 $\omega_0 L = (6.28)(200 \times 10^3)(0.957 \times 10^{-3}) = 1{,}200\ \Omega$

$$Q_o = \frac{\omega_0 L}{R} = \frac{1{,}200}{40} = 30$$

3 $\omega_0 L = (6.28)(200 \times 10^3)(0.319 \times 10^{-3}) = 400\ \Omega$

$$Q_o = \frac{\omega_0 L}{R} = \frac{400}{40} = 10$$

////

The voltage across the capacitance at resonance can be derived in terms of Q_o. In Fig. 12-9, at resonance,

$$\mathbf{I} = \frac{\mathbf{E}}{R}$$

$$\mathbf{E}_C = -jX_C\mathbf{I} = -j\frac{X_C}{R}\mathbf{E}$$

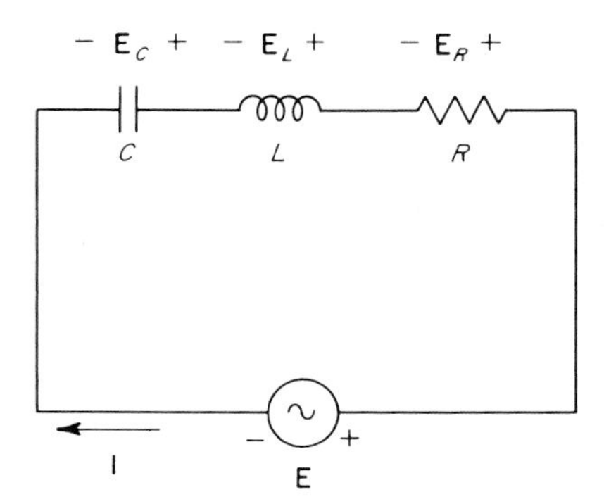

FIGURE 12-9
Circuit for Example 12-6.

Since $X_L = X_C$ at resonance,

$$Q_o = \frac{X_L}{R} = \frac{X_C}{R}$$

Therefore,

$$\mathbf{E}_C = -jQ_o\,\mathbf{E} \tag{12-6}$$

Equation (12-6) indicates a voltage-gain factor of Q_o for $\mathbf{E}_C$ at resonance. Example 12-7 demonstrates voltage gain in a series-resonant circuit.

The reader will note that the magnitude of the voltage drop across the inductance equals the drop across the capacitance. *In all series-resonant circuits, the magnitude of the voltage drop across either is Q_o times the applied voltage.*

EXAMPLE 12-7 Calculate $\mathbf{E}_C$ at resonance for the circuit elements given in Example 12-6. The voltage applied to the circuit of Fig. 12-6 is $40 \times 10^{-3} \angle 0°$ V.

SOLUTION

1 The current at resonance is

$$\mathbf{I} = \frac{\mathbf{E}}{R} = \frac{40 \times 10^{-3} \angle 0°}{40} = 1 \times 10^{-3} \angle 0° \text{ A}$$

When $C = 332$ pF,

$$X_C = \frac{1}{(6.28)(200 \times 10^3)(332 \times 10^{-12})} = 2{,}400\ \Omega$$

Therefore,

$$\mathbf{E}_C = -jX_C\mathbf{I} = (2.4 \times 10^3 \angle -90°)(1 \times 10^{-3}) = 2.4 \angle -90° \text{ V}$$

But $\mathbf{E}_C$ can be calculated by Eq. (12-6).

$$\mathbf{E}_C = -jQ_o\mathbf{E}$$

Q_o was calculated in Example 12-6 to be 60. Therefore,

$$\mathbf{E}_C = -j(60)(40 \times 10^{-3} \angle 0°) = 2.4 \angle -90° \text{ V}$$

2 When $C = 663$ pF,

$$X_C = \frac{1}{(6.28)(200 \times 10^3)(663 \times 10^{-12})} = 1{,}200\ \Omega$$

$$\mathbf{E}_C = -jX_C\mathbf{I} = (1.2 \times 10^3 \angle -90°)(1 \times 10^{-3}) = 1.2 \angle -90° \text{ V}$$

By Eq. (12-6),

$$\mathbf{E}_C = -jQ_o\mathbf{E} = 1.2 \angle -90° \text{ V}$$

3 When $C = 1{,}900$ pF,

$$X_C = \frac{1}{(6.28)(200 \times 10)(1{,}990 \times 10^{-12})} = 400\ \Omega$$

$$\mathbf{E}_C = -jX_C\mathbf{I} = (0.4 \times 10^3) \angle -90°(1 \times 10^{-3}) = 0.4 \angle -90° \text{ V}$$

By Eq. (12-6),

$$\mathbf{E}_C = -jQ_o\mathbf{E} = 0.4 \angle -90° \text{ V}$$ ////

12-4 BANDWIDTH

The frequency-response curve of the series-resonant circuit of Fig. 12-9 shows maximum current at resonance. Maximum power is delivered by the source at resonant frequency. As current drops off on either side of resonant frequency, there are frequencies above and below resonant frequency at which the current is 0.707 times the current magnitude at resonance. The frequency above resonant frequency at which the current $I = 0.707I_{\text{max}}$ is shown as f_2 in Fig. 12-10. The frequency below resonant frequency at which the current $I = 0.707I_{\text{max}}$ is shown as f_1. Recall that $I = |\mathbf{I}|$.

The frequencies at which current is $0.707I_{\text{max}}$ are termed the *half-power points.* Thus f_1 is the *lower half-power point*, and f_2 is the *upper half-power point*. The half-power relationship is derived as follows:

$$\begin{aligned} P_{\text{max}} &= (I_{\text{max}})^2 R \\ P_{f_1} &= (0.707I_{\text{max}})^2 R \\ &= 0.500(I_{\text{max}})^2 R \\ &= 0.5P_{\text{max}} \end{aligned}$$

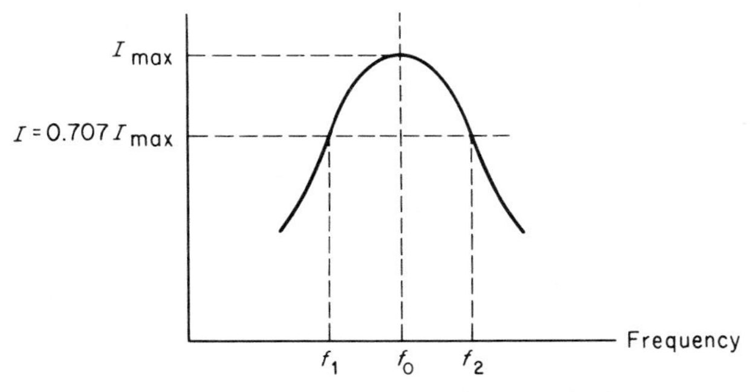

FIGURE 12-10
Half-power points of a series-resonant circuit.

The frequency span between the lower half-power point f_1 and the upper half-power point f_2 is defined as the *bandwidth*. The bandwidth is given by Eq. (12-7).

$$\text{BW} = f_2 - f_1 \tag{12-7}$$

The bandwidth can be derived as a function of circuit constants as follows.

The current at f_1 is I_1, and the current at f_2 is I_2. It must be noted that in this derivation magnitudes only are dealt with.

The impedance at the half-power points must be 1.414 times as great as the impedance at resonance, since the current at these points is 0.707 of current at resonance. Since the impedance at resonance is R, it follows that

$$1.414R = |\mathbf{Z}_1| = |\mathbf{Z}_2|$$

$$\mathbf{Z} = R + j\omega L - j\frac{1}{\omega c} = R + j\left(\omega L - \frac{1}{\omega C}\right)$$

$$|\mathbf{Z}| = \sqrt{R^2 + \left(\omega L - \frac{1}{\omega C}\right)^2}$$

$$|\mathbf{Z}|^2 = R^2 + \left(\omega L - \frac{1}{\omega C}\right)^2$$

At f_1 and f_2,

$$|\mathbf{Z}|^2 = (1.414\ R)^2 = 2\ R^2$$

$$2R^2 = R^2 + \left(\omega_x L - \frac{1}{\omega_x C}\right)^2$$

$$R = \pm\left(\omega_x L - \frac{1}{\omega_x C}\right) \tag{12-8}$$

where

$$\omega_x = 2\pi f_1 \quad \text{or} \quad \omega_x = 2\pi f_2$$

Rearranging Eq. (12-8) and using poth positive and negative values of the reactance yields Eqs. (12-9) and (12-10).

$$\omega_x^2 LC - \omega_x RC - 1 = 0 \tag{12-9}$$

$$\omega_x^2 LC + \omega_x RC - 1 = 0 \tag{12-10}$$

Solving Eq. (12-9) for ω_x,

$$\omega_x = \frac{RC \pm \sqrt{R^2C^2 + 4LC}}{2LC}$$

$$= \frac{R}{2L} \pm \sqrt{\frac{R^2}{4L^2} + \frac{1}{LC}} \tag{12-11}$$

Solving Eq. (12-10) for ω_x,

$$\omega_x = \frac{-RC \pm \sqrt{R^2C^2 + 4LC}}{2LC}$$

$$= \frac{-R}{2L} \pm \sqrt{\frac{R^2}{4L^2} + \frac{1}{LC}} \tag{12-12}$$

Since

$$\frac{1}{LC} \gg \frac{R^2}{4L^2} \quad \text{because} \quad \frac{1}{C} \gg \frac{R^2}{4L}$$

we have

$$\omega_x = \frac{R}{2L} \pm \sqrt{\frac{1}{LC}} \tag{12-13}$$

and

$$\omega_x = -\frac{R}{2L} \pm \sqrt{\frac{1}{LC}} \tag{12-14}$$

But

$$f_o = \frac{1}{2\pi\sqrt{LC}}$$

Therefore,

$$2\pi f_o = \frac{1}{\sqrt{LC}}$$

$$\omega_0 = \frac{1}{\sqrt{LC}} \tag{12-15}$$

Substitute Eq. (12-15) into Eqs. (12-13) and (12-14).

$$\omega_x = \frac{R}{2L} \pm \omega_0 \tag{12-16}$$

$$\omega_x = -\frac{R}{2L} \pm \omega_0 \tag{12-17}$$

With the positive value of ω_0, ω_x of Eq. (12-16) must be

$$2\pi f_2 = \frac{R}{2L} + 2\pi f_o \tag{12-18}$$

and ω_x of Eq. (12-17) must be

$$2\pi f_1 = -\frac{R}{2L} + 2\pi f_o \tag{12-19}$$

Subtracting Eq. (12-19) from Eq. (12-18),

$$2\pi f_2 - 2\pi f_1 = \frac{R}{2L} + \frac{R}{2L}$$

$$2\pi(f_2 - f_1) = \frac{R}{L}$$

$$f_2 - f_1 = \frac{R}{2\pi L}$$

$$\text{BW} = \frac{R}{2\pi L} \tag{12-20}$$

The bandwidth can be given in terms of resonant frequency and Q_o. From Eq. (12-20), and multiplying by 1 as f_o/f_o

$$\text{BW} = \frac{R}{2\pi L}\frac{f_o}{f_o}$$

$$= \frac{R}{\omega_0 L} f_o$$

$$= \frac{f_o}{Q_o} \tag{12-21}$$

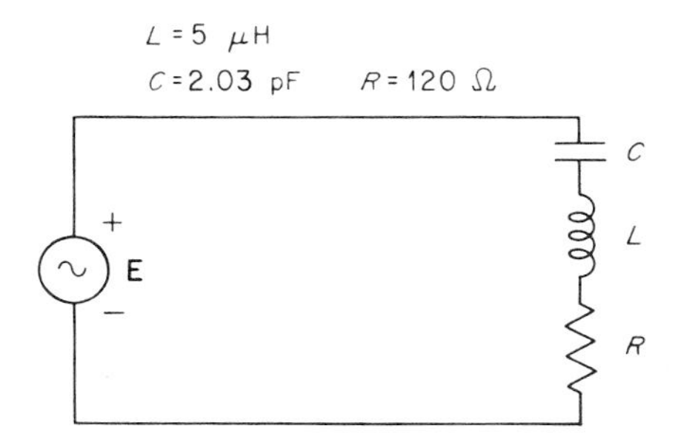

FIGURE 12-11
Series-LCR circuit for Example 12-8.

EXAMPLE 12-8 Calculate the bandwidth at resonance for the circuit of Fig. 12-11.

SOLUTION

$$f_o = \frac{1}{2\pi\sqrt{LC}} = \frac{1}{6.28\sqrt{5 \times 10^{-6} \times 2.03 \times 10^{-12}}} = 50 \text{ MHz}$$

$$X_L = 2\pi f_o L = (6.28)(50 \times 10^6)(5 \times 10^{-6}) = 1{,}570 \ \Omega$$

$$Q_o = \frac{X_L}{R} = \frac{1{,}570}{120} = 13.1$$

$$\text{BW} = \frac{f_o}{Q_o} = \frac{50 \times 10^6}{13.1} = 3.82 \times 10^6 = 3.82 \text{ MHz}$$

////

The change in circuit current between $I_{\max}$ at f_o and $0.707I_{\max}$ at f_1 or f_2 is commonly expressed in decibels.

$$\begin{aligned} \text{dB} &= 20 \log \frac{I \text{ at } f_1 \text{ or } f_2}{I \text{ at } f_o} \\ &= 20 \log \frac{0.707 I_{\max}}{I_{\max}} = 20 \log 0.707 \\ &= -3 \end{aligned}$$

Thus at either half-power frequency, the signal magnitude is 3 dB less than that at resonance.

12-5 THE IDEAL PARALLEL-RESONANT CIRCUIT

Figure 12-12 shows pure capacitance in parallel with pure inductance. According to general definition, the parallel circuit of Fig. 12-12 appears resistive when $\mathbf{I}_C$ is equal in magnitude, to $\mathbf{I}_L$. For $\mathbf{I}_L$ and $\mathbf{I}_C$ to be equal, the reactances must be equal. The

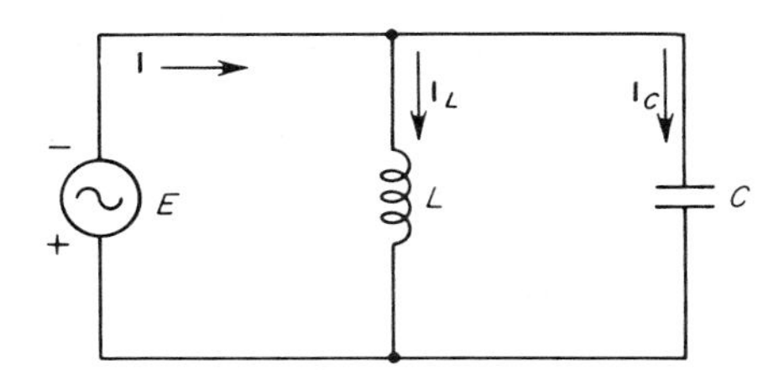

FIGURE 12-12
The ideal parallel-resonant circuit.

parallel-resonant frequency is sometimes referred to as the *antiresonant frequency*, and f_o is used as before to indicate the resonant frequency.

For the ideal circuit of Fig. 12-12, the resonant frequency is

$$f_o = \frac{1}{2\pi\sqrt{LC}} \tag{12-22}$$

Converting the reactances to susceptances at parallel resonance,

$$B_L = B_C \tag{12-23}$$

where

$$B_L = \frac{1}{\omega L} \quad \text{and} \quad B_C = \omega C$$

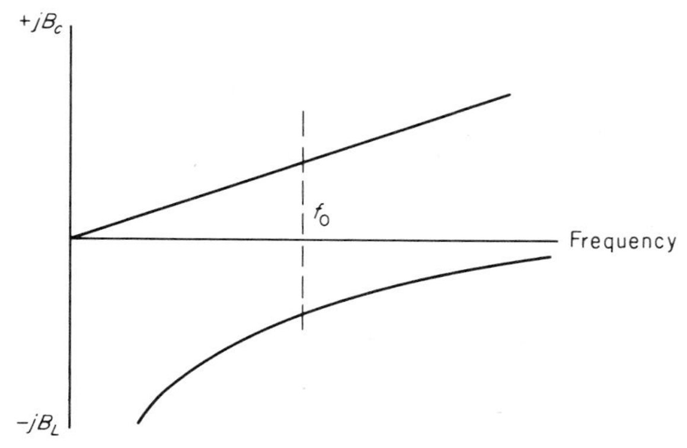

FIGURE 12-13
B_L and B_C plotted versus frequency.

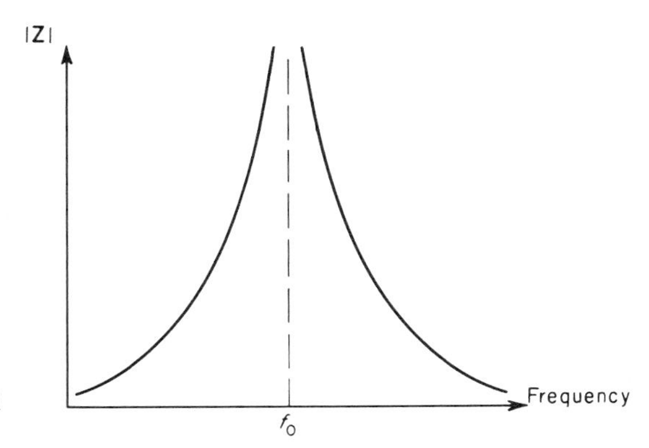

FIGURE 12-14
Frequency versus impedance in the ideal parallel resonant circuit.

B_L and B_C are plotted versus frequency in Fig. 12-13. As shown, at frequencies below resonance the net susceptance is inductive susceptance, and the parallel circuit appears inductive.

At frequencies above resonance, the net susceptance is capacitive susceptance, and the parallel circuit appears capacitive. At f_o, the inductive susceptance and capacitive susceptance are equal in magnitude. Therefore,

$$\mathbf{Z}\Big|_{f=f_o} = \frac{1}{jB_C - jB_L} = \frac{1}{0} = \infty$$

The impedance magnitude of the ideal parallel circuit is plotted versus frequency in Fig. 12-14. Also, the total current magnitude versus frequency is shown in Fig. 12-15.

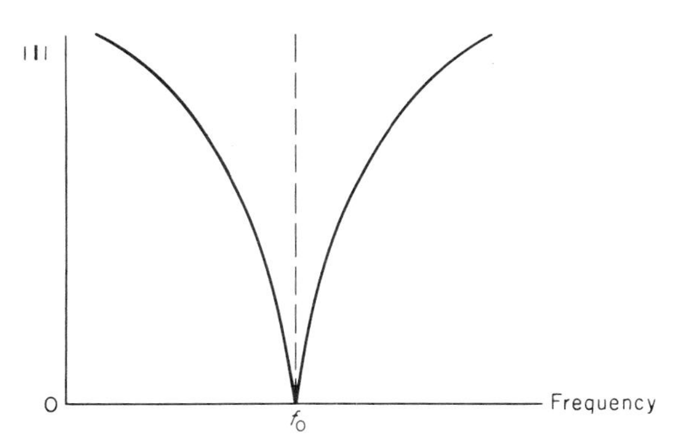

FIGURE 12-15
Total current magnitude versus frequency for the ideal parallel circuit.

12-6 RESONANCE OF A PRACTICAL PARALLEL-LCR CIRCUIT

It is impossible to wind a coil that does not have some resistance. The ideal circuit of Fig. 12-12 is converted to the more practical circuit of Fig. 12-16.

The conditions under which the circuit of Fig. 12-16 appears purely resistive may now be derived in general.

$$\mathbf{Z}_t = \frac{(R + jX_L)(-jX_C)}{R + j(X_L - X_C)} \tag{12-24}$$

$$= \frac{X_L X_C - jRX_C}{R + j(X_L - X_C)} \frac{R - j(X_L - X_C)}{R - j(X_L - X_C)}$$

$$= \frac{RX_L X_C - jR^2 X_C - jX_L X_C(X_L - X_C) - RX_C(X_L - X_C)}{R^2 + (X_L - X_C)^2}$$

$$= \frac{RX_C{}^2}{R^2 + (X_L - X_C)^2} + \frac{-j(R^2 X_C + X_L{}^2 X_C - X_L X_C{}^2)}{R^2 + (X_L - X_C)^2} \tag{12-25}$$

For $\mathbf{Z}_t$ to be purely resistive, the numerator of the right-hand term of Eq. (12-25) must be zero.

$$R^2 X_C + X_L{}^2 X_C - X_L X_C{}^2 = 0$$

$$\frac{R^2}{2\pi fC} + \frac{2\pi fL^2}{C} - \frac{L}{2\pi fC^2} = 0 \tag{12-26}$$

Solving Eq. (12-26) for frequency, multiplying both sides by fC

$$\frac{R^2}{2\pi} + 2\pi f^2 L^2 - \frac{L}{2\pi C} = 0$$

$$2\pi f^2 L^2 = \frac{L}{2\pi C} - \frac{R^2}{2\pi}$$

$$f^2 = \frac{L}{4\pi^2 L^2 C} - \frac{R^2}{4\pi^2 L^2}$$

$$f = \sqrt{\frac{1}{4\pi^2 L^2}\left(\frac{L}{C} - R^2\right)}$$

$$f_o = \frac{1}{2\pi L}\sqrt{\frac{L}{C} - R^2} \tag{12-27}$$

Equation (12-27) yields the frequency at which the circuit of Fig. 12-16 appears purely resistive. Notice that if R^2 is larger than L/C, the radical is negative, and

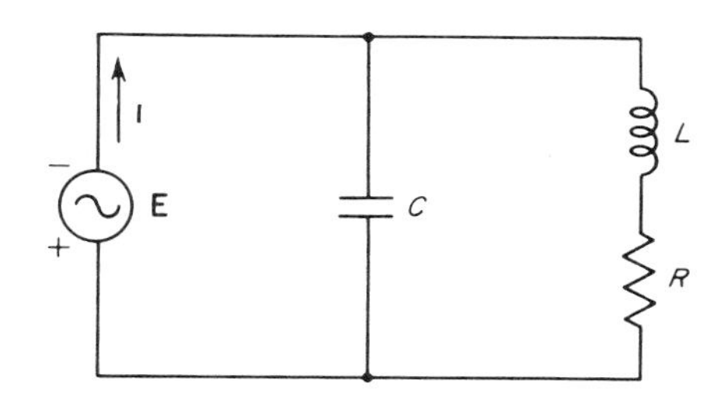

FIGURE 12-16
A practical parallel-LCR circuit.

parallel resonance does not occur. When L/C is much greater than R^2, Eq. (12-26) reduces to

$$f_o = \frac{1}{2\pi L}\sqrt{\frac{L}{C}}$$

$$= \frac{1}{2\pi\sqrt{LC}} \quad \text{for} \quad \frac{L}{C} \gg R^2 \qquad (12\text{-}28)$$

Equation (12-28) is identical to Eq. (12-22) for the ideal parallel circuit. When f_o is taken as in Eq. (12-28), Eq. (12-24) becomes

$$\mathbf{Z}_t = \frac{X_L X_C - jRX_C}{R}$$

Since the magnitude of the inductive reactance equals the magnitude of the capacitive reactance,

$$\mathbf{Z}_t = \frac{X_L X_L}{R} - jX_L \qquad (12\text{-}29)$$

but

$$\frac{X_L}{R} = Q_o$$

Therefore,

$$\mathbf{Z}_t = X_L Q_o - jX_L \qquad (12\text{-}30)$$

If Q_o is greater than 10, for all practical purposes

$$\mathbf{Z}_t = X_L Q_o$$

$$\mathbf{Z}_t = \omega L Q_o \qquad (12\text{-}31)$$

Since $\mathbf{Z}_t$ is resistive,

$$R_p = \omega L Q_o \qquad (12\text{-}31a)$$

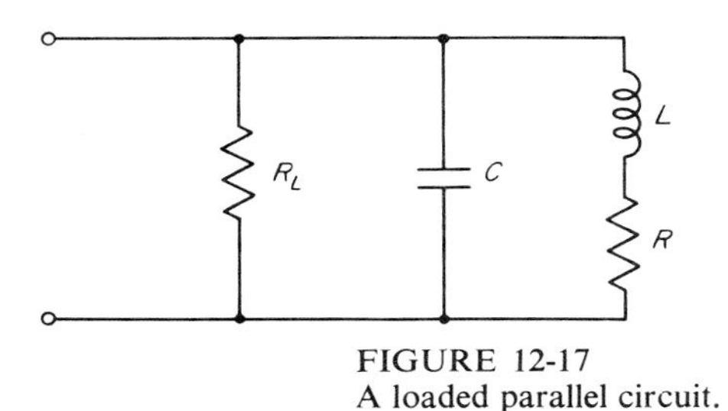

FIGURE 12-17
A loaded parallel circuit.

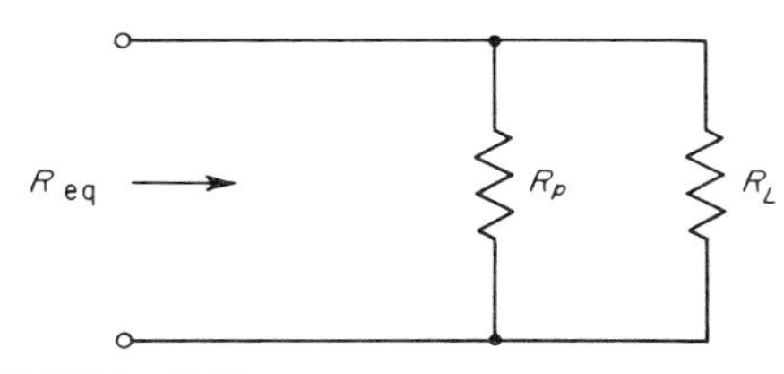

FIGURE 12-18
The effective circuit for Fig. 12-17 at resonance.

In Eq. (12-31*a*), R_p is the effective parallel resistance of the circuit of Fig. 12-16 at resonance. If a load resistance is connected across the parallel circuit of Fig. 12-16, the circuit of Fig. 12-17 results.

In Fig. 12-18, R_{eq} is

$$R_{eq} = \frac{R_L \times R_p}{R_L + R_p}$$

A new Q results from the conditions in Fig. 12-18, where the new Q (Q_o') is the effective Q of the resonant circuit under load.

$$R_{eq} = \omega L Q_o' \qquad (12\text{-}32)$$

As has been shown for the ideal parallel circuit, the maximum impedance condition occurs at the resonant frequency and is equal to R_p or R_{eq} depending upon the circuit. Above resonance, the parallel circuit appears capacitive, and below resonance the parallel circuit appears inductive. Figure 12-19 shows the frequency response for the magnitude of the impedance and the phase angle of the impedance.

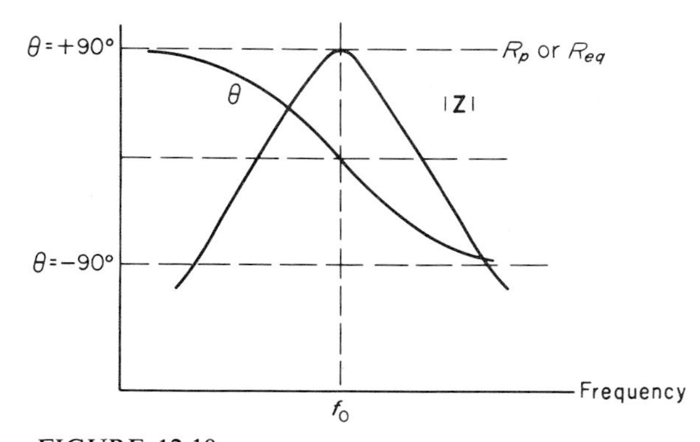

FIGURE 12-19
Graph of magnitude and phase angle of the impedance of a parallel circuit.

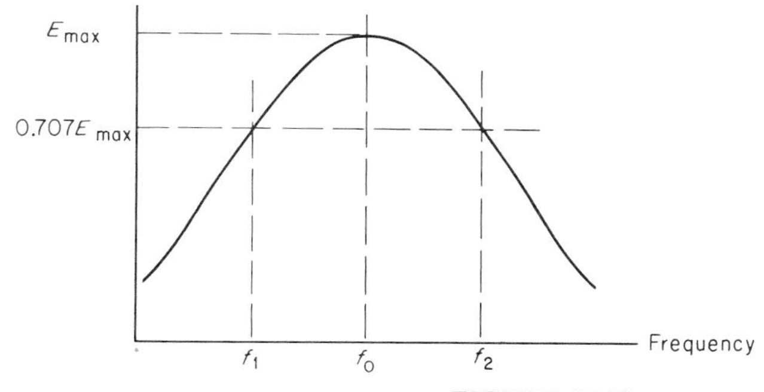

FIGURE 12-20
Half-power points of a parallel-*LCR* circuit.

If a constant current source is applied to the parallel circuit, the voltage drop across the circuit versus frequency is shown in Fig. 12-20. Recall that $E = |\mathbf{E}|$.

The half-power (-3 DB) points are shown in Fig. 12-20 as they were for the series-resonant circuit. The bandwidth equations are applied to the parallel-resonant circuit as they were applied to the series-resonant circuit.

The effect of load resistance R_L in Fig. 12-17 is to reduce Q_o to a new value Q_o' and thus increase the bandwidth.

EXAMPLE 12-9 Calculate Q_o and the resonant frequency. Use Eqs. (12-27) and (12-28) to calculate f_o for Fig. 12-21 when (1) $R = 10\ \Omega$, (2) $R = 50\ \Omega$, (3) $R = 150\ \Omega$, and (4) $R = 200\ \Omega$.

SOLUTION

1 By Eq. (12-28),

$$f_o = \frac{1}{2\pi\sqrt{LC}}$$

$$= \frac{1}{6.28\sqrt{(0.373 \times 10^{-3})(0.00114 \times 10^{-6})}} = 245\ \text{kHz}$$

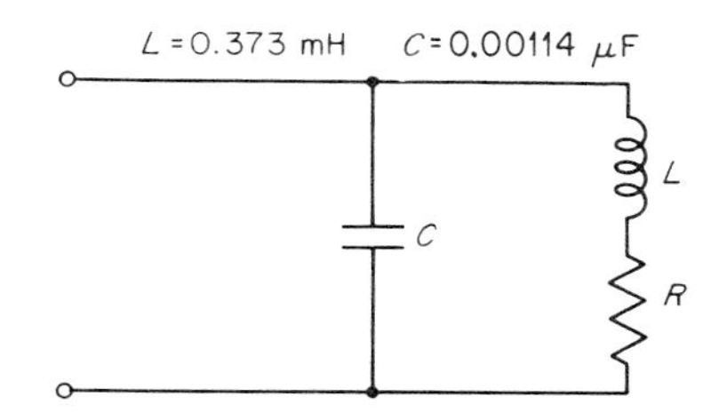

FIGURE 12-21
Circuit for Example 12-9.

By Eq. (12-27),

$$f_o = \frac{1}{2\pi L}\sqrt{\frac{L}{C} - R^2}$$

$$= \frac{1}{(6.28)(0.373 \times 10^{-3})}\sqrt{\frac{0.373 \times 10^{-3}}{0.00114 \times 10^{-6}} - (10)^2}$$

$$= (0.428 \times 10^3)\sqrt{3.27 \times 10^5 - 100}$$

For all practical purposes,

$$f_o = (0.428 \times 10^3)\sqrt{3.27 \times 10^5} = 245 \text{ kHz}$$

$$Q_o = \frac{X_L}{R} = \frac{(6.28)(245 \times 10^3)(0.373 \times 10^{-3})}{10} = 57.4$$

2 The resonant frequency from Eq. (12-28) will be unchanged.

$$f_o = \frac{1}{2\pi L}\sqrt{\frac{L}{C} - R^2}$$

$$= (0.428 \times 10)^3\sqrt{327 \times 10^3 - 50^2} = (0.428 \times 10^3)\sqrt{327 \times 10^3 - 2.5 \times 10^3}$$

$$= (0.428 \times 10^3)\sqrt{324.5 \times 10^3} = 244 \text{ kHz}$$

$$Q_o = \frac{(6.28)(244 \times 10^3)(0.373 \times 10^{-3})}{50} = 11.43$$

3

$$f_o = \frac{1}{2\pi L}\sqrt{\frac{L}{C} - R^2}$$

$$= (0.428 \times 10^3)\sqrt{32.7 \times 10^4 - (150)^2}$$

$$= (0.428 \times 10^3)\sqrt{32.7 \times 10^4 - 2.25 \times 10^4}$$

$$= (0.428 \times 10^3)\sqrt{30.45 \times 10^4} = 235 \text{ kHz}$$

$$Q_o = \frac{(6.28)(235 \times 10^3)(0.373 \times 10^{-3})}{150} = 3.7$$

4

$$f_o = \frac{1}{2\pi L}\sqrt{\frac{L}{C} - R^2}$$

$$= (0.428 \times 10^3)\sqrt{32.7 \times 10^4 - (200)^2}$$

$$= (0.428 \times 10^3)\sqrt{32.7 \times 10^4 - 4 \times 10^4)}$$

$$= (0.428 \times 10^3)\sqrt{28.7 \times 10^4} = 229 \text{ kHz}$$

$$Q_o = \frac{(6.28)(229 \times 10^3)(0.373 \times 10^{-3})}{200} = 2.7$$

////

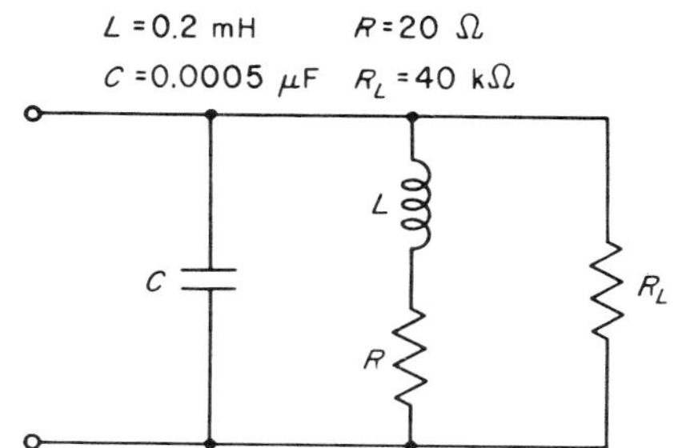

FIGURE 12-22
Circuit for Example 12-10.

Example 12-9 demonstrates two important points: (1) As resistance is increased so that Q_o becomes less than 10, Eq. (12-28) does not yield the true parallel-resonant frequency. (2) The parallel-resonant frequency is affected by resistance.

EXAMPLE 12-10 Calculate the parallel-resonant frequency, Q_o, and the bandwidth of the circuit of Fig. 12-22. Calculate the effective Q and the bandwidth if a 40,000 Ω resistor is shunted across the circuit.

SOLUTION

Without R_L,

$$f_o = \frac{1}{2\pi L}\sqrt{\frac{L}{C} - R^2}$$

$$= \frac{1}{(6.28)(2 \times 10^{-4})}\sqrt{\frac{0.2 \times 10^{-3}}{5 \times 10^{-10}} - 20^2}$$

$$= (0.0794 \times 10^4)\sqrt{4 \times 10^5 - 400} = 504 \text{ kHz}$$

$$Q_o = \frac{2\pi f L}{R} = \frac{(6.28)(504 \times 10^3)(0.2 \times 10^{-3})}{20} = 31.6$$

$$BW = \frac{f_o}{Q_o} = \frac{504 \times 10^3}{31.6} = 15.9 \text{ kHz}$$

At resonance, when $Q_o \geq 10$,

$$\mathbf{Z}_t = R_p = Q_o X_L = (31.6)(628) = 19.9 \text{ k}\Omega$$

When R_L is shunted across the parallel-resonant circuit,

$$R_{eq} = \frac{R_L R_p}{R_L + R_p} = \frac{(40 \times 10^3)(19.9 \times 10^3)}{59.9 \times 10^3} = 13.3 \text{ k}\Omega$$

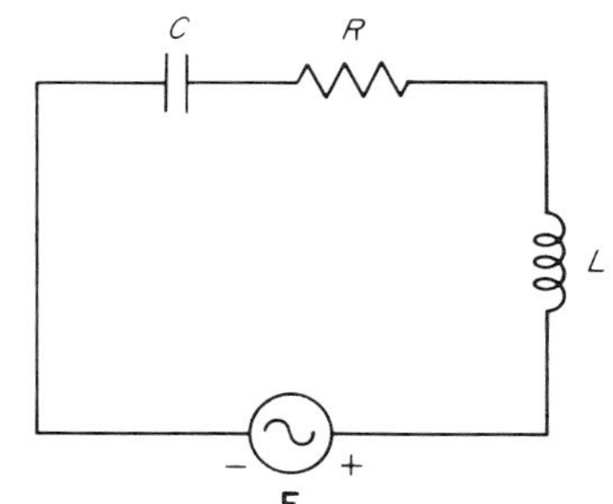

FIGURE 12-23
Circuit for Probs. 12-1 to 12-17.

Therefore,

$$R_{eq} = \omega L Q'_o$$

$$Q'_o = \frac{R_{eq}}{X_L} = \frac{13.3 \times 10^3}{628} = 21.2$$

$$BW = \frac{f_o}{Q'_o} = \frac{504 \times 10^3}{21.2} = 23.8 \text{ kHz}$$

////

As shown by Example 12-10, the bandwidth of a parallel-resonant circuit is increased by a shunting resistance. Since a shunting resistance is a power-dissipating load, Example 12-10 demonstrates the effect of loading a parallel-resonant circuit.

PROBLEMS

12-1 In Fig. 12-23, $C = 0.005\ \mu\text{F}$, $L = 0.2$ mH, $R = 30\ \Omega$, and $\mathbf{E} = 2\angle 0°$ mV. Calculate (*a*) resonant frequency, (*b*) $\mathbf{E}_R$ at resonance, and (*c*) $\mathbf{E}_C$ at resonance.

12-2 Repeat Prob. 12-1 with the following components: $C = 0.002\ \mu\text{F}$, $L = 0.9$ mH, $R = 70\ \Omega$, and $\mathbf{E} = 2\angle 40°$ mV.

12-3 Repeat Prob. 12-1 with the following components: $C = 0.08\ \mu\text{F}$, $L = 23\ \mu\text{H}$, $R = 25\ \Omega$, and $\mathbf{E} = 4\angle 135°$ mV.

12-4 In Fig. 12-23, if inductance L is 0.05 mH and the resonant frequency is 275 kHz, calculate the required capacitance.

12-5 Repeat Prob. 12-4 when L is 2.4 mH and the resonant frequency is 120 kHz.

12-6 In Fig. 12-23, if the capacitance is 20 pF and the resonant frequency is 20 MHz, calculate the inductance.

12-7 In Fig. 12-23, if the capacitance is 50 pF and the resonant frequency is 20 MHz, calculate the inductance.

12-8 Calculate Q_o for Prob. 12-1.

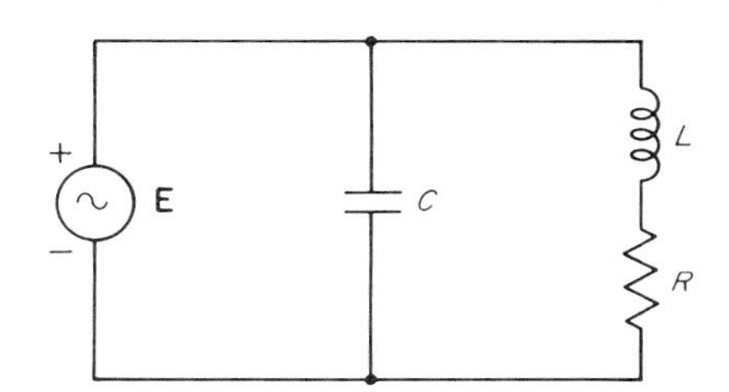

FIGURE 12-24
Circuit for Probs. 12-18 to 12-23.

12-9 Calculate Q_o for Prob. 12-2.

12-10 Calculate Q_o for Prob. 12-3.

12-11 A Q_o of 30 is desired for Prob. 12-4. Calculate the required value of R.

12-12 A Q_o of 15 is desired for Prob. 12-5. Calculate the required value of R.

12-13 A Q_o of 40 is desired for Prob. 12-6. Calculate the required value of R.

12-14 A Q_o of 90 is desired for Prob. 12-7. Calculate the required value of R.

12-15 For Prob. 12-1, calculate (*a*) bandwidth and (*b*) upper and lower half-power frequencies.

12-16 For Prob. 12-2, calculate (*a*) bandwidth and (*b*) upper and lower half-power frequencies.

12-17 For Prob. 12-3, calculate (*a*) bandwidth, and (*b*) upper and lower half-power frequencies.

12-18 Calculate the parallel-resonant frequency if one exists for the circuit of Fig. 12-24 by Eqs. (12-27) and (12-28) when $L = 0.01$ mH, $C = 0.0045\ \mu$F, and (*a*) $R = 10\ \Omega$, (*b*) $R = 15\ \Omega$, (*c*) $R = 50\ \Omega$, (*d*) $R = 200\ \Omega$, and (*e*) $R = 175\ \Omega$.

12-19 Calculate the parallel-resonant frequency of Fig. 12-24 by Eqs. (12-27) and (12-28) when $L = 0.3$ mH, $C = 0.001\ \mu$F, and (*a*) $R = 12\ \Omega$, (*b*) $R = 25\ \Omega$, (*c*) $R = 50\ \Omega$, and (*d*) $R = 220\ \Omega$.

12-20 In Fig. 12-24, $C = 0.0005\ \mu$F. If a bandwidth of 50 kHz is desired for a resonant frequency of 1.2 MHz, calculate the required R and L.

12-21 What resistance must be shunted across C of Fig. 12-24 to increase the bandwidth of Prob. 12.20 to 80 kHz?

12-22 In Fig. 12-24, $C = 0.008\ \mu$F. A bandwidth of 5 kHz is desired for a resonant frequency of 220 kHz. Calculate the required R and L.

12-23 What resistance must be shunted across C of Fig. 12-24 to increase the bandwidth of Prob. 12-22 to 9 kHz?

13

COUPLED CIRCUITS

13-1 INTRODUCTION

All circuits are coupled circuits in the sense that there is coupling between a source and a load. However, the term has a special meaning in electronics and electric power distribution. *Circuits are said to be coupled when there is mutual impedance between them.* This mutual impedance may take the form of pure resistance, any combination of resistance and reactance, mutual inductance, or a combination of mutual inductance and impedances. Because of the mutual impedance, any change in one of the circuits affects current and voltage in the other circuits. Figure 13-1 shows three circuits coupled to each other by mutual impedances.

The coupling network between circuits usually has two functions: (1) It acts as an impedance modifier to make a low impedance at one pair of terminals appear as a high impedance at some other pair of terminals or, conversely, to make a high impedance at one pair of terminals appear as a low impedance at another pair of terminals. (2) A coupling circuit may offer impedance characteristics that vary with frequency, so that the coupling circuit acts as a filter. The circuit of Fig. 13-2 is a coupling network used for impedance transformation. In this circuit, it is desired that

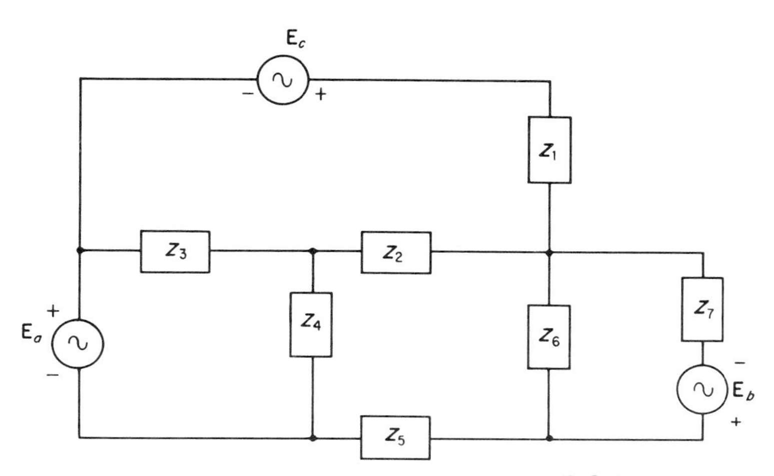

FIGURE 13-1
Three mutually coupled circuits.

resistor R_L appear to be equal to resistance R_1 to obtain maximum power transfer to terminals *a-a*. The coupling network "reflects" resistance R_L, modified by the factor k, back into the input circuit so that kR_L equals R_1.

13-2 THE TWO-PORT NETWORK

Consider the circuit of Fig. 13-3. The word *port* represents a terminal pair. A set of general loop equations will be developed for any two-port network, regardless of the components within the terminals of the network. Terminal currents are assumed to flow into the network. Terminal voltages are assumed negative at the upper terminal with respect to the lower terminal.

The loop equations at the input and output are commonly referred to as *open-circuit impedance equations.* z_{11}, z_{12}, z_{21}, and z_{22} are the *open-circuit impedance parameters:*

$$\mathbf{E}_1 = z_{11}\mathbf{I}_1 + z_{12}\mathbf{I}_2 \tag{13-1}$$

$$\mathbf{E}_2 = z_{21}\mathbf{I}_1 + z_{22}\mathbf{I}_2 \tag{13-2}$$

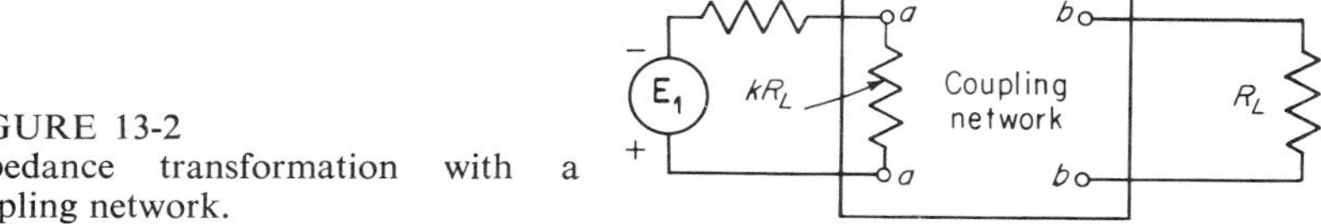

FIGURE 13-2
Impedance transformation with a coupling network.

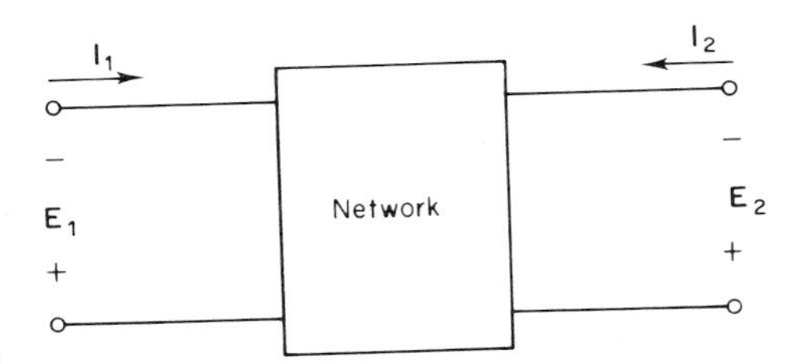

FIGURE 13-3
Two-port network.

where z_{11} is the input impedance with output open-circuited

$$z_{11} = \frac{\mathbf{E}_1}{\mathbf{I}_1} \qquad (\mathbf{I}_2 = 0)$$

where z_{12} is the reverse-transfer impedance with input open-circuited

$$z_{12} = \frac{\mathbf{E}_1}{\mathbf{I}_2} \qquad (\mathbf{I}_1 = 0)$$

where z_{21} is the forward-transfer impedance with output open-circuited

$$z_{21} = \frac{\mathbf{E}_2}{\mathbf{I}_1} \qquad (\mathbf{I}_2 = 0)$$

and where z_{22} is the output impedance with input open-circuited

$$z_{22} = \frac{\mathbf{E}_2}{\mathbf{I}_2} \qquad (\mathbf{I}_1 = 0)$$

Figure 13-4 illustrates the various setups required to obtain the impedance parameters z_{11}, z_{12}, z_{21}, and z_{22}.

Networks may be divided into two general groups, passive and active. An *active network* is one with internal sources of energy, such as amplifying tubes or transistors. One identifying feature of an active network is that z_{12} differs from z_{21}. A *passive network* is any combination of circuit elements, such as resistors, capacitors, and inductors. In a passive network, z_{12} always equals z_{21} if no initial voltages or currents exist. This follows directly from the reciprocity theorem.

To determine the reflected impedance at the input terminals, the first step is to find the driving-point impedance seen by source $\mathbf{E}_1$. The circuit of Fig. 13-3 is terminated in the output with load impedance $\mathbf{Z}_L$ as shown in Fig. 13-5.

The driving-point impedance is

$$\mathbf{Z}_{11} = \frac{\mathbf{E}_1}{\mathbf{I}_1} \tag{13-3}$$

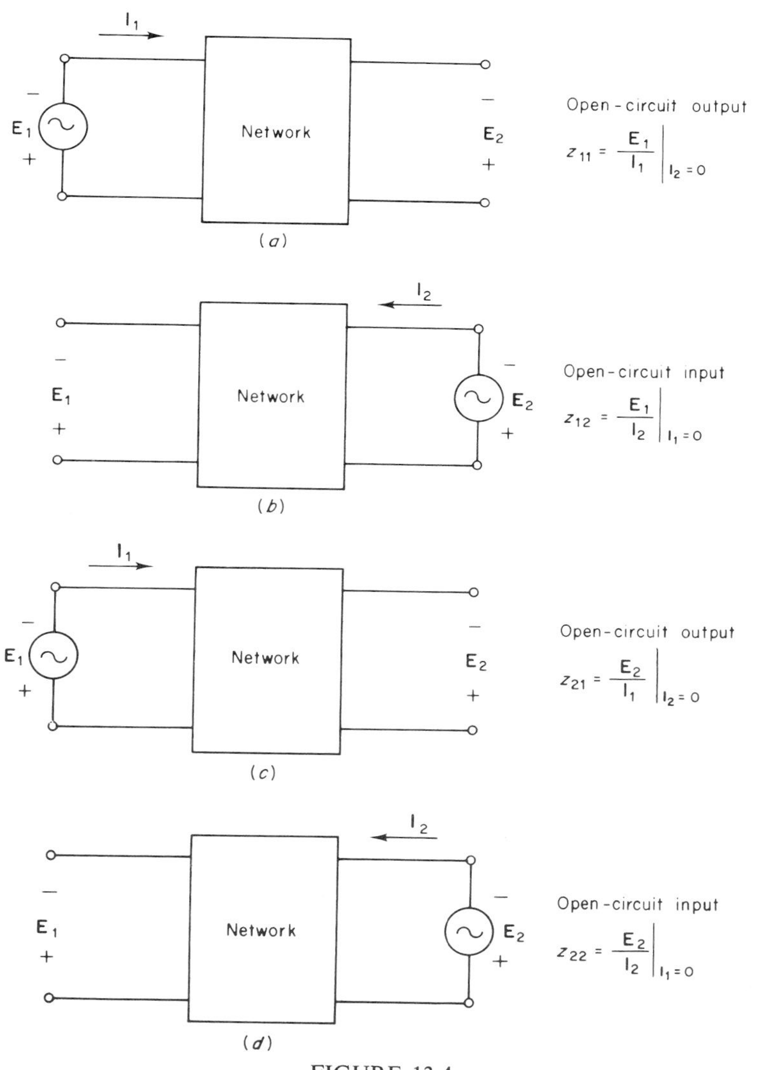

FIGURE 13-4
Test setup to determine (*a*) z_{11}, (*b*) z_{12}, (*c*) z_{21}, (*d*) z_{22}.

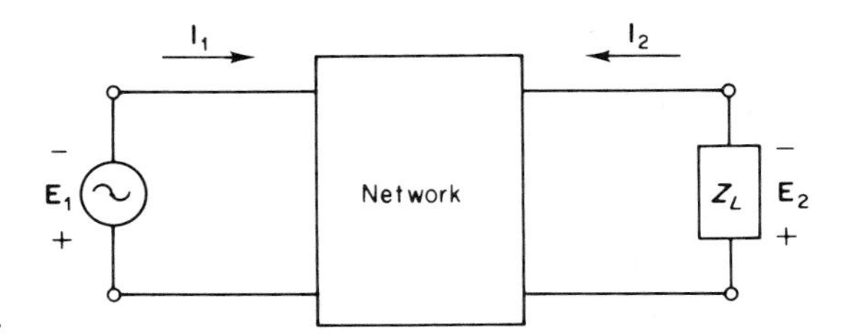

FIGURE 13-5
Two-port terminated in $\mathbf{Z}_L$.

for a specified passive load termination $\mathbf{Z}_L$. Equation (13-3) is not equivalent to the impedance parameter z_{11}. Recall that z_{11} is defined under the condition of an open-circuit output.

The output voltage is equal to

$$\mathbf{E}_2 = -\mathbf{Z}_L \mathbf{I}_2 \tag{13-4}$$

Rewriting Eq. (13-1) and substituting Eq. (13-4) into Eq. (13-2) yields the input and output impedance equations.

$$\mathbf{E}_1 = z_{11}\mathbf{I}_1 + z_{12}\mathbf{I}_2 \tag{13-5}$$

$$0 = z_{21}\mathbf{I}_1 + (z_{22} + \mathbf{Z}_L)\mathbf{I}_2 \tag{13-6}$$

Solving Eqs. (13-5) and (13-6) for $\mathbf{I}_1$ gives

$$\mathbf{I}_1 = \frac{\begin{vmatrix} \mathbf{E}_1 & z_{12} \\ 0 & z_{22} + \mathbf{Z}_L \end{vmatrix}}{\begin{vmatrix} z_{11} & z_{12} \\ z_{21} & z_{22} + \mathbf{Z}_L \end{vmatrix}} = \frac{\mathbf{E}_1(z_{22} + \mathbf{Z}_L)}{z_{11}(z_{22} + \mathbf{Z}_L) - z_{12}z_{21}}$$

Since $z_{12} = z_{21}$, let $z_M = z_{12} = z_{21}$ represent the mutual impedance between input and ouput. Thus, the driving-point impedance is

$$\mathbf{Z}_{11} = \frac{\mathbf{E}_1}{\mathbf{I}_1}$$

$$= z_{11} - \frac{z_M^2}{z_{22} + \mathbf{Z}_L} \tag{13-7}$$

The reflected impedance is the difference between the input impedance with no load, z_{11}, and the driving-point impedance with load, $\mathbf{Z}_{11}$.

$$\mathbf{Z}_{\text{ref}} = \frac{z_M^2}{z_{22} + \mathbf{Z}_L} \tag{13-8}$$

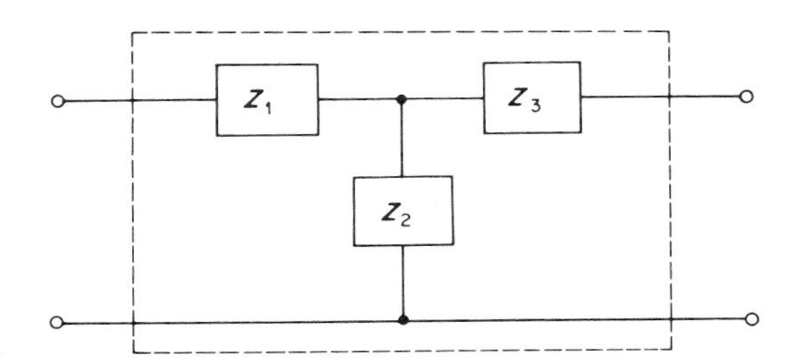

FIGURE 13-6
T network as a two-port.

If the two-port network is a T network, as shown in Fig. 13-6, determination of the z parameters is straightforward. Using the definitions of z_{11}, z_{12}, z_{21}, and z_{22}, by inspection

$$z_{11} = \mathbf{Z}_1 + \mathbf{Z}_2$$
$$z_M = \mathbf{Z}_2$$
$$z_{22} = \mathbf{Z}_2 + \mathbf{Z}_3$$

Arriving at the z parameters for a pi network in a two-port as shown in Fig. 13-7 is not as simple as for the T network.

$$z_{11} = \mathbf{Z}_1 \parallel (\mathbf{Z}_2 + \mathbf{Z}_3)$$
$$= \frac{\mathbf{Z}_1(\mathbf{Z}_2 + \mathbf{Z}_3)}{\mathbf{Z}_1 + \mathbf{Z}_2 + \mathbf{Z}_3}$$

z_M can be found with the aid of the circuit in Fig. 13-8.

$$z_{12} = z_M = \left.\frac{\mathbf{E}_1}{\mathbf{I}_2}\right|_{\mathbf{I}_1 = 0}$$

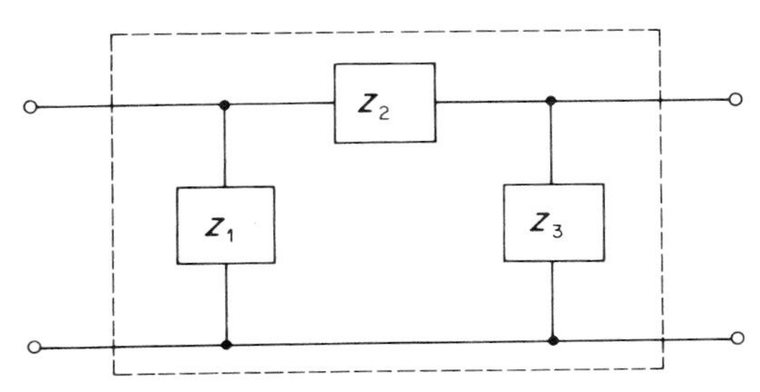

FIGURE 13-7
Pi network as a two-port.

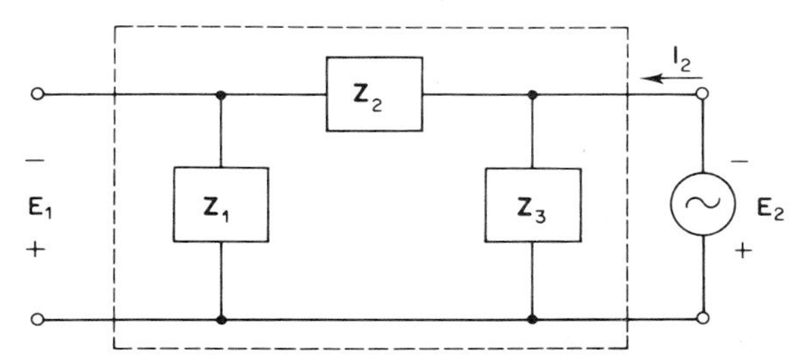

FIGURE 13-8
Condition to find z_{12}.

Using the VDR for $\mathbf{E}_1$,

$$\mathbf{E}_1 = \mathbf{E}_2 \frac{\mathbf{Z}_1}{\mathbf{Z}_1 + \mathbf{Z}_2}$$

$$\mathbf{I}_2 = \frac{\mathbf{E}_2}{\dfrac{\mathbf{Z}_3(\mathbf{Z}_1 + \mathbf{Z}_2)}{\mathbf{Z}_1 + \mathbf{Z}_2 + \mathbf{Z}_3}} = \frac{\mathbf{E}_2(\mathbf{Z}_1 + \mathbf{Z}_2 + \mathbf{Z}_3)}{\mathbf{Z}_3(\mathbf{Z}_1 + \mathbf{Z}_2)}$$

$$z_M = \frac{\mathbf{Z}_1\mathbf{Z}_3}{\mathbf{Z}_1 + \mathbf{Z}_2 + \mathbf{Z}_3}$$

By inspection,

$$z_{22} = \mathbf{Z}_3 \parallel (\mathbf{Z}_1 + \mathbf{Z}_2)$$

$$= \frac{\mathbf{Z}_3(\mathbf{Z}_1 + \mathbf{Z}_2)}{\mathbf{Z}_1 + \mathbf{Z}_2 + \mathbf{Z}_3}$$

EXAMPLE 13-1 Given the circuit of Fig. 13-9, find (*a*) the driving-point impedance, (*b*) the reflected impedance, and (*c*) the input current.

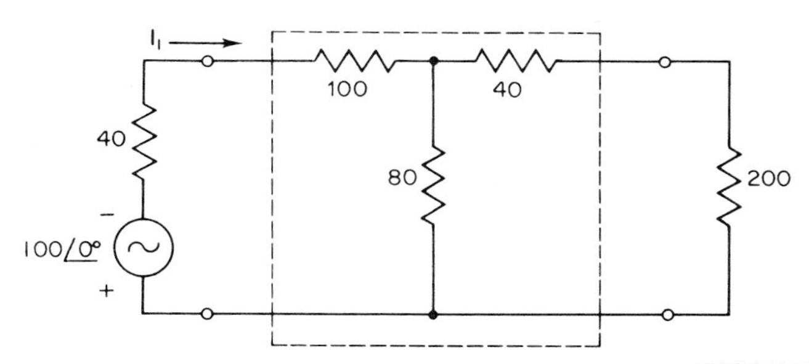

FIGURE 13-9
Circuit for Example 13-1.

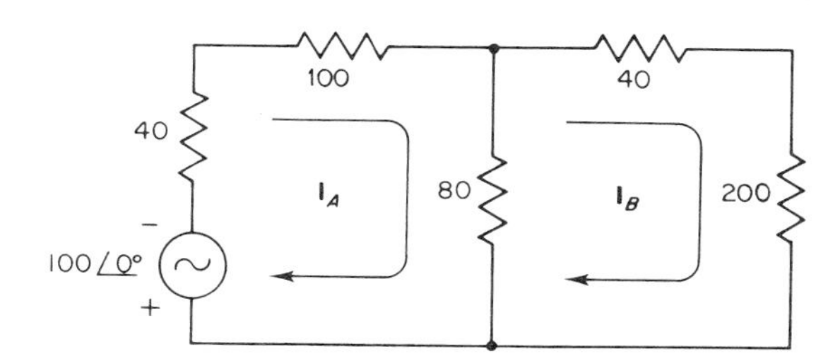

FIGURE 13-10

SOLUTION The z parameters for the two-port are

$$z_{11} = \mathbf{Z}_1 + \mathbf{Z}_2 = 180\ \Omega$$

$$z_{12} = z_{21} = z_M = \mathbf{Z}_2 = 80\ \Omega$$

$$z_{22} = \mathbf{Z}_2 + \mathbf{Z}_3 = 120\ \Omega$$

Driving-point impedance is equal to

$$\mathbf{Z}_{11} = z_{11} - \frac{z_M{}^2}{z_{22} + \mathbf{Z}_L}$$

$$= 180 - \frac{6{,}400}{320} = 160\ \Omega$$

$$\mathbf{Z}_{\text{ref}} = \frac{z_M{}^2}{z_{22} + \mathbf{Z}_L} = 20\ \Omega$$

Input current is found by the division

$$\mathbf{I}_1 = \frac{100\angle 0^\circ}{40 + \mathbf{Z}_{11}} = 0.5\ \angle 0^\circ\ \text{A}$$

Since the circuit within the two-port is known, $\mathbf{I}_1$ can be verified using mesh equations (see Fig. 13-10).

A-mesh equation: $\quad 100\angle 0^\circ = 220\mathbf{I}_A - 80\mathbf{I}_B$

B-mesh equation: $\quad 0 = -80\mathbf{I}_A + 320\mathbf{I}_B$

Solving for $\mathbf{I}_A$,

$$\mathbf{I}_A = \frac{\begin{vmatrix} 100\ \angle 0^\circ & -80 \\ 0 & 320 \end{vmatrix}}{\begin{vmatrix} 220 & -80 \\ -80 & 320 \end{vmatrix}} = \frac{32{,}000\ \angle 0^\circ}{64{,}000}$$

$$\mathbf{I}_A = \mathbf{I}_1 = 0.5\ \angle 0^\circ\ \text{A} \qquad \text{(checks)}$$

////

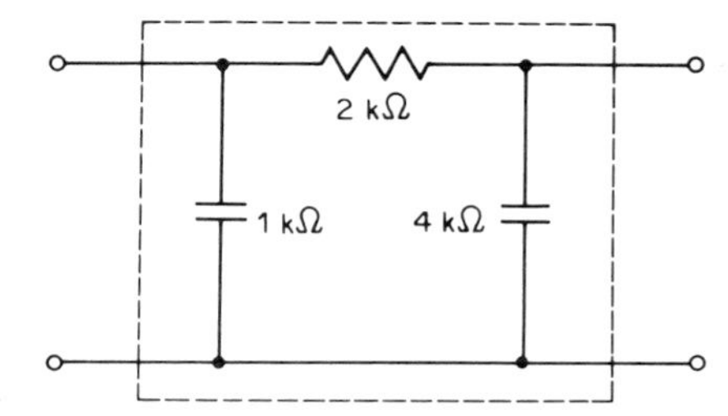

FIGURE 13-11
Circuit for Example 13-2.

EXAMPLE 13-2 Determine open-circuit impedance parameters for the circuit in Fig. 13-11.

SOLUTION

$$\mathbf{Z}_1 = -j(1\ \text{k}\Omega) = 1\ \text{k}\Omega \angle -90^\circ$$

$$\mathbf{Z}_2 = 2\ \text{k}\Omega$$

$$\mathbf{Z}_3 = -j(4\ \text{k}\Omega) = 4\ \text{k}\Omega \angle -90^\circ$$

z parameters have been expressed in terms of $\mathbf{Z}_1$, $\mathbf{Z}_2$, and $\mathbf{Z}_3$ for the pi network.

$$z_{11} = \frac{\mathbf{Z}_1(\mathbf{Z}_2 + \mathbf{Z}_3)}{\mathbf{Z}_1 + \mathbf{Z}_2 + \mathbf{Z}_3} = \frac{1\ \text{k}\Omega \angle -90^\circ[2\ \text{k}\Omega - j(4\ \text{k}\Omega)]}{2\ \text{k}\Omega - j(5\ \text{k}\Omega)}$$

$$= 827 \angle -85^\circ\ \Omega$$

$$z_M = \frac{\mathbf{Z}_1\mathbf{Z}_3}{\mathbf{Z}_1 + \mathbf{Z}_2 + \mathbf{Z}_3} = \frac{4 \times 10^6 \angle -180^\circ}{2\ \text{k}\Omega - j(4\ \text{k}\Omega)}$$

$$= 740 \angle -112^\circ\ \Omega$$

$$z_{22} = \frac{\mathbf{Z}_3(\mathbf{Z}_1 + \mathbf{Z}_2)}{\mathbf{Z}_1 + \mathbf{Z}_2 + \mathbf{Z}_3} = \frac{4\ \text{k}\Omega \angle -90^\circ[2\ \text{k}\Omega - j(1\ \text{k}\Omega)]}{2\ \text{k}\Omega - j(5\ \text{k}\Omega)}$$

$$= 1.66\ \text{k}\Omega \angle -48^\circ\ \Omega$$

////

Short-circuit Admittance Parameters

The coupling network of Fig. 13-3 can be described by the following equations:

$$\mathbf{I}_1 = y_{11}\mathbf{E}_1 + y_{12}\mathbf{E}_2 \tag{13-9}$$

$$\mathbf{I}_2 = y_{21}\mathbf{E}_1 + y_{22}\mathbf{E}_2 \tag{13-10}$$

where y_{11} is the input admittance with output terminals shorted

$$y_{11} = \frac{\mathbf{I}_1}{\mathbf{E}_1} \qquad (\mathbf{E}_2 = 0)$$

where y_{12} is the reverse-transfer admittance with input terminals shorted

$$y_{12} = \frac{\mathbf{I}_1}{\mathbf{E}_2} \qquad (\mathbf{E}_1 = 0)$$

where y_{21} is the forward transfer admittance with output terminals shorted

$$y_{21} = \frac{\mathbf{I}_2}{\mathbf{E}_1} \qquad (\mathbf{E}_2 = 0)$$

and where y_{22} is the output admittance with input terminals shorted

$$y_{22} = \frac{\mathbf{I}_2}{\mathbf{E}_2} \qquad (\mathbf{E}_1 = 0)$$

Figure 13-12 illustrates the various test setups needed to measure the short-circuit admittance parameters. y_{12} and y_{21} are equal in all passive networks without initial voltages or currents. This condition is easily proved using the reciprocity theorem.

The admittance seen by the source $\mathbf{E}_1$ when the two-port is terminated with load admittance $\mathbf{Y}_L$ is referred to as the *driving-point admittance*. Figure 13-13 shows a two-port terminated in load admittance $\mathbf{Y}_L$.

The driving-point admittance is

$$\mathbf{Y}_{11} = \frac{\mathbf{I}_1}{\mathbf{E}_1} \tag{13-11}$$

for a specified passive load termination $\mathbf{Y}_L$. Equation (13-11) is not equivalent to the admittance parameter y_{11}. Recall that y_{11} is defined under the condition of a short-circuit output.

The output current is equal to

$$\mathbf{I}_2 = -\mathbf{Y}_L \mathbf{E}_2 \tag{13-12}$$

Rewriting Eq. (13-9) and substituting Eq. (13-12) into Eq. (13-10) yields the input and output admittance equations.

$$\mathbf{I}_1 = y_{11}\mathbf{E}_1 + y_{12}\mathbf{E}_2 \tag{13-13}$$

$$0 = y_{21}\mathbf{E}_1 + (y_{22} + \mathbf{Y}_L)\mathbf{E}_2 \tag{13-14}$$

Solving Eqs. (13-13) and (13-14) for $\mathbf{E}_1$ gives

$$\mathbf{E}_1 = \frac{\begin{vmatrix} \mathbf{I}_1 & y_{12} \\ 0 & y_{22} + \mathbf{Y}_L \end{vmatrix}}{\begin{vmatrix} y_{11} & y_{12} \\ y_{21} & y_{22} + \mathbf{Y}_L \end{vmatrix}} = \frac{\mathbf{I}_1(y_{22} + \mathbf{Y}_L)}{y_{11}(y_{22} + \mathbf{Y}_L) - y_{12}y_{21}}$$

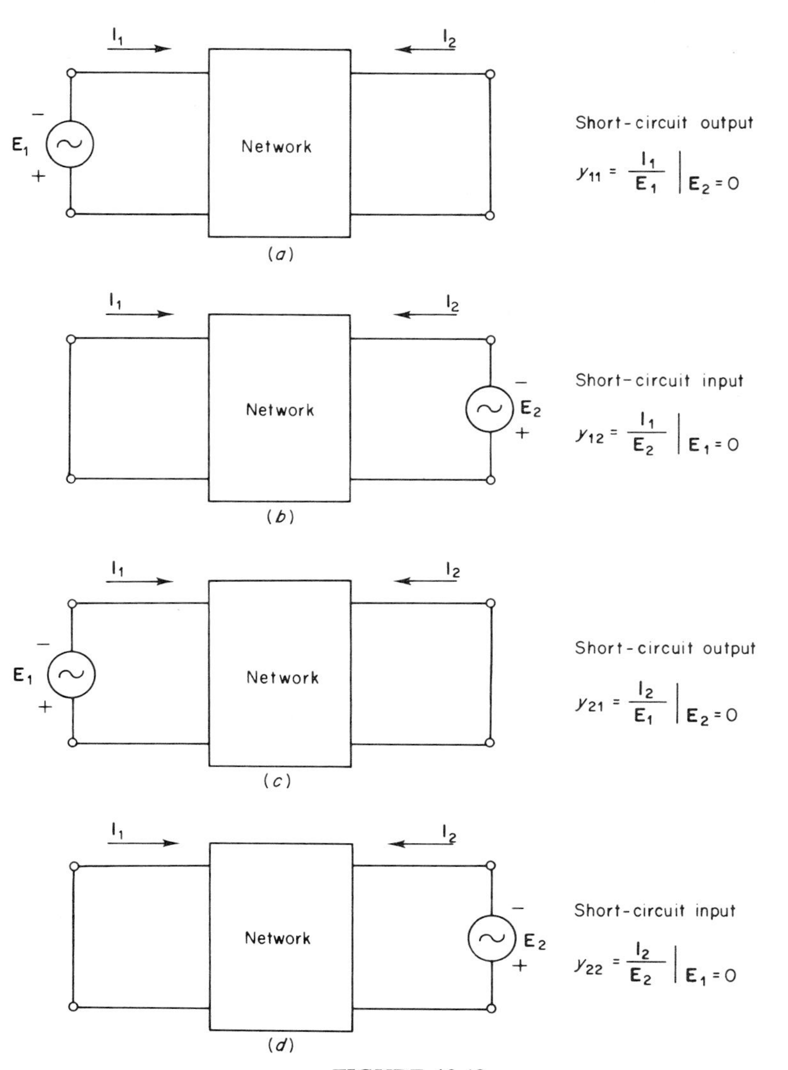

FIGURE 13-12
Test setup to determine (*a*) y_{11}, (*b*) y_{12}, (*c*) y_{21}, (*d*) y_{22}.

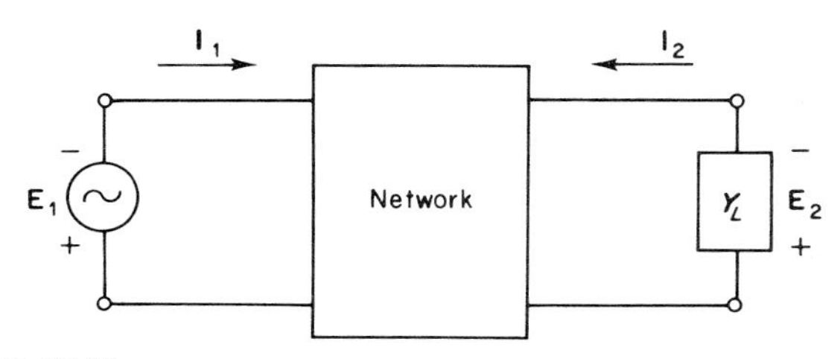

FIGURE 13-13
Two-port terminated in Y_L.

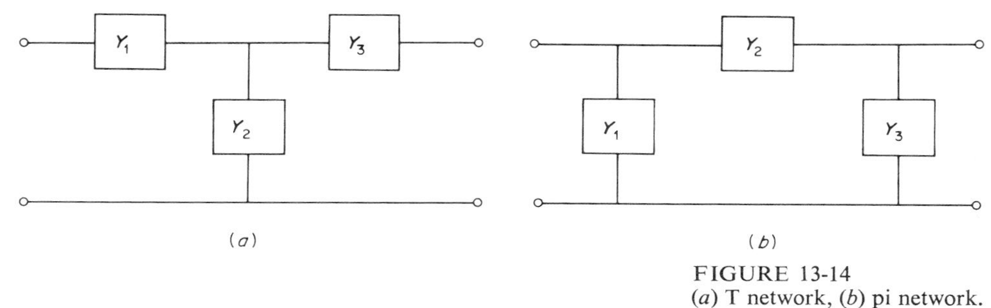

FIGURE 13-14
(*a*) T network, (*b*) pi network.

Since $y_{12} = y_{21}$, let $y_M = y_{12} = y_{21}$ represent the mutual admittance between input and output. Thus, the driving-point admittance is

$$\mathbf{Y}_{11} = \frac{\mathbf{I}_1}{\mathbf{E}_1}$$

$$= y_{11} - \frac{{y_M}^2}{y_{22} + \mathbf{Y}_L} \tag{13-15}$$

The reflected admittance is the difference between the input admittance with a short, y_{11}, and the driving-point admittance with load, $\mathbf{Y}_{11}$.

$$\mathbf{Y}_{\text{ref}} = \frac{{y_M}^2}{y_{22} + \mathbf{Y}_L} \tag{13-16}$$

If the two-port is a T network or a pi network, as shown in Fig. 13-14, the resulting admittance parameters are those listed in Table 13-1.

Table 13-1 ADMITTANCE PARAMETERS FOR T AND PI NETWORKS

Admittance parameter	T network	Pi network
y_{11}	$\frac{\mathbf{Y}_1(\mathbf{Y}_2 + \mathbf{Y}_3)}{\mathbf{Y}_1 + \mathbf{Y}_2 + \mathbf{Y}_3}$	$\mathbf{Y}_1 + \mathbf{Y}_2$
$y_M = y_{12} = y_{21}$	$\frac{-\mathbf{Y}_1\mathbf{Y}_3}{\mathbf{Y}_1 + \mathbf{Y}_2 + \mathbf{Y}_3}$	$-\mathbf{Y}_2$
y_{22}	$\frac{\mathbf{Y}_3(\mathbf{Y}_1 + \mathbf{Y}_2)}{\mathbf{Y}_1 + \mathbf{Y}_2 + \mathbf{Y}_3}$	$\mathbf{Y}_2 + \mathbf{Y}_3$

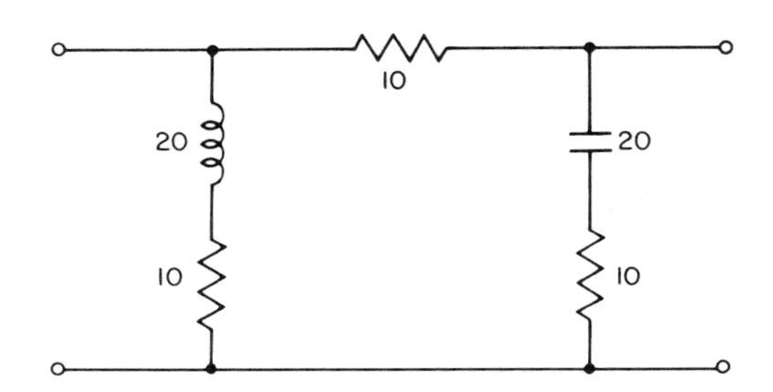

FIGURE 13-15
Circuit for Example 13-3.

EXAMPLE 13-3 Determine the short-circuit admittance parameters for the pi network shown in Fig. 13-15.

SOLUTION Using the equations from Table 13-1,

$$y_{11} = \mathbf{Y}_1 + \mathbf{Y}_2$$

$$= \frac{1}{10 + j20} + \frac{1}{10} = 0.12 - j0.04 \text{ mho}$$

$$y_M = -0.1 \text{ mho}$$

$$y_{22} = \mathbf{Y}_2 + \mathbf{Y}_3$$

$$= \frac{1}{10} + \frac{1}{10 - j20} = 0.12 + j0.04 \text{ mho}$$

////

EXAMPLE 13-4 For the circuit in Fig. 13-15, calculate the driving-point admittance and reflected admittance with a 25-Ω load termination.

SOLUTION y parameters for the network are

$$y_{11} = 0.12 - j0.04 = 0.127 \angle -18.4°$$

$$y_M = -0.1$$

$$y_{22} = 0.12 + j0.04 = 0.127 \angle 18.4°$$

Using Eq. (13-15),

$$\mathbf{Y}_{11} = y_{11} - \frac{{y_M}^2}{y_{22} + \mathbf{Y}_L}$$

$$= 0.12 - j0.04 - \frac{1 \times 10^{-2}}{0.16 + j0.04}$$

$$= 0.061 - j0.025 \text{ mho}$$

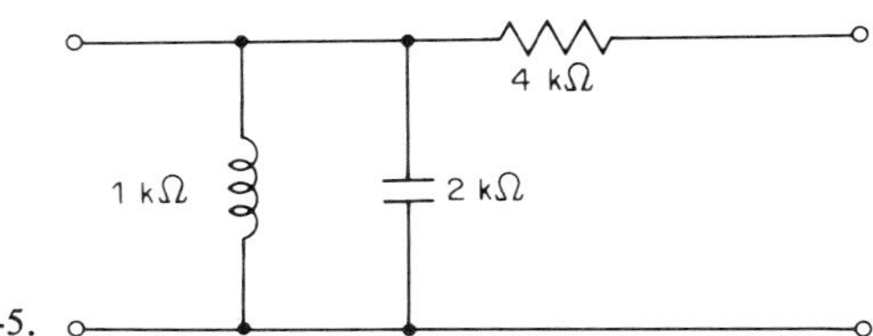

FIGURE 13-16
Circuit for Example 13-5.

Using Eq. (13-16),

$$\mathbf{Y}_{\text{ref}} = \frac{{y_M}^2}{y_{22} + \mathbf{Y}_L} = y_{11} - \mathbf{Y}_{11} = 0.12 - j0.04 - 0.061 + j0.025$$

$$= 0.059 - j0.015 \text{ mho}$$

////

EXAMPLE 13-5 Determine the y parameters for the circuit in Fig. 13-16.

SOLUTION The circuit is similar to a pi network where $\mathbf{Y}_3 = 0$.

$$\mathbf{Y}_2 = \frac{1}{4\ \text{k}\Omega} = 2.5 \times 10^{-4}$$

$$\mathbf{Y}_1 = -j1 \times 10^{-3} + j0.5 \times 10^{-3} = -j5 \times 10^{-4}$$

Thus,

$$y_{11} = \mathbf{Y}_1 + \mathbf{Y}_2 = 2.5 \times 10^{-4} - j5 \times 10^{-4} \text{ mho}$$

$$y_M = -\mathbf{Y}_2 = -2.5 \times 10^{-4} \text{ mho}$$

$$y_{22} = \mathbf{Y}_2 + \mathbf{Y}_3 = 2.5 \times 10^{-4} \text{ mho}$$

////

Up to this point, the reader has been subjected to an array of equations and formulas. Now it would be appropriate to state the basis of the open-circuit and short-circuit parameters. In any linear circuit, current and voltage are variables that depend upon one another. The equation $\mathbf{E} = R\mathbf{I}$ simply makes voltage the dependent and current the independent variable. The same is true of the open-circuit impedance parameters. In Eqs. (13-1) and (13-2), terminal voltages are dependent variables and currents at the terminals are independent variables. On the other hand, currents can be the dependent variables and terminal voltages the independent variables. Then the equations yield the short-circuit admittance parameters.

In this section on general analysis of circuits, frequent use has been made of the expressions "open-circuit" and "short-circuit." Just what is a short circuit? Obviously, the ideal short circuit would be one with zero impedance. But this is impossible to achieve. *A portion of a circuit is considered to be a short when its impedance is*

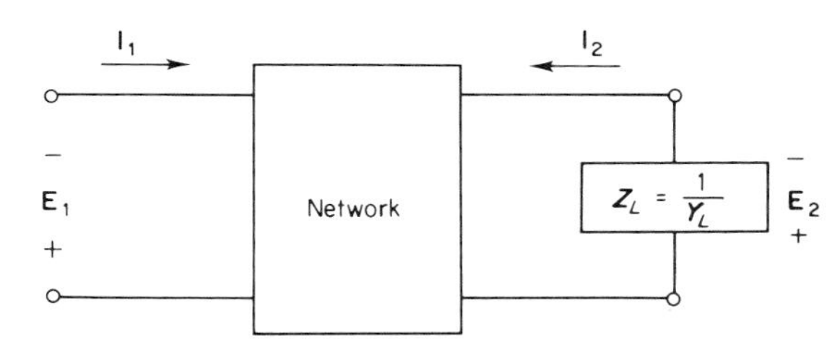

FIGURE 13-17
Output-terminated two-port.

so small in comparison to the rest of the circuit that assuming its impedance to be zero does not significantly change the total impedance of the circuit.

Along the same line of reasoning, it might be asked: What is an open circuit? An ideal open would have infinite impedance. This, too, is impractical, and it is necessary to define the practical open circuit. *A portion of a circuit is considered to be open when its impedance is so high that assuming it to be infinite does not significantly change the total impedance of the circuit.* For example, if a 150-Ω resistor is connected in series with a reactance of 10,000 Ω, the total impedance for all practical purposes is still 10,000 Ω. The 150-Ω resistor looks like a "short" in comparison to the 10,000-Ω reactance. On the other hand, if the two components were connected in parallel, the total impedance would be $150 \angle 0° \ \Omega$. The reactance could be ignored and could be considered to be "open" in comparison to the 150-Ω resistor.

Up to this point only the driving-point impedance and driving-point admittance have been evaluated for a two-port represented by z or y parameters with an external load termination. Several other functional relations exist between the four terminal responses. Referring to Fig. 13-17, a list of transfer functions for the terminated two-port is as follows:

Voltage ratio: $$G_{12} = \frac{\mathbf{E}_2}{\mathbf{E}_1} \quad \text{no units}$$

Current ratio: $$\alpha_{12} = \frac{-\mathbf{I}_2}{\mathbf{I}_1} \quad \text{no units}$$

Transfer impedance: $$\mathbf{Z}_{12} = \frac{\mathbf{E}_2}{\mathbf{I}_1} \quad \Omega$$

Transfer admittance: $$\mathbf{Y}_{12} = \frac{-\mathbf{I}_2}{\mathbf{E}_1} \quad \text{mhos}$$

The current ratio α_{12} is negative because the actual direction of $\mathbf{I}_2$ is opposite the assumed direction with $\mathbf{I}_1$ flowing into the network. A similar effect causes $\mathbf{Y}_{12}$ to be negative since $\mathbf{I}_2$ would flow out of the network with $\mathbf{E}_1$'s polarity orientation.

Considering a two-port represented by z parameters, some of the network calculations for the circuit in Fig. 13-17 were done previously. Referring to Eqs. (13-5) and (13-6),

$$\mathbf{E}_1 = z_{11}\mathbf{I}_1 + z_{12}\mathbf{I}_2 \tag{13-17}$$

$$0 = z_{21}\mathbf{I}_1 + (z_{22} + \mathbf{Z}_L)\mathbf{I}_2 \tag{13-18}$$

$\mathbf{I}_1$ was evaluated previously.

$$\mathbf{I}_1 = \frac{\mathbf{E}_1(z_{22} + \mathbf{Z}_L)}{z_{11}(z_{22} + \mathbf{Z}_L) - z_M{}^2}$$

Let

$$\Delta z = z_{11} z_{22} - z_M{}^2$$

then

$$\mathbf{I}_1 = \frac{\mathbf{E}_1(z_{22} + \mathbf{Z}_L)}{\Delta z + z_{11}\mathbf{Z}_L} \tag{13-19}$$

Solving Eqs. (13-17) and (13-18) for $\mathbf{I}_2$,

$$\mathbf{I}_2 = \frac{\begin{vmatrix} z_{11} & \mathbf{E}_1 \\ z_{21} & 0 \end{vmatrix}}{\Delta z + z_{11}\mathbf{Z}_L} = \frac{-\mathbf{E}_1 z_M}{\Delta z + z_{11}\mathbf{Z}_L} \tag{13-20}$$

The list of transfer functions can be solved in terms of the z parameters.

$$G_{12} = \frac{\mathbf{E}_2}{\mathbf{E}_1} = \frac{-\mathbf{I}_2\mathbf{Z}_L}{\mathbf{E}_1}$$

$$= \frac{z_M\mathbf{Z}_L}{\Delta z + z_{11}\mathbf{Z}_L} \tag{13-21}$$

$$\alpha_{12} = \frac{-\mathbf{I}_2}{\mathbf{I}_1} = \frac{\dfrac{\mathbf{E}_1 z_M}{\Delta z + z_{11}\mathbf{Z}_L}}{\dfrac{\mathbf{E}_1(z_{22} + \mathbf{Z}_L)}{\Delta z + z_{11}\mathbf{Z}_L}}$$

$$= \frac{z_M}{z_{22} + \mathbf{Z}_L} \tag{13-22}$$

$$\mathbf{Z}_{12} = \frac{\mathbf{E}_2}{\mathbf{I}_1} = \frac{-\mathbf{I}_2\mathbf{Z}_L}{\mathbf{I}_1}$$

$$= \frac{z_M\mathbf{Z}_L}{z_{22} + \mathbf{Z}_L} \tag{13-23}$$

$$\mathbf{Y}_{12} = \frac{-\mathbf{I}_2}{\mathbf{E}_1} = \frac{\dfrac{\mathbf{E}_1 z_M}{\Delta z + z_{11}\mathbf{Z}_L}}{\mathbf{E}_1}$$

$$= \frac{z_M}{\Delta z + z_{11}\mathbf{Z}_L} \tag{13-24}$$

A set of transfer functions can be derived in terms of y parameters. Referring to Eqs. (13-13) and (13-14),

$$\mathbf{I}_1 = y_{11}\mathbf{E}_1 + y_{12}\mathbf{E}_2 \tag{13-25}$$

$$0 = y_{21}\mathbf{E}_1 + (y_{22} + \mathbf{Y}_L)\mathbf{E}_2 \tag{13-26}$$

$\mathbf{E}_1$ was evaluated previously.

$$\mathbf{E}_1 = \frac{\mathbf{I}_1(y_{22} + \mathbf{Y}_L)}{y_{11}(y_{22} + \mathbf{Y}_L) - y_M^2}$$

Let

$$\Delta y = y_{11}y_{22} - y_M^2$$

then

$$\mathbf{E}_1 = \frac{\mathbf{I}_1(y_{22} + \mathbf{Y}_L)}{\Delta y + y_{11}\mathbf{Y}_L} \tag{13-27}$$

Solving Eqs. (13-25) and (13-26) for $\mathbf{E}_2$,

$$\mathbf{E}_2 = \frac{\begin{vmatrix} y_{11} & \mathbf{I}_1 \\ y_{21} & 0 \end{vmatrix}}{\Delta y + y_{11}\mathbf{Y}_L} = \frac{-\mathbf{I}_1 y_M}{\Delta y + y_{11}\mathbf{Y}_L} \tag{13-28}$$

The list of transfer functions can be solved in terms of the y parameters.

$$G_{12} = \frac{\mathbf{E}_2}{\mathbf{E}_1} = \frac{-\dfrac{\mathbf{I}_1 y_M}{\Delta y + y_{11}\mathbf{Y}_L}}{\dfrac{\mathbf{I}_1(y_{22} + \mathbf{Y}_L)}{\Delta y + y_{11}\mathbf{Y}_L}}$$

$$= -\frac{y_M}{y_{22} + \mathbf{Y}_L} \tag{13-29}$$

$$\alpha_{12} = \frac{-\mathbf{I}_2}{\mathbf{I}_1} = \frac{\mathbf{E}_2\mathbf{Y}_L}{\mathbf{I}_1}$$

$$= -\frac{y_M\mathbf{Y}_L}{\Delta y + y_{11}\mathbf{Y}_L} \tag{13-30}$$

$$\mathbf{Z}_{12} = \frac{\mathbf{E}_2}{\mathbf{I}_1} = \frac{\dfrac{-\mathbf{I}_1 y_M}{\Delta y + y_{11}\mathbf{Y}_L}}{\mathbf{I}_1}$$

$$= -\frac{y_M}{\Delta y + y_{11}\mathbf{Y}_L} \tag{13-31}$$

$$\mathbf{Y}_{12} = \frac{-\mathbf{I}_2}{\mathbf{E}_1} = \frac{\mathbf{E}_2\mathbf{Y}_L}{\mathbf{E}_1} = \frac{\dfrac{-\mathbf{I}_1 y_M \mathbf{Y}_L}{\Delta y + y_{11}\mathbf{Y}_L}}{\dfrac{\mathbf{I}_1(y_{22} + \mathbf{Y}_L)}{\Delta y + y_{11}\mathbf{Y}_L}}$$

$$= \frac{-y_M \mathbf{Y}_L}{y_{22} + \mathbf{Y}_L} \tag{13-32}$$

Table 13-2 summarizes the transfer function for both z and y parameters.

EXAMPLE 13-6 The z parameters for the two-port network in Fig. 13-18 are

$$z_{11} = 2 - j3 \qquad z_M = 1 + j1 \qquad z_{22} = 4 + j3$$

Calculate (a) G_{12}, (b) α_{12}, (c) $\mathbf{Z}_{12}$, (d) $\mathbf{Y}_{12}$, (e) $\mathbf{Z}_{11}$.
(f) Find the values $\mathbf{Z}_1$, $\mathbf{Z}_2$, and $\mathbf{Z}_3$ in an equivalent T network.
(g) If $\mathbf{E}_1 = 4\angle 30°$, find $\mathbf{I}_1$, $\mathbf{E}_2$, and $\mathbf{I}_2$.

SOLUTION Δz must be calculated for transfer-function calculations.

$$\Delta z = z_{11}z_{22} - z_M^2$$
$$z_{11} = 2 - j3 = 3.6\angle -56.4°$$
$$z_M = 1 + j1 = 1.41\angle 45°$$
$$z_{22} = 4 + j3 = 5\angle 36.8°$$

Table 13-2 TRANSFER FUNCTIONS OF A TWO-PORT

Transfer function		z parameters	y parameters
Voltage ratio	G_{12}	$\dfrac{z_M \mathbf{Z}_L}{\Delta z + z_{11}\mathbf{Z}_L}$	$-\dfrac{y_M}{y_{22} + \mathbf{Y}_L}$
Current ratio	α_{12}	$\dfrac{z_M}{z_{22} + \mathbf{Z}_L}$	$-\dfrac{y_M \mathbf{Y}_L}{\Delta y + y_{11}\mathbf{Y}_L}$
Transfer impedance	$\mathbf{Z}_{12}$	$\dfrac{z_M \mathbf{Z}_L}{z_{22} + \mathbf{Z}_L}$	$-\dfrac{y_M}{\Delta y + y_{11}\mathbf{Y}_L}$
Transfer admittance	$\mathbf{Y}_{12}$	$\dfrac{z_M}{\Delta z + z_{11}\mathbf{Z}_L}$	$-\dfrac{y_M \mathbf{Y}_L}{y_{22} + \mathbf{Y}_L}$
		$\Delta z = z_{11}z_{22} - z_M^2$	$\Delta y = y_{11}y_{22} - y_M^2$

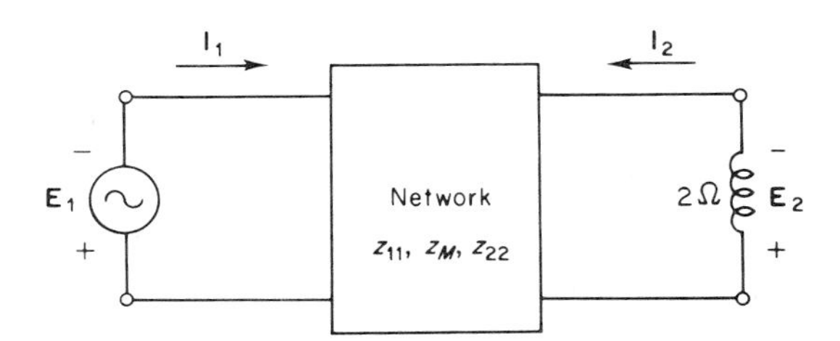

FIGURE 13-18
Circuit for Example 13-6.

Thus,

$$\Delta z = (3.6 \angle -56.4^\circ)(5 \angle 36.8^\circ) - (1.41 \angle 45^\circ)^2$$
$$= 17 - j8 = 18.8 \angle -25.2^\circ$$

(*a*)
$$G_{12} = \frac{z_M \mathbf{Z}_L}{\Delta z + z_{11}\mathbf{Z}_L} = \frac{(1.41 \angle 45^\circ)(2 \angle 90^\circ)}{17 - j8 + (3.6 \angle -56.4^\circ)(2 \angle 90^\circ)}$$
$$= \frac{2.82 \angle 135^\circ}{23.4 \angle -10^\circ} = 0.12 \angle 145^\circ$$

(*b*)
$$\alpha_{12} = \frac{z_M}{z_{22} + \mathbf{Z}_L} = \frac{1.41 \angle 45^\circ}{6.4 \angle 51.4^\circ}$$
$$= 0.22 \angle -6.4^\circ$$

(*c*)
$$\mathbf{Z}_{12} = \frac{z_M \mathbf{Z}_L}{z_{22} + \mathbf{Z}_L} = \alpha_{12}\mathbf{Z}_L = (0.22 \angle -6.4^\circ)(2 \angle 90^\circ)$$
$$= 0.44 \angle 83.6^\circ\ \Omega$$

(*d*)
$$\mathbf{Y}_{12} = \frac{z_M}{\Delta z + z_{11}\mathbf{Z}_L} = \frac{G_{12}}{\mathbf{Z}_L} = \frac{0.12 \angle 145^\circ}{2 \angle 90^\circ}$$
$$= 0.06 \angle 55^\circ \text{ mho}$$

(*e*)
$$\mathbf{Z}_{11} = z_{11} - \frac{z_M{}^2}{z_{22} + \mathbf{Z}_L} = 2 - j3 - (0.22 \angle -6.4^\circ)(1.41 \angle 45^\circ)$$
$$= 1.76 - j3.2 = 3.65 \angle -61.2^\circ\ \Omega$$

(*f*) The z parameters in terms of T-network impedances are

$$z_{11} = \mathbf{Z}_1 + \mathbf{Z}_2$$
$$z_M = \mathbf{Z}_2$$
$$z_{22} = \mathbf{Z}_2 + \mathbf{Z}_3$$

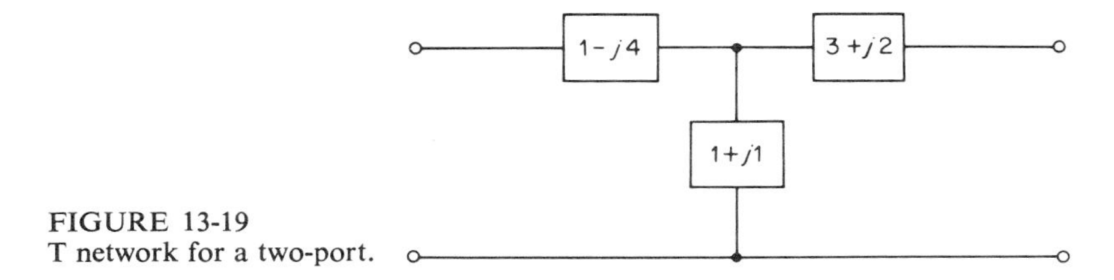

FIGURE 13-19
T network for a two-port.

These three equations can be solved for $\mathbf{Z}_1$, $\mathbf{Z}_2$, and $\mathbf{Z}_3$.

$$\mathbf{Z}_1 = z_{11} - z_M$$

$$\mathbf{Z}_2 = z_M$$

$$\mathbf{Z}_3 = z_{22} - z_M$$

Thus,

$$\mathbf{Z}_1 = 2 - j3 - 1 - j1 = 1 - j4$$

$$\mathbf{Z}_2 = 1 + j1$$

$$\mathbf{Z}_3 = 4 + j3 - 1 - j1 = 3 + j2$$

The resulting T network is shown in Fig. 13-19.

(*g*)

$$\mathbf{I}_1 = \frac{\mathbf{E}_1}{\mathbf{Z}_{11}} = \frac{4 \angle 30^\circ}{3.65 \angle -61.2^\circ} = 1.1 \angle 91.2^\circ \text{ A}$$

$$\mathbf{E}_2 = G_{12}\,\mathbf{E}_1 = (0.12 \angle 145^\circ)(4 \angle 30^\circ) = 0.48 \angle 175^\circ \text{ V}$$

$$\mathbf{I}_2 = -\alpha_{12}\,\mathbf{I}_1 = (-0.22 \angle -6.4^\circ)(1.1 \angle 91.2^\circ)$$

$$= -0.242 \angle 84.8^\circ \text{ A}$$

or

$$\mathbf{I}_2 = -\mathbf{Y}_{12}\,\mathbf{E}_1 = (-0.06 \angle 55^\circ)(4 \angle 30^\circ)$$

$$= -0.24 \angle 85^\circ \text{ A}$$

////

Since the two-port network has been expressed in terms of z and y parameters, it should be evident that there is some means of conversion between the two parameter systems. A common mistake made by neophytes is the assumption that $y_{11} = 1/z_{11}$, $y_M = 1/z_M$, and $y_{22} = 1/z_{22}$. The correct conversion, which can be determined from Eqs. (13-1), (13-2), (13-9), and (13-10), is shown in Table 13-3.

There are other sets of general circuit equations that can be written. In one set, input voltage and output current are the dependent variables, and input current and output voltage are the independent variables. The parameters obtained with these equations are called *hybrid parameters* and are widely used in the analysis of transistor circuits.

$$\mathbf{E}_1 = h_{11}\mathbf{I}_1 + h_{12}\mathbf{E}_2 \tag{13-33}$$

$$\mathbf{I}_2 = h_{21}\mathbf{I}_1 + h_{22}\mathbf{E}_2 \tag{13-34}$$

where h_{11} is the input impedance with output shorted

$$h_{11} = \frac{\mathbf{E}_1}{\mathbf{I}_1} \qquad (\mathbf{E}_2 = 0)$$

where h_{12} is the reverse voltage-transfer ratio with input open

$$h_{12} = \frac{\mathbf{E}_1}{\mathbf{E}_2} \qquad (\mathbf{I}_1 = 0)$$

where h_{21} is the forward current-transfer ratio with output shorted

$$h_{21} = \frac{\mathbf{I}_2}{\mathbf{I}_1} \qquad (\mathbf{E}_2 = 0)$$

and where h_{22} is the output admittance with input open

$$h_{22} = \frac{\mathbf{I}_2}{\mathbf{E}_2} \qquad (\mathbf{I}_1 = 0)$$

Hybrid parameters are useful with active circuits but are of no practical value in the analysis of passive networks. No further work will be done with them in this text. They have been mentioned merely to illustrate to the reader the variety of methods available for network analysis.

Table 13-3 CONVERSION BETWEEN z AND y

Impedance to admittance	Admittance to impedance
$y_{11} = \frac{z_{22}}{\Delta z}$	$z_{11} = \frac{y_{22}}{\Delta y}$
$y_M = \frac{-z_M}{\Delta z}$	$z_M = \frac{-y_M}{\Delta y}$
$y_{22} = \frac{z_{11}}{\Delta z}$	$z_{22} = \frac{y_{11}}{\Delta y}$

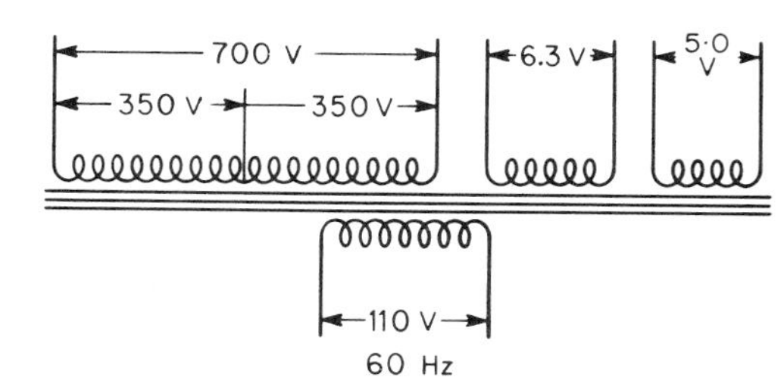

FIGURE 13-20
Typical power transformer for a TV receiver.

13-3 TRANSFORMERS

In Chap. 7, we discussed mutual inductance and its ability to transfer energy from one circuit to another. A transformer is a device that couples circuits through mutual inductance. Transformers are classified according to their application and frequency of operation. Hence, there are audio transformers, power transformers, radio-frequency transformers, output transformers, pulse transformers, as well as many others.

Transformers for use at power-line frequencies and at audio frequencies are constructed on laminated cores of high permeability, and are characterized by very high coefficients of coupling. It is common practice to assume that a transformer wound on a closed high-permeability core has a coefficient of coupling of 1. On the other hand, transformers for use at radio frequencies are wound on cores of very low permeability, and their coefficients of coupling are low, ranging from as little as 0.002 to about 0.7. The windings of transformers used at radio frequencies are often made part of resonant circuits, while transformers used at audio frequencies must have broadband characteristics that minimize discrimination against frequencies within the desired range.

It is common practice to refer to the input side of a transformer as the primary winding and the output side or sides as the secondary windings. Figure 13-20 illustrates the arrangement of the windings for a typical power transformer used in electronic equipment. Note that the transformer has several secondaries, one of which works as a step-up unit and the others as step-down units.

A secondary circuit does not have to be insulated from the primary circuit. A type of transformer that is widely used in electric power applications is the *autotransformer*. In this type of transformer, the primary and secondary windings are connected in series-aiding. Voltage is applied to the primary, and the magnetic field that is set up induces a voltage across the remaining turns of the winding. The secondary voltage is the sum of the applied primary voltage and the voltage induced in the secondary. Figure 13-21 shows a typical autotransformer. The illustration shows a step-up arrangement, although step-down operation is equally possible.

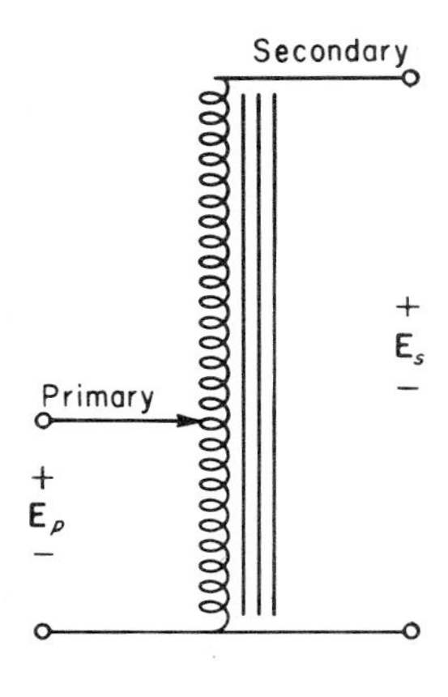

FIGURE 13-21
Autotransformer.

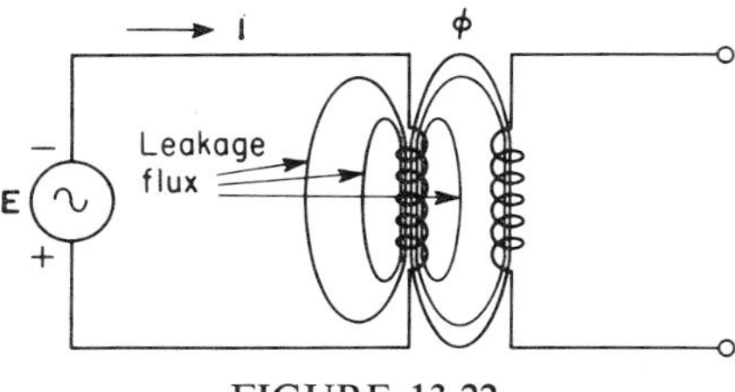

FIGURE 13-22
Leakage flux in a transformer.

All practical transformers have power losses. There are hysteresis and eddy-current losses in the core and I^2R losses in the windings. In addition, there is always some leakage flux, even in a transformer with a very high coefficient of coupling. Leakage loss occurs when the flux from one winding fails to link the turns of another. Figure 13-22 illustrates a transformer with leakage flux. Leakage flux represents inductance that is not participating in the transfer of energy from primary to secondary of the transformer. In the equivalent circuit of a transformer, leakage flux is represented by *leakage inductance* in series with the *incremental inductance* (that portion of inductance which participates in the transformer action).

There is capacitance between the windings and between the windings and the core. At high frequencies, these capacitances have great effect on the performance of the transformer, and therefore they must be considered in transformers used at audio and radio frequencies. Figure 13-23 shows the equivalent circuit of a typical audio-frequency transformer when all losses are taken into consideration.

Radio-frequency transformers have very high leakage inductance and little or no hysteresis or eddy-current losses because their cores are usually of nonmagnetic material. In some instances coils used at radio frequencies are wound on cores made of powdered magnetic materials suspended in a ceramic binder.

13-4 ANALYSIS OF CLOSELY COUPLED TRANSFORMERS

Power transformers, audio-frequency transformers, and other so-called iron-core transformers are considered to be *closely coupled.* For the purposes of this analysis, it will be assumed that the transformer under consideration is an *ideal* transformer. It is therefore assumed that the reluctances of the magnetic paths are the same for

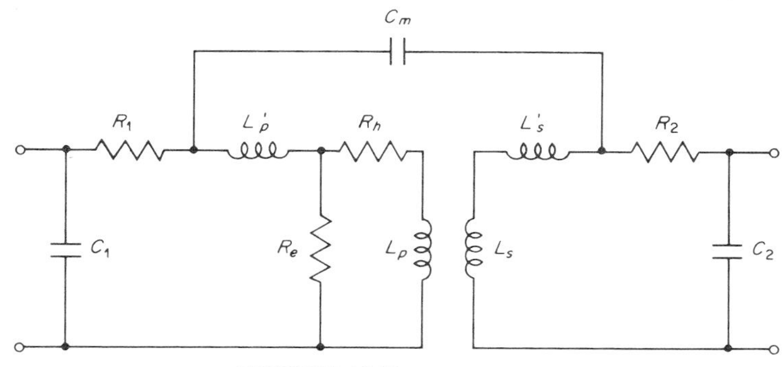

FIGURE 13-23
Equivalent circuit of an audio-frequency transformer.
R_1 = resistance of primary winding
R_h = resistance that represents power loss due to hysteresis
R_e = resistance that represents power loss due to eddy currents
R_2 = resistance of secondary winding
C_m = capacitance between windings
C_1 = capacitance from primary to core
C_2 = capacitance from secondary to core
L_p' = leakage inductance of primary winding
L_p = incremental inductance of primary winding
L_s' = leakage inductance of secondary winding
L_s = incremental inductance of secondary winding

both primary and secondary. It is also assumed that there is negligible leakage loss and core loss. It is further assumed that the reactance of the incremental inductance of the primary is very much greater than the resistance of the primary. On the basis of these assumptions, the primary and secondary voltages will be determined for the transformer of Fig. 13-24. Equations will be derived for the currents of primary and secondary circuits and for the effect of the secondary circuit on the primary circuit.

In accordance with Faraday's law the voltage induced in a winding is proportional to the rate of change of flux linkages.

$$e = N\frac{d\Phi}{dt}$$

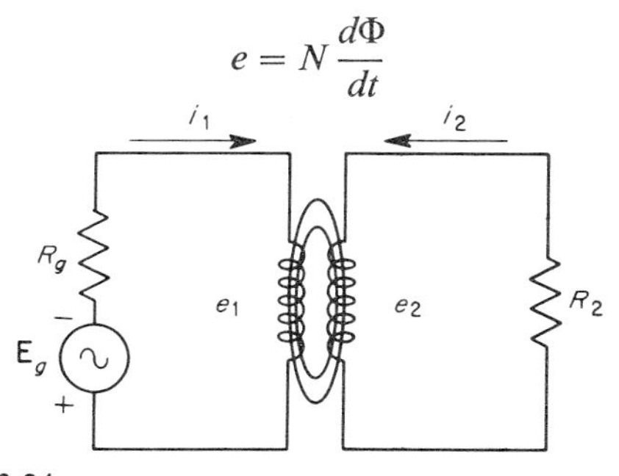

FIGURE 13-24
Circuit for analysis of impedance reflection in an ideal transformer.

Then in the circuit of Fig. 13-24 the voltage induced across the primary and secondary windings by primary flux can be found.

$$e_1 = N_1 \frac{d\Phi}{dt} \tag{13-35}$$

$$e_2 = N_2 \frac{d\Phi}{dt} \tag{13-36}$$

If the *turns ratio* of the transformer is defined as the ratio of primary turns to secondary turns,

$$a = \frac{N_1}{N_2} \tag{13-37}$$

then, taking the ratio of e_1/e_2 from Eqs. (13-35) and (13-36),

$$a = \frac{e_1}{e_2} \tag{13-38}$$

This relationship makes it possible to define the secondary voltage in terms of the primary voltage and the turns ratio of the transformer.

$$e_2 = \frac{e_1}{a} \tag{13-39}$$

Thus a step-up transformer has more secondary than primary turns, and a is less than 1. A step-down transformer has a turns ratio greater than 1.

Since effective values and phasor quantities are proportional to their instantaneous magnitudes, Eq. (13-38) can be written in effective or phasor form.

$$a = \frac{E_{1,\,\mathrm{eff}}}{E_{2,\,\mathrm{eff}}} = \frac{|\mathbf{E}_1|}{|\mathbf{E}_2|} \tag{13-40}$$

The relationship between primary and secondary currents is shown by one of the qualifying assumptions of this discussion. It has been assumed that core losses are negligible, and therefore apparent primary power equals apparent secondary power.

$$P_{1,\mathrm{appar}} = E_{1,\mathrm{eff}}\, I_{1,\mathrm{eff}} \qquad P_{2,\mathrm{appar}} = E_{2,\mathrm{eff}}\, I$$

$$E_{1,\mathrm{eff}}\, I_{1,\mathrm{eff}} = E_{2,\mathrm{eff}}\, I_{2,\mathrm{eff}}$$

$$\frac{I_{1,\mathrm{eff}}}{I_{2,\mathrm{eff}}} = \frac{E_{2,\mathrm{eff}}}{E_{1,\mathrm{eff}}} \tag{13-41}$$

Since $a = \dfrac{E_{1,\,\mathrm{eff}}}{E_{2,\,\mathrm{eff}}}$,

$$I_{1,\,\mathrm{eff}} = \frac{I_{2,\,\mathrm{eff}}}{a} \tag{13-42}$$

Equation (13-42) indicates that primary current is a function of secondary current and therefore of secondary loading. This effect of the secondary on the primary is most certainly due to the fact that flux of the secondary links primary turns. However, the effect of the secondary on the primary is represented as a reflected impedance appearing in shunt in the primary circuit. It has already been assumed that primary and secondary powers are equal. Since power can be dissipated only in a resistance, there must be resistance present in the primary circuit that dissipates power equal to that dissipated in the secondary. In the circuit of Fig. 13-24, R_g represents the internal resistance of the source. The reactance of the primary inductance is a constant quantity of ωL_p. If an increase in primary current with a decrease in impedance in the secondary is to be accounted for, it is apparent that any impedance reflection from the secondary into the primary must be in shunt with the incremental inductance of the primary. The reflected impedance will be derived with the aid of Fig. 13-24. Since $P_{1,\text{appar}} = P_{2,\text{appar}}$, it follows that

$$\frac{E_{1,\text{eff}}^2}{R_1} = \frac{E_{2,\text{eff}}^2}{R_2} \tag{13-43}$$

where R_2 is the secondary load resistance and R_1 is the resistance reflected from secondary into primary. Solving for R_1,

$$R_1 = R_2 \left(\frac{E_{1,\text{eff}}}{E_{2,\text{eff}}} \right)^2 \tag{13-44}$$

Since $a = E_{1,\text{eff}}/E_{2,\text{eff}}$, Eq. (13-44) can be written

$$R_1 = a^2 R_2 \tag{13-45}$$

Equation (13-45) indicates that a resistance connected across the secondary of a closely coupled transformer appears across the primary modified by the square of the turns ratio. This relationship is of great importance in audio circuitry and in the matching of impedances for power distribution.

The equations in this section are based on an ideal transformer. Their validity will be made clear to the reader by a more rigorous analysis of loosely coupled transformers.

EXAMPLE 13-7 A transformer has a step-up ratio of 1:4. If the primary voltage is 110 V, what is the secondary voltage?

$$a = \frac{N_1}{N_2} = 0.25$$

SOLUTION

$$E_{2,\text{eff}} = \frac{E_{1,\text{eff}}}{a} = \frac{100}{0.25}$$
$$= 400 \text{ V}$$

////

EXAMPLE 13.8 An ideal transformer has 100 turns in the primary and 8,000 turns in the secondary. If 100 V is applied to the primary and a resistance of 2 MΩ is connected across the secondary, what is the primary current?

SOLUTION

$$a = \frac{N_1}{N_2} = \frac{100}{8{,}000} = 1.25 \times 10^{-2}$$

$$E_{2,\text{eff}} = \frac{E_{1,\text{eff}}}{a} = \frac{100}{1.25 \times 10^{-2}}$$
$$= 8 \times 10^3 \text{ V}$$

$$I_{2,\text{eff}} = \frac{E_{2,\text{eff}}}{R_2} = \frac{8 \times 10^3}{2 \times 10^6}$$
$$= 4 \text{ mA}$$

$$I_{1,\text{eff}} = \frac{I_{2,\text{eff}}}{a} = \frac{4 \text{ mA}}{1.25 \times 10^{-2}}$$
$$= 320 \text{ mA}$$

An alternate method involves dividing the primary voltage by the impedance reflected to the primary.

$$R_1 = a^2 R_2 = (1.25 \times 10^{-2})^2 (2 \times 10^6)$$
$$= 3.12 \times 10^2 \ \Omega$$

$$I_{1,\text{eff}} = \frac{E_{1,\text{eff}}}{R_1} = \frac{100}{3.12 \times 10^2}$$
$$= 320 \text{ mA.}$$

////

EXAMPLE 13-9 In the circuit of Fig. 13-24, it is desired to match an 8-Ω speaker to the primary impedance of 2 kΩ. Which of the following transformers would be best for the purpose?

(*a*)	$N_1 = 600$	$N_2 = 20$
(*b*)	$N_1 = 320$	$N_2 = 20$
(*c*)	$N_1 = 500$	$N_2 = 20$
(*d*)	$N_1 = 800$	$N_2 = 20$

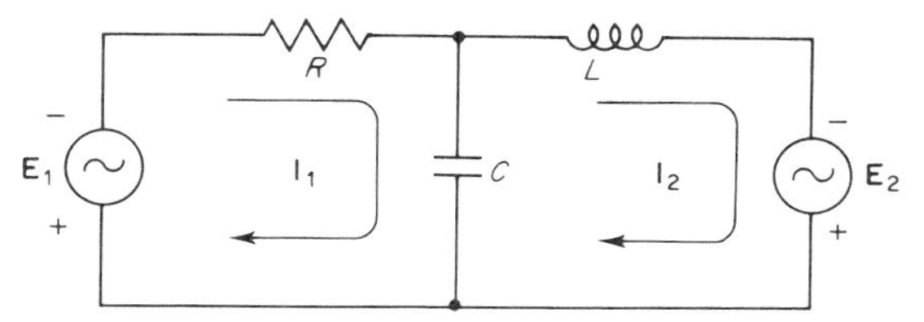

FIGURE 13-25
General conductive coupled circuit.

SOLUTION

$$R_1 = a^2 R_2$$

$$a = \sqrt{\frac{R_1}{R_2}} = \sqrt{\frac{2{,}000}{8}}$$

$$a = 15.8$$

All the transformer choices have $N_2 = 20$.

$$a = \frac{N_1}{N_2}$$

$$N_1 = aN_2 = (15.8)(20) = 316 \text{ turns}$$

Therefore select choice (*b*). ////

13-5 ANALYSIS OF LOOSELY COUPLED TRANSFORMERS

All transformers whose coefficients of coupling are less than 1 are classed as loosely coupled transformers. In such transformers, the permeabilities of the magnetic paths are not the same for primary and secondary, and in addition they have large leakage inductances. Therefore, analyses based on turns ratios are impossible.

Mesh equations could be written directly for a conductively coupled circuit, as shown in Fig. 13-25. The impedance common to the meshes appears negative in the mesh equations when the assumed mesh currents are taken in a clockwise direction. Mesh equations for Fig. 13-25 are

First mesh: $$\mathbf{E}_1 = \left(R + \frac{1}{j\omega C}\right)\mathbf{I}_1 - \frac{1}{j\omega C}\mathbf{I}_2$$

Second mesh: $$-\mathbf{E}_2 = -\frac{1}{j\omega C}\mathbf{I}_1 + \left(j\omega L + \frac{1}{j\omega C}\right)\mathbf{I}_2$$

Mesh equations can also be written for circuits that have magnetic coupling rather than conductive coupling between meshes. Once again the mesh currents

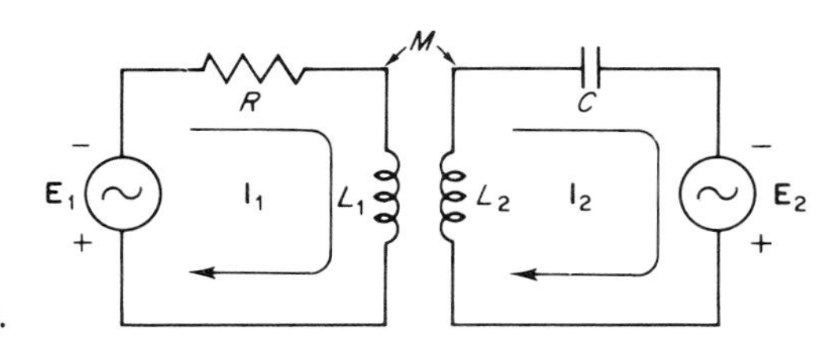

FIGURE 13-26
Circuit with magnetic coupling.

are assumed to flow in a clockwise direction. Figure 13-26 shows a magnetic coupling between mesh 1 and mesh 2. The two meshes are coupled through the mutual inductance M. A voltage is induced in mesh 1 due to M and the current in mesh 2. Likewise, a voltage is induced in mesh 2 due to M and the current in mesh 1. Of course these induced voltages must be considered when writing the mesh equations. However, are the induced voltages positive or negative in the mesh equations?

One of two methods may be used in determining the polarity of the induced voltage. The transformer can be physically examined, if possible, to determine the winding sense of the turns on primary and secondary. This method is not very practical since the transformer might require dissection or a great deal of time might be spent in determining the winding sense.

A far better method is the use of dot notation on the transformer. Figure 13-27 shows a transformer with dot notation. The dot notation indicates whether the induced voltage through mutual inductance will aid or oppose the voltage from self-inductance. Two basic rules for dot notation indicate the polarity of the induced voltage.

Rule 1 If the two assumed currents enter the dot terminals or terminals without dots, the polarity of the mutual coupled term is positive.

Rule 2 If one assumed current enters a dot and the other assumed current leaves a dot, the polarity of the mutual coupled term is negative.

Dot notation will be used in writing the mesh equations for the circuit in Fig. 13-28. Both $\mathbf{I}_1$ and $\mathbf{I}_2$ enter the dot terminals on the transformer. The polarity of $jM\mathbf{I}$ is positive.

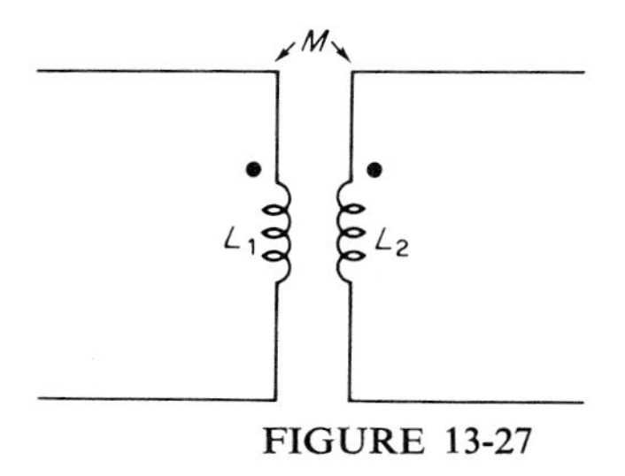

FIGURE 13-27

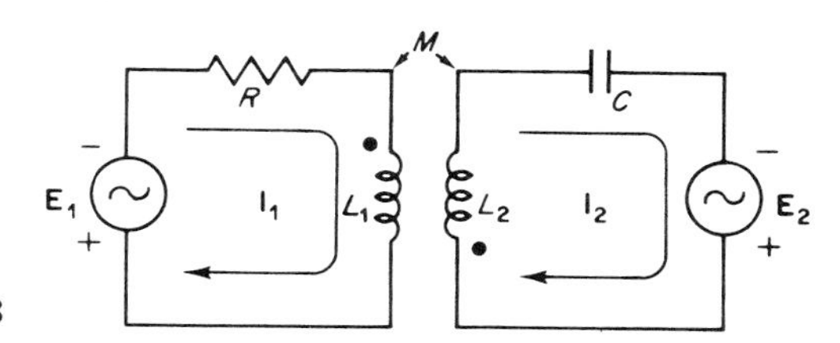

FIGURE 13-28

First mesh: $$E_1 = (R + j\omega L_1)\mathbf{I}_1 + (j\omega M)\mathbf{I}_2$$

Second mesh: $$-\mathbf{E}_2 = (j\omega M)\mathbf{I}_1 + \left(j\omega L_1 + \frac{1}{j\omega C}\right)\mathbf{I}_2$$

If the secondary of a transformer is terminated in a load impedance $\mathbf{Z}_L$, the reflected impedance can be found through a mesh-current analysis. Consider the circuit in Fig. 13-29. The two mesh equations are:

Primary: $$\mathbf{E}_p = (j\omega L_1)\mathbf{I}_p + (j\omega M)\mathbf{I}_s$$

Secondary: $$0 = (j\omega M)\mathbf{I}_p + (j\omega L_2 + \mathbf{Z}_L)\mathbf{I}_s$$

The total impedance at the primary terminals is

$$\mathbf{Z}_{\text{tot}} = \frac{\mathbf{E}_p}{\mathbf{I}_p}$$

solving mesh equations for $\mathbf{I}_p$,

$$\mathbf{I}_p = \frac{\begin{vmatrix} \mathbf{E}_p & j\omega M \\ 0 & j\omega L_2 + \mathbf{Z}_L \end{vmatrix}}{\begin{vmatrix} j\omega L_1 & j\omega M \\ j\omega M & j\omega L_2 + \mathbf{Z}_L \end{vmatrix}} = \frac{\mathbf{E}_p(j\omega L_2 + \mathbf{Z}_L)}{j\omega L_1(j\omega L_2 + \mathbf{Z}_L) + (\omega M)^2}$$

Let $\mathbf{Z}_s = j\omega L_2 + \mathbf{Z}_L$ represent the total impedance in the secondary. Then

$$\mathbf{I}_p = \frac{\mathbf{E}_p(\mathbf{Z}_s)}{j\omega L_1(\mathbf{Z}_s) + (\omega M)^2}$$

$$\mathbf{Z}_{\text{tot}} = \frac{\mathbf{E}_p}{\mathbf{I}_p}$$

Therefore: $$\mathbf{Z}_{\text{tot}} = j\omega L_1 + \frac{(\omega M)^2}{\mathbf{Z}_s} \qquad (13\text{-}46)$$

Since $j\omega L_1$ is associated with the primary, the reflected impedance from the secondary is

$$\mathbf{Z}_{\text{ref}} = \frac{(\omega M)^2}{\mathbf{Z}_s} \qquad (13\text{-}47)$$

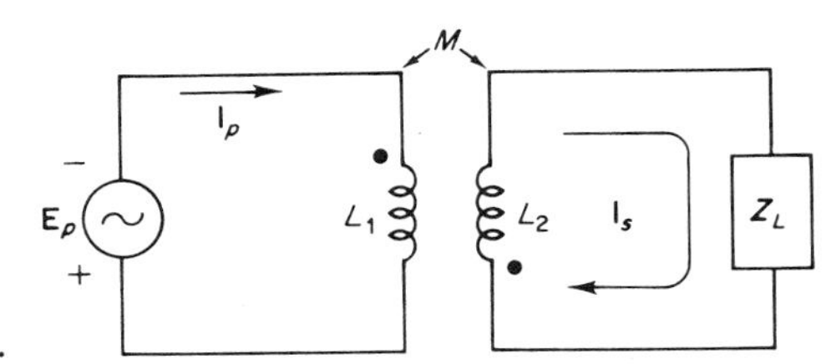

FIGURE 13-29
Secondary terminated in $\mathbf{Z}_L$.

Both Eqs. (13-46) and (13-47) are independent of dot orientation on the transformer. The reader can validate this by placing the dot of L_2 in Fig. 13-29 at the top and solving the resulting mesh equations. Thus, the reflected impedance is not simply the product of turns ratio squared and secondary impedance as was found in closely coupled ($k \approx 1$) transformers.

EXAMPLE 13-10 Given the circuit of Fig. 13-30, find (1) the primary current, (2) the secondary current, and (3) the voltage drop at R_L.

SOLUTION

First mesh:
$$2 \angle 0° = \left(R_1 + j\omega L_1 + \frac{1}{j\omega C_1}\right)\mathbf{I}_1 - (j\omega M)\mathbf{I}_2$$

Second mesh:
$$0 = -(j\omega M)\mathbf{I}_1 + \left(R_L + j\omega L_2 + \frac{1}{j\omega C_2}\right)\mathbf{I}_2$$

$$M = k\sqrt{L_1 L_2} = 0.6\sqrt{300 \times 10^{-6}} = 10.4 \text{ mH}$$

$$\omega L_1 = (2\pi)(3.8 \times 10^5)(1.5 \times 10^{-2}) = 3.58 \times 10^4$$

$$\frac{1}{\omega C_1} = \frac{1}{\omega C_2} = \frac{1}{(2\pi)(0.38 \times 10^6)(0.2 \times 10^{-10})} = 2.1 \times 10^4$$

$$\omega L_2 = (2\pi)(3.8 \times 10^5)(2.0 \times 10^{-2}) = 4.76 \times 10^4$$

$$\omega M = (2\pi)(3.8 \times 10^5)(1.04 \times 10^{-2}) = 2.48 \times 10^4$$

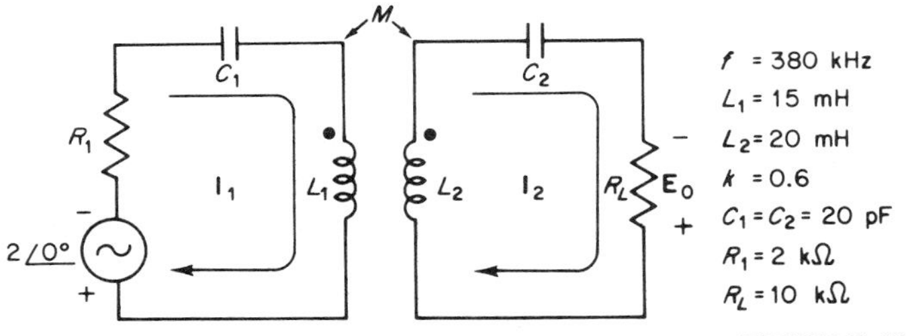

FIGURE 13-30
Circuit for Example 13-10.

Substituting values in the mesh equations,

First mesh: $2\angle 0° = 15\ \text{k}\Omega\ \angle 82.3°\ \mathbf{I}_1 - 24.8\ \text{k}\Omega\ \angle 90°\ \mathbf{I}_2$

Second mesh: $0 = -24.8\ \text{k}\Omega\ \angle 90°\ \mathbf{I}_1 + 28.4\ \text{k}\Omega\ \angle 69.5°\ \mathbf{I}_2$

1 Solve for $\mathbf{I}_1$.

$$\mathbf{I}_1 = \frac{\begin{vmatrix} 2\angle 0° & -24.8\ \text{k}\Omega\ \angle 90° \\ 0 & 28.4\ \text{k}\Omega\ \angle 69.5° \end{vmatrix}}{\begin{vmatrix} 15\ \text{k}\Omega\ \angle 82.3° & -24.8\ \text{k}\Omega\ \angle 90_{\circ} \\ -24.8\ \text{k}\Omega\ \angle 90° & 28.4\ \text{k}\Omega\ \angle 69.5° \end{vmatrix}} = \frac{56.8\ \text{k}\Omega\ \angle 69.5°}{31.4 \times 10^7\ \angle 40°}$$

$$= 181\ \angle 29.5°\ \mu\text{A}$$

2 Solve for $\mathbf{I}_2$.

$$\mathbf{I}_2 = \frac{\begin{vmatrix} 15\ \text{k}\Omega\ \angle 82.3° & 2\angle 0° \\ -24.8\ \text{k}\Omega\ \angle 90° & 0 \end{vmatrix}}{31.4 \times 10^7\ \angle 40°} = \frac{49.6\ \text{k}\Omega\ \angle 90°}{31.4 \times 10^7\ \angle 40°}$$

$$= 158\ \angle 50°\ \mu\text{A}$$

3 Solve for $\mathbf{E}_0$.

$$\mathbf{E}_0 = R_L \mathbf{I}_2 = (10\ \text{k}\Omega)(0.158 \times 10^{-3}\ \angle 50°)$$

$$= 1.58\ \angle 50°\ \text{V}$$

////

EXAMPLE 13-11 Given the circuit of Fig. 13-30, find (1) the total impedance seen at primary terminals, (2) the reflected impedance, and (3) the primary current.

SOLUTION

1

$$\mathbf{Z}_{\text{tot}} = j\omega L_1 + \frac{(\omega M)^2}{\mathbf{Z}_s}$$

$$\mathbf{Z}_s = R_L + j\omega L_2 - \frac{1}{j\omega C_2}$$

$$= 28.4\ \text{k}\Omega\ \angle 69.5°$$

$$\mathbf{Z}_{\text{tot}} = j35.8\ \text{k}\Omega + \frac{6.15 \times 10^8}{28.4\ \text{k}\Omega\ \angle 69.5°}$$

$$= 7.55\ \text{k}\Omega + j15.6\ \text{k}\Omega$$

2

$$\mathbf{Z}_{\text{ref}} = \frac{(\omega M)^2}{\mathbf{Z}_s} = 7.55\ \text{k}\Omega - j20.2\ \text{k}\Omega$$

3

$$\mathbf{I}_1 = \frac{2\angle 0°}{R_1 + 1/j\omega C_1 + \mathbf{Z}_{\text{tot}}} = \frac{2\angle 0°}{9.55\ \text{k}\Omega - j5.4\text{k}\Omega}$$

$$= 182\ \angle 29.5°\ \mu\text{A}$$

////

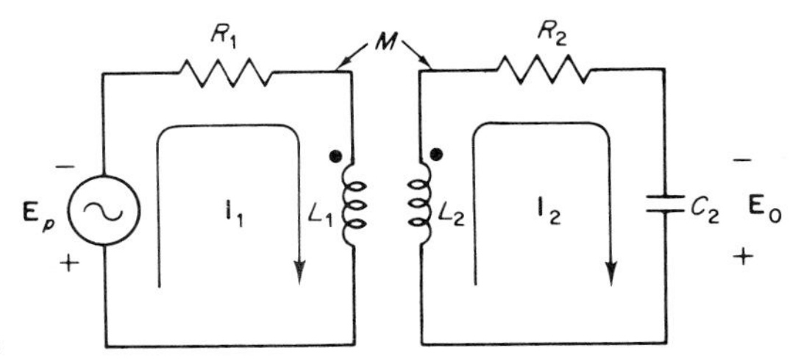

FIGURE 13-31
Transformer with tuned secondary.

Note this value corresponds to $\mathbf{I}_1$ found in the previous example.

Because loosely coupled transformers are used at frequencies well above the audio range, the properties of resonant circuits are commonly added to the transformer-coupled circuits. There are two general types of transformer-coupled circuits with low coefficients of coupling: the tuned-secondary–untuned-primary circuit and the tuned-secondary–tuned-primary (bandpass-coupled) circuit.

A transformer-coupled circuit with tuned secondary and untuned primary is shown in Fig. 13-31. R_2 represents the winding resistance of the secondary inductance L_2. At resonance, the inductive reactance of L_2 equals the capacitive reactance of C_2. The mesh equations at resonance are

First mesh:
$$\mathbf{E}_p = (R_1 + j\omega_0 L_1)\mathbf{I}_1 - (j\omega_0 M)\mathbf{I}_2$$

Second mesh:
$$0 = -(j\omega_0 M)\mathbf{I}_1 + \left(R_2 + j\omega_0 L_2 - j\frac{1}{\omega_0 C_2}\right)\mathbf{I}_2$$

since $\omega_0 L_2 = 1/\omega_0 C_2$ at resonance,

$$\mathbf{E}_p = (R_1 + j\omega_0 L_1)\mathbf{I}_1 - (j\omega_0 M)\mathbf{I}_2$$
$$0 = -(j\omega_0 M)\mathbf{I}_1 + (R_2)\mathbf{I}_2$$

Output voltage:
$$\mathbf{E}_0 = \left(-j\frac{1}{\omega_0 C_2}\right)\mathbf{I}_2 = (-j\omega_0 L_2)\mathbf{I}_2$$

Solve for $\mathbf{I}_2$.

$$\mathbf{I}_2 = \frac{\begin{vmatrix} R_1 + j\omega_0 L_1 & \mathbf{E}_p \\ -j\omega_0 M & 0 \end{vmatrix}}{\begin{vmatrix} R_1 + j\omega_0 L_1 & -j\omega_0 M \\ -j\omega_0 M & R_2 \end{vmatrix}} = \frac{j\omega_0 M\mathbf{E}_p}{R_2(R_1 + j\omega_0 L_1) + (\omega_0 M)^2}$$

Solve for $\mathbf{I}_1$.

$$\mathbf{I}_1 = \frac{\begin{vmatrix} \mathbf{E}_p & -j\omega_0 M \\ 0 & R_2 \end{vmatrix}}{R_2(R_1 + j\omega_0 L_1) + (\omega_0 M)^2} = \frac{R_2\mathbf{E}_p}{R_2(R_1 + j\omega_0 L_1) + (\omega_0 M)^2}$$

Expressing $\mathbf{E}_p$ as a function of $\mathbf{I}_1$,

$$\mathbf{E}_p = \frac{R_2(R_1 + j\omega_0 L_1) + (\omega_0 M)^2}{R_2} \mathbf{I}_1$$

Substituting $\mathbf{E}_p$ into the equation for $\mathbf{I}_2$ gives

$$\mathbf{I}_2 = \frac{j\omega_0 M}{R_2} \mathbf{I}_1$$

Output voltage is given by

$$\mathbf{E}_0 = (-j\omega_0 L_2)\frac{j\omega_0 M}{R_2} \mathbf{I}_1$$

$$= \frac{\omega_0 M \omega_0 L_2}{R_2} \mathbf{I}_1$$

The unloaded Q of the secondary coil is

$$Q_2 = \frac{\omega_0 L_2}{R_2}$$

Therefore, $\mathbf{E}_0 = \omega_0 M Q_2 \mathbf{I}_1$ adjacent dots (13-48)

If the transformer dots are staggered,

$\mathbf{E}_0 = -\omega_0 M Q_2 \mathbf{I}_1$ staggered dots (13-49)

It is of utmost importance for the reader to note the significance of Eqs. (13-48) *and* (13-49). They state that *the voltage developed across the tuning capacitance in a tuned-secondary transformer-coupled circuit is the Q of the secondary times the voltage induced from primary to secondary.*

EXAMPLE 13-12 In the circuit of Fig. 13-32, find the voltage ($\mathbf{E}_0$) across the output terminals of the secondary. $f_o = 8.65$ MHz, $L_1 = 24$ μH, $L_2 = 18$ μH, $k = 0.09$, $R_2 = 5$ Ω, and C_2 is adjusted to resonance. Assume $\mathbf{I}_1 = 3\angle 0°$ mA.

$$M = k\sqrt{L_1 L_2} = 0.09\sqrt{432 \times 10^{-12}}$$

$$= 1.87\ \mu\text{H}$$

Secondary Q is

$$Q_2 = \frac{\omega_0 L_2}{R_2} = \frac{(2\pi)(8.65 \times 10^6)(0.18 \times 10^{-4})}{5}$$

$$= 196$$

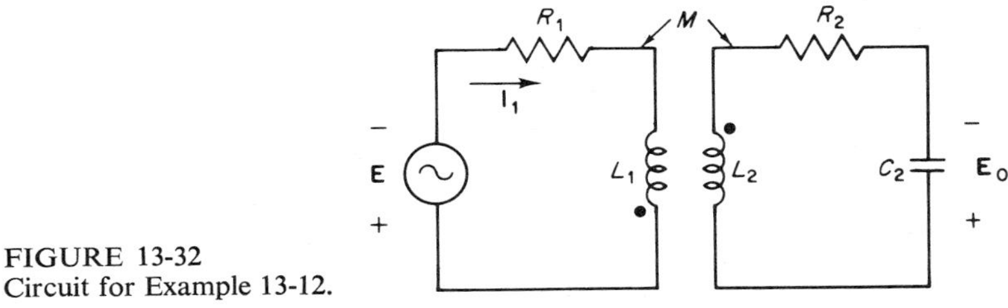

FIGURE 13-32
Circuit for Example 13-12.

Since the dots are staggered,

$$\mathbf{E}_0 = -\omega_0 M Q_2 \mathbf{I}_1$$
$$= -(2\pi)(8.65 \times 10^6)(1.87 \times 10^{-6})(0.196 \times 10^3)(3 \times 10^{-3} \angle 0°)$$
$$= 59.6 \angle 180° \text{ V}$$

If $\mathbf{E}$ instead of $\mathbf{I}_1$ is known, $\mathbf{I}_1$ can be calculated from

$$\mathbf{I}_1 = \frac{\mathbf{E}}{R_1 + \mathbf{Z}_{tot}} \qquad \text{where } \mathbf{Z}_s = R_2 \text{ at resonance} \qquad ////$$

The output voltage of the coupling circuit of Fig. 13-32 has frequency-response characteristics like those of any resonant circuit. The magnitude of the output voltage is directly proportional to the Q of the secondary, and the bandwidth is inversely proportional to the secondary Q. Figure 13-33 illustrates the output characteristics of the circuit of Fig. 13-32 for various values of Q.

Figure 13-33 illustrates one of the problems associated with untuned-primary–tuned-secondary coupling. To obtain broadband frequency-response characteristics,

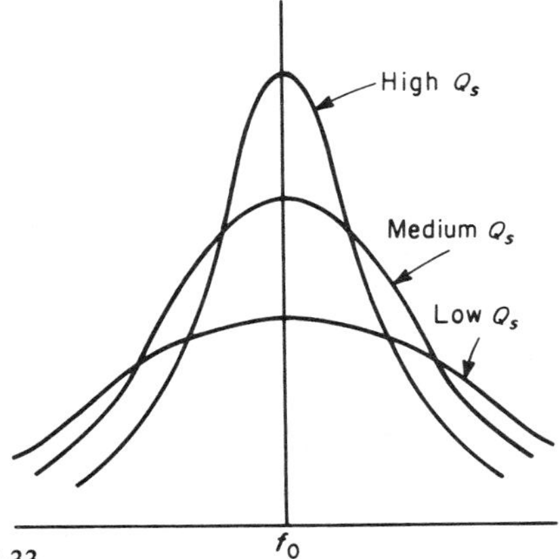

FIGURE 13-33
Variation of frequency response with secondary Q for the circuit of Fig. 13-32.

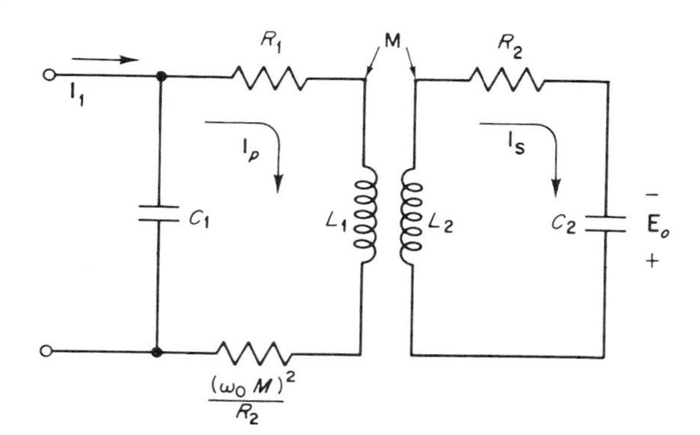

FIGURE 13-34
Loosely coupled transformer with tuned primary and tuned secondary.

the Q of the secondary must be low. This reduces the output amplitude and causes the circuit to develop appreciable output voltages at frequencies outside the desired bandwidth. The solution to the problem is found in tuning both primary and secondary. A transformer-coupled circuit of this type is often referred to as a *bandpass-coupled circuit*. Figure 13-34 illustrates such a circuit. This circuit is analyzed by reflecting the secondary impedance into the primary. $(\omega_0 M)^2/R_2$ represents the reflected impedance at resonance. The primary current can be determined from the CDR.

$$\mathbf{I}_p = \mathbf{I}_1 \frac{\dfrac{1}{R_1 + (\omega_0 M)^2/R_2 + j\omega_0 L_1}}{j\omega_0 C_1 + \dfrac{1}{R_1 + (\omega_0 M)^2/R_2 + j\omega_0 L_1}} \tag{13-50}$$

Rearranging Eq. (13-50) gives

$$\mathbf{I}_p = \mathbf{I}_1 \frac{1}{1 + j\omega_0 C_1[R_1 + (\omega_0 M)^2/R_2 + j\omega_0 L_1]} \tag{13-51}$$

$$\omega_0 C_1 = \frac{1}{\omega_0 L_1} \qquad \text{and} \qquad M = k\sqrt{L_1 L_2}$$

Substituting into Eq. (13-51),

$$\mathbf{I}_p = \mathbf{I}_1 \frac{1}{1 + j\dfrac{R_1}{\omega_0 L_1} + j\dfrac{k^2 \omega_0 L_1 \omega_0 L_2}{\omega_0 L_1 R_2} - 1} \tag{13-52}$$

$$Q_1 = \frac{\omega_0 L_1}{R_1} \qquad \text{and} \qquad Q_2 = \frac{\omega_0 L_2}{R_2}$$

Substituting into Eq. (13-52),

$$\mathbf{I}_p = \mathbf{I}_1 \frac{1}{j\left(\dfrac{1}{Q_1} + k^2 Q_2\right)} \tag{13-53}$$

Write the second mesh equation,

$$0 = -j\omega_0 M \mathbf{I}_p + R_2 \mathbf{I}_s \qquad \text{at resonance} \tag{13-54}$$

Solve for secondary current.

$$\mathbf{I}_s = \frac{j\omega_0 M}{R_2} \mathbf{I}_p \tag{13-55}$$

Substitute Eq. (13-53) into Eq. (13-55).

$$\mathbf{I}_s = \frac{\omega_0 M}{R_2\left(\dfrac{1}{Q_1} + k^2 Q_2\right)} \mathbf{I}_1$$

$$\mathbf{E}_0 = \left(-j\frac{1}{\omega_0 C_2}\right)\mathbf{I}_s$$

$$= -j\frac{\omega_0 M}{\omega_0 C_2 R_2\left(\dfrac{1}{Q_1} + k^2 Q_2\right)} \mathbf{I}_1$$

$$\omega_0 C_2 = \frac{1}{\omega_0 L_2} \qquad \text{and} \qquad \frac{R_2}{\omega_0 L_2} = \frac{1}{Q_2}$$

$$\mathbf{E}_0 = -j\frac{\omega_0 M}{k^2 + \dfrac{1}{Q_1 Q_2}} \mathbf{I}_1 \qquad \text{dots adjacent} \tag{13-56}$$

$$\mathbf{E}_0 = j\frac{\omega_0 M}{k^2 + \dfrac{1}{Q_1 Q_2}} \mathbf{I}_1 \qquad \text{dots staggered} \tag{13-57}$$

Equations (13-56) and (13-57) give the magnitude of the output voltage at the resonant frequency only. It is of interest to note that the voltage developed across the output capacitor in the bandpass-coupled circuit is displaced 90° from the voltage across the output terminals of the tuning capacitor in the singly tuned transformer-coupled circuit. *It is of utmost importance to note that the current* $\mathbf{I}_1$ *in Eqs.* (13-56) *and* (13-57) *is the current flowing into the primary circuit from an external source and is not the current through the primary.*

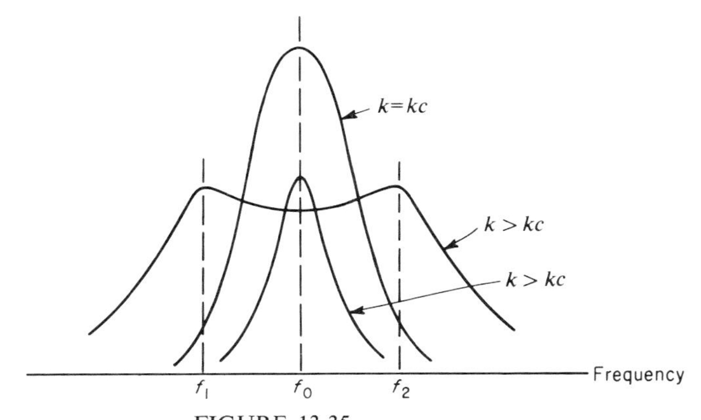

FIGURE 13-35
Bandwidth as a function of coupling in a bandpass-coupled circuit.

The frequency-response characteristics of the bandpass-coupled circuit are shown in Fig. 13-35. In all three curves, primary and secondary circuits are unchanged, loading is constant, and the respective Q's are constant. *The only change is in the coefficient of coupling. At this point, critical coupling must be defined. Critical coupling is that coefficient of coupling which yields a reflected resistance exactly equal to the primary-circuit resistance.*

$$k_c = \frac{1}{\sqrt{Q_1 Q_2}} \tag{13-58}$$

The equation is derived as follows:

$$R_1 = \frac{(\omega_0 M)^2}{R_2}$$

$$R_1 R_2 = k_c^2 \omega_0 L_1 \omega_0 L_2$$

$$k_c^2 = \frac{R_1 R_2}{\omega_0 L_1 \omega_0 L_2}$$

$$k_c^2 = \frac{1}{Q_1 Q_2}$$

$$k_c = \frac{1}{\sqrt{Q_1 Q_2}}$$

For coefficients of coupling less than critical coupling, the bandwidth is narrow, less than the bandwidth of a singly tuned circuit. This is because the reflected impedance is small and has little effect on the primary circuit. The entire circuit behaves

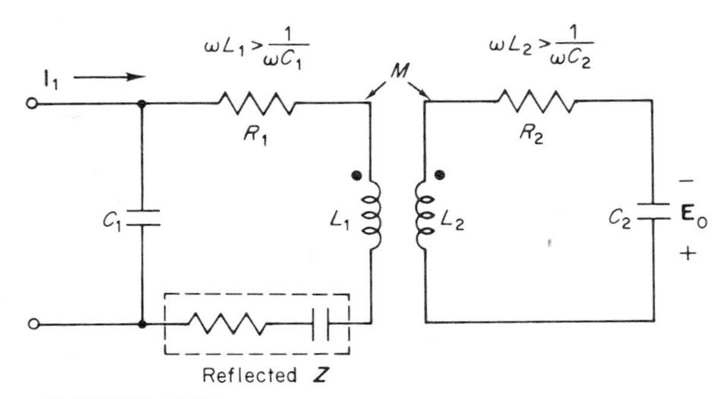

FIGURE 13-36
Equivalent circuit for Fig. 13-34 when frequency of operation is slightly less than the resonant frequency of the tuned circuits.

like two separate tuned circuits and has overall bandpass characteristics of two separate tuned circuits. The output voltage at resonance is less than the output at critical coupling because the voltage induced from primary to secondary by the low mutual inductance is so small.

At critical coupling, the output voltage at resonance is more than for any other degree of coupling. Critical coupling is the optimum compromise between high reflected impedance and high mutual inductance. High voltages are induced from primary to secondary. When coupling exceeds critical coupling, the output voltage at resonance decreases because the reflected resistance reduces primary current and voltage induced in the secondary.

Critical coupling also gives a frequency-response characteristic with fairly uniform output at frequencies slightly off the resonant frequency of the circuit. This can be explained with the aid of Fig. 13-36. In this illustration, consider the frequency of operation to be slightly higher than resonant frequency. Both circuits then appear slightly inductive. The reflected impedance is capacitive, and this reflected capacitive reactance helps to make the primary circuit resonant at this frequency. Thus primary current remains high, inducing high voltage in the secondary and maintaining an output voltage that is almost as great as the voltage at resonance. A similar analysis applies to frequencies slightly below the resonant frequency of the circuit.

The curve of the overcoupled bandpass-coupled circuit (Fig. 13-35) shows two peaks in the output. This would lead one to suspect that resonance occurs at two frequencies other than the original resonant frequency of the system, and this is indeed the case. At the original resonant frequency of the circuit, output is low because of the high reflected resistance. The primary current is very low, and therefore

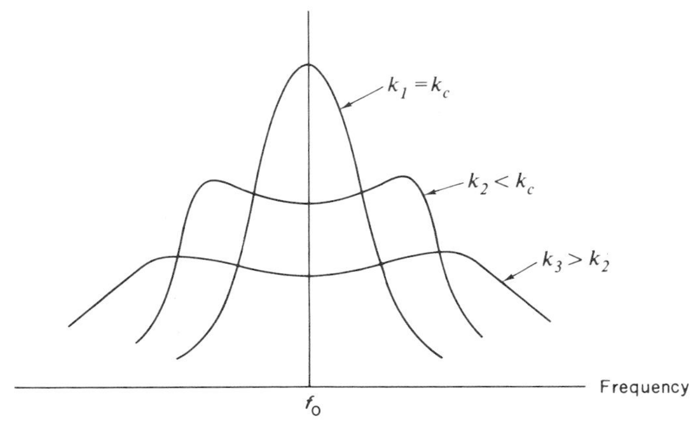

FIGURE 13-37
Bandwidth at critical coupling, slight overcoupling, and very great overcoupling.

the voltage induced from primary to secondary is low. At frequencies above and below the original resonant frequency, secondary impedance is high and reactive. This gives rise to reflected impedance that is low and of opposite sign to the reactance of the primary. At f_1 and f_2 the reflected reactance has canceled the primary reactance, yielding resonance at these points. The resulting high primary currents induce high voltages in the secondary circuit. If the coefficient of coupling is reduced, but remains more than k_c, the two peaks move closer together. When the coefficient of coupling is increased, the peaks move farther apart. This is shown in Fig. 13-37.

EXAMPLE 13-13 Given the circuit of Fig. 13-38, calculate (1) the output voltage at critical coupling and at a coefficient of coupling five times critical coupling for $f = f_o$,

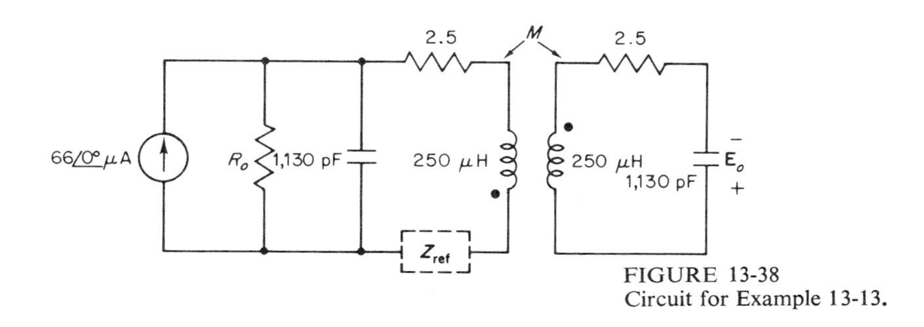

FIGURE 13-38
Circuit for Example 13-13.

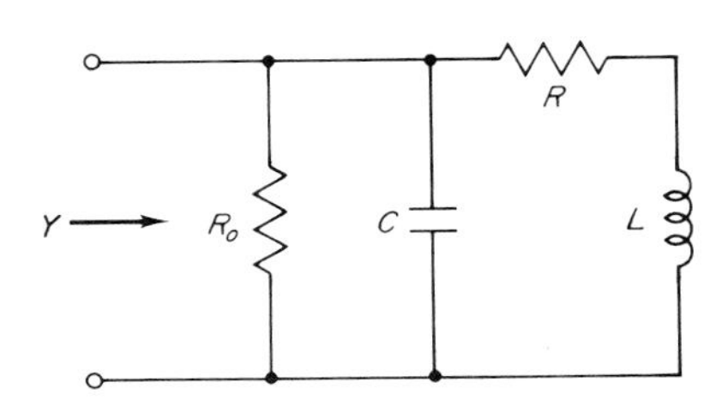

FIGURE 13-39

and (2) the output voltage at 295 kHz for each value of coupling; $f_o = 300$ kHz; $R_o = 63$ kΩ.

SOLUTION

1 The output resistance of the source shunts the primary and thus affects the Q. R_o may be lumped with other components in the primary circuit. Consider the circuit in Fig. 13-39.

$$\mathbf{Y} = \frac{1}{R_o} + j\omega_0 C + \frac{1}{R + j\omega_0 L}$$

$$= \frac{1}{R_o} + j\omega_0 C + \frac{R - j\omega_0 L}{R^2 + (\omega_0 L)^2}$$

Q's of inductors in tuned circuits are larger than 10.

$$Q = \frac{\omega_0 L}{R} > 10$$

Thus
$$(\omega_0 L)^2 \gg R^2$$

$$\mathbf{Y} = \frac{1}{R_o} + j\omega_0 C + \frac{R}{(\omega_0 L)^2} - j\frac{1}{\omega_0 L}$$

$$\omega_0 C = \frac{1}{\omega_0 L}$$

Therefore,
$$\mathbf{Y} = \frac{1}{R_o} + \frac{R}{(\omega_0 L)^2} = \frac{1}{R_o} + \frac{1}{R_p}$$

where $R_p = (\omega_0 L)^2/R$ represents the parallel equivalent resistance of the series resistance in L. The total parallel resistance of the primary is

$$R_{eq} = R_o \parallel R_p = \frac{R_o R_p}{R_o + R_p}$$

Since
$$R_p = (\omega_0 L)^2/R, \; R = (\omega_0 L)^2/R_p.$$

Thus the series equivalent of R_{eq} is

$$R' = \frac{(\omega_0 L)^2}{R_{eq}}$$

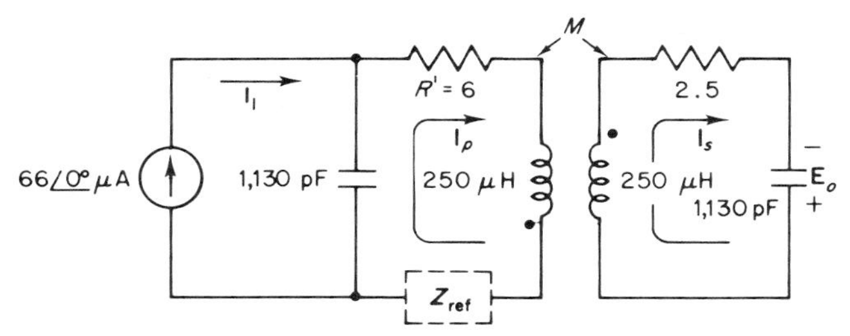

FIGURE 13-40

Returning to the example,

$$R_p = \frac{(\omega_0 L_1)^2}{R} = \frac{(470)^2}{2.5} = 88.4\ \text{k}\Omega$$

$$R_{eq} = R_o \parallel R_p = 63\ \text{k}\Omega \parallel 88.4\ \text{k}\Omega = 36.7\ \text{k}\Omega$$

$$R' = \frac{(\omega_0 L_1)^2}{R_{eq}} = \frac{(470)^2}{36.7\ \text{k}\Omega} = 6\ \Omega$$

Figure 13-40 shows the circuit equivalent to the one in Fig. 13-38.

$$Q_1 = \frac{\omega_0 L_1}{R'} = \frac{470}{6} = 78.5$$

$$Q_2 = \frac{\omega_0 L_2}{R_2} = \frac{470}{2.5} = 188$$

Note how R_o affected Q_1 compared to Q_2.

$$k_c = \frac{1}{\sqrt{Q_1 Q_2}} = 0.0082$$

$$M = k\sqrt{L_1 L_2} = 2.05\ \mu\text{H}$$

$$\mathbf{Z}_{\text{ref}} = \frac{(\omega_0 M)^2}{R_2} = 6\ \Omega$$

Since the dots are staggered, refer to Eq. (13-57).

$$\mathbf{E}_o = j\frac{\omega_0 M}{k^2 + \dfrac{1}{Q_1 Q_2}}\mathbf{I}_1$$

$$= j\frac{3.86}{1.35 \times 10^{-4}}(66 \times 10^{-6})$$

$$= j1.88\ \text{V} \qquad \frac{k = k_c}{f = f_o}$$

When $k = 5k_c$,

$$M = k\sqrt{L_1 L_2} = 10.25\ \mu\text{H}$$

$$\mathbf{Z}_{\text{ref}} = \frac{(\omega_0 M)^2}{R_2} = 150\ \Omega$$

$$\mathbf{E}_o = j\frac{\omega_0 M}{k^2 + \dfrac{1}{Q_1 Q_2}}\,\mathbf{I}_1 = j\frac{19.3}{17.5 \times 10^{-4}}(66 \times 10^{-6})$$

$$= j0.73\ \text{V} \qquad \frac{k = 5k_c}{f = f_o}$$

2 Output voltage at $f = 295$ kHz.

$$k = k_c = 0.0082$$

$$M = 2.05\ \mu\text{H}$$

$$\mathbf{Z}_{\text{ref}} = \frac{(\omega M)^2}{\mathbf{Z}_s}$$

$\mathbf{Z}_{\text{ref}}$ is not a resistance since $\mathbf{Z}_s$ is not a pure resistance ($f \neq f_o$). Thus, Eqs. (13.56) and (13.57) cannot be used because $(\omega M)^2/\mathbf{Z}_s$ was resistive in the derivation.

$$\mathbf{Z}_s = j\omega L_2 - j\frac{1}{\omega C_2} + R_2 = j463 - j478 + 2.5$$

$$= 15.2\ \angle -80.5^\circ$$

$$\mathbf{Z}_{\text{ref}} = \frac{(\omega M)^2}{Z_S} = 0.95\ \angle 80.5^\circ = 0.156 + j0.936$$

Using the CDR,

$$\mathbf{I}_p = \mathbf{I}_1 \frac{\dfrac{1}{R' + j\omega L_1 + \mathbf{Z}_{\text{ref}}}}{j\omega C_1 + \dfrac{1}{R' + j\omega L_1 + \mathbf{Z}_{\text{ref}}}}$$

$$= 66\ \mu\text{A}\ \frac{\dfrac{1}{6.156 + j463.936}}{j2.09 \times 10^{-3} + \dfrac{1}{6.156 + j463.936}}$$

$$= 2.02\ \angle -23.3^\circ\ \text{mA}$$

The mesh equation for the secondary gives

$$0 = (j\omega M)\mathbf{I}_p + \left(R_2 + j\omega L_2 - j\frac{1}{\omega C_2}\right)\mathbf{I}_s$$

$$\mathbf{I}_s = \frac{-j\omega M \mathbf{I}_p}{R_2 + j\omega L_2 - j\dfrac{1}{\omega C_2}} = \frac{(3.8\angle -90^\circ)(2.02\angle -23.3^\circ \text{ mA})}{15.2\angle -80.5^\circ}$$

$$= 0.505\angle -32.8^\circ \text{ mA}$$

$$\mathbf{E}_o = \left(-j\frac{1}{\omega C_2}\right)\mathbf{I}_s = (478\angle -90^\circ)(0.505\angle -32.8^\circ \text{ mA})$$

$$= 0.242\angle -122.8^\circ \text{ V} \qquad \frac{k = k_c}{f = 295 \text{ kHz} \neq f_o}$$

when $k = 5k_c$,

$$k = 0.041$$

$$M = 10.25\ \mu\text{H}$$

$$\mathbf{Z}_{\text{ref}} = \frac{(\omega M)^2}{\mathbf{Z}_s} = \frac{361}{15.2\angle -80.5^\circ}$$

$$= 23.8\angle 80.5^\circ = 3.93 + j23.5$$

Using the CDR,

$$\mathbf{I}_p = \mathbf{I}_1 \frac{\dfrac{1}{R' + j\omega L_1 + \mathbf{Z}_{\text{ref}}}}{j\omega C_1 + \dfrac{1}{R' + j\omega L_1 + \mathbf{Z}_{\text{ref}}}}$$

$$= 66\ \mu\text{A}\ \frac{1}{0.0268\angle 129^\circ} = 2.46\angle -129^\circ \text{ mA}$$

$$\mathbf{I}_s = \frac{-j\omega\ \ \mathbf{I}_p}{R_2 + j\omega L_2 - j\dfrac{1}{\omega C_2}} = \frac{19\angle -90\ 2.46\angle -129^\circ \text{ mA}}{15.2\angle -80.5^\circ}$$

$$= 3.08\angle -138.5^\circ \text{ mA}$$

$$\mathbf{E}_o = \left(-j\frac{1}{\omega C_2}\right)\mathbf{I}_s = (478\angle -90^\circ)(3.08\angle -138.5^\circ \text{ mA})$$

$$= 1.47\angle 131.5^\circ \text{ V} \qquad \frac{k = 5k_c}{f = 295 \text{ kHz} \neq f_o}$$

////

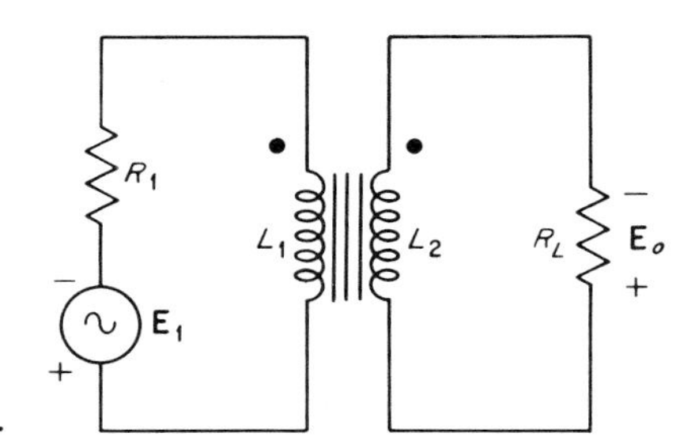

FIGURE 13-41
Iron-core transformer with load.

The solution of Example 13-13 was certainly laborious and involved many steps. Primarily, this example demonstrates to the reader the effect that increasing the coupling has upon the frequency response of a bandpass-coupled circuit. This type of coupled circuit is very widely used, especially in superheterodyne radio receivers.

13-6 THE FREQUENCY RESPONSE OF IRON-CORE TRANSFORMERS

In the previous section, the behavior of loosely coupled transformers was discussed. The manner in which output voltage varies with frequency was covered, and we saw that resonant circuits limit useful output to a relatively narrow range of frequencies. In the so-called iron-core type of transformer, output voltages are available over a wide range of frequencies. In the analysis of transformer response over this wide frequency range, the frequency range will be divided into three sections. The middle range will be defined as that range of frequencies over which the output of the coupling circuit is constant. Low frequencies are those below middle range, and highs are those above middle-range frequencies. Note that the middle frequencies have no definite range, but may vary from circuit to circuit.

Consider the circuit of Fig. 13-41. It is desired to use input signals over a frequency range from 100 Hz to 15 kHz and for this range of frequencies to write

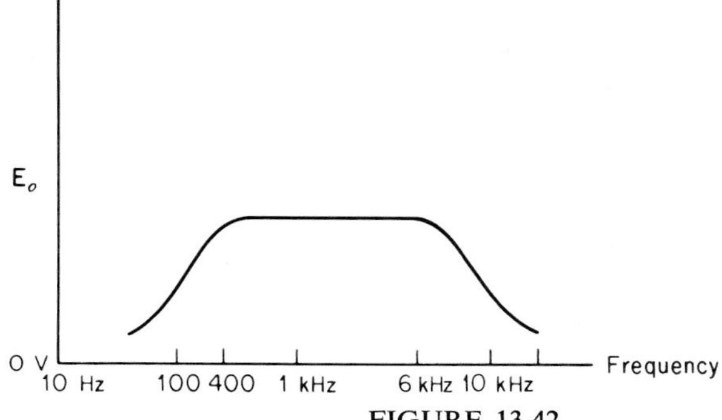

FIGURE 13-42
Frequency response of the circuit of Fig. 13-41.

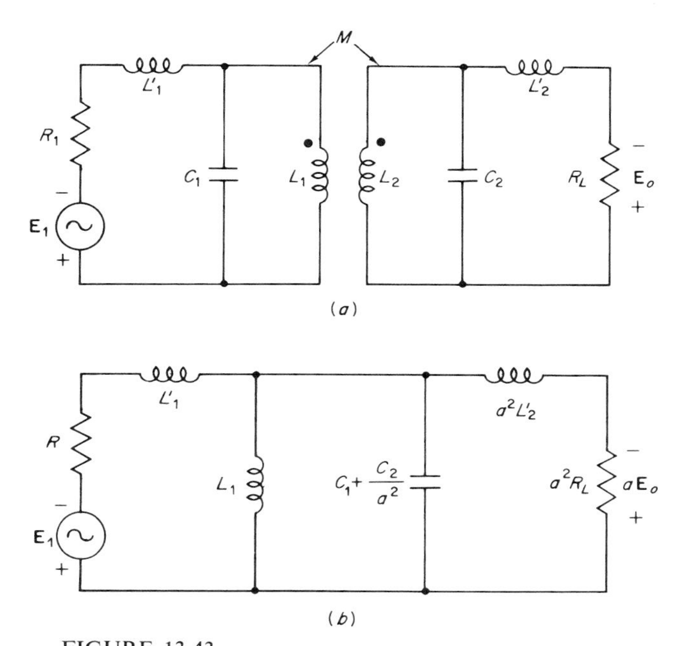

FIGURE 13-43
(*a*) Equivalent circuit of Fig. 13-41. (*b*) T equivalent circuit of Fig. 13-41 with all secondary quantities referred back to the primary.

equations for the output voltage. Measurements of output voltage versus frequency yield the results shown in Fig. 13-42. In this graph, it is at once apparent that the middle frequencies range from 400 Hz to 6 kHz. To analyze the behavior of the coupling circuit, an equivalent circuit must be developed in which all of the circuit appears on the input, or primary, side. This equivalent circuit is shown in Fig. 13-43. The load, multiplied by the square of the turns ratio, is reflected back into the primary. The secondary leakage inductance has been included after being properly modified by the square of the turns ratio. The output voltage, with proper modification, is also shown on the primary side. In the figure,

$$a = \frac{N_1}{N_2} = \sqrt{\frac{L_1}{L_2}} \qquad \text{since } k \approx 1.0$$

L_1 = incremental primary inductance
L_2 = incremental secondary inductance
L'_1 = primary leakage inductance
L'_2 = secondary leakage inductance
C_1, C_2 = capacitance between turns

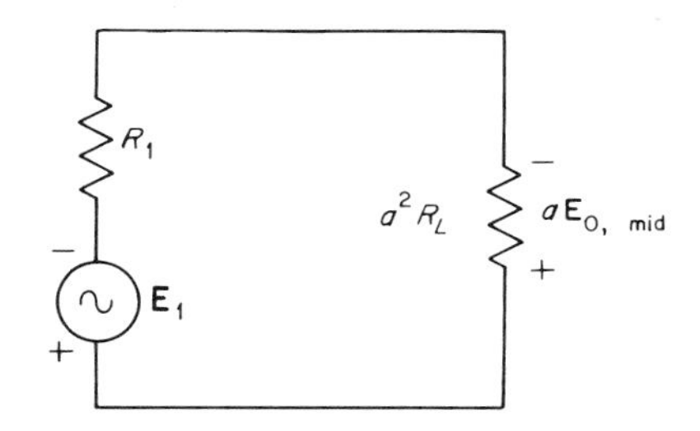

FIGURE 13-44
Mid-frequency equivalent circuit of Fig. 13-41.

Middle-frequency Analysis

Inspection of the circuit of Fig. 13-43*b* shows that the circuit contains many reactive components, yet the output, as shown in the graph of Fig. 13-42, is constant over a range of frequencies from 400 Hz to 6 kHz. This must mean that over this range of frequencies the circuit behaves as if it consisted of pure resistance. This is indeed true. At mid-frequencies, the leakage inductance has too low a reactance to be considered part of the circuit. On the other hand, the shunt capacitance has no effect over the entire range of the amplifier because of its high reactance in comparison to $a^2 R_L$. The reactance of the primary incremental inductance L_1 is so great as to appear as an open in this range of frequencies. The equivalent circuit then reduces to the very simple circuit of Fig. 13-44, and the equation for the output voltage can be written immediately.

$$a\mathbf{E}_{o,\text{mid}} = \mathbf{E}_1 \frac{a^2 R_L}{R_1 + a^2 R_L}$$

$$\mathbf{E}_{o,\text{mid}} = \frac{aR_L}{R_1 + a^2 R_L} \mathbf{E}_1 \qquad (13\text{-}59)$$

Low-frequency Analysis

At frequencies below mid-range, the leakage inductance still looks like a short circuit in comparison to the rest of the circuit. However, the reactance of the incremental inductance is decreasing and can no longer be considered to be an open circuit. The equivalent circuit accurate for the low frequencies is shown in Fig. 13-45.

Application of Thévenin's theorem to the circuit of Fig. 13-45 yields the circuit of Fig. 13-46. The reader will note that the open-circuit voltage (with L_1 removed) is the same as the mid-frequency voltage across the reflected load resistance.

$$a\mathbf{E}_{o,\text{low}} = a\mathbf{E}_{o,\text{mid}} \frac{j\omega L_1}{R_e + j\omega L_1} \qquad (13\text{-}60)$$

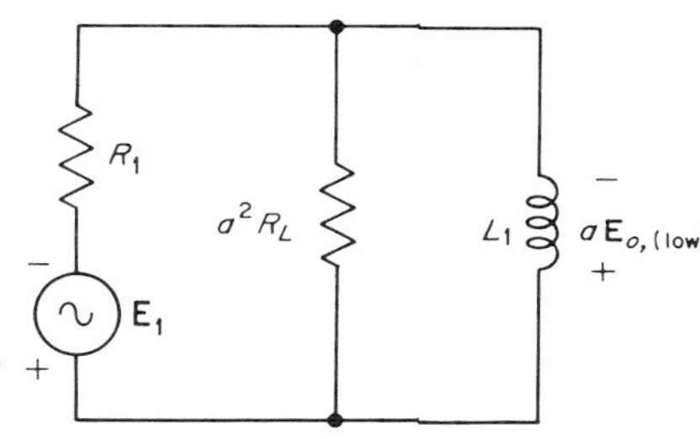

FIGURE 13-45
Low-frequency equivalent circuit of Fig. 13-41.

where R_e is the parallel combination of R_1 and a^2R_L.

$$\mathbf{E}_{o,\text{low}} = \frac{j\omega L_1}{R_e + j\omega L_1}\mathbf{E}_{o,\text{mid}} \tag{13-61}$$

at $f = f_1$

$$|\mathbf{E}_{o,\text{low}}| = 0.707|\mathbf{E}_{o,\text{mid}}|$$

Therefore,

$$0.707 = \left|\frac{j\omega_1 L_1}{R_e + j\omega_1 L_1}\right| = \frac{\omega_1 L_1}{\sqrt{R_e^{\,2} + (\omega_1 L_1)^2}}$$

Squaring both sides gives

$$(0.707)^2[R_e^{\,2} + (\omega_1 L_1)^2] = (\omega_1 L_1)^2$$

$$0.5[R_e^{\,2} + (\omega_1 L_1)^2] = (\omega_1 L_1)^2$$

Thus,

$$R_e = \omega_1 L_1$$

$$f_1 = \frac{R_e}{2\pi L_1} \tag{13-62}$$

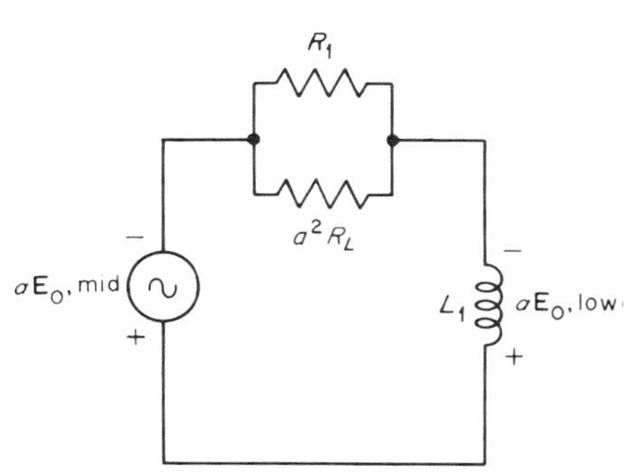

FIGURE 13-46
Thévenin equivalent circuit of Fig. 13-45.

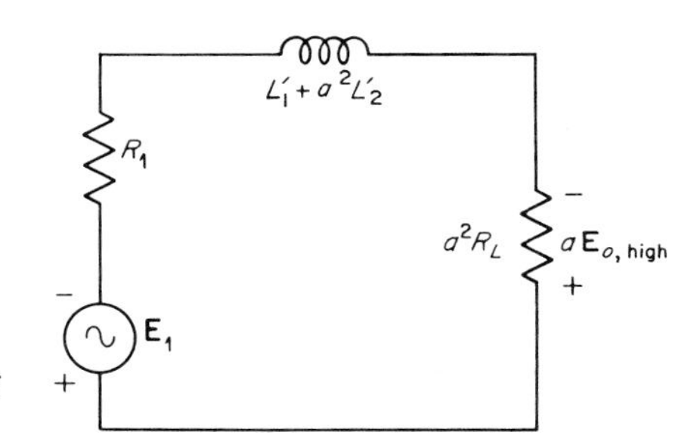

FIGURE 13-47
High-frequency equivalent circuit of Fig. 13-41.

Rewriting Eq. (13-61),

$$\mathbf{E}_{o,\text{low}} = \frac{1}{1 - j\dfrac{R_e}{2\pi f L_1}}\,\mathbf{E}_{o,\text{mid}}$$

Substituting

$$\frac{R_e}{2\pi L_1} = f_1$$

$$\mathbf{E}_{o,\text{low}} = \frac{1}{1 - j\dfrac{f_1}{f}}\,\mathbf{E}_{o,\text{mid}} \tag{13-63}$$

High-frequency Analysis

At high frequencies, the leakage inductance no longer looks like a short in comparison to the remainder of the circuit. The equivalent circuit accurate for the highs is shown in Fig. 13-47.

Let

$$L_t' = L_1' + a^2L_2'$$

$$a\mathbf{E}_{o,\text{high}} = \frac{a^2R_L}{a^2R_L + R_1 + j\omega L_t'}\,\mathbf{E}_1$$

$$\mathbf{E}_{o,\text{high}} = \frac{\dfrac{aR_L}{R_1 + a^2R_L}\,\mathbf{E}_1}{1 + j\dfrac{\omega L_t'}{R_1 + a^2R_L}}$$

$$= \frac{1}{1 + j\dfrac{\omega L_t'}{R_1 + a^2R_L}}\,\mathbf{E}_{o,\text{mid}} \tag{13-64}$$

at $f = f_2$

$$|\mathbf{E}_{o,\text{high}}| = 0.707|\mathbf{E}_{o,\text{mid}}|$$

Therefore

$$0.707 = \left|\frac{1}{1 + j\dfrac{\omega_2 L_t'}{R_1 + a^2 R_L}}\right| = \frac{1}{\sqrt{1 + \left(\dfrac{\omega_2 L_t'}{R_1 + a^2 R_L}\right)^2}}$$

$$0.707 = \frac{1}{\sqrt{2}} = \frac{1}{\sqrt{1 + 1}}$$

Thus

$$\frac{\omega_2 L_t'}{R_1 + a^2 R_L} = 1$$

$$f_2 = \frac{R_1 + a^2 R_L}{2\pi L_t'} \tag{13-65}$$

Substituting Eq. (13-65) into Eq. (13-64),

$$\mathbf{E}_{o,\text{high}} = \frac{1}{1 + j\dfrac{f}{f_2}} \mathbf{E}_{o,\text{mid}} \tag{13-66}$$

EXAMPLE 13-14 Given the circuit of Fig. 13-41, where $R_L = 20\ \Omega$, $L_1' = 20$ mH, $L_2' = 0.15$ mH, $R_1 = 3$ kΩ, $L_1 = 14$ H, $a = 15$, and $\mathbf{E}_1 = 205\ \angle 0°$ V. Determine f_1 and f_2. Then find the output voltage at (1) 10 Hz, (2) 3 kHz, and (3) 40 kHz.

SOLUTION

$$f_1 = \frac{R_e}{2\pi L_1}$$

$$R_e = R_1 \parallel a^2 R_L = 3\text{k}\Omega \parallel 4.5\text{ k}\Omega = 1.8\text{ k}\Omega$$

$$f_1 = \frac{1{,}800}{2\pi \times 14} = 20.4\text{ Hz}$$

$$f_2 = \frac{R_1 + a^2 R_L}{2\pi(L_1' + a^2 L_2')} = \frac{7.5\text{ k}\Omega}{2\pi \times 0.538 \times 10^{-1}}$$

$$= 22.2\text{ kHz}$$

At the mid-frequency range,

$$\mathbf{E}_{o,\text{mid}} = \frac{aR_L}{R_1 + a^2 R_L}\mathbf{E}_1$$

$$= \frac{300}{7{,}500} 205\ \angle 0°$$

$$= 8.2\ \angle 0°\text{ V}$$

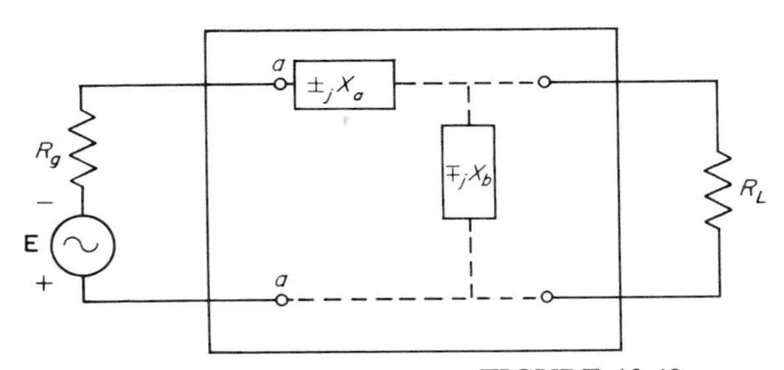

FIGURE 13-48
Impedance matching with reactive networks.

1 $f = 10$ Hz. This frequency is in the low-frequency range since $f_1 = 20.4$ Hz.

$$\mathbf{E}_{o,\text{low}} = \frac{1}{1 - j\dfrac{f_1}{f}}\,\mathbf{E}_{o,\text{mid}} = \frac{1}{1 - j2.04}\,8.2\,\angle 0^\circ$$

$$= 3.62\,\angle 64^\circ \text{ V}$$

2 $f = 3$ kHz. Response is flat between $10f_1 = 204$ Hz and $f_2/10 = 2.2$ kHz. Thus 3 kHz is in the high-frequency range.

$$\mathbf{E}_{o,\text{high}} = \frac{1}{1 + j\dfrac{f}{f_2}}\,\mathbf{E}_{o,\text{mid}} = \frac{1}{1 + j0.135}\,8.2\,\angle 0^\circ$$

$$= 8.1\,\angle -7.7^\circ \text{ V}$$

3 $f = 40$ kHz. This is in the high-frequency range since $f_2 = 22.2$ kHz.

$$\mathbf{E}_{o,\text{high}} = \frac{1}{1 + j\dfrac{f}{f_2}}\,\mathbf{E}_{o,\text{mid}} = \frac{1}{1 + j1.8}\,8.2\,\angle 0^\circ$$

$$= 3.98\,\angle -61^\circ \text{ V}$$

////

13-7 IMPEDANCE TRANSFORMATION

In this chapter, a great deal of time has been spent on transformers and their application. It has been shown how transformers reflect impedance from secondary to primary, thus permitting impedance transformation. In addition, the first sections of this chapter discussed general four-terminal network analysis and introduced the concept of impedance transformation. In those sections, T and pi networks were discussed and analyzed. This section will be devoted to discussion of the *L* section for impedance-matching purposes.

In Fig. 13-48, the load R_L is to be connected to the source through a lossless network of reactances so that at the terminals *a-a* there will be a resistance equal to R_g.

In the first step of the analysis, consider the parallel combination of R_L and $\pm jX_b$. Let this impedance be $\mathbf{Z}_1$.

$$\mathbf{Z}_1 = \frac{\pm jR_L X_b}{R_L \pm jX_b} \tag{13-67}$$

$$= \frac{jR_L X_b}{R_L \pm jX_b} \frac{R_L \mp jX_b}{R_L \mp jX_b}$$

$$= \frac{R_L X_b^2 \pm jR_L^2 X_b}{R_L^2 + X_b^2}$$

$$= \frac{R_L X_b^2}{R_L^2 + X_b^2} \pm j \frac{R_L^2 X_b}{R_L^2 + X_b^2} \tag{13-68}$$

Equation (13-68) consists of two parts: the first term is the resistive portion of the $\mathbf{Z}_1$ impedance, and the second term is the reactive portion of the $\mathbf{Z}_1$ impedance. To cause a resistance equal to R_g to appear at terminals *a-a*, two things must be done. First, reactance X_b must be selected so that the resistance term in Eq. (13-68) equals R_g. Then impedance $\pm jX_a$ must be selected so that it is equal in magnitude but opposite in sign to the reactance term in Eq. (13-68)

$$R_g = \frac{R_L X_b^2}{R_L^2 + X_b^2} \tag{13-69}$$

Solving for X_b,

$$X_b = R_L \sqrt{\frac{R_g}{R_L - R_g}} \tag{13-70}$$

To solve for X_a, first set X_a equal to the magnitude of the reactive term of Eq. (13-68).

$$X_a = \frac{R_L^2 X_b}{R_L^2 + X_b^2} \tag{13-71}$$

Substituting Eq. (13-70) for X_b in Eq. (13-71),

$$X_a = \frac{R_L^2 \left(R_L \sqrt{\dfrac{R_g}{R_L - R_g}} \right)}{R_L^2 + \left(R_L \sqrt{\dfrac{R_g}{R_L - R_g}} \right)^2}$$

Upon proper algebraic manipulation,

$$X_a = \sqrt{R_g (R_L - R_g)} \tag{13-72}$$

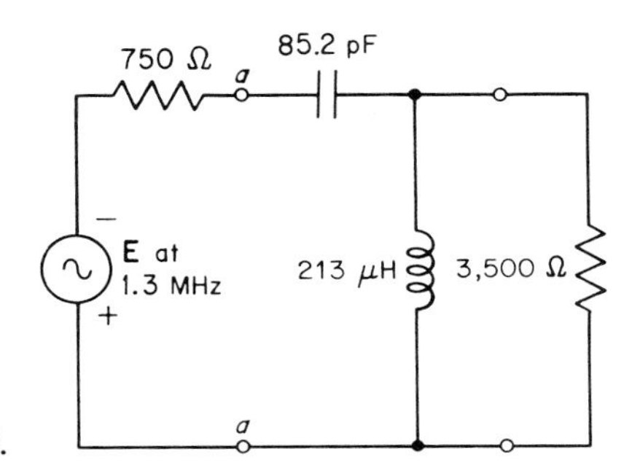

FIGURE 13-49
Solution for Example 13-15.

EXAMPLE 13-15 In the circuit of Fig. 13-48, $R_g = 750\ \Omega$, and $R_L = 3{,}500\ \Omega$. The frequency of operation is 1.3 MHz. Design two L matching networks that will cause the generator to work into a 750-Ω resistive load.

SOLUTION

1 With Eq. (13-70), X_b can be calculated.

$$X_b = 3{,}500\sqrt{\frac{750}{2{,}750}} = 1{,}825\ \Omega$$

2 With Eq. (13-72), X_a can be calculated.

$$X_a = \sqrt{750 \times 2{,}750} = 1{,}436\ \Omega$$

3 One solution would be $-jX_a = -j1{,}436\ \Omega$ and $jX_b = j1{,}825\ \Omega$ where X_a is a capacitor and X_b is an inductor.

$$C_a = \frac{1}{\omega X_a}$$

$$= \frac{1}{6.28 \times 1.3 \times 10^6 \times 1.436 \times 10^3} = 85.2\ \text{pF}$$

$$L_b = \frac{X_b}{\omega}$$

$$= \frac{1.825 \times 10^3}{6.28 \times 1.3 \times 10^6} = 213\ \mu\text{H}$$

4 Figure 13-49 illustrates this solution to the problem. ////

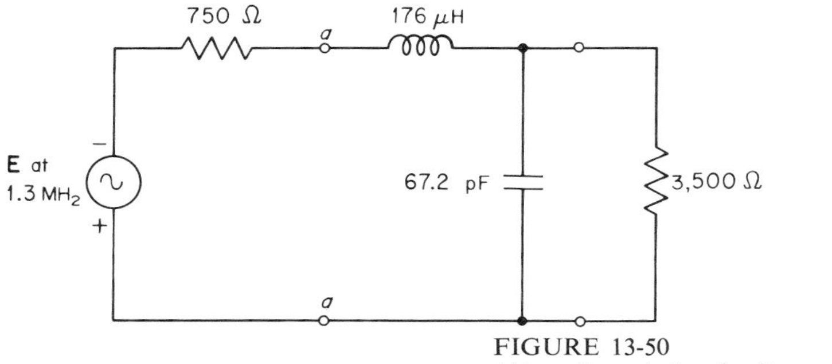

FIGURE 13-50
Alternative solution for Example 13-15.

ALTERNATIVE SOLUTION

1 The alternative solution would be $jX_a = j1{,}436\ \Omega$ and $-jX_b = -j1{,}825\ \Omega$ where X_a is an inductor and X_b is a capacitor.

$$L_a = \frac{X_a}{\omega}$$

$$= \frac{1.436 \times 10^3}{6.28 \times 1.3 \times 10^6} = 176\ \mu\text{H}$$

$$C_b = \frac{1}{\omega X_b}$$

$$= \frac{1}{6.28 \times 1.3 \times 10^6 \times 1.825 \times 10^3} = 67.2\ \text{pF}$$

2 Figure 13-50 illustrates this alternative solution. ////

It is important to note that either solution provides a resistance match at the operating frequency. It is even more important for the reader to realize that *this impedance match is at the operating frequency only*. At all other frequencies there will be a mismatch. The direction of the change in characteristics of the circuit when off frequency depends upon which of the solutions is actually used. If the solution shown in Fig. 13-49 is used, the loading on the source will rise with frequency above operating frequency. On the other hand, the alternative solution will cause the loading to rise as frequency drops below operating frequency. In addition, either solution will have a certain series-resonant frequency, and at this frequency maximum current will be drawn from the source. This can be destructive in small electronic equipment, so the *L*-section impedance transformation must be used with care, and only within a restricted range of frequencies.

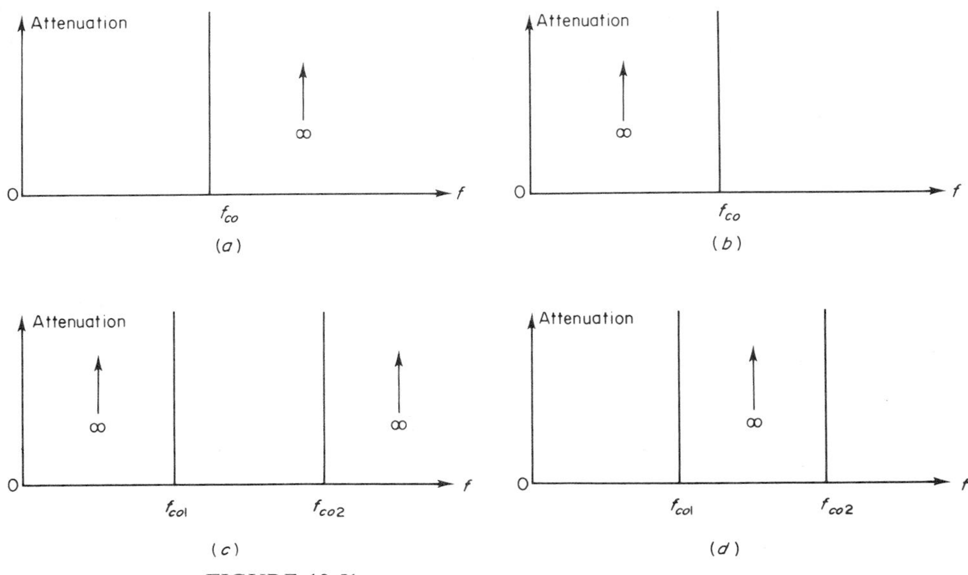

FIGURE 13-51
Attenuation curves of ideal filters: (*a*) lowpass, (*b*) highpass (*c*) bandpass, (*d*) band-reject.

13-8 IDEAL FILTER

A filter is basically a coupling network that allows some frequencies from the source to reach the load while other frequencies from the input have their magnitudes attenuated before reaching the load. This text will concern itself with passive filters as opposed to active filters. The passive filters are considered to be lossless, which implies that only L, C components appear in a filter.

The ideal filter allows all signals with frequencies in the passband to reach the load with zero attenuation. Signals with frequencies outside the passband, those in the stopband, receive infinite attenuation, and they do not appear at the load. Location of the passband in the frequency spectrum dictates the name given to the filter's pass region.

Lowpass filters attenuate signal frequencies above a specified cutoff frequency. Highpass filters allow signal frequencies above the cutoff frequency to reach the load. Bandpass filters have two cutoff frequencies. Frequencies are passed between the cutoff frequencies. Band-reject filters attenuate signal frequencies between two cutoff frequencies. Fig. 13-51 shows the attenuation curves of ideal filters in the various pass regions.

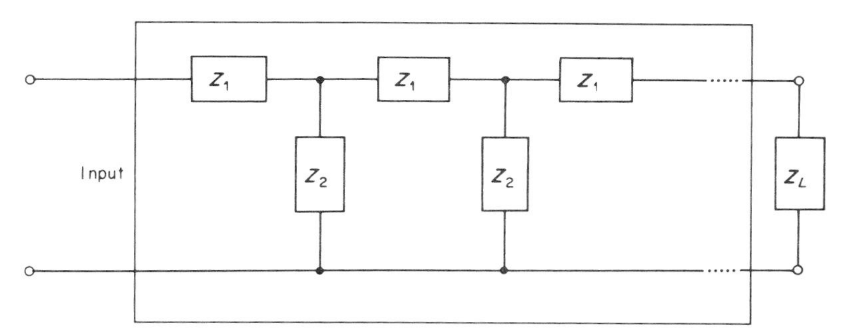

FIGURE 13-52
General filter network.

13-9 CHARACTERISTIC IMPEDANCE

A filter network has a pair of input terminals and a pair of output terminals. Thus it comes under the definition of a two-port network. From previous discussion on two-port networks, the reader is aware that any two-port may be represented by an equivalent T network or an equivalent pi network. The filter may be represented by the network in Fig. 13-52. $\mathbf{Z}_1$ is the series impedance. $\mathbf{Z}_2$ is the shunt impedance. Cascading a series of T networks or pi networks produces the same network orientation shown in Fig. 13-52. Figure 13-53 shows how T networks are connected in cascade, and Fig. 13-54 shows how pi networks are connected to duplicate the circuit in Fig. 13-52. The characteristic impedance $\mathbf{Z}_0$ is the value of load impedance $\mathbf{Z}_L$ terminating the filter that causes the input impedance $\mathbf{Z}_{\text{in}}$ to equal the load impedance. The circuit of Fig. 13-55 is used in determining the characteristic impedance of a T section.

$$\mathbf{Z}_{\text{in}} = \frac{\mathbf{Z}_1}{2} + \frac{\mathbf{Z}_2(\mathbf{Z}_1/2 + \mathbf{Z}_L)}{\mathbf{Z}_2 + \mathbf{Z}_1/2 + \mathbf{Z}_L} \tag{13-73}$$

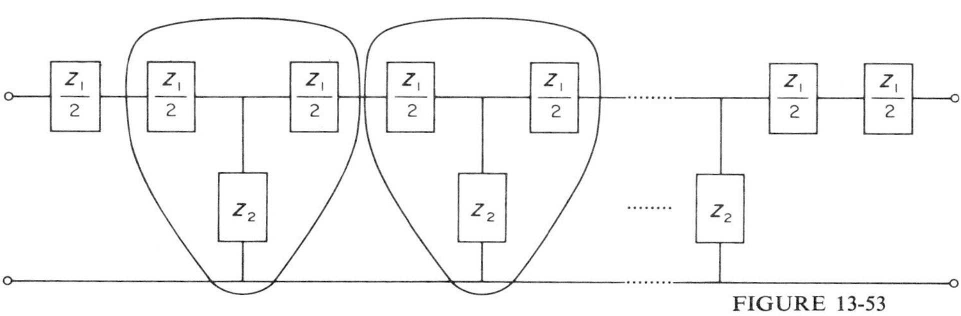

FIGURE 13-53
Cascade of T networks.

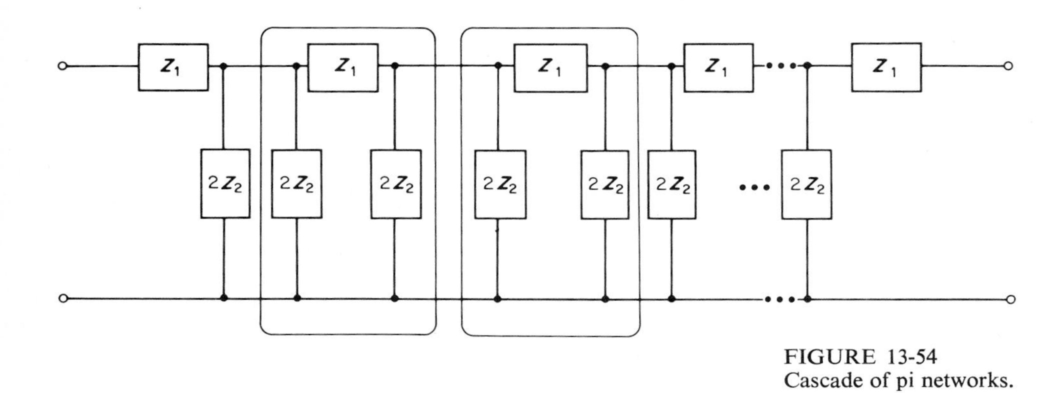

FIGURE 13-54
Cascade of pi networks.

The characteristic impedance is found under the assumption that

$$\mathbf{Z}_{0T} = \mathbf{Z}_{\text{in}} = \mathbf{Z}_L \qquad \text{T section}$$

Substituting in Eq. (13-73),

$$\mathbf{Z}_{0T} = \frac{\mathbf{Z}_1}{2} + \frac{\mathbf{Z}_2(\mathbf{Z}_1/2 + \mathbf{Z}_{0T})}{\mathbf{Z}_2 + \mathbf{Z}_1/2 + \mathbf{Z}_{0T}}$$

$$\mathbf{Z}_{0T} = \sqrt{\mathbf{Z}_1\mathbf{Z}_2 + \frac{\mathbf{Z}_1^{\ 2}}{4}} \tag{13-74}$$

The characteristic impedance of a pi section is found from the circuit in Fig. 13-56.

$$\mathbf{Z}_{\text{in}} = \frac{2\mathbf{Z}_2\left(\mathbf{Z}_1 + \dfrac{2\mathbf{Z}_2\,\mathbf{Z}_L}{2\mathbf{Z}_2 + \mathbf{Z}_L}\right)}{2\mathbf{Z}_2 + \mathbf{Z}_1 + \dfrac{2\mathbf{Z}_2\,\mathbf{Z}_L}{2\mathbf{Z}_2 + \mathbf{Z}_L}} \tag{13-75}$$

$$\mathbf{Z}_{0\pi} = \mathbf{Z}_{\text{in}} = \mathbf{Z}_L \qquad \text{pi section}$$

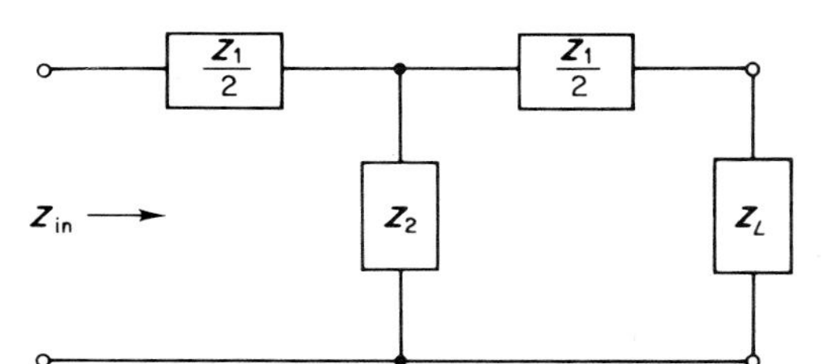

FIGURE 13-55

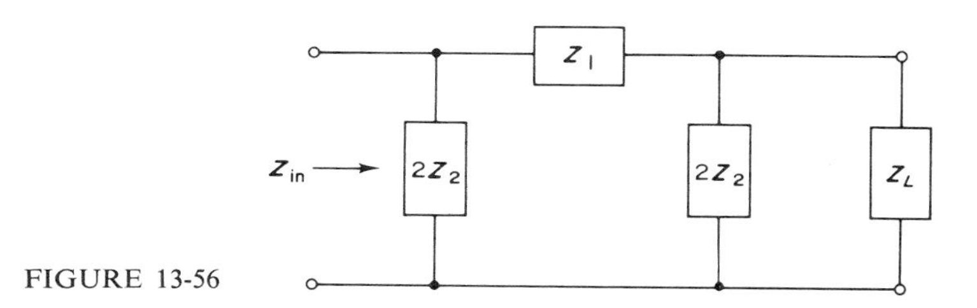

FIGURE 13-56

Substituting in Eq. (13-75),

$$\mathbf{Z}_{0\pi} = \frac{2\mathbf{Z}_2\left(\mathbf{Z}_1 + \dfrac{2\mathbf{Z}_2\mathbf{Z}_{0\pi}}{2\mathbf{Z}_2 + \mathbf{Z}_{0\pi}}\right)}{2\mathbf{Z}_2 + \mathbf{Z}_1 + \dfrac{2\mathbf{Z}_2\mathbf{Z}_{0\pi}}{2\mathbf{Z}_2 + \mathbf{Z}_{0\pi}}} \tag{13.76}$$

$$\mathbf{Z}_{0\pi} = \sqrt{\frac{\mathbf{Z}_1\mathbf{Z}_2}{\mathbf{Z}_1\mathbf{Z}_2 + \dfrac{\mathbf{Z}_1{}^2}{4}}}$$

EXAMPLE 13-16 Determine the characteristic impedance of the T section in Fig. 13-57; $f = 1$ kHz.

SOLUTION

$$\mathbf{Z}_{0T} = \sqrt{\mathbf{Z}_1\mathbf{Z}_2 + \frac{\mathbf{Z}_1{}^2}{4}}$$

$$\frac{\mathbf{Z}_1}{2} = j\omega L = j2\pi(1 \times 10^3)(0.7 \times 10^{-1}) = j440$$

$$\mathbf{Z}_1 = j880\ \Omega$$

$$\mathbf{Z}_2 = -j\frac{1}{\omega C} = -j\frac{1}{(2\pi)(1 \times 10^3)(0.8 \times 10^{-6})} = -j199$$

$$\mathbf{Z}_{0T} = \sqrt{1.9 \times 10^4\ \angle -180^\circ} \text{ or } \sqrt{1.9 \times 10^4\ \angle 180^\circ}$$

$$= -j138 \text{ or } j138$$

If the T section were terminated in the impedance $-j138$, the input impedance would be $-j138$. Likewise, a $j138$ termination provides an input impedance of $j138$. ////

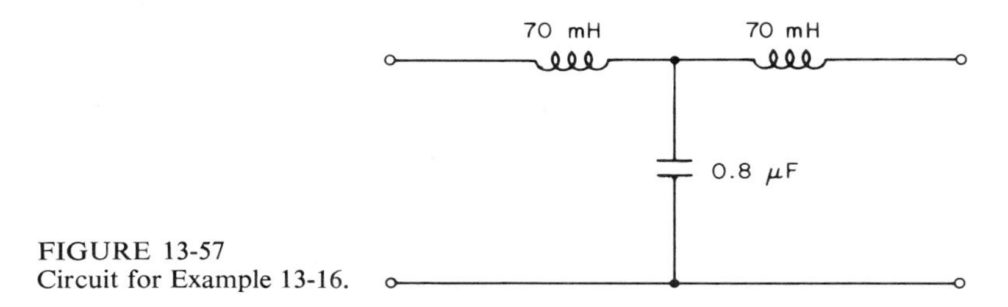

FIGURE 13-57
Circuit for Example 13-16.

13-10 CONSTANT-K FILTER

Filter sections in which the product $\mathbf{Z}_1\mathbf{Z}_2 = K^2$ are called *constant-K filters.* K represents a constant which is independent of frequency. This type of filter permits maximum power transfer to a constant load within the passband.

Figure 13-58 shows a lowpass T filter. With inductor values $L/2$, $\mathbf{Z}_1 = j\omega L$. The shunt capacitance yields $\mathbf{Z}_2 = 1/j\omega C$.

$$K^2 = \mathbf{Z}_1\mathbf{Z}_2 = j\omega L \frac{1}{j\omega C} = \frac{L}{C}$$

$$K = R_o = \sqrt{\frac{L}{C}} \tag{13-77}$$

R_o represents the desired constant-load termination. Therefore,

$$\mathbf{Z}_{OT} = R_o = \sqrt{\frac{L}{C}} \tag{13-78}$$

The cutoff frequency is found through analysis of the characteristic impedance equation.

$$\mathbf{Z}_{OT} = \sqrt{\mathbf{Z}_1\mathbf{Z}_2 + \frac{\mathbf{Z}_1^{\,2}}{4}}$$

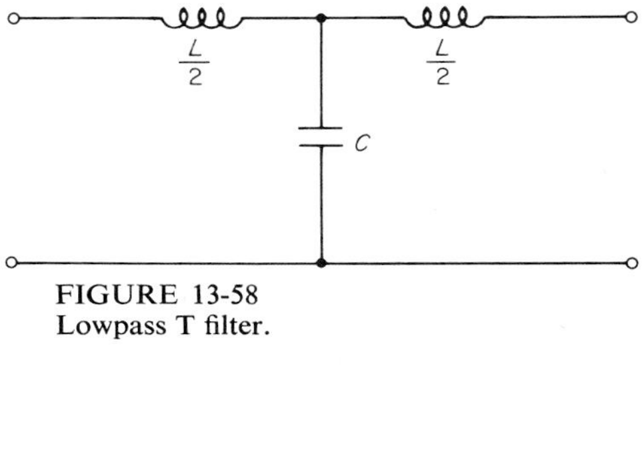

FIGURE 13-58
Lowpass T filter.

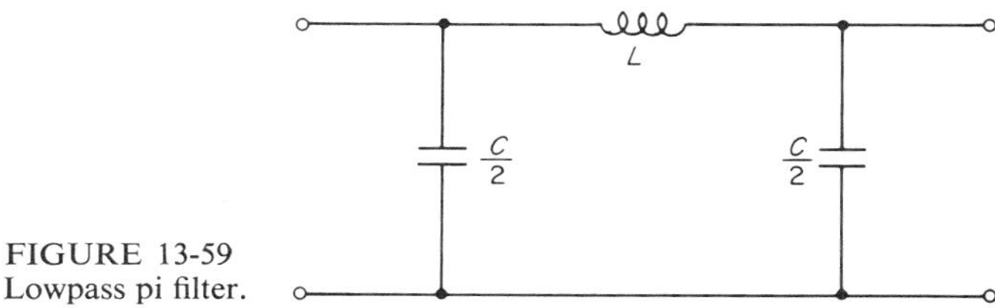

FIGURE 13-59
Lowpass pi filter.

At $f = f_{co}$, $\mathbf{Z}_{0T} = 0$ (signals shunted to ground).

$$\mathbf{Z}_{0T} = \sqrt{\mathbf{Z}_1\left(\mathbf{Z}_2 + \frac{\mathbf{Z}_1}{4}\right)}$$

$$= 0 \qquad \text{if } \mathbf{Z}_2 + \frac{\mathbf{Z}_1}{4} = 0$$

$$\frac{1}{j\omega_{co} C} + \frac{j\omega_{co} L}{4} = 0$$

Therefore,

$$\omega_{co} = \frac{2}{\sqrt{LC}} \tag{13-79}$$

Using Eqs. (13-78) and (13-79),

$$L = \frac{R_o}{\pi f_{co}} \qquad C = \frac{1}{\pi R_o f_{co}} \tag{13-80}$$

Through a similar analysis, the filter-design equations for the lowpass pi section in Fig. 13-59 are easily obtained. The required L and C are the same as stated in Eq. (13-80).

A highpass T filter is shown in Fig. 13-60. Component positions are reversed when compared with the lowpass T filter.

$$\frac{\mathbf{Z}_1}{2} = \frac{1}{j\omega 2C}$$

$$\mathbf{Z}_1 = \frac{1}{j\omega C} \qquad \mathbf{Z}_2 = j\omega L$$

$$K = \sqrt{\frac{L}{C}}$$

$$\mathbf{Z}_{0T} = R_o = \sqrt{\frac{L}{C}} \tag{13-81}$$

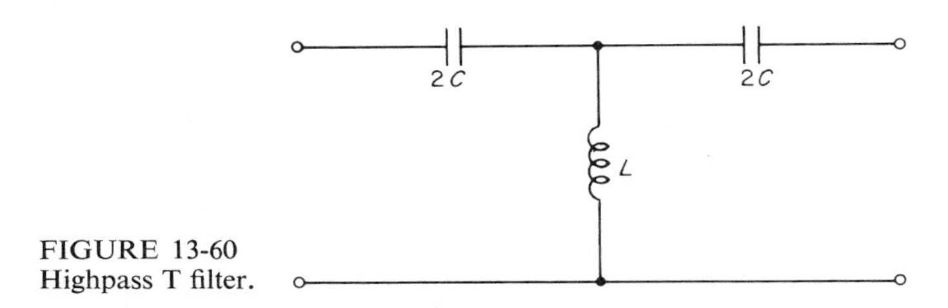

FIGURE 13-60
Highpass T filter.

Once again $\mathbf{Z}_{0T}$ is investigated to find the cutoff frequency. At

$$f = f_{co}$$

$$\mathbf{Z}_{0T} = 0$$

$$\mathbf{Z}_{0T} = \sqrt{\mathbf{Z}_1\left(\mathbf{Z}_2 + \frac{\mathbf{Z}_1}{4}\right)}$$

$$= 0 \qquad \text{if } Z_2 + \frac{Z_1}{4} = 0$$

$$j\omega_{co}L + \frac{1}{j4\omega_{co}C} = 0$$

Therefore,

$$\omega_{co} = \frac{1}{2\sqrt{LC}} \tag{13-82}$$

Using Eqs. (13-81) and (13-82),

$$L = \frac{R_o}{4\pi f_{co}} \qquad C = \frac{1}{4\pi f_{co} R_o} \tag{13-83}$$

The values of L and C in the highpass pi filter of Fig. 13-61 are the same as those in Eq. (13-83).

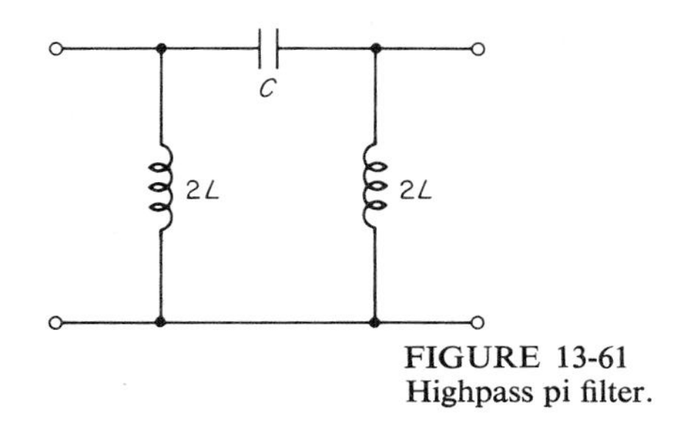

FIGURE 13-61
Highpass pi filter.

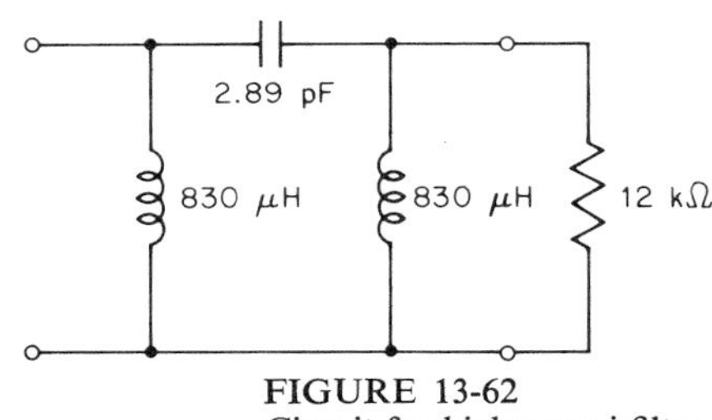

FIGURE 13-62
Circuit for highpass pi filter.

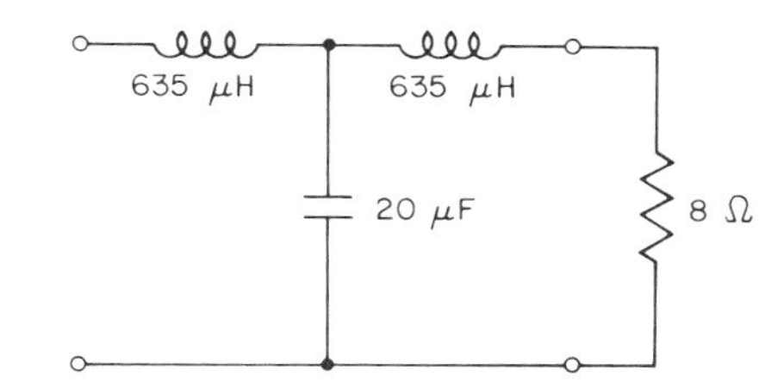

FIGURE 13-63
Circuit for lowpass T filter.

EXAMPLE 13-17 Design a highpass pi filter with a cutoff frequency of 2.3 MHz. Terminate the filter in a constant load of 12 kΩ. Sketch the resulting design.

SOLUTION

$$R_o = 12\ \text{k}\Omega \qquad f_{co} = 2.3\ \text{MHz}$$

$$L = \frac{R_o}{4\pi f_{co}} = 415\ \mu\text{H}$$

$$C = \frac{1}{4\pi f_{co} R_o} = 2.89\ \text{pF}$$

The resulting design is shown in Fig. 13-62. ////

EXAMPLE 13-18 Design a lowpass T filter that has a load termination of 8 Ω. The cutoff frequency is 2 kHz. Sketch the resulting design.

SOLUTION

$$R_o = 8\ \Omega \qquad f_{co} = 2\ \text{kHz}$$

$$L = \frac{R_o}{\pi f_{co}} = 1.27\ \text{mH}$$

$$C = \frac{1}{\pi R_o f_{co}} = 20\ \mu\text{F}$$

The resulting design is shown in Fig. 13-63. ////

The circuit of a bandpass T filter is shown in Fig. 13-64. This circuit is equivalent to cascading a lowpass T filter and a highpass T filter. The cutoff frequency of the lowpass section becomes the upper frequency limit in the passband, f_2. The lower frequency in the passband, f_1, is due to the cutoff frequency in the highpass section.

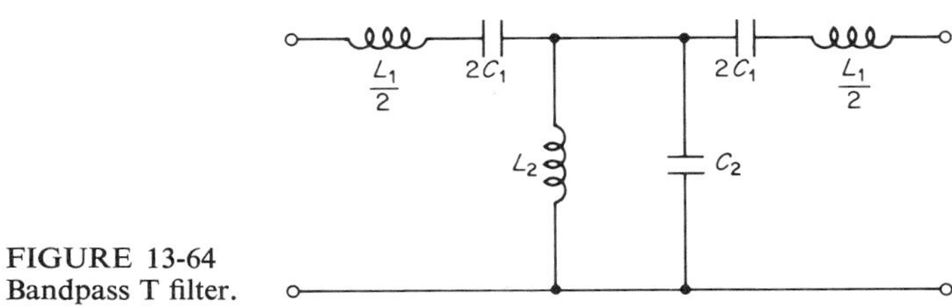

FIGURE 13-64
Bandpass T filter.

An analysis similar to preceding ones will be used to obtain design equations for the bandpass filter.

$$\frac{\mathbf{Z}_1}{2} = j\omega\frac{L_1}{2} + \frac{1}{j\omega 2C_1} = \frac{1-\omega^2 L_1 C_1}{j\omega 2C_1}$$

$$\mathbf{Z}_1 = \frac{1-\omega^2 L_1 C_1}{j\omega C_1}$$

$$\mathbf{Z}_2 = \frac{(j\omega L_2)(1/j\omega C_2)}{j\omega L_2 + 1/j\omega C_2} = \frac{j\omega L_2}{1-\omega^2 L_2 C_2}$$

$$K^2 = \mathbf{Z}_1\mathbf{Z}_2 = \frac{L_2}{C_1}\frac{1-\omega^2 L_1 C_1}{1-\omega^2 L_2 C_2}$$

For K to be independent of frequency, we must have

$$L_1 C_1 = L_2 C_2$$

Therefore,

$$K = R_o = \frac{L_2}{C_1} = \frac{L_1}{C_2} \tag{13-84}$$

At the two cutoff frequencies f_2 and f_1, $\mathbf{Z}_{0T}$ must equal zero to stop frequencies above f_2 and below f_1.

$$\mathbf{Z}_{0T} = \sqrt{\mathbf{Z}_1\mathbf{Z}_2 + \frac{\mathbf{Z}_1^{\,2}}{4}} = \sqrt{\mathbf{Z}_1\left(\mathbf{Z}_2 + \frac{\mathbf{Z}_1}{4}\right)}$$

at f_2 or f_1

$$\mathbf{Z}_{0T} = 0$$

Therefore,

$$\mathbf{Z}_2 + \frac{\mathbf{Z}_1}{4} = 0$$

Multiply by $\mathbf{Z}_1$ and rearrange

$$\mathbf{Z}_1\mathbf{Z}_2 = -\frac{\mathbf{Z}_1^{\,2}}{4}$$

$$\mathbf{Z}_1^{\,2} = -4\mathbf{Z}_1\mathbf{Z}_2 \qquad \mathbf{Z}_1\mathbf{Z}_2 = R_o^{\,2}$$

$$\mathbf{Z}_1 = \pm j2R_o$$

Since

$$\mathbf{Z}_1 = \frac{1 - \omega^2 L_1 C_1}{j\omega C_1}$$

$$\frac{1 - \omega^2 L_1 C_1}{j\omega C_1} = \pm j2R_0$$

multiplying by $j\omega C_1$ and arranging

$$\omega^2 L_1 C_1 \mp \omega 2C_1 R_o - 1 = 0$$

With the ω term being $(-\omega 2C_1 R_o)$,

$$\omega^2 L_1 C_1 - \omega 2C_1 R_0 - 1 = 0$$

$$\omega = \frac{2C_1 R_o \pm \sqrt{4C_1^{\,2} R_o^{\,2} + 4L_1 C_1}}{2L_1 C_1}$$

Only the positive square root is allowed since $\omega > 0$. With the ω term being $(+\omega 2C_1 R_o)$,

$$\omega^2 L_1 C_1 + \omega 2C_1 R_o - 1 = 0$$

$$\omega = \frac{-2C_1 R_o \pm \sqrt{4C_1^{\,2} R_o^{\,2} + 4L_1 C_1}}{2L_1 C_1}$$

Only the positive square root is allowed since $\omega > 0$. Therefore

$$\omega_2 = \frac{2C_1 R_o + \sqrt{4C_1^{\,2} R_o^{\,2} + 4L_1 C_1}}{2L_1 C_1}$$

$$\omega_1 = \frac{-2C_1 R_o + \sqrt{4C_1^{\,2} R_o^{\,2} + 4L_1 C_1}}{2L_1 C_1}$$

The frequency range of the passband is

$$\omega_2 - \omega_1 = \frac{2R_o}{L_1} \tag{13-85}$$

From Eq. (13-85)

$$L_1 = \frac{R_o}{\pi(f_2 - f_1)} \tag{13-86}$$

The product of the two cutoff frequencies yields

$$\omega_2 \omega_1 = \frac{1}{L_1 C_1} \tag{13-87}$$

C_1 can be found from Eqs. (13-86) and (13-87).

$$C_1 = \frac{f_2 - f_1}{4\pi f_2 f_1 R_o} \tag{13-88}$$

From Eqs. (13-84), (13-86), and (13-88) the other components can be found.

$$L_2 = \frac{R_o(f_2 - f_1)}{4\pi f_2 f_1} \tag{13-89}$$

$$C_2 = \frac{1}{\pi(f_2 - f_1)R_o} \tag{13-90}$$

EXAMPLE 13-19 Design a bandpass T filter which has a passband from 5 to 15 kHz. The filter will be terminated in a resistance of 1 kΩ. Sketch the resulting circuit.

SOLUTION

$$f_2 = 15\text{ kHz} \qquad f_1 = 5\text{ kHz}$$

$$L_1 = \frac{R_o}{\pi(f_2 - f_1)} = \frac{1 \times 10^3}{\pi(1 \times 10^4)}$$

$$= 31.8\text{ mH}$$

$$C_1 = \frac{f_2 - f_1}{4\pi f_2 f_1 R_o} = \frac{10 \times 10^3}{(4\pi)(1.5 \times 10^4)(0.5 \times 10^4)(1 \times 10^3)}$$

$$= 0.0106\ \mu\text{F}$$

$$L_2 = \frac{R_o(f_2 - f_1)}{4\pi f_2 f_1} = \frac{(1 \times 10^3)(10 \times 10^3)}{(4\pi)(1.5 \times 10^4)(0.5 \times 10^4)}$$

$$= 10.6\text{ mH}$$

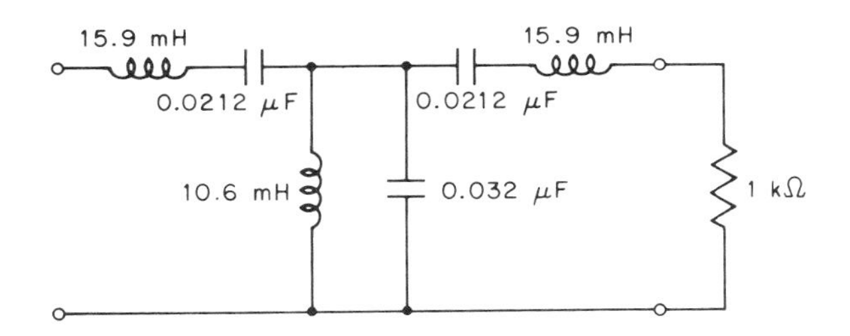

FIGURE 13-65
Bandpass (5- to 15-kHz) T filter.

$$C_2 = \frac{1}{\pi(f_2 - f_1)R_o} = \frac{1}{\pi(0.1 \times 10^5)(1 \times 10^3)}$$
$$= 0.032\ \mu\text{F}$$

The resulting circuit is shown in Fig. 13-65.

A band-reject filter is a cascade connection of a lowpass and a highpass filter. The cutoff frequency of the lowpass is the low-frequency side of the rejection band. Thus the high-frequency side of the rejection band is the cutoff frequency for the highpass filter. A band-reject T filter is shown in Fig. 13-66. The approach used in previous filters will yield the specified design equations.

$$L_1 = \frac{R_o(f_2 - f_1)}{\pi f_2 f_1} \tag{13-91}$$

$$C_1 = \frac{1}{4\pi(f_2 - f_1)R_o} \tag{13-92}$$

$$L_2 = \frac{R_o}{4\pi(f_2 - f_1)} \tag{13-93}$$

$$C_2 = \frac{f_2 - f_1}{\pi f_2 f_1 R_o} \tag{13-94}$$

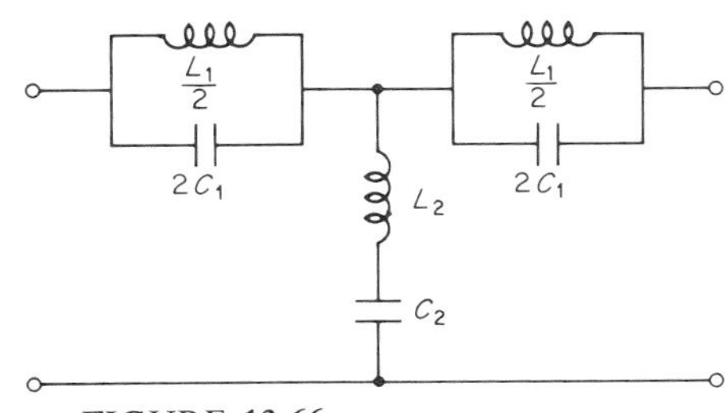

FIGURE 13-66
Band-reject T filter.

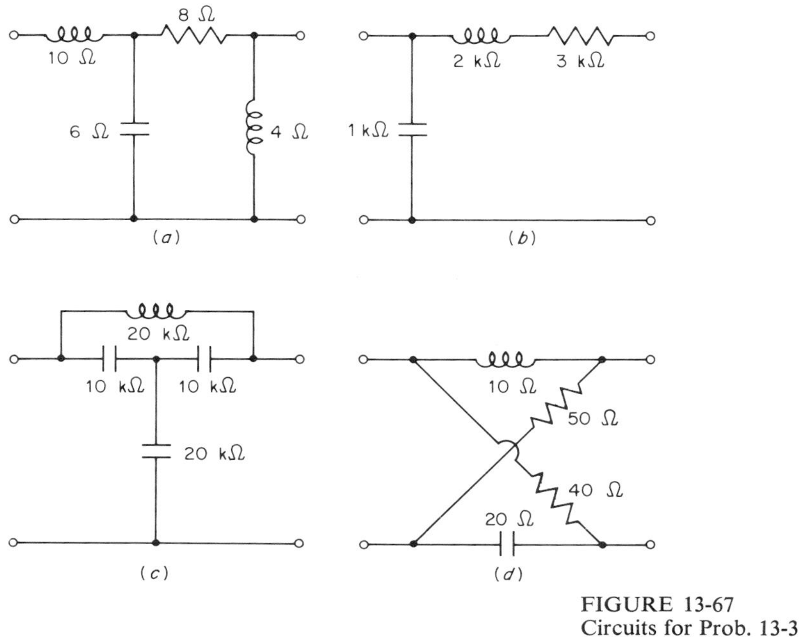

FIGURE 13-67
Circuits for Prob. 13-3.

PROBLEMS

13-1 The open-circuit impedance parameters of a two-port are $z_{11} = 2\ \text{k}\Omega\ \angle 30°$, $z_{22} = 1.6\ \text{k}\Omega\ \angle -70°$, and $z_{12} = 4\ \text{k}\Omega\ \angle 120°$. If $\mathbf{E}_1 = 0$ V and $\mathbf{E}_2 = 14\ \angle 20°$, determine the currents $\mathbf{I}_1$ and $\mathbf{I}_2$. Input is shorted.

13-2 The z parameters for a two-port are $z_{11} = 10\ \angle 0°$, $z_{22} = 6\ \angle 40°$, and $z_{12} = 15\ \angle 50°$. If $\mathbf{E}_1 = 31\ \angle 70°$ and $\mathbf{E}_2 = 19\ \angle -50°$, determine the currents $\mathbf{I}_1$ and $\mathbf{I}_2$.

13-3 Find the open-circuit parameters for the circuits in Fig. 13-67.

13-4 Determine driving-point impedance for two-ports with the following sets of z parameters. $\mathbf{Z}_L = j800\ \Omega$:

(*a*)	$z_{11} = 400$	$z_{22} = 1\ \text{k}\Omega$	$z_M = 200$
(*b*)	$z_{11} = 200\ \angle 40°$	$z_{22} = 700\ \angle -40°$	$z_M = 1\ \text{k}\Omega$
(*c*)	$z_{11} = 1.2\ \text{k}\Omega\ \angle 20°$	$z_{22} = 1\ \text{k}\Omega\ \angle 0°$	$z_M = 500\ \angle 0°$
(*d*)	$z_{11} = 600\ \angle 80°$	$z_{22} = 400\ \angle 90°$	$z_M = 100\ \angle 20°$

13-5 A two-port is terminated in a load of $28 - j60$. Calculate the reflected impedance if $z_{11} = 14\ \angle 80°$, $z_{22} = 20\ \angle -30°$, and $z_M = 12\ \angle 25°$.

13-6 For the T networks in Fig. 13-68, calculate the open-circuit parameters.

13-7 For the pi networks in Fig. 13-69, calculate the open-circuit parameters.

13-8 The short-circuit admittance parameters of a two-port are $y_{11} = 1 \times 10^{-3}\ \angle 70°$, $y_{22} = 0.8 \times 10^{-3}\ \angle -40°$, and $y_{12} = 2 \times 10^{-3}\ \angle 0°$. If $\mathbf{I}_1 = 2\ \angle 0°$ mA and $\mathbf{I}_2 = 0$ A, determine the voltages $\mathbf{E}_1$ and $\mathbf{E}_2$.

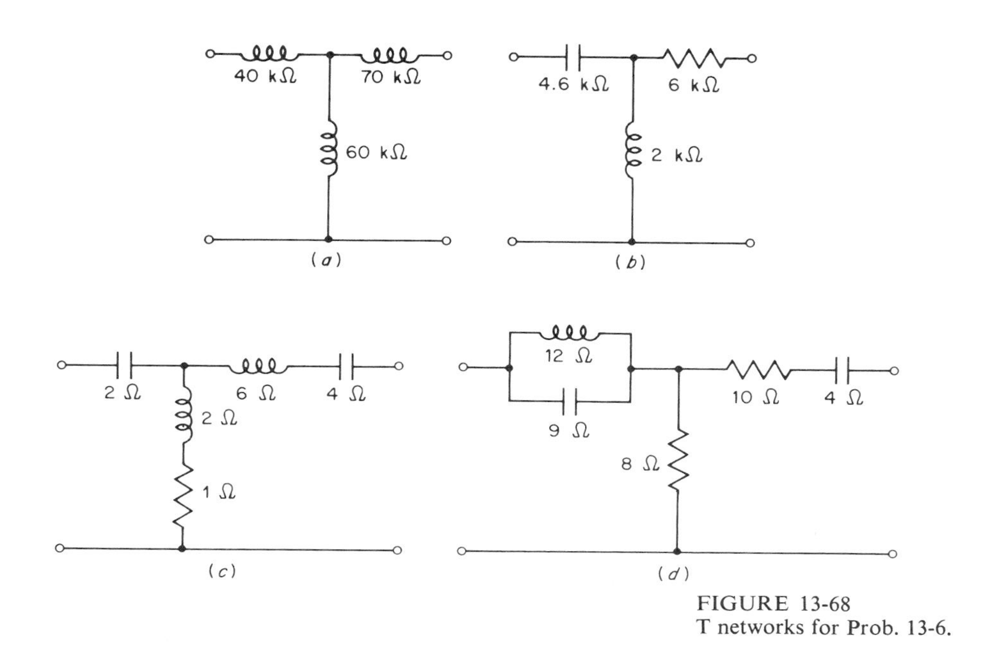

FIGURE 13-68
T networks for Prob. 13-6.

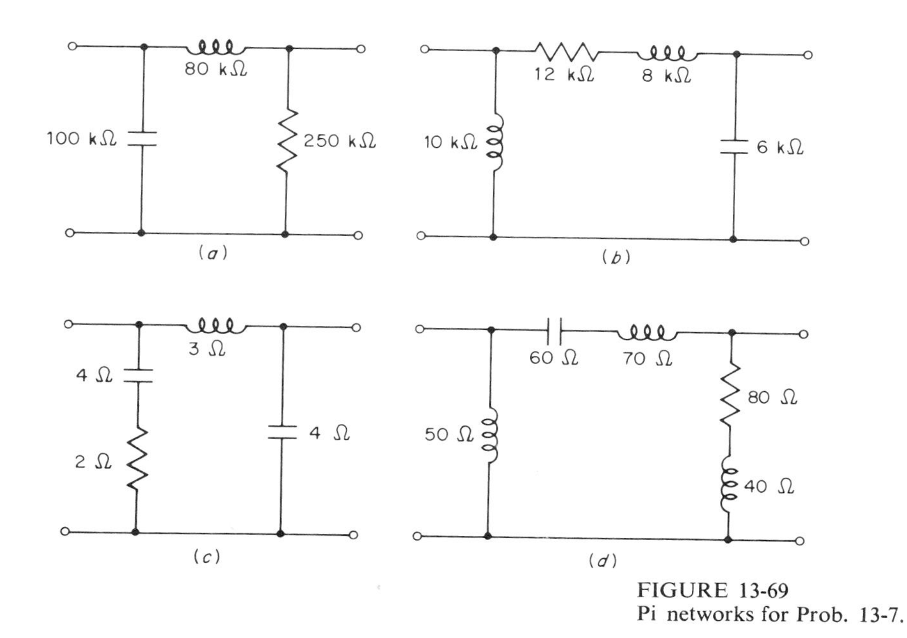

FIGURE 13-69
Pi networks for Prob. 13-7.

13-9 The y parameters for a two-port are $y_{11} = 0.5 \angle 0°$, $y_{22} = 2 \angle -20°$, and $y_{12} = 3 \angle 40°$. If $\mathbf{I}_1 = 6.4 \angle 65°$ and $\mathbf{I}_2 = 5 \angle -30°$, determine the voltages $\mathbf{E}_1$ and $\mathbf{E}_2$.

13-10 Find the short-circuit parameters for the circuits in Fig. 13-70.

13-11 Determine the driving-point admittance for two-ports with the following sets of y parameters; $\mathbf{Z}_L = j1\ \text{k}\Omega$:

(*a*) $y_{11} = 1 \times 10^{-3} \angle 0°$ $\quad y_{22} = 1.2 \times 10^{-3} \angle 30°$ $\quad y_M = 2.5 \times 10^{-3} \angle 0°$
(*b*) $y_{11} = 4.8 \times 10^{-4} \angle -20°$ $\quad y_{22} = 8.9 \times 10^{-4} \angle 35°$ $\quad y_M = 1.2 \times 10^{-3} \angle 60°$
(*c*) $y_{11} = 20 \times 10^{-4} \angle 0°$ $\quad y_{22} = 8 \times 10^{-4} \angle 50°$ $\quad y_M = 3 \times 10^{-3} \angle -20°$
(*d*) $y_{11} = 14 \times 10^{-3} \angle -100°$ $\quad y_{22} = 18 \times 10^{-3} \angle 70°$ $\quad y_M = 3.6 \times 10^{-2} \angle 45°$

13-12 A two-port is terminated in a load of $10 + j30\ \Omega$. Calculate the reflected impedance if $y_{11} = 5 \times 10^{-2} \angle 0°$, $y_{22} = 8 \times 10^{-2} \angle 50°$, and $y_M = 2 \times 10^{-2} \angle 20°$.

13-13 For the T networks in Fig. 13-71, calculate the short-circuit parameters.

13-14 For the pi networks in Fig. 13-72, calculate the short-circuit parameters.

13-15 For each circuit in Fig. 13-73, determine the (*a*) z parameters, (*b*) driving-point impedance $\mathbf{Z}_{11}$, (*c*) voltage-transfer ratio G_{12}, (*d*) current-transfer ratio α_{12}, (*e*) transfer impedance $\mathbf{Z}_{12}$, (*f*) transfer admittance $\mathbf{Y}_{12}$, (*g*) values of $\mathbf{Z}_1$, $\mathbf{Z}_2$, and $\mathbf{Z}_3$ in the equivalent T network, (*h*) values of $\mathbf{Z}_1$, $\mathbf{Z}_2$, and $\mathbf{Z}_3$ in the equivalent pi network.

13-16 For each circuit in Fig. 13-74, determine the (*a*) y parameters, (*b*) driving-point impedance $\mathbf{Z}_{11}$, (*c*) voltage-transfer ratio G_{12}, (*d*) current-transfer ratio α_{12}, (*e*) transfer impedance $\mathbf{Z}_{12}$, (*f*) transfer admittance $\mathbf{Y}_{12}$, (*g*) values of $\mathbf{Y}_1$, $\mathbf{Y}_2$, and $\mathbf{Y}_3$ in the equivalent T network, (*h*) values of $\mathbf{Y}_1$, $\mathbf{Y}_2$, and $\mathbf{Y}_3$ in the equivalent pi network.

13-17 Determine z parameters from the given y parameters.

(*a*) $y_{11} = 0.1 \angle 45°$ $\quad y_{22} = 0.6 \angle -36°$ $\quad y_M = 0.25 \angle -80°$
(*b*) $y_{11} = 1 \times 10^{-6} \angle 0°$ $\quad y_{22} = 4 \times 10^{-6} \angle 90°$ $\quad y_M = 2.6 \times 10^{-6} \angle -42°$
(*c*) $y_{11} = 4 \times 10^{-4} \angle 10°$ $\quad y_{22} = 7 \times 10^{-4} \angle 70°$ $\quad y_M = 3.8 \times 10^{-4} \angle -140°$

13-18 Determine y parameters from the given z parameters.

(*a*) $z_{11} = 6 \angle 78°$ $\quad z_{22} = 10 \angle 0°$ $\quad z_M = 3 \angle -45°$
(*b*) $z_{11} = 1\ \text{K} \angle -20°$ $\quad z_{22} = 4.3\ \text{K} \angle 66°$ $\quad z_M = 2\ \text{K} \angle 80°$
(*c*) $z_{11} = 42\ \text{K} \angle 0°$ $\quad z_{22} = 56\ \text{K} \angle 34°$ $\quad z_M = 18\ \text{K} \angle 70°$

13-19 Determine the turns ratio for a closely coupled transformer that couples a 4-Ω load to a 200-Ω primary resistance.

13-20 An effective ac voltage of 36 V is attached to the primary of a closely coupled transformer. If the transformer has a turns ratio of $a = 6$, calculate the primary current, secondary current, and secondary voltage with a 800-Ω resistor connected on the output.

13-21 Write mesh equations and solve for the mesh currents in the circuits of Fig. 13-75.

13-22 Given the circuits of Fig. 13-76, determine the (*a*) impedance seen at primary terminals, (*b*) reflected impedance, (*c*) primary current, assumed direction shown.

13-23 For each of the following cases:

(*a*) $f_o = 100$ kHz $\quad \mathbf{E} = 4.5 \angle -41°$ $\quad k = 0.13$ $\quad L_1 = 1.6$ mH $\quad L_2 = 2$ mH
$\quad Q_2 = 400$ $\quad Q_1 = 150$
(*b*) $f_o = 3.4$ MHz $\quad \mathbf{E} = 600 \angle 66°$ mV $\quad k = 0.07$ $\quad L_1 = 5\ \mu$H $\quad L_2 = 13\ \mu$H
$\quad Q_2 = 80$ $\quad Q_1 = 100$

determine C_2, primary current $\mathbf{I}_1$, and output voltage $\mathbf{E}_o$ in the circuit of Fig. 13-77.

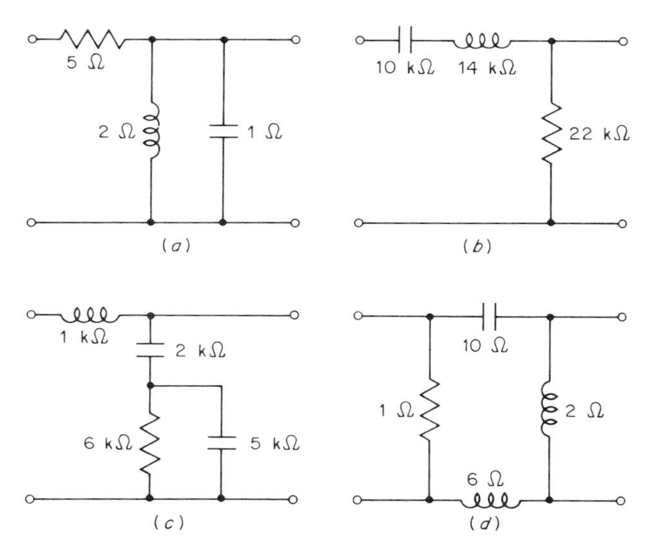

FIGURE 13-70
Circuits for Prob. 13-10.

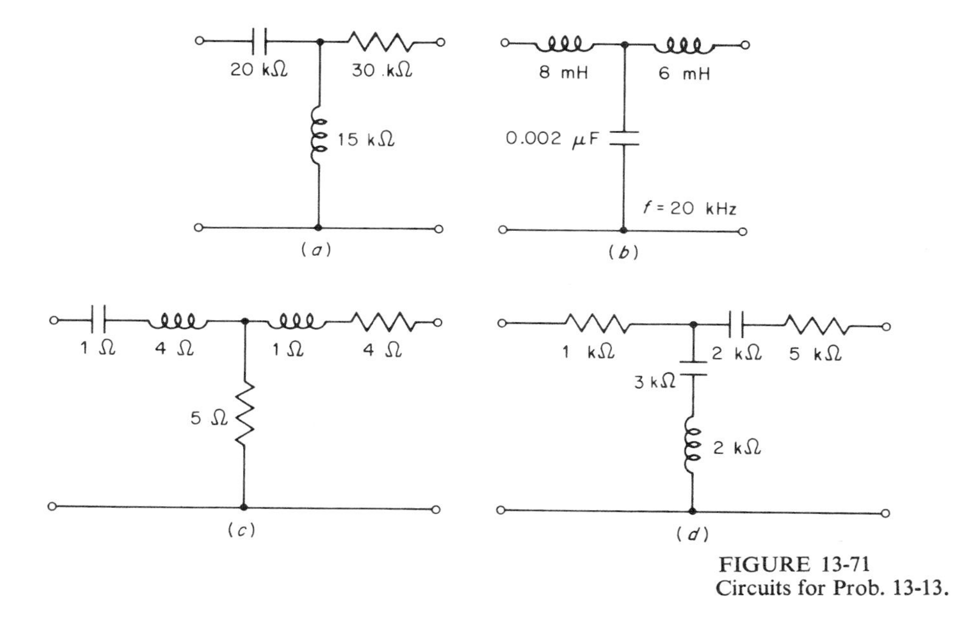

FIGURE 13-71
Circuits for Prob. 13-13.

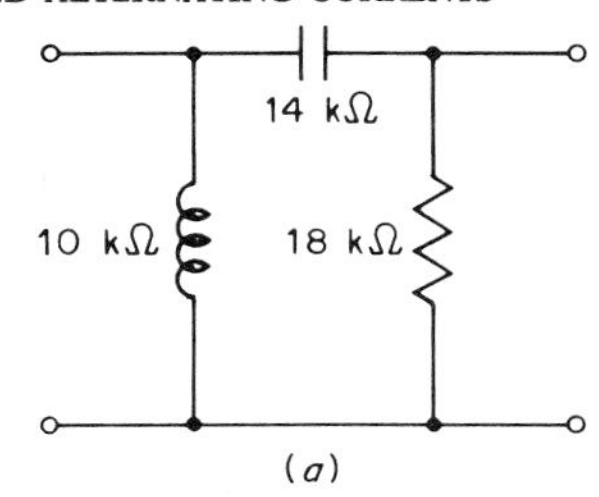

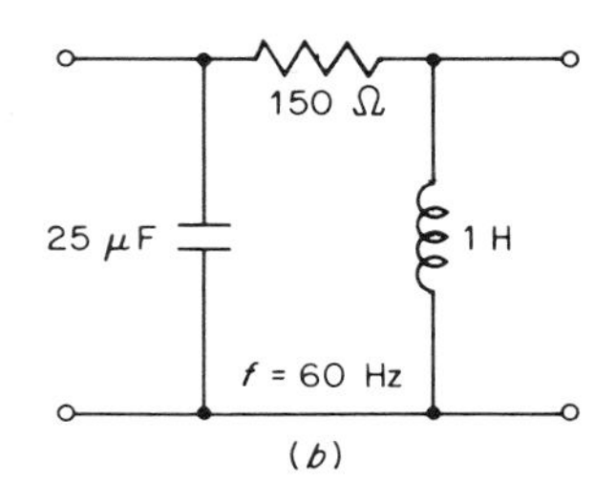

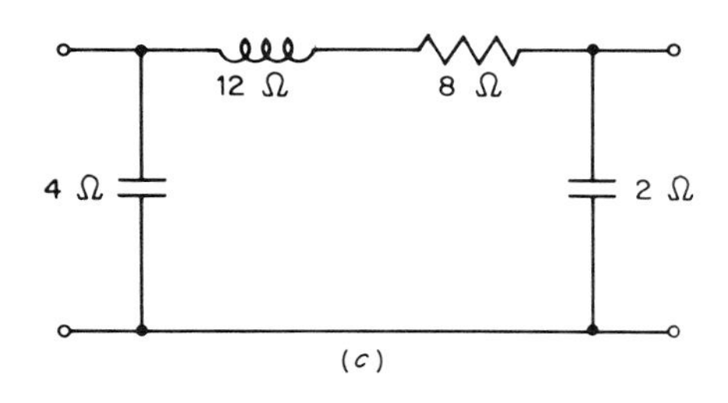

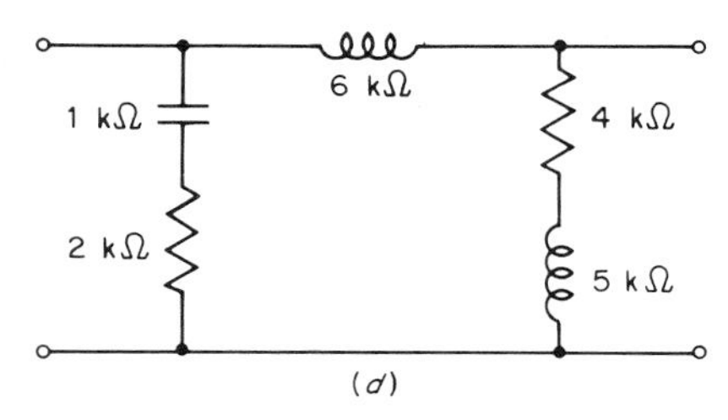

FIGURE 13-72
Circuits for Prob. 13-14.

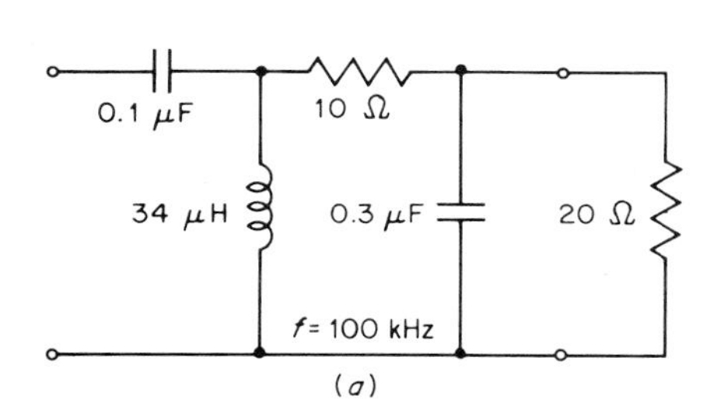

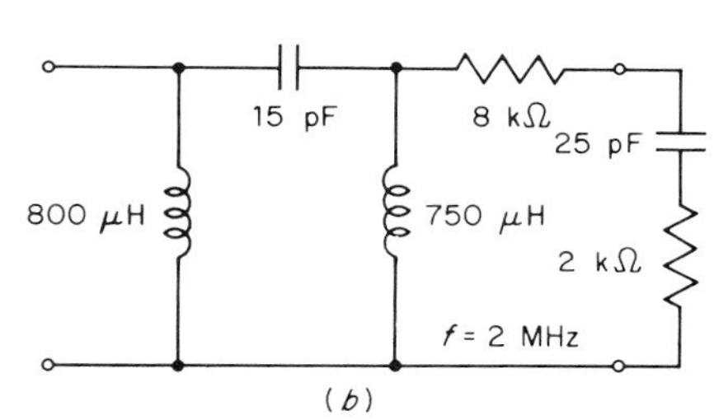

FIGURE 13-73
Circuits for Prob. 13-15.

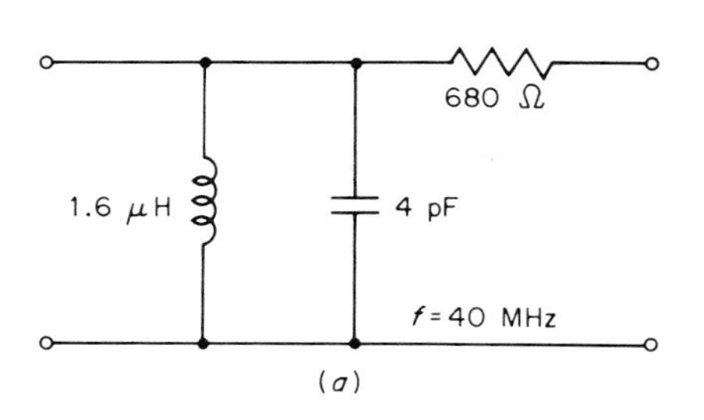

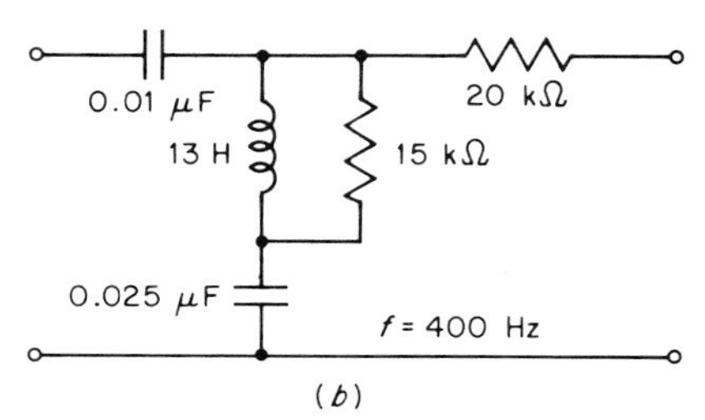

FIGURE 13-74
Circuits for Prob. 13-16.

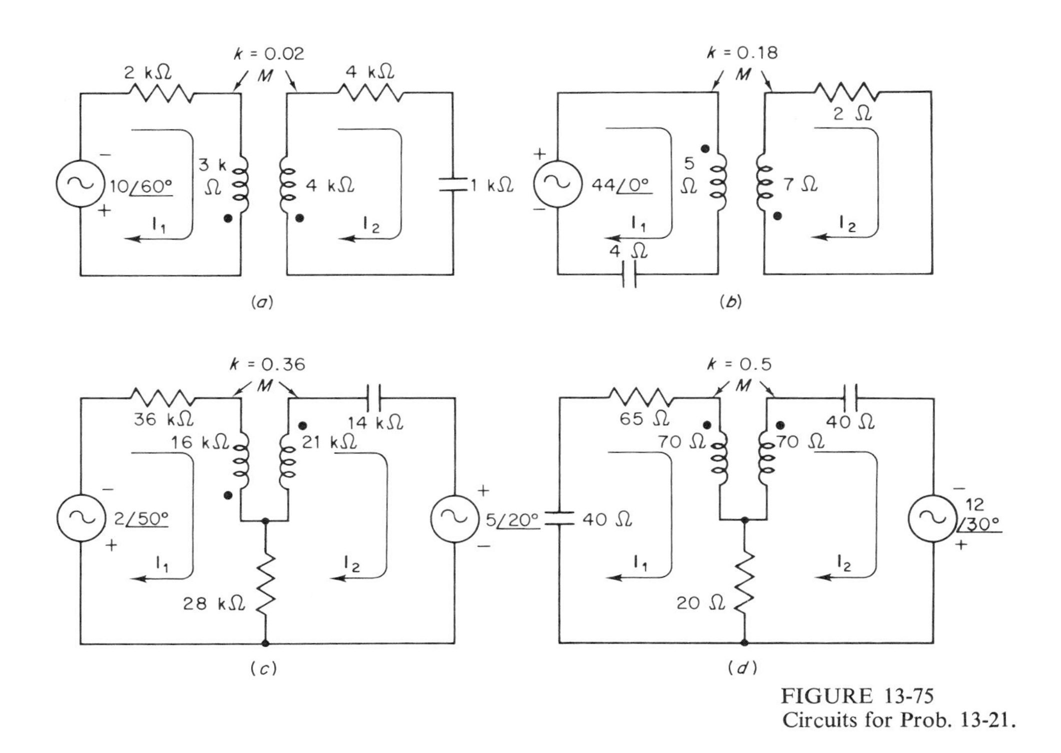

FIGURE 13-75
Circuits for Prob. 13-21.

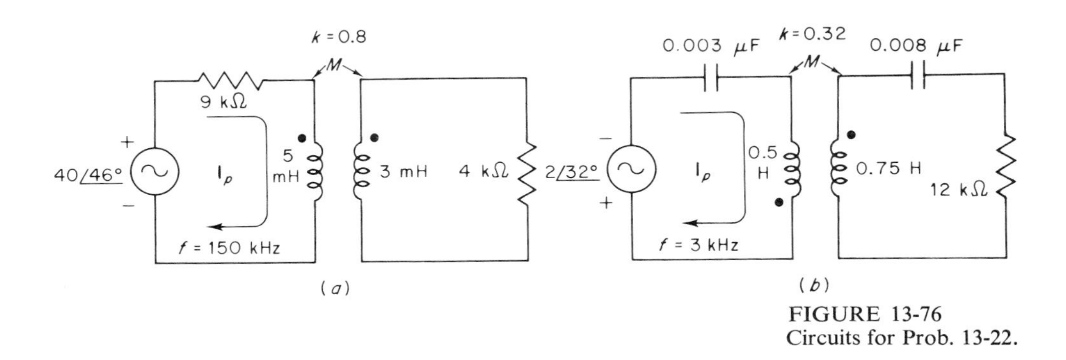

FIGURE 13-76
Circuits for Prob. 13-22.

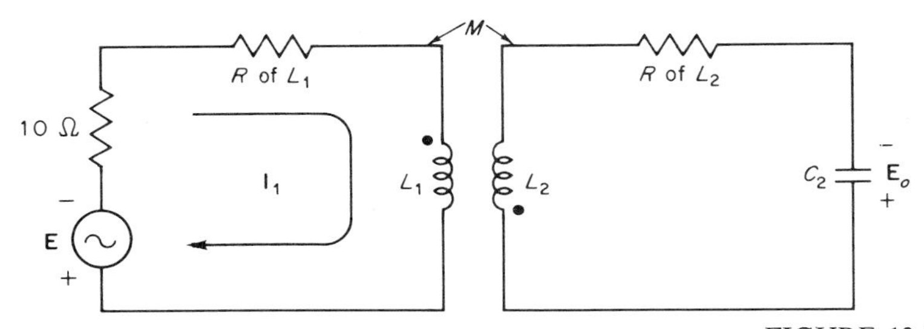

FIGURE 13-77
Circuit for Prob. 13-23.

13-24 Determine the output voltage in the circuit of Fig. 13-78 for the following cases. Also find C_1 and C_2.

(*a*) $f_o = 455$ kHz $\quad R_o = 20$ kΩ $\quad L_1 = 100\ \mu$H $\quad L_2 = 150\ \mu$H $\quad k = k_c$
$Q_1 = 140 \quad Q_2 = 160$

(*b*) $f_o = 32.7$ MHz $\quad R_o = 15$ kΩ $\quad L_1 = 5\ \mu$H $\quad L_2 = 8\ \mu$H $\quad k = 1.2\ k_c$
$Q_1 = 80 \quad Q_2 = 60$

13-25 A portion of a vacuum-tube cascade amplifier with transformer coupling is shown in Fig. 13-79. The output voltage $\mathbf{E}_o$ is the input voltage to the second stage. Figure 13-80 is the ac equivalent circuit of the amplifier. The turns ratio is selected so the reflected secondary resistance equals r_p. Design the circuit to provide $f_1 = 100$ Hz and $f_2 = 15$ kHz. Determine:

(*a*) turns ratio (*b*) L_1 (*c*) L_2
(*d*) M (*e*) L'_t (*f*) $\mathbf{E}_{o,\ \mathrm{mid}}$
(*g*) $\mathbf{E}_o$ at $f = 400$ Hz (*h*) $\mathbf{E}_o$ at $f = 20$ Hz (*i*) $\mathbf{E}_o$ at $f = 10$ kHz
(*j*) $\mathbf{E}_o$ at $f = 20$ kHz (*k*) $\mathbf{E}_{o,\ \mathrm{mid}}$ if $a = 0.03$ (*l*) f_1 if $a = 0.3$

13-26 Design two L matching networks for each of the following frequency conditions (see Fig. 13-81):

(*a*) $f = 10$ Hz (*b*) $f = 8$ MHz (*c*) $f = 1.2$ MHz
(*d*) $f = 49$ kHz (*e*) $f = 5$ kHz (*f*) $f = 1$ kHz
(*g*) $f = 50$ MHz (*h*) $f = 100$ Hz

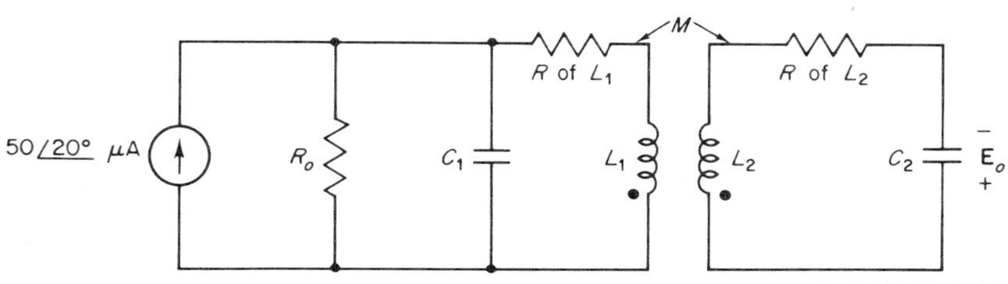

FIGURE 13-78
Circuit for Prob. 13-24.

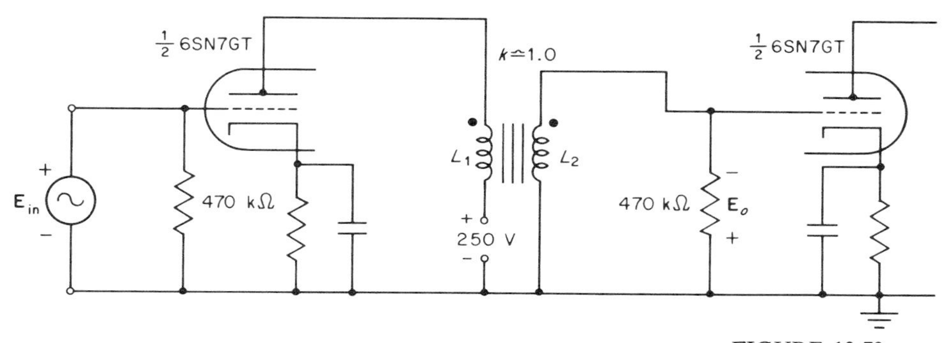

FIGURE 13-79
Circuit for Prob. 13-25.

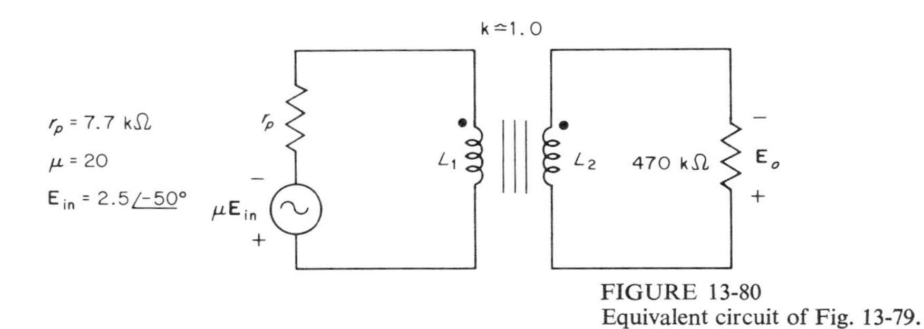

FIGURE 13-80
Equivalent circuit of Fig. 13-79.

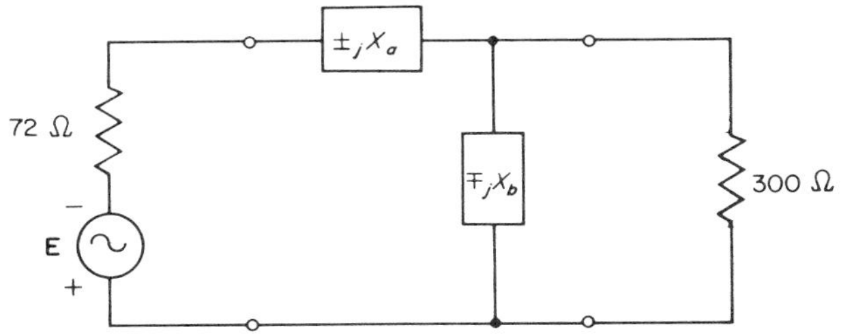

$\pm jX_a$
72 Ω
$\mp jX_b$
300 Ω
−
E
+

FIGURE 13-81
Circuit for Prob. 13-26.

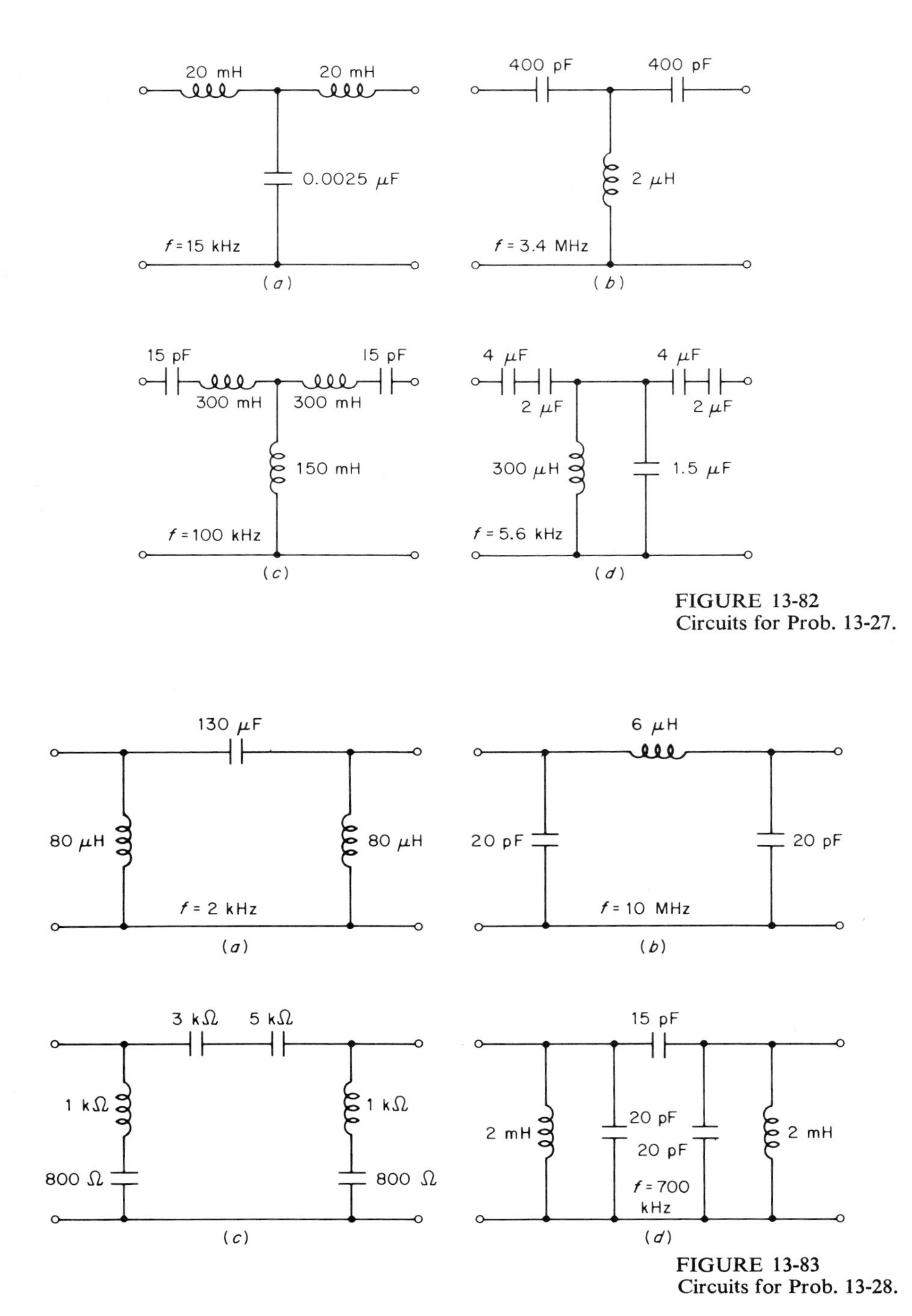

FIGURE 13-82
Circuits for Prob. 13-27.

FIGURE 13-83
Circuits for Prob. 13-28.

13-27 Determine the characteristic impedance of the T sections in Fig. 13-82.

13-28 Determine the characteristic impedance of the pi sections in Fig. 13-83.

13-29 Design both a lowpass T filter and lowpass pi filter for the following specifications. Sketch the resulting circuits.

(*a*) Cutoff at 400 Hz, load of 1 kΩ
(*b*) Cutoff at 400 Hz, load of 14 kΩ
(*c*) Cutoff at 1 MHz, load of 2 kΩ
(*d*) Cutoff at 20 MHz, load of 100 kΩ
(*e*) Cutoff at 100 kHz, load of 1 MΩ

13-30 Design both a highpass T filter and highpass pi filter for the following specifications. Sketch the resulting circuits.

(*a*) Cutoff at 40 Hz, load of 100 Ω
(*b*) Cutoff at 8 MHz, load of 20 kΩ
(*c*) Cutoff at 30 kHz, load of 1 kΩ
(*d*) Cutoff at 18 kHz, load of 4 kΩ
(*e*) Cutoff at 2.8 MHz, load of 86 kΩ

13-31 Design both a bandpass filter and band-reject filter for the following specifications. Sketch the resulting circuits.

(*a*) $f_1 = 800$ Hz, $f_2 = 1{,}300$ Hz, load of 400 Ω
(*b*) $f_1 = 260$ kHz, $f_2 = 740$ kHz, load of 7 kΩ
(*c*) $f_1 = 1.4$ MHz, $f_2 = 9.8$ MHz, load of 1 kΩ
(*d*) $f_1 = 100$ Hz, $f_2 = 200$ Hz, load of 70 Ω
(*e*) $f_1 = 10$ MHz, $f_2 = 12$ MHz, load of 12 kΩ

14

NONSINUSOIDAL VOLTAGES AND CURRENTS

14-1 INTRODUCTION

It was pointed out in earlier chapters that any nonsinusoidal alternating voltage or current can be shown to consist of a combination of sinusoidal waveforms, each sinusoid having a different frequency from the others. Where the overall wave is periodic (repeating in both shape and time of period), the various sinusoids have definite frequency relationships to one another. The basic frequency of the nonsinusoidal wave is called the *fundamental*, and the frequency of the fundamental is determined by the period of the wave. The other sinusoidal components of the wave are called *harmonics* and are invariably whole-number multiples of the fundamental frequency. *It is impossible to develop a periodic nonsinusoidal wave by combining frequencies that are not integral multiples of the fundamental frequency.* The harmonic frequency that is twice the fundamental frequency is the second harmonic, the frequency that is three times the fundamental is the third harmonic, and so on.

If the waveform is pulsating (that is, the average value from 0 to π is not the same as the average value from π to 2π), then the wave contains a dc component in addition to the fundamental and harmonics. Figure 14-1 illustrates the waveform that results when a fundamental and its third harmonic are combined. In this particular sketch, the two waves start in phase.

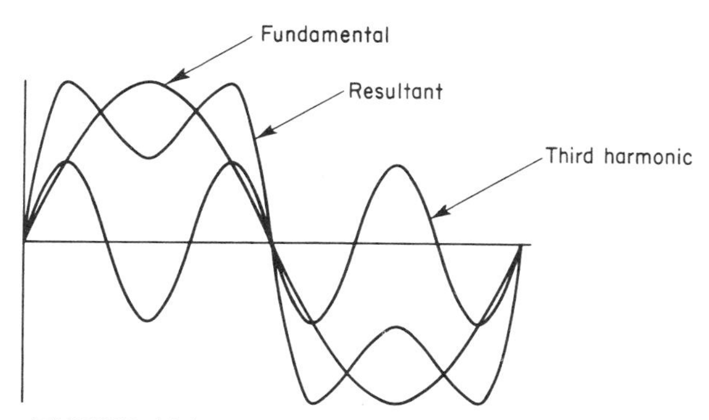

FIGURE 14-1
Combination of a fundamental with its third harmonic. The harmonic is in phase with the fundamental.

Figure 14-2 shows the resultant waveform when the frequency components of Fig. 14-1 are combined, but the third harmonic is shifted in phase by 180°.

Figure 14-3 illustrates a combination of fundamental and second-harmonic components. Note that while the two halves of the resultant waveform are not symmetrical, their average values are equal and therefore there is no dc component. In Fig. 14-4, the same combination of frequencies is shown, but the second-harmonic component has been shifted in phase by $-90°$, and now there is a dc component in the resultant waveform.

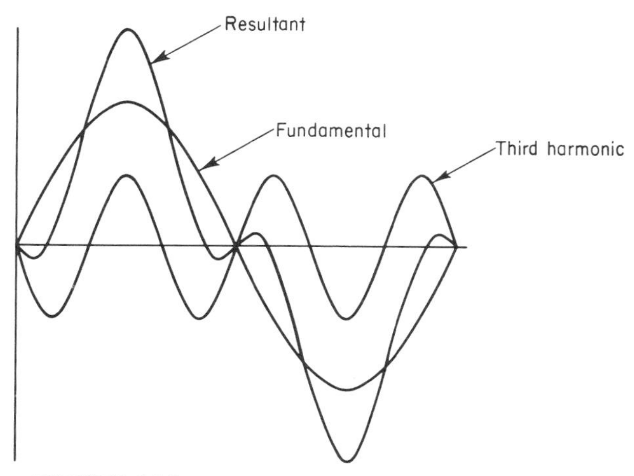

FIGURE 14-2
Combination of a fundamental with its third harmonic. The harmonic is 180° out of phase with the fundamental.

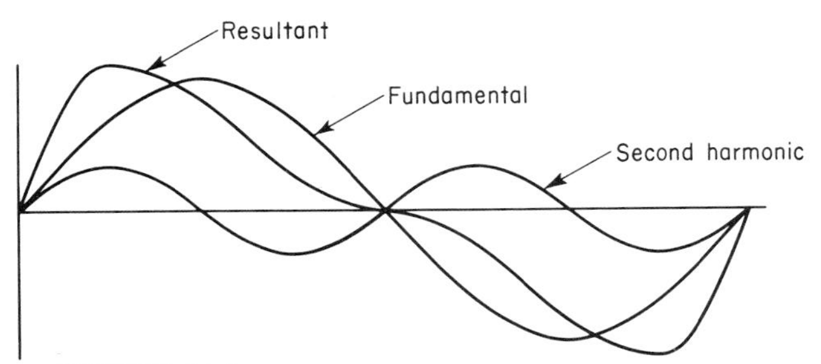

FIGURE 14-3
Combination of a fundamental with its second harmonic. The second harmonic is in phase.

14-2 SYMMETRICAL AND ASYMMETRICAL WAVES

A waveform is symmetrical if its form in the interval from π to 2π is identical with its form from 0 to π. A very simple way to detect symmetry is by visualizing the wave from π to 2π as rectified and then comparing this wave with the wave from 0 to π. If the waveshapes of these two intervals are identical, the wave is symmetrical. A symmetrical waveshape can result only from a combination of fundamental and odd-harmonic frequencies. If the waveform contains any even-order harmonics, the waveform is asymmetrical. These statements can be verified by examining the waveforms

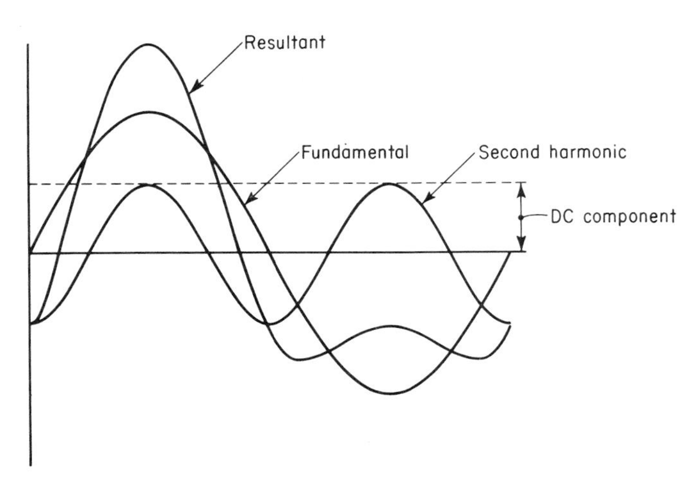

FIGURE 14-4
Combination of a fundamental with its second harmonic. The second harmonic lags by 90°.

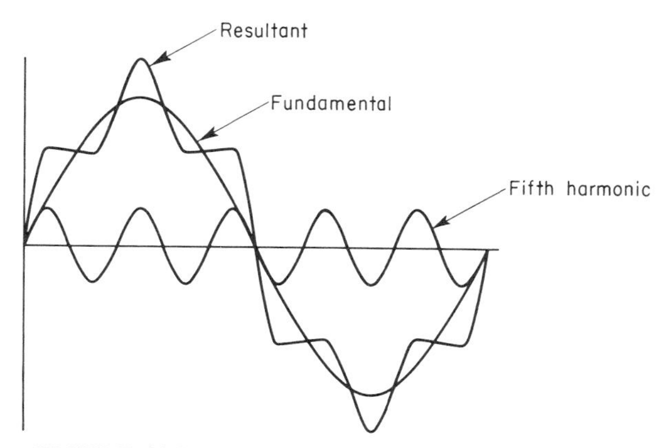

FIGURE 14-5
Combination of a fundamental and its fifth harmonic. The harmonic is in phase.

illustrated in Figs. 14-1 to 14-4. As further visual proof, Fig. 14-5 shows a combination of fundamental and fifth harmonic in a waveform that is symmetrical. Figure 14-6 illustrates a combination of a fundamental frequency and its fourth harmonic. The resultant waveform is nonsymmetrical. Figure 14-7 shows a combination of a fundamental and its third harmonic, with the third harmonic shifted in phase by $+90°$.

14-3 HARMONIC CONTENT OF NONSINUSOIDAL WAVES

In the previous section, it was shown that any periodic nonsinusoidal wave consists of a fundamental frequency and additional frequency components that are harmonics of the fundamental frequency. These harmonically related frequency components can have any phase relationship with respect to each other and to the fundamental. The phase relationships remain constant in each cycle because the harmonics are integral multiples of the fundamental.

The phase relation between a harmonic and its fundamental is always measured in degrees of the cycle of the harmonic. The point of reference is the 0° point on the cycle for the fundamental. For example, if a second harmonic lags the fundamental by 90°, the phase displacement is one-fourth the cycle of the harmonic, not one-fourth the cycle of the fundamental. This is illustrated in Fig. 14-8. The reader will note that 90° in the cycle of the second harmonic is the same as 45° in the cycle of the fundamental.

The equation of the resultant wave is always written with the angular velocity of the fundamental as the reference. It follows, therefore, that all harmonic frequencies have angular velocities that are multiples of the fundamental. In the equation

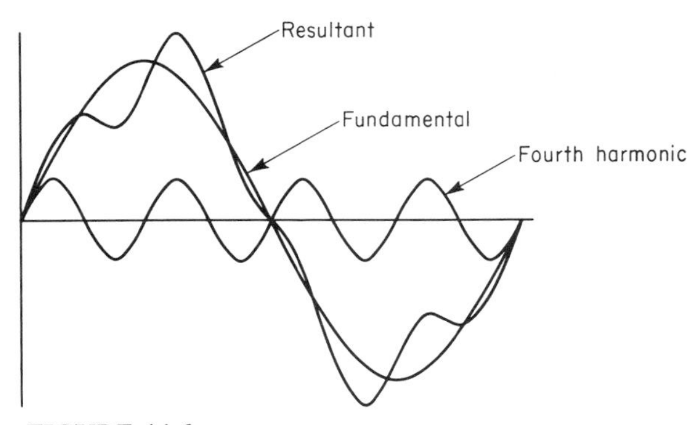

FIGURE 14-6
Combination of a fundamental and its fourth harmonic. The harmonic is in phase.

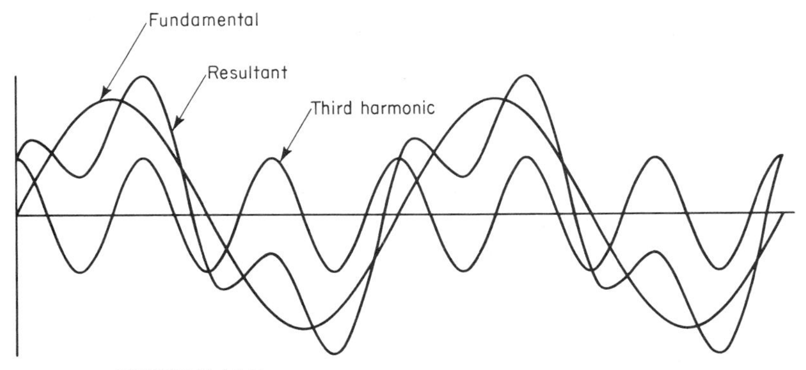

FIGURE 14-7
Combination of a fundamental and its third harmonic. The harmonic leads by 90°.

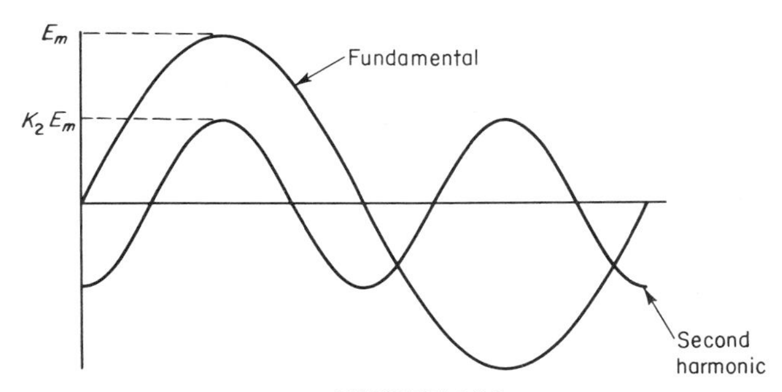

FIGURE 14-8
Fundamental with its second harmonic lagging by 90°.

of the resultant wave, the peak amplitude of the fundamental is used for each component of the wave, and the various peak amplitudes of the harmonic components are stated in terms of the peak amplitude of the fundamental. In the following equations E_m is the peak value of the fundamental, k_2 is the ratio of the second harmonic to the fundamental, k_3 is the ratio of the third harmonic to the fundamental, and k_n is the ratio of the nth harmonic to the fundamental.

$$e_1 = E_m \sin \omega_1 t \tag{14-1}$$

$$e_2 = k_2 E_m \sin 2\omega_1 t \tag{14-2}$$

$$e_3 = k_3 E_m \sin 3\omega_1 t \tag{14-3}$$

$$e_n = k_n E_m \sin n\omega_1 t \tag{14-4}$$

The value of the complex wave at any instant is the algebraic sum of the fundamental and the harmonics. A single equation for the complex wave can be written.

$$e = e_1 + e_2 + e_3 + \cdots + e_n \tag{14-5}$$

$$e = E_m(\sin \omega_1 t + k_2 \sin 2\omega_1 t + k_3 \sin 3\omega_1 t + \cdots + k_n \sin n\omega_1 t) \tag{14-6}$$

EXAMPLE 14-1 A complex wave has a fundamental frequency of 200 Hz. The peak amplitude of the fundamental current is 45 mA. The wave consists of a third harmonic whose peak amplitude is 22.5 mA and a sixth harmonic whose peak amplitude is 15 mA. Write the equations for the frequency components. Write the equation for the complex wave, assuming that the fundamental and the harmonics start in phase. Draw a sketch showing the fundamental, the harmonics, and the resultant wave.

SOLUTION

$$\omega = 2\pi(200) = 1{,}256 \text{ rad/s}$$

Fundamental: $i_1 = 45 \times 10^{-3} \sin 1{,}256t$
Third harmonic: $i_3 = 22.5 \times 10^{-3} \sin 3{,}768t$
Sixth harmonic: $i_6 = 15 \times 10^{-3} \sin 7{,}536t$
Complex wave: $i = 45 \times 10^{-3} \sin 1{,}256t + 0.5 \sin 3{,}768t + 0.333 \sin 7{,}536t$

Figure 14-9 shows the fundamental, the harmonics, and the resultant complex wave. ////

Up to this point all discussion has been about the sine wave and various combinations of sine waves. The cosine wave is a sinusoidal wave that is displaced from a sine wave by $\pm 90°$. Equation 14-6 can be written in terms of cosine waves as well as

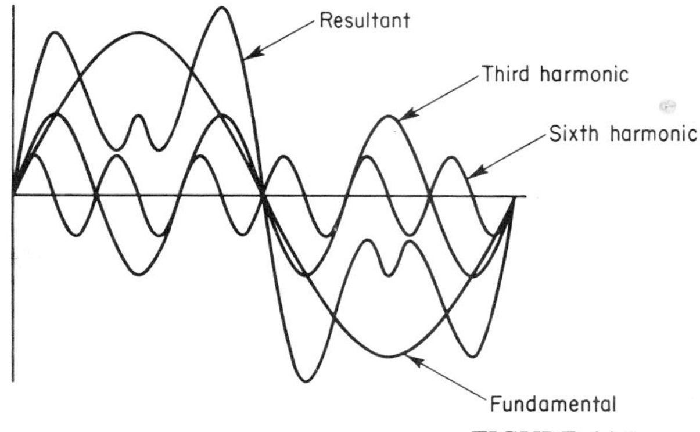

FIGURE 14-9
Resultant waveforms for Example 14-1.

sine waves. As a matter of fact, the usual equation for a complex wave will contain both sine and cosine terms. For example, the equation for the wave in Fig. 14-8 may be written in terms of sine waves only, or both sine and cosine waves.

In sine waves,

$$e = E_m[\sin \omega_1 t + k_2 \sin(2\omega_1 t - 90°)]$$

In sine and cosine terms,

$$e = E_m(\sin \omega_1 t - k_2 \cos 2\omega_1 t)$$

Figure 14-10 illustrates the relationship between a sine wave and leading or lagging waves.

The reader can see from this discussion that the instantaneous value of a nonsinusoidal voltage or current is the algebraic sum of the individual frequency components of voltage or current at any instant in time.

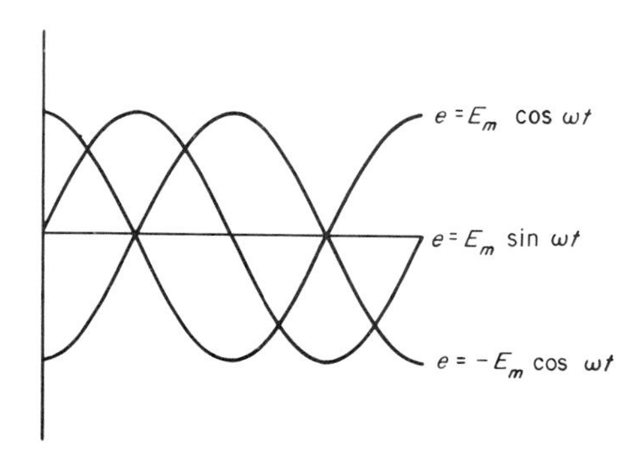

FIGURE 14-10
Leading and lagging cosine waves.

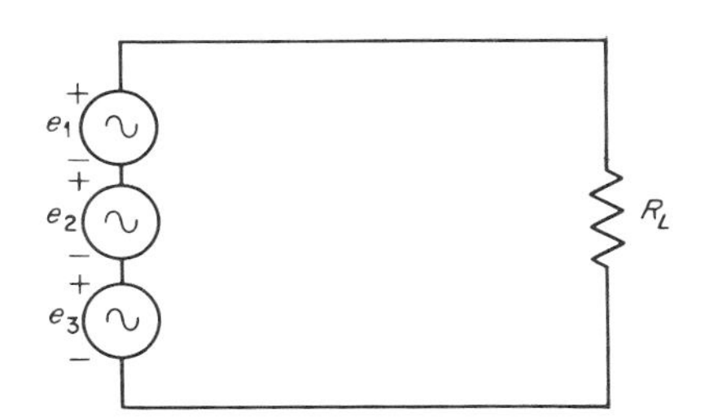

FIGURE 14-11
Treating a nonsinusoidal wave as three separate sources.

14-4 EFFECTIVE VALUE

The principle of superposition enables us to determine the effective value of the nonsinusoid. If a nonsinusoidal voltage is applied across a resistance, the power dissipated results from all the frequency components causing current flow in the resistor. By the superposition theorem, we can visualize a fundamental and its harmonics as individual sources of voltage, all developing power in the resistance. The total power developed in the resistance is then the sum of the powers developed by each of the frequency components of the wave. Figure 14-11 illustrates this concept for a nonsinusoidal wave consisting of a fundamental and its second and third harmonics.

Solving for the power in the resistor, we have

$$P_t = P_1 + P_2 + P_3 \tag{14-7}$$

where P_t = total power dissipated in resistor

P_1 = power dissipation in resistor due to fundamental

P_2 = power dissipation in resistor due to second harmonic

P_3 = power dissipation in resistor due to third harmonic

In terms of voltage, the total power is

$$P_t = \frac{E_{1,\text{eff}}^2}{R_L} + \frac{E_{2,\text{eff}}^2}{R_L} + \frac{E_{3,\text{eff}}^2}{R_L}$$

$$= \frac{E_{1,\text{eff}}^2 + E_{2,\text{eff}}^2 + E_{3,\text{eff}}^2}{R_L} \tag{14-8}$$

where $E_{1,\text{eff}}$, $E_{2,\text{eff}}$, and $E_{3,\text{eff}}$ are effective values of the fundamental, the second harmonic, and the third harmonic.

$$P_t = \frac{E_m^2 + (k_2 E_m)^2 + (k_3 E_m)^2}{2R_L} \tag{14-9}$$

where the voltages are specified in terms of peak values of the fundamental.

In terms of current, the total power is

$$P_t = I_{1,\text{eff}}^2 R_L + I_{2,\text{eff}}^2 R_L + I_{3,\text{eff}}^2 R_L$$
$$= (I_{1,\text{eff}}^2 + I_{2,\text{eff}}^2 + I_{3,\text{eff}}^2) R_L \qquad (14\text{-}10)$$

where the currents are specified in effective values.

When the currents are written in terms of the peak value of the fundamental, the total power is

$$P_t = \frac{[I_m^2 + (k_2 I_m)^2 + (k_3 I_m)^2] R_L}{2} \qquad (14\text{-}11)$$

Previous sections have shown how to determine the instantaneous value of a nonsinusoidal current or voltage. With the aid of Eqs. (14-8) to (14-11), it is possible to solve for the effective value of a nonsinusoidal current or voltage.

The total power in any resistance is the square of the effective voltage across the resistance divided by the resistance. If this relationship is solved for the effective voltage,

$$E_{t,\text{eff}} = \sqrt{P_t R_L}$$

From Eq. (14-8), it is at once apparent that the effective value of a nonsinusoidal voltage is the square root of the sum of the squares of the effective voltages of the frequency components of the wave.

$$E_{t,\text{eff}} = \sqrt{E_{1,\text{eff}}^2 + E_{2,\text{eff}}^2 + \cdots + E_{n,\text{eff}}^2} \qquad (14\text{-}12)$$

From Eq. (14-9),

$$E_{t,\text{eff}} = \sqrt{\frac{E_m^2 + (k_2 E_m)^2 + (k_3 E_m)^2 + \cdots + (k_n E_m)^2}{2}} \qquad (14\text{-}13)$$

The total power dissipated in any resistance is the product of the square of the effective value of current and the resistance. The current found from this relationship is

$$I_{t,\text{eff}} = \sqrt{\frac{P_t}{R_L}}$$

From Eq. (14-10), it is apparent that the effective current of a nonsinusoidal wave is found in a manner similar to the solution for the effective voltage of the wave.

$$I_{t,\text{eff}} = \sqrt{I_{1,\text{eff}}^2 + I_{2,\text{eff}}^2 + \cdots + I_{n,\text{eff}}^2} \qquad (14\text{-}14)$$

From Eq. (14-11),

$$I_{t,\text{eff}} = \sqrt{\frac{I_m^2 + (k_2 I_m)^2 + (k_3 I_m)^2 + \cdots + (k_n I_m)^2}{2}} \qquad (14\text{-}15)$$

EXAMPLE 14-2 A nonsinusoidal wave whose equation of voltage is given below is applied to a series circuit consisting of a 1,000-Ω resistor and a 4,700-Ω resistor. What is the effective voltage across the 4,700-Ω resistor, and what is the effective current in the circuit?

$$e = 0.45(\sin \omega_1 t + 0.4 \sin 2\omega_1 t)$$

SOLUTION

1 The effective voltage of the wave is found.

$$E_{t,\text{eff}} = \sqrt{\frac{0.45^2 + (0.4 \times 0.45)^2}{2}} = \sqrt{\frac{0.235}{2}} = 0.343 \text{ V}$$

2 The effective current is found by Ohm's law.

$$I_{t,\text{eff}} = \frac{0.343}{5.7 \times 10^3} = 60.1\ \mu\text{A}$$

3 The effective drop across the 4,700-Ω resistor is found.

$$E_{\text{eff}} = (60.1 \times 10^{-6})(4.7 \times 10^3) = 0.238 \text{ V}$$ ////

14-5 CIRCUIT ANALYSIS WITH NONSINUSOIDAL WAVES

It should be clear that the principle of superposition permits the treatment of each frequency component of a nonsinusoidal wave as an individual source of voltage or current in any circuit. In solving circuits, some very interesting results are obtained when nonsinusoids are applied to circuits containing reactance and resistance.

In any circuit containing reactances, the impedances seen by the various frequency components of the input wave will differ. This changes the waveshapes of voltage and current through circuit components to forms different from the waveshapes of the input. In addition, the various ratios between harmonics and the fundamental are changed.

EXAMPLE 14-3 In the circuit shown in Fig. 14-12, the applied voltage is given by

$$e_{\text{in}} = 150(\sin 6{,}280t + 0.3 \sin 12{,}560t + 0.1 \sin 31{,}400t)$$

(1) Write the equation for the voltage across the 2,700-Ω resistor. (2) Draw the input waveform. (3) Draw the output waveform across the resistor.

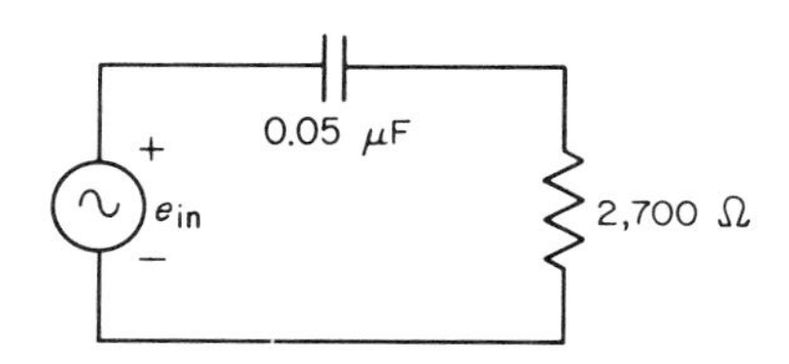

FIGURE 14-12
Circuit for Example 14-3.

SOLUTION

1 The frequency components are found by dividing each term of the input voltage by 2π.

$$\text{Fundamental} = 1{,}000 \text{ Hz}$$
$$\text{Second harmonic} = 2{,}000 \text{ Hz}$$
$$\text{Fifth harmonic} = 5{,}000 \text{ Hz}$$

2 The input-voltage equation indicates that all frequency components are in phase.

3 By the superposition theorem, the current through the circuit is found for each frequency component.

Fundamental:

$$X_c = \frac{1}{\omega_1 C} = 3{,}180\ \Omega$$

$$\mathbf{Z}_1 = 2{,}700 - j3{,}180 = 4{,}160 \angle -49.6^\circ\ \Omega$$

$$\mathbf{I}_1 = \frac{150}{4{,}160 \angle -49.6^\circ} = 36.1 \angle 49.6^\circ \text{ mA}$$

Second harmonic:

$$X_c = \frac{1}{\omega_2 C} = 1{,}590\ \Omega$$

$$\mathbf{Z}_2 = 2{,}700 - j1{,}590 = 3{,}140 \angle -30.5^\circ\ \Omega$$

$$\mathbf{I}_2 = \frac{0.3 \times 150}{3{,}140 \angle -30.5^\circ} = 14.3 \angle 30.5^\circ \text{ mA}$$

Fifth harmonic:

$$X_c = \frac{1}{\omega_5 C} = 636\ \Omega$$

$$\mathbf{Z}_5 = 2{,}700 - j636 = 2{,}770 \angle -13.3^\circ\ \Omega$$

$$\mathbf{I}_5 = \frac{0.1 \times 150}{2{,}770 \angle 13.3^\circ} = 5.4 \angle 13.3^\circ \text{ mA}$$

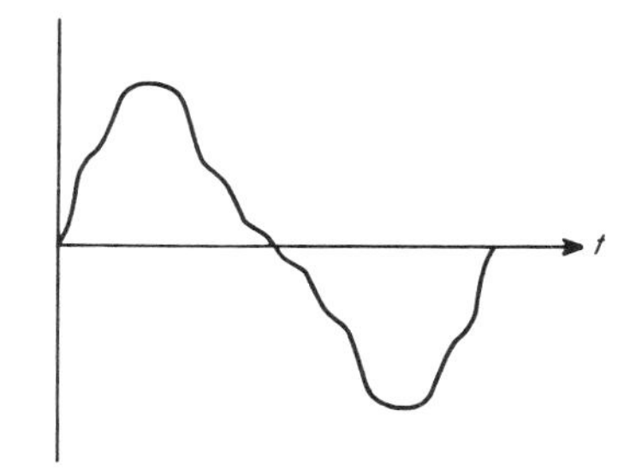

FIGURE 14-13
Input waveform of Example 14-3. One cycle.

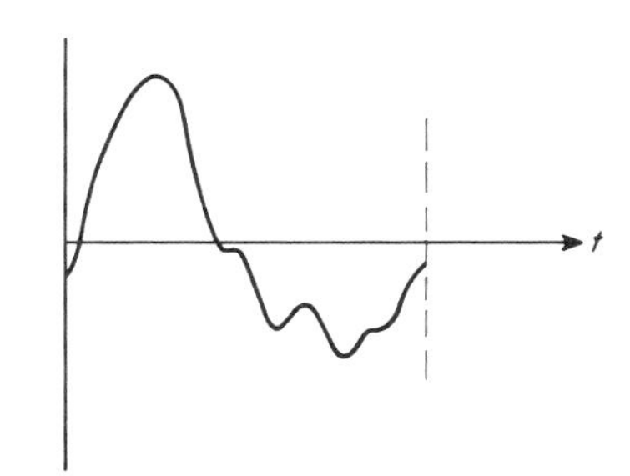

FIGURE 14-14
Output waveform of Example 14-3. One cycle.

4 The equation of the current waveform can now be found.

$$k_2 = \frac{|\mathbf{I}_2|}{|\mathbf{I}_1|} = \frac{14.3}{36.1} = 0.396$$

$$k_5 = \frac{|\mathbf{I}_5|}{|\mathbf{I}_1|} = \frac{5.4}{36.1} = 0.15$$

The second-harmonic component of the current lags the fundamental by 19.1° (49.6 − 30.5°), and the fifth-harmonic component of the current lags the fundamental by 36.3° (49.6 − 13.3°).

$$i = 36.1[\sin 6{,}280t + 0.396 \sin(12{,}560t - 19.1°) + 0.15 \sin(31{,}400t - 36.3°)] \text{ mA}$$

5 The equation of the voltage across the resistor can now be written.

$$e = iR = 97.5[\sin 6{,}280t + 0.396 \sin(12{,}560t - 19.1°) + 0.15 \sin(31{,}400t - 36.3°)]$$

6 The input-voltage waveform is shown in Fig. 14-13.

7 The output-voltage waveform is shown in Fig. 14-14. ////

EXAMPLE 14-4 In the circuit of Fig. 14-15, the total current is

$$i = 210(\sin 37.68 \times 10^4 t + 0.04 \sin 113.04 \times 10^4 t)\ \mu\text{A}$$

Write the equation for the voltage between terminals a and b.

SOLUTION The equation of the current indicates a waveform with a fundamental frequency of 60 kHz and a third harmonic at 180 kHz. The voltage between terminals a and b is the sum of the drops due to the components of the current.

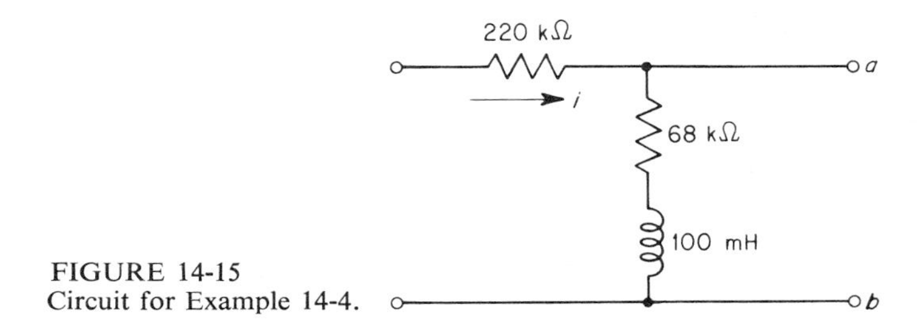

FIGURE 14-15
Circuit for Example 14-4.

1 Solve for the impedance at each frequency.
At 60 kHz:

$$\mathbf{Z}_1 = 68 \times 10^3 + j37.6 \times 10^3 = 77.8 \times 10^3 \angle 29^\circ\ \Omega$$

At 180 kHz:

$$\mathbf{Z}_3 = 68 \times 10^3 + j113 \times 10^3 = 132 \times 10^3 \angle 58.9^\circ\ \Omega$$

2 The voltage drop at the fundamental is the product of the impedance at the fundamental frequency and the current at the fundamental.

$$\mathbf{E}_1 = \mathbf{Z}_1 \mathbf{I}_1 = (77.8 \times 10^3 \angle 29^\circ)(210 \times 10^{-6}) = 16.3 \angle 29^\circ \text{ V}$$

The voltage drop at the third harmonic is the product of the third-harmonic current and the impedance at the third harmonic.

$$\mathbf{E}_3 = \mathbf{Z}_3 \mathbf{I}_3 = (132 \times 10^3 \angle 58.9^\circ)(0.04)(210 \times 10^{-6}) = 1.11 \angle 58.9^\circ \text{ V}$$

3 The results found in step 2 show that the voltage component at the third harmonic leads the fundamental by 29.9° (58.9° − 29°). The magnitude of the third harmonic is related to the fundamental by the factor 0.068 (1.11/16.3). The equation of the voltage across terminals *a* and *b* may now be written.

$$e = 16.3[\sin 37.68 \times 10^4 t + 0.068 \sin(113.04 \times 10^4 + 29.9^\circ)t] \quad ////$$

PROBLEMS

14-1 Given the circuit of Fig. 14-15, where the voltage drop across the 68-kΩ resistor is 3.6 ($\sin 62{,}800t + 0.1 \sin 188{,}400t$), (*a*) find the current in the circuit and (*b*) find the voltage drop across the inductance.

14-2 In the circuit of Fig. 14-16, the applied voltage is

$$e = 19[\sin 3{,}104t + 0.2 \sin(6{,}208t + 30^\circ) + 0.05 \sin(9{,}312t + 80^\circ)]$$

(*a*) Find the voltage drop across terminals *x*-*y* at the fundamental frequency.
(*b*) Find the voltage drop across *x*-*y* for the second harmonic.
(*c*) Find the voltage drop across *x*-*y* for the third harmonic.

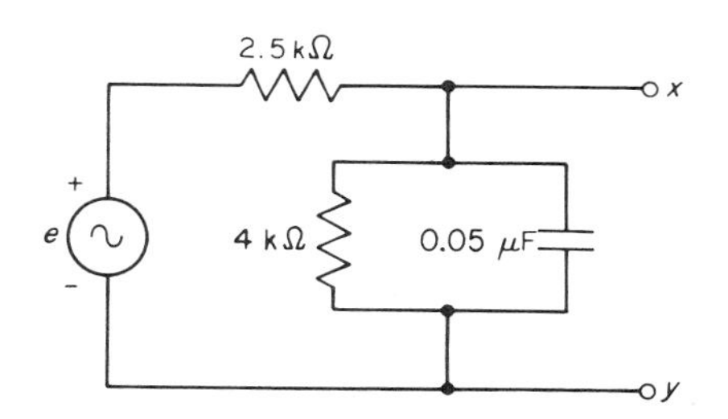

FIGURE 14-16
Circuit for Prob. 14-2.

14-3 What is the effective value of the current in Prob. 14-1?

14-4 Find the power dissipated in the 220-kΩ resistor of Fig. 14-15 under the operating conditions of Prob. 14-1.

14-5 What is the power dissipated in the 2,500-Ω resistor of Prob. 14-2?

14-6 An excitation of $e = 9[\sin(6 \times 10^5)t - 0.4 \cos(12 \times 10^5)t + 0.8 \sin(24 \times 10^5)t]$ is applied to the circuit of Fig. 14-17.

(*a*) Find the current in the circuit.
(*b*) Determine the voltage drop across the capacitor.
(*c*) Find the effective value of the applied voltage.
(*d*) Calculate the effective value of the voltage drop.

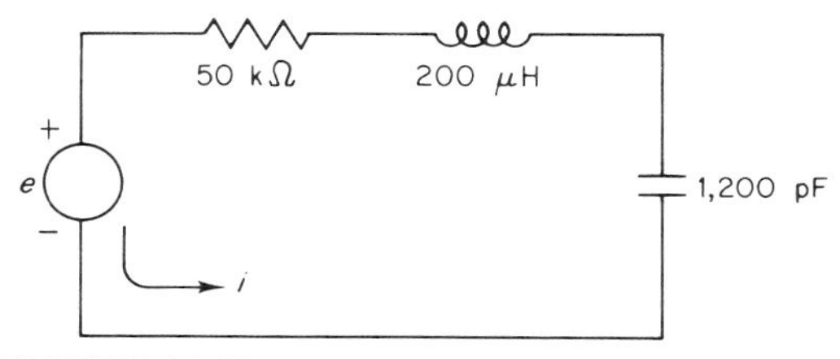

FIGURE 14-17
Circuit for Prob. 14-6.

15

POLYPHASE CIRCUITS

15-1 INTRODUCTION

Any circuit with a single source of alternating current is simply referred to as single-phase alternating current. A *polyphase* circuit or system consists of several single-phase circuits that are displaced from one another in phase and are interconnected in some manner.

The polyphase circuit can deliver electric energy to a single load, as in a polyphase rectifier circuit and three-phase motors and transformers, or to several independent single-phase loads, as in house wiring. Single-phase sources are inefficient compared to three-phase sources, especially for the distribution of power over long distances. In addition, a single-phase source is too "pulsating" for effective operation of large motors. From the standpoint of distribution, it will be shown that a three-phase system requires less copper than a single-phase system. Three-phase motors and generators are lighter, more rugged, and less expensive than single-phase machines that deliver the same power. For these reasons single-phase generators are seldom used, and all commercial single-phase sources are simply one phase of a polyphase system.

Theoretically a polyphase system may have any number of phases. It will be shown that a three-phase system has advantages over two- or four-phase systems, and

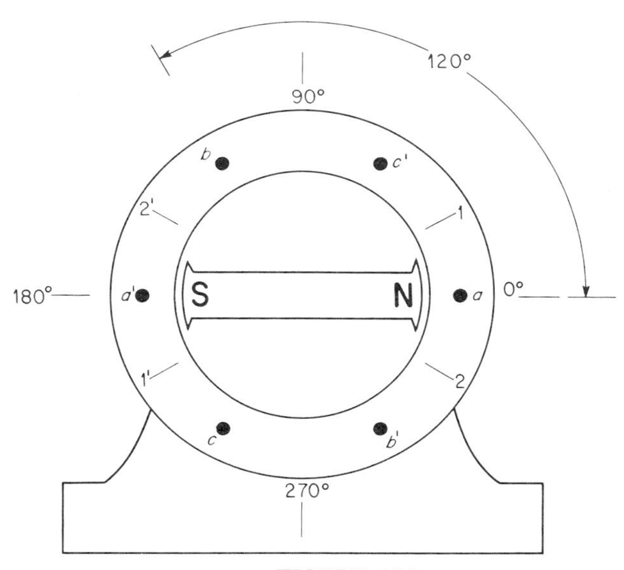

FIGURE 15-1
Cross section of an elementary three-phase alternator.

therefore nearly all polyphase systems are three-phase (or multiples of three-phase) systems. Discussion in this chapter will include two-phase systems as well as three-phase systems, primarily so that an intelligent comparison can be made between them.

15-2 THE GENERATION OF POLYPHASE VOLTAGES

Polyphase voltages are generated in the same manner as single-phase voltages. Consider the cross section of the elementary three-phase generator shown in Fig. 15-1. Assume that the direction of rotation of the rotor is clockwise. Assume that when the north pole of the rotor is passing under a stator winding, the polarity of the voltage induced in the winding is positive. The stator coils each consist of two parts; their halves are identified by the same letter (a-a', b-b', c-c'). All the windings are electrically isolated from one another, and three pairs of terminals are available on the generator panel so that the outputs of the stator windings can be connected to one another in several ways.

The voltages generated in the stator coils during one cycle of operation are shown in Fig. 15-2. For the sake of simplicity, terminals a, b, and c will be referred to as *starts*, and a', b', and c' as *finishes*. The graph in Fig. 15-2 represents the phase voltages

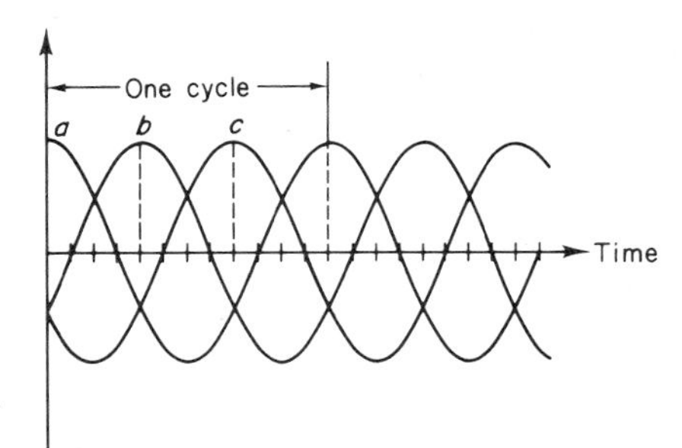

FIGURE 15-2
Graph of output voltage of alternator of Fig. 15-1 with phase sequence *abc*.

at the start terminals with the finishes as the reference points. When the rotor is at 0° and passing under coils a-a', there will be a maximum voltage generated in the coils, with start a positive with respect to finish a'. At the same instant, there will be equal voltages developed in coils b-b' and c-c'. These two voltages will be negative with respect to their finishes, since the south pole of the rotor is passing their respective starts. As the rotor is turned, the voltages in the phases change, and the reader will note that the next phase winding to go maximum positive is b, then c. This sequence of voltage generation is called the *phase sequence*, and in this case the phase sequence is *abc*. If the direction of rotation is reversed, the phase sequence is *acb*. The phase sequence can be changed externally by the order in which the phases are connected to a load. Obviously, if the loads on the phases are isolated, phase sequence is of no importance. However, when the generator is connected to a three-phase load, the sequence is of great importance. Changing the phase sequence to a motor causes it to rotate in the opposite direction. Changing the phase sequence to an unbalanced system changes the line currents and completely changes the characteristics of a polyphase system.

For the diagram of phase voltages in Fig. 15-2, the phase-voltage equations can be written.

$$e_a = E_{\max} \cos \omega t \tag{15-1}$$

$$e_b = E_{\max} \cos(\omega t - 120°) \tag{15-2}$$

$$e_c = E_{\max} \cos(\omega t - 240°) \tag{15-3}$$

It can be seen from Fig. 15-2 and from Eqs. (15-1) to (15-3) that the *sum of the phase voltages at any instant is zero.*

$$e_a + e_b + e_c = 0 \tag{15-4}$$

This can be demonstrated by calculating the voltages at several instants. When ωt is zero,

$$e_a = E_{\max}$$

$$e_b = E_{\max} \cos(-120°) = -0.5E_{\max}$$

$$e_c = E_{\max} \cos(-240°) = -0.5E_{\max}$$

The sum of the three voltages is zero. When $\omega t = 90°$,

$$e_a = 0$$

$$e_b = E_{\max} \cos(-30°) = 0.866E_{\max}$$

$$e_c = E_{\max} \cos(-150°) = -0.866E_{\max}$$

Again the sum of the voltages is zero. The student will find it instructive to try various values of ωt and demonstrate to himself that the algebraic sum of the phase voltages is always zero.

The discussion of the generation of polyphase voltages included a simple three-phase alternator with single coils for each phase. This is a simplification. In actual practice, there are coils around the entire periphery of the stator frame. In the sketch in Fig. 15-1, the numbers 1, 2, 1′, and 2′ are shown. In the space between 1 and 2, and 1′ and 2′ would be coils that are part of phase a. The distance between 1 and 2 and between 1′ and 2′ is called a *phase belt*. There are similar phase belts for phases b and c.

Note that the angular displacement between phases is 120°. In general, the phase difference in degrees can be found by dividing 360° by the number of phases.

$$a = \frac{360°}{n} \tag{15-5}$$

where a is the phase displacement in degrees between phases and n is the number of phases generated.

It follows, then, that a two-phase system should have phase voltages 180° apart. This is not the case, however, because the two-phase system is in reality four phases with 90° between phases. Figure 15-3 illustrates a simple two-phase alternator.

The method of internal connection of starts and finishes determines whether the generator output is two- or four-phase. Consider the phasor diagram in Fig. 15-4. If all the midpoints of the stator coils are connected to a common terminal, there will be four voltages available at the output terminals of the generator. Each voltage will be displaced from each of the others by 90 or 180°. The resultant is a four-phase

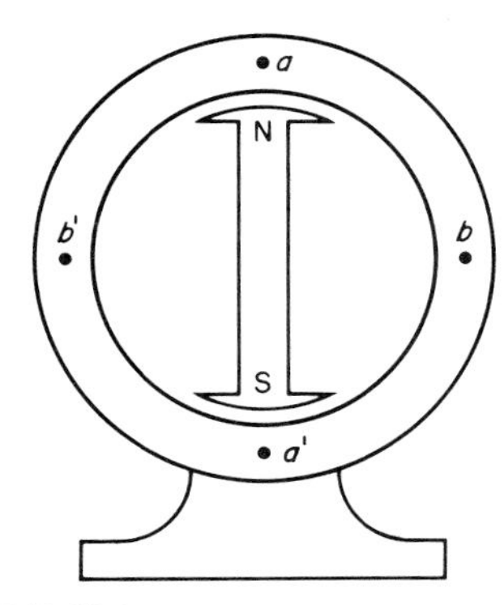

FIGURE 15-3
Cross section of elementary two-phase alternator.

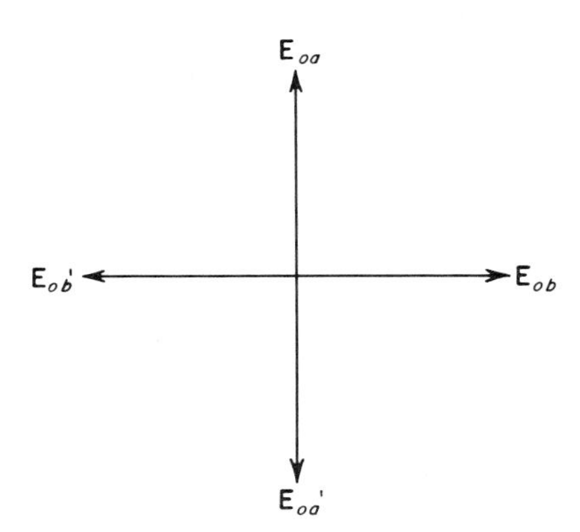

FIGURE 15-4
Phase relations in a four-phase system.

system with 90° between adjacent phases. If the total voltage across each winding is used as output voltage, only two voltages are available and the system is a two-phase system. However, the angular displacement between phases is still 90°, as shown in the phasor diagram of Fig. 15-5.

15-3 DOUBLE-SUBSCRIPT NOTATION

It will be noted that in Figs. 15-4 and 15-5 the voltages are identified by double subscripts. This notation is used to indicate the sense of a voltage (or current). The first subscript is the reference point, and the second is the point of measurement. For example, $\mathbf{E}_{oa}$ means the voltage at terminal a with respect to o. Another way of saying

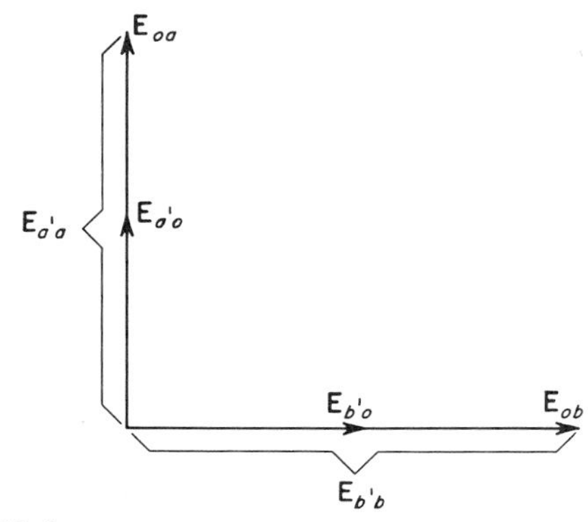

FIGURE 15-5
Phase relations in a two-phase system.

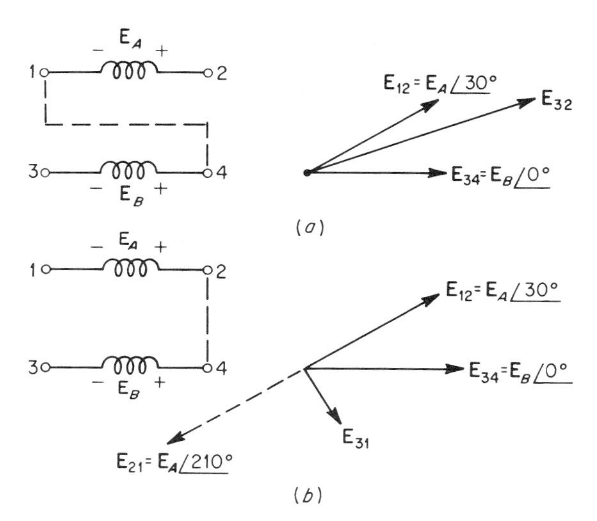

FIGURE 15-6
(*a*) Windings connected so that E_A and E_B are in series-aiding. The resultant voltage is measured from terminal 3 to terminal 2. (*b*) Windings connected so the E_A and E_B are in series-opposing. The resultant voltage is measured from terminal 3 to terminal 1.

the same thing is that $\mathbf{E}_{oa}$ is the voltage measured from o (as an origin) to a. Voltage $\mathbf{E}_{ao}$ is simply the negative of $\mathbf{E}_{oa}$.

This double-subscript notation is necessitated in polyphase circuits by the fact that phase voltages must be interconnected. It should be apparent that various phase voltages can be interconnected to be either additive or subtractive, hence the need for some means of identifying the method of interconnection.

For example, suppose that two phase windings are to be connected to yield a voltage that is the additive sum of the individual phase voltages. Without double-subscript notation, there would be no way of determining which terminals must be tied together to combine the voltages in series-aiding. The only alternative method would be trial and error. On the other hand, double-subscript notation shows immediately which terminals must be connected. This problem is illustrated in Fig. 15-6. Note that if the terminals of either coil are reversed, the voltages are 150° out of phase, rather than 30°.

15-4 TWO-PHASE CIRCUITS

It is common practice for all polyphase alternators to keep the individual phase windings electrically insulated from one another within the machine. Interconnections between phases are made at the terminal board of the alternator. In Sec. 15-2, it was pointed out that two-phase voltages can be generated by an alternator with two sets of

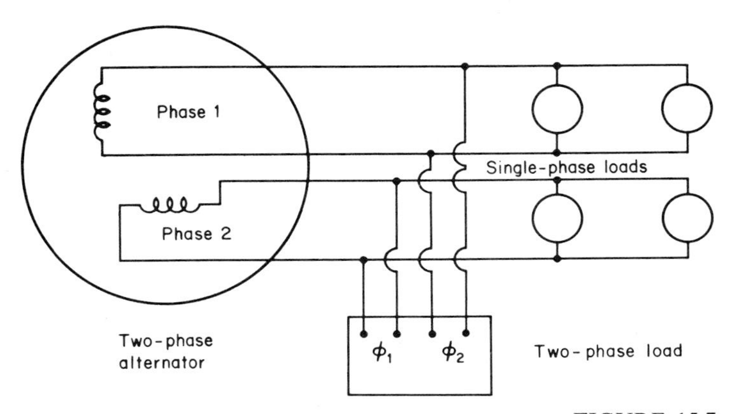

FIGURE 15-7
Four-wire two-phase system.

stator windings arranged to deliver two voltages 90° apart at the terminals. When the output voltages of an alternator are maintained as two separate single-phase circuits to the loads, the polyphase system is referred to as a *four-wire two-phase system*. Figure 15-7 illustrates such a system. The reader will note the common method of representing the stators of a two-phase alternator.

A *three-wire two-phase system* is established when two-phase voltages are interconnected with a common lead. In this method, loads may be connected to either phase as single-phase loads, but there is no longer complete isolation of phases.

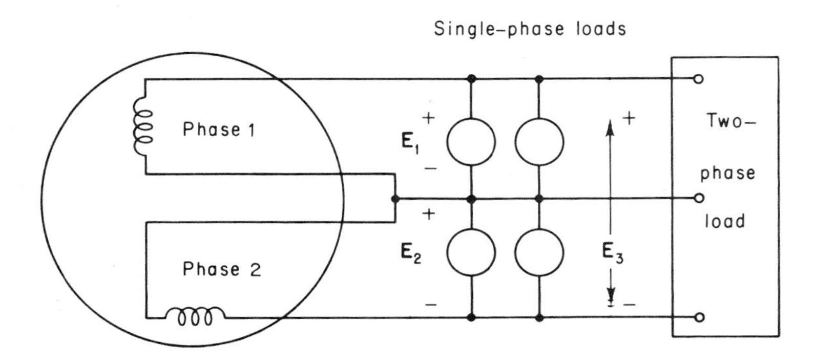

FIGURE 15-8
Three-wire two-phase system.

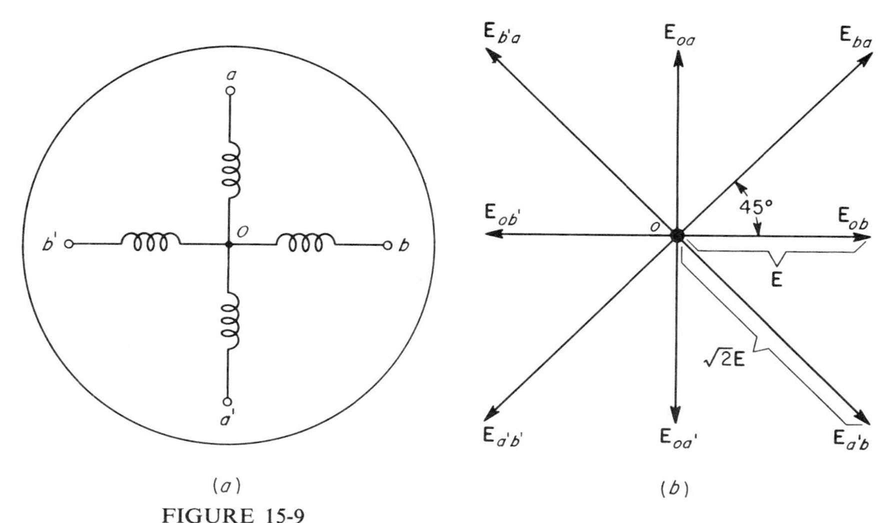

FIGURE 15-9
(*a*) Star-connected stators of a four-phase alternator. (*b*) Phasor diagram of phase and line voltages available at the terminals of the four-phase alternator.

The advantages of the three-wire system are a reduction in copper requirements and simplification of the wiring. One disadvantage of the three-wire system is that the voltage between the outer two wires of the system is 1.414 times the voltage of a single phase. This requires more insulation than the single-phase system. Figure 15-8 illustrates the two-phase three-wire system.

The two-phase system is never used to distribute electric power because three-phase systems are more efficient. The main application of two-phase systems is in the operation of motors, in particular, small servo motors of the type used in aircraft. In any two-phase system, the voltage between phases is always 1.414 $\sqrt{(2)}$ times the voltage of a single phase. This is because two-phase voltages are 90° out of phase and equal in magnitude. The phasor sum of such voltages is always 1.414 times either voltage. In Fig. 15-8, $|\mathbf{E}_1| = |\mathbf{E}_2|$, and $|\mathbf{E}_3| = 1.414|\mathbf{E}_1|$. The current in the common line is the phasor sum of the load currents. In a balanced system, the current in the return line is $\sqrt{2}$ times a single-phase load current.

The four-phase system is an extension of the two-phase system. If an alternator is arranged as shown in Fig. 15-9, there will be four voltages available at the terminals. The phasor diagram of this alternator is shown in Fig. 15-4. It can be seen from this diagram that the sum of all the voltages is zero. The connection shown in Fig. 15-9 is the *four-phase star*.

$$\mathbf{E}_{oa} + \mathbf{E}_{ob} + \mathbf{E}_{oa'} + \mathbf{E}_{ob'} = 0 \tag{15-6}$$

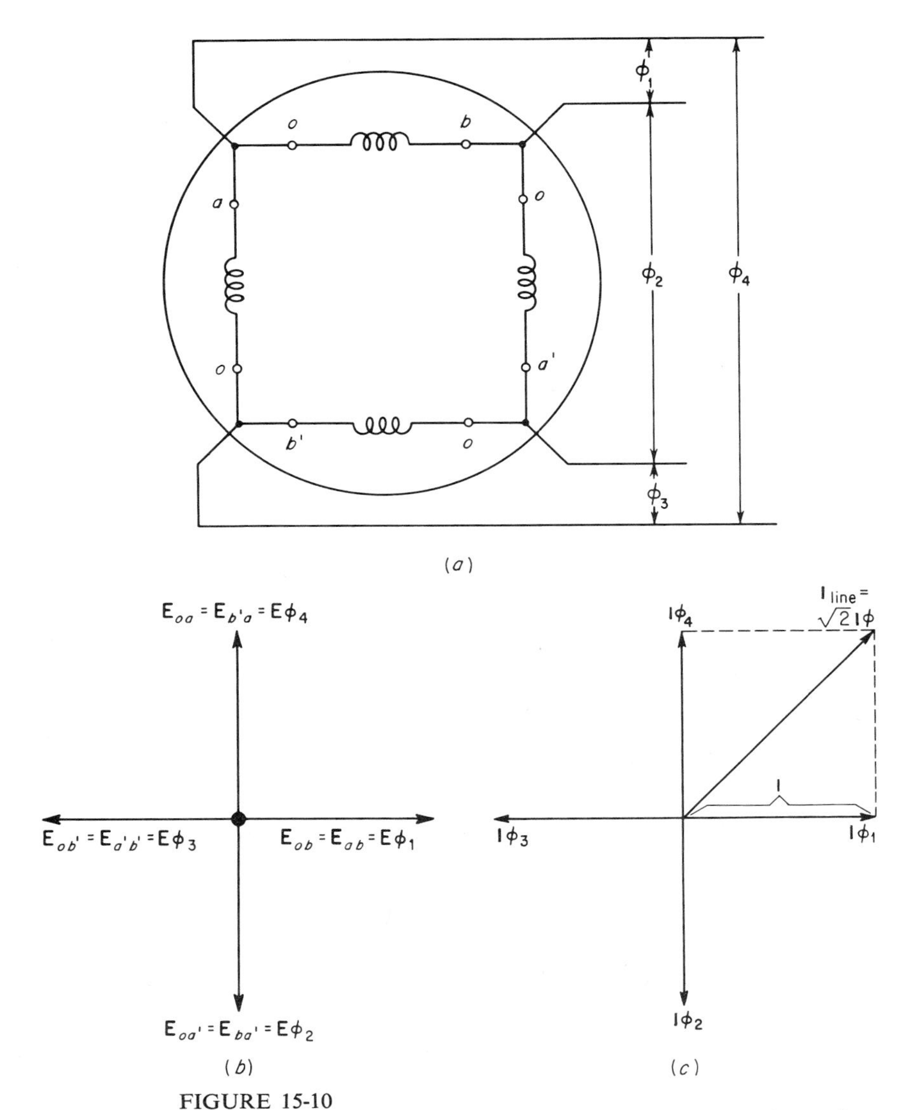

FIGURE 15-10
(*a*) Mesh-connected stators of a four-phase alternator. (*b*) Phasor diagram of the voltages of a four-phase mesh. (*c*) Phasor diagram of the currents of a four-phase mesh.

The star connection makes it possible to operate many single-phase and two-phase loads off the same alternator.

Because the sum of the phase voltages of the four-phase alternator is zero, it is possible to connect the four coils in the form of a mesh instead of a star. The star offers an advantage over the mesh in that a voltage $\sqrt{2}$ times a phase voltage is available. However, the mesh requires less insulation. Figure 15-10 shows the arrangement of stator coils for a mesh connection and the phasor diagram of the mesh voltages. The line current of a four-phase mesh is $\sqrt{2}$ times phase current.

15-5 THREE-PHASE Y SYSTEMS

The generation of three-phase voltages was discussed in Sec. 15-2. It has been shown that three separate single-phase voltages are developed by a three-phase alternator. It is therefore possible to operate three separate single-phase loads from one source. But this is not common practice. Polyphase generators usually operate polyphase loads in addition to single-phase loads; hence they must be connected so that other voltages as well as the individual phase voltages are available.

If either the starts or the finishes of the stator windings of the alternator of Fig. 15-1 are connected together, there will be three sets of voltages available at the terminals in addition to the individual single-phase voltages. This is shown in Fig. 15-11, where all finishes are connected together. The stator windings are drawn so that the 120° displacement between stators can be seen. The individual phase voltages are available between the line connecting all finishes and any of the individual start terminals. The line connecting all finishes is the neutral line, and in all discussions the common termination will be called the *neutral*.

The connection shown in Fig. 15-11*a* is the star, or Y, connection. The voltage from each line to the neutral is the *phase voltage*. Voltage between lines will simply be referred to as *line voltage*. Line voltage is found by taking the *difference* of potential between the lines. Therefore, to find the line voltage between phases, the diagram of Fig. 15-12 will be followed.

The reader will note from the phasor diagram of Fig. 15-12 that the line voltages are larger than the phase voltages. The numerical relationship between a phase voltage and a line voltage is found in the following manner.

Line voltage is the difference between two equal voltages 120° apart, or it is the sum of two equal voltages 60° apart. These statements are clearly illustrated in Fig. 15-12. Line voltage $\mathbf{E}_{12}$ is the difference between phase voltages E_{na} and E_{nb}. The negative of $\mathbf{E}_{na}$ appears in the diagram as $\mathbf{E}_{an}$. Hence $\mathbf{E}_{nb} - \mathbf{E}_{na} = \mathbf{E}_{nb} + \mathbf{E}_{an}$. The angular difference between $\mathbf{E}_{nb}$ and $\mathbf{E}_{an}$ is 60°.

$$\mathbf{E}_{12} = \mathbf{E}_{nb} + \mathbf{E}_{an}$$

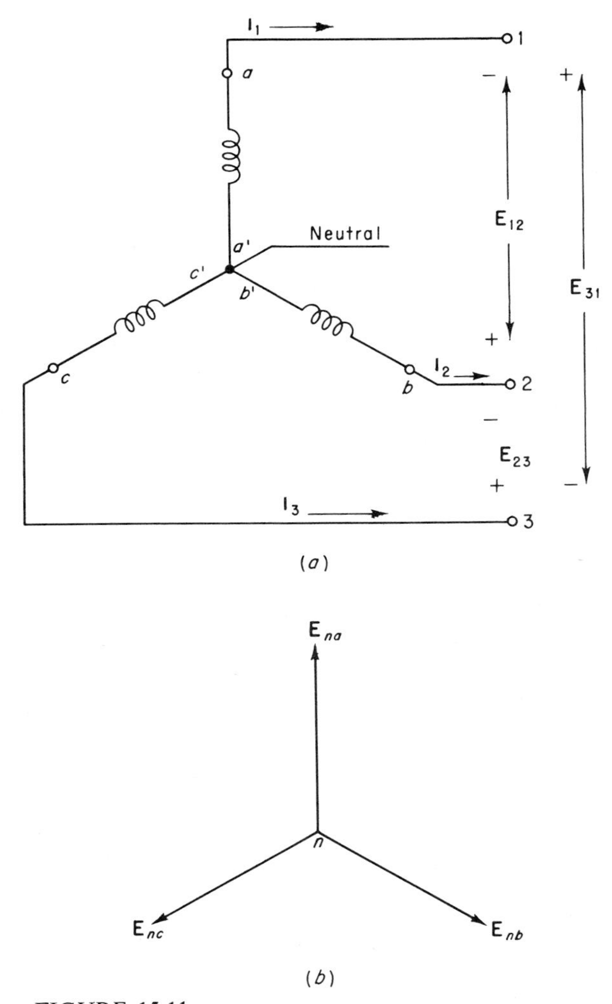

FIGURE 15-11
(*a*) Y-connected stator coils of a three-phase alternator. (*b*) Phasor diagram of the voltages in a three-phase Y connection.

$\mathbf{E}_{nb}$ is shown in the diagram at $-30°$, where it equals $0.866 - j0.5$. $\mathbf{E}_{an}$ is shown at $-90°$, where it equals $0 - j1.0$. $\mathbf{E}_{nb}$ and $\mathbf{E}_{an}$ are assumed to be unity-magnitude phasors for ease of calculation.

The sum of $\mathbf{E}_{nb}$ and $\mathbf{E}_{an}$ is $0.866 - j1.5$. The polar resultant of these rectangular quantities is $1.732 \angle -60°$, which is exactly where $\mathbf{E}_{12}$ is drawn in Fig. 15-12. Figure 15-13 illustrates this solution in detail. Since

$$1.732 = \sqrt{3}$$

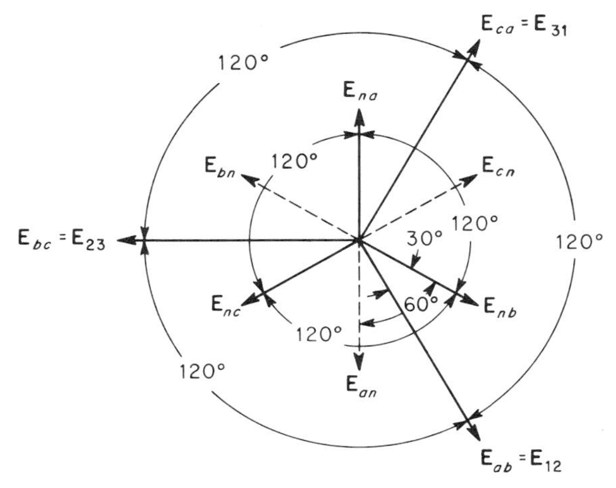

FIGURE 15-12
Phasor diagram of phase and line voltages for a three-phase Y-connected system.

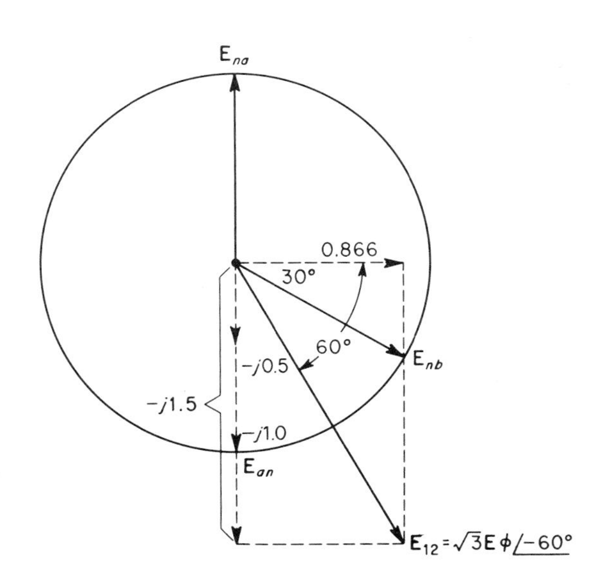

FIGURE 15-13
Phasor diagram showing the addition of two phases of a three-phase source. The resultant is 1.732 times a phase voltage.

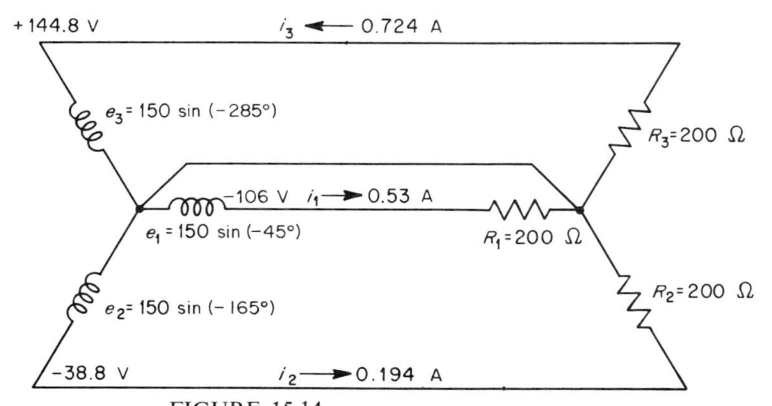

FIGURE 15-14
Three-phase Y-connected circuit and balanced load with a neutral line.

the following rule applies to the relationship between line voltage and phase voltage in the three-phase Y system:

$$E_{\text{line,eff}} = \sqrt{3}E_{\text{phase,eff}} \tag{15-7}$$

The currents in the Y system are displaced from one another by 120°. If the system is balanced (all loads equal), the currents are equal. The phase relationships between the currents and line voltages depend on the characteristics of the load circuit, as in any system. It should be apparent from inspection of Fig. 15-11*a* that the line currents are the same as phase currents.

Consider the circuit of Fig. 15-14. It is desired to determine the current per phase, the current in the lines, and the current in the neutral.

It has already been stated that current in the neutral is zero when the load is balanced. This statement will now be proved with the aid of Fig. 15-14. Voltages e_1, e_2, and e_3 are taken at some instant such that $\omega t = -45°$.

$$e_1 = 150\sin(-45°) = -106$$

$$e_2 = 150\sin(-45° - 120°) = 150\sin(-165°) = -38.8$$

$$e_3 = 150\sin(-45° - 240°) = 150\sin(-285°) = 144.8$$

The polarities of the voltages at this instant are shown in the figure. They are considered with respect to the neutral. The currents are found by Ohm's law.

$$i_1 = 0.53\text{ A}$$

$$i_2 = 0.194\text{ A}$$

$$i_3 = 0.724\text{ A}$$

The 45° angle was selected for this example. However, any other angle would have been equally suitable.

The currents are drawn on the circuit of Fig. 15-14. It is obvious that the current in the neutral is zero, since the sum of the currents entering the junction of the loads exactly equals current i_3. Kirchhoff's law of currents dictates that there be no current in the neutral. It should be clear that the only time there is current in the neutral is when the Y circuit operates into an unbalanced load. The circuit of Fig. 15-14 is referred to as a *four-wire three-phase Y*. The neutral is needed in circuits with unbalanced loads to make the voltages across all sections of the load equal. When the Y has a neutral, the voltage across each section of the line is the voltage of a phase, whether the load is balanced or not.

When the Y-connected source works into a balanced load, the neutral can be removed without affecting the behavior of the circuit. The current in each line and the voltage drop across each phase of the load will be precisely the same as if the neutral were in the circuit. A Y-connected circuit without a neutral is referred to as a *three-wire* three-phase Y system. A three-wire three-phase Y can be used with unbalanced loads, but under such conditions the neutral point of the load shifts so that the voltage drops across the various legs of the load are not the same. In a later section of this chapter, the performance of a three-wire Y circuit with unbalanced loads will be analyzed. Figure 15-15 illustrates a three-phase Y operating single-phase and polyphase loads. The neutral is effective only with unbalanced single-phase loads; the three-phase motor acting as a polyphase load is unaffected by the neutral line.

When a three-phase load is Y-connected, as in Fig. 15-15, the voltage per leg of the load equals a phase voltage, even though the line voltage is $\sqrt{3}$ times a phase voltage. This is because the three-phase load is a balanced load, and the circuit operates in the same manner as if a neutral line were connected between source and load. On the other hand, the three-phase load can be connected in delta (Sec. 15-6). When the load is delta-connected, the voltage across each leg of the load equals line voltage rather than phase voltage. This method of connecting a three-phase source to a three-phase load is called a *Y-delta* connection and is illustrated in Fig. 15-16.

EXAMPLE 15-1 In the circuit of Fig. 15-16, calculate the current in each load.

SOLUTION

1 To find the current in each leg of the delta-connected load, it is necessary to find the line voltage of the Y-connected source and the impedance per leg of the load.

$$E_{\text{line,eff}} = \sqrt{3}E_{\text{phase,eff}} = 208 \text{ V}$$

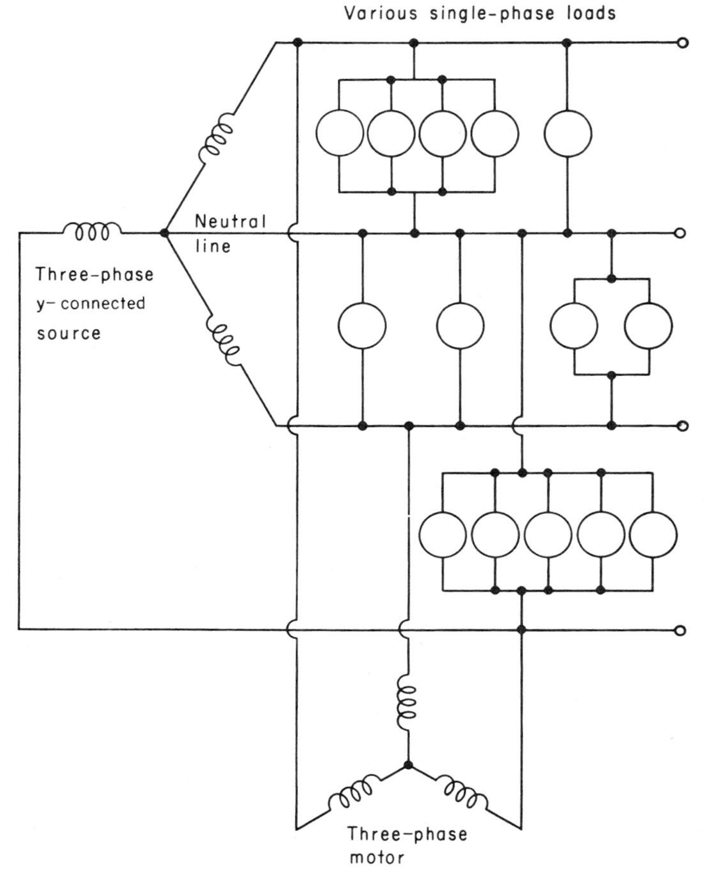

FIGURE 15-15
Three-phase Y-connected source with balanced three-phase loads and differing single-phase loads.

Impedance per leg

$$\mathbf{Z} = 40 + j25 = 47.3 \angle 32^\circ \ \Omega$$

Current per leg

$$I_{\text{line, eff}} = \frac{E_{\text{line,eff}}}{|\mathbf{Z}_{\text{leg}}|} = \frac{208}{|47.3 \angle 32^\circ|} = 4.41 \text{ A}$$

(The angle merely indicates the angle of lag between current and line voltage. The currents in the delta load will be 120° apart, just as the line currents are.)

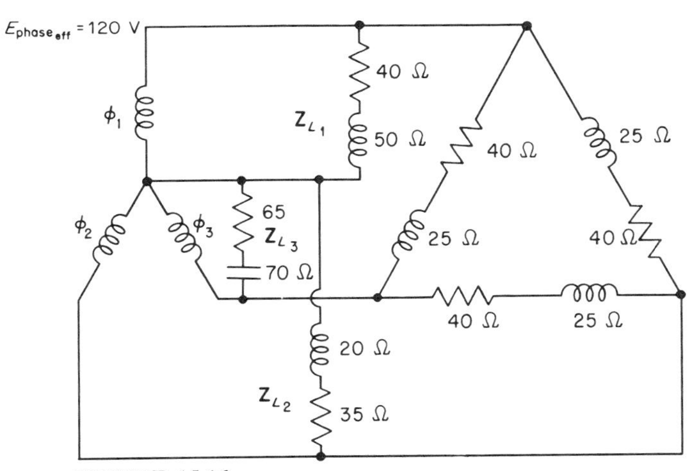

FIGURE 15-16
Three-phase Y-connected source with balanced delta load and differing single-phase loads. (Circuit for Example 15-1.)

2 If it is assumed that phase 1 is at $\angle 0°$ and the phase sequence is 123, then phase 2 is at $-120°$, and phase 3 is at $-240° \, (+120°)$.

$$\mathbf{I}_3 = \frac{E_{\text{phase,eff}} \angle 120°}{\mathbf{Z}_{L_3}} = \frac{120 \angle 120°}{95.5 \angle -47.1°} = 1.26 \angle 167.1° \text{ A}$$

$$\mathbf{I}_1 = \frac{E_{\text{phase,eff}} \angle 0°}{\mathbf{Z}_{L_1}} = \frac{120 \angle 0°}{64 \angle 51.3°} = 1.88 \angle -51.3° \text{ A}$$

$$\mathbf{I}_2 = \frac{E_{\text{phase,eff}} \angle -120°}{\mathbf{Z}_{L_2}} = \frac{120 \angle -120°}{40.2 \angle 29.8°} = 2.98 \angle -149.8° \text{ A}$$

////

15-6 THREE-PHASE DELTA SYSTEMS

It was pointed out in Sec. 15-2 that the phasor sum of the voltages of a three-phase system is zero [Eq. (15-4)]. It is therefore possible to interconnect the stator windings of the alternator so that the start of one winding is connected to the finish of another winding, as in Fig. 15-17. (Throughout these discussions an alternator is used as the source of three-phase voltages for the circuits. This is a matter of convenience. It

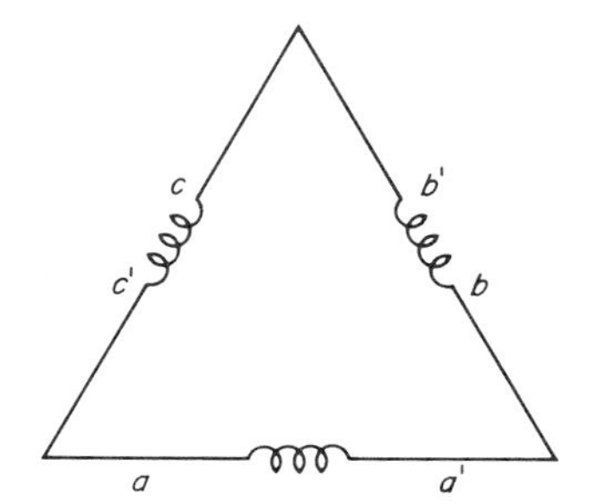

FIGURE 15-17
Delta-connected source, showing interconnection of "starts" and "finishes."

would be just as simple to use the secondary windings of a three-phase transformer as the voltage source for the terminals of the mesh.) This method of connection is called the *delta*, or *mesh*, connection. There will be no loop current in the mesh because the total voltage of the mesh is zero. However, loads can be connected to the terminals of the mesh. Each load has a voltage across it equal to a phase voltage, and the current per phase is determined by the individual loads. When the loads per phase are unequal, the delta system is unbalanced and a resultant current flows in the mesh. This will be analyzed in the section dealing with unbalanced three-wire systems.

When the delta-connected system operates into a balanced load, the currents per phase are equal and current per line is $\sqrt{3}$ times current per phase. Figure 15-19 is a phasor diagram of the currents in the circuit of Fig. 15-18. In the circuit of Fig. 15-18, currents are shown leaving the source and entering the load. This simplifies the drawing. It is obvious that at any instant the direction of the current is a function of the polarity of the source voltage, and at no time can all the currents be flowing in the

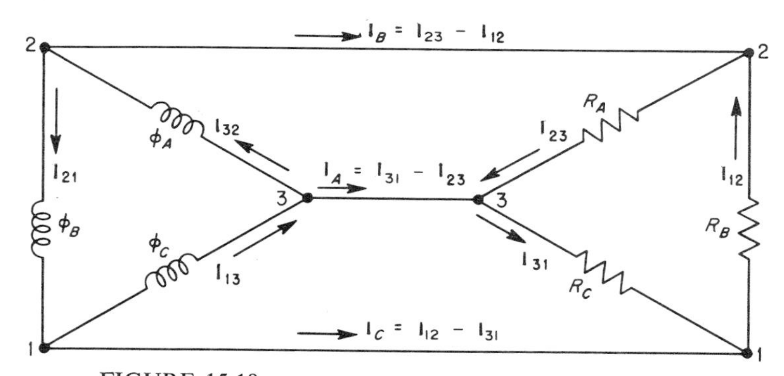

FIGURE 15-18
Delta-connected source and load. Voltage per phase is 220 V. (Circuit for Example 15-2.)

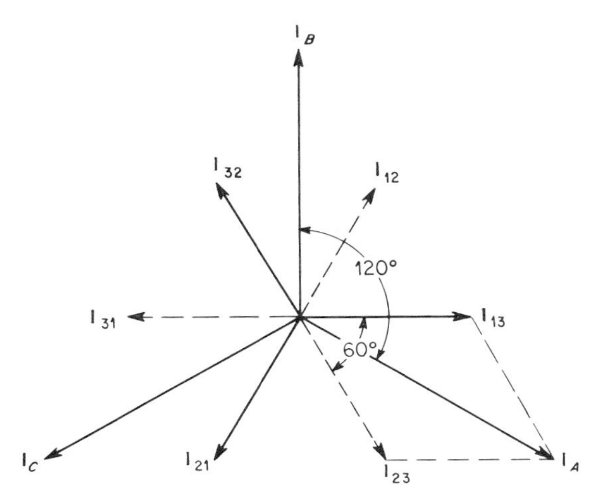

FIGURE 15-19
Phasor diagram of the currents in the circuit of Fig. 15-18.

same direction. Since line current is the phasor sum of two currents 60° apart, line current is 1.732 ($\sqrt{3}$) times phase current.

$$I_{\text{line,eff}} = \sqrt{3} I_{\text{phase,eff}} \tag{15-8}$$

EXAMPLE 15-2 In the circuit of Fig. 15-18, the voltage per phase is 220 V at a frequency of 60 Hz. Load resistors R_A, R_B, and R_C are each 35 Ω. The phase sequence is *CBA*. (1) Find the effective current in each line. (2) Find the instantaneous current per line when $t = 0.51$ s.

SOLUTION

1 The rms current per line is 1.732 times rms current per branch of the delta load.

$$I_{\text{load,eff}} = \frac{E_{\text{phase,eff}}}{R_{\text{load}}} = \frac{220}{35} = 6.29 \text{ A}$$

$$I_{A,\text{eff}} = I_{B,\text{eff}} = I_{C,\text{eff}} = 1.732 \times 6.29 = 10.9 \text{ A}$$

2 Find the instantaneous voltage per phase in the order of the phase sequence.

$$e_C = 311 \sin 377t$$

$$e_B = 311 \sin(377t - 120°)$$

$$e_A = 311 \sin(377t - 240°)$$

3 Find the angular displacement α of phase C at the instant $t = 0.51$ s. Find the time of one cycle.

$$T = \frac{1}{f} = 0.0166 \text{ s}$$

Find the number of cycles during 0.51 s.

$$N = \frac{5.1 \times 10^{-1}}{1.66 \times 10^{-2}} = 30.6 \text{ cycles}$$

4 From the phase sequence, it is assumed that at time $t = 0$, phase C starts a cycle. Therefore, α must be 0.6 cycle, 0.1 cycle = 36°, and it follows that

$$\alpha = 216°$$

5 Solve for the voltage per phase.

$$E_C = 311 \sin 216° = 182.8 \text{ V}$$

$$E_B = 311 \sin 96° = 309 \text{ V}$$

$$E_A = 311 \sin -24° = -126.5 \text{ V}$$

6 Solve for the currents by Ohm's law.

$$I_{31} = 5.22 \text{ A}$$

$$I_{12} = 8.84 \text{ A}$$

$$I_{23} = 3.62 \text{ A}$$

7 Solve for line currents.

$$I_C = I_{12} - I_{31}$$

$$I_C = 8.84 - 5.22$$

$$= 3.62A$$

$$I_B = I_{23} - I_{12}$$

$$I_B = 3.62 - 8.84$$

$$= -5.22 \text{ A}$$

$$I_A = I_{31} - I_{23}$$

$$I_A = 5.22 - 3.62$$

$$= 1.6 \text{ A}$$

////

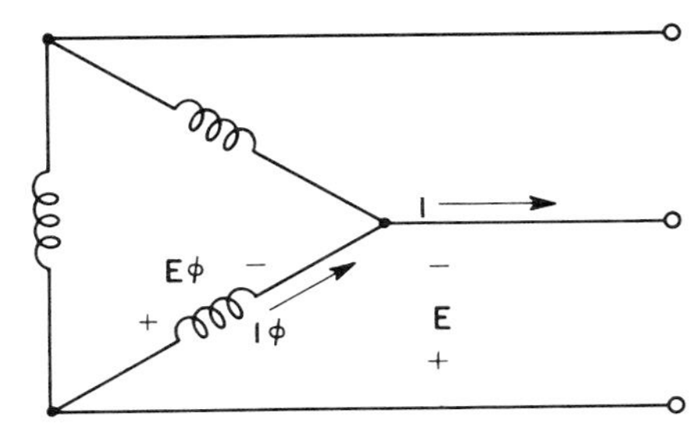

FIGURE 15-20
Delta source used to illustrate power computations in three-phase systems.

Solving for the line currents in Example 15-2 makes this an appropriate point to call attention to a statement made in the first part of this section. It was stated that no current flows around the mesh. The reader will note that the sum of currents I_C, I_B, and I_A is zero, as stated.

$$I_A + I_B + I_C = 0 \tag{15-9}$$

15-7 POWER IN BALANCED THREE-PHASE SYSTEMS

The total power in any polyphase system is the sum of the powers of the individual phases. This section will deal only with balanced loads, so that the characteristics of all phases are identical.

Consider the delta-connected source of Fig. 15-20. Power per phase is the product of phase voltage, current, and the power factor.

$$P_\phi = E_{\phi,\text{eff}} I_{\phi,\text{eff}} \cos\theta \tag{15-10}$$

The total power in the circuit is three times the power per phase.

$$P_t = 3E_{\phi,\text{eff}} I_{\phi,\text{eff}} \cos\theta \tag{15-11}$$

Equation (15-8) stated that line current is $\sqrt{3}$ times phase current. Line voltage is the same as phase voltage in a delta-connected circuit. Substituting line voltage and line current into Eq. (15-11),

$$P_t = \sqrt{3} E_{\text{eff}} I_{\text{eff}} \cos\theta \tag{15-12}$$

where E_{eff} = line voltage

I_{eff} = line current

$\cos\theta$ = power factor

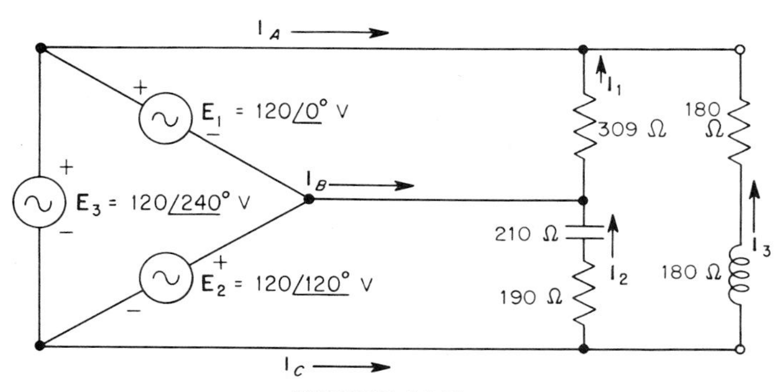

FIGURE 15-21
Unbalanced delta-delta system. (Circuit for Example 15-3.)

When the source is Y-connected, the same equation yields total power. Power per phase is found with Eq. (15-10). In the Y-connected circuit, line current is the same as a phase current, and line voltage is $\sqrt{3}$ times phase voltage. Substituting these values into Eq. (15-11), we get a total-power equation that is the same as for a delta circuit.

In either a delta- or a Y-connected balanced circuit, the total power is three times the power of an individual phase.

15-8 UNBALANCED THREE-PHASE SYSTEMS

In an unbalanced system, one or more of the load impedances differs from the other load impedances. In circuits of this type, some of the rules established in earlier sections of this chapter still apply; however, load currents and/or voltages will differ from one another in unbalanced systems. These systems may be delta-delta, Y-delta, delta-Y, or Y-Y.

Delta-Delta

In a three-phase system with a delta-connected source and delta-connected load, the voltages across all legs of the load are the same and equal the voltage per phase of the source. The load currents are inverse functions of the load impedances, and the line currents are the phasor sums of the load currents.

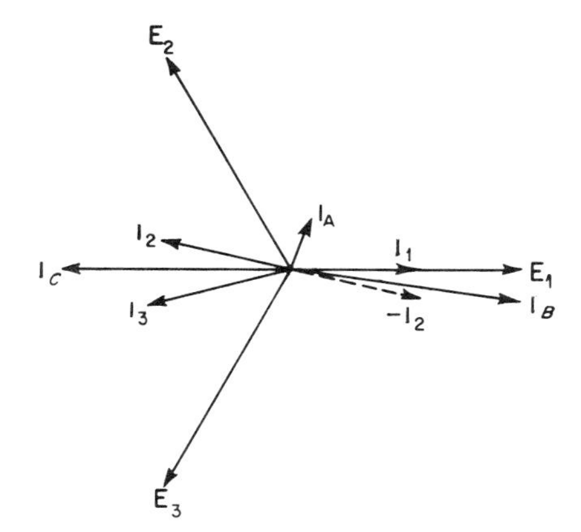

FIGURE 15-22
Phasor diagram of currents and voltages for the circuit of Example 15-3.

EXAMPLE 15-3 Given the circuit of Fig. 15-21, determine the current through the load impedances and the line currents.

SOLUTION

1
$$\mathbf{I}_1 = \frac{\mathbf{E}_1}{\mathbf{Z}_1} = \frac{120 \angle 0^\circ}{309 \angle 0^\circ} = 0.389 \angle 0^\circ \text{ A}$$

2
$$\mathbf{I}_2 = \frac{\mathbf{E}_2}{\mathbf{Z}_2} = \frac{120 \angle 120^\circ}{190 - j210} = 0.424 \angle 167.8^\circ \text{ A}$$

3
$$\mathbf{I}_3 = \frac{\mathbf{E}_3}{\mathbf{Z}_3} = \frac{120 \angle 240^\circ}{180 + j180} = 0.473 \angle 195^\circ \text{ A}$$

4
$$\begin{aligned}\mathbf{I}_A &= -\mathbf{I}_1 - \mathbf{I}_3 \\ &= (-0.389 - j0) - (-0.456 - j0.122) \\ &= +0.067 + j0.122 = 0.139 \angle 61^\circ \text{ A}\end{aligned}$$

5
$$\begin{aligned}\mathbf{I}_B &= \mathbf{I}_1 - \mathbf{I}_2 \\ &= (0.389 + j0) - (-0.415 + j0.0882) \\ &= 0.804 - j0.0882 = 0.811 \angle -6.24^\circ \text{ A}\end{aligned}$$

6
$$\begin{aligned}\mathbf{I}_C &= \mathbf{I}_2 + \mathbf{I}_3 \\ &= (-0.415 + j0.0882) + (-0.456 - j0.122) \\ &= -0.871 - j0.0338 = 0.871 \angle 180^\circ \text{ A} \qquad \text{approx.}\end{aligned}$$

Figure 15-22 is a phasor diagram illustrating all of the voltages and currents in the example. ////

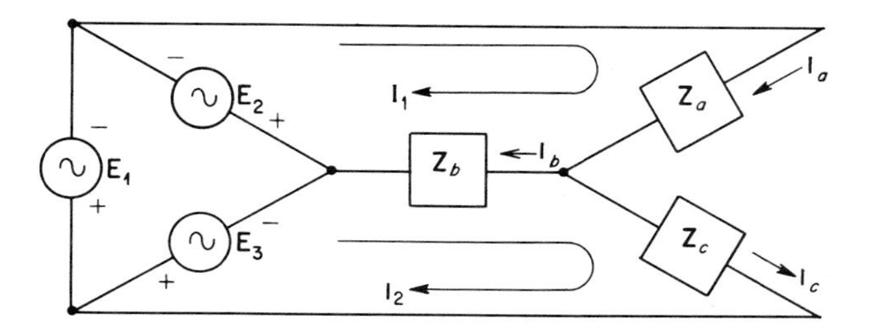

FIGURE 15-23
Delta-Y circuit for Example 15-4.

Y-Delta

In solving this type of circuit, the currents in the legs of the load and the line currents are found in the same manner as in the delta-delta circuit. The only difference is that the voltage per leg is the line voltage of a Y-connected source and is therefore 1.732 times a phase voltage.

Delta-Y

Solution of unbalanced Y-connected loads can be tedious. The usual procedure is to use Kirchhoff's laws or to change the Y-connected load to an equivalent delta load (Chap. 4) and solve for currents. Then the currents found in the equivalent delta are the currents through the actual Y-connected load. However, another approach to this problem is possible. It is necessary to write only two loop equations in which two currents traverse all parts of the load circuits. The loop equations take the standard form discussed in Chap. 11. In the circuit of Fig. 15-23, two loop currents are drawn so that current flows through each of the load. The directions assigned the currents are arbitrary.

The following equations express the results of tracing the two current loops shown in Fig. 15-23:

$$\mathbf{E}_2 = (\mathbf{Z}_a + \mathbf{Z}_b)\mathbf{I}_1 - \mathbf{Z}_b\mathbf{I}_2 \qquad (15\text{-}13)$$

$$\mathbf{E}_3 = -\mathbf{Z}_b\mathbf{I}_1 + (\mathbf{Z}_b + \mathbf{Z}_c)\mathbf{I}_2 \qquad (15\text{-}14)$$

The simultaneous equations (15-13) and (15-14) can be solved for currents $\mathbf{I}_1$ and $\mathbf{I}_2$.

$$\mathbf{I}_1 = \frac{(\mathbf{Z}_b + \mathbf{Z}_c)\mathbf{E}_2 + \mathbf{Z}_b\mathbf{E}_3}{(\mathbf{Z}_a + \mathbf{Z}_b)(\mathbf{Z}_b + \mathbf{Z}_c) - \mathbf{Z}_b^2} \qquad (15\text{-}15)$$

$$\mathbf{I}_2 = \frac{(\mathbf{Z}_a + \mathbf{Z}_b)\mathbf{E}_3 + \mathbf{Z}_b\mathbf{E}_2}{(\mathbf{Z}_a + \mathbf{Z}_b)(\mathbf{Z}_b + \mathbf{Z}_c) - \mathbf{Z}_b^2} \qquad (15\text{-}16)$$

EXAMPLE 15-4 Given the circuit of Fig. 15-23, where $\mathbf{E}_1 = 150 \angle 30° \text{ V}$, $\mathbf{E}_2 = 150 \angle 150° \text{ V}$, $\mathbf{E}_3 = 150 \angle 270° \text{ V}$, $\mathbf{Z}_a = 100 \angle 90° \ \Omega$, $\mathbf{Z}_b = 100 \angle 0° \ \Omega$, and $\mathbf{Z}_c = 90 \angle 45° \ \Omega$, find currents $\mathbf{I}_a$, $\mathbf{I}_b$, and $\mathbf{I}_c$.

SOLUTION

1 Assume loop currents as drawn in the figure.

2 Determine

$$\mathbf{Z}_a + \mathbf{Z}_b = (0 + j100) + (100 + j0) = 100 + j100 = 141 \angle 45°$$

$$\mathbf{Z}_b = 100 \angle 0° \ \Omega$$

$$\mathbf{Z}_b + \mathbf{Z}_c = (100 + j0) + (63.4 + j63.4)$$
$$= 163.4 + j63.4 = 175 \angle 21.2° \ \Omega$$

3 Use Eqs. (15-15) and (15-16).

$$\mathbf{I}_1 = \frac{(150 \angle 150°)(175 \angle 21.2°) + (150 \angle 270°)(100 \angle 0°)}{(141 \angle 45°)(175 \angle 21.2°) - (100 \angle 0°)(100 \angle 0°)}$$

$$= \frac{26{,}250 \angle 171.2° + 15{,}000 \angle 270°}{24{,}700 \angle 66.2° - 10{,}000 \angle 0°}$$

$$= \frac{(-25{,}950 - j4{,}000) + (0 - j15{,}000)}{(9{,}950 + j22{,}600) - (10{,}000 + j0)}$$

$$= \frac{-25{,}950 - j19{,}000}{-50 + j22{,}600}$$

$$= \frac{32{,}200 \angle 233.9°}{22{,}600 \angle 90°}$$

$$= 1.42 \angle 143.9° = -1.15 + j0.84 \text{ A}$$

$$\mathbf{I}_2 = \frac{(150 \angle 270°)(141 \angle 45°) + (150 \angle 150°)(100 \angle 0°)}{(141 \angle 45°)(175 \angle 21.2°) - (100 \angle 0°)(100 \angle 0°)}$$

$$= \frac{21{,}150 \angle 315° + 15{,}000 \angle 150°}{24{,}700 \angle 66.2° - 10{,}000 \angle 0°}$$

$$= \frac{(14{,}950 - j4{,}950) + (-13{,}000 + j7{,}500)}{(9{,}950 + j22{,}600) - (10{,}000 + j0)}$$

$$= \frac{1{,}950 - j7{,}450}{-50 + j22{,}600}$$

$$= \frac{7{,}700 \angle -76.6°}{22{,}600 \angle 90°}$$

$$= 0.34 \angle -165.6° = -0.33 - j0.0846 \text{ A}$$

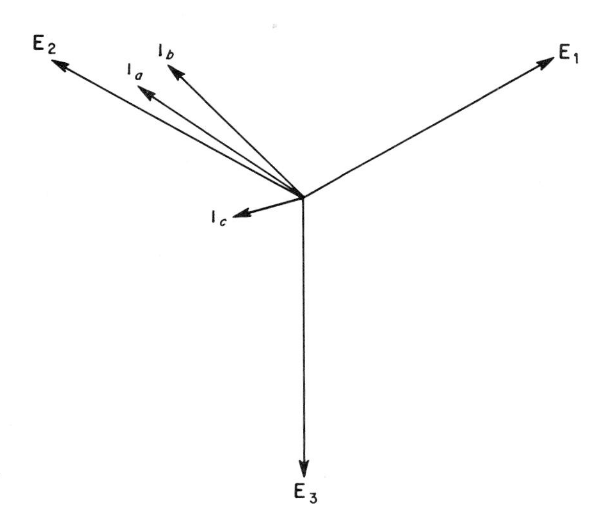

FIGURE 15-24
Phasor diagram of currents and voltages in the circuit of Example 15-4.

4 Inspection of Fig. 15-23 shows that $\mathbf{I}_1 = \mathbf{I}_a$, $\mathbf{I}_2 = \mathbf{I}_c$, and $\mathbf{I}_b = \mathbf{I}_1 - \mathbf{I}_2$.

$$\mathbf{I}_c = 0.34 \angle -165.6° \text{ A}$$

$$\mathbf{I}_a = 1.42 \angle 143.9° \text{ A}$$

$$\mathbf{I}_b = (-1.15 + j0.84) - (-0.33 - j0.0846)$$

$$= -0.82 + j0.925 = 1.23 \angle 131.5° \text{ A} \qquad ////$$

Figure 15-24 is a phasor diagram of the currents and voltages in the circuit of Example 15-4.

Y-Y

The solution of Y-Y circuits with unbalanced loads is similar to that for the delta-Y circuit. The only difference is that the line voltage of the Y-connected source is 1.732 times a phase voltage. All other computations are handled as in Example 15-4.

15-9 COPPER REQUIREMENTS FOR POLYPHASE CIRCUITS

One of the important advantages of three-phase circuits for power distribution systems is that they require less copper to deliver the same power as single-phase or two-phase systems.

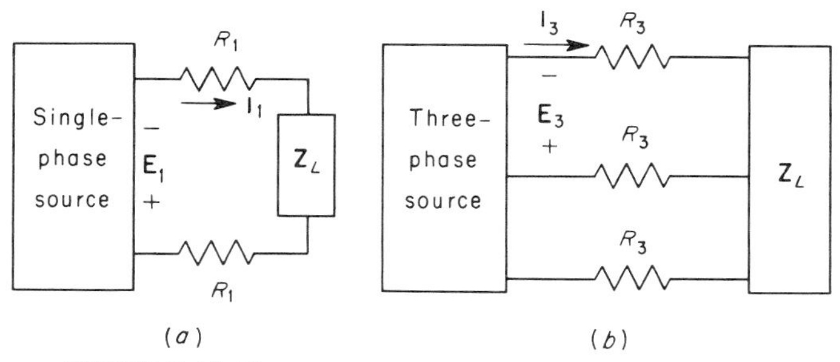

FIGURE 15-25
(*a*) Single-phase source connected to a load. (*b*) Three-phase source connected to three-phase load. The power requirements of both loads are the same.

Figure 15-25 shows two circuits connected to loads that require the same power. In Fig. 15-25*a*, a single-phase source supplies power to a load. Resistances R_1 represent the connecting lines. Part *b* of the same figure shows a three-phase source supplying the same power as the single-phase source. Resistors R_3 represent the lines. Assume that the polyphase source is delta-connected and that the voltages in both systems are equal. The line losses in both systems can now be investigated.

$$P_1 = P_3$$

$$E_{1,\text{eff}} I_{1,\text{eff}} \cos \theta = \sqrt{3}\, E_{3,\text{eff}} I_{3,\text{eff}}$$

Since $E_{1,\text{eff}} = E_{3,\text{eff}}$, and $P_1 = P_3$,

$$I_{1,\text{eff}} = \sqrt{3}\, I_{3,\text{eff}}$$

The line losses are set equal to each other to determine the amount of copper needed for each system.

Single-phase line loss: $\quad 2I_{1,\text{eff}}^2 R_1 = 2(\sqrt{3}\, I_{3,\text{eff}})^2 R_1$

Three-phase line loss: $\quad 3I_{3,\text{eff}}^2 R_3$

Setting the losses equal,

$$2(3I_{3,\text{eff}}^2)R_1 = 3I_{3,\text{eff}}^2 R_3$$

Solve for R_1.

$$R_1 = \frac{R_3}{2}$$

It should be apparent from this equation that the area of wire required in each line of the single-phase system is twice the area required in each line of the three-phase

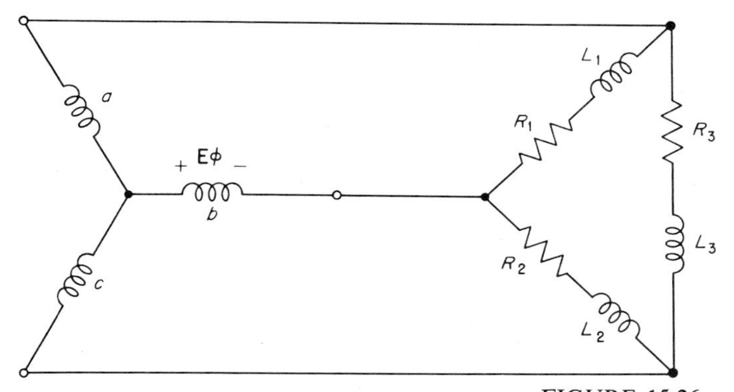

FIGURE 15-26
Circuit for Probs. 15-1 and 15-2.

system. Since there are two lines in the single-phase system and three lines in the three-phase system, it is at once apparent that the total copper requirement for a single-phase system is four-thirds the requirement for a three-phase system.

In the preceding discussion, a delta-connected source was considered. The discussion is equally valid for a Y-connected source if we assume that no neutral is required. If a neutral line is added, the copper requirements for both systems are the same.

It can be demonstrated that the three-phase or multiple three-phase system is the most economical of all polyphase systems insofar as copper requirements are concerned.

PROBLEMS

15-1 In the circuit of Fig. 15-26, find (*a*) the voltage and currents per phase in the load, and (*b*) the line currents when the phase sequence is *abc*. Given $E_{\phi,\text{eff}} = 115$ V, 60 Hz, $L_1 = L_2 = L_3 = 0.8$ H, and $R_1 = R_2 = R_3 = 292\ \Omega$.

15-2 In the circuit of Fig. 15-26, find the current per phase and the line currents when the phase sequence is *cba*. Given: $E_{\phi,\text{eff}} = 150$ V, 60 Hz, $L_1 = 0.6$ H, $L_2 = 0.9$ H, $L_3 = 1.1$ H, $R_1 = 310\ \Omega$, $R_2 = 180\ \Omega$, and $R_3 = 450\ \Omega$.

15-3 In the circuit of Fig. 15-27, the line frequency is 60 Hz. The line current is 1.08 A. Find the line voltage. Given: $R_1 = R_2 = R_3 = 190\ \Omega$, and $C_1 = C_2 = C_3 = 8\ \mu$F.

15-4 Find the total power input to the circuit of Prob. 15-3.

15-5 In the circuit of Fig. 15-28, solve for currents $\mathbf{I}_a$, $\mathbf{I}_b$, and $\mathbf{I}_c$. Given $E_{\phi,\text{eff}} = 120$ V, 60 Hz; $R_1 = 175\ \Omega$; $R_2 = 230\ \Omega$; $R_3 = 95\ \Omega$; $L_3 = 0.4$ H; $C_1 = 15\ \mu$F; and $C_2 = 8\ \mu$F. Phase sequence is *cab*.

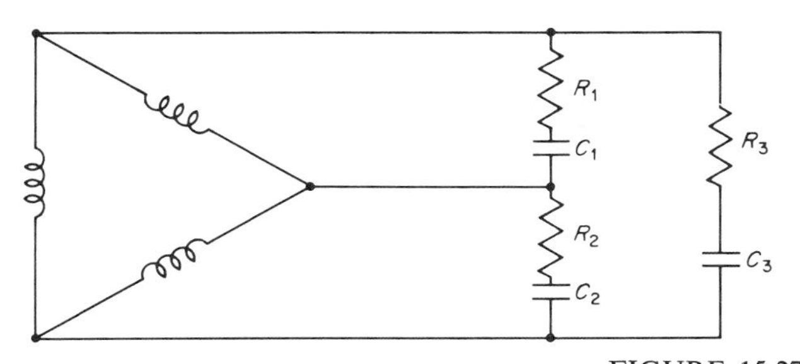

FIGURE 15-27
Circuit for Probs. 15-3 and 15-4.

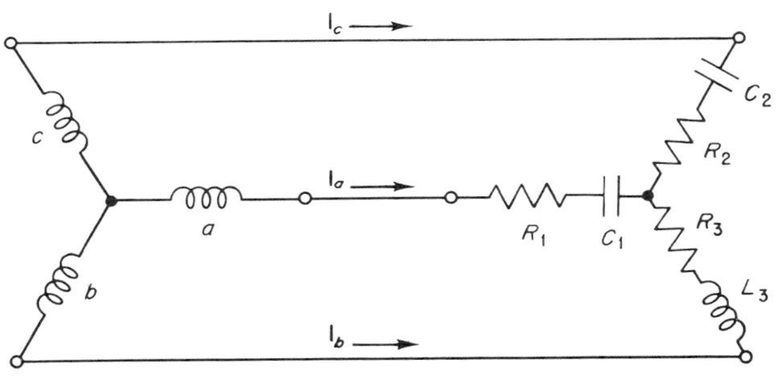

FIGURE 15-28
Circuit for Prob. 15-5.

APPENDIX

Table A-1 ATOMIC NUMBERS AND ATOMIC MASSES*

Element	Symbol	Atomic number	Average atomic mass
Actinium	Ac	89	227
Aluminum	Al	13	26.98
Americium	Am	95	243
Antimony	Sb	51	121.75
Argon	Ar	18	39.95
Arsenic	As	33	74.92
Astatine	At	85	210
Barium	Ba	56	137.34
Berkelium	Bk	97	249
Beryllium	Be	4	9.012
Bismuth	Bi	83	208.98
Boron	B	5	10.81
Bromine	Br	35	79.91
Cadmium	Cd	48	112.40
Calcium	Ca	20	40.08
Californium	Cf	98	251
Carbon	C	6	12.011
Cerium	Ce	58	140.12
Cesium	Cs	55	132.91
Chlorine	Cl	17	35.45
Chromium	Cr	24	52.00
Cobalt	Co	27	58.93
Copper	Cu	29	63.54
Curium	Cm	96	247
Dysprosium	Dy	66	162.50
Einsteinium	Es	99	254
Erbium	Er	68	167.26
Europium	Eu	63	151.96
Fermium	Fm	100	253
Fluorine	F	9	19.00

* Based on $^{12}C = 12.000$.

Table A-1 *Continued*

Element	Symbol	Atomic number	Average atomic mass
Francium	Fr	87	223
Gadolinium	Gd	64	157.25
Gallium	Ga	31	69.72
Germanium	Ge	32	72.59
Gold	Au	79	196.97
Hafnium	Hf	72	178.49
Helium	He	2	4.003
Holmium	Ho	67	164.93
Hydrogen	H	1	1.0080
Indium	In	49	114.82
Iodine	I	53	126.90
Iridium	Ir	77	192.2
Iron	Fe	26	55.85
Krypton	Kr	36	83.80
Lanthanum	La	57	138.91
Lawrencium	Lw	103	257
Lead	Pb	82	207.19
Lithium	Li	33	6.939
Lutetium	Lu	71	174.04
Magnesium	Mg	12	24.31
Manganese	Mn	25	54.94
Mendelevium	Md	101	256
Mercury	Hg	80	200.59
Molybdenum	Mo	42	95.94
Neodymium	Nd	60	144.24
Neon	Ne	10	20.18
Neptunium	Np	93	237
Nickel	Ni	28	58.71
Niobium	Nb	41	92.91
Nitrogen	N	7	14.007
Nobelium	No	102	253
Osmium	Os	76	190.2
Oxygen	O	8	15.999
Palladium	Pd	46	106.4
Phosphorus	P	15	30.97

Table A-1 *Continued*

Element	Symbol	Atomic number	Average atomic mass
Platinum	Pt	78	195.09
Plutonium	Pu	94	242
Polonium	Po	84	210
Potassium	K	19	39.10
Praseodymium	Pr	59	140.91
Promethium	Pm	61	147
Protactinium	Pa	91	231
Radium	Ra	88	226
Radon	Rn	86	222
Rhenium	Re	75	186.23
Rhodium	Rh	45	102.91
Rubidium	Rb	37	85.47
Ruthenium	Ru	44	101.1
Samarium	Sm	62	150.35
Scandium	Sc	21	44.96
Selenium	Se	34	78.96
Silicon	Si	14	28.09
Silver	Ag	47	107.870
Sodium	Na	11	22.990
Strontium	Sr	38	87.62
Sulfur	S	16	32.06
Tantalum	Ta	73	180.95
Technetium	Tc	43	99
Tellurium	Te	52	127.60
Terbium	Tb	65	158.92
Thallium	Tl	81	204.37
Thorium	Th	90	232.04
Thulium	Tm	69	168.93
Tin	Sn	50	118.69
Titanium	Ti	22	47.90
Tungsten	W	74	183.85
Uranium	U	92	238.03
Vanadium	V	23	50.94
Xenon	Xe	54	131.30
Ytterbium	Yb	70	173.04
Yttrium	Y	39	88.91
Zinc	Zn	30	65.37
Zirconium	Zr	40	91.22

Table A-2 PERIODIC TABLE OF THE ELEMENTS

H 1																	**He** 2
Li 3	**Be** 4											**B** 5	**C** 6	**N** 7	**O** 8	**F** 9	**Ne** 10
Na 11	**Mg** 12											**Al** 13	**Si** 14	**P** 15	**S** 16	**Cl** 17	**Ar** 18
K 19	**Ca** 20	**Sc** 21	**Ti** 22	**V** 23	**Cr** 24	**Mn** 25	**Fe** 26	**Co** 27	**Ni** 28	**Cu** 29	**Zn** 30	**Ga** 31	**Ge** 32	**As** 33	**Se** 34	**Br** 35	**Kr** 36
Rb 37	**Sr** 38	**Y** 39	**Zr** 40	**Nb** 41	**Mo** 42	**Tc** 43	**Ru** 44	**Rh** 45	**Pd** 46	**Ag** 47	**Cd** 48	**In** 49	**Sn** 50	**Sb** 51	**Te** 52	**I** 53	**Xe** 54
Cs 55	**Ba** 56	°	**Hf** 72	**Ta** 73	**W** 74	**Re** 75	**Os** 76	**Ir** 77	**Pt** 78	**Au** 79	**Hg** 80	**Tl** 81	**Pb** 82	**Bi** 83	**Po** 84	**At** 85	**Rn** 86
Fr 87	**Ra** 88	†	**Ku** 104	**Ha** 105													

° **Lanthanides**	**La** 57	**Ce** 58	**Pr** 59	**Nd** 60	**Pm** 61	**Sm** 62	**Eu** 63	**Gd** 64	**Tb** 65	**Dy** 66	**Ho** 67	**Er** 68	**Tm** 69	**Yb** 70	**Lu** 71
† **Actinides**	**Ac** 89	**Th** 90	**Pa** 91	**U** 92	**Np** 93	**Pu** 94	**Am** 95	**Cm** 96	**Bk** 97	**Cf** 98	**Es** 99	**Fm** 100	**Md** 101	**No** 102	**Lr** 103

From *Chemistry*, 4/e, by Sienko and Plane. Copyright © 1971 by McGraw-Hill, Inc. Used by permission of McGraw-Hill Book Company.

Table A-3 THE GREEK ALPHABET AND COMMON MEANINGS

Name	Capital	Lower case	Commonly used to designate
Alpha	A	α	Angles, area, coefficients
Beta	B	β	Angles, flux density, coefficients
Gamma	Γ	γ	Conductivity, specific gravity
Delta	Δ	δ	Variation, density
Epsilon	E	ϵ	Base of natural logarithms
Zeta	Z	ζ	Impedance, coefficients, coordinates
Eta	H	η	Hysteresis, coefficients, efficiency
Theta	Θ	θ	Temperature, phase angle
Iota	I	ι	
Kappa	K	κ	Dielectric constant, susceptibility
Lambda	Λ	λ	Wavelength
Mu	M	μ	Micro, amplification factor, permeability
Nu	N	ν	Reluctivity
Xi	Ξ	ξ	
Omicron	O	o	
Pi	Π	π	Ratio of circumference to diameter = 3.1416
Rho	P	ρ	Resistivity
Sigma	Σ	σ	Sign of summation
Tau	T	τ	Time constant, time phase displacement
Upsilon	ϒ	υ	
Phi	Φ	ϕ	Magnetic flux, angles
Chi	X	χ	
Psi	Ψ	ψ	Dielectric flux, phase difference
Omega	Ω	ω	Capital, ohms; lowercase, angular velocity

Table A-4 DIELECTRIC MATERIALS

Dielectric	Relative permittivity
Air	1.0006
Amber	2.9
Barium-strontium titanate (ceramic)	7,500
Bakelite	7.0
Ebonite	2.86
Fiber	4.52
Glass, crown	6.0
flint	7.0
Pyrex	4.5
Mica	5.0
Oil (transformer)	4.0
Paper	2.5
Paraffin	4.0
Porcelain	6.0
Quartz	4.2
Rubber	3.0
Slate	7.0
Styrene	2.6
Teflon	2.0
Vinyl resins	4.0
Water (pure)	81

Table A-5 RESISTIVITY

Material	Resistivity, Ω·cmil/ft at 20°C
Silver	9.9
Copper, annealed	10.37
Aluminum	17.0
Tungsten	33.0
Nickel	47.0
Platinum	60.0
Iron, commercial	75.0
Lead	132.0
Mercury	577.0
Nichrome	600.0
Germanium	2.73×10^8
Silicon	1.35×10^{11}

Table A-6 ELECTROMOTIVE SERIES

Element	Potential, V
Lithium	−3.03
Potassium	−2.92
Rubidium	−2.92
Barium	−2.90
Strontium	−2.89
Calcium	−2.87
Sodium	−2.71
Magnesium	−2.34
Aluminum	−2.30
Titanium	−1.63
Zinc	−0.76
Iron	−0.44
Cadmium	−0.40
Nickel	−0.25
Tin	−0.14
Lead	−0.13
Hydrogen	0
Copper	+0.34
Silver	+0.80
Mercury	+0.80
Palladium	+0.99
Gold	+1.68

Table A-7 AMERICAN WIRE GAUGE (B & S)

Gauge No.	Diameter, mils at (20°C)	Gauge No.	Diameter, mils at 20°C
0000	460.0	19	35.89
000	409.6	20	31.96
00	364.8	21	28.46
0	324.9	22	25.35
1	289.3	23	22.57
2	257.6	24	20.10
3	229.4	25	17.90
4	204.3	26	15.94
5	181.9	27	14.20
6	162.0	28	12.64
7	144.3	29	11.26
8	128.5	30	10.03
9	114.4	31	8.928
10	101.9	32	7.950
11	90.74	33	7.080
12	80.81	34	6.305
13	71.96	35	5.615
14	64.08	36	5.000
15	57.07	37	4.453
16	50.82	38	3.965
17	45.26	39	3.531
18	40.30	40	3.145

Table A-8 CONVERSION FACTORS

Length	cm	m	km	ft
1 centimeter	1	10^{-2}	10^{-5}	3.281×10^{-2}
1 meter	100	1	10^{-3}	3.281
1 kilometer	10^{5}	1,000	1	3,281.0
1 inch	2.540	2.54×10^{-2}	2.54×10^{-5}	8.333×10^{-2}
1 foot	30.48	0.3048	3.048×10^{-4}	1
1 statute mile	1.609×10^{5}	1,609	1.609	5,280

Mass	g	lb	slug	ton
1 gram	1	2.205×10^{-3}	6.582×10^{-5}	1.102×10^{-6}
1 kilogram	1,000	2.205	6.852×10^{-2}	1.102×10^{-3}
1 ounce (avdp)	28.35	6.250×10^{-2}	1.943×10^{-3}	3.125×10^{-5}
1 pound (avdp)	453.6	1	3.108×10^{-2}	5×10^{-4}
1 slug	1.459×10^{4}	32.17	1	1.609×10^{-2}
1 ton	9.072×10^{5}	2,000.0	62.16	1

Time	day	h	min	s
1 year	365.2	8.766×10^{3}	5.259×10^{5}	3.156×10^{7}
1 day	1	24	1,440	8.64×10^{4}
1 hour	4.167×10^{-2}	1	60	3,600
1 minute	6.944×10^{-4}	1.667×10^{-2}	1	60
1 second	1.157×10^{-5}	2.778×10^{-4}	1.667×10^{-2}	1

Electric charge	ah	C	f
1 ampere-hour	1	3,600	3.730×10^{-2}
1 coulomb	2.778×10^{-4}	1	1.036×10^{-5}
1 farad	26.81	9.652×10^{4}	1

Energy	ft lb	J	Wh
1 foot-pound	1	1.356	3.766×10^{-7}
1 joule	0.7376	1	2.778×10^{-7}
1 kilowatt-hour	2.655×10^{6}	3.6×10^{6}	1

Power	ft lb/s	W
1 foot-pound/second	1	1.356
1 watt	0.7376	1

Table A-9 EXPONENTIALS* ϵ^x and ϵ^{-x}

x	ϵ^x	Diff.	x	ϵ^x	Diff.	x	ϵ^x	x	ϵ^{-x}	Diff.	x	ϵ^{-x}	x	ϵ^{-x}
0.00	1.000	10	0.50	1.649	16	1.0	2.718†	0.00	1.000	−10	0.50	0.607	1.0	0.368
0.01	1.010	10	0.51	1.665	17	1.1	3.004	0.01	0.990	−10	0.51	0.600	1.1	0.333
0.02	1.020	10	0.51	1.682	17	1.2	3.320	0.02	0.980	−10	0.52	0.595	1.2	0.301
0.03	1.030	11	0.53	1.699	17	1.3	3.669	0.03	0.970	−9	0.53	0.589	1.3	0.273
0.04	1.041	10	0.54	1.716	17	1.4	4.055	0.04	0.961	−10	0.54	0.583	1.4	0.247
0.05	1.051	11	0.55	1.733	18	1.5	4.482	0.05	0.951	−9	0.55	0.577	1.5	0.223
0.06	1.062	11	0.56	1.751	17	1.6	4.953	0.06	0.942	−10	0.56	0.571	1.6	0.202
0.07	1.073	10	0.57	1.768	18	1.7	5.474	0.07	0.932	−9	0.57	0.566	1.7	0.183
0.08	1.083	11	0.58	1.786	18	1.8	6.050	0.08	0.923	−9	0.58	0.560	1.8	0.165
0.09	1.094	11	0.59	1.804	18	1.9	6.686	0.09	0.914	−9	0.59	0.554	1.9	0.150
0.10	1.105	11	0.60	1.822	18	2.0	7.389	0.10	0.905	−9	0.60	0.549	2.0	0.135
0.11	1.116	11	0.61	1.840	19	2.1	8.166	0.11	0.896	−9	0.61	0.543	2.1	0.122
0.12	1.127	12	0.62	1.859	19	2.2	9.025	0.12	0.887	−9	0.62	0.538	2.2	0.111
0.13	1.139	11	0.63	1.878	18	2.3	9.974	0.13	0.878	−9	0.63	0.533	2.3	0.100
0.14	1.150	12	0.64	1.896	20	2.4	11.02	0.14	0.869	−8	0.64	0.527	2.4	0.0907
0.15	1.162	12	0.65	1.916	19	2.5	12.18	0.15	0.861	−9	0.65	0.522	2.5	0.0821
0.16	1.174	11	0.66	1.935	19	2.6	13.46	0.16	0.852	−8	0.66	0.517	2.6	0.0743
0.17	1.185	12	0.67	1.954	20	2.7	14.88	0.17	0.844	−9	0.67	0.512	2.7	0.0672
0.18	1.197	12	0.68	1.974	20	2.8	16.44	0.18	0.835	−8	0.68	0.507	2.8	0.0608
0.19	1.209	12	0.69	1.994	20	2.9	18.17	0.19	0.827	−8	0.69	0.502	2.9	0.0550
0.20	1.221	13	0.70	2.014	20	3.0	20.09	0.20	0.819	−8	0.70	0.497	3.0	0.0498
0.21	1.234	12	0.71	2.034	20	3.1	22.20	0.21	0.811	−8	0.71	0.492	3.1	0.0450
0.22	1.246	13	0.72	2.054	21	3.2	24.53	0.22	0.803	−8	0.72	0.487	3.2	0.0408
0.23	1.259	12	0.73	2.075	21	3.3	27.11	0.23	0.795	−8	0.73	0.482	3.3	0.0369
0.24	1.271	13	0.74	2.096	21	3.4	29.96	0.24	0.787	−8	0.74	0.477	3.4	0.0334
0.25	1.284	13	0.75	2.117	21	3.5	33.12	0.25	0.779	−8	0.75	0.472	3.5	0.0302
0.26	1.297	13	0.76	2.138	22	3.6	36.60	0.26	0.771	−8	0.76	0.468	3.6	0.0273
0.27	1.310	13	0.77	2.160	21	3.7	40.45	0.27	0.763	−7	0.77	0.463	3.7	0.0247
0.28	1.323	13	0.78	2.181	22	3.8	44.70	0.28	0.756	−8	0.78	0.458	3.8	0.0224
0.29	1.336	14	0.79	2.203	23	3.9	49.40	0.29	0.748	−7	0.79	0.454	3.9	0.0202
0.30	1.350	13	0.80	2.226	22	4.0	54.60	0.30	0.741	−8	0.80	0.449	4.0	0.0183
0.31	1.363	14	0.81	2.248	22	4.1	60.34	0.31	0.733	−7	0.81	0.445	4.1	0.0166
0.32	1.377	14	0.82	2.270	23	4.2	66.69	0.32	0.726	−7	0.82	0.440	4.2	0.0150
0.33	1.391	14	0.83	2.293	23	4.3	73.70	0.33	0.719	−7	0.83	0.436	4.3	0.0136
0.34	1.405	14	0.84	2.316	24	4.4	81.45	0.34	0.712	−7	0.84	0.432	4.4	0.0123
0.35	1.419	14	0.85	2.340	23	4.5	90.02	0.35	0.705	−7	0.85	0.427	4.5	0.0111
0.36	1.433	15	0.86	2.363	24			0.36	0.698	−7	0.86	0.423		
0.37	1.448	14	0.87	2.387	24	5.0	148.4	0.37	0.691	−7	0.87	0.419	5.0	0.00674
0.38	1.462	15	0.88	2.411	24	6.0	403.4	0.38	0.684	−7	0.88	0.415	6.0	0.00248
0.39	1.477	15	0.89	2.435	25	7.0	1097.	0.39	0.677	−7	0.89	0.411	7.0	0.000912
0.40	1.492	15	0.90	2.460	24	8.0	2981.	0.40	0.670	−6	0.90	0.407	8.0	0.000335
0.41	1.507	15	0.91	2.484	25	9.0	8103.	0.41	0.664	−7	0.91	0.403	9.0	0.000123
0.42	1.522	15	0.92	2.509	26	10.0	22026.	0.42	0.657	−6	0.92	0.399	10.0	0.000045
0.43	1.537	16	0.93	2.535	25	$\pi/2$	4.810	0.43	0.651	−7	0.93	0.395		
0.44	1.553	15	0.94	2.560	26	$2\pi/2$	23.14	0.44	0.644	−6	0.94	0.391		
0.45	1.568	16	0.95	2.586	26	$3\pi/2$	111.3	0.45	0.638	−7	0.95	0.387	$\pi/2$	0.208
0.46	1.584	16	0.96	2.612	26	$4\pi/2$	535.5	0.46	0.631	−6	0.96	0.383	$2\pi/2$	0.0432
0.47	1.600	16	0.97	2.638	26	$5\pi/2$	2576	0.47	0.625	−6	0.97	0.379	$3\pi/2$	0.00898
0.48	1.616	16	0.98	2.664	27	$6\pi/2$	12392	0.48	0.619	−6	0.98	0.375	$4\pi/2$	0.00187
0.49	1.632	17	0.99	2.691	27	$7\pi/2$	59610	0.49	0.613	−6	0.99	0.372	$5\pi/2$	0.000388
0.50	1.649		1.00	2.718		$8\pi/2$	286751	0.50	0.607		1.00	0.386	$6\pi/2$	0.000081

$\varepsilon = 2.71828$. $1/\epsilon = 0.367879$. $\log \epsilon = 0.4343$. $1/0.4343 = 2.3026$.

* From Lionel S. Marks, "Mechanical Engineers' Handbook."

† NOTE: Do not interpolate in this column.

Table A-10 COMMON LOGARITHMS

No.	0	1	2	3	4	5	6	7	8	9
10	0000	0043	0086	0128	0170	0212	0253	0294	0334	0374
11	0414	0453	0492	0531	0569	0607	0645	0682	0719	0755
12	0792	0828	0864	0899	0934	0969	1004	1038	1072	1106
13	1139	1173	1206	1239	1271	1303	1335	1367	1399	1430
14	1461	1492	1523	1553	1584	1614	1644	1673	1703	1732
15	1761	1790	1818	1847	1875	1903	1931	1959	1987	2014
16	2041	2068	2095	2122	2148	2175	2201	2227	2253	2279
17	2304	2330	2355	2380	2405	2430	2455	2480	2504	2529
18	2553	2577	2601	2625	2648	2672	2695	2718	2742	2765
19	2788	2810	2833	2856	2878	2900	2923	2945	2967	2989
20	3010	3032	3054	3075	3096	3118	3139	3160	3181	3201
21	3222	3243	3263	3284	3304	3324	3345	3365	3385	3404
22	3424	3444	3464	3483	3502	3522	3541	3560	3579	3598
23	3617	3636	3655	3674	3692	3711	3729	3747	3766	3784
24	3802	3820	3838	3856	3874	3892	3909	3927	3945	3962
25	3979	3997	4014	4031	4048	4065	4082	4099	4116	4133
26	4150	4166	4183	4200	4216	4232	4249	4265	4281	4298
27	4314	4330	4346	4362	4378	4393	4409	4425	4440	4456
28	4472	4487	4502	4518	4533	4548	4564	4579	4594	4609
29	4624	4639	4654	4669	4683	4698	4713	4728	4742	4757
30	4771	4786	4800	4814	4829	4843	4857	4871	4886	4900
31	4914	4928	4942	4955	4969	4983	4997	5011	5024	5038
32	5051	5065	5079	5092	5105	5119	5132	5145	5159	5172
33	5185	5198	5211	5224	5237	5250	5263	5276	5289	5302
34	5315	5328	5340	5353	5366	5378	5391	5403	5416	5428
35	5441	5453	5465	5478	5490	5502	5514	5527	5539	5551
36	5563	5575	5587	5599	5611	5623	5635	5647	5658	5670
37	5682	5694	5705	5717	5729	5740	5752	5763	5775	5786
38	5798	5809	5821	5832	5843	5855	5866	5877	5888	5899
39	5911	5922	5933	5944	5955	5966	5977	5988	5999	6010
40	6021	6031	6042	6053	6064	6075	6085	6096	6107	6117
41	6128	6138	6149	6160	6170	6180	6191	6201	6212	6222
42	6232	6243	6253	6263	6274	6284	6294	6304	6314	6325
43	6335	6345	6355	6365	6375	6385	6395	6405	6415	6425
44	6435	6444	6454	6464	6474	6484	6493	6503	6513	6522
45	6532	6542	6551	6561	6571	6580	6590	6599	6609	6618
46	6628	6637	6646	6656	6665	6675	6684	6693	6702	6712
47	6721	6730	6739	6749	6758	6767	6776	6785	6794	6803
48	6812	6821	6830	6839	6848	6857	6866	6875	6884	6893
49	6902	6911	6920	6928	6937	6946	6955	6964	6972	6981
50	6990	6998	7007	7016	7024	7033	7042	7050	7059	7067
51	7076	7084	7093	7101	7110	7118	7126	7135	7143	7152
52	7160	7168	7177	7185	7193	7202	7210	7218	7226	7235
53	7243	7251	7259	7267	7275	7284	7292	7300	7308	7316
54	7324	7332	7340	7348	7356	7364	7372	7380	7388	7396
No.	0	1	2	3	4	5	6	7	8	9

Table A-10 (*Continued*)

No	0	1	2	3	4	5	6	7	8	9
55	7404	7412	7419	7427	7435	7443	7451	7459	7466	7474
56	7482	7490	7497	7505	7513	7520	7528	7536	7543	7551
57	7559	7566	7574	7582	7589	7597	7604	7612	7619	7627
58	7634	7642	7649	7657	7664	7672	7679	7686	7694	7701
59	7709	7716	7723	7731	7738	7745	7752	7760	7767	7774
60	7782	7789	7796	7803	7810	7818	7825	7832	7839	7846
61	7853	7860	7868	7875	7882	7889	7896	7903	7910	7917
62	7924	7931	7938	7945	7952	7959	7966	7973	7980	7987
63	7993	8000	8007	8014	8021	8028	8035	8041	8048	8055
64	8062	8069	8075	8082	8089	8096	8102	8109	8116	8122
65	8129	8136	8142	8149	8156	8162	8169	8176	8182	8189
66	8195	8202	8209	8215	8222	8228	8235	8241	8248	8254
67	8261	8267	8274	8280	8287	8293	8299	8306	8312	8319
68	8325	8331	8338	8344	8351	8357	8363	8370	8376	8382
69	8388	8395	8401	8407	8414	8420	8426	8432	8439	8445
70	8451	8457	8463	8470	8476	8482	8488	8494	8500	8506
71	8513	8519	8525	8531	8537	8543	8549	8555	8561	8567
72	8573	8579	8585	8591	8597	8603	8609	8615	8621	8627
73	8633	8639	8645	8651	8657	8663	8669	8675	8681	8686
74	8692	8698	8704	8710	8716	8722	8727	8733	8739	8745
75	8751	8756	8762	8768	8774	8779	8785	8791	8797	8802
76	8808	8814	8820	8825	8831	8837	8842	8848	8854	8859
77	8865	8871	8876	8882	8887	8893	8899	8904	8910	8915
78	8921	8927	8932	8938	8943	8949	8954	8960	8965	8971
79	8976	8982	8987	8993	8998	9004	9009	9015	9020	9025
80	9031	9036	9042	9047	9053	9058	9063	9069	9074	9079
81	9085	9090	9096	9101	9106	9112	9117	9122	9128	9133
82	9138	9143	9149	9154	9159	9165	9170	9175	9180	9186
83	9191	9196	9201	9206	9212	9217	9222	9227	9232	9238
84	9243	9248	9253	9258	9263	9269	9274	9279	9284	9289
85	9294	9299	9304	9309	9315	9320	9325	9330	9335	9340
86	9345	9350	9355	9360	9365	9370	9375	9380	9383	9390
87	9395	9400	9405	9410	9415	9420	9425	9430	9435	9440
88	9445	9450	9455	9460	9465	9469	9474	9479	9484	9489
89	9494	9499	9504	9509	9513	9518	9523	9528	9533	9538
90	9542	9547	9552	9557	9562	9566	9571	9576	9581	9586
91	9590	9595	9600	9605	9609	9614	9619	9624	9628	9633
92	9638	9643	9647	9652	9657	9661	9666	9671	9675	9680
93	9685	9689	9694	9699	9703	9708	9713	9717	9722	9727
94	9731	9736	9741	9745	9750	9754	9759	9763	9768	9773
95	9777	9782	9786	9791	9795	9800	9805	9809	9814	9818
96	9823	9827	9832	9836	9841	9845	9850	9854	9859	9863
97	9868	9872	9877	9881	9886	9890	9894	9899	9903	9908
98	9912	9917	9921	9926	9930	9934	9939	9943	9948	9952
99	9956	9961	9965	9969	9974	9978	9983	9987	9991	9996
No	0	1	2	3	4	5	6	7	8	9

Table A-11 NATURAL TRIGONOMETRIC FUNCTIONS

Angle°	sin	tan	cot	cos	Angle°	Angle°	sin	tan	cot	cos	Angle°
0.0	.00000	.00000	∞	1.00000	**90.0**	**4.5**	.07846	.07870	12.706	.99692	**85.5**
.1	.00175	.00175	572.96	1.00000	.9	.6	.08020	.08046	12.429	.99678	.4
.2	.00349	.00349	286.48	0.99999	.8	.7	.08194	.08221	12.163	.99664	.3
.3	.00524	.00524	190.98	.99999	.7	.8	.08368	.08397	11.909	.99649	.2
.4	.00698	.00698	143.24	.99998	.6	.9	.08542	.08573	11.664	.99635	.1
.5	.00873	.00873	114.59	.99996	.5	**5.0**	.08716	.08749	11.430	.99619	**85.0**
.6	.01047	.01047	95.489	.99995	.4	.1	.08889	.08925	11.205	.99604	.9
.7	.01222	.01222	81.847	.99993	.3	.2	.09063	.09101	10.988	.99588	.8
.8	.01396	.01396	71.615	.99990	.2	.3	.09237	.09277	10.780	.99572	.7
.9	.01571	.01571	63.657	.99988	.1	.4	.09411	.09453	10.579	.99556	.6
1.0	.01745	.01746	57.290	.99985	**89.0**	.5	.09585	.09629	10.385	.99540	.5
.1	.01920	.01920	52.081	.99982	.9	.6	.09758	.09805	10.199	.99523	.4
.2	.02094	.02095	47.740	.99978	.8	.7	.09932	.09981	10.019	.99506	.3
.3	.02269	.02269	44.066	.99974	.7	.8	.10106	.10158	9.8448	.99488	.2
.4	.02443	.02444	40.917	.99790	.6	.9	.10279	.10334	9.6768	.99470	.1
.5	.02618	.02619	38.188	.99966	.5	**6.0**	.10453	.10510	9.5144	.99452	**84.0**
.6	.02792	.02793	35.801	.99961	.4	.1	.10626	.10687	9.3572	.99434	.9
.7	.02967	.02968	33.694	.99956	.3	.2	.10800	.10863	9.2052	.99415	.8
.8	.03141	.03143	31.821	.99951	.2	.3	.10973	.11040	9.0579	.99396	.7
.9	.03316	.03317	30.145	.99945	.1	.4	.11147	.11217	8.9152	.99377	.6
2.0	.03490	.03492	28.636	.99939	**88.0**	.5	.11320	.11394	8.7769	.99357	.5
.1	.03664	.03667	27.271	.99933	.9	.6	.11494	.11570	8.6427	.99337	.4
.2	.03839	.03842	26.031	.99926	.8	.7	.11667	.11747	8.5126	.99317	.3
.3	.04013	.04016	24.898	.99919	.7	.8	.11840	.11924	8.3863	.99297	.2
.4	.04188	.04191	23.859	.99912	.6	.9	.12014	.12101	8.2636	.99276	.1
.5	.04362	.04366	22.904	.99905	.5	**7.0**	.12187	.12278	8.1443	.99255	**83.0**
.6	.04536	.04541	22.022	.99897	.5	.1	.12360	.12456	8.0285	.99233	.9
.7	.04711	.04716	21.205	.99889	.3	.2	.12533	.12633	7.9158	.99211	.8
.8	.04885	.04891	20.446	.99881	.2	.3	.12706	.12810	7.8062	.99189	.7
.9	.05059	.05066	19.740	.99872	.1	.4	.12880	.12988	7.6996	.99167	.6
3.0	.05234	.05241	19.081	.99863	**87.0**	.5	.13053	.13165	7.5958	.99144	.5
.1	.05408	.05416	18.464	.99854	.9	.6	.13226	.13343	7.4947	.99122	.4
.2	.05582	.05591	17.886	.99844	.8	.7	.13399	.13521	7.3962	.99098	.3
.3	.05756	.05766	17.343	.99834	.7	.8	.13572	.13698	7.3002	.99075	.2
.4	.05931	.05941	16.832	.99824	.6	.9	.13744	.13876	7.2066	.99051	.1
.5	.06105	.06116	16.350	.99813	.5	**8.0**	.13917	.14054	7.1154	.99027	**82.0**
.6	.06279	.06291	15.895	.99803	.4	.1	.14090	.14232	7.0264	.99002	.9
.7	.06453	.06467	15.464	.99792	.3	.2	.14263	.14410	6.9395	.98978	.8
.8	.06627	.06642	15.056	.99780	.2	.3	.14436	.14588	6.8548	.98953	.7
.9	.06802	.06817	14.669	.99767	.1	.4	.14608	.14767	6.7720	.98927	.6
4.0	.06976	.06993	14.301	.99756	**86.0**	.5	.14781	.14945	6.6912	.98902	.5
.1	.07150	.07168	13.951	.99744	.9	.6	.14954	.15124	6.6122	.98876	.4
.2	.07324	.07344	13.617	.99731	.8	.7	.15126	.15302	6.5350	.98849	.3
.3	.07498	.07519	13.300	.99719	.7	.8	.15299	.15481	6.4596	.98823	.2
.4	.07672	.07695	12.996	.99705	.6	.9	.15471	.15660	6.3859	.98796	.1
4.5	.07846	.07870	12.706	.99692	**85.5**	**9.0**	.15643	.15838	6.3138	.98769	**81.0**
Angle°	cos	cot	tan	sin	Angle°	Angle°	cos	cot	tan	sin	Angle°

Table A-11 *(Continued)*

Angle°	sin	tan	cot	cos	Angle°	Angle°	sin	tan	cot	cos	Angle°
9.0	.15643	.15838	6.3138	.98769	**81.0**	**13.5**	.23345	.24008	4.1653	.97237	.5
.1	.15816	.16017	6.2432	.98741	.9	.6	.23514	.24193	4.1335	.97196	.4
.2	.15988	.16196	6.1742	.98714	.8	.7	.23684	.24377	4.1022	.97155	.3
.3	.16160	.16376	6.1066	.98686	.7	.8	.23853	.24562	4.0713	.97113	.2
.4	.16333	.16555	6.0405	.98657	.6	.9	.24023	.24747	4.0408	.97072	.1
.5	.16505	.16734	5.9758	.98629	.5	**14.0**	.24192	.24933	4.0108	.97030	**76.0**
.6	.16677	.16914	5.9124	.98600	.4	.1	.24362	.25118	3.9812	.96987	.9
.7	.16849	.17093	5.8502	.98570	.3	.2	.24531	.25304	3.9520	.96945	.8
.8	.17021	.17273	5.7894	.98541	.2	.3	.24700	.25490	3.9232	.96902	.7
.9	.17193	.17453	5.7297	.98511	.1	.4	.24869	.25676	3.8947	.96858	.6
10.0	.17365	.17633	5.6713	.98481	**80.0**	.5	.25038	.25862	3.8667	.96815	.5
.1	.17537	.17813	5.6140	.98450	.9	.6	.25207	.26048	3.8391	.96771	.4
.2	.17708	.17993	5.5578	.98420	.8	.7	.25376	.26235	3.8118	.96727	.3
.3	.17880	.18173	5.5026	.98389	.7	.8	.25545	.26421	3.7848	.96682	.2
.4	.18052	.18353	5.4486	.98357	.6	.9	.25713	.26608	3.7583	.96638	.1
.5	.18224	.18534	5.3955	.98325	.5	**15.0**	.25882	.26795	3.7321	.96593	**75.0**
.6	.18395	.18714	5.3435	.98294	.4	.1	.26050	.26982	3.7062	.96547	.9
.7	.18567	.18895	5.2924	.98261	.3	.2	.26219	.27169	3.6806	.96502	.8
.8	.18738	.19076	5.2422	.98229	.2	.3	.26387	.27357	3.6554	.96456	.7
.9	.18910	.19257	5.1929	.98196	.1	.4	.26556	.27545	3.6305	.96410	.6
11.0	.19081	.19438	5.1446	.98163	**79.0**	.5	.26724	.27732	3.6059	.96363	.5
.1	.19252	.19619	5.0970	.98129	.9	.6	.26892	.27921	3.5816	.96316	.4
.2	.19423	.19801	5.0504	.98096	.8	.7	.27060	.28109	3.5576	.96269	.3
.3	.19595	.19982	5.0045	.98061	.7	.8	.27228	.28297	3.5339	.96222	.2
.4	.19766	.20164	4.9594	.98027	.6	.9	.27396	.28486	3.5105	.96174	.1
.5	.19937	.20345	4.9152	.97992	.5	**16.0**	.27564	.28675	3.4874	.96126	**74.0**
.6	.20108	.20527	4.8716	.97958	.4	.1	.27731	.28864	3.4646	.96078	.9
.7	.20279	.20709	4.8288	.97922	.3	.2	.27899	.29053	3.4420	.96029	.8
.8	.20450	.20891	4.7867	.97887	.2	.3	.28067	.29242	3.4197	.95981	.7
.9	.20620	.21073	4.7453	.97851	.1	.4	.28234	.29432	3.3977	.95931	.6
12.0	.20791	.21256	4.7046	.97815	**78.0**	.5	.28402	.29621	3.3759	.95882	.5
.1	.20962	.21438	4.6646	.97778	.9	.6	.28569	.29811	3.3544	.95832	.4
.2	.21132	.21621	4.6252	.97742	.8	.7	.28736	.30001	3.3332	.95782	.8
.3	.21303	.21804	4.5864	.97705	.7	.8	.28903	.30192	3.3122	.95732	.2
.4	.21474	.21986	4.5483	.97667	.6	.9	.29070	.30382	3.2914	.95681	.1
.5	.21644	.22169	4.5107	.97630	.5	**17.0**	.29237	.30573	3.2709	.95630	**73.0**
.6	.21814	.22353	4.4747	.97592	.4	.1	.29404	.30764	3.2506	.95579	.9
.7	.21985	.22536	4.4373	.97553	.3	.2	.29571	.30955	3.2305	.95528	.8
.8	.22155	.22719	4.4015	.97515	.2	.3	.29737	.31147	3.2106	.95476	.7
.9	.22325	.22903	4.3662	.97476	.1	.4	.29904	.31338	3.1910	.95424	.6
13.0	.22495	.23087	4.3315	.97437	**77.0**	.5	.30071	.31530	3.1716	.95372	.5
.1	.22665	.23271	4.2972	.97398	.9	.6	.30237	.31722	3.1524	.95319	.4
.2	.22835	.23455	4.2635	.97358	.8	.7	.30403	.31914	3.1334	.95266	.3
.3	.23005	.23639	4.2303	.97318	.7	.8	.30570	.32106	3.1146	.95213	.2
.4	.23175	.23823	4.1976	.97278	.6	.9	.30736	.32299	3.0961	.95159	.1
13.5	.23345	.24008	4.1653	.97237	.5	**18.0**	.30902	.32492	3.0777	.95106	**72.0**
Angle°	**cos**	**cot**	**tan**	**sin**	**Angle°**	**Angle°**	**cos**	**cot**	**tan**	**sin**	**Angle°**

Table A-11 *(Continued)*

Angle°	sin	tan	cot	cos	Angle°	Angle°	sin	tan	cot	cos	Angle°
18.0	.30902	.32492	3.0777	.95106	**72.0**	**22.5**	.38268	.41421	2.4142	.92388	.5
.1	.31068	.32685	3.0595	.95052	.9	.6	.38430	.41626	2.4023	.92321	.4
.2	.31233	.32878	3.0415	.94997	.8	.7	.38591	.41831	2.3906	.92254	.3
.3	.31399	.33072	3.0237	.94943	.7	.8	.38752	.42036	2.3789	.92186	.2
.4	.31565	.33266	3.0061	.94888	.6	.9	.38912	.42242	2.3673	.92119	.1
.5	.31730	.33460	2.9887	.94832	.5	**23.0**	.39073	.42447	2.3559	.92050	**67.0**
.6	.31896	.33654	2.9714	.94777	.4	.1	.39234	.42654	2.3445	.91982	.9
.7	.32061	.33848	2.9544	.94721	.3	.2	.39394	.42860	2.3332	.91914	.8
.8	.32227	.34043	2.9375	.94665	.2	.3	.39555	.43067	2.3220	.91845	.7
.9	.32392	.34238	2.9208	.94609	.1	.4	.39715	.43274	2.3109	.91775	.6
19.0	.32557	.34433	2.9042	.94552	**71.0**	.5	.39875	.43481	2.2998	.91706	.5
.1	.32722	.34628	2.8878	.94495	.9	.6	.40035	.43689	2.2889	.91636	.4
.2	.32887	.34824	2.8716	.94438	.8	.7	.40195	.43897	2.2781	.91566	.3
.3	.33051	.35020	2.8556	.94380	.7	.8	.40355	.44105	2.2673	.91496	.2
.4	.33216	.35216	2.8397	.94322	.6	.9	.40514	.44314	2.2566	.91425	.1
.5	.33381	.35412	2.8239	.94264	.5	**24.0**	.40674	.44523	2.2460	.91355	**66.0**
.6	.33545	.35608	2.8083	.94206	.4	.1	.40833	.44732	2.2355	.91283	.9
.7	.33710	.35805	2.7929	.94147	.3	.2	.40992	.44942	2.2251	.91212	.8
.8	.33874	.36002	2.7776	.94088	.2	.3	.41151	.45152	2.2148	.91140	.7
.9	.34038	.36199	2.7625	.94029	.1	.4	.41310	.45362	2.2045	.91068	.6
20.0	.34202	.36397	2.7475	.93969	**70.0**	.5	.41469	.45573	2.1943	.90996	.5
.1	.34366	.36595	2.7326	.93909	.9	.6	.41628	.45784	2.1842	.90924	.4
.2	.34530	.36793	2.7179	.93849	.8	.7	.41787	.45995	2.1742	.90851	.3
.3	.34694	.36991	2.7034	.93789	.7	.8	.41945	.46206	2.1624	.90778	.2
.4	.34857	.37190	2.6889	.93728	.6	.9	.42104	.46418	2.1543	.90704	.1
.5	.35021	.37388	2.6746	.93667	.5	**25.0**	.42262	.46631	2.1445	.90631	**65.0**
.6	.35184	.37588	2.6605	.93606	.4	.1	.42420	.46843	2.1348	.90557	.9
.7	.35347	.37787	2.6464	.93544	.3	.2	.42578	.47056	2.1251	.90483	.8
.8	.35511	.37986	2.6325	.93483	.2	.3	.42736	.47270	2.1155	.90408	.7
.9	.35674	.38186	2.6187	.93420	.1	.4	.42894	.47483	2.1060	.90334	.6
21.0	.35837	.38386	2.6051	.93358	**69.0**	.5	.43051	.47698	2.0965	.90259	.5
.1	.36000	.38587	2.5916	.93295	.9	.6	.43209	.47912	2.0872	.90183	.4
.2	.36162	.38787	2.5782	.93232	.8	.7	.43366	.48127	2.0778	.90108	.3
.3	.36325	.38988	2.5649	.93169	.7	.8	.43523	.48342	2.0682	.90032	.2
.4	.36488	.39190	2.5517	.93106	.6	.9	.43680	.48557	2.0594	.89956	.1
.5	.36650	.39391	2.5386	.93042	.5	**26.0**	.43837	.48773	2.0503	.89879	**64.0**
.6	.36812	.39593	2.5257	.92978	.4	.1	.43994	.48989	2.0413	.89803	.9
.7	.36975	.39795	2.5129	.92913	.3	.2	.44151	.49206	2.0323	.89726	.8
.8	.37137	.39997	2.5002	.92849	.2	.3	.44307	.49423	2.0233	.89649	.7
.9	.37299	.40200	2.4876	.92784	.1	.4	.44464	.49640	2.0145	.89571	.6
22.0	.37461	.40403	2.4751	.92718	**68.0**	.5	.44620	.49858	2.0057	.89493	.5
.1	.37622	.40606	2.4627	.92653	.9	.6	.44776	.50076	1.9970	.89415	.4
.2	.37784	.40809	2.4504	.92587	.8	.7	.44932	.50295	1.9883	.89337	.3
.3	.37946	.41013	2.4383	.92521	.7	.8	.45088	.50514	1.9797	.89259	.2
.4	.38107	.41217	2.4262	.92455	.6	.9	.45243	.50733	1.9711	.89180	.1
22.5	.38268	.41421	2.4142	.92388	.5	**27.0**	.45399	.50953	1.9626	.89101	**63.0**
Angle°	cos	cot	tan	sin	Angle°	Angle°	cos	cot	tan	sin	Angle°

Table A-11 *(Continued)*

Angle°	sin	tan	cot	cos	Angle°	Angle°	sin	tan	cot	cos	Angle°
27.0	.45399	.50953	1.9626	.89101	**63.0**	**31.5**	.52250	.61280	1.6319	.85264	**59.5**
.1	.45554	.51173	1.9542	.89021	.9	.6	.52399	.61520	1.6255	.85173	.4
.2	.45710	.51393	1.9458	.88942	.8	.7	.52547	.61761	1.6191	.85081	.3
.3	.45865	.51614	1.9375	.88862	.7	.8	.52696	.62003	1.6128	.84989	.2
.4	.46020	.51835	1.9292	.88782	.6	.9	.52844	.62245	1.6066	.84897	.1
.5	.46175	.52057	1.9210	.88701	.5	**32.0**	.52992	.62487	1.6003	.84805	**58.0**
.6	.46330	.52279	1.9128	.88620	.4	.1	.53140	.62730	1.5941	.84712	.9
.7	.46484	.52501	1.9047	.88539	.3	.2	.53288	.62973	1.5880	.84619	.8
.8	.46639	.52724	1.8967	.88458	.2	.3	.53435	.63217	1.5818	.84526	.7
.9	.46793	.52947	1.8887	.88377	.1	.4	.53583	.63462	1.5757	.84433	.6
28.0	.46947	.53171	1.8807	.88295	**62.0**	.5	.53730	.63707	1.5697	.84339	.5
.1	.47101	.53395	1.8728	.88213	.9	.6	.53877	.63953	1.5637	.84245	.4
.2	.47255	.53620	1.8650	.88130	.8	.7	.54024	.64199	1.5577	.84151	.3
.3	.47409	.53844	1.8572	.88048	.7	.8	.54171	.64446	1.5517	.84057	.2
.4	.47562	.54070	1.8495	.87965	.6	.9	.54317	.64693	1.5458	.83962	.1
.5	.47716	.54296	1.8418	.87882	.5	**33.0**	.54464	.64941	1.5399	.83867	**57.0**
.6	.47869	.54522	1.8341	.87798	.4	.1	.54610	.65189	1.5340	.83772	.9
.7	.48022	.54748	1.8265	.87715	.3	.2	.54756	.65438	1.5282	.83676	.8
.8	.48175	.54975	1.8190	.87631	.2	.3	.54902	.65688	1.5224	.83581	.7
.9	.48328	.55203	1.8115	.87546	.1	.4	.55048	.65938	1.5166	.83485	.6
29.0	.48481	.55431	1.8040	.87462	**61.0**	.5	.55194	.66189	1.5108	.83389	.5
.1	.48634	.55659	1.7966	.87377	.9	.6	.55339	.66440	1.5051	.83292	.4
.2	.48786	.55888	1.7893	.87292	.8	.7	.55484	.66692	1.4994	.83195	.3
.3	.48938	.56117	1.7820	.87207	.7	.8	.55630	.66944	1.4938	.83098	.2
.4	.49090	.56347	1.7747	.87121	.6	.9	.55775	.67197	1.4882	.83001	.1
.5	.49242	.56577	1.7675	.87036	.5	**34.0**	.55919	.67451	1.4826	.82904	**56.0**
.6	.49394	.56808	1.7603	.86949	.4	.1	.56064	.67705	1.4770	.82806	.9
.7	.49546	.57039	1.7532	.86863	.3	.2	.56208	.67960	1.4715	.82708	.8
.8	.49697	.57271	1.7461	.86777	.2	.3	.56353	.68215	1.4659	.82610	.7
.9	.49849	.57503	1.7391	.86690	.1	.4	.56497	.68471	1.4605	.82511	.6
30.0	.50000	.57735	1.7321	.86603	**60.0**	.5	.56641	.68728	1.4550	.82413	.5
.1	.50151	.57968	1.7251	.86515	.9	.6	.56784	.68985	1.4496	.82314	.4
.2	.50302	.58201	1.7182	.86427	.8	.7	.56928	.69243	1.4442	.82214	.3
.3	.50453	.58435	1.7113	.86340	.7	.8	.57071	.69502	1.4388	.82115	.2
.4	.50603	.58670	1.7045	.86251	.6	.9	.57215	.69761	1.4335	.82015	.1
.5	.50754	.58905	1.6977	.86163	.5	**35.0**	.57358	.70021	1.4281	.81915	**55.0**
.6	.50904	.59140	1.6909	.86074	.4	.1	.57501	.70281	1.4229	.81815	.9
.7	.51054	.59376	1.6842	.85985	.3	.2	.57643	.70542	1.4176	.81714	.8
.8	.51204	.59612	1.6775	.85895	.2	.3	.57786	.70804	1.4124	.81614	.7
.9	.51354	.59849	1.6709	.85806	.1	.4	.57928	.71066	1.4071	.71513	.6
31.0	.51504	.60086	1.6643	.85717	**59.0**	.5	.58070	.71329	.14019	.81412	.5
.1	.51653	.60324	1.6577	.85627	.9	.6	.58212	.71593	1.3968	.81310	.4
.2	.51803	.60562	1.6512	.85536	.8	.7	.58354	.71857	1.3916	.81208	.3
.3	.51952	.60801	1.6447	.85446	.7	.8	.58496	.72122	1.3865	.81106	.2
.4	.52101	.61040	1.6383	.85355	.6	.9	.58637	.72388	1.3814	.81004	.1
31.5	.52250	.61280	1.6319	.85264	**59.5**	**36.0**	.58779	.72654	1.3764	.80902	**54.0**
Angle°	cos	cot	tan	sin	Angle°	Angle°	cos	cot	tan	sin	Angle°

Table A-11 *(Concluded)*

Angle°	sin	tan	cot	cos	Angle°	Angle°	sin	tan	cot	cos	Angle°
36.0	.58779	.72654	1.3764	.80902	**54.0**	**40.5**	.64945	.85408	1.1708	.76041	**49.5**
.1	.58920	.72921	1.3713	.80799	.9	.6	.65077	.85710	1.1667	.75927	.4
.2	.59061	.73189	1.3663	.80696	.8	.7	.65210	.86014	1.1626	.75813	.3
.3	.59201	.73457	1.3613	.80593	.7	.8	.65342	.86318	1.1585	.75700	.2
.4	.59342	.73726	1.3564	.80489	.6	.9	.65474	.86623	1.1544	.75585	.1
.5	.59482	.73996	1.3514	.80386	.5	**41.0**	.65606	.86929	1.1504	.75471	**49.0**
.6	.59622	.74267	1.3465	.80282	.4	.1	.65738	.87236	1.1463	.75356	.9
.7	.59763	.74538	1.3416	.80178	.3	.2	.65869	.87543	1.1423	.75241	.8
.8	.59902	.74810	1.3367	.80073	.2	.3	.66000	.87852	1.1383	.75126	.7
.9	.60042	.75082	1.3319	.79968	.1	.4	.66131	.88162	1.1343	.75011	.6
37.0	.60182	.75355	1.3270	.79864	**53.0**	.5	.66262	.88473	1.1303	.74896	.5
.1	.60321	.75629	1.3222	.79758	.9	.6	.66393	.88784	1.1263	.74780	.4
.2	.60460	.75904	1.3175	.79653	.8	.7	.66523	.89097	1.1224	.74664	.3
.3	.60599	.76180	1.3127	.79547	.7	.8	.66653	.89410	1.1184	.74548	.2
.4	.60738	.76456	1.3079	.79441	.6	.9	.66783	.89725	1.1145	.74431	.1
.5	.60876	.76733	1.3032	.79335	.5	**42.0**	.66913	.90040	1.1106	.74314	**48.0**
.6	.61015	.77010	1.2985	.79229	.4	.1	.67043	.90357	1.1067	.74198	.9
.7	.61153	.77289	1.2938	.79122	.3	.2	.67172	.90674	1.1028	.74080	.8
.8	.61291	.77568	1.2892	.79016	.2	.3	.67301	.90993	1.0990	.73963	.7
.9	.61429	.77848	1.2846	.78908	.1	.4	.67430	.91313	1.0951	.73846	.6
38.0	.61566	.78129	1.2799	.78801	**52.0**	.5	.67559	.91633	1.0913	.73728	.5
.1	.61704	.78410	1.2753	.78694	.9	.6	.67688	.91955	1.0875	.73610	.4
.2	.61841	.78692	1.2708	.78586	.8	.7	.67816	.92277	1.0837	.73491	.3
.3	.61978	.78975	1.2662	.78478	.7	.8	.67944	.92601	1.0799	.73373	.2
.4	.62115	.79259	1.2617	.78369	.6	.9	.68072	.92926	1.0761	.73254	.1
.5	.62251	.79544	1.2572	.78261	.5	**43.0**	.68200	.93252	1.0724	.73135	**47.0**
.6	.62388	.79829	1.2527	.78152	.4	.1	.68327	.93578	1.0686	.73016	.9
.7	.62524	.80115	1.2482	.78043	.3	.2	.68455	.93906	1.0649	.72897	.8
.8	.62660	.80402	1.2437	.77934	.2	.3	.68582	.94235	1.0612	.72777	.7
.9	.62796	.80690	1.2393	.77824	.1	.4	.68709	.94565	1.0575	.72657	.6
39.0	.62932	.80978	1.2349	.77715	**51.0**	.5	.68835	.94896	1.0538	.72537	.5
.1	.63068	.81268	1.2305	.77605	.9	.6	.68962	.95229	1.0501	.72417	.4
.2	.63203	.81558	1.2261	.77494	.8	.7	.69088	.95562	1.0464	.72297	.3
.3	.63338	.81849	1.2218	.77384	.7	.8	.69214	.95897	1.0428	.72176	.2
.4	.63473	.82141	1.2174	.77273	.6	.9	.69340	.96232	1.0392	.72055	.1
.5	.63608	.82434	1.2131	.77162	.5	**44.0**	.69466	.96569	1.0355	.71934	**46.0**
.6	.63742	.82727	1.2088	.77051	.4	.1	.69591	.96907	1.0319	.71813	.9
.7	.63877	.83022	1.2045	.76940	.3	.2	.69717	.97246	1.0283	.71691	.8
.8	.64011	.83317	1.2002	.76828	.2	.3	.69842	.97586	1.0247	.71569	.7
.9	.64145	.83613	1.1960	.76717	.1	.4	.69966	.97927	1.0212	.71447	.6
40.0	.64279	.83910	1.1918	.76604	**50.0**	.5	.70091	.98270	1.0176	.71325	.5
.1	.64412	.84208	1.1875	.76492	.9	.6	.70215	.98613	1.0141	.71203	.4
.2	.64546	.84507	1.1833	.76380	.8	.7	.70339	.98958	1.0105	.71080	.3
.3	.64679	.84806	1.1792	.76267	.7	.8	.70463	.99304	1.0070	.70957	.2
.4	.64812	.85107	1.1750	.76154	.6	.9	.70587	.99652	1.0035	.70834	.1
40.5	.64945	.85408	1.1708	.76041	**49.5**	**45.0**	.70711	1.00000	1.0000	.70711	**45.0**
Angle°	cos	cot	tan	sin	Angle°	Angle°	cos	cot	tan	sin	Angle°

ANSWERS TO SELECTED PROBLEMS

Chapter 2

2-2. 838.2 m
2-4. 0.038 g
2-6. 4,500 s
2-8. 0.10 Ω
2-10. 0.002 in.
2-12. 8,140 ft.
2-14. Brass
2-16. 0.1313 Ω
2-18. (*a*) 121.6°C
(*b*) −100.8°C
2-20. (*a*) 1,440 Ω
(*b*) 1,451 Ω

Chapter 3

3-1. 6.91 mA
3-3. 48 mA
3-5. 62.5 kΩ
3-7. 625 mA
3-9. 714 Ω
3-11. 0.4A
3-13. 214 V
3-15. 20 V
3-17. 81.3 Ω
3-19. 56.2 Ω
3-21. 94.3 Ω
3-23. $E_{load} = 189$ V
$E_{lines} = 30.4$ F
3-25. 12 mA
3-27. 1.55 A
3-29. 1 A
3-31. $R = 3.5$ Ω
$I = 1.8$ A
3-33. $I_{total} = 1.18$ A
$R = 35.6$ Ω
3-35. 211 Ω
3-37. 2 kΩ
3-39. 485 Ω
3-41. $I_{total} = 0.412$ A
3-43. 160 V
3-45. 50 mV
3-47. 29 V
3-49. 213 Ω
3-51. $I_{total} = 3.3$ A
$E_{applied} = 439$ V
3-53. 305 Ω
3-55. 17.5 V
3-57. 1 kΩ
3-59. 1,180 V
3-61. 264 Ω
3-63. 107 mA
3-65. 222.5 V
3-67. 247 mA
3-69. 92 V
3-71. 3,450 J
3-73. 40.3 C
3-75. 4.655 kW
3-77. 36.4 W: 10.3 kJ
33.2 W: 9.36 kJ
26 W: 7.33 kJ
3-79. 800 mW
3-81. 37.3 Ω: 131 W
3-85. 39%
3-91. $R_1 = 107$ Ω
$E_s = 78.2$ V
3-97. −60.8 dB
3-101. 114 dB

Chapter 4

4-2. (*b*) 32.75 V
4-4. (*c*) $-6 = (7\text{ k}\Omega)I$
$I = -.857$ mA
4-6. (*b*) -14.4 mA
4-8. (*b*) $I_1 = 13$ mA
$I_2 = -3$ mA
4-10. $-100 = 1{,}350I_1 - 1{,}150I_2 - 200I_3$
$40 = -1{,}150I_1 + 1{,}450I_2 - 300I_3$
$0 = -200I_1 - 300I_2 + 1{,}500I_3$
$I_1 = -215$ mA
4-16. $I_1 = -163$ mA
$I_2 = -100$ mA
$I_3 = -30.4$ mA
4-18. Four
4-20. $E_1 = 16.5$ V
$E_2 = 16.7$ V
4-24. $E_1 = -4.02$ V
$E_2 = -8.05$ V
4-32. (*b*) 4.12 mA
4-38. (*c*)

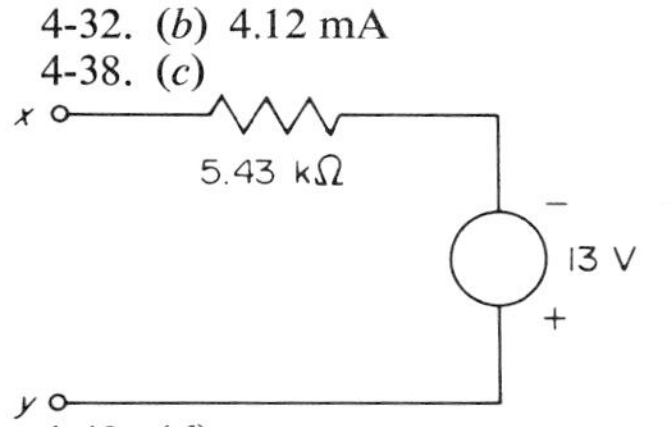

4-40. (*d*)

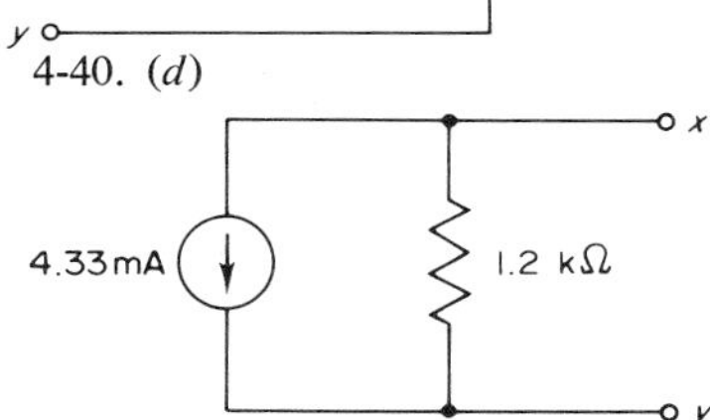

4-42. $E_{th} = 5$ V
$R_{th} = R_n = 99.8\ \Omega$
$I_n = 50.1$ mA
4-46. (*a*) 236 μmhos
4-48. (*b*)

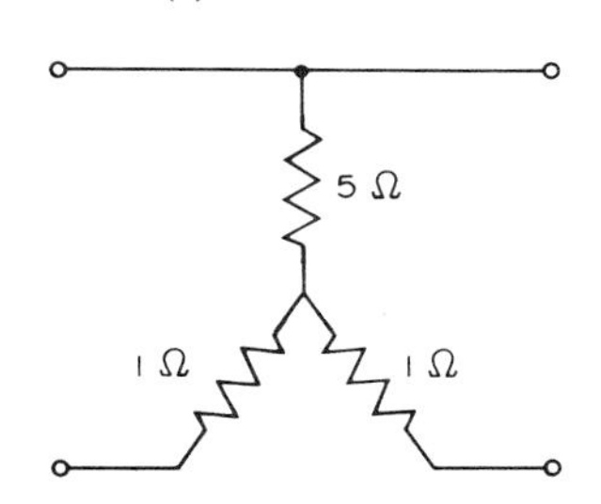

Chapter 5

5-1. 0.066×10^{-6} F
0.066 μF
5-3. 1×10^{-6} C
5-5. 0.014 μF
5-7. 157,000 V
94,200 V
5-9. 2 μF
5-11. 0.064 μF
5-15. 0.782 mA
5-19. (*a*) $e_R = 187.7$ V
$e_c = 62.3$ V
(*b*) 5 mA
(*c*) 1.75 s
(*d*) 1.79 s
5-23. (*a*) 19.8 V
(*b*) 2.5 s
(*c*) 439 ms
5-25. (*a*) 105 μA
(*b*) 2.12 V
(*c*) 1.46 s
5-27. (*a*) 5.47 ms
(*b*) 6.03 ms

Chapter 6

6-2. $F = NI = \mathscr{R}\Phi = Bl/\mu_0$
$I = 37$ A
6-4. 3.73 At/m
6-6. 534 At/Wb
6-8. 0.62 Wb/m^2
6,200 G
40 kilolines/in^2
6-10. 1.35×10^4 G
6-14. 40 T
6-16. Branch legs contain flux $\Phi/2$ whereas center section has flux Φ. Center area is twice the branch areas. Thus, B and H are the same at all points. From Amperes Circuital Law
$NI = H(l_{ab} + l_{bc} + l_{cd} + l_{da})$
$H = 625$ At/m
$B = 1.28$ Wb/m^2
6-18. 0.82×10^{-2} Wb
6-20. 220 T

Chapter 7

7-1. 20×10^{-3} Wb
7-5. 445 T
7-7. 88.5 V
7-9. 9.28 mH
7-11. 39.3 mH
7-13. 623 mH
$\mathbf{K} = 0.5$
7-15. 147 mH
7-17. 0.57 s
7-21. 238 Ω
7-23. 0.789 J
7-25. (*a*) 160 mH

Chapter 8

8-2. (*a*) 60 Hz
(*b*) 138.5 V
(*c*) 0 V
8-4. $e = 45 \sin (203 \times 10^3 t)$ mV
8-6. (*a*) 8.333 kHz
(*b*) 120 μs
8-8. (*a*) 0 A
(*b*) 10.6 mA
(*c*) −6.34 mA
8-14. 222°
8-16. 945 mA
8-18. 6.25 kΩ
8-20. 0.237 mA

Chapter 9

9-1. 49.5 mA
9-3. $E_1 = 77.4$ V
$E_2 = E_3 = 64$ V
9-5. $i_1 = 0.1245$ A
$i_2 = 0.0685$ A
$i_3 = 0.056$ A
9-7. $E_4 = 905$ V
9-9. 650 mW
9-11. −17.2 mA
9-13. (*b*) 90 Ω
9-15. (*b*) 20 mH
9-17. (*e*) 32 Ω
9-19. (*c*) 300 μF
9-21. (*c*) 1. 27 kΩ
2. 0.667 mA
3. $0.667 \sin (6{,}500t + 90^\circ)$ mA
9-23. (*d*) 1. 73.5 mA
2. 52 mA
3. $i = 73.5 \sin (2.2 \times 10^6 t + 72^\circ)$ mA
$e = 12 \sin (2.2 \times 10^6 t - 18^\circ)$ V
9-25. 33.3 A
$e = 282.8 \sin(377t)$V
$i = 47 \sin(377t - 90^\circ)$A
9-27. (*f*) 1. 1.33 mA
2. 200 mH
3. $1.33 \sin (48{,}000t - 170^\circ)$mA
9-29. $207 \angle 41.7^\circ$ V
9-33. $\mathbf{I} = 0.67 \angle -50^\circ$ A
186 kHz
9-35. $1.56 \angle -90^\circ$ kΩ
9-37. (*c*) $R = 1.57$ kΩ

Chapter 10

10-2. (*g*) $500 + j314$ Ω
$590 \angle 32.2^\circ$ Ω
10-4. (*c*) $R = 85.5$ Ω
$C = 0.0016$ μF
10-8. $429 \angle 30^\circ$ Ω
10-10. $30.2 \angle 125.2^\circ$ V
10-14. (*a*) $278 \angle 52.3^\circ$ Ω
(*b*)
$\mathbf{E}_c = 186 \angle -142^\circ$ V
$\mathbf{E}_{R1} = 16.4 \angle -52.3^\circ$ V
$\mathbf{E}_L = 213 \angle 37.7^\circ$ V
$\mathbf{E}_{R2} = 5 \angle 52.3^\circ$ V
10-18. 0.00336 μF
10-20. 509 μH
10-22. (*a*) 6.63 pF
(*b*) 13.5 μH
(*c*) $378 \angle -84^\circ$ V
(*d*) $5.4 \angle -64^\circ$ kΩ
10-24. $2.75 \angle -82^\circ$ kΩ
10-26. (*d*) $0.668 \angle -49^\circ$ A
(*f*)
$5.04 \angle -89^\circ$ mmhos
10-30. (*c*) $94 \angle -38^\circ$ Ω
10-32. $\mathbf{I}_2 = 0.406 \angle 96^\circ$ A
$\mathbf{I}_4 = 1.0 \angle 73^\circ$ A
10-36. $167 \angle -57.7^\circ$ Ω
10-40. $205 \angle -10.4^\circ$ Ω
10-42. (*a*) $36.5 \angle -36^\circ$ V
(*b*) $27.5 \angle 129^\circ$ V
(*c*) $15.1 \angle 42.9^\circ$ V
10-44. (*a*) $362 \angle 112^\circ$ V
(*b*) $166 \angle -17^\circ$ V
(*c*) $0.975 \angle 36.2^\circ$ A
10-52. $640 \angle 1.3^\circ$ Ω
10-56. $44.3 \angle -28^\circ$ V
10-60. $77.9 \angle -145^\circ$ mA
10-64. $P_{true} = 1.06$ W
$P_{reac} = 1.83$ VARS
$P_{appar} = 2.12$VA
10-68. $P_{true} = 2.94$ W
$P_{reac} = 26.2$ VARS
$P_{appar} = 26.4$ VA
10-76. $P_1 = 48.8$ W
$P_2 = 14.5$ W
$P_3 = 18.4$ W
$P_4 = 15.8$ W
$P_{total} = 97.5$ W

Chapter 11

11-1. (*c*)

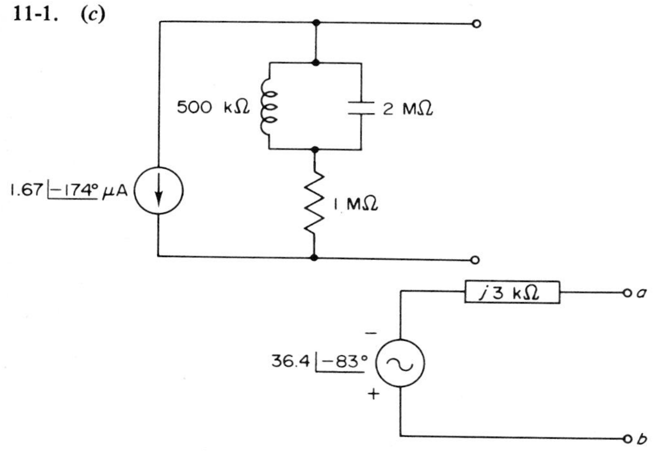

11-3. (*d*) $6.7 \angle 168.2°$ V

11-5. (*c*) $\mathbf{I}_1 = .545 \angle 51°$ mA
$\mathbf{I}_2 = .513 \angle 90°$ mA

11-7. (*c*)
$\mathbf{E}_1 = 27 \angle 23.5°$ V
$\mathbf{E}_2 = 12.6 \angle -11.5°$ V

11-9. (*a*)
$\mathbf{I}_E = 18.7 \angle 68.6°$ mA
$\mathbf{I}_I = 11.7 \angle -81.4°$ mA
$\mathbf{I} = 10.3 \angle 34°$ mA

11-11. (*d*)

11-13. $2.08 \angle -175°$ A

11-15. (*d*) $R_x = 40$ kΩ
$L_x = 2.25\ \mu$H

Chapter 12

12-2. (*a*) 119 kHz
(*b*) $2 \angle 40°$ mV
(*c*) $19.2 \angle 50°$ mV

12-4. 6.73 nF

12-6. 3.2 μH

12-8. 6.67

12-10. 0.684

12-12. 1.2 Ω

12-14. 1.78 Ω

12-16. (*a*) 12.45 kHz
(*b*) 112.8 kHz,
125.2 kHz

12-18. By Eq. (12-28),
$f_0 = 750$ kHz
By Eq. (12-27),
(*a*) $f_0 - 730$ kHz
(*d*) no f_0

12-20. $L = 33\ \mu$H
$R = 10\ \Omega$

12-22. $L = 60\ \mu$H
$R = 1.82\ \Omega$

Chapter 13

13-1. $\mathbf{I}_1 = 3.56 \angle -88.5°$ mA

13-3. (*a*)
$z_{11} = 6.62 \angle 50°\ \Omega$
$z_{22} = 4.84 \angle 67.2°\ \Omega$
$z_M = 2.91 \angle 13.8°\ \Omega$

13-5. $1.73 \angle 107°\ \Omega$

13-7. (*a*)
$z_{11} = 105 \angle -67.7°$ kΩ
$z_{22} = 20 \angle -85.4°$ kΩ
$z_M = 100 \angle -85.4°$ kΩ

13-9. $\mathbf{E}_2 = 2.28 \angle 24°$ V

13-11. (*a*)
$4.64 \angle -154.5°$ mmhos

13-13. (*a*)
$y_{11} = 100 \angle 53°\ \mu$mhos
$y_{22} = 14.9 \angle -63.5°$ μmhos
$y_M = 44.5 \angle -63.5°$ μmhos

13-17. (*b*)
$z_{11} = 370 \angle -3.8°$ kΩ
$z_{22} = 92.5 \angle -93.8°$kΩ
$z_M = 241 \angle 44.2°$ kΩ

13-19. $a = 7.07$

13-21. (*a*) $\mathbf{I}_1 = 2.76 \angle 3.7°$ mA

13-23. (*a*)
$C_2 = 1{,}270$ ph
$\mathbf{I}_1 = .657 \angle -49.4°$ mA
$\mathbf{E}_0 = 38.4 \angle 130.6°$ V

13-25. (*a*) $a = .128$
(*f*)
$\mathbf{E}_{0,\text{mid}} = 195 \angle -50°$ V

13-27. (*c*) $\pm$j149 kΩ

13-29. (*e*)

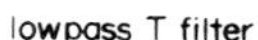

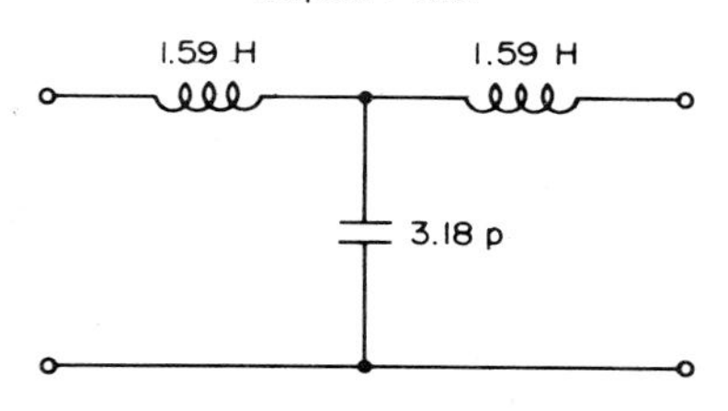

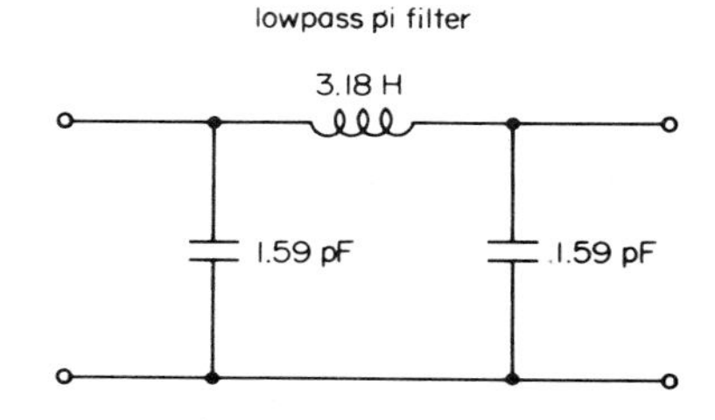

13-31. (*e*)

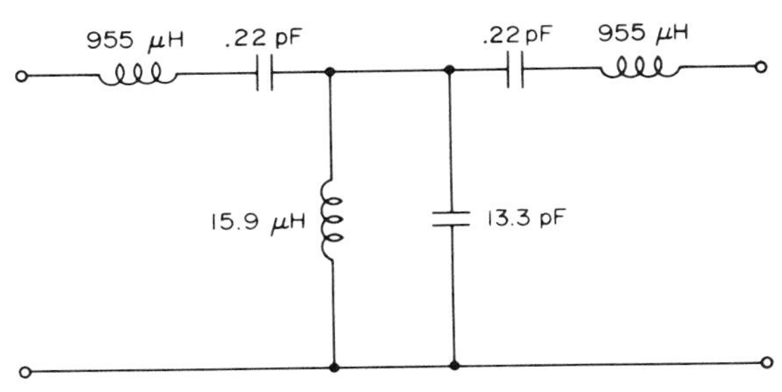

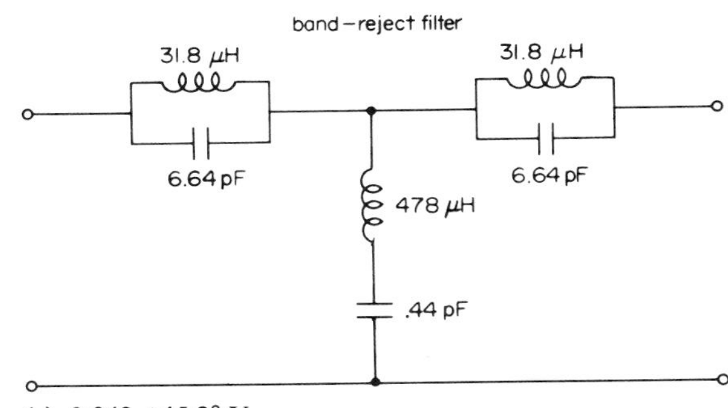

Chapter 14

14-2. (*a*) $8.08\angle -50.3°$ V
(*b*) $1.5\angle 4.5°$ V
(*c*) $0.342\angle 45.2°$ V

14-4. 371 μW

Chapter 15

15-1. (*a*)
$\mathbf{E}_1 = 198\angle -150°$ V
$\mathbf{E}_2 = 198\angle 90°$ V
$\mathbf{E}_3 = 198\angle -30°$ V
$\mathbf{I}_1 = 0.471\angle -196°$ A
$\mathbf{I}_2 = 0.471\angle 44°$ A
$\mathbf{I}_3 = 0.471\angle -76°$ A
(*b*)
$\mathbf{I}_a = 0.471\angle -136°$ A
$\mathbf{I}_b = 0.471\angle 104°$ A
$\mathbf{I}_c = 0.471\angle -16°$ A

15-3. 208 V

15-5. $\mathbf{I}_a = 0.975\angle 147°$ A
$\mathbf{I}_b = 0.941\angle -209°$ A
$\mathbf{I}_c =$
$0.0948\angle -131.6°$ A